Manfred Gottwald
Thomas Kenkmann
Wolf Uwe Reimold

TERRESTRIAL IMPACT STRUCTURES

The TanDEM-X Atlas

1

Africa, North/Central America, South America

"There is to be found in the almost level plain country, about five miles almost due south of Sunshine Station, on the Atchinson, Topeka & Santa Fé Ry., in Coconino county, Arizona, a very remarkable and almost perfectly round crater, differing in many respects, as will be hereafter seen, from any crater on the earth's surface with which I, at least, am familiar."

"Having observed all these things, containing as they do many arguments in favor of the theory that this great hole in the plain was produced by the impact of a body falling out of space, and against the theory that it was produced by either volcanic action or by a steam explosion, it naturally suggested itself to us to endeavor to prove absolute synchronism of the two events namely, the falling of a very great meteor on this particular spot and the formation of this crater."

Daniel Moreau Barringer (1906),
in: *Coon Mountain and its Crater*.
Proceedings of the Academy of Natural Science of Philadelphia,
Vol. 57, pp. 861–886.

Meteor Crater in the early 1930's (photo: U.S. Army Air Service).

Manfred Gottwald
Thomas Kenkmann
Wolf Uwe Reimold

TERRESTRIAL IMPACT STRUCTURES

The TanDEM-X Atlas

1

Africa, North/Central America, South America

Verlag Dr. Friedrich Pfeil • München 2020
ISBN 978-3-89937-261-8

Bibliografische Information Der Deutschen Bibliothek

Die Deutsche Bibliothek verzeichnet diese Publikation in der Deutschen Nationalbibliografie;
detaillierte bibliografische Daten sind im Internet über
http://dnb.dnb.de abrufbar.

front cover
Vredefort Dome, Meteor Crater, Serra da Cangalha (from left to right).
Top row: TanDEM-X topographic maps; bottom row Sentinel-2 RGB images,
fused with the TanDEM-X digital elevation model.

back cover
Top: Meteor Crater (photo: T. Kenkmann);
bottom: B.P. structure (photo: W. U. Reimold).

print support by UNI FREIBURG and Wolf Uwe Reimold
and Manfred Gottwald

Druckvorstufe: Verlag Dr. Friedrich Pfeil, München
Druck: PBtisk a.s., Příbram I – Balonka

ISBN 978-3-89937-261-8

Printed in the European Union

Verlag Dr. Friedrich Pfeil, Wolfratshauser Straße 27, 81379 München, Germany
Tel.: + 49 (0)89 5528600-0 • Fax: + 49 (0)89 5528600-4 • E-Mail: info@pfeil-verlag.de
www.pfeil-verlag.de

Contents

Preface

The idea for this atlas was born already several years ago. Then, a precise topographic atlas presenting terrestrial impact craters, the scars of the impacts of solid bodies onto the Earth's surface, had not been established yet. This was astonishing, as such impacts had long been considered as the "most fundamental process on the surfaces of the terrestrial planets", as the late Eugene Shoemaker, one of the pioneers of impact geology, had put it. But when the first digital elevation data had been processed from the early phase of the TanDEM-X mission, they immediately showed their great potential for mapping applications. We were particularly excited, because the objective of this space-borne undertaking of generating a precise high-resolution digital elevation model for the entire solid surface of Earth would allow, for the first time, to map the morphology of all terrestrial impact structures with a topographic expression, in detail. What had been the limiting factor of the existing datasets for such an exercise in the past – no global coverage, data gaps, or data artefacts – would no longer hamper the cartographic work. Only very small impact craters below the resolution of the TanDEM-X data would escape recognition in the TanDEM-X maps.

Patience, however, was required before the final TanDEM-X Digital Elevation Model (DEM) was released and access to the data for science applications was granted. In the meantime, we had investigated the workflow to present the topography of an impact structure and its environs by using Raw DEM scenes. Because our intention was not to merely publish a sequence of high-quality maps, our atlas concept foresaw to additionally provide short and concise, illustrated text sections for each impact structure. As two of us, Thomas Kenkmann and Wolf Uwe Reimold, have throughout their careers as impact geologists confirmed the impact origin of a considerable number of structures and participated in the studies of many more, these so-called "fact sheets" also reflect personal experience. Clearly this is the reason why these texts sometimes differ in style – this is indeed intended. What is more, the fact sheets also reflect the varied degrees of detailed study that different impact structures have permitted to date.

The individual texts also provide information about the location of a structure and how to access it. For those structures in very remote parts of the globe, the access notes can, however, be limited. A list of selected references completes each text. These are the main references that were used by the authors of the individual fact sheets – who are acknowledged at the top of each structure's first page. These references are also intended to provide the layperson with a strategic entry into the literature pertaining to any of these impact structures.

The atlas consists of three parts: introductory chapters, the physical maps with corresponding texts for all impact structures covered herein, and an annexure with supplementary information. The first introductory chapter briefly explains why interplanetary space is filled with small bodies and why they sometimes approach Earth's orbit. Chapter 2 goes a bit deeper into the principles of hypervelocity impact. It explains crater formation, shock metamorphism, and specific impact-related lithologies. This chapter introduces impact related concepts that are addressed in more detail in the subsequent impact structure-related fact sheets. The third chapter describes the terrestrial impact crater record. A short account on its completeness – or rather lack thereof – is followed by its status at the time of the manuscript deadline as of February 2020. As the last two years have been very productive in identifying previously unknown impact structures, we have thrived towards a status that is as complete as possible. Our atlas, however, has a differentiated view on some alleged impact structures. Where we feel that the evidence for an impact origin is still incomplete or ambiguous, we treat these structures separately, even though other impact data bases may list these cases as confirmed – without further analysis of the evidence.

The remaining introductory chapters explain the remote sensing principles behind data acquisition and processing, together with how the physical maps were generated. Chapter 4 illustrates radar remote sensing, particularly the concept of Synthetic Aperture Radar (SAR) that allows high-resolution space-borne imaging in the microwave spectral domain. While SAR yields 2D imagery, it is a pre-condition for SAR interferometry, which exploits elevation and therefore delivers a 3D view. This concept has been successfully implemented in the TanDEM-X mission, which is the topic of chapter 5. This section not only summarizes the characteristics of both radar satellites, but it also explains how the challenging task of interferometric data takes was pursued with subsequent data processing. Chapter 6 explains how the maps were produced. This demanded a stepwise approach with observation of various requirements, such as map projection, illumination, and dealing with the presence of water bodies.

With chapter 7 the map-oriented part of the atlas – by far the major portion of our book – begins: Africa, North/Central America and South America are covered in Part 1, Asia, Australia and Europe in Part 2. All impact structures on a continent are listed alphabetically. Each section comprises a one-page physical map together with the illustrated fact sheet. Those structures where confirmation as impact structures has remained ambiguous or incomplete appear separately at the end of each continent chapter.

An annex with the complete list of selected references, a glossary, a geologic timescale, and lists of acronyms and abbreviations completes the atlas.

For whom, which readership, was this atlas compiled? Our atlas is intended to provide the professional expert with a reference book summarizing the terrestrial impact crater record as of February 2020 in a complete and coherent, as well as concise manner. For the interested layman our book can be a source of impact cratering information at different levels – basic in the introductory chapters and more specific in the fact sheets. The maps together with the indication how to access an individual impact site may even trigger visits to some of the more conveniently located structures.

In this atlas our knowledge and expertise from different research areas – impact geology, a bit of astronomy, and remote sensing – has been combined. Thomas Kenkmann and Wolf Uwe Reimold assembled the fact sheets for the currently confirmed impact structures. Their ongoing work in the past years even helped to add several new impact structures to the terrestrial impact record – and to the atlas content. We believe that this work provides a status that is as up-to-date as possible. During their field work, they obtained a large repository of imagery, some of which is shared here in the book. Manfred Gottwald dealt with the remote sensing related tasks, formulated the introductory chapters, and took care of the graphic artwork. This included the processing of the TanDEM-X digital elevation data, the production of the physical maps, and, where appropriate, the fusion of the TanDEM-X data with Sentinel-2 data.

What we present in this atlas about the terrestrial impact structure record – its content and parameterization – reflects the status that, as we feel, provides a complete and consistent view as of February 2020. For several impacts, however, different results can be found in literature. As the study of impact structures is an ongoing and vivid research topic, for these structures only future work will reveal the final truth.

Considerable effort has been put into the making of this reference work. We hope that our result will provide assistance to many, and enjoyment as well. It may be a reference volume for years to come.

Manfred Gottwald, Garching bei München
Thomas Kenkmann, Albert-Ludwigs-Universität, Freiburg
Wolf Uwe Reimold, University of Brasília, Brasília
February 2020

Acknowledgments

Many individuals have supported us in the course of compiling this atlas; they all deserve explicit and grateful acknowledgment.

Manfred Gottwald acknowledges the wealth of support he did receive from experts in radar remote sensing at DLR in Oberpfaffenhofen, from the Earth Observation Center, and from the Microwaves and Radar Institute. This brought him quickly into a position to understand and discuss the specifics of the TanDEM-X mission and its data. Dana Floricioiu, Birgit Schättler, Helko Breit, Thomas Fritz, Achim Roth and Stefan Buckreuß were always patient discussion partners and teachers. Silke Kerkhoff and Andreas Niedermeier provided access to the raw digital elevation data and allowed him to develop the workflow for topographic map generation well in advance of the availability of the final TanDEM-X datasets. Katja Reissig took care of the necessary software, and she was always prepared to assist with troubleshooting. Three colleagues at DLR deserve particular acknowledgment regarding the making of this book. They are Thomas Busche, who could be approached whenever help was needed concerning data access, Wolfgang Balzer who, as a constant supporter of this undertaking, helped to move forward when project management support was required, and Thomas Trautmann, who, although as section head of atmospheric remote sensing not involved in TanDEM-X, provided the necessary backing for Manfred Gottwald to accomplish this yearlong work.

We received support, either detailed information for specific impact structures or imagery, from many colleagues: Luis Calçada from the European Southern Observatory contributed his view of the early solar system, and Marco Langbroek from The Netherlands captured the two TanDEM-X satellites when passing over the city of Leiden. Ralf-Thomas Schmitt from the Museum für Naturkunde in Berlin produced excellent photos of impact lithologies from the collection at the museum. Charaf Chabou from Ferhat Abbas University in Sétif, Algeria provided access information for impact craters in Algeria. Photos and know-how for Luizi came from Ludovic Ferriére from the Natural History Museum in Vienna, Austria, and Luigi Folco from the University in Pisa, Italy, contributed images for Kamil crater. Lutz Hecht from the Museum für Naturkunde in Berlin offered images of impact melt rock from the Tenoumer crater. Christian Koeberl from the Natural History Museum in Vienna, Austria shared his extensive knowledge about African impact structures. Michael Poelchau from the University of Freiburg, Germany provided images from the Chicxulub drilling campaign, and for the Santa Fe and Karla impact structures. The drone view of the Douglas crater field came from Kent Sundell at the Casper College, Casper in Wyoming. The University of Arizona's Mineral Museum contributed images of the Brenham and the Sikhote Alin meteorites. Gordon Osinski from the Western University in London, Canada, permitted the use of photos from Mistastin, and Chris Herd from the University of Alberta, Edmonton in Canada offered two images of the Whitecourt crater. Chuck O'Dale from Canada added the photos of the giant shatter cones on Slate Islands and the aerial view showing the location of the Bloody Creek structure. Several of the photos for Manicouagan came from Martin Schmieder from the University of Applied Sciences in Neu-Ulm, Germany, who made them available to the public. The texts about South American impact structures were reviewed by Alvaro Crósta from the State University in Campinas, Brazil. He also contributed photos for several impact structures on this continent (Araguainha, Cerro do Jarau, Santa Marta, Vista Alegre, Vargeão Dome).

For the impact structures in India, Jayanta Pati from the University of Allahabad, India reviewed our text about Dhala, and Amar Agarwal from the University of Freiburg, Germany, provided the photos for Lonar. The panorama of El'gygytgyn comes from Ulli Raschke, formerly of the Museum für Naturkunde in Berlin. Edwin Gnos from the Natural History Museum in Geneva contributed figures for the Wabar crater field. Ming Chen from the Guangzhou Institute of Geochemistry at the Chinese Academy of Sciences supplied us with photos of the Xiuyan impact crater, and Jean-Luc Dauvergne, with the help of Sylvain Bouley, provided the images of the Kara-Kul structure.

Peter Haines from the Geological Survey of Western Australia in Perth added images for the Australian impact structures Acraman, Boxhole, Foelsche and Goyder. The panoramic views of Kärdla and Morasko were contributed by Jakob Wilk from the University of Freiburg and Kai Wünnemann from the Museum für Naturkunde in Berlin, respectively. Jüri Plado from the University of Tartu, Estonia, shared the image of the Kärdla drill core, and Satu Hietala from the Geological Survey of Finland contributed the Summanen photo. Alexander Zubov and Tatyana Shumilova from the Institute of Geology of Kami Scientific Center in Syktyvkar offered their drone views of Kara. Laurence Garvie from the Center for Meteorite Studies at the Arizona State University provided the photo of the Sterlitamak fragment. We also greatly value the open access to the Copernicus Sentinel-2 data that ensured their availability for generating the combination of TanDEM-X and multispectral RGB-images.

Thomas Kenkmann acknowledges the persistent funding by the German Research Foundation (DFG), as well as financial support by the University of Freiburg and, before 2010, by the Museum of Natural History Berlin. This support made the study of a number of terrestrial impact craters possible. He would like to express his gratitude to the many colleagues accompanying him in the field. Special thanks go to Gerwin Wulf and Michael Poelchau from the University of Freiburg for ongoing fruitful discussions on various topics related to impact craters and impact processes, and to Alex Deutsch from the University of Münster for allocating a large literature collection on impact craters to his group in Freiburg.

Wolf Uwe Reimold's impact research benefitted greatly from grants from the National Research Foundation of South Africa (prior to 2006) and from the German Research Foundation (DFG) until 2017. He was also generously supported by the Museum für Naturkunde in Berlin, until that year. Since then he has been supported by the Brazilian National Council for Scientific and Technological Development (CNPq) through fellowship grant 305761/2019-6. This work was financed in part by the Coordenação de Aperfeiçoamento de Pessoal de Nível Superior – Brasil (CAPES) – Finance Code 001. In particular, Uwe Reimold would like to thank the numerous colleagues and friends, with whom he could share decades of exciting impact research.

As the compilation of such an extensive atlas is a rather time-consuming endeavor, the authors also would like to particularly thank their families and partners for their support and abundance of patience. Our thanks also go to the teams of scientists, engineers and managers at the German Aerospace Center (DLR) and at Airbus Defence & Space, who were responsible for the TanDEM-X mission. It was their efforts that made this mission such a huge success and, ultimately, allowed us to achieve this atlas. If the authors have omitted to acknowledge another contribution to this work, they would like to apologize here. To compile such an encyclopedic work is no simple feat; but they obviously assume full responsibility for any such omission.

Last but not least, we, the authors, also contributed to the financial support that permitted printing the atlas at quite reasonable conditions. Thomas Kenkmann is greatly indebted to the University of Freiburg for providing a substantial lump sum, and both Wolf Uwe Reimold and Manfred Gottwald made considerable contributions from their personal funds.

1. Small Bodies in the Solar System

The Beginning

About 4.6 billion years ago the solar system began to form when a dense molecular cloud of interstellar gas and dust collapsed, presumably triggered by a nearby supernova explosion. Driven by gravity, most of the cloud matter condensed in a central core, whilst the rest accumulated in a surrounding disk. Once pressure and temperature were sufficiently high in the central core, hydrogen fusion set in and our home star, the Sun, was born.

The gas and dust disk became the nursery of the solar system planets. Its dimension was much smaller than that of the originating molecular cloud. Therefore, conservation of angular momentum had caused this disk to rotate at high velocities. In the inner part of the protoplanetary disk, close to the forming Sun, temperatures were high, whereas at its edge it remained cold. In the beginning, the composition of the disk matter resembled that of interstellar space but at higher densities than in the molecular cloud. Because a particle wind, orders of magnitudes stronger than today's solar wind, pushed the light elements in the disk outward, a radial gradient developed. Only heavy elements could condense close to the protosun. Dust grains, owing to their increased density and rapid turbulent motion, collided and began to coalesce from sub-micrometer particles to larger and larger particles, then bodies, finally reaching the size of *planetesimals* of diameters of several kilometers.

Planetesimals are considered one of the subsequent steps in planet formation. Well after the formation of the protoplanetary disk, there must have existed countless planetesimals, providing a huge probability for collisions between them. Often in these collisions the two impactors were destroyed – thereby seeding the disk with smaller objects again, sometimes even going through a melting and solidifying stage. However, in the case of suitable collision parameters, both planetesimals could have survived. Eventually multiple collisions may have led to the formation of planetary embryos, which finally evolved by merging into planets.

The radial gradient in the protoplanetary disk resulted in planetesimals and planetary embryos of different compositions. Close to the Sun only elements with high melting points condensed, whereas at larger distances ice formed and dusty ice cores developed. When these bodies had collected sufficient mass, their gravitational pull began to accumulate gas, which at that time was still abundant at distances further out in the protoplanetary disk. It is believed that in a relatively short time one of such solid cores of rock and ice was embedded in a thick atmosphere of hydrogen and helium. This is how the gas-planet Jupiter could have formed as the first large planet. Because this process left Jupiter's environs devoid of gas and dust, other gas and ice planets in the outer solar system, i. e., Saturn, Uranus and Neptune, emerged much more slowly and without reaching Jupiter's enormous mass.

Closer to the young Sun, the protoplanetary disk no longer contained a large amount of volatiles. Rocky planetesimals and planetary embryos were the only progenitors of the terrestrial planets. Consequently, their development took longer. Mass accumulation, collisional heating and energy from radioactive decay may have led to a complete-melting stage in the evolution of these planets, with subsequent differentiation, i. e., segregation of lighter and heavier elements into the outer and inner parts of the proto-planets, resulting in the formation of an iron core.

These ideas are currently our best understanding of the solar system's origin and early evolution of the planets. It also provides the clue why our home planet has – first frequently and then later only occasionally – collided with small bodies in the course of Earth's history.

Artist's view of a protoplanetary disk around a young star (not drawn to scale). Large solid bodies had, at this stage, already formed. Gaps within the concentric rings of gas suggest planet formation (image: ESO/L. Calçada).

Relics of Planetary Formation

Not all planetesimals and planetary embryos ended up in one of the solar system planets. Gravitational perturbations caused by the large planets, in particular Jupiter, hampered some small bodies to coalesce to larger objects. They remained in their infant state and occasional collisions shattered them to even smaller pieces. Today we find most of such rocky remnants of early accretion in the *Asteroid Belt* between the orbits of Mars and Jupiter.

Many icy planetesimals were expelled from the solar system in the early phase of its evolution, again by gravitational interaction with the young gas giants. Today countless such objects form the *Oort Cloud*, a region at a distance of up to one lightyear from the Sun but still under its gravitational influence. The Oort Cloud is believed to be the source of long-period comets that sometimes traverse the inner solar system on their orbital journey. Other, not so distant regions, hosting small bodies with a size of up to a few hundred kilometers, are the *Kuiper Belt* beyond the orbit of Neptune and the so-called *Scattered Disk*, a region even further out and populated with objects, whose orbits may exhibit extreme eccentricities and very long periods. The Kuiper Belt is considered the source of the short-period comets.

In summary, interplanetary space between the large planets is not empty. Instead, still today, 4.5–4.6 billion years after the solar system was formed, it is host to innumerable small bodies. Their sizes range from dust grains to minor planets.

One hypothesis suggests that when the planets were still young, the orbital migration of the gas and ice planets in the outer solar system may have sent asteroids and comets in large numbers into the inner solar system, where they eventually collided with the terrestrial planets. This phase has been long over. But even today, orbit perturbations or occasional collisions can cause small bodies to visit the inner solar system. Their perihelions bring them close to or within the orbits of the terrestrial planets.

Of particular interest are those planetary bodies that could come close to Earth, the so-called *Near Earth Objects (NEOs)*. They are defined as asteroids and comets having a perihelion distance of less than 1.3 times the orbital radius of Earth, i. e., 1.3 AU. In the case of a comet its orbital period must not exceed 200 years. Currently more than 23 000 asteroids are classified as Near Earth Asteroids, supplemented by slightly more than one hundred Near Earth Comets. However, this is only a small fraction of the estimated total number of small bodies entering the inner solar system. Among the NEOs are objects that may approach Earth even further. These *Potentially Hazardous Objects (PHOs)* have a minimum orbit intersection distance with Earth of less than 7.5 Million kilometers and their brightness can be translated into an approximate diameter of at least 140 meters. At the time of writing we know of more than 2000 PHOs. Ongoing sky surveys continuously increase the numbers of NEOs and PHOs. None of the presently identified objects is considered a threat for our home planet.

In the evolution of Earth, however, there have indeed been really hazardous asteroids and comets. They collided with Earth at a rate that seems to have been more or less stable since at least three billion years. Each collision has left an impact structure on the Earth's surface, which remained obvious for a certain period of time. Understanding the processes during and after an impact is the key to understanding today's terrestrial impact crater record.

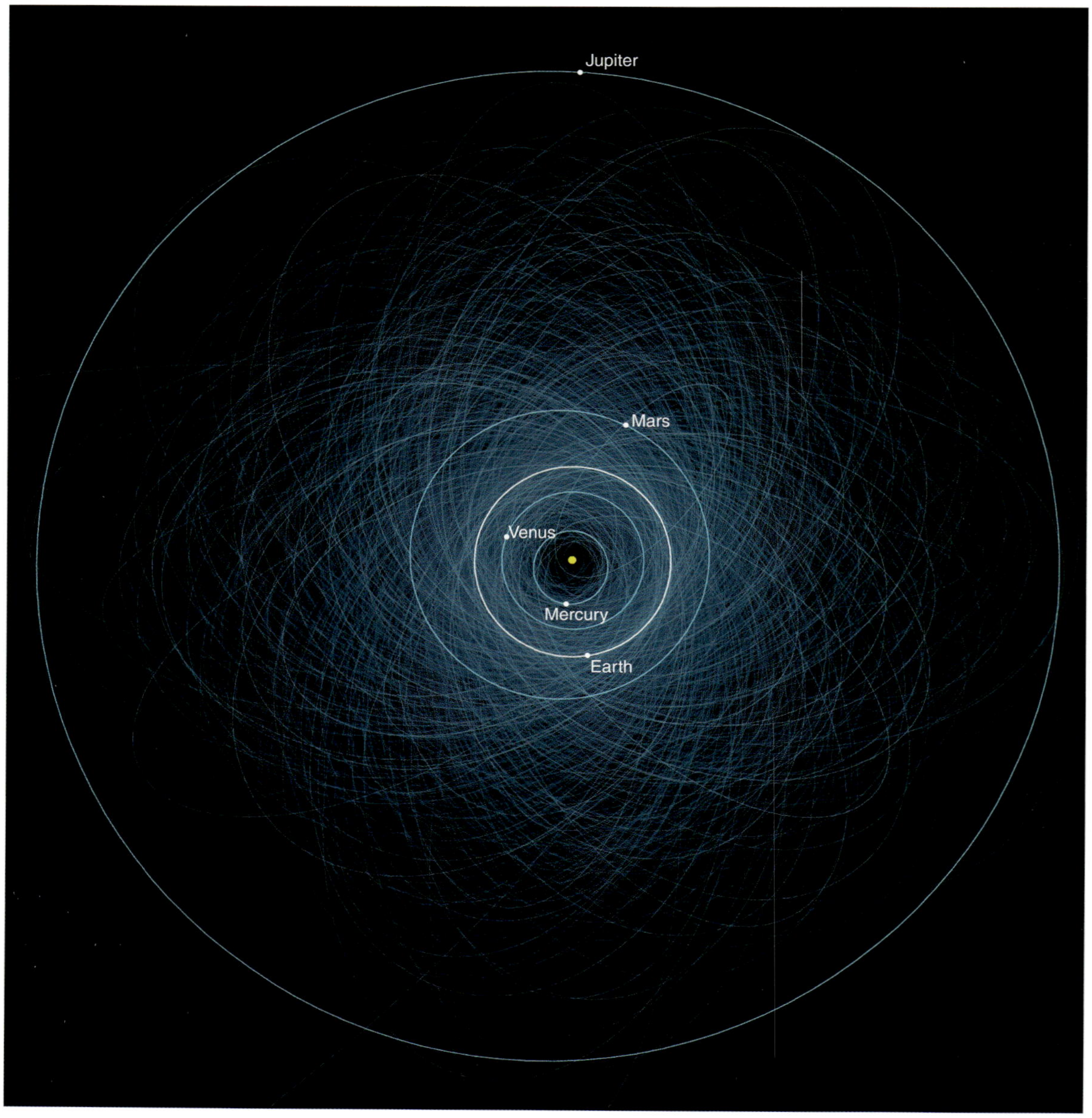

Orbits of the Potentially Hazardous Objects (PHOs) as of early 2013, together with the orbits of the terrestrial planets and Jupiter (image: NASA/JPL-Caltech).

Selected References

Beatty, J. K., Collins Petersen, C., and Chaikin, A. (eds.) (1999). The New Solar System. Sky Publishing Corporation and Cambridge University Press, 4th edition, 421 pp.

Minor Planet Center: http://www.minorplanetcenter.net/

Rothery, D. A., McBride, N., and Gilmour, I. (eds.) (2018). An Introduction to the Solar System. Cambridge University Press, 3rd edition, 440 pp.

2. Impacts

Small bodies orbit in interplanetary space under the Sun's gravitational force. When one of them enters the Earth's atmosphere, they arrive with a speed of at least 11.2 km/s. This speed is equivalent to the velocity that would be required for an object, launched from Earth, to overcome the terrestrial gravitational pull to reach interplanetary space. The average impact velocity with Earth, however, is higher than 11.2 km/s because of orbital velocity differences between the asteroid and the Earth, and amounts to roughly 17 km/s. In reality, the dense atmosphere shields the Earth's surface to some degree. Atmospheric drag sets in already high above the ground and increases towards lower altitudes. Depending on mass, shape and composition, interplanetary intruders might disintegrate or become slowed down to eventually reach the Earth's surface at much lower speed.

At altitudes of more than 100 km, small meteoroids, not much larger than interplanetary dust grains, collide with air molecules. Along their paths they become ionized and excite atmospheric atoms, which causes the well-known meteor flash of light. At the end of its path, however, the interplanetary particle will be stopped and disintegrates. The remaining dust can be traced in the high layers of Earth's atmosphere. The higher and stronger the mass of a particle, the further the projectile may penetrate into the atmosphere. Some reach altitudes of only a few tens of kilometers before they are completely destroyed. Such cases provide the brilliant fireballs, sometimes even accompanied by sound. When a large meteoroid, also referred as a small asteroid, survives down to a height of 10–20 km above ground, it might either stay intact or fragment into smaller pieces, with a wake forming behind. The atmosphere has drastically reduced its interplanetary speed such that the meteoroids, either as an isolated fall or a meteorite shower, reach the surface with terminal velocity of just a few hundred km/h. Besides a mere impact pit, no further morphological structure is formed. In current view, this fate occurs to stony asteroids below a size of about 50 m. Disintegration can occur in a violent explosion, the so-called "airburst". Because of the low altitudes, widespread damage on the ground may result. The Tunguska event of 1908 is the most prominent example for this. More than 2000 km^2 of Siberian boreal forests were destroyed by that event. Another extreme airburst was witnessed on February 15, 2013 above the city of Chelyabinsk in Siberia. Over a thousand people were injured, mostly from flying glass debris, and numerous buildings were damaged. Only under very specific circumstances (shallow entry angle) can a small stony asteroid survive along its impact path with retaining a fraction of its original speed. In such a case, a small crater may result. This was the case for the Carancas impact in Peru, where on September 15, 2007 a meter-sized stony asteroid impacted with a speed of about 350–600 m/s, which produced a 14.2 m wide crater.

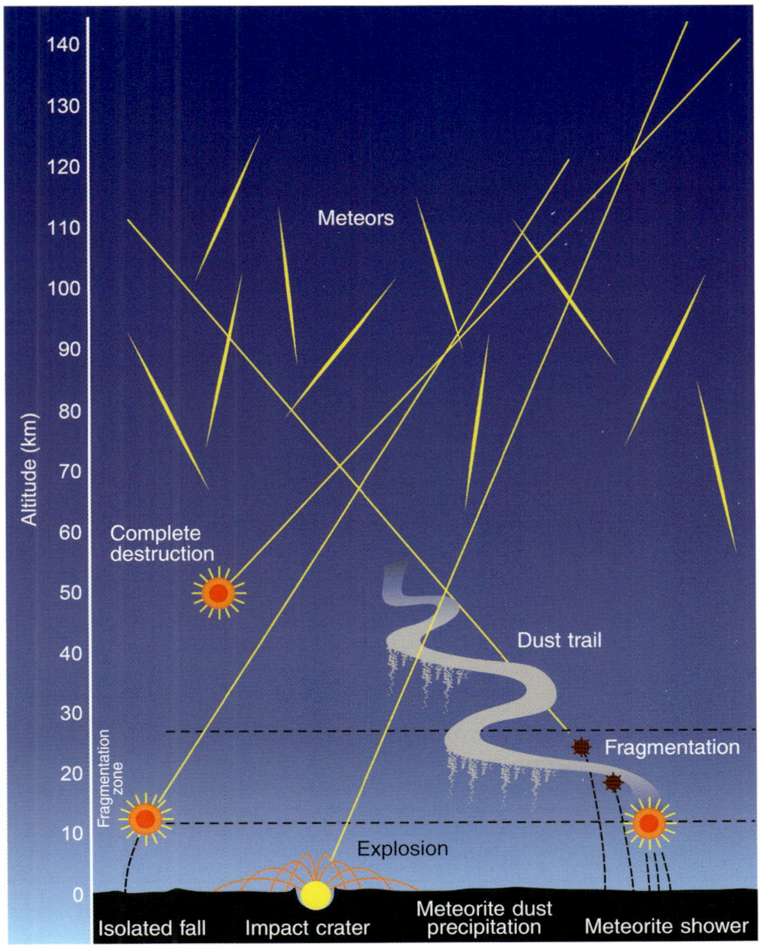

Small bodies colliding with Earth and its atmosphere (modified after Lunar and Planetary Information Bulletin, issue 133, May 2013).

Hypervelocity Impacts

When asteroids of stony composition larger than about 50 m enter a collision course to Earth, the atmosphere cannot appreciably slow them down. This limit is even smaller for projectiles composed of iron-nickel, but they constitute only a very small fraction of the small bodies in the Solar System. These objects are more robust than stony projectiles of similar size, and can withstand disintegration much better. They hit the surface with their incoming, extremely high speed. This so-called "hypervelocity" can range from the minimum 11 km/s to a maximum of more than 70 km/s in the case of an object coming from beyond the Solar System and colliding with Earth head-on. Hypervelocity impacts are in any case a very special phenomenon that has no rival with respect to its ability to generate enormous energy. Such events are essentially instantaneous, and the kinetic energy of these projectiles is concentrated in a small area where they may cause extreme physical conditions at the impact site. These conditions modify the target rocks and minerals in a unique manner – unknown from any other geological events. Fortunately, hypervelocity impacts occur at a very low rate.

The kinetic energy for an impactor of only 50 m diameter and a "moderate" speed of 20 km/s amounts to an equivalent of 11 million tons of TNT. For a bolide with a diameter of 10 km and the same speed, this value increases to almost a tremendous 90 trillion tons of TNT. The energy carried by a projectile is only limited by its mass and its velocity, and not by terrestrial constraints. Impact energy is released within a very short time. At impact, the projectile's kinetic energy is immediately transformed into an intense shock wave that radiates outward from the point of impact. The shock wave causes extremely high pressures to the target rocks, with material acceleration to a few kilometers per second. The affected target rock suffers shattering, brittle deformation, mineral transformations, melting, and vaporization. This all occurs within seconds for a crater with a diameter of 1 km, and even the largest structures known on Earth, about 200–250 km wide, required less than 10–15 minutes to reach their final crater shapes. The release of energy not only occurs instantly but it is also concentrated onto a very small area. There the energy is expended into the target rocks and into the atmosphere, and subsequently can cause environmental effects on regional or even global scales.

The intense shock wave produced upon impact travels through the target rocks and back into the projectile with supersonic speed and provides extreme stresses, much higher than rocks in the Earth's crust usually experience. In a "normal" geological environment rocks either deform slowly or break when the stresses exceed a limit. When impact-deformed, the target rocks cannot respond similarly when a shock wave races through them. Then pressures may reach several hundred GPa beneath the impact site, and further away large volumes of target rocks can still be exposed to several tens of GPa. In addition, high shock pressures induce high temperatures of up to several 1000 °C. Affected rocks immediately melt or even vaporize. For a short while, within a very small volume of the terrestrial surface, pressures and temperatures become comparable to those of the inner core of the Earth. These extreme conditions have severe consequences for the target rocks and their minerals. They are modified in a unique way, unknown from other geological phenomena. This deformation and transformation process is known as **shock metamorphism**, or as **impact metamorphism**. Some of these shock-metamorphic effects are considered distinct and diagnostic proof for the impact origin of a suspected structure or of mineral grains found in crustal deposits. The discovery of meteoritic material at an assumed site may provide a clear hint for the extraterrestrial origin of a structure, if it is possible to link this occurrence with the crater-forming event. However, only for a very small number of impact crater structures fragments of the projectile could be recovered, because the impactor's mass is most likely, and in its majority, either vaporized or melted, or weathering has erased such material from the surface. Shock-metamorphic effects in rocks and minerals persist for a long time. They can survive to tell the impact story millions or even billions of years after a projectile collided with Earth.

Large impacts are very rare events. For what is known as historic time, none have been recorded. The small Carancas crater in Peru and the even smaller but meanwhile vanished Sterlitamak crater in Russia are the only examples where modern impact geology has had access to a contemporary impact structure, even though they were produced by an impactor with a speed lower than hypervelocity. For our civilization the collision with a large asteroid and its consequences is still without experience – although earlier hominids in all likelihood could have witnessed some impact events over the past hundreds of thousands of years. This has clearly made it difficult for many to accept impact cratering as a fundamental process in the evolution of Earth, which has contributed significantly to the shaping of the planet's surface, together with the other, much less violent, geological phenomena.

Impact Crater Formation

Although the formation of an impact crater is a continuous and very rapid process, for convenience, it can be subdivided into several successive stages, each addressing a certain aspect of crater formation: *contact and compression*, *excavation*, and *modification* that eventually leads to the completion of the final crater.

Contact and Compression

When the front of a projectile comes into contact with the target surface, it starts to decelerate and penetrates only a short distance of 1–2 times its diameter into the target – within a mere fraction of a second. The predominant part of the kinetic energy is transferred to the target material through a shock wave radiating outward from the impact point at speeds of possibly higher than 10 km/s – thus exceeding the speed of sound in the affected rocks. While the wave rushes outward, it generates extreme pressures in the rocks. The produced stresses are far higher than what can be usually experienced in a geological environment on Earth. The rocks undergo unique deformations triggered by the high-pressure shock wave. When the shock wave spreads out hemispherically through the target rocks, the

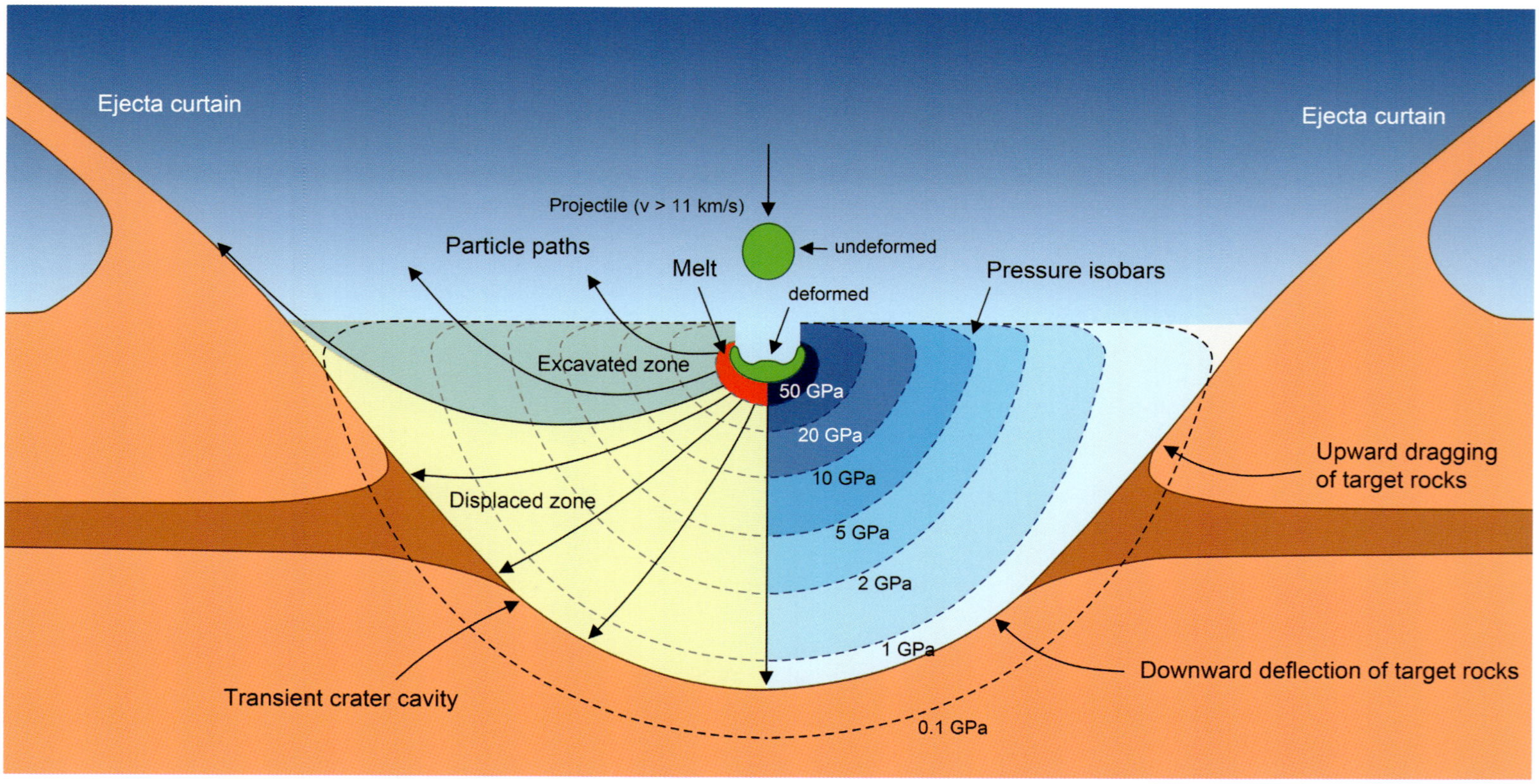

Schematic diagram of the initial impact and subsequent formation of the transient crater cavity during the excavation stage. As a direct consequence of the impact, a shock wave passes through the target rocks. Close to the point of impact, rocks are melted or turned to vapor. With increasing distance, the pressure of the shock wave decays and deforms the rock to different levels of shock metamorphism. The passage of the shock wave sets the target material in motion, which either leads to excavation and ejection of the material, or to displacement and downward deflection of target rocks of the transient crater (modified after Kenkmann et al., 2014).

wave's energy density decreases because the affected area becomes larger with increasing distance from the impact point. Further energy loss is attributed to heating, deformation, and acceleration of target rocks during excavation of the crater. Pressure in the target rock falls off with distance from the impact point. Depending on the transferred energy, peak values of several 100 GPa are found close to where the projectile had hit the ground. Such shock pressures cause vaporization and melting in large volumes of the target material upon pressure release. This also occurs within the projectile, as a shock wave propagates backward into it from the impact point. Thus, it will be transformed, almost entirely, to vapor, with the remainder becoming incorporated into impact-generated melt.

If a kilometer-sized impactor hits the surface, shock pressures can remain at levels of several 10 GPa over a distance of several kilometers from the impact point. Under these conditions, post-shock temperatures are not high enough anymore to cause melting; but target rocks are subject to various shock deformation effects. At distances even further out from the impact point the shock wave has slowed down to the speed of sound in rocks and has attained the character of seismic waves. The associated pressures have dropped to a few GPa, a value where characteristic shock deformation no longer occurs. In this wider region of the target, brecciation, faulting and fracturing are common deformation processes.

Pressures are eventually released, when the shock wave has reached the back-end of the projectile and is reflected back as a rarefaction wave, which travels slightly faster through the compressed target material than the initial shock wave. This causes unloading of the pressures in the rocks, whereby much of the projectile's kinetic energy is transferred into the target as thermal energy.

The end of the compression stage is usually attributed to the moment when the rarefaction wave enters the target rocks and initiates their decompression. Even for very large impact magnitudes, this will be achieved in only a few seconds after contact. For small objects of a few hundred meter size the backward-travelling shock wave and the subsequent forward-travelling rarefaction wave travel through the projectile in a fraction of a second.

Excavation

The excavation stage is defined by the opening of the crater cavity while the shock wave expands hemispherically through the target rocks, thereby becoming weaker with increasing distance from the impact point. A rarefaction wave follows the shock wave. It has its origin at the target surface where the upward travelling shock wave is reflected downward. The combination of shock wave and trailing rarefaction wave sets the target in motion in the form of a symmetrical flow to quickly excavate a volume around the impact point. This excavation flow is directed downward and outward. Target materials from the deeper parts of the affected target volume (displaced zone) are displaced downward into the ground. Only material from upper levels (excavation zone) leaves the cavity. Neglecting atmospheric effects, the debris ejected from the growing crater follows ballistic trajectories. Innermost ejecta are launched first, are ejected to high altitudes, and at highest velocities. Ejecta originating further from the center are launched later, move more slowly, and are deposited in the ejecta blanket nearer to the crater rim. This behavior leads to the formation of an ejecta curtain that has the shape of an inverted cone. The ejecta cone, developed over the growing crater, expands outward, and the ejecta debris from this cone is deposited onto the target surface at increasing distances from the crater.

Above the opening transient cavity, the impact plume, a mixture of vapor with other debris phases, rises upward. Depending on the size of the projectile, the plume can reach beyond the top of the atmosphere.

The excavation flow stops when the remaining kinetic energy is insufficient to displace the target or to overcome the strength of the target material. The resulting, unstable cavity at the end of the excavation stage is called the *transient cavity*, a bowl-shaped depression with an uplifted rim and a depth of about 1/3 of its diameter. This diameter is multiple times larger than the size of the impacting

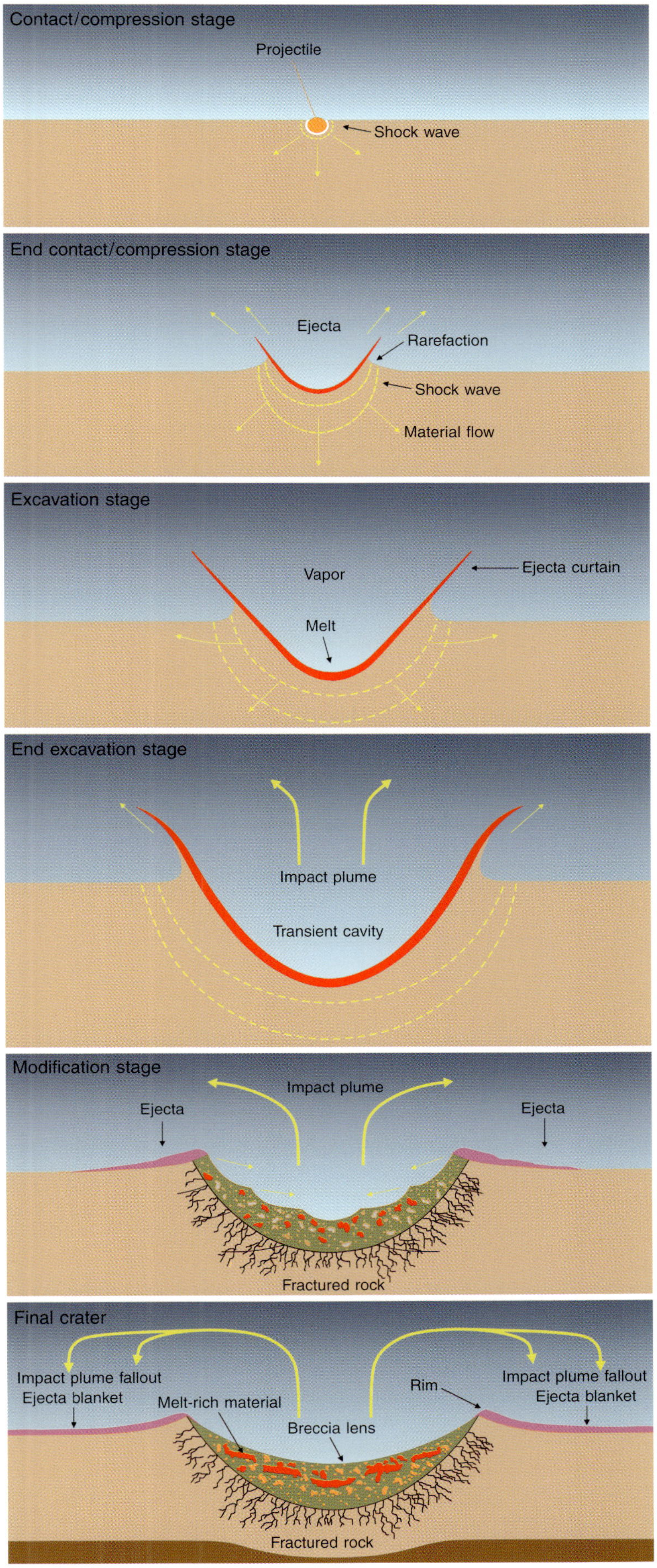

Formation of a simple impact crater: The sequence starts with the contact and compression stage, which is followed by the development of the transient cavity in the excavation stage. The formation of the crater structure is complete once the final crater has been formed after modification of the transient cavity (modified after David Kring, Lunar and Planetary Institute, Houston).

projectile. Velocity and size of the projectile are crucial parameters defining how large the transient cavity can grow. Transient cavity formation takes longer than the initial contact and compression stage but is still very short on geological timescales. It lasts only a few seconds to excavate a crater of about 1 km. The transient cavities of the largest known terrestrial impact structures of some 200–250 km final crater size formed in only 1–2 minutes.

Modification

The modification stage begins when the excavation flow comes to a halt and the transient cavity reaches its largest horizontal extent at the level of the target surface. The shock wave has expanded considerably and only provides low stresses to the target rocks beyond the crater rim. Gravity is now the principal force that drives the collapse of the transient cavity. Depending on the degree of modification, the final crater is classified either as a *simple* or *complex* structure. Simple and complex impact craters have fundamental differences in morphology and structure; both these aspects are a function of size. Simple craters generally have a bowl-shaped morphology, while complex craters have terraced rims and an uplifted central portion of the crater floor. The crater diameter at which the simple-to-complex transition occurs depends mainly on surface gravity. On Earth the largest simple craters were formed in crystalline targets and have diameters of up to 4 km, whereas in sedimentary targets that are weaker in strength, this upper limit is only about 2 km.

Two mechanisms compete after transient cavity formation: An upward directed force (buoyancy) attempting to close the crater versus the strength of the target that resists movements. If the size of the transient cavity exceeds a certain threshold, buoyancy forces exceed the strength of the target and the cavity can completely or partly collapse leading to formation of a complex crater. In the other case, a simple crater of bowl-shaped morphology, similar in shape to the transient cavity, results. Then, modification is mainly restricted to mass movements along the steep transient cavity rim and the deposition of brecciated material that fills the cavity. The visible floor of the crater is underlain by a lens of allochthonous breccias composed of unshocked and shocked target rock. Slumping of the crater rim increases the crater diameter slightly. Together with the infill of the cavity, the final modified simple crater will have a somewhat shallower shape with respect to the initial transient cavity with a depth-to-diameter ratio of about 0.2.

Morphologies of complex impact craters differ significantly from those of their transient cavities. The strong modification is the result of an extensive gravity-driven collapse. The collapse occurs first at the deepest point of the transient cavity with the cavity floor starting to rise upward and inward – pushing the cavity floor upward. The upward and inward flow creates a mass deficit in the subsurface beneath the cavity rim, which ultimately results in a down-sagging of the steep crater walls and causes an enlargement of the cavity's rim. With increasing diameter, the gravitationally-driven growth of the central uplift can progress fast, and the uplift can grow high enough so that it itself becomes gravitationally

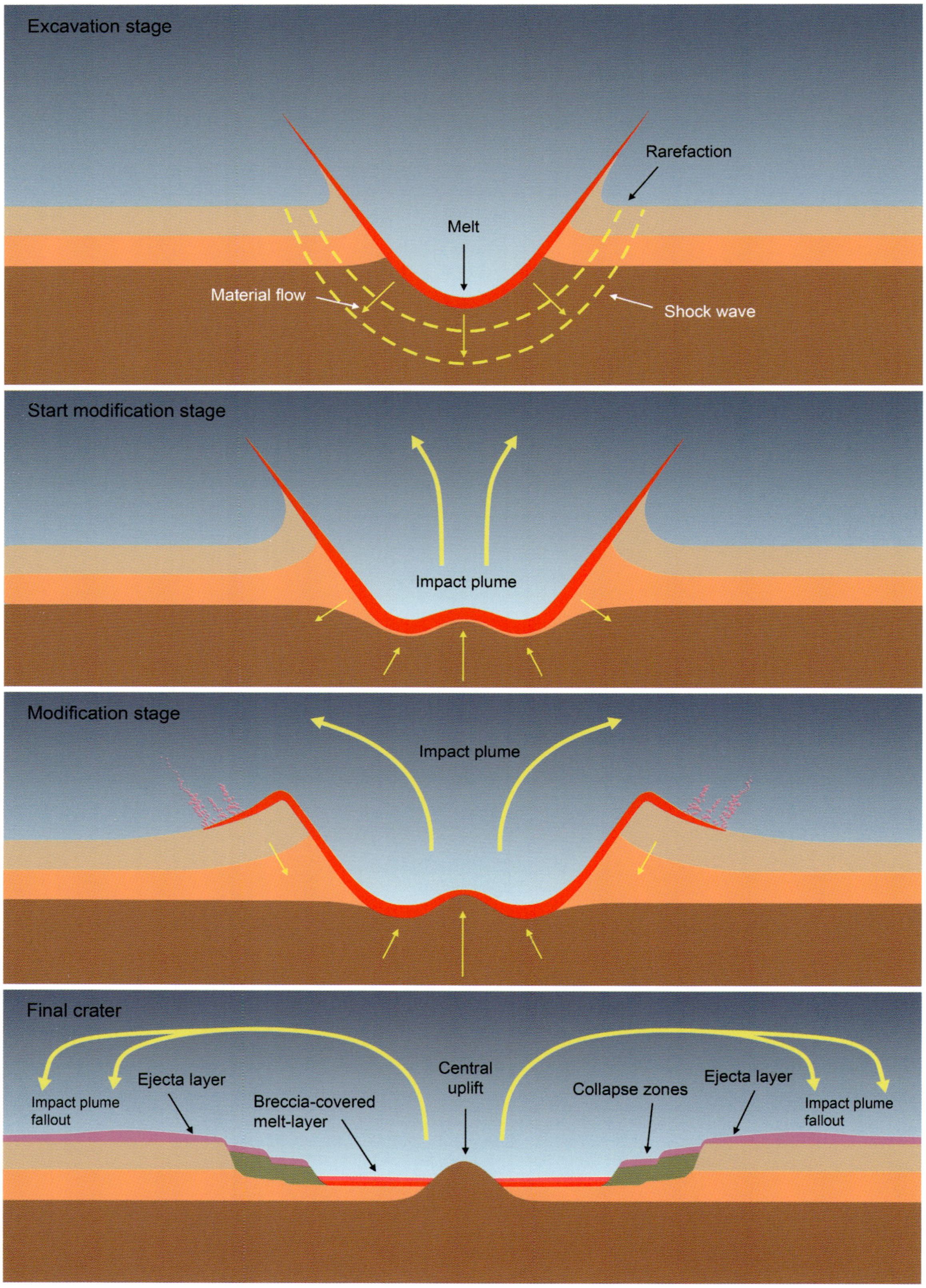

Formation of a large complex impact structure: This sequence starts at the end of the contact and compression stage. At the beginning of the modification stage, the central uplift begins to develop (modified after David Kring, Lunar and Planetary Institute, Houston).

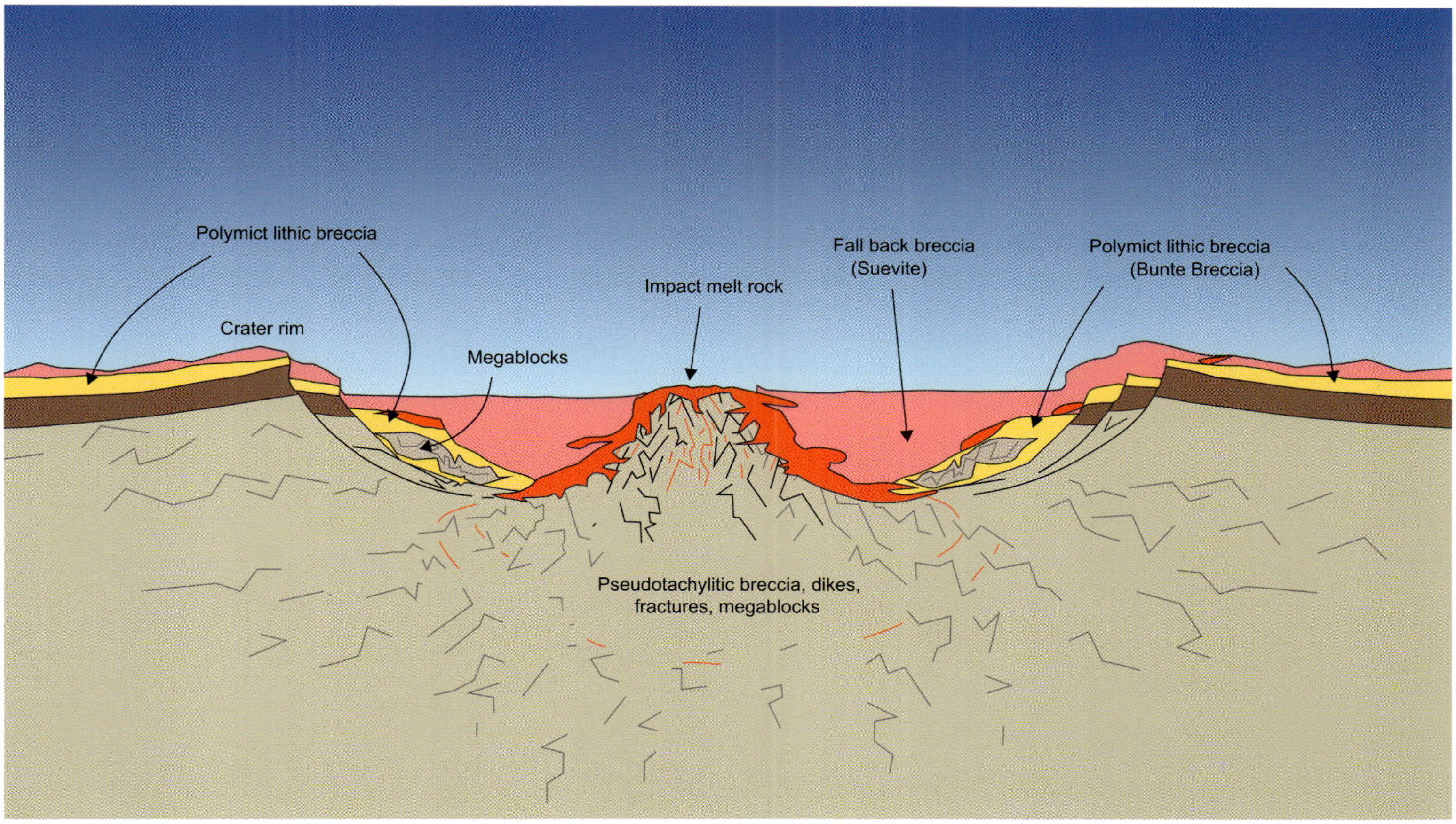

Schematic cross-section through a mid-sized complex impact crater, illustrating the settings of lithologies formed through the impact cratering process. Near-surface or ejected lithologies include lithic breccia that can be either monomict or polymict but only contains clastic components. If a polymict breccia with a clastic matrix contains melt fragments, it is referred to as a "suevite". If the matrix itself is composed of "melt", the breccia is referred as impact melt rock. At the base of the crater, monomict breccias are commonly generated through fracturing, faulting, and the formation of megablocks (modified after Kenkmann et al., 2014).

unstable and collapses under its own weight. The material flow from this process interacts with, and superposes, the material that is simultaneously moving from the cavity rim towards the crater center. The resulting impact structures are called "peak-ring craters". The central uplift or peak ring rocks are often more resistant to weathering than the surrounding crater fill of impact breccias and may, thus, be the only remnant of deeply eroded impact structures. Complex craters have a depth-to-diameter ratio of the order of only 1/10. The most remarkable morphological difference between simple and complex impact craters is that complex impact craters contain an uplifted crater floor of deformed and shocked rocks that either form a central uplift, a patchy distribution of hills and hummocks, a peak ring, or a flat crater floor. The stratigraphic uplift of the central crater floor systematically increases with increasing final crater size. The crater floor that surrounds the central uplift typically appears flat. There the autochthonous crater floor is overlain by allochthonous breccias with materials of various shock degrees and, possibly, incorporating impact melt. The crater rim region of complex impact structures is characterized by terracing, whereby the terraces are separated by steep scarps.

The end of the modification stage is reached when all significant motion of the target has come to an end. The duration of crater modification depends on impact energy, gravity, and strength properties of the target material. Its duration can take up to a few minutes for a 25 km sized impact crater. After the formation process has ended, the resulting final, more or less circular crater structure will be covered by crater-fill deposits and surrounded by an ejecta blanket.

The final impact structure becomes immediately subject to the usual geological processes, such as erosion, sedimentation, or tectonic deformation. This phase is no longer part of the modification stage but represents the post-impact history.

Shock Metamorphic Effects

Impact cratering is a unique geological process because of the extreme conditions caused by the shock wave expanding from the impact point. The energy deposited into the target material can raise its temperature to melting or even vaporization points. Compared to normal geological processes, the pressures induced by the shock wave provide a very sudden and very high stress on the target rocks. This stress is a transient phenomenon occurring on extremely short timescales, which results in strain rates unrivaled by other geological events/processes such as earthquakes, volcanic eruptions, or tectonics. Rocks can withstand shock compression without plastic deformation and internal rearrangement up to a certain limit. Above this limit, which is known as the *Hugoniot elastic limit*, they undergo a *shock-metamorphic* overprint; below this limit, brittle deformation is dominant. The type of shock deformation effects developed in a given target rock sample depends on shock pressure. As this decreases hemispherically from the impact point, a series of different shock metamorphic effects can be observed.

Shock metamorphism sets in at shock pressures of about 2 GPa. At still lower pressures, the target material is only subject to fracturing and cataclasis. In the regime from 2 to 10 GPa, and maybe higher, shatter cones are produced. These are the only macroscopic shock features known to be formed by impact. Sometimes they can become several meters in size. Shatter cones exhibit a conical shape with striations radiating from its top (apex). Microscopic attributes of shock-deformed minerals are planar fractures (PFs) or feather features (FFs) that are generated in the shock domain of 5–10 GPa.

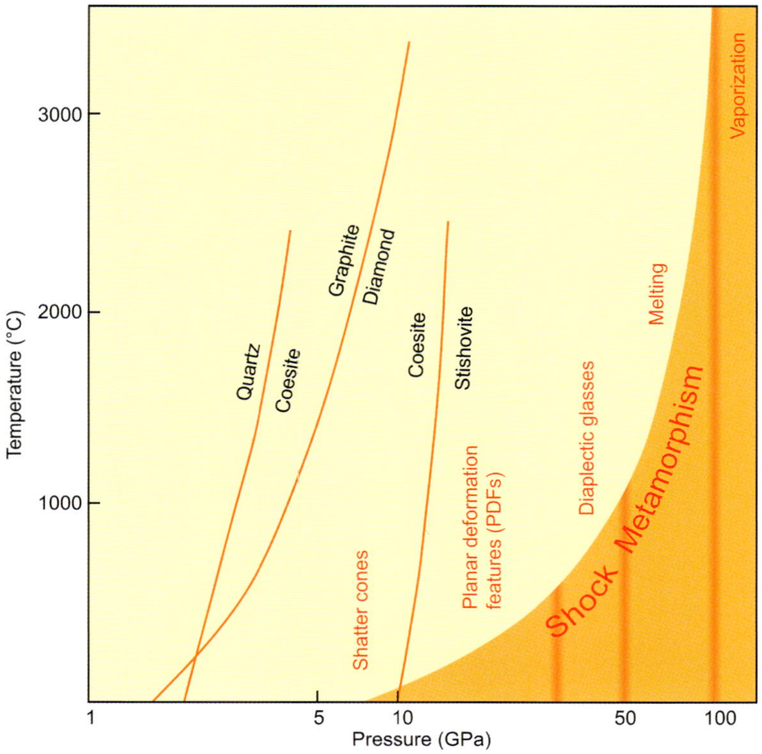

Shock metamorphism illustrated as a function of pressure and temperature. Shock metamorphism occurs at shock pressures above about 7 GPa and extends to pressures exceeding 100 GPa. Post-shock temperatures are given for granitic crystalline rock. Stability curves for the transitions quartz/coesite, graphite/diamond, and coesite/stishovite are illustrated for static equilibrium conditions, and differ from the non-static impact case (modified after French, 1998).

Planar fractures are parallel sets of multiple (sub)planar cracks in a mineral grain, whereas feather features are short microstructures that appear like the fine feather structures on the quill of a bird's feather. The feather features extend variably inclined from a longer planar fracture and indicate a shear displacement on the planar fracture.

When the pressure exceeds about 10 GPa, so-called "planar deformation features (PDFs)" can be formed. Other than planar fractures, PDFs are not open cracks. Instead, they are narrow-spaced (less than 10 µm) and thin (less than 2 µm) amorphous lamellae within the crystal lattice that occur in sets of parallel features. Their orientations are strictly parallel to specific crystallographic directions. In quartz, amongst the most abundant minerals in the upper crust, they develop at shock pressures between about 5–10 and 30–35 GPa. PDFs are arguably the most important shock criterion, as they are distinct from deformation related to all other geological processes. They are also most prominent in the Earth's crust most abundant minerals – quartz and feldspar. And particularly important, they can be readily identified optically – requiring only a polarizing light microscope, together with some experience.

At even higher pressures exceeding 30–35 GPa, for quartz and feldspar minerals, the shock wave deposits even more energy into the mineral grains. The mineral is transformed under solid-state conditions, i.e., without melt production, into an isotropic, quasi-glass-like phase, known as diaplectic glass. This type of glass differs from glass that is produced by melting and subsequent solidification.

Closer to the point of impact, target material is transformed to a liquid melt phase. This sudden melting of considerable amounts of target material is a unique consequence of the impact process, due to the shock wave depositing a fraction of the impactor's kinetic energy as thermal energy into the target rocks. When shock pressures go beyond 40–50 GPa, the associated shock temperatures exceed 1000 °C, and rock-forming minerals start to melt, whereas between 60 and 100 GPa the affected target rocks undergo wholesale melting. Finally, at highest shock pressures above 100 GPa vaporization becomes the fate of the target material.

The mineral zircon, with the chemical formula $ZrSiO_4$, is also important as a shock diagnostic mineral. It is highly stable not only under weathering conditions but, because of its very high melting point, also under metamorphic conditions. Planar deformation bands and twinning can be produced by shock pressures of the order, or in excess of 20 GPa, and the mineral is converted to reidite, a high shock-pressure polymorph at pressures between 20 and 40 GPa. At even higher shock pressures the mineral dissociates into the assemblage silica plus baddeleyite (zirconia, ZrO_2). This conversion requires temperatures in excess of 1700 °C, which can only be produced by impact generated shock in rocks from the upper crust of the Earth.

High-Pressure Polymorphs

Minerals, when exposed to high pressures, can change their crystal structure. Such so-called "high-pressure polymorphs" are produced in nature under static conditions at sufficient depths in the Earth's crust or mantle. Tectonic processes or volcanic eruptions may transport them to the terrestrial surface. The abundant mineral quartz, for example, may be statically transformed into coesite or stishovite, and graphite may become diamond.

Upon impact, the high shock pressures can also trigger the formation of such polymorphs. However, the impact environment differs considerably from static equilibrium. The pressure and temperature field required for the formation of high-pressure phases is transiently reached within a shock wave but it depends on the kinetics of the mineral conversion and the heat attained whether or

A delicate shatter cone formed in Upper Jurassic limestone of the Steinheim impact structure at the periphery of the central uplift. Image width is 30 cm (photo: T. Kenkmann).

Quartz grain of Pentecost Sandstone of the Spider impact crater with parallel Planar Fractures (PFs) from which Feather Feature (FF) lamellae emanate. Note shearing along the PF lamellae offsets the quartz grain boundary (sample: T. Kenkmann).

One of the rare quartz grains in sandstone samples from the Gweni-Fada structure with multiple sets of PDFs. Width of field of view about 560 µm wide. Cross polarized light image (photo: W. U. Reimold).

not the crystals of the high-pressure phase will be preserved upon unloading. Stishovite, for example, was found in quartz containing PDFs indicating shock pressures of 12–15 GPa, and coesite was discovered in diaplectic quartz glass that had experienced shock pressures of more than 30 GPa. The detection of coesite, stishovite and diamond provided the first mineralogical evidence for the impact origin of suspected structures, years before planar deformation features became a well-established criterion.

Impactites

Within and around impact structures, a series of lithologies, formed as a direct consequence of the impact event, occurs. These materials are generally known as *impactites*. They can occur either proximal or distal to an impact structure. The latter comprises ejecta that can be distributed regionally and discontinuously around the impact structure, or on continental and even global scales. In contrast, proximal impactites are found within and just around a crater.

Amongst the proximal impactites, one distinguishes *shocked rocks* and *impact breccias*. The former are rocks that have generally maintained their pre-impact texture and fabric but contain shock metamorphosed minerals. Among the impact breccias, one distinguishes *monomict* breccias and *polymict* breccias. The former comprises breccias from a single lithology of the target material; they are nothing else but shattered target rock, perhaps carrying shocked minerals. Should shock metamorphism not be in evidence, such rocks may resemble tectonically produced cataclasites, and it would be very difficult to recognize the impact origin only with this material. Polymict impact breccias are classified according to their clast content and groundmass composition. Four major types can be distinguished. *Lithic breccias* are composed of clastic components only, i.e., the result of mechanical disintegration. *Suevite* is a distinct rock type comprising mineral and lithic clasts of varied shock metamorphic degree but that also includes a melt fragment component, embedded in a matrix of both melted and unmelted material. *Impact melt rock* is composed of a coherent melt matrix with variable amounts of mineral and lithic clasts of variable shock metamorphic overprint. All these impact breccia types can occur as local formations or as melt injections in the crater floor. In this latter case, they are labelled breccia dikes. They can also be found injected into the crater walls. Besides impact melt, there is another type of melt vein or melt body, also containing clastic components, known as *pseudotachylitic breccia*. Large volumes of this rock occur in the largest known terrestrial impact structures, Vredefort and Sudbury. The origin of this material, which is locally formed in crater floors and central uplifts, is still strongly debated.

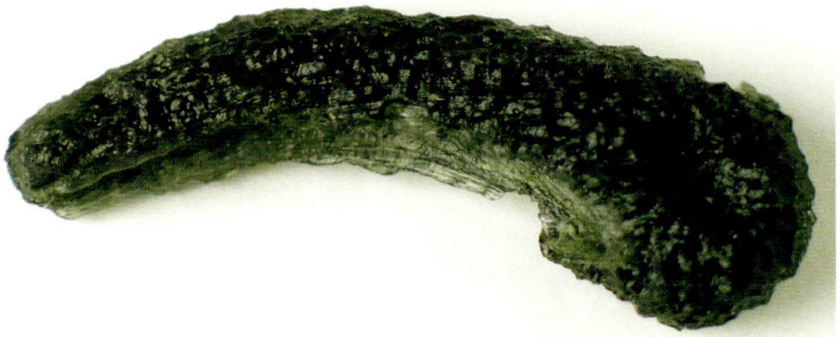

A 2 cm long translucent greenish tektite, known as a Moldavite, which originates from the Ries crater (photo: Ralf-Thomas Schmitt, Museum für Naturkunde, Berlin).

Impact glass differs from impact melt rock. This material is either fully glassy or only semihyaline. *Tektites* and *microtektites*, found in strewn fields with an extent of many hundreds to thousands of kilometers, represent a unique form of impact melted-material. They are the result of the very early ejection of melt from an impact structure. These melt bodies cooled quickly during their atmospheric trajectories to form glass, which then was deposited outside of the source crater. In the case of the Ries impact of southern Germany, the associated tektites are known as moldavites, named after the Vltava (Moldau) river in the Czech Republic where the first samples were found 200 years ago, at a distance of about 300 km from the impact site.

Suevite sample from the Aumühle quarry outside of the northeastern rim of the Ries crater. The sample contains a large, dark grey melt lumb (photo courtesy: Ralf-Thomas Schmitt, Museum für Naturkunde Berlin).

Tracing the Meteoritic Projectile in Impact Breccias and Impact Melt Rock

A further criterion valid for the diagnosis of an impact structure or impact-affected material is a geochemical signature characteristic for a meteoritic component imprinted on impact breccias and impact melt rock. This chemical tracing is possible as the extraterrestrial projectile may have a very different composition from the rock types normally making up the upper crust of the Earth. In particular, the so-called "siderophile" and "highly siderophile" elements are strongly enriched in many types of meteorites in comparison to their abundances in terrestrial crustal rocks. Thus, analysis of the concentrations of elements such as Ni, Co, and Cr in impact breccias – preferably impact melt rock – and all available country rocks that constituted the target may show an enrichment in the impact lithology.

However, there is some ambiguity in this technique. It must be truthfully demonstrated that there are no rocks derived from the Earth's mantle involved in the make-up of the target, as mantle material is also enriched in these elements. Thus, in recent years several other techniques, of both chemical and isotope chemical nature, have been preferred. First, there is the analysis of concentrations of the platinum group elements, such as Ir, Ru, Re, or Os, to check for possible enrichment. In comparison with the analytical databases for different types of meteorites, the interelement ratios of such elements may then reveal to what kind of meteoritic class a projectile likely belonged. Isotope ratios between meteorite types and terrestrial material may also be very different, so that the Cr isotope and the Re-Os isotope method have been successfully applied to the detection and/or classification of traces of projectiles. Chromium isotopes are not as sensitive for the detection of meteoritic contamination, but they have the advantage that this method can be used to distinguish between carbonaceous and ordinary chondrite projectiles. The Re-Os technique is more sensitive and may trace projectile contributions to melt rock as low as 0.2 %.

Selected References

French, B. M. (1998). Traces of Catastrophe: A Handbook of Shock-Metamorphic Effects in Terrestrial Meteorite Impact Structures. LPI Contribution No. 954, Lunar and Planetary Institute, Houston. 120 pp.

Kenkmann, T., Poelchau, M. H., and Wulf, G. (2014). Structural geology of impact craters. Journal of Structural Geology, Vol. 62, pp. 156–182.

Koeberl, C. (2014). The geochemistry and cosmochemistry of impacts. In: Planets, Asteriods, Comets and The Solar System, Davies, A. M. (ed.), 2nd edition, Treatise on Geochemistry, Elsevier, Amsterdam, Vol. 2, pp. 73–118.

Reimold, W. U., and Koeberl, C. (2014). Impact structures in Africa: A review. Journal of African Earth Sciences, Vol. 93, pp. 57–175.

3. Terrestrial Impact Structures

Earth's Impact Crater Record

Our Solar System neighbors with solid surfaces – planets, moons, and asteroids – are littered with impact craters. Mars exhibits more than 300 000 impact structures with diameters of more than 1 km, the Moon's highlands are densely covered with innumerable impact structures, and for Mercury, the innermost planet, crater counts have yielded an impact density similar to that of the Moon. All three bodies are characterized by surfaces with ages of several billion years, although Mars had occasional volcanic activity and exogenic processes such as wind erosion and accumulation or glacial drift that renewed parts of its surface, but at a slow pace. Impact structures are well preserved in those environments. Earth is different. Its surface has been, and is still, undergoing permanent modification. Weathering, erosion, tectonic activity, volcanic resurfacing, plate subduction, and sedimentation contribute to surface renewal. The oceanic basins account for two thirds of the terrestrial crust, with seafloor generally significantly younger than 200 Myr. Thus, Earth's face is a mosaic of geological provinces of different ages with a large fraction being less than 200 Myr old. Only some limited portions of Earth's crust date back to Archean or early Proterozoic times. This terrestrial "youth" is the reason why so many impact structures of old age have vanished – erased by the "normal" geological processes over millions and billions of years.

Today, more than a century after the impact of an extraterrestrial projectile was first proposed for Meteor Crater, more than 200 geological structures of confirmed impact origin are known. Their impact evidence is based on the established criteria in impact geology – the proof of unambiguous shock-induced deformation in rock, which demonstrates that the associated structure was formed by an impact. These criteria are (1) shock-metamorphic evidence such as macroscopic shatter cones or microscopic shock deformation features, and (2) relics of the meteoritic projectile or chemical traces of it left in impact melt rock. The proof of at least one of these criteria requires proper field study or core drilling in case of buried structures. Therefore, impact structure identification always requires detailed "ground-truthing".

On average, about 2–3 new confirmed impact structures can be expected to be added to Earth's impact record per year. Either field studies finally found the "smoking gun" in shock-metamorphosed rock, after a candidate had been considered a potential impact for a long time, or a previously unknown structure can be added to the list when impact-affected rock was discovered accidentally. How many more impact craters can we expect to find? Studies taking erosion rates into account have shown that for diameters larger than 6 km, the record of exposed terrestrial impacts is rather complete. New discoveries in that range are still likely, however, simply because of statistical variations and uncertainty. For diameters of less than 6 km a pronounced deficiency has been found for the terrestrial impact crater database. Estimates yield about 350 craters that can still be expected in the range from 0.25 to 6 km, with 90 of these to have diameters of more than 1 km. Adding the already confirmed impact structures and considering statistical aspects, the Earth's impact structure record

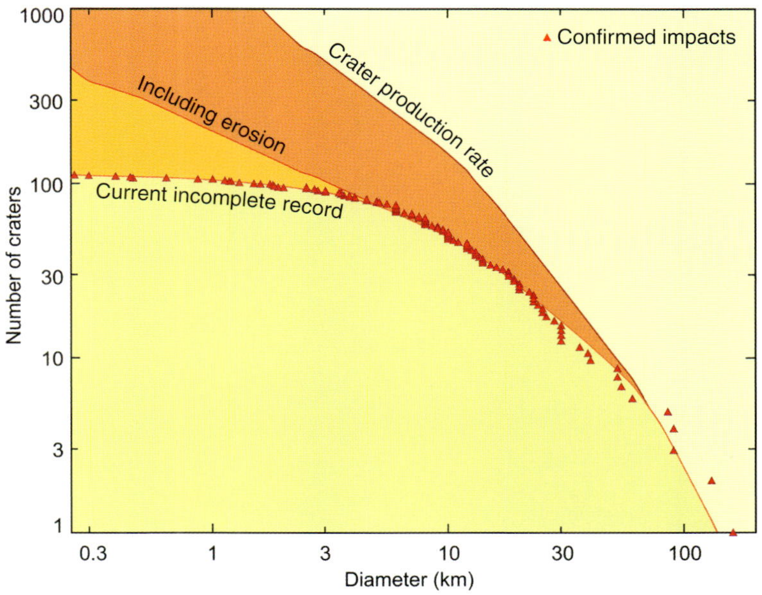

Cumulative crater-size distributions. The confirmed impact structures used for these computations are only the exposed crater structures listed in the Earth Impact Database. The crater production rate was computed without consideration of erosion. It was rescaled under the assumption that the inventory of impacts larger than 85 km is complete (modified after Hergarten and Kenkmann, 2015).

may actually count 550–600 structures, much more than the current limited record but still only a fraction of what is known from other Solar System bodies with, on average, much older surfaces.

Various impact databases are maintained by science institutions or laypersons, and are made available via websites. They differ quite widely in content. Some only comprise confirmed impact structures, others mix confirmed and proposed structures, and there are even some sites that are populated without applying any established criteria of impact geology. Because the process of impact cratering has meanwhile received considerable public attention, this situation can become rather confusing for the interested and abundantly lay readership.

A widely appreciated and accepted database has been the **Earth Impact Database (EID)**. It has slowly grown out of a Canadian effort to identify impact structures on the Canadian Shield. Since 2001 it has been hosted and maintained by the Planetary and Space Science Center (PASSC) of the University of New Brunswick, Canada. The EID only lists impact structures – 190 entries at the time of writing – for which it is claimed that the impact origin has been confirmed by the stated diagnostic criteria. However, even this careful approach does not prevent the EID from containing occasional ambiguous or even erroneous entries. Careful research indicates that for several EID entries the final proof of shock-metamorphic effects is indeed still lacking.

The Actual Impact Structure Record

Our atlas uses the EID as a starting point but adds all new impact structures which have been claimed *"confirmed based on the established criteria of shock metamorphism"* in recent years. In addition, a relatively unknown meteorite fall producing a small crater in Russia has also been considered. This adds another 18 impact structures bringing the total record to 208. Because for a small number of these a final, unambiguous evidence for a shock overprint in the affected rock still has to be presented, we consider two types of "samples" in this atlas: One where the quality of the studied shock features justifies to label them unequivocally *"confirmed"*. This is by far the majority. The second sample includes all those structures where confirmation, even though alleged, can be doubted or has been achieved only

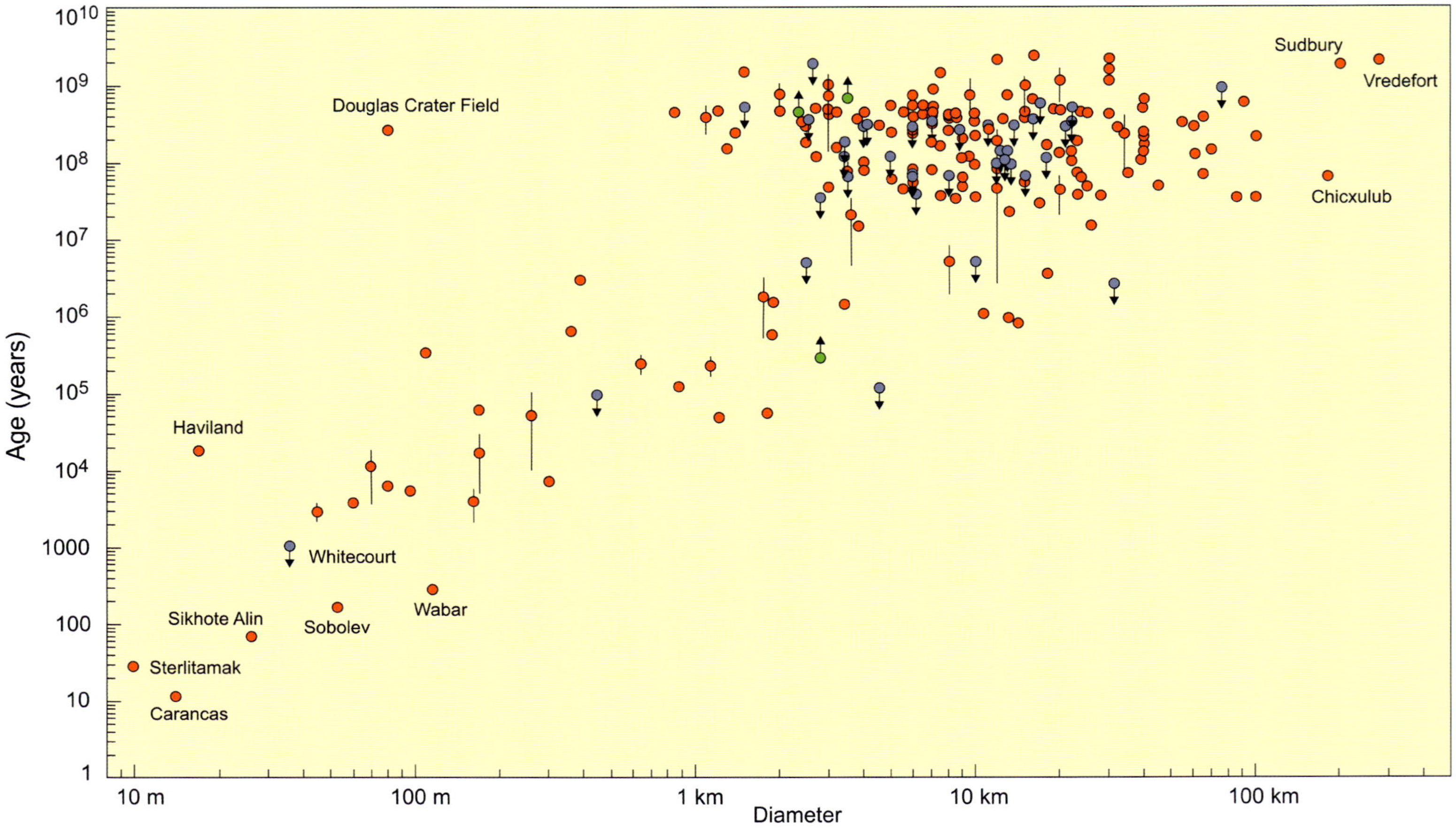

The relation between diameters and ages of those impact structures with firm confirmation from this atlas, with errors and upper age (blue dots) or lower age (green dots) limits, where applicable.

very recently such that additional studies are recommended. This may be due to sometimes decade-long studies without finding shock effects but where an origin other than an impact seems beyond imagination. Other structures are lacking elaborate field studies or detailed publications with proper documentation of evidence. And even very recent findings may claim the proof of shock overprint but the circumstances of their discovery would warrant a second, much closer look. All structures requiring still an unambiguous confirmation are not listed in alphabetical order in each continent chapter but added to its end – clearly separating them from the "safe" entries. The questionable/debated ones include Bloody Creek (Nova Scotia, Canada), Colônia Basin and Nova Colinas (both Brazil), Connolly Basin (Western Australia), Crawford and Flaxman (South Australia), Hiawatha (Greenland), Kara-Kul (Tajikistan), Yilan (China), Mount Toondina (South Australia), Ouarkziz (Algeria), Pantasma (Nicaragua), Piccaninny (Western Australia), and the Rio Cuarto Crater Field (Argentina).

Nine entries refer to impact crater strewn fields. These are Ilumetsä, Kaalijärv and Morasko in Europe, Campo del Cielo and Rio Cuarto in South America, Sikhote Alin in Asia, Henbury in Australia, and Douglas and Odessa in North America. The Campo del Cielo field extends over an area of 18.5×3 km, the debated Rio Cuarto structures occur over at least 50×15 km, and the "fossilized" small craters of Douglas cover a region with a minimum size of 13×7 km in southeast-northwest direction. In all other strewn fields the small craters crouch closely around the main crater. At Sikhote Alin more than 120 very small crater-like pits have been identified.

All atlas entries belong to one of the following categories: (1) Impact structure at surface and topographically visible, (2) surface topography completely erased but exposures still present, (3) impact structures buried underground but identified by geophysical methods and accessed via drilling, and (4) impacts submerged. In all four cases it is ensured that the position of the impact structure is known, even when no longer visible in the topography at the terrestrial surface.

It is known, however, that other impacts must have occurred in recent geological times where the search for the crater imprint has still been unsuccessful. The tracers for such impacts can be tektite strewn fields, ejecta blankets, or a local increase of the amount of meteoritic material at a certain site. Well-known cases of that kind are:

- The Australasian tektite strewn field covers about 30 % of the Earth's surface. The tektites are about 0.8 Myr old but no source crater, which must be of considerable size larger than 10 km, has been discovered so far. A favored location is Southeast Asia.
- The Cambrian Vakkejokk Breccia in Sweden, the Alamo Breccia in Nevada, and Stak Fada in Scotland are impact ejecta in which shock overprints in quartz grains have been found. Positions for the associated impact craters have been proposed but no buried structures have been unveiled yet.
- Eltanin in the Bellinghausen Sea, about 1500 km off the coast of Chile, is the only known impact in a deep ocean basin. It is only slightly older than 2 Myr. Its discovery is based on a strong iridium anomaly in sediment drill cores; later, a high concentration of meteoritic ejecta was found. No crater structure exists.

None of those impacts without a defined crater location has been included in this atlas.

Impact Structure Parameterization

The most obvious parameters characterizing an impact structure are "diameter", "position", and "age". For well-preserved simple craters, the diameters are those that can be directly measured morphologically. When the structure has already been heavily eroded or is even buried, the apparent, present-day, crater diameter can be determined, from which the original diameter can be inferred if the amount of erosion can be quantified. In those cases circumferential crater rim faults or the diameter of geophysical disturbance are used to determine the crater size.

The smallest listed single crater structures are Sterlitamak in Russia with 9.4 m, which does no longer exist, and Carancas in Peru with a diameter of 14.2 m. On the opposite side of the scale resides Vredefort in South Africa, as the largest confirmed impact structure on Earth with an original diameter of 250–300 km.

When the impact structure shows an imprint in topography, preferentially with a circular morphology, the position can be precisely determined up to an accuracy of *seconds* in geographic latitude and longitude. The exact position of buried craters can be a matter of debate. This applies particularly to those located in the former Soviet Union, where publications, often in Russian, only provide vague or even contradicting locations. For all atlas entries where the position could not be directly extracted from the TanDEM-X maps, we used locations from the literature.

The age of impact structures relies on stratigraphic and paleontological data or radiometric dating. Methods applied may yield a relative, or in the case of radiometric dating, an absolute time value. Notably when only stratigraphy or paleontology can be used to answer the question when an impact had occurred, uncertainties can be quite considerable or only upper or lower limits can be given. The youngest impact event occurred in 2007 in Peru forming the Carancas meteorite impact crater, whilst at the opposite end of the age scale several structures date back to Paleoproterozoic times, including Vredefort in Africa with an age of 2023 ± 4 Myr and Yarrabubba in Australia, which has recently been dated to 2246 ± 17 Myr. Some of the ages currently referred in the literature can be seriously questioned, as the evidence at hand does not suffice to evaluate whether the given age really relates to the impact age – or, for example, relates to the age of a target rock.

The lifetime of a crater, that is the period until a crater is degraded to a level where no confirmation is possible anymore, depends on its size. Therefore a lot of small craters are amongst the youngest in the record. Once formed, geologic processes erase them quickly. The larger the structure, the higher the probability that the crater has remained preserved, even long after the impact. One exception is the Douglas impact crater strewn field, which consists of dozens of small craters only 16–100 m in size but which dates back to the Permian. These craterlings still exist because the impacts occurred in an environment that preserved them from destruction by rapid burial.

A fourth parameter used in the atlas refers to the topographic visibility of an impact structure. It defines whether a particular impact has left an imprint in the local topography, either with a clear crater morphology (simple or complex) or with what is left after eon-long erosion and tectonic activity but which is still recognizable as an anomalous feature in the area. When the topography of non-buried craters has been erased beyond recognition, the particular impact structure may still be recognizable at the surface. This occurs through a variety of impact lithologies which is still preserved and exposed at the surface.

A large number of impact structures is buried underground. They can be revealed by exploiting geophysical methods, such as, e.g., gravity or magnetic surveys. When drilled, rocks from the extracted cores may finally show the effects of shock metamorphism, or a central stratigraphic uplift may be revealed through comparison of the stratigraphic column in the central part of the structure and the sequence of undisturbed strata in its environs. Over a small number of buried impact structures the local surface does provide a hint at what is hidden underneath. This can either be a subdued depression, as in the case of Holleford in Canada, or a series of sinkholes (cenotes) arranged in a ring reflecting the location of Chicxulub in Mexico.

Impact craters can also be shielded from view because they are submerged. Three impact structures lie on the continental shelf. Two of these, Mjølnir and Montagnais, are entirely covered by the Barents Sea and the Atlantic Ocean, respectively. The third, Chesapeake Bay straddles the Delmarva Peninsula but is mainly submerged in shallow sea as well. In Scandinavia several structures are "drowned" by the Baltic Sea, and still others at northern latitudes are covered by, often eponymous, lakes.

	Area (10^6 km^2)	Number of impact structures	Density (impact/10^6 km^2)
Africa	30.37	21 (20)	0.69 (0.66)
Asia	44.58	23 (21)	0.52 (0.47)
Australia	7.69	31 (26)	4.03 (3.38)
Europe	10.18	54	5.30
North/Central America	24.71	65 (62)	2.63 (2.51)
South America	17.84	14 (11)	0.78 (0.62)

Terrestrial impact structures (figures in brackets refer to impact structures, for which firm confirmation has been obtained).

The majority of confirmed impact structures is found in North America, Europe, and Australia. Comparatively far less have been identified in Asia, Africa, and South America. One reason for this is that on continents with high impact numbers, countries have pursued vivid space research and geological exploration already in the past. There, the search for impact structures dates back to the middle of the 20th century. In addition, distinguished individuals have promoted this work and were actively involved. This is particularly true for Australia where Eugene M. Shoemaker and his wife Carolyn contributed considerably to the impressive Australian impact record. Contrary to this, in large regions with an obvious impact structure deficiency, studies have been hindered by vast areas of rainforest or a large number of countries with enduring civil strife.

However, all field work would have been in vain if the geological histories of individual regions had not supported the preservation of impact structures. The accumulated impact inventory on old, stable geological provinces certainly differs from areas which are dominated by young crust. Nonetheless, when future campaigns will slowly unveil the still missing part of the terrestrial impact crater record, we expect that continents which are presently underrepresented will receive their share. With Greenland and Antarctica, the surfaces of two large landmasses are hidden from our view by massive ice shields. Their total area of about 16 million km^2 may host another significant number of impact structures. How many of these will be discovered in future has to be seen. There is growing interest in the status of the ice sheets of Greenland and Antarctica in the context of global warming studies. The increasing cryospheric research exploiting sophisticated remote sensing methods may find possible impact candidates presently hidden under thick layers of ice. In such cases, it will be the task of laborious field work to identify the shock metamorphic effects in *in situ* rock, which will then permit an unambiguous confirmation of the impact origin.

Selected References

Earth Impact Database: http://www.passc.net/EarthImpactDatabase/New website_05-2018/Index.html

French, B. M. (1998). Traces of Catastrophe: A Handbook of Shock-Metamorphic Effects in Terrestrial Meteorite Impact Structures. LPI Contribution No. 954, Lunar and Planetary Institute, Houston, 120 pp. Available at http://www.lpi.usra.edu/publications/books/CB-954/CB-954.pdf

Hergarten, S., and Kenkmann, T. (2015). The number of impact craters on Earth: Any room for further discoveries? Earth and Planetary Science Letters, Vol. 425, pp. 187–192.

Schmieder, M., and Kring, D. A. (2020). Earth's impact events through geologic time: A list of recommended ages for terrestrial impact structures and deposits. Astrobiology, Vol. 20, pp. 1–51, DOI: 10.1089/ast.2019.2085.

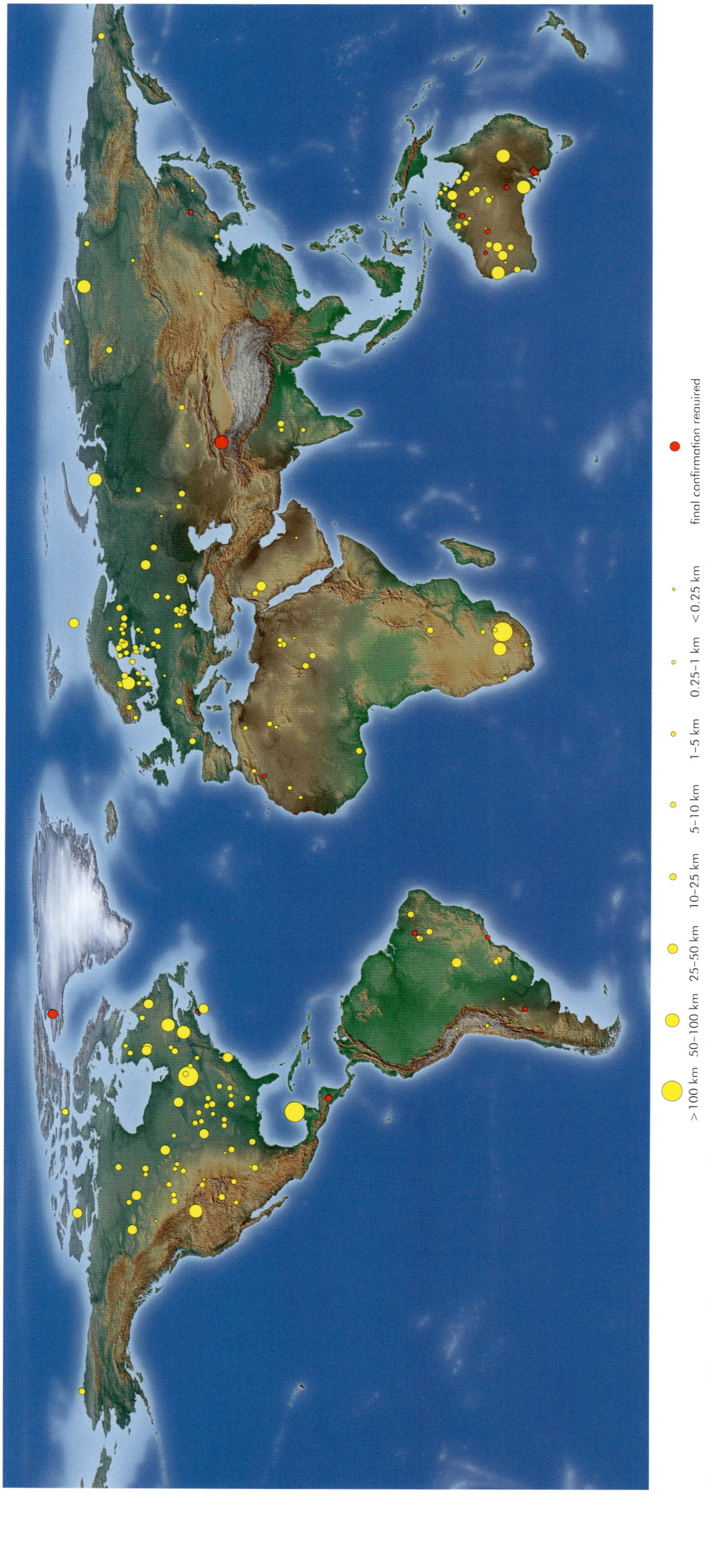

The inventory of terrestrial impact structures. The size of a symbol is a measure for the structure's diameter. Red dots indicate those impact structures where an unambiguous confirmation is still required. Made with Natural Earth. Free vector and raster map data @ naturalearthdata.com.

4. Radar Remote Sensing

Interplanetary space exploration over the past 50 years has shown that impact craters are abundant on all large and small solid bodies throughout the Solar System. Orbiting platforms or spacecraft on fly-by missions, equipped with remote sensing instruments, are the only means for gathering information about impacts on extraterrestrial solid bodies on a large scale. Earth is different. Here, candidate impact craters can be studied in situ. However, even the understanding of our home planet can benefit vastly from remote-sensing studies.

The view from space-borne sensors allows access to all environments – from the tropical rainforest to arid deserts, or to the cold polar regions. This opens up new opportunities. For impact structures, where field work is hampered by unfavorable conditions, satellite data from low-Earth orbit can provide first geological context. Large structures, often considerably eroded, may unveil their morphologies from an outpost several hundred kilometers above the terrestrial surface, whilst it may be tricky, or even impossible, to identify such features from the ground. Remotely sensed data may be useful in the search for new impact crater candidates. In addition, the geological analysis of impact structures or impact structure candidates can be supported by high-resolution remote-sensing imagery.

Remote sensing instruments aboard satellite platforms exploit regions of the electromagnetic spectrum. They can measure scattered and reflected sunlight or radiation emitted by Earth, or another target body, as a consequence of its thermal status. Such instruments are termed *passive*. Another type of sensors generates electromagnetic radiation itself, transmits it down to Earth, and receives the backscattered signal. These are *active* instruments.

The narrow visible part of the electromagnetic spectrum and the region from microwaves to radiowaves can penetrate the atmosphere. With the presence of small particles – dust, water droplets – light between 400 and 700 nanometers also becomes blocked. Only a cloud-free, clean atmosphere provides unobscured views. Mapping a whole surface with a remote sensing instrument working in the visible spectrum is possible for bodies with only a thin or without an atmosphere. Therefore our solar system neighbors – Moon, Mars, Mercury and others – can easily be studied by such techniques. For Earth, however, an extensive charting task is always challenged by the presence of a cloudy atmosphere.

By using microwaves, particularly those from the radar (radio detection and ranging) range, such difficulties can be avoided. The property to penetrate clouds makes the atmosphere transparent for this type of radiation. Commonly used wavelengths comprise the X-band (8–12.5 GHz, $\lambda = 2.5–4$ cm), C-band (3.95–5.8 GHz, $\lambda = 4–8$ cm) and L-band (1–2.6 GHz, $\lambda = 15–30$ cm).

Synthetic Aperture Radar

When imaging Earth's surface from orbit with an optical device, the height of the satellite together with the properties of the sensor's optics and focal plane detector determine what detail can be resolved. Today's cameras are far advanced, providing meanwhile high-resolution views in the sub-meter domain. This does not apply to an active radar instrument. The microwave radiation transmitted downwards by a typical radar antenna of a few meter length from an altitude of several 100 kilometers illuminates on ground a beam several kilometers wide, equivalent to the achievable maximum resolution in such a real aperture configuration – far less than what is required in high-resolution imaging.

By exploiting Synthetic Aperture Radar (SAR) this shortcoming can be avoided. In SAR configuration the radar antenna aboard the satellite platform moves at an altitude of a few hundred kilometers with the orbital velocity of slightly more than 7 km/s, and transmits radar pulses downward. They illuminate a footprint sideways of the satellite ground track. The pulses have known frequency and phase relations. Therefore, the radar waves are referred to as *coherent*. All objects lying in the footprint are sequentially illuminated as the satellite advances in orbit – thus defining a continuous swath. Because of the orbital motion of the antenna, an object enters the footprint on its leading side, passes through its middle, and leaves the footprint at its trailing edge. For this entire time period, the radar antenna receives the backscattered pulses from the object. As the antenna first moves towards, and then off, the object, the frequencies of the backscattered pulses are shifted, caused by the *Doppler effect*. The backscattered radar raw data cannot be visualized; they do not exhibit any recognizable image content. However, evaluation of the *Doppler history* permits to derive a high-resolution image of the scene. The achievable spatial resolution is defined by the distance the antenna has moved in orbit from footprint ingress to egress of the object. This distance is equivalent to the virtual (synthetic) aperture of the sensing radar antenna, which lies in the order of several kilometers. An antenna of this length allows to obtain resolutions of meters or less.

How the transmitted microwaves are backscattered depends on the roughness and local slopes of the sensed surface, together with

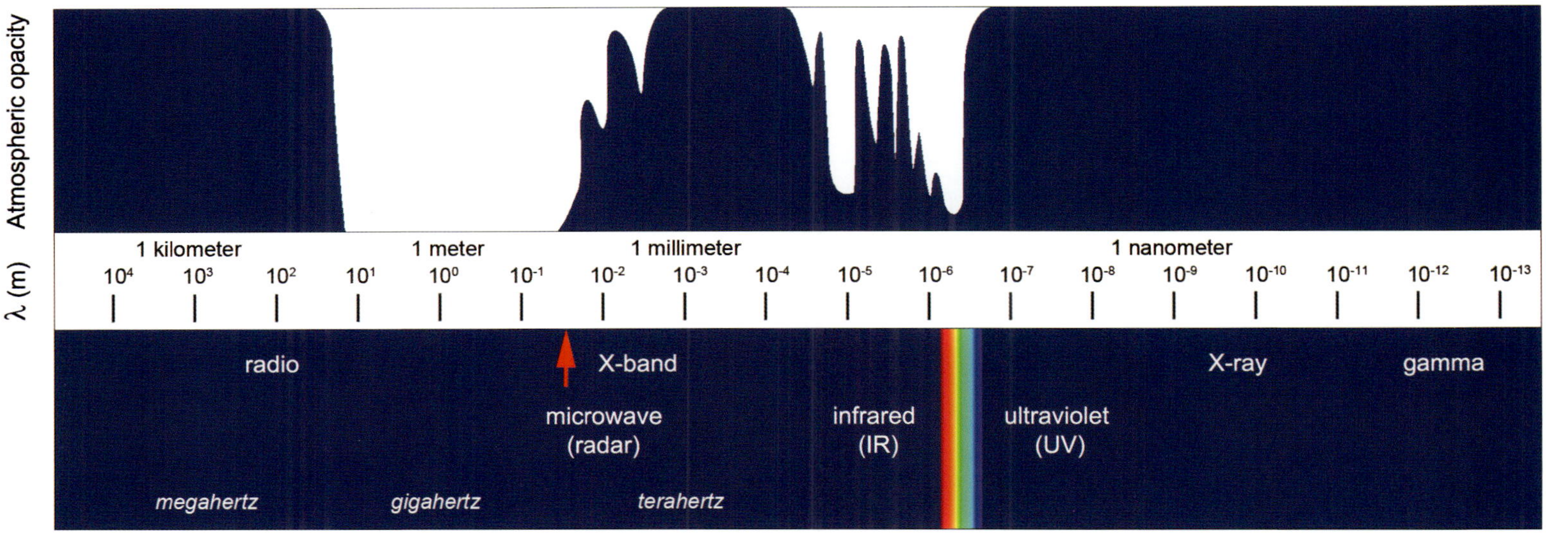

The electromagnetic spectrum from gamma rays to radio waves, with the visible light part indicated between the ultraviolet and infrared sections.

the electric properties of the illuminated matter. A smooth surface tends to reflect away from the instrument passing overhead in-orbit – thus resulting in low radar signals. The opposite is found for rough ground where considerable backscatter towards the satellite is generated. Areas facing the radar antenna return more backscattered radiation than those pointing away or even lying within the radar shadow. High water content in material at solid surfaces favors increased backscatter, while dry soil causes reduced radar returns. All these effects contribute to the radar image which finally is composed of a wealth of grey levels. These values represent the strength of the radar signal as received from the individual ground pixels. The imaged scene depicts the layers where backscatter occurs. This is not always "bare" Earth. Vegetation, particularly dense tree canopy, acts as an obstacle for X-band microwaves, whilst L-band pulses are not blocked by leaves and penetrate down to the ground, with the C-band falling in between these cases.

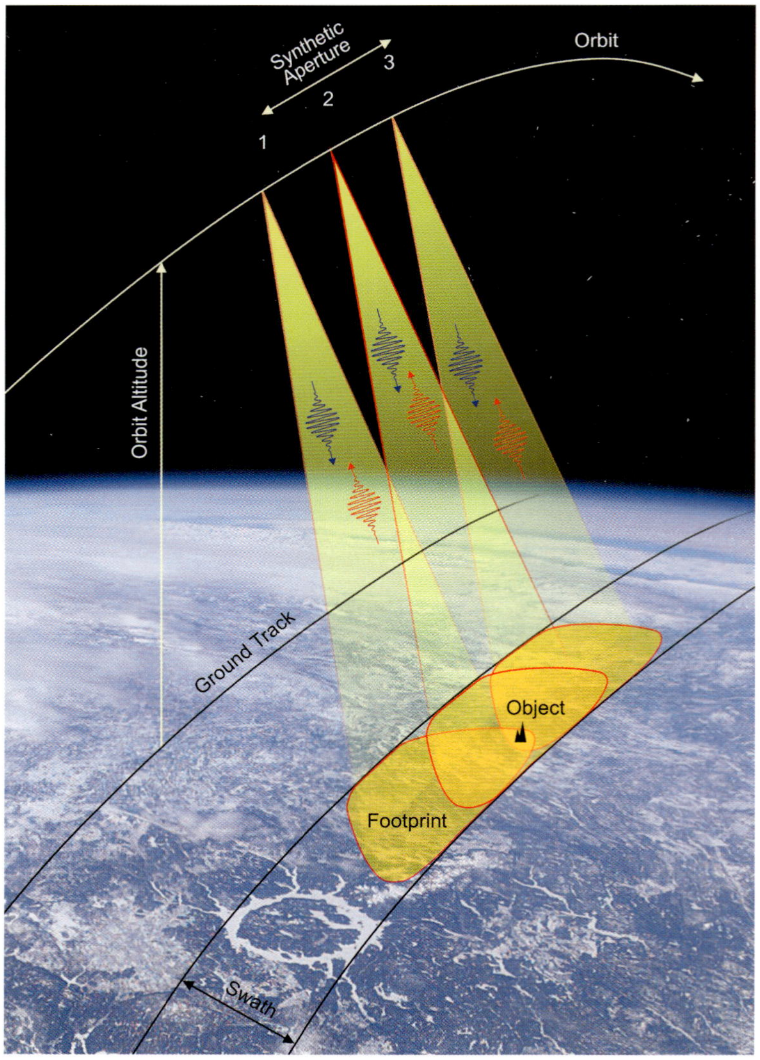

Principle of a Synthetic Aperture Radar: Between the orbital positions 1 and 3 the object is within the footprint. The distance between 1 and 3 defines the length of the synthetic aperture.

SAR Interferometry

SAR images provide high-resolution displays in 2D. Elevation, the third dimension, is required to fully characterize the morphology of impact structures. On a global scale only space-borne remote sensing can provide accurate elevation information. It can be derived from optical stereo pairs, altimeter measurements, or radar data obtained in interferometric mode. Whilst optical sensors require a cloud free, transparent atmosphere, and altimeters may lack high resolution or fail to provide sufficient coverage, Synthetic Aperture Radar (SAR) interferometry is not compromised by such effects. Features introduced by the SAR method and its interferometric application can be minimized through appropriate mission design, together with careful calibration and elevation extraction during data processing.

A single SAR sensor measures the radar signal amplitude together with the incoming phase of the backscattered microwave pulse. Operating a second receiving antenna in close vicinity of the transmitting antenna allows to "see" the same scene simultaneously – only from a slightly different position. Because all measurements occur in a single orbit using two spacecraft, the method is known as *single-pass interferometry*.

Radar pulses backscattered from ground have to travel different distances until they arrive at one of the two antennas. As a consequence, their phases at arrival differ as well. The measured phase difference is a measure for the path length, i.e., the difference in distances between ground and both antennas. It can be derived for each ground pixel in the scene. The resulting two-dimensional image is the *interferogram*. As phase can only be measured between 0 and 2π, whilst the path length is usually of the order of a few 100 m, the phase difference information is ambiguous.

The challenging computational step of phase unwrapping removes that ambiguity. Once the orbital altitude and position of both satellites is known with high precision, the unwrapped phase allows calculation of the true elevations above a reference sphere in the imaged scene. Their accuracies not only depend on the exact knowledge of the orbital parameters but also on how far both radar antennas have been separated in space during measurements, as given by their baseline. All derived elevation data constitute the Digital Elevation Model (DEM).

SAR interferometry is a rather new field. The method received increasing attention in the 1990's when first tests in orbit proved the

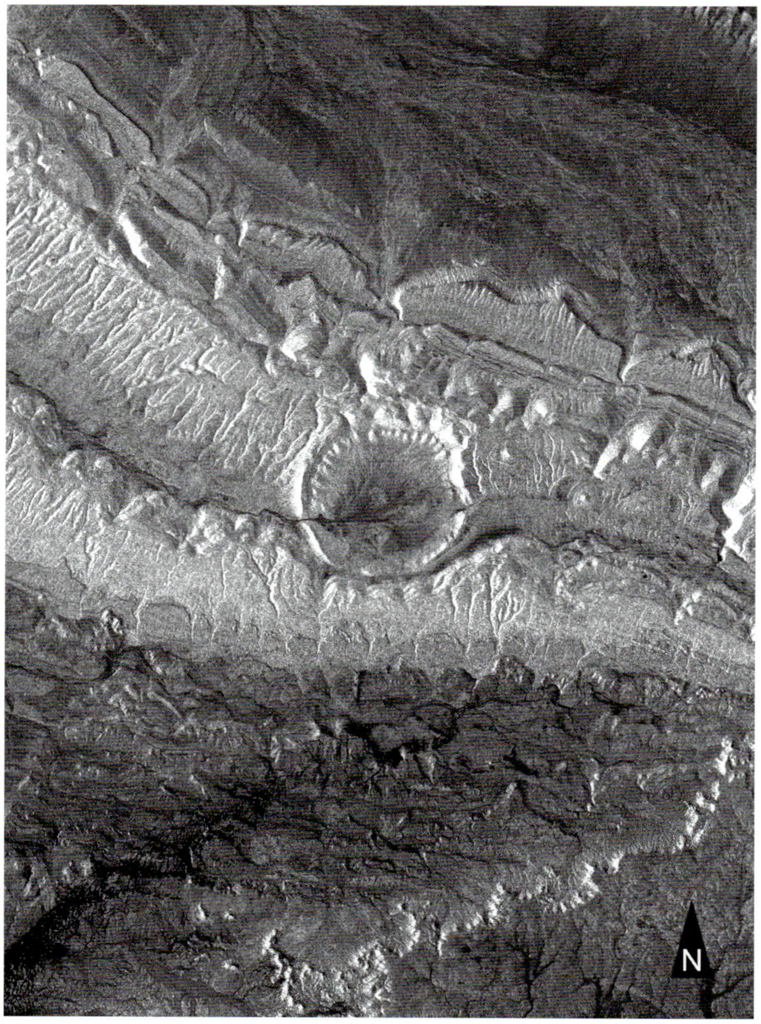

The 3.5 km wide Ouarkziz structure in Algeria, as it appears in a TanDEM-X SAR image. The satellite passed the scene west of the impact crater. Therefore, slopes facing the radar antenna's ground track appear bright. Dark patches are related to sandy areas with low backscatter.

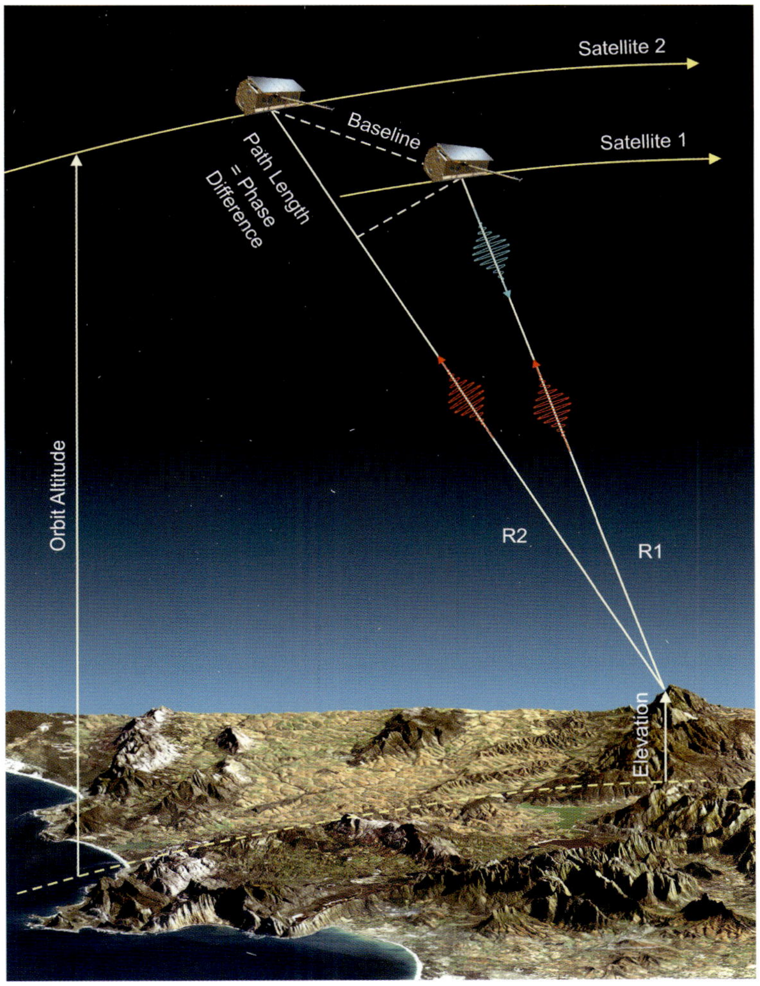

◁ Principle of single-pass SAR interferometry: The radar antenna on satellite 1 transmits radar pulses, whereas both antennas receive the backscattered signals. The phase difference is a measure for the path length. Knowing the geometry allows deriving the elevation.

validity of the concept. Until recently, before the TanDEM-X era, the most accurate space-borne DEM originated from the Shuttle Radar Topography Mission (SRTM), a joint project of NASA, Germany's Aerospace Center (DLR), and the Italian Space Agency. From February 11–22, 2000, Space Shuttle *Endeavour* carried an X-band and a C-band radar sensor in its payload bay. With a second receiving antenna for both bands at the end of a 60 m boom for the duration of the mission, SRTM operated as a single-pass interferometer. The data from both sensors were processed to yield two DEMs between the 56° S and 60° N latitudes, the region accessible from the Shuttle orbit. Their pixels were spaced by 1", equivalent to 30 m, at the equator.

Selected Reference

Moreira, A., Prats-Iraola, P., Younis, M., Krieger, G., Hajnsek, I., and Papathanassiou, K. P. (2013). A tutorial on synthetic aperture radar. IEEE Geoscience and Remote Sensing Magazine, Vol. 1, pp. 6–43.

Bottom left: The interferogram of the Ouarkziz scene shown on the previous page and a greyscale digital elevation map derived thereof (bottom right). In the interferogram the cyclic rainbow colours indicate sequences of phase differences from 0–2π. The elevation map displays altitudes between 380 m and 690 m, with dark hues indicating the lowest, and bright ones the highest elevations.

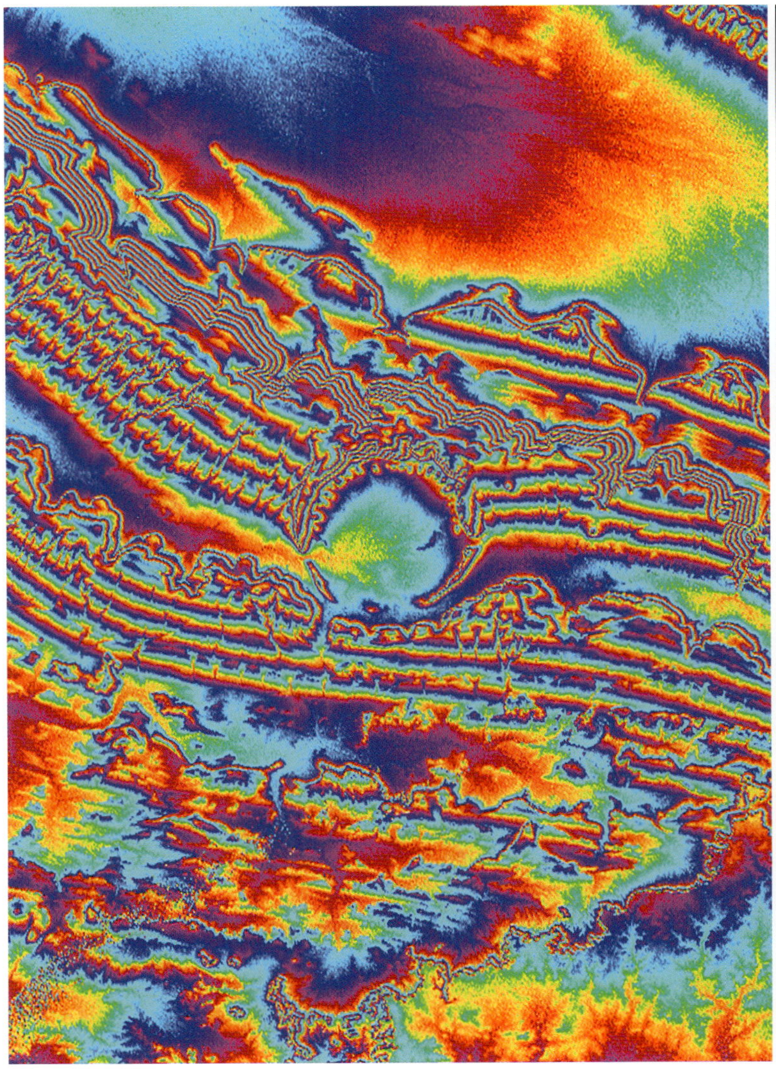

5. The TanDEM-X Mission

With the success of the SRTM digital elevation data it became apparent that a pole-to-pole DEM with even higher accuracy could indeed be accomplished by a dedicated space-borne radar mission. Both the geoscience community and commercial or public users were in urgent need of it. Therefore, shortly after the SRTM flight, the German Aerospace Center (DLR) initiated the *TanDEM-X*, short for *TerraSAR-X add-on for Digital Elevation Measurement,* mission for that purpose. Its goal was to process – from radar X-band data using the technique of SAR interferometry – a global terrestrial elevation model with a significant improvement in spatial sampling and elevation accuracy as compared to SRTM.

The TanDEM-X mission consists of two identical X-band radar satellites, TerraSAR-X and TanDEM-X. The first, TerraSAR-X, was launched on June 15, 2007. For about 3 years it acquired data as an individual X-band mission until it was joined in orbit by the TanDEM-X platform on June 21, 2010. After a 6-month long successful commissioning of TanDEM-X, both satellites began to operate in close formation with combined data acquisitions performed as the world's first single-pass interferometer in space. The configuration was primarily operated in a mode where one satellite provided the microwave illumination of the Earth's surface, and both satellites simultaneously received the radar echoes. A novel approach was a dedicated inter-satellite link for the synchronization of the phase of the two independent oscillators, required for modulation and demodulation of the radar pulses. This was a prerequisite for a precise calibration of the sensors, which reduced errors in interferometric processing considerably.

Launch of the TanDEM-X satellite from Bajkonur/Kazakhstan with a Dnjepr-1 rocket on June 21, 2010 (photo: DLR, CC-BY 3.0).

DEM requirements	
Independent pixel spacing	12 m (at equator)
Absolute vertical accuracy	< 10 m
Relative vertical accuracy	< 2 m (flat terrain)
	< 4 m (steep terrain)
Platform dimensions	
Mass	1209 kg
Power	800 W
Size	5 m × 2.4 m (length × diameter)
Platform orbit	
Type	polar sun-synchronous
Local time ascending equator crossing	18.00 (dusk/dawn)
Altitude	515 km
Orbital period	95 min
Inclination	97.44°

TanDEM-X mission facts.

The challenging DEM requirements could only be achieved by selecting inter-satellite distances of the order of 200–500 m, sometimes even reaching a minimum of only 120 m. These distances had to be known with a precision of at most a few millimeters. Although both spacecraft moved with a speed of more than 25 000 km/h, only separated by at most several hundred meters, the formation flying of both satellites, known as *helix formation,* ensured safe operations. Over both poles vertical separation and over the equator horizontal separation of the satellites was highest. The helix formation was fully configurable from ground. This permitted data acquisition in various viewing geometries as required for DEM generation. Particularly, the size of the satellite's baseline can be translated in SAR interferometry into the *height of ambiguity*, a measure for the sensitivity of SAR interferometry to the ground relief, which is then related to the achievable height accuracy. A longer baseline causes a smaller. height of ambiguity, equivalent to a larger topographic sensitivity of the interferogram. On the other hand, step-like elevation changes can only be resolved when the height of ambiguity is larger than the step. Therefore, the mission concept foresaw a combination of different TanDEM-X and TerraSAR-X configurations in subsequent acquisition phases.

TerraSAR-X and TanDEM-X in flight. The satellite tracks were taken on December 8, 2010 over Leiden, The Netherlands. Flight direction is from left to right with TerraSAR-X being the leading spacecraft (photo: M. Langbroek, Leiden, The Netherlands).

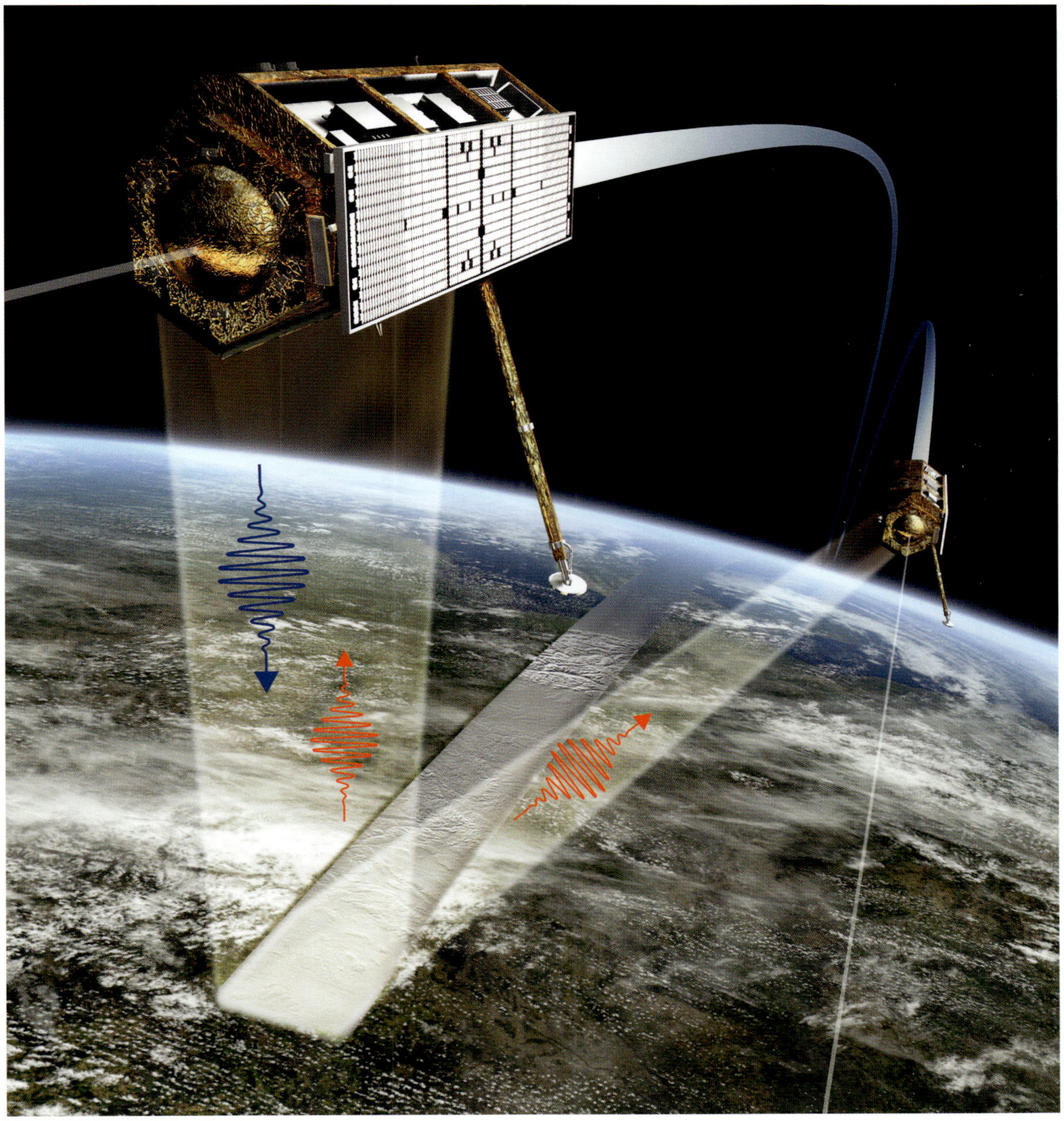

Artist's impression of the two TanDEM-X radar satellites in single-pass interferometric configuration. The leading antenna transmits radar pulses (blue) while both receive the backscattered signals (red) (modified after DLR, CC-BY 3.0).

TanDEM-X Data Acquisition

Even though the purpose of TanDEM-X was the delivery of a high-resolution elevation model, both radar antennas could not continuously transmit X-band pulses for interferometric measurements. Several limits imposed hard constraints: The observation geometry had to be suitable for SAR interferometry, in certain orbital positions the close vicinity of the satellites hampered radar transmissions, or the measurement program requested operation as a single imaging SAR mission. In fact, data acquisition was possible only for up to 300 s of the entire orbital period of 95 minutes.

Measurements for DEM data takes occurred in a phased approach over a period of almost 6 years. First, between December 2010 and March 2013 two global coverages of the entire terrestrial solid surface were achieved. These were followed by two mappings of Antarctica, interleaved with activities to fill gaps where previous acquisitions proved unsuccessful, and a 9-month phase during which the satellite configuration was re-adjusted to allow imaging of difficult terrain with opposite viewing geometry. By end 2015, all acquisitions had been accomplished. Since then, TerraSAR-X and TanDEM-X continue their missions as SAR imagers with further scientific experiments being scheduled due to the special nature of the formation flight.

TanDEM-X DEM Generation

The maximum time devoted to interferometric acquisitions of 300 s per orbit corresponded to a swath length of 2250 km with a width of 30 km according to the antenna's footprint. This whole data set was divided into individual scenes of about 30 km × 50 km extent, each of which was processed to yield a *Raw DEM*. About 450 000 overlapping Raw DEMs, generated from all acquisitions in single-pass interferometer configuration, were finally found suitable to be used in the processing of the TanDEM-X digital elevation model.

Each Raw DEM was the result of the specified interferometric processing chain. This includes focusing of the SAR image, co-registration of the two channels, generation of the interferogram, and unwrapping the ambiguous phase with the conversion of phase to height. Despite being the result of a single measurement, the precision achieved in a Raw DEM came already very close to that required for the final TanDEM-X DEM. However, several systematic inaccuracies, for example limited regions with incorrect heights due to phase unwrapping errors, tilts across the scene of at most a few centimeters per km, or

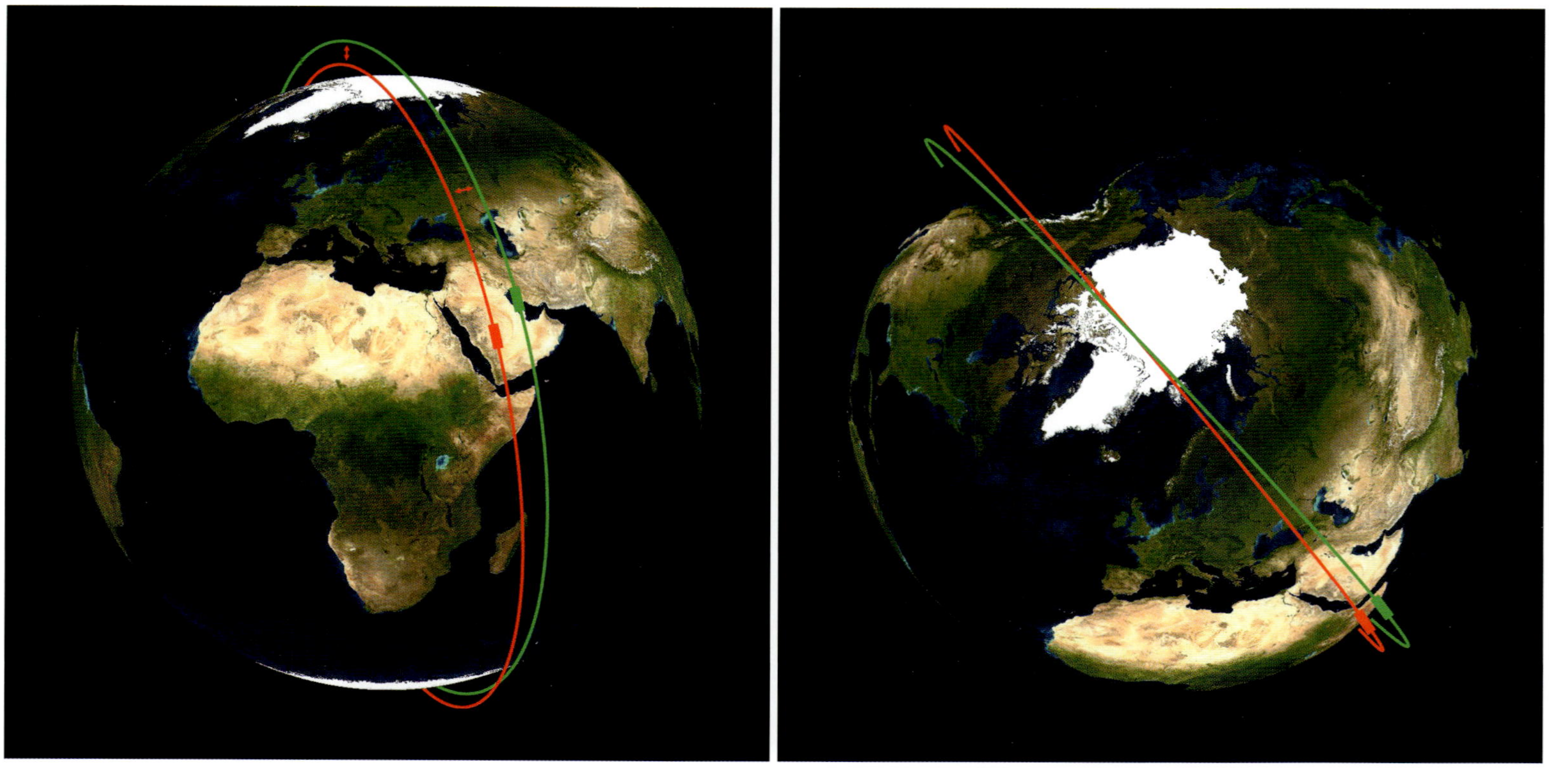

Schematic view of the helix formation. The red and green boxes indicate the positions of the two satellites. The helix formation permitted formation flying with short inter-satellite distances (modified after DLR, CC-BY 3.0).

occasional height offsets of up to several tens of meters still persisted. Such inconsistencies were eliminated or reduced to a minimum in the mosaicking of the individual Raw DEMs to form the final DEM. This mosaicking turned out to be a rather challenging processing step. Not only had the sheer multitude of Raw DEMs from different acquisition phases to be combined. Over difficult terrain sensing occurred with crossing acquisitions. For achieving high accuracies over sandy desert areas the antennas had to transmit their pulses at steeper angles, while over large forests or mountainous regions the antenna baseline needed to be adjusted to facilitate the unwrapping of the ambiguous phase.

By fall 2016 the Tandem-X DEM had been completed. It covers approximately 150 million square kilometers of land surface with unprecedented accuracy. The quality of the global elevation model has surpassed all expectations. Exceeding the required 10-meter accuracy, the topographic map has an elevation accuracy of only a single meter over large areas.

TanDEM-X was implemented on behalf of DLR using funds from the Federal Ministry for Economic Affairs and Energy (Bundesministerium für Wirtschaft und Energie), with Airbus Defence and Space as the industrial partner. DLR's responsibility covers radar operation and calibration, control of the two satellites, generation of the digital elevation model, and scientific coordination via DLR's TanDEM-X Science Office. This occurs at the facilities at Oberpfaffenhofen/Germany, namely at the Microwaves and Radar Institute, the Earth Observation Center, and the Space Operations Center. Airbus Defence & Space built the satellites and is sharing the development and operating costs. Access to the TanDEM-X digital elevation data is restricted. DLR supplies the data free of charge for approved scientific projects. Commercial users can purchase the data from Airbus Defence & Space, the industrial partner.

Selected References

Gottwald, M., Fritz, T. Helko, B., Schättler, B., and Harris, A. (2017). Remote Sensing of terrestrial impact craters: The TanDEM-X digital elevation model. Meteoritics and Planetary Science, Vol. 52, pp. 1412–1427.

Gruber, A., Wessel, B., Martone, M., and Roth, A. (2016). The TanDEM-X DEM mosaicking: Fusion of multiple acquisitions using InSAR quality parameters. IEEE Journal of Selected Topics in Applied Earth Observations and Remote Sensing, Vol. 9, pp. 1047–1057.

Zink, M. (2015). TanDEM-X mission status. The International Archives of the Photogrammetry, Remote Sensing and Spatial Information Sciences XL-7/W3, pp. 1345–1352.

Zink, M., Bachmann, M., Bräutigam, B., Fritz, T., Hajnsek, I., Krieger, G., Moreira, A., and Wessel, B. (2014). TanDEM-X: The new global DEM takes shape. IEEE Geoscience and Remote Sensing Magazine, Vol. 2, pp. 8–23.

On January 11, 2013 TanDEM-X orbited northwards over southern Germany. Between 17:08:52 UTC and 17:09:22 UTC data were acquired for interferometric processing. The 210 km long swath was divided into 5 Raw DEMs, displayed here in color-coded hill-shaded form. The second one of these covers the Ries crater. The background image shows the scene in a Landsat-8 panchromatic mosaic.

6. TanDEM-X Maps of Impact Structures

This atlas is aimed at providing views of the terrestrial impact structures in identical formatting, irrespective of their location, size, or preservation status. One prerequisite for this goal is the fact that the TanDEM-X digital elevation model is global, has a high-resolution pixel spacing of 12 m, and is of uniform excellent data quality for about 99 % of the Earth's landmass. This ensures that all currently known impact structures can be mapped with high precision. The other precondition is the approach to apply the same workflow to all structures, whereby the final maps comply with established cartographic standards.

Digital elevation data deliver a height value for a location in a scene defined by geographic latitude and longitude, and spaced as defined by the independent pixel posting. When they are displayed as a grey-scaled or color-coded image, they appear "flat" (see chapter 4). For transferring them into a physical map, two steps are required. In order to highlight the topography of the terrain, the flat scene needs to be transformed into a 3D-model. This is achieved by adding shadows, which requires artificial illumination of the elevation map. For all TanDEM-X maps we have used an artificial light source at the standard values of elevation = 45°/azimuth = 315° and applied these parameters to all structures. This ensured a consistent visual impression, although the selected azimuth does, usually, not represent the illumination conditions for the real Sun, when observing impact structures with space-borne sensors. How the azimuth of the light source affects the hill-shaded scene is illustrated for the Matt Wilson structure in Australia. For obtaining the physical map with terrain information as presented in the atlas, the grey-scale hill-shaded scene had to be finally combined with the color-coded elevation map.

The final TanDEM-X DEM is organized in tiles of 1° width in latitude and 1°, 2° or 4° width in longitude, depending on latitude.

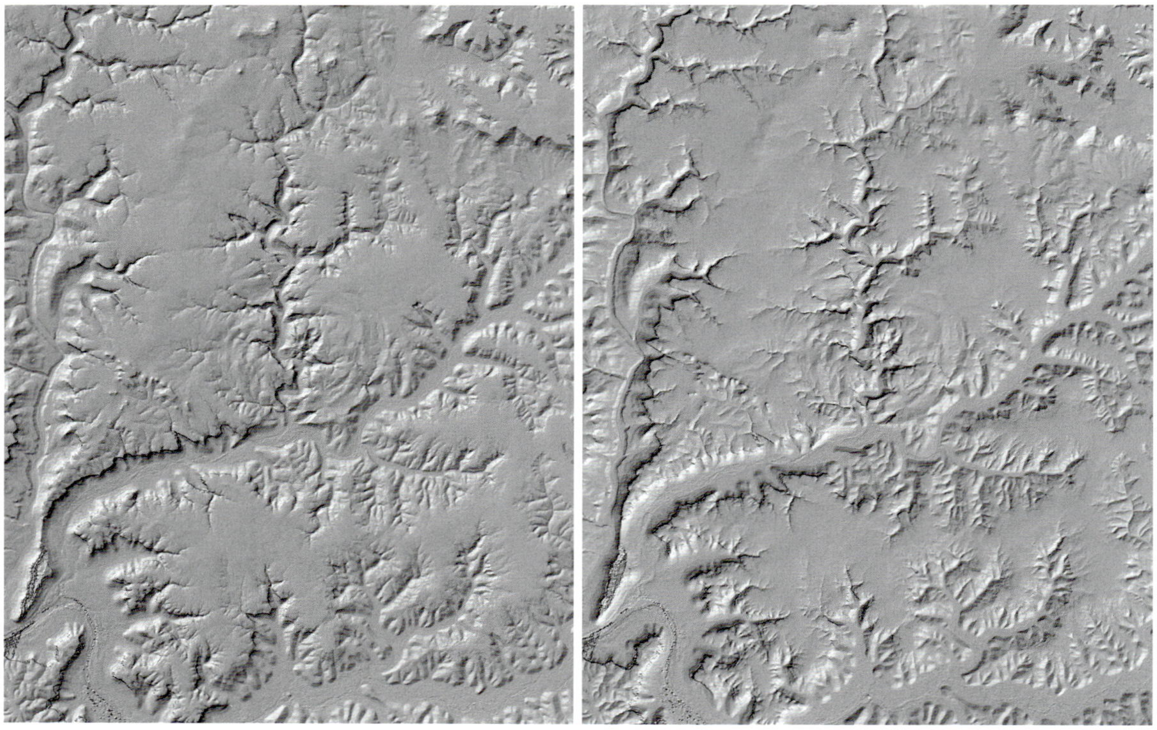

The elliptical Matt Wilson structure in Australia is located at the center of this 18 × 24 km wide scene. The image on the left illustrates illumination with the standard elevation and azimuth of 45° and 315° used throughout the atlas. For the right panel the light source was moved to the opposite side at an azimuth of 135°. The appearance of the canyon-like structures changes considerably.

The Ouarkziz structure in a color-coded TanDEM-X DEM (left), the grey-scaled hill-shaded version after artificial illumination (middle), and the final physical map derived therefrom (right). This scene is 13.5 × 18.5 km wide.

The southwestern sector of the rim of the Ries impact structure, in a map derived from the SRTM DEM (left) and from the TanDEM-X DEM (right). The high resolution of the TanDEM-X map shows details down to the scale of roads and buildings, which are still missing in the SRTM image. The TanDEM-X map is also devoid of noticeable elevation noise. This scene is 21 × 23 km wide.

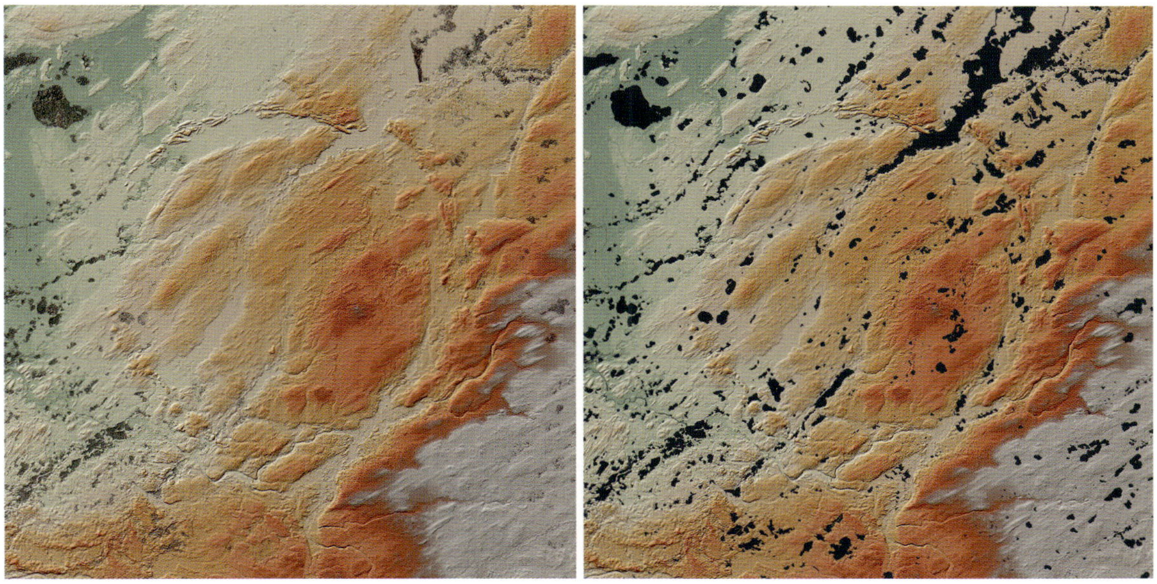

The Carswell impact structure in Saskatchewan/Canada in a TanDEM-X map without (on the left) and with (on the right) water mask. The digital elevation information in the central swaths was derived from data acquisitions when the lakes were frozen – the water bodies appear smooth. The swaths east and west of Carswell were recorded at times with open water surfaces and elevation noise is obvious. The water-masked map takes both situations into account. This scene is 52 × 54 km wide.

About 20 000 tiles cover the entire terrestrial landmass of about 150×10^6 km^2. TanDEM-X used a geographic latitude/longitude projection with the WGS84 ellipsoid giving the horizontal and vertical datum. For generating the atlas maps, the tiles were transformed into the *Universal Transverse Mercator* (UTM) coordinate system with equal linear pixel sizes in north-south and east-west directions. The elevations were transferred to heights above the EGM96 system, which approximates the mean sea level.

From the transformed tiles, mosaicked in the case of large impact structures or craters lying at the boundaries of tiles, the impacts were extracted with the aims to (1) provide sufficient high-resolution detail of the crater's morphology and (2) to illustrate the environs for providing the context with the local topography. The size of an impact structure was the dominating parameter that defined the map scale. However, the local surface texture had to be taken into account as well. The specifics of the radar X-band, like sensing the vegetation top or becoming sketchier over sandy areas, required to adapt scales for certain impact sites to achieve suitable mapping performance. The quality that could be achieved with the TanDEM-X DEM becomes obvious when comparing a crater map with one extracted from the same workflow using the 30 m DEM from SRTM, one of the best available open access near-global space-borne elevation data prior to TanDEM-X. Only mapping of impact structures in hilly terrain with steep slopes may be hampered by reduced elevation data quality. This is the case for a few scenes only. Upheaval Dome resides in the Canyonlands National Park. For some of the cliffs along the canyons unreliable elevation data had to be masked out. In the corresponding atlas map we filled them with data from the 30 m DEM which were obtained from optical stereo pairs taken with the panchromatic stereo mapping sensor PRISM aboard the Advanced Land Observing Satellite (ALOS). The Ritland impact lies at Norway's west coast shaped by steep fjords. Along such fjords the Ritland map displays some elevation noise. This is also the case for Piccaninny in Australia's Purnululu National Park, where creeks cut into the Bungle Bungle Range.

Water surfaces display a high degree of elevation noise because the radar pulse coherence, a precondition for SAR interferometry, is lost during data collection. In order to avoid mapping of such areas with erroneous elevation information, a water mask is added in scenes where required. This mask is derived from the X-band amplitude signal. Because of the low backscatter over water, lakes as well as rivers and ocean can be traced via the very low X-band amplitude. Over small water bodies the TanDEM-X DEM exhibits smooth elevations. This also occurred in northern regions when the digital elevation data had been derived in a season with ice- and snow-covered water surfaces. In order to obtain maps of generally homogeneous appearance, in both cases the water mask had to be applied as well. Identification of small open water bodies also exploited the low X-band amplitude. Frozen, snow-covered water surfaces are characterized, in contrast, by a high backscatter. Therefore such areas could be identified via the very high X-band amplitude. The DEM for large northern impact structures was sometimes mosaicked from summer and winter data takes. Then the water mask had to be combined, using both approaches. In the final impact maps water surfaces received a dark-blue hue, similar to what is obtained when imaging the Earth's surface from orbit. Australia with its ephemeral lakes is a special case. To highlight the transient and shallow nature of such lakes, the water mask displays a bright greyish hue.

TanDEM-X provides digital elevation data with a pixel spacing of 12 m. Very small craters with a diameter of the order of this spacing cannot be resolved by the DEM. The smallest impact crater that can be identified in the color-coded, hill-shaded DEM is Kamil with a

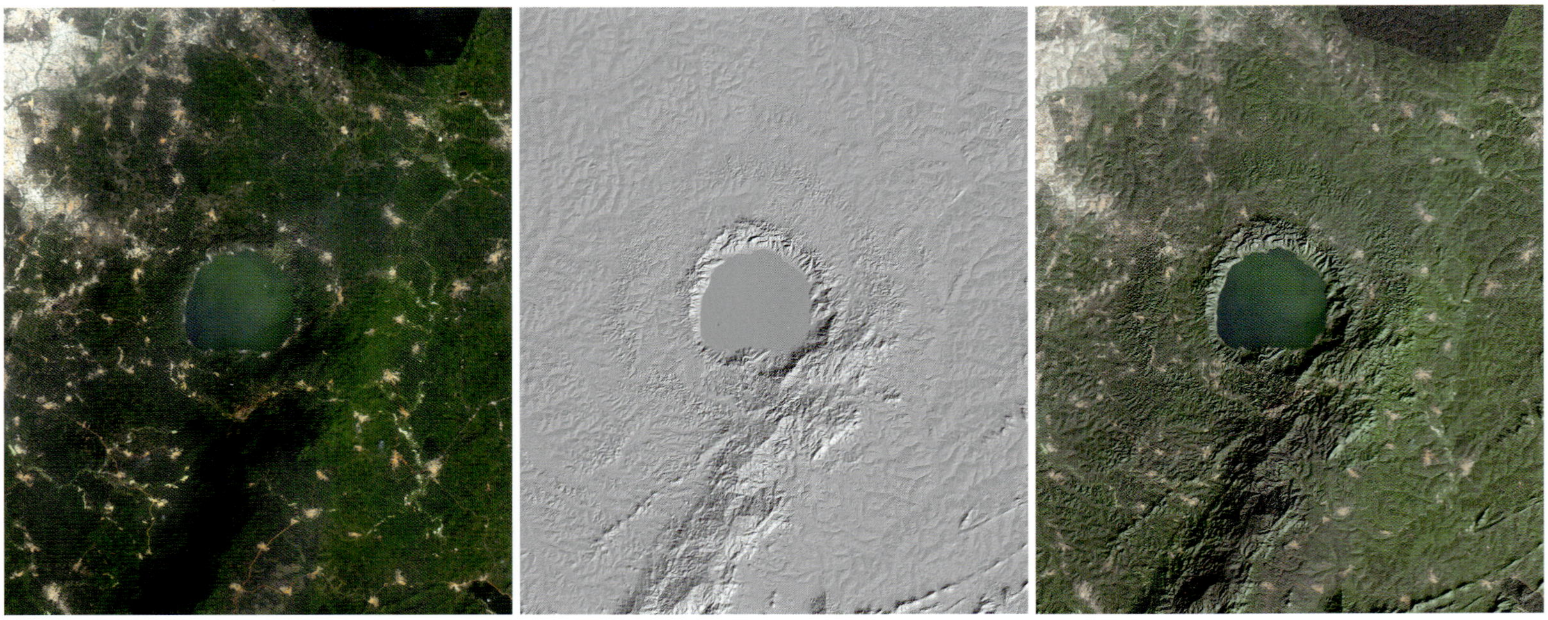

The Bosumtwi impact structure in a Sentinel-2 RGB-composite (left), the corresponding TanDEM-X hill-shaded DEM for identical illumination conditions (middle) and the combination of both (right). The hill-shaded DEM adds light and shadow such that the topography becomes visible, even in a scene with dense vegetation. This scene is 37 x 45 km wide.

diameter of 45 m. Craters of such size are only recognizable in dry, reasonably rough regions. Strong vegetation hampered the sensing of the bare ground with TanDEM-X. Also large amounts of fine-grained sand spoiled SAR interferometry. Then even craters of about 100 m diameter could escape identification in the digital maps.

Data Fusion TanDEM-X / Multispectral Sensor

The atlas also displays visible imagery for those impact structures, which are obvious in the TanDEM-X topographic maps. These images are based on RGB-composites of the European Copernicus Sentinel-2 mission, or, when no cloud-free scenes were available from Sentinel-2 – as was the case for Kara-Kul – from Landsat-8, a collaborative Earth Observation mission of NASA and the United States Geological Survey (USGS). The composites use the corresponding sensor's channel 2 (blue), channel 3 (green) and channel 4 (red). Sentinel-2 provides a resolution of 10 m for the RGB channel, close to that of TanDEM-X with 12 m, whereas for Landsat-8 this value amounts to 30 m. However, the panchromatic channel 8 of Landsat-8 is available with a resolution of 15 m, so that the method of "pansharpening" delivers a final resolution at this level.

Even when the multispectral image acquisitions occurred with oblique Sun illumination at solar elevations of 60° and lower, topography is usually not obvious in the RGB images. Therefore the RGB composites are enhanced by fusion with a hill-shaded TanDEM-X digital elevation model, which was illuminated with a light source at an elevation of 45° and at an azimuth as for the multispectral data take. This provides pronounced light and shadow to the RGB image, thus highlighting any topography in the scene.

How a Sentinel-2 RGB-composite is enhanced by adding the TanDEM-X DEM is demonstrated for the region around the Bosumtwi impact structure in Ghana/Africa. The Sentinel-2 acquisition took place on January 22, 2018, with a solar azimuth of 138° and an elevation of 54°.

All TanDEM-X/Sentinel-2 and TanDEM-X/Landsat-8 data-fused images in the atlas have the same orientation and projection as the physical (topographic) TanDEM-X maps.

Selected Reference

Gottwald, M., Fritz, T. Helko, B., Schättler, B., and Harris, A. (2017). Remote Sensing of terrestrial impact craters: The TanDEM-X digital elevation model. Meteoritics and Planetary Science, Vol. 52, pp. 1412–1427.

7. The Atlas

The atlas displays topographic (physical) maps for 208 terrestrial structures, either of confirmed impact origin or where additional work is required to prove their claimed formation as a result of a hypervelocity impact. These maps are supplemented by one map for a recent airburst (Tunguska). The cover areas of the maps reflect the wide range in impact crater diameters. Minimum extent occurs for Kamil, the smallest detectable impact crater with slightly more than 10 × 13 km (longitude × latitude) map area, whereas the maps for Vredefort, Chicxulub, and Acraman display scenes of about 165 × 220 km size.

The position of an impact structure that has a visible topography but that is difficult to recognize on the map because of its small diameter and/or the local texture of the scene, is emphasized by a cross-hair. Cross hairs also mark the location of an impact structure that is exposed but where the topography has been completely erased, which is buried underground, or which is submerged. In all these cases, the proposed diameter of the structure is displayed as a dashed circle. The scale of maps displaying buried impact structures on the territory of the former Soviet Union has been selected so that alternative locations in the vicinity, as listed in other publications, can also be found on these maps.

More than 50 % of all maps have at least 1–2 landmarks labelled. These are usually cities or villages, rivers, or lakes. The maps of impact structures in remote settings, such as sites in Australia's outback or in northern hemisphere regions beyond the Arctic Circle, often lack such markings.

Chapters 7-1 to 7-6 describe the terrestrial impact structures continent by continent. Each chapter 7-1 to 7-6 comprises also a map displaying the corresponding impact content per continent and a table listing the impact features by the parameters *name*, *country/region*, *latitude*, *longitude*, *diameter* (km), *age* (Myr), and *topographic visibility*. For impact structures where the position cannot be inferred from their topography, the table lists the most up to date publication that provided the listed location.

The table is followed by the sequence of impact structures that includes, for each of these, the annotated topographic map with cartographic information and a brief, illustrated description of the structure highlighting its geological setting, impact related morphology and lithology, together with a short account of how the structure was discovered and identified. Notes about the location and how to access the impact site, together with a list of selected references, supplements each impact entry. All entries with unequivocal confirmation are listed alphabetically. At the end of each continent chapter follow those structures where an impact origin is widely considered "confirmed" but where we feel additional field and laboratory work is required to unambiguously prove this based on the established criteria of shock metamorphism or by finding traces of the meteoritic projectile.

7.1 Africa

Bosumtwi in January 2018 – TanDEM-X/Copernicus Sentinel-2.

African Impact Structures – Overview

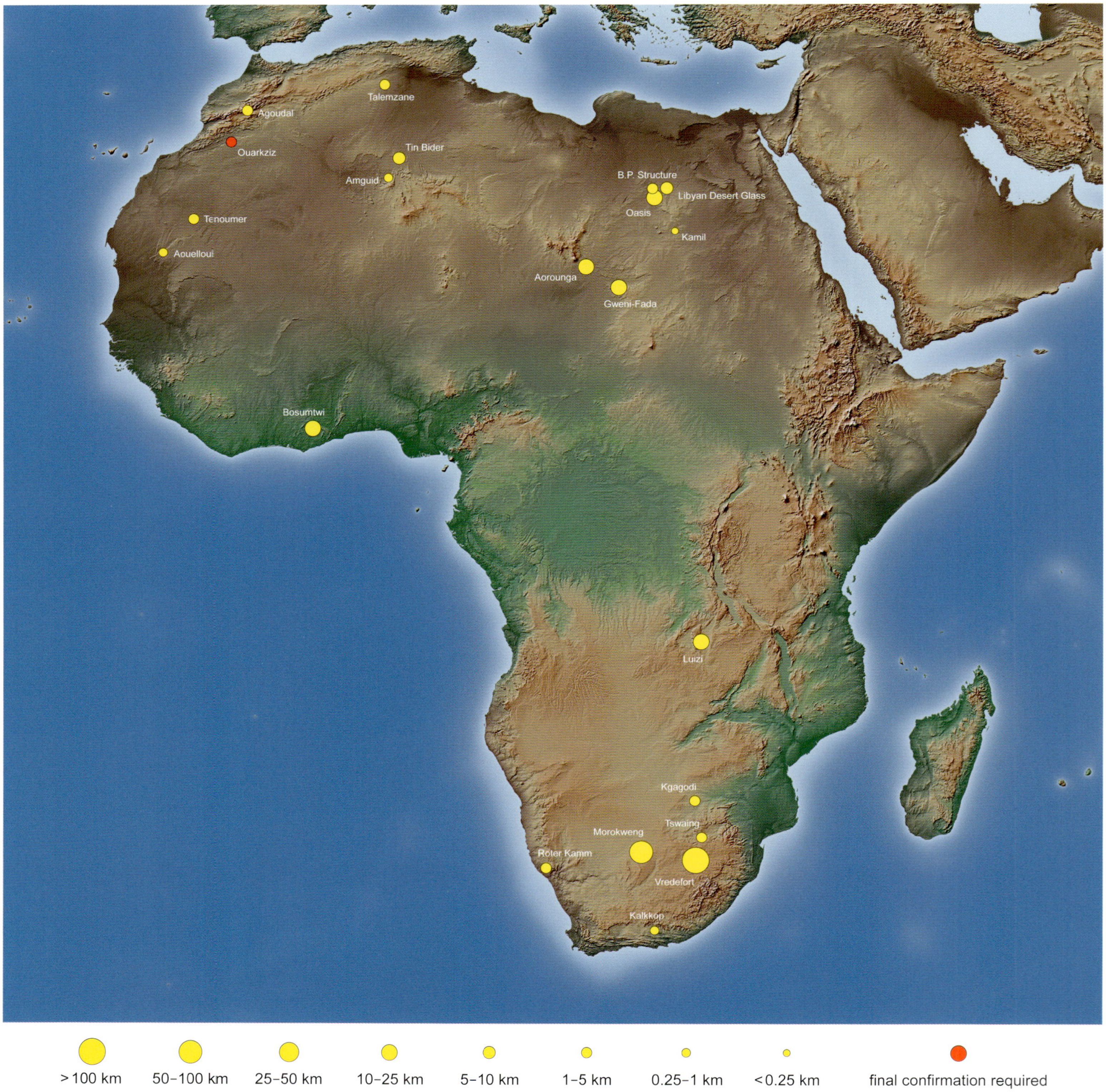

Table: African Impact Structures – Overview

ID	Name	Country/Region	Latitude	Longitude	Diameter (km)	Age (Myr)	Topographic Visibility
1	Agoudal[1]	Morocco	31°59'12"N	05°30'57"W	1.4–4.2	> 0.3	no
2	Amguid	Algeria	26°05'15"N	04°23'42"E	0.45	< 0.1	yes
3	Aorounga	Chad	19°05'31"N	19°14'37"E	16	< 355	yes
4	Aouelloul	Mauritania	20°14'28"N	12°40'28"W	0.39	3.1 ± 0.3	yes
5	Bosumtwi	Ghana	06°30'09"N	01°24'27"W	10.5	1.07	yes
6	B.P.	Libya	25°19'08"N	24°18'37"E	3.4	< 120	yes
7	Gweni-Fada	Chad	17°25'07"N	21°45'17"E	22	< 355	yes
8	Kalkkop	South Africa	32°42'31"S	24°25'56"E	0.64	0.25 ± 0.05	yes
9	Kamil	Egypt	22°01'06"N	26°05'15"E	0.045	0.00302 ± 0.00060	yes
10	Kgagodi Basin[2]	Botswana	22°28'28"S	27°34'48"E	3.4	< 180	no (buried)
11	Libyan Desert Glass	Egypt	25°25'N	25°30'E	unknown	29 ± 1	no
12	Luizi	DR Congo	10°10'09"S	28°00'21"E	17	< 573	yes
13	Morokweng[3]	South Africa	26°28'S	23°32'E	70	145 ± 2	no (buried)
14	Oasis	Libya	24°34'33"N	24°24'43"E	18	< 120	yes
15	Roter Kamm	Namibia	27°45'58"S	16°17'21"E	2.5	< 5	yes
16	Talemzane	Algeria	33°18'52"N	04°02'05"E	1.75	0.5–3	yes
17	Tenoumer	Mauritania	22°55'07"N	10°24'20"W	1.9	1.52 ± 0.14	yes
18	Tin Bider	Algeria	27°36'07"N	05°06'44"E	6	50–66	yes
19	Tswaing	South Africa	25°24'31"S	28°04'57"E	1.13	0.220 ± 0.052	yes
20	Vredefort	South Africa	27°00'S	27°30'E	250–300	2023 ± 4	yes
Further confirmation required							
21	Ouarkziz	Algeria	29°00'38"N	07°33'02"W	3.5	> 66	yes

Position references for selected impact structures:

1 Sadilenko, D. A., Lorenz, C. A. Ivanova, M. A., Roshchina, I. A., and Korochantsev, A. V. (2013). A new small impact crater in the High Atlas, in the Agoudal iron strewn field. 76th Annual Meeting of The Meteoritical Society, Edmonton, Canada, Abstract # 5215.

2 Brandt, D., Holmes, H., Reimold, W. U., Paya, B. K., Koeberl, C., and Hancox, P. J. (2002). Kgagodi Basin: the first impact structure recognized in Botswana. Meteoritics and Planetary Science, Vol. 37, pp. 1765–1779.

3 Koeberl, C., and Reimold, W. U. (2003). Geochemistry and petrography of impact breccias and target rocks from the 145 Ma Morokweng impact structure, South Africa. Geochimica et Cosmochimica Acta, Vol. 67, pp. 1837–1862.

Agoudal
Morocco

Thomas Kenkmann

Agoudal, the only known site with impact relevance in Morocco, is situated in the High Atlas. The site is unusual and neither characterized by a circular signature nor any topographic pecularity. However, there is a confirmed presence of well-preserved, seemingly in situ shatter cones in mid-Jurassic marly limestones. Shatter cones occur over an area of 700 × 400 m.

Geological mapping revealed that this area is also structurally disturbed. Strata are locally in upright position and strike irregularly and partly perpendicular to the general trend in the High Atlas range. Local brecciation and intense in situ jointing go along with faulting of these rocks at Agoudal. The highly disturbed strata are surrounded by undeformed strata of the same stratigraphic unit. The size of the disturbed zone roughly coincides with the area where shatter cones were found. As a general trend known from other structures, shatter cones occur within an area corresponding to 1/6 to 1/2 of the estimated original diameter of an impact crater. Based on this rough assumption, an original crater size of 1.4–4.2 km could recently be inferred for Agoudal.

Erosion rates estimated from cosmogenic nuclides (^{36}Cl) range from 220–430 m per 1 Myr in this region. These erosion rates may remove the topographic expression of such a crater and its ejecta in a time period of about 0.3 to 1.9 Myr. This age is older than the Agoudal iron meteorite. Agoudal was discovered in 2013 by chance in the course of investigating the Agoudal iron meteorite strewn field. The fall of this meteorite with an estimated mass of more than 2 tons occurred 0.105 ± 0.040 Myr ago. The overlap of the strewn field of the Agoudal iron meteorite with the area of shatter cone occurrence led to a debate regarding a possible genetic link between the meteorite shower and the formation of the shatter cones, perhaps by the formation of several small, less than 100 m impact craters. According to the recent age estimates for the perceived impact structure, the current interpretation suggests that both events are not linked with each other and that the impact event significantly predates the fall of the meteorite.

Shatter cone in marly limestone from the discovery site at Agoudal, just south of Agoudal village. The length of the specimen is about 10 cm (photo courtesy: Ralf-Thomas Schmitt, Museum für Naturkunde Berlin, sample from W. U. Reimold).

Location/Access

Agoudal is situated about 180 km northeast of the city of Ouarzazate and 20 km southeast of the town of Imilchil. Agoudal village is located close to the shatter cone occurrence. This village is connected to Imilchil via Highway R317 and to Ouarzazate via R704. The impact site can be reached from Agoudal village by driving along regional road R704 in southwesterly direction for about 3 km. The disturbed area is located in the foothills of the mountains south of the road.

Selected References

Chennaoui-Aoudjehane, H., El Kerni, H., Reimold, W. U., Baratoux, D., Koeberl, C., Bouley, S., and Aoudjehane, M. (2016). The Agoudal (High Atlas Mountains, Morocco) shatter cone conundrum: A recent meteorite fall onto the remnant of an impact site. Meteoritics and Planetary Science, Vol. 51, pp. 1497–1518.

El Kerni, H., Chennaoui-Aoudjehane, H., Baratoux, D., Aoudjehane, M., Charrière, A. Ibouh, H., Rochette, P., Quesnel, Y., Uehara, M., Kenkmann, T., Wulf, G., Poelchau, M. H., Nguyen, V. B., Aboulahris, M., Makhoukhi, S., Aumaître, G., Bourlès, D., and Keddadouche, K. (2019). Geological and geophysical studies of the Agoudal impact structure (Central High Atlas, Morocco): New evidence for crater size and age. Meteoritics and Planetary Science, Vol. 54, pp. 2483–2509.

Agoudal

Morocco

latitude = 31°59'12" N / longitude = 05°30'57" W / diameter = 1.4 - 4.2 km / age = > 0.3 Myr

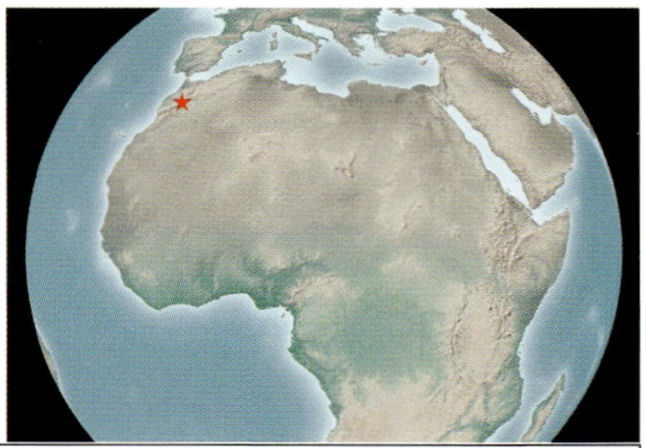

scale = 1 : 144 000 / scene = 25.9 × 34.5 km²
0 1.44 7.20 km
N
1860 meter 3195

Amguid
Algeria

Wolf Uwe Reimold

Amguid is a mere 450 m diameter crater structure in Algeria, which was first recognized from an airplane in 1954. The small crater is exposed in Lower Devonian sandstones. An up to 50 m high, elevated rim surrounds the central depression of the simple, bowl-shaped crater. The structure is filled with coarse-grained alluvial and eolian deposits that are covered in the central part of the crater by fine-grained eolian silts. The dips of the sandstone beds of the crater rim become progressively steeper in the upper part of the rim wall. Overturned beds have been described from the north-northwest and south-southeast parts of the rim.

Limited field investigation has taken place at Amguid. Evidence for impact was reported in the form of irregular fractures and undulatory extinction in quartz from the crater wall. Less than 1 % of the quartz grains studied displayed planar deformation features (PDFs), with one or two orientations per host grain.

A very young age has been estimated for this structure, at less than 0.1 Myr, on the basis of the excellent preservation of the crater. This evidence includes a continuous ejecta blanket outside of the crater, extending up to about 100 m from the crater rim.

In 1969 Jean-Phillippe Lefranc was the first to suggest an impact origin based on the regular circular shape, the central depression, a raised rim, and the regional peculiarity of this crater feature.

A 5.5 × 4.5 × 2 cm piece of Amguid polymict impact breccia (photo courtesy: Ralf-Thomas Schmitt, Museum für Naturkunde Berlin, sample from Serra Romano, Bologna).

Location/Access

The Amguid meteorite crater is difficult to visit. It can be reached with a four-wheel-drive vehicle on a desert track from the town of In Salah, in a south-easterly direction over a distance of 200 km. The track is very rough over a stretch of some 140 km. Final access to the site is impossible by vehicle. It will be necessary to walk about 20 km to reach the crater.

Selected References

Lambert, P., McHone, J. F., Dietz, R. S., and Houfani, M. (1980). Impact and impact-like structures in Algeria, Part 1. Four bowl-shaped depressions. Meteoritics, Vol. 15, pp. 157–179.

Lambert, P., and Lamali, A. (2009). Impact structures in Algeria (abstract). In: 1st Arab Impact Cratering and Astrogeology Conference, Amman, pp. 50–53.

Lefranc, J. P. (1969). Reconnaissance du cratère météoritique d'Amguid (Mouydir, Sahara Central). Comptes Rendus de l'Académie des Sciences, Paris, Vol. 268, pp. 900–902.

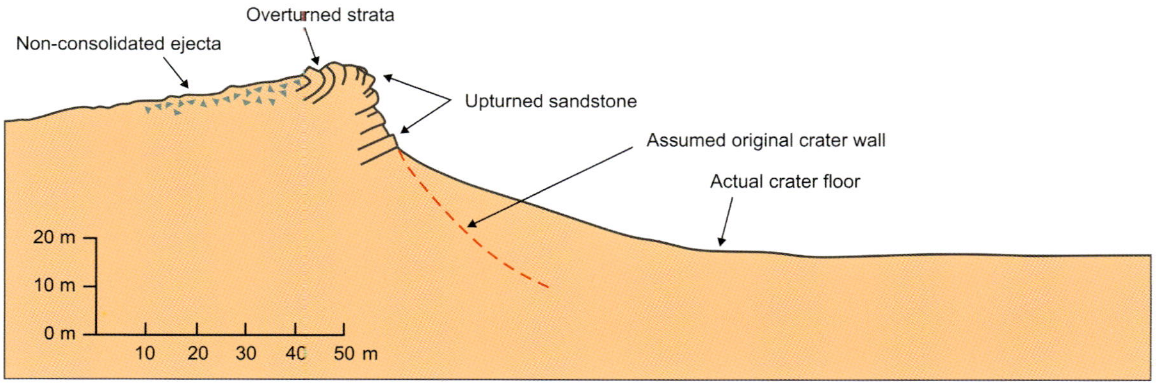

Schematic half cross-section through the Amguid impact crater, from north-northwest (at left) towards the center of the structure (modified after Lambert et al., 1980).

Amguid

Algeria

latitude = 26°05'15" N / longitude = 04°23'42" E / diameter = 450 m / age < 0.1 Myr

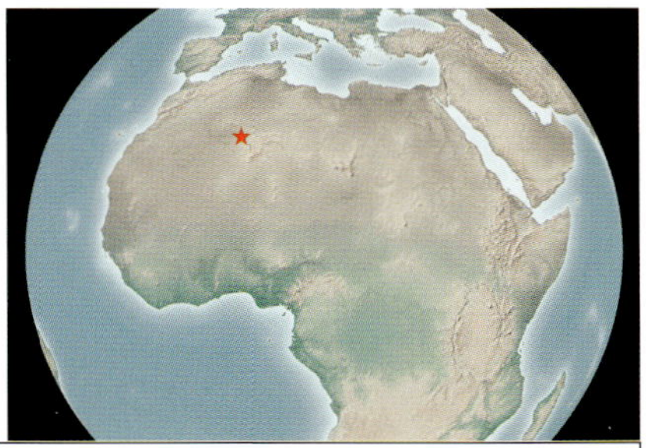

scale = 1:84 000 / scene = 15.1 × 20.2 km²

0 0.84 4.20 km

N

445 meter 1040

The Amguid impact crater and environs on September 6, 2018 as a composite image based on TanDEM-X DEM and Copernicus Sentinel-2 (channel 2/3/4) data.

Aorounga
Chad

Wolf Uwe Reimold

Aorounga is a quite deeply eroded impact structure in northern Chad, about 110 km southeast of the prominent Emi Koussi volcano of the Tibesti Massif. The entire area of the structure is transected by more or less parallel longitudinal yardangs, parallel ridges with sharp crests sculptured by wind abrasion. Remote sensing imagery demonstrates the considerable erosion of the structure.

The Aorounga structure was formed in fine-grained, well-sorted, sandstone of Upper Devonian age, which has a minor carbonate component. Early studies by Jean-Francois Becq-Giraudon and his team distinguished an outer and an inner ring wall at 7 and 11 km distance from the center, separated by an annular depression. Apparently, both ring structures rise up to 100 m above the environs. The innermost part of the structure is formed by a hilly terrain of about 1.5 km width. Also a breccia comprising fine-grained clasts and characterized by a fluidal texture had been found on top of the inner ring wall.

More recent remote sensing, petrographic and geochemical results were published by Christian Koeberl and co-workers in 2005. They found that Aorounga comprises a 9 km wide, structurally complex central zone, which is surrounded by a 2–3 km wide, ring structure of rather low topography, and then a second, outer ring of 3.5 km width. It was proposed that the latter could represent the crater rim zone, which would make the diameter of the Aorounga impact structure as large as 16 km, more than a diameter of about 13 km which had been quoted in older literature. In addition, new evidence of shock deformation for monomict breccia samples of quartzite and sandstone was found. In one sample quartz grains with up to 5 sets of planar deformation features (PDFs) were observed. In several other samples abundant grains with planar fractures (PFs) and rare grains with single sets of PDFs could be noted. These results provided unambiguous support for Aorounga being an impact structure, a conclusion which already had been put forward by Becq-Giraudon and his team but had not been considered convincing at that time.

The first time when Aorounga had been investigated with aerial and satellite imagery was in 1976 by Norbert Roland. He suggested that the structure could either represent a granite "diapir" or was formed as an impact crater, and then decided to favor the "diapir" hypothesis. In contrast, twelve years later, Richard Grieve and co-workers supported an impact origin.

The age of the structure was estimated by the French expedition of 1992 based on ^{14}C dating at 3500–12 000 years, which in the light of the eroded appearance of the structure appears to be much too young. Unfortunately, to date no melt-bearing samples have been found in the structure, which might have provided a basis for dating attempts with the more promising ^{40}Ar-^{39}Ar and U-Pb zircon chronometers. Cosmogenic nuclide exposure age dating was attempted to obtain an exposure age. Also thermoluminescence and electron spin resonance dating of quartz from Aorounga was exploited as an alternative method. These efforts yielded 0.5–1 Myr ages that Wolf Uwe Reimold and C. Koeberl in their review of African impact structures regarded as likely minimum age values for this impact event. Thus, currently the assumed Upper Devonian age of the target rock is the only – maximum – age limit available for the Aorounga impact event.

Aorounga

Chad

latitude = 19°05'31" N / longitude = 19°14'37" E / diameter = 16 km / age = < 355 Myr

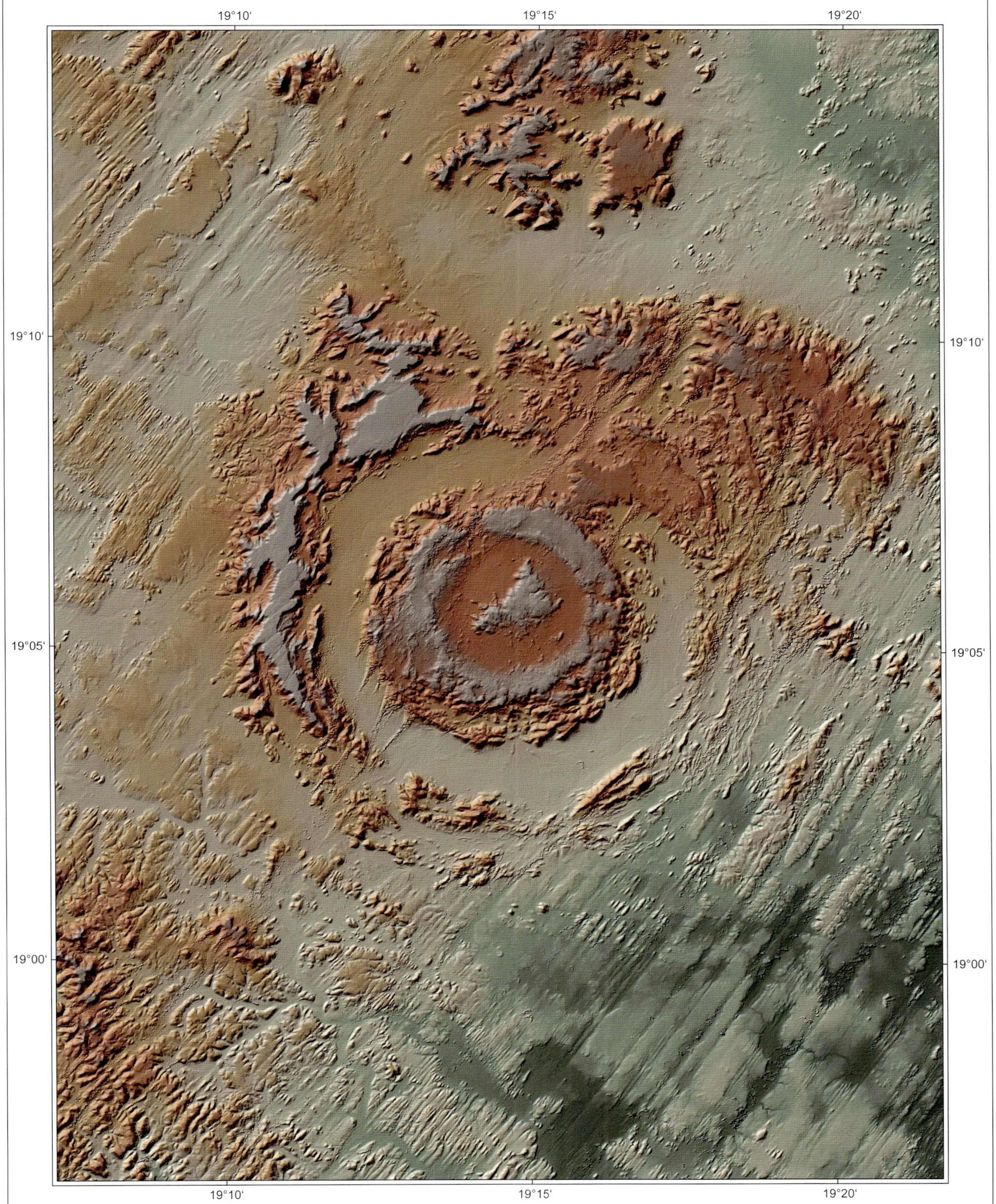

scale = 1 : 141 000 / scene = 25.5 × 34.0 km²

0 1.41 7.05 km

N

380 meter 705

The Aorounga impact structure on July 25, 2018 as a composite image based on TanDEM-X DEM and Copernicus Sentinel-2 (channel 2/3/4) data.

Location/Access

There is much scope for further detailed ground analysis of this impact structure. However, like the rest of northern and western Africa, Chad and adjacent territories have been subject to intense political unrest for the past decades, so that any attempt to venture towards Aorounga must be strongly discouraged.

Relatively widely spaced planar fractures (PFs) and closer spaced planar deformation features (PDFs) in a quartz grain from a sandstone specimen from the Aorounga structure (Chad). Width of field of view about 0.8 mm. Plane polarized light image (photo courtesy: W. U. Reimold).

Selected References

Becq-Giraudon, J. F., Rouzeau, O., Goachet, E., and Solages, S. (1992). Impact hyperveloce d'une météorite géante a l'origine de la dépression circulaire d'Aorounga au Tchad (Afrique). Comptes Rendus de l'Académie des Sciences, Paris, Vol. 315, pp. 83–88.

Grieve, R. A. F., Wood, C. A., Garvin, J. B., Mclaughlin, G., and McHone, J. F. (1988). Astronaut's Guide to Terrestrial Impact Craters. Lunar and Planetary Institute Technical Report, 88-03, Lunar and Planetary Institute, Houston, 89 pp.

Koeberl, C., Reimold, W. U., Cooper, G., Cowan, D., and Vincent, P. (2005). Aorounga and Gweni Fada impact structures, Chad: remote sensing, and petrography and geochemistry of target rocks. Meteoritics and Planetary Science, Vol. 41, pp. 1455–1471.

Reimold, W. U., and Koeberl, C. (2014). Impact structures in Africa: A review. Journal of African Earth Sciences, Vol. 93, pp. 57–175.

Roland, N. W. (1976). Die Ringstruktur Aorounga (Borkou, Süd-Sahara). Geologisches Jahrbuch, Vol. 33, pp. 117–131.

Aouelloul

Mauretania

Wolf Uwe Reimold

Aouelloul is a small, simple bowl-shaped crater located in the Adrar region of Mauritania. Its rim-to-rim diameter amounts to only 390 m. The crater was formed mainly in sandstone, besides some quartzite; both formations are of Ordovician age. A well-preserved rim rises some 15–25 m above the surrounding desert, and 53 m above the crater floor. The rim shows a distinct, overturned sequence of strata, an observation that is also in line with limited erosion of the crater. The crater fill comprises poorly-sorted, sandy silt that is overlain by eolian sand. Gravity analysis demonstrated that the thickness of the sedimentary fill was about 23 m. Below this occurs a breccia lens down to the crater floor at 100–130 m depth.

The crater rim is covered by eolian deposits. Locally loose blocks of sandstone are exposed that could possibly represent ejecta from the growing crater. In rim outcrops, strata are upturned, and dip at 35–60° towards the crater center. A lot of glass fragments could be collected on the outer flank of the southern, southeastern, and northern crater rim.

This small, well-preserved impact crater was originally discovered from the air by André Pourquié in 1938. It took another twelve years until he, together with Théodore Monod, visited the crater for the first time on the ground, in 1950. They were also the first to suggest an impact origin, from the discovery of Ni-rich glass. Shortly thereafter, Walter Campbell-Smith and Max Hey found that the glass was silica glass of a composition similar to that of Darwin and Wabar glasses. They determined a mixture of local sandstone and remnants of a meteoritic projectile with Ni-rich Fe spherules possessing 1.7–9 wt% Ni, which was thought to be from the projectile. Aouelloul glass was shown to be heterogeneous in composition and to have a well-developed schlieren texture. Individual schlieren were of different chemical compositions, and the glass contained digested quartz and feldspar grains. The glass chemistry resembled the composition of the target sandstone but with some enrichment of siderophile elements. The high-temperature mineral phases lechatelierite and baddeleyite were identified in the glass. Later studies by Anton Beran and Christian Koeberl determined a very low water content of the glass, which also represents evidence for a high-temperature formation. These analytical observations together were considered conclusive evidence for an impact origin of Aouelloul. The age of this impact event was determined by K-Ar and fission track dating of impact glass to 3.1 ± 0.3 Myr.

However, definite evidence for impact in the form of shock metamorphosed minerals could not be detected so far, only planar fluid inclusion trails in quartz that were interpreted as possible healed remnants of original planar deformation features (PDFs). In 1998, Koeberl and co-workers detected widespread fracturing and cataclasis of quartz grains in sandstone samples, also with abundant sub-planar and planar fractures (PFs). They also described several grains of quartz with two sets of relatively straight and closely spaced fluid inclusion trails, but did not consider this as impact deformation

Aoulleul impact melt rock pieces of about 3 cm size each (photo courtesy: Ralf-Thomas Schmitt, Museum für Naturkunde Berlin, sample from D. Stöffler, originally supplied by R. B. Dietz).

but rather as healed pre-impact features. Their discussion addressed the fact that the smaller an impact structure is, the less energetic the event would have been, so that only a very limited zone of the target would have been subjected to shock deformation in excess of the *Hugoniot Elastic Limits* of major rock-forming minerals. For quartz, this amounts to 8–10 GPa as the value for onset of formation of planar deformation features. When applying the Re-Os isotope method to Aouelloul glass, they convincingly demonstrated that it contained a definite meteoritic component – thus providing the ultimate proof that this small crater was of impact origin.

Location/Access

As in all of North and West Africa, extreme caution is demanded when planning to travel. In the past a trek to Aouelloul involved a flight to the country's capital, Nouakchott, and then overland travel by four-wheel transport to Chinguetti. The condition of these roads to date is not known to us. From Chinguetti one would take smaller tracks until finally cross-desert driving would be demanded, all the way to the crater. As with all North and West African travel, it is also advised to rely on local collaborators or at least responsible guides.

Selected References

Beran, A., and Koeberl, C. (1997). Water in tektites and impact glasses by FTIR spectrometry. Meteoritics and Planetary Science, Vol. 32, pp. 211–216.

Campbell-Smith, W., and Hey, M. H. (1952). The silica glass from the crater of Aouelloul (Adrar, western Sahara). Bulletin de l'Institut fondamental d'Afrique noire, Vol. 14, pp. 762–776.

Fudali, R. F., and Cassidy, W. A. (1972). Gravity reconnaissance at three Mauretanian craters of explosive origin. Meteoritics, Vol. 7, pp. 51–70.

Grieve, R. A. F., Garvin, J. B., Coderre, J. M., and Rupert, J. (1989). Test of a geometric model for the modification stage of simple impact crater development. Meteoritics, Vol. 24, pp. 83–88.

Koeberl, C., Reimold, W. U., and Shirey, S. B. (1998). The Aouelloul crater, Mauritania: on the problem of confirming the impact origin of a small crater. Meteoritics and Planetary Science, Vol. 33, pp. 513–517.

The Aouelloul impact crater and environs on September 27, 2018 as a composite image based on TanDEM-X DEM and Copernicus Sentinel-2 (channel 2/3/4) data.

Aouelloul

Mauritania

latitude = 20°14‘28“ N / longitude = 12°40‘28“ W / diameter = 390 m / age = 3.1 ± 0.3 Myr

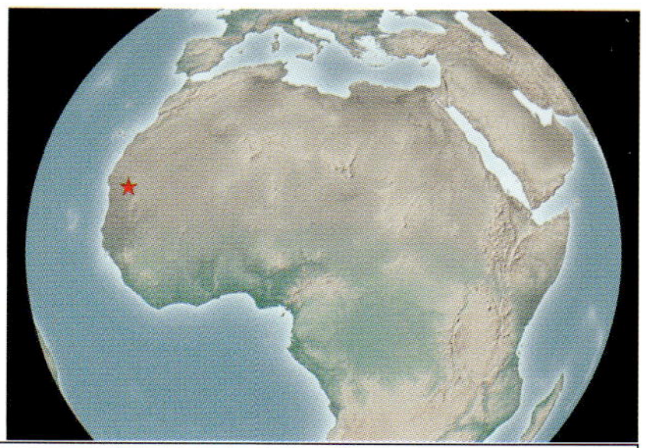

scale = 1 : 86 000 / scene = 15.5 × 20.6 km²

0 0.86 4.30 km

N

425 meter 775

Bosumtwi
Ghana

Wolf Uwe Reimold

The Bosumtwi impact structure is located in the Ashanti region of Ghana, at about 32 km east of the provincial capital city of Kumasi. Bosumtwi ranks amongst the best-studied impact structures in the world. The impact structure is only 1.07 Myr old and is arguably the youngest and best preserved terrestrial impact structure larger than 6 km in diameter. Bosumtwi's somewhat asymmetrical structure is marked by a pronounced rim, with a rim-to-rim diameter of 10.5 km. The rather straight southern margin must have been caused when the impact occurred against the flank of the up to 800 m high Obuom Mountain Range that trends northeast-southwest tangentially to the impact structure. The range represents a major regional watershed. The crater structure is to a large extent filled by Lake Bosumtwi with a diameter of 8.5 km and a maximum depth of about 80 m. The crater rim rises 250–300 m above the lake level. Bosumtwi's location in the equatorial region, much of which is covered by tropical forest, in a rather warm and humid climate with high rainfall, favors strong chemical weathering. In the entire area, lateritic soils and pediments have been developed; in places, they are tens of meters thick.

The Bosumtwi impact crater was excavated in lower greenschist facies metasediments of the 2100–2200 Myr Birimian Supergroup – mainly sandstones, schists, meta-graywackes, and shales. Birimian metavolcanic rocks occur to the southeast of the crater. Several Proterozoic granite intrusions can be found in the area of the structure, and some strongly weathered granitic dikes were mapped along the inner crater rim. Granitic complexes and dikes probably mainly belong to the 2000–2200 Myr Kumasi-type granite intrusions.

There are three types of breccias at Bosumtwi: the apparent monomict breccia that is thought to be autochthonous, a lithic polymict breccia that is thought to be allochthonous, and allochthonous breccia with glass/melt fragments, i. e., suevite. The Bosumtwi suevite is a glass/melt-bearing breccia that is similar to the suevite of the Ries crater. It contains up to 40 cm sized clasts of the target rocks that display all stages of shock metamorphism, including the transformation to vitreous and devitrified impact glasses. Suevite samples were analyzed as early as 1961 when coesite was documented; later Ni-rich iron spherules and baddeleyite were described from vesicular glass. Planar deformation features (PDFs) have been described from target rock clasts in suevite as well. At Bosumtwi, suevite occurs in exposures of up to several meter size within a 2.5 km wide zone around the crater. Shallow drilling has proven the presence of this impact breccia beyond this limit of exposure.

Seismic reflection data identified a central uplift with a diameter of almost 2 km, a height of more than 100 m and an estimated stratigraphic uplift of at least 800 m. However, this central peak is completely submerged below Lake Bosumtwi. The apparent crater depth amounts to 550 m. Already in the 1980s a weak negative gravity anomaly had been noted over the crater structure. In 1997, a high-resolution aero-geophysical survey was flown to better characterize the subsurface structure below and around the lake. This was the basis for a magnetic model for the Bosumtwi structure that showed a magnetic halo at ca. 6 km from the center, around the structure. The suevites were shown to have comparatively stronger magnetization, lower densities, and higher porosities than the target

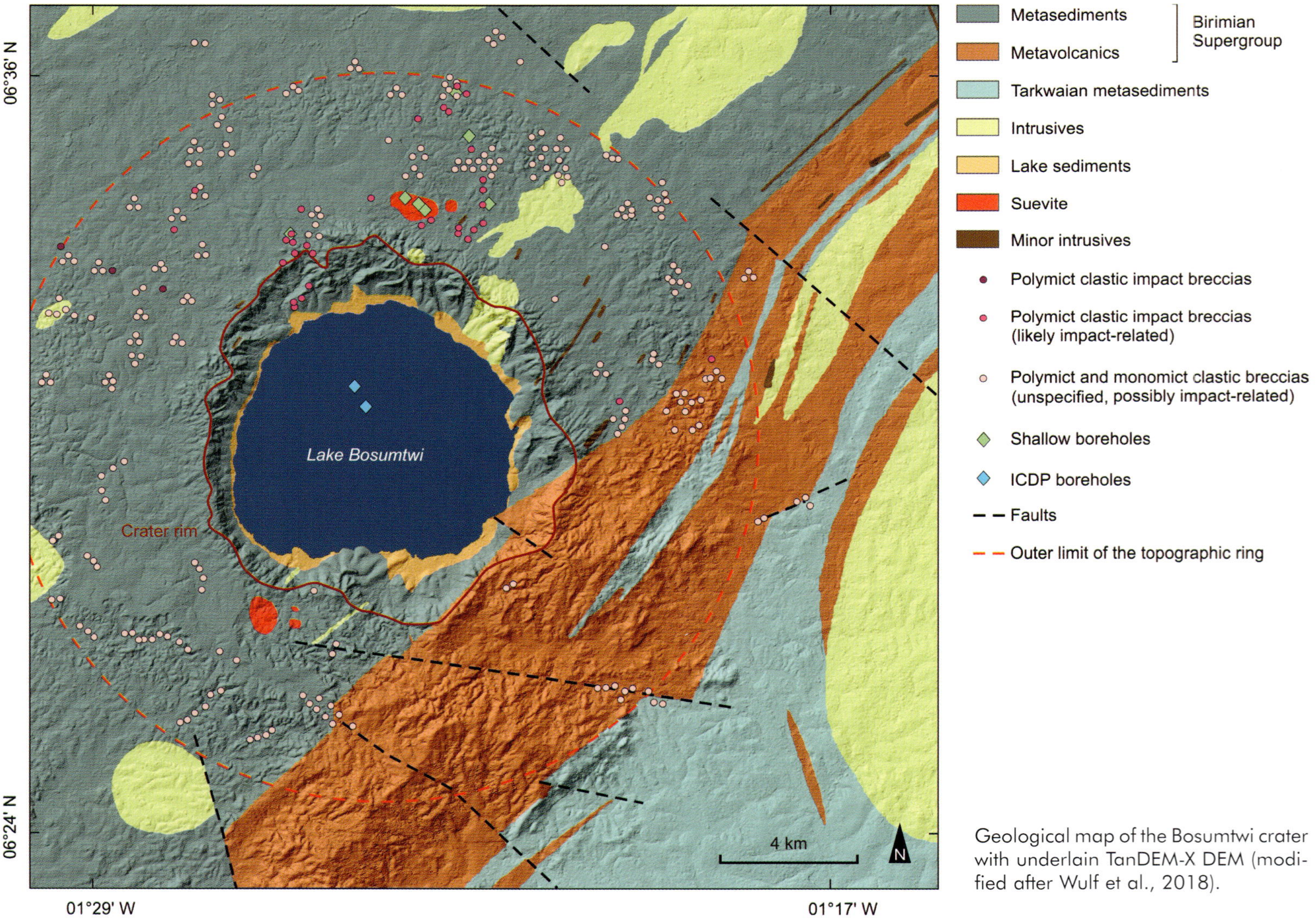

Geological map of the Bosumtwi crater with underlain TanDEM-X DEM (modified after Wulf et al., 2018).

Bosumtwi

Ghana

latitude = 06°30'09" N / longitude = 01°24'27" W / diameter = 10.5 km / age = 1.07 Myr

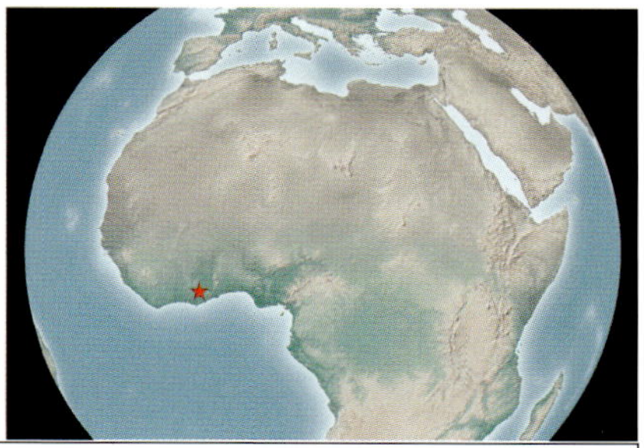

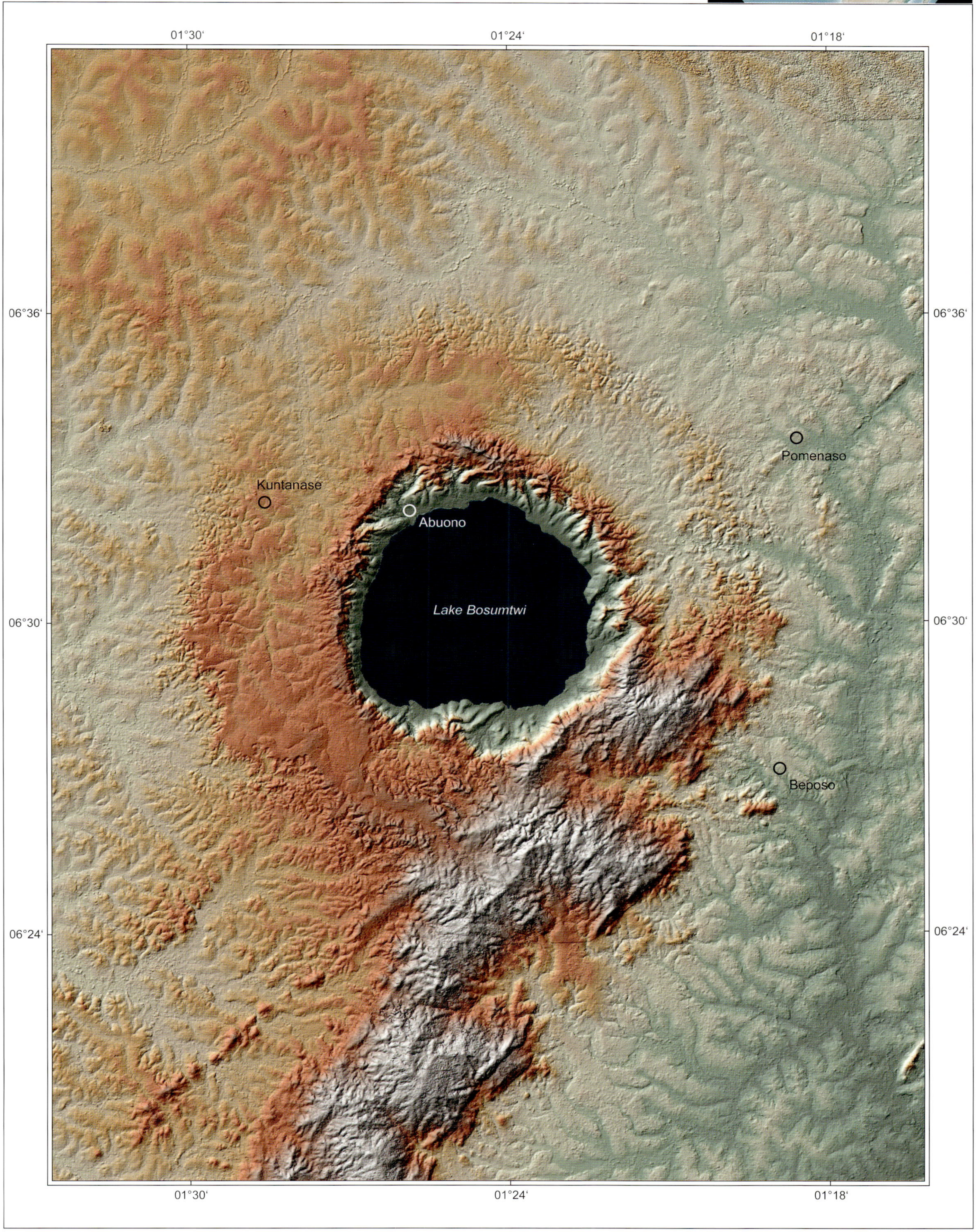

scale = 1 : 168 000 / scene = 30.2 × 40.2 km²

0 1.68 8.40 km

N

100 meter 735

A large block of suevite impact breccia on the northern crater rim of the Bosumtwi structure. The polymict clast population includes mainly Birimian metasedimentary lithologies (image courtesy: C. Koeberl, Naturhistorisches Museum Wien, Austria).

Bosumtwi suevite, 9 × 5 × 3 cm in size, from the southern part of the structure (photo courtesy: Ralf-Thomas Schmitt, Museum für Naturkunde Berlin, sample from E. W. Sharp, University of South Carolina).

Sparse outcrops outside the Bosumtwi crater exhume polymict breccia of the crater's ejecta blanket (photo: C. Koeberl, Naturhistorisches Museum Wien, Austria)

Intensely brecciated metasedimentary rock of the mid-section of the Bosumtwi crater rim, along the tar road to the village of Abuono at the lakeshore. A large light-colored leucogranitic dike injection into the Birimian metasedimentary rocks has been dismembered. The three parts of this megablock are displaced by faults. A younger co-author (WUR) for scale (photo: C. Koeberl, Naturhistorisches Museum Wien, Austria).

rocks. A near-circular, slight depression was already noted in the early 1980s at 7–8.5 km from the center of the crater, and a shallow outer topographic ring feature at 18–20 km from the center. Especially this outer feature is evident in spaceborne radar images as well as in aero-radiometry data. Combining remote sensing and geomorphological analyses with landform evolution modeling, Gerwin Wulf and co-workers could show in 2018 that the morphological and spectral characteristics of the ejecta outside the crater exhibit striking similarities to those of Martian rampart craters, especially double-layered ejecta craters. In the same year, the group of David Baratoux arrived at the same conclusions regarding ejecta fluidization and rampart formation, from radiometric data analysis and geomorphological constraints.

In 2004 a drilling project was launched at Bosumtwi by the International Continental Scientific Drilling Program (ICDP). The objectives were primarily to make use of the fact that this equatorial crater represents a closed basin with presumably undisturbed environment, which would have the prospect of delivering a unique, 1 Myr paleoclimatic record for the equatorial region. In addition, the pre-impact geophysical studies and core-based analysis would provide a wealth of information about such a mid-sized, complex impact structure, including a complete section through the impact breccia deposits in the central parts of the crater. At the top of one of these drill cores, the transition between the underlying impact breccias and the post-impact sediments was completely recovered. This includes a 30 cm thick layer representing the basal, fine-grained impact fallback from the ejecta plume. It contains in its top 10 cm accretionary lapilli, microtektite-like glass spherules, and shocked quartz grains. The composition of these spherules is very similar to that of Ivory Coast tektites and microtektites, with the exception of elevated CaO in the spherules. The presence of this fallback layer proved that the complete impactite sequence was recovered at Bosumtwi, and that Bosumtwi is a well-preserved impact structure.

A magnetic survey that had been carried out prior to drilling seemingly indicated the presence of a large amount of melt within the crater fill. However, drill core analysis revealed that this interpretation was wrong, in that much more melt had been predicted than was actually observed. Also interesting, much more melt had been incorporated into the suevite sampled outside of the crater, in comparison with the very low amount observed in suevite from the drill cores from the crater interior.

As Lake Bosumtwi is a hydrologically closed lake beneath the path of seasonal migration of the Intertropical Convergence Zone (ITCZ), it was anticipated from earlier shallow drilling that the sedimentary record of the lake beds would be a unique repository of information about the past changes of monsoon activity in West Africa. The continuous 300 m long drilling record of lake sediments from Lake

Aerial view of Lake Bosumtwi with its prominent crater rim. On the left, a photo taken from a helicopter and, on the right, a close-up of the drilling-barge of the ICDP drilling project (images courtesy: C. Koeberl, Naturhistorisches Mueum, Wien).

Bosumtwi provides the longest, continuous lacustrine sequences obtained from an extant lake on equatorial Earth. The detailed investigation of these drillings showed that Bosumtwi contains a unique record of Late Quaternary climate change for West Africa.

The origin of the Bosumtwi crater structure was first discussed in the 1930s, when geologists such as Herbert Rohleder or Norman Junner related it to "cryptovolcanic" activity. In the 1960s multiple definitive evidence for impact such as coesite, evidence pointing at a meteoritic component or baddeleyite was reported, and in the 1980s other shock deformation in clasts within suevite was added to this. In the mid-1990s further work on site and in the laboratory on impactites from Bosumtwi was carried out, and around that time the idea of

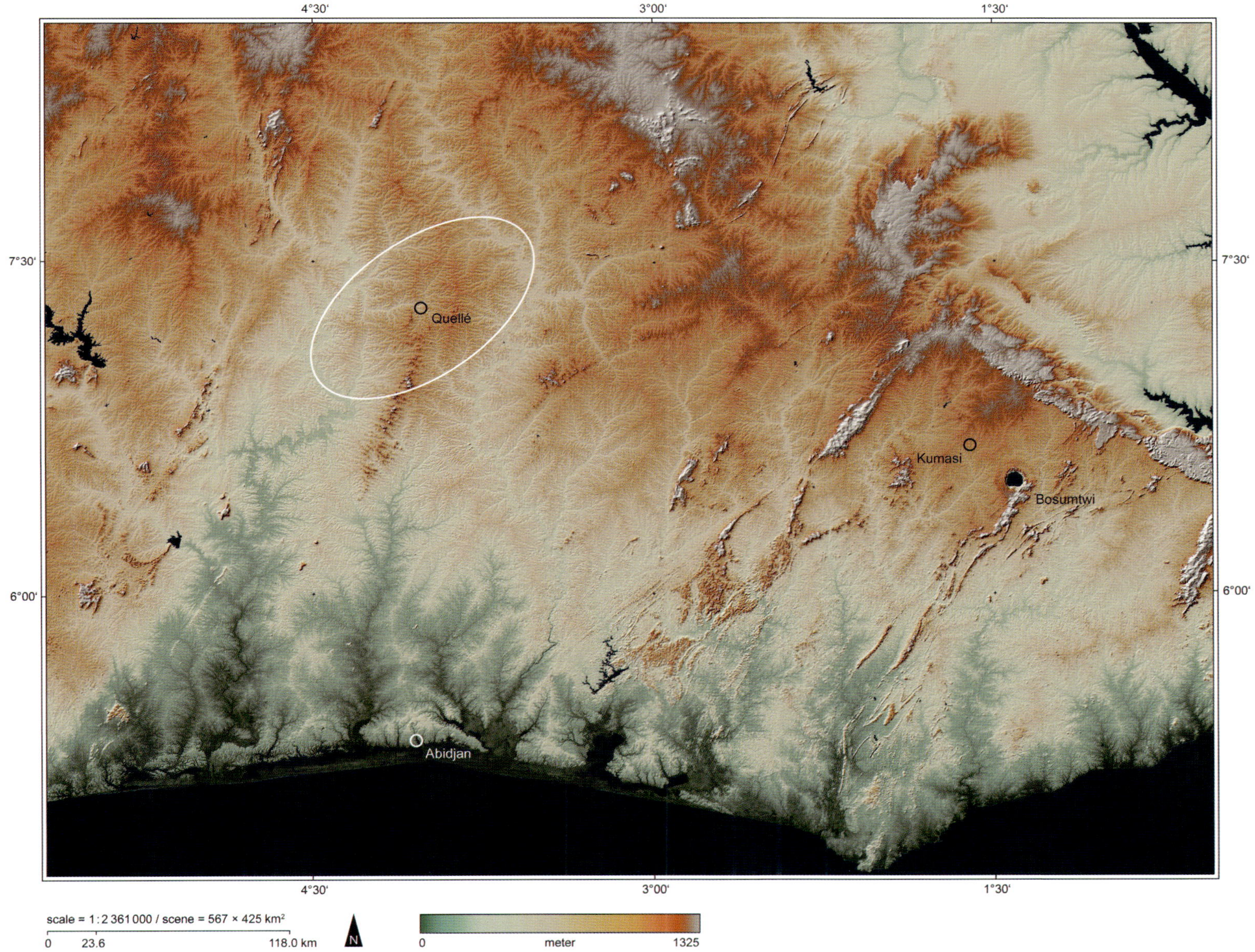

The location of the Ivory Coast tektite strewn field (white ellipse) and Bosumtwi, the source impact crater, in a physical map generated from the TanDEM-X 90m DEM.

a drilling project into Lake Bosumtwi was born – which brought the culmination of research at this intriguing and well-preserved structure. The findings of this drilling campaign were published in 2007 by Christian Koeberl and his team.

Bosumtwi is also known as the source crater for the **Ivory Coast tektites**. These were first reported in 1934 from an area of about 40 km radius close to the town of Quellé in the central Ivory Coast. Since then they have only yielded a limited number of tektites, much less than the samples found from the Australasian or Central European strewn fields. The link between Bosumtwi and the Ivory Coast tektites is based on related chemical and isotopic compositions of the target rock sequence at Bosumtwi and of the tektites, and because of the similar ages for impact melt from Bosumtwi suevite and for Ivory Coast tektites. Microtektites were found in deep-sea cores off the coast of Western Africa and are understood to be related to the tektites found on land. These findings extended the geographical distribution of the Ivory Coast tektite strewn field out into the Atlantic Ocean to an area more than 20° (latitude) × 30° (longitude) wide.

Location/Access

Starting point to access Lake Bosumtwi is the provincial capital, the city of Kumasi. Kumasi can be reached from Accra by road – some 250 km, with a minimum of 5 hours driving time on this very busy and quite dangerous road – or by air. From Kumasi take the paved Lake Road to Kuntanase, and from there the winding road down the crater rim to Abuono at the lakeshore (on aggregate 35 km). There is a reasonable hotel on the northwest lakeshore. Most roads around the lake are not paved and a four-wheel drive car is recommended.

Selected References

Baratoux, D., Niang, C. A. B., Reimold, W. U., Salah, M. S., Jessell, M. W., Boamah, D., Faye, G., Bouley, S., and Vanderheaghe, O. (2019). Bosumtwi impact structure, Ghana: Evidence for fluidized emplacement of the ejecta. Meteoritics and Planetary Science, Vol. 54, pp. 2541–2556.

Koeberl, C. and Reimold, W. U. (2005). Bosumtwi impact crater, Ghana (West Africa): an updated and revised geological map, with explanations. Jahrbuch der Geologischen Bundesanstalt Vol. 145, Heft 1, pp. 31–70 (one map at 1:50,000).

Koeberl, C., Milkereit, B., Overpeck, J. T., Scholz, C. A., Amoako, P. Y. O., Boamah, D., Danuor, S. K., Karp, T., Kueck, J., Hecky, R. E., King, J., and Peck, J. A. (2007). An international and multidisciplinary drilling project into a young complex impact structure: the 2004 ICDP Bosumtwi impact crater, Ghana, drilling project – an overview. Meteoritics and Planetary Science, Vol. 42, pp. 483–511.

Koeberl, C., Milkereit, B., and Reimold, W. U. (2007). The ICDP Drilling Project in Bosumtwi, Ghana. Meteoritics and Planetary Science, Vol. 42, pp. 477–896.

Reimold, W. U., and Koeberl, C. (2014). Impact structures in Africa – A review. Journal of African Earth Sciences, Vol. 93, pp. 57–175.

Wulf, G., Hergarten, S., and Kenkmann, T. (2018). Combined remote sensing analyses and landform evolution modeling reveal the terrestrial Bosumtwi impact structure as a Mars-like rampart crater. Earth and Planetary Science Letters, Vol. 506, pp. 209–220.

B.P.
Libya

Wolf Uwe Reimold

Close to the intersection of the borders of Libya, Egypt and Sudan, in southeast Libya there are two long-known impact structures. Both the **Oasis** and B.P. structures were discovered by petroleum exploration geologists and named after their respective oil companies. The B.P. structure is a small, but complex impact structure. In its interior, it has a small central uplift structure of strongly folded sandstone, much of it with steeply upturned bedding. At 1 km from the center, there is a notable ring of up to 50 m high hills that was interpreted to represent the remains of the crater rim. Sandstone dips at 30–50° outward, and parts of the rim are folded and faulted. Further out at about 1.4 km and at 1.7 km radius from the center of the structure, there are more, subdued, no more than 20 m high ring features preserved over some sector of the circumference. This outermost ring structure likely represents a shallowly inward dipping fault structure. It is suggested that it may have developed during collapse of the crater structure, when the material was directed inward in the course of central uplift formation. The overall not very prominent morphology of this crater structure is strongly suggestive of advanced erosion. Although a diameter for B.P. of 2 km is often cited, the morphology and structure indicate a diameter of perhaps as much as 3.4 km.

No impact breccia has been found in this crater structure to date. Country rocks in the B.P. area are sandstones that are more or less enriched in iron. There are also some intercalated siltstones and, locally, conglomerate exposures. These rocks are considered to belong to the Nubia Sandstone Formation, for which an Early Cretaceous age is often given in the literature. However, care has to be taken, as the stratigraphy and age of the Nubia Formation are not straightforward, with contrasting ideas given in the respective Libyan and Egyptian mapping reports. In the environs of the B.P. structure, ferruginous sandstones form the caps on many of the ranges and inselberge. These exposures show generally (sub)horizontal attitudes, with at most gentle warping of the layers. In contrast, the rocks at B.P. must certainly be considered deformed, at both the macroscopic and mesoscopic scales. Strata in the small central uplift area show highly varied orientations of bedding, from horizontal to vertical attitudes. The age of B.P. is only constrained by the assumed age of the Nubian Sandstone Formation (90–120 Myr) – giving an upper age limit for the impact. No materials susceptible to radiometric dating, i.e., melt breccia or authigenic minerals formed after impact, have been discovered to date.

A first indication of this structure was given in 1967 by Truman Kohman and co-workers on the basis of space and aerial photography, but it was A. Martin two years later who first discussed B.P. as a possible impact structure. Initial field analysis was carried out by James Underwood in the mid-1970s, with Underwood and Edward Fisk adding further details about the geology of the structure in 1980. Bevan French and his team carried out a detailed petrographic study of medium- to coarse-grained ortho-quartzite samples from the B.P. structure in 1974. Quartz grains displaying multiple sets of planar deformation features (PDFs, "planar elements" was the term applied then) were discovered, which confirmed the impact origin for B.P.

The samples referred by Christian Koeberl and co-workers in 2005 have so far not exhibited any planar deformation features, only planar fractures (PFs). Wolf Uwe Reimold in 2010 collected a further sample suite from the central uplift. This produced a few thin sections with shock metamorphosed quartz. One or two sets of PDFs occur in – again rare – quartz grains. PDFs are generally short and occur in very small patches within, or at the edges of, the host grains. In addition, PFs occur, in up to three sets per host grain, and feather features (FFs) as well. This visit was the last to the structure before Libya skittered into the throes of civil war.

Location/Access

The B.P. impact structure occurs in the remote southeastern part of Libya, within the triangle between southeast Libya, southwest Egypt, and northwest Sudan. Provided the civil unrest in this country will abate at some stage, a traveler to B.P. will in all likelihood begin the journey in the capital city of Tripoli in order to ensure that all travel permits have been obtained and are in order. The route from there

B.P.

Libya

latitude = 25°19'08" N / longitude = 24°18'37" E / diameter = 3.4 km / age = < 120 Myr

scale = 1 : 111 000 / scene = 20.0 × 26.7 km²

0 1.11 5.55 km

N

335 meter 500

A panoramic view onto the ca. 100 m wide central uplift of the 2 km diameter B.P. impact structure. The sedimentary strata in the central uplift are strongly deformed by faulting and fracturing, as well as brecciation. They also locally display effects of ductile deformation in the form of more or less pronounced folding (photo: W. U. Reimold).

Approaching the B.P. structure from the west – only the highest hills of the crater rim are visible, and the scenery much resembles the inselberge dominated environs of the impact structure. Only with the help of a GPS may it be possible to hit the target (photo: W. U. Reimold).

will follow the main road along the Mediterranean Sea in direction of Benghazi. At Ahbadija one takes the equally paved road southwards towards Al Kufrah Oasis. This town is the last option to replenish supplies before heading off into the desert. Al Kufrah does have an airport. Already in 2010 it was a dicey undertaking to go from there into the desert towards the east, as those areas were heavily patrolled by army, and the threat that smugglers might want to harm visitors was mentioned all the time. The crater is located about 165 km northeast of Al Kufrah Oasis, and about 80 km due north of the Oasis impact structure. There were a large number of vehicle tracks in the desert between Al Kufrah and Oasis in 2010. Even in politically quiet times, it is recommended to have local guides and guards. Equipment including multiple GPS and a satellite telephone is a must.

Selected References

French, B. M., Underwood, J. R., and Fisk, E. P. (1974). Shock metamorphic features in two meteorite impact structures, south-eastern Libya. Geological Society of America Bulletin, Vol. 85, pp. 1425–1428.

Koeberl, C., Reimold, W. U., and Plescia, J. (2005). BP and Oasis impact structures, Libya: remote sensing and field studies. In: Impact Tectonics, Impact Studies Series Series, Koeberl, C., and Henkel, H. (eds.), Springer, Berlin Heidelberg, pp. 161–190.

Martin, A. J. (1969). Possible impact structure in Southern Cyrenaica, Libya. Nature, Vol. 223, pp. 940–941.

Underwood, J. R. (1975). Reconnaissance geology of meteorite impact structures in SE Libya. Abstracts and Programs of the Geological Society of America, Vol. 7, p. 242.

Underwood, J. R. (1976). Impact structures in the Libyan Sahara: some comparisons with Mars. Proceedings of the International Colloquium on Planetary Geology. Geologica Romana, Vol. 15, pp. 337–340.

Underwood, J. R., and Fisk, E. P. (1980). Meteorite impact structures, Southeast Libya. 2nd Symposium on the Geology of Libya, Tripoli, Libya, pp. 893–900.

The B.P. (indicated by cross hairs) and Oasis (bottom) impact structures in September 2018 as a composite image based on TanDEM-X DEM and Copernicus Sentinel-2 (channel 2/3/4) data.

Gweni Fada
Chad

Wolf Uwe Reimold

The Gweni Fada structure in northern Chad appears strongly eroded and is somewhat asymmetric: the northwest-southeast diameter is slightly longer than the diametrically opposite one. The crater structure comprises a broad depression of about 12 km diameter that forms a crescent around two-thirds of the structurally complex innermost zone. The broad ring syncline is delimited on its outside by apparently steeply dipping limestones. To the north, the syncline is delimited by an elevated outer ring of outward-dipping sandstones. The southern part of Gweni Fada lacks the broad depression and exhibits tilted/folded sandstones. In the innermost rugged terrain, hills of up to several hundred meters elevation are composed of sandstones of varied dips. Gweni Fada – like **Aorounga** 320 km to the northwest – is located in Upper Devonian strata. This represents the only (maximum) age constraint for this structure.

Landsat 5 satellite imagery and Shuttle Radar Topography remote sensing data indicated a near-circular structure with a slight offset of a central rugged area from its center to the south that is surrounded in the western, northern and eastern sectors by an apparently flattish terrain. The crater seems to be truncated in its southern sector by a west-southwest/east-northeast trending fault that was active after the impact event and must represent the reason for the offset on the crater rim. In contrast to earlier workers, who favored a diameter of 12–14 km, more recent research considered the actual diameter of the Gweni Fada structure to be perhaps as large as 21–23 km, which would require a central uplift area of 7–9 km in diameter. This was indeed shown for Gweni Fada by Christian Koeberl and co-workers in 2005.

The interesting Gweni Fada structure was first reported in 1996 from Landsat imagery and aerial photography by Pierre Vincent and Alain Beauvilain. They visited the structure and returned sandstone samples with alleged shock metamorphic effects in quartz. Later, Koeberl and colleagues reported shock evidence for a number of quartzite, sandstone, and quartz conglomerate samples from the central area and from the outer margin of the ring depression. Several samples exhibited – altogether rare – quartz grains with 1 or 2 sets of planar deformation features (PDFs). These two petrographic investigations provided definite evidence for impact. However, further work at both Gweni Fada and **Aorounga** might still be rewarding with regard to a better understanding of shock metamorphism of sandstones and perhaps unearthing of datable material.

The Gweni Fada impact structure on May 18, 2018 as a composite image based on TanDEM-X DEM and Copernicus Sentinel-2 (channel 2/3/4) data.

Location/Access

The Gweni Fada structure is located some 320 km southeast of the Aorounga impact structure and 30 km northeast of the Fada palm grove, in the Ennedi district of northern Chad. There is need for further detailed ground analysis of this impact structure. However, like the rest of northern and western Africa, Chad and adjacent territories have been subject to intense political unrest for the past decades, so that any attempt to venture towards Gweni Fada must be strongly discouraged at this time.

Selected References

Koeberl, C., Reimold, W. U., Cooper, G., Cowan, D., and Vincent, P. (2005). Aorounga and Gweni Fada impact structures, Chad: remote sensing, and petrography and geochemistry of target rocks. Meteoritics and Planetary Science, Vol. 41, pp. 1455–1471.

Vincent, P. M., and Beauvilain, A. (1996). Découverte d'un nouveau cratère d'impact météoritique en Afrique: l'astroblème de Gweni Fada (Ennedi, Sahara de Tchad). Comptes Rendus de l'Académie des Sciences, Paris, Vol. 323, pp. 987–997.

Wacrenier, P., Hudeley, H., and Vincent, P. M. (1958). Carte géologique du Borkou-Ennedi-Tibesti a 1/100000. Brazzaville, Direction des Mines et la Géologie.

Gweni-Fada

Chad

latitude = 17°25'07" N / longitude = 21°45'17" E / diameter = 22 km / age = < 355 Myr

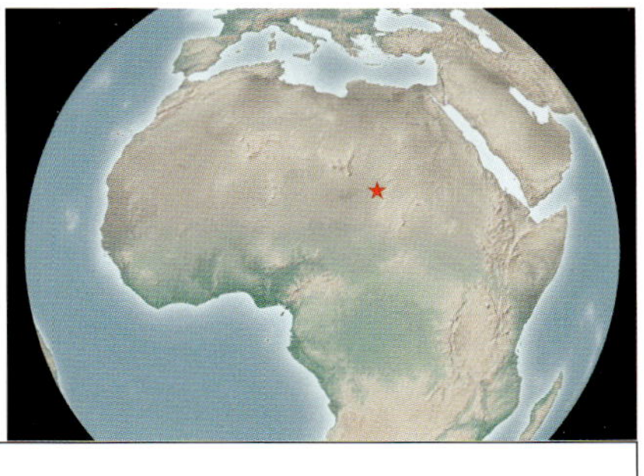

scale = 1 : 197 000 / scene = 35.5 × 47.3 km^2

0 1.97 9.85 km

N

Kalkkop
South Africa

Wolf Uwe Reimold

The Kalkkop, or *Kalkkoppie* in Afrikaans, impact crater measures only 640 m in diameter. This nearly circular structure is located in South Africa's Eastern Cape Province between the towns of Graaff-Reinet and Jansenville, actually about 50 km south of Graaff-Reinet. The structure is obvious on aerial photographs as a circular, light-colored area surrounded by a darker annulus. The interior of the crater is filled with light-colored limestone. The outer ring feature comprises upturned sandstone and shale of the Koonap Formation of the Beaufort Group of the Karoo Supergroup, which are rather poorly exposed. The best exposure known is located just southwest of the road along the northern side of the structure.

Non-impact diagnostic fracturing is widespread in clasts of brecciated bedrock. However, the critical evidence was obtained when the crater was drilled in 1992. The core transected the crater sediment fill to a depth of 89 m, then impact breccia to 160 m, and finally to 380 m depth the sedimentary basement. Subsequently, rare planar deformation features (PDFs) were discovered in quartz from this polymict impact breccia. Some samples of this breccia also contained a very small component of impact melt fragments, partially altered by carbonate, which indicates that these breccias are suevite. Re-Os isotopic analysis of breccia, as well as of some Beaufort Group country rocks (sandstone and mudstone) showed that the breccia has a meteoritic component, obviously from the projectile. This represented one of the first successful applications of this isotope technique for the confirmation of the presence of a meteoritic component.

U-Th series dating of samples from the upper and lower crater fill indicated that the impact age is of the order of 0.250 ± 0.050 Myr. This is – within error limits – similar to the age of the **Tswaing** impact, and it has been suggested that, maybe, Kalkkop and Tswaing could be the results of a dual impact. One way to try to proof – or disproof – this idea might be to compare the compositions of the two projectiles that caused the formation of the Kalkkop and Tswaing craters. Both impact breccias contain a small meteoritic component – but analysis of the highly siderophile elements in further samples would be required and is still pending.

Kalkkop was drilled in the 1940s, in the cause of energy-resource evaluation of the wider Karoo Basin. The limestone crater fill was also investigated in the 1980s when the Geological Survey of South Africa, now the Council for Geoscience, commissioned a survey of the country's limestone resources. Further drilling with extensive core recovery was carried out in 1992 and proofed unambiguously the impact origin, when Wolf Uwe Reimold and his team published the results of their detailed petrographic and geochemical studies on the core in 1998.

Location/Access

The crater is located about 50 km south of the town of Graaff-Reinet, on the old, unpaved road to the small town of Jansenville. Kalkkop is located just south of the crossing of this track over the Bulsrivier stream. The track skirts the crater on its north-northeast margin, where there is a low outcrop of seemingly upturned bedrock in contact with limestone of the crater fill. A sturdy vehicle is a must on this track. The nearest airport from which to stage an expedition to the crater is in the city of Port Elisabeth.

Selected References

Koeberl, C., Reimold, W. U., Shirey, S. B., and Le Roux, F. G. (1994). Kalkkop crater, Cape Province, South Africa: confirmation of impact origin using osmium isotope systematics. Geochimica et Cosmochimica Acta, Vol. 58, pp. 1229–1234.

Reimold, W. U., Koeberl, C., and Reddering, J. S. V. (1998). The Kalkkop impact crater, Eastern Cape Province, South Africa: stratigraphy, petrography, and geochemistry of the 1992 drill core, and age. Journal of African Earth Sciences, Vol. 26, pp. 573–592.

Reimold, W. U., and Koeberl, C. (2014). Impact structures in Africa – A review. Journal of African Earth Sciences, Vol. 93, pp. 57–175.

Kalkkop suevite sample, $6 \times 3.5 \times 3$ cm in size, from the 1992 drill core by the Council for Geoscience, Pretoria (photo courtesy: Ralf-Thomas Schmitt, Museum für Naturkunde Berlin, sample from W. U. Reimold).

Kalkkop
South Africa

latitude = 32°42‘31“ S / longitude = 24°25‘56“ E / diameter = 640 m / age = 0.25 ± 0.05 Myr

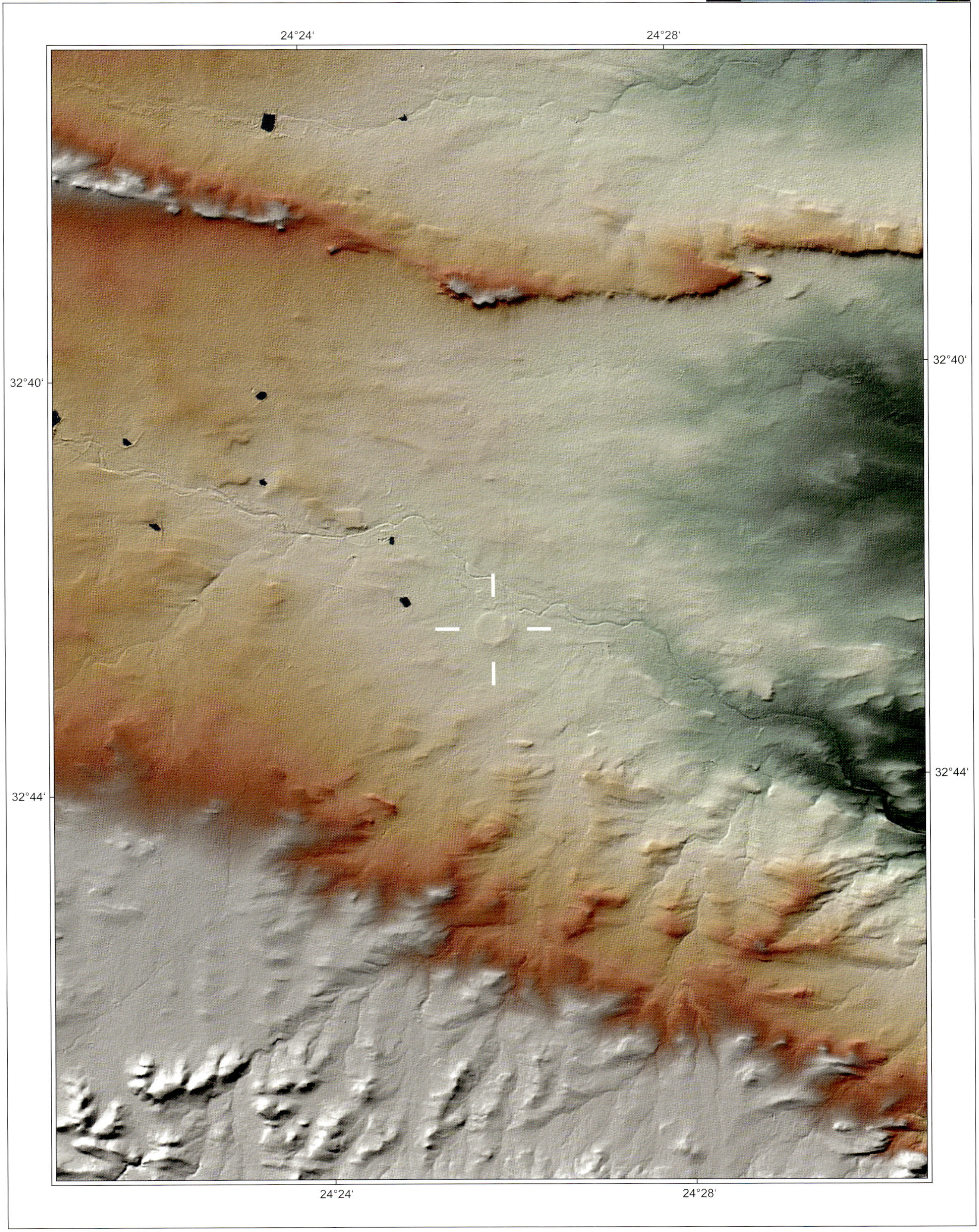

scale = 1:84 000 / scene = 15.1 × 20.1 km^2

0 0.84 4.20 km

N

575 meter 915

Kamil
Egypt

Wolf Uwe Reimold

In the Libyan Desert of the East Uweinat District of southwestern Egypt occurs a merely 45 m wide, perfectly preserved impact crater – known as Kamil. This little crater marvel is preserved including a pristine ejecta blanket, with extensive ejecta ray structures. Kamil, to date, represents the only confirmed impact structure in Egypt.

The crater was formed in pale sandstones, mainly quartz arenite in exposure, of the Early Cretaceous Gilf Kebir Formation. Locally this formation is overlain by some loose soil. The sandstone is sub-horizontally bedded and forms part of the sedimentary cover that uncomformably overlies the Precambrian crystalline basement.

The small "craterling" has a typical simple bowl shape. The rim is raised by about 3 m above the pre-impact surface. The true crater floor is located about 16 m below the rim crest. It is overlain by an about 6 m thick crater fill composed of blocks and stratified fallout. Along the crater walls, the bedding of the sandstone is upturned by up to 40°, and dips are oriented radially outward. An overturned flap of ejected material was locally observed on the rim crest. One of the striking features of this crater is a continuous ejecta blanket that extends radially for about 50 m from the crater rim in northerly, easterly, and southerly directions. Three ejecta rays extend up to 350 m from the crater rim into northerly, southeasterly, and southwesterly directions.

Systematic field work has been conducted in the Kamil area by a joint Italian and Egyptian expedition team. More than 5000 iron meteorite specimens, to a total amount of 1.7 tons, were identified within the crater and in its environs. The meteorite type has been identified as a Ni-rich ataxite, and the meteorite has been named *Gebel Kamil*. The total mass of the impactor has been estimated from the multidisciplinary geological-geophysical field work to 10 tons – but noble gas and radionuclide analysis even suggest that the original meteoroid may have had a mass of as much as 20 tons or more. Meteorite fragments all have the characteristics of shrapnel that was generated when the extraterrestrial impactor was shredded upon impact. Therefore, it is suggested that the Kamil projectile may have impacted essentially intact, instead of having been ripped apart while traversing through the atmosphere. As meteoritic debris was found to be strongly concentrated to the southeast of the crater and is more or less absent from the northwestern environs, the investigators presumed that the impactor approached Earth obliquely from the northwest.

View of Kamil crater from the west (image courtesy: Luigi Folco, Università di Pisa, Italy).

The 83 kg regmaglypted individual found 230 m due north of the crater (image courtesy: Mario di Martino, Istituto Nazionale di Astrofisica, Torino, Italy).

Shock metamorphic features have been found that suggest a wide range of shock pressures (locally in some samples up to about 60 GPa) and temperatures (even exceeding 1700 °C). This includes the possible occurrence of shatter cones on sandstone – although these features are not very well formed, very thin melt veinlets in sandstone samples from ejecta, and planar deformation features (PDFs) in quartz. Also the high-pressure silica polymorphs coesite and stishovite, silica glass (lechatelierite), and baddeleyite from the shock-induced decomposition of zircon in shocked sandstone have been detected. In addition, impact melt lapilli and bombs, as well as microscopic impact spherules have been reported to occur scattered around the crater.

The Kamil crater was found only in 2008 by Vincenzo de Michele, former curator of the Natural History Museum in Milan,

Kamil

Egypt

latitude = 22°01‘06“ N / longitude = 26°05‘15“ E / diameter = 45 m / age = 3020 ± 600 yr

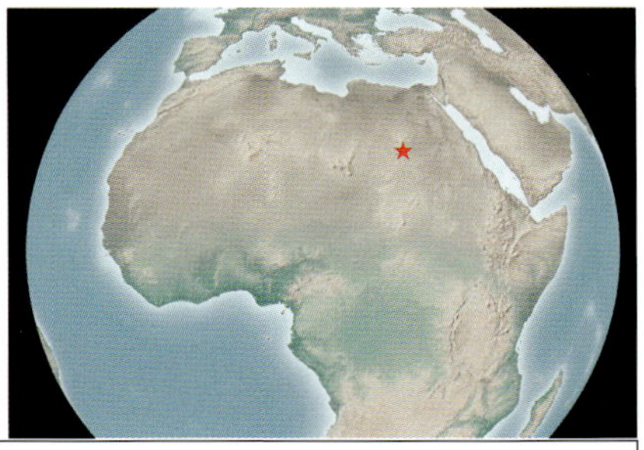

scale = 1 : 58 000 / scene = 10.4 × 13.8 km²

0 0.58 2.90 km

N

560 meter 830

during a Google Earth™ survey. Geological-geophysical field work was then carried out in the Kamil area as part of the 2010 Italian-Egyptian geophysical investigation by Luigi Folco and co-workers, which confirmed the impact origin of the crater. Subsequent studies by Folco and his team using some geo-archeological reasoning, such as the chronological relationship between the impact crater and a number of other features, e.g., trails, or settlements attesting to prehistoric human occupation of this area, suggested that the impact took place at less than 5000 years ago. In 2015 this quite recent age was confirmed when Gian Paolo Sighinolfi and co-workers arrived at a preferred age interval between 400 and 1600 before current epoch (BCE) by thermoluminescence dating of quartz in shocked target rocks.

Location/Access

Travelling in Egypt is not that easy. There are permits to acquire. First of all, the location of Kamil crater is near the Egyptian-Sudanese border. This means that for military reasons, this area is generally off-limits. It is advisable to make a first approach through an Egyptian embassy. It may also help to have Egyptian cooperation partners, but in all likelihood there are still several permits from ministries to be obtained along the way. Access from Cairo would then involve driving on tar via Farafra and Dakhla to Bir Tarfaui. From there to Kamil one would cross the Great Sand Sea desert. Travel duration would be more than two days.

Selected References

D'Orazio, M., Folco, L., Zeoli, A., and Cordier, C. (2011). Gebel Kamil: the iron meteorite that formed the Kamil Crater (Egypt). Meteoritics and Planetary Science, Vol. 46, pp. 1179–1196.

Fazio, A., Folco, L., D'Orazio, M., Frezzotti, M. L., and Cordier, C. (2014). Shock metamorphism and impact melting in small impact craters on Earth: Evidence from Kamil Crater, Egypt. Meteoritics and Planetary Science, Vol. 49, pp. 2175–2200.

Folco, L., Di Martino, M., El Barkooky, A., D'Orazio, M., Lethy, A., Urbini, S., Nicolosi, I., Hafez, M., Cordier, C., van Ginneken, M., Zeoli, A., Radwan, A. M., El Khrepy, S., El Gabry, M., Gomaa, M., Barakat, A. A., Serra, R., and El Sharkawi, M. (2010). The Kamil Crater in Egypt. Science, Vol. 329, p. 804.

Folco, L., Di Martino, M., El Barkooky, A., D'Orazio, M., Lethy, A., Urbini, S., Nicolosi, I., Hafez, M., Cordier, C., van Ginneken, M., Zeoli, A., Radwan, A. M., El Khrepy, S., El Gabry, M., Gomaa, M., Barakat, A. A., Serra, R., El Sharkawi, M. (2011). Kamil Crater (Egypt): ground truth for small scale meteorite impact on Earth. Geology, Vol. 39, pp. 179–182.

Folco, L., Mugnaioli, E., Gemelli, M., Masotta, M., and Campanale, F. (2018). Direct quartz-coesite transformation in shocked porous sandstone from Kamil Crater (Egypt). Geology, Vol. 46, pp. 739–742.

Folco, L., Reimold, W. U., and El Barkooky, A. (2019). Impact craters and meteorites: the Egyptian record. In: The Geology of Egypt, Hamimi, Z., El Barkooky, A., Martínez-Frías, J., Fritz, H., and Abdelrahman, Y. M. (eds.), Springer Publishers, Regional Geology Reviews, pp. 415–444.

Sighinolfi, G. P., Sibilia, E., Contini, G., and Martini, M. (2015). Thermoluminescence dating of the Kamil impact crater (Egypt). Meteoritics and Planetary Science, Vol. 50, pp. 204–213.

Urbini, S., Nicolosi, I., Zeoli, A., El Khrepy, S., Lethy, A., Hafez, M., El Gabry, M., El Barkooky, A., Barakat, A., Gomaa, M., Radwan, A., El Sharkawi, M., D'Orazio, M., and Folco, L. (2012). Geological and geophysical investigation of Kamil Crater, Egypt. Meteoritics and Planetary Science, Vol. 47, pp. 1842–1868.

Kgagodi Basin

Botswana

Wolf Uwe Reimold

The 3.4 km diameter Kgagodi Basin impact structure, located approximately 7 km south of Kgagodi village in southeastern Botswana, does not have any significant surface geological expression. It is recognizable by a weak annular pattern on satellite imagery. This is not surprising, as the large southern African country of Botswana is covered to 80 % by Cretaceous to Holocene deposits of the Kalahari Group. Therefore, it was considered likely that first hints at the presence of a possible impact structure in that country would come predominantly from geophysical surveys. For the initial discovery of Botswana's so far only confirmed impact structure, extensive drilling for water resources was key in this semi-arid to arid terrane.

Quartz grain with multiple sets of PDFs in a granitic clast from a suevite sample in drill core from the Kgagodi impact structure. The field of view is 0.9 mm wide. Plane polarized light image (photo: W. U. Reimold).

The geology in the area comprises Archean granitoid basement, with local occurrence of Karoo dolerite with an age of about 180 Myr. A large outcrop of Karoo dolerite occurs just to the southeast of the crater structure. Thus, it is thought that the impact occurred into Karoo strata overlying granitic basement. The only surface outcrops in this area are a few small calcrete exposures, most of which were actually shown to delineate the subsurface crater rim.

The only firm age constraint for the impact event is provided by the age of the Karoo dolerite, which provides a maximum age limit. Palynological evidence in the form of thermally altered pollen from impact breccia suggests an Upper Cretaceous to Early Tertiary age for the impact event.

The presence of an anomalous structure was first recognized in 1997/1998 in the course of a water drilling project. The perceived structure was drilled by the Geological Survey of Botswana to a depth of 274 m. Tentative indications of the presence of a small basin structure were supported by interpretation of aerial photography and satellite imagery. B. K. Paya and co-workers showed that the drill core contained a breccia, which lead them to propose Kgagodi Basin as an impact structure. They offered a gravity model consistent with a 3.4 km wide, simple bowl-shaped crater. According to this model, the drill core was drawn at about 400 m inside of the crater's edge. However, their initial petrographic analysis did not reveal the necessary proof in the form of shock metamorphic evidence. That was finally reported by Dion Brandt and his team in 2002, after they had resampled the drill core. The core comprises crater-fill sediment to a depth of 158 m, which is followed by impact breccia to a depth of 165 m, below which locally brecciated basement rocks were drilled that gradually show reduced deformation to 250 m depth, where undeformed basement is reached. Shock deformation could only be

Kgagodi Basin
Botswana

latitude = 22°28'28" S / longitude = 27°34'48" E / diameter = 3.4 km / age = < 180 Myr

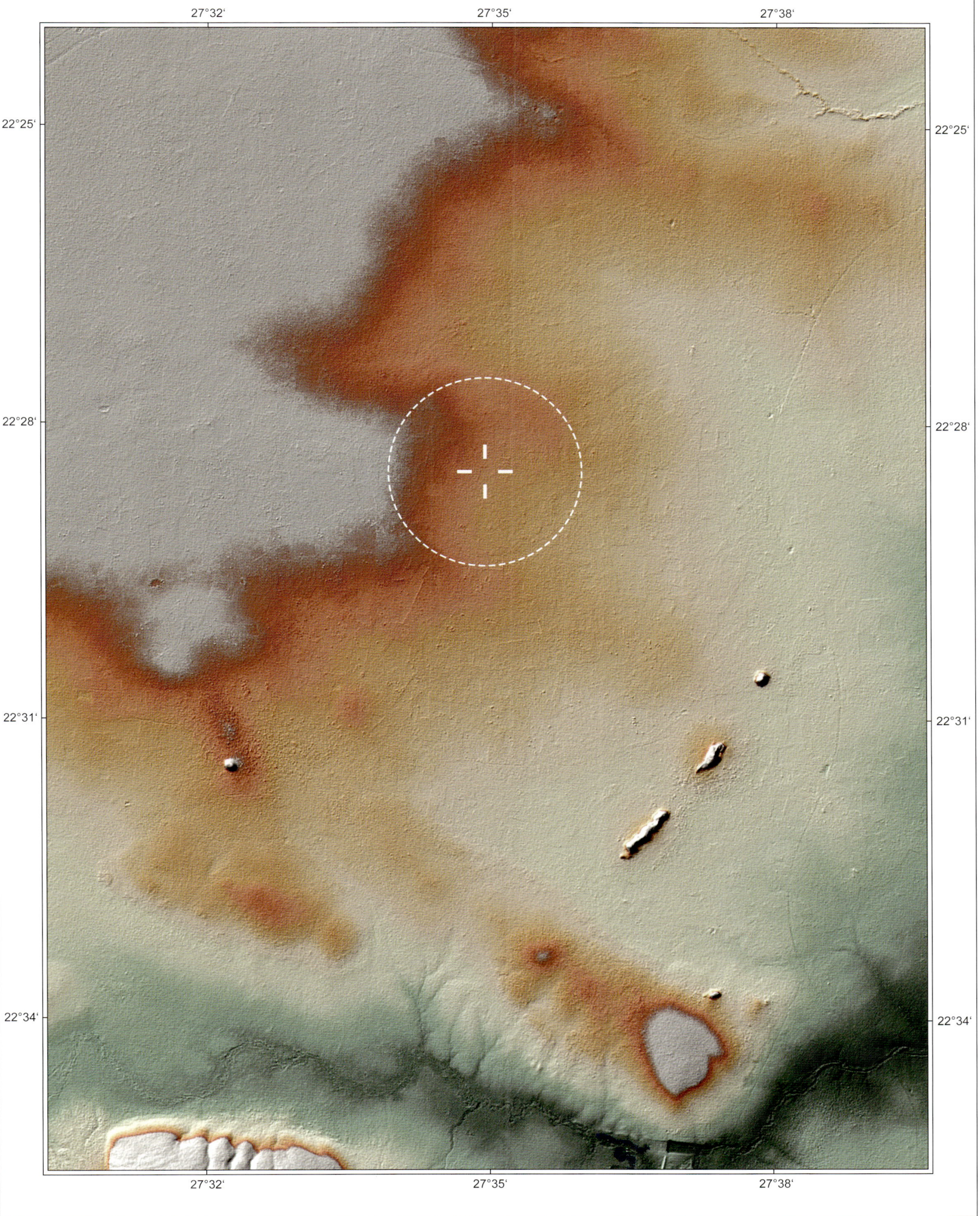

scale = 1:89 000 / scene = 15.9 × 21.3 km²

0 0.89 4.45 km

N

830 meter 1030

Polymict lithic impact breccia from the drill core obtained from the Kgagodi impact structure, Botswana (photo: W. U. Reimold).

observed in granitic clasts from the impact breccia interval. Multiple sets of planar deformation features (PDFs) in quartz and feldspar, diaplectic quartz glass, and partially to completely isotropized feldspar were reported. They also observed rare melt fragments and consequently classified the impact breccia as suevite. It was also established that the abundances of some siderophile elements and of iridium were enhanced in breccia samples, compared to the background levels in the target rocks. They interpreted this as evidence for a small projectile component in the suevite.

In recent years several efforts have been undertaken to solicit support from the Council for Geoscience (Geological Survey) in South Africa and the Geological Survey of Botswana for a new drilling project in the center of the structure. The argument is that this small impact crater has enormous potential to provide a unique, long-term paleo-environmental record for this part of the southern hemisphere. From gravity modeling it is assumed that drilling in the center of the crater to a maximum drilling depth of about 800 m might be required – unless additional geophysical surveying would reveal that there is a small central uplift hidden at depth.

Location/Access

From Gaborone follow the A1 national road in the direction of Francistown, for some 300 km via the towns of Mpokane, Dinokwe, Mahalapye and Palapye. Some 40 km after Palapya leave the A1 towards the east and follow the road to Swartwater (South Africa). After about 40 km you will drive south of the village of Kgagodi, where a subsidiary road turns off to the south-southeast that will take you after 7 km directly into the Kgagodi Basin structure. The whole tour should take perhaps 4 hours. Should you go about 2 km too far, you will be confronted by a large dolerite outcrop telling you to "turn around!"

The structure can also be accessed from Johannesburg, via the N1 past Pretoria, and on to Bela-Bela, where the R33 to Groblersdal and into Botswana should be taken. Subsidiary roads allow to reach the structure from the southern side.

Selected References

Brandt, D., Holmes, H., Reimold, W. U., Paya, B. K., Koeberl, C., and Hancox, P. J. (2002). Kgagodi Basin: the first impact structure recognized in Botswana. Meteoritics and Planetary Sciences, Vol. 37, pp. 1765–1779.

Paya, B. K., Holmes, H., Reimold, W. U., and Farr, J. (1999). Kgagodi Basin: a possible impact structure (abstract). 62nd Annual Meeting of The Meteoritical Society, Johannesburg, South Africa.

Reimold, W. U., and Koeberl, C. (2014). Impact structures in Africa – A review. Journal of African Earth Sciences, Vol. 93, pp. 57–175.

Libyan Desert Glass

Egypt

Wolf Uwe Reimold

The origin of the enigmatic Libyan Desert Glass (LDG), a natural high-silica glass that occurs in corridors between the high sand dunes of the southwestern-most part of the Great Sand Sea of Egypt, has been debated ever since its first scientific investigation by Patrick Clayton and L. Spencer in 1934. Since then, not less than hundreds of scientific papers have dealt with the physical-mineralogical-chemical characteristics of LDG. Most of the authors have agreed that this must be an impact-generated material, so-called "impact glass". However, a source crater has remained missing.

LDG is found as irregular shaped masses up to several tens of centimeters in size. They occur on exposed sandstone of the Cretaceous Nubia Formation in rather linear corridors between some 100 m high and 3–5 km apart, roughly north-south-trending sand dunes, so-called "seifs", of the Great Sand Sea. The glasses are found within a more or less oval region of approximately 95 × 30 km. A major concentration of finds is centered around 25°25'N, 25°30'E. The total mass of LDG has been estimated at roughly 10^3 t. However, due to loss by weathering and burial in the soil, the original mass could have been much greater. Primitive people used it for tool-making, and in Egyptian dynastic times it was also used as a semi-precious stone. The largest LDG specimen so far found has a mass of about 26 kg and is curated in the Muséum National d'Histoire Naturelle in Paris, France.

Color of LDG varies from bright yellow-green to nearly colorless, pale yellow, to grey-green. Some specimens contain dark bands and schlieren, and others may be opaque to milky white due to the presence of abundant vesicles. Pale and dark, millimeter-sized inclusions can be discerned with the naked eye in some samples.

Mineralogical analysis has shown that LDG contains microscopic quartz, cristobalite, lechatelierite (pure silica glass), zircon and its thermal decomposition product baddeleyite, anatase, rutile and decomposition products of this mineral, Ti-magnetite, corundum, Al-rich orthopyroxene, and mullite inclusions. The occurrences of lechatelierite, cristobalite, baddeleyite, decomposed Ti-magnetite, and mullite indicate that melting took place at temperatures exceeding 1670 °C. Also emulsion textures in the dark bands have been revealed by transmission electron microscopy: nanometer-scale spherules of glass enriched in Al, Fe and Mg in liquid immiscibility with the surrounding silica glass. Recently evidence for former reidite, the high-pressure polymorph of the mineral zircon, has been detected in granular zircon from LDG.

Many geologists have investigated the chemical composition of LDG. It is extremely silica-rich, at 96.5 to 99 wt%. Lithophile elements are inversely correlated with silica, and the LDG rare earth element pattern is typical of upper continental crust materials – suggestive of a dominant quartz-rich sedimentary source. Strontium and Nd isotopic abundances and oxygen isotopic compositions point at a mature Cretaceous sandstone (or sands), most likely derived from the erosion of intrusive rocks of Neoproterozoic age as the dominant precursor material for LDG. Several studies have suggested that the precursor rock of LDG was most probably quartz sand, perhaps from weathering of Cretaceous sandstones from the Gilf Kebir Plateau, which had been deposited in a high-energy, fluviatile environment.

The fact that dark bands in glass specimens are enriched in Cr, Co, Ni and Ir, and as the contents of these elements are positively

Libyan Desert Glass (Strewn Field)

Egypt

latitude = ~ 25°25‘ N / longitude = ~ 25°30‘ E / diameter = unknown / age = 29 ± 1 Myr

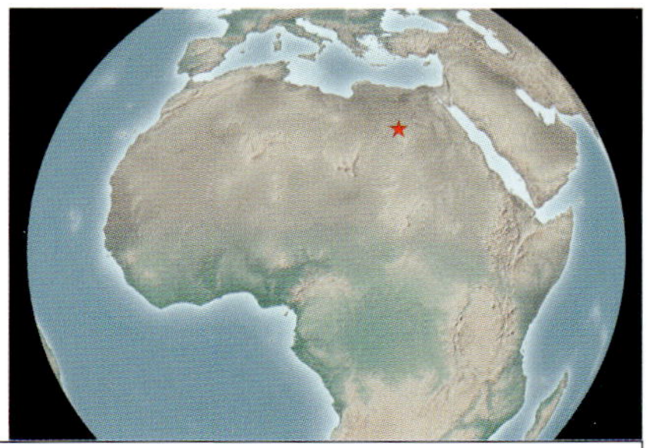

scale = 1 : 535 000 / scene = 96.3 × 128.4 km²

0 5.35 26.75 km

N

310 meter 690

Two specimens of Libyan Desert Glass (3.5 and 2 cm long), from the collections of the Museum für Naturkunde Berlin (photo courtesy: Ralf-Thomas Schmitt, Museum für Naturkunde Berlin).

correlated with those of Fe and Mg, has been interpreted that these bands contain a contribution from a meteoritic component. Several studies have favored a chondritic meteoritic projectile type. Re-Os isotopic analysis, and some radiogenic isotope work confirmed the presence of an extraterrestrial component, and showed that only a negligible mantle-derived component may have been involved.

Two independent fission-track ages obtained for the LDG have been reported and suggest an age of 29 ± 1 Myr.

The very high formation temperatures for a range of mineral inclusion types and the presence of a chondritic component in the dark bands are strong evidence that Libyan Desert Glass represents impact glass. In the past both the **B.P.** and the **Oasis** impact structures located some 150 km west-southwest of the LDG strewn field in Libya have been implicated as possible source craters. However, the ages of these two impact craters are still unknown, so that proper correlation has remained elusive. Some workers have resorted to a hypothesis that the LDG was formed locally, by radiative heating/melting caused by a low-altitude atmospheric airburst of a cometary or asteroidal object without forming a crater in the ground.

A further development in the long history of LDG analysis dates to 2013, when Jan Kramers and co-workers reported about a dark, $3.5 \times 3.2 \times 2.1$ cm, carbon-dominated, diamond-bearing rock fragment from the vicinity of the LDG strewn field. They called this specimen "Hypatia". Their analysis of the Ar isotopic composition of LDG glass and non-terrestrial $\partial^{13}C$ values lead them to the conclusion that the fragment was most likely of extraterrestrial origin. They speculated that it could be a fragment from a comet nucleus that had exploded in the terrestrial atmosphere. Two years later Guillaume Avice and his team confirmed the extraterrestrial nature of Hypatia with noble gas and nitrogen isotopic analysis.

Other complications regarding the local strewn field relate to isotopic evidence that suggests that different specimens of glass could have had very different exposure histories, suggesting that major redistribution of glass could have occurred since its formation. And then there is evidence for the loss of some 400 m of sand and rock deposit since formation of the LDG, which opens the question what this may mean in terms of erosion of a possible impact crater and redistribution of glass. Already 20 years ago, several float specimens of shocked sandstone from the area of the LDG strewn field were discussed as possible evidence for an actual impact cratering event. And recently, shock-produced planar micro-deformation features, namely planar fractures (PFs), planar deformation features (PDFs), and feather features (FFs), were found in quartz grains from sandstone sampled in situ in this area. This strongly supports that there was a physical impact event, not just an airburst, at this site. Although it remains difficult to understand that such an impact crater would be nearly entirely eroded but that evidence in the form of essentially fresh impact glass is still present widely, some of it in the form of large blocks. Perhaps the high silica content of the glass may have served its preservation.

Location/Access

Travelling in Egypt is not that easy. There are permits to acquire. First of all, the location of the LDG area is near the Egyptian-Libyan border. This means that for military reasons, this area is generally off-limits. It is advisable to make a first approach through an Egyptian embassy. It may also help to have Egyptian cooperation partners, but in all likelihood there are still several permits from ministries to be obtained along the way.

Selected References

Avice, G., Meier, M. M. M., Marty, B., Wieler, R., Kramers, J. D., Langenhorst, F., Cartigny, P., Maden, C., Zimmermann, L., and Andreoli, M. A. G. (2015). A comprehensive study of noble gases and nitrogen in "Hypatia", a diamond-rich pebble from SW Egypt. Earth and Planetary Science Letters, Vol. 432, pp. 243–253.

Belyanin, G. A., Kramers, J. D., Andreoli, M. A. G., Greco, F., Gucsik, A., Makhubela, T. V., Przybylowicz, W. J., and Wiedenbeck, M. (2018). Petrography of the carbonaceous, diamond-bearing stone "Hypatia" from southwest Egypt: A contribution to the debate on its origin. Geochimica et Cosmochimica Acta, Vol. 223, pp. 462–492.

Clayton, P. A., and Spencer, L .J. (1934). Silica Glass from the Libyan Desert. Lecture presented on the 9 November 1933 at the Annual Meeting of the Mineralogical Society. Mineralogical Magazine, Vol. 23, pp. 501–508.

Folco, L., Reimold, W. U., and El Barkooky, A. (2019). Impact Craters and Meteorites: the Egyptian Record. In: The Geology of Egypt, Hamimi, Z., El Barkooky, A., Martínez-Frías, J., Fritz, H., and Abdelrahman, Y. M. (eds.), Springer Publishers, Regional Geology Reviews, in press.

Koeberl, C., and Ferrière, L. (2019). Libyan Desert Glass area in western Egypt: Shocked quartz in bedrock points to a possible deeply eroded impact structure in the region. Meteoritics and Planetary Science, Vol. 54, pp. 2398–2408.

Kramers, J. D., Andreoli, M. A. G., Atanasova, M., Belyanin, G. A., Block, D., Franklyn, C., Harris, C., Lekgoathi, M., Montross, C. S., Ntsoane, T., Pischedda, V., Segonyane, P., Viljoen, K. S., and Westraadt, J. E. (2013). Unique chemistry of a diamond-bearing pebble from the Libyan Desert Glass strewnfield, SW Egypt: evidence for a shocked comet fragment. Earth and Planetary Science Letters, Vol. 382, pp. 21–31.

Luizi
Democratic Republic of Congo

Wolf Uwe Reimold

The central African region between the Sahara desert and the southern African subcontinent is essentially devoid of known impact structures. Large parts of central Africa are thickly covered with rain forest that hinders remote sensing analysis and does not really allow ground-based exploration of large areas. In addition, especially in the Democratic Republic of Congo (DRC), impact crater search has been severely hampered because of decades long civil war. Only a single impact structure, Luizi, has been confirmed in this huge terrane so far. The Luizi impact structure is located in a rather unexplored region on the Kundelungu Plateau of Katanga Province in the southeastern DRC. Some observations from satellite imagery were initially available and suggested an impact origin of this quite circular structure. Luizi is a 17 km diameter, complex crater structure, with an intermediate ring at ca. 5.2 km from the center, and an inner ring structure at 1 km from the center, around a central depression. This morphology is very similar to that of the Brazilian **Serra da Cangalha** structure. The outer rim at Luizi is elevated by 300–350 m above the crater interior. The topography over the structure is rugged, with a difference of elevation between the deeper part of the central depression and the rim, i. e., to the level of the plateau in the environs of the structure, of up to 350 m. The hills in the interior parts are between 50 and 125 m high. The structure superposes a north-northwest/south-southeast trending regional-scale shear zone on its eastern side. In addition, a pronounced semi-annular ring of hills at about 5 km from the center is observed. It is tempting to interpret this as the edge of the central uplift feature, which, in turn, is surrounded by a ring of subdued topography before the outer rim is reached.

Shatter cones in sandstone from the Luizi impact structure (image courtesy: Ludovic Ferriére, Natural History Museum in Vienna, Austria).

Outcrop of clast-rich impact melt rock from the Luizi impact structure in the Democratic Republic of Congo (image courtesy: Ludovic Ferriére, Natural History Museum in Vienna, Austria).

The Luizi structure was formed in massive, tabular sandstone with intercalated argillaceous sandstones of the Biano Subgroup. This facies is of late Neoproterozoic age and forms the uppermost part of the Kundelungu Group. In the outer parts of the structure, these strata lie subhorizontal, but in the inner parts they have steeply dipping to vertical attitudes. Abundant, well-developed shatter cones of up to 40 cm size in the inner, 3.2 km wide part of the structure can be noted. Within this zone occur also up to 2 m thick dikes of monomict lithic breccia, samples of which did not show any shock deformation. However, planar deformation features (PDFs) are abundant in quartz and rare in feldspar in samples of arkosic sandstone, with and without shatter cones, from the inner part of the structure. Apparently, the samples from the innermost 2 km of the structure were shock metamorphosed more strongly than the parts further out. Recently, several other breccia types have been noted. They include impact breccia dikes that intrude sandstone beds in the inner 1 km of the structure. These dikes consist of rock clasts and shocked mineral clasts that are in part or totally melted, showing flow texture and irregular shapes, in a fine-grained matrix with possibly local melt. In addition, there are large occurrences of clast-rich impact melt rock.

It does appear from the little that is known about this structure that Luizi has been eroded to a significant degree. The entire crater fill has been removed, and impact breccias have so far only been observed in the form of dike injections into bedrock. An age for the Luizi impact is only poorly constrained by stratigraphic considerations. The maximum age of the sedimentary strata of the Kundelungu Group in this region is ca. 573 Myr, which thus provides a maximum age for the impact. There is another somewhat controversial idea that the fault zone straddled by the impact structure could be of Neogene age, which would require that a youngish structure superposing this fault zone became deeply eroded in a geologically very short interval. Though, hardly anything is known about this regional-scale fault zone.

The Luizi structure was first mentioned in a geological report by E. Grosse back in 1919. More than 70 years later, Paul Dumont and Philippe Claeys with co-workers reported remote sensing information, and finally the impact origin of Luizi was confirmed in 2011 by ground-truthing by Ludovic Ferrière of the Natural History Museum in Vienna.

Location/Access

The easiest way to the Luizi structure, now that some bridges have been reconstructed, is to drive from the city of Lubumbashi about 200 km towards the northwest on National Road No. 5. A large part of this stretch is gravel road of reasonable condition. From the village of Lukeka one has to proceed on foot to cover the remaining 15 km towards the center of the structure in due west direction. Note, on this march through the bush you have to carry all your supplies for the duration of your trip, including water.

The DRC has been a hotbed for civil strife for a number of decades. Any visits to this country should be carefully planned, and any signs of civil war activity along the way towards Luizi should be carefully evaluated.

Selected References

Claeys, P., Chan, J. Ch.-W., and Dujardin, R. (2008). The Luizi structure: remote sensing study of a possible impact crater in Central Africa. 39th Lunar and Planetary Science Conference, Lunar and Planetary Institute, Houston, Texas, Abstract #1720.

Dumont, P. (1990). Plaidoyer pour une revalorisation des photographies aériennes. Bulletin de la Société belge de Géologie, Vol. 99, pp. 57–65.

Ferrière, L., Lubala, F. R. T., Osinski, G. R., and Kaseti, P. K. (2011). The newly confirmed Luizi impact structure, Democratic Republic of Congo – insights into central uplift formation and post-impact erosion. Geology, Vol. 39, pp. 851–854.

Grosse, E. (1919). Grundlinien der Geologie und Petrographie des östlichen Katanga. Neues Jahrbuch für Mineralogie, Geologie und Palaontologie, Beilage-Band, Vol. 42, pp. 272–419.

Master, S., Rainaud, C., Armstrong, R. A., Phillips, D., and Robb, L. J. (2005). Provenance ages of the Neoproterozoic Katanga Supergroup (Central African Copperbelt), with implications for basin evolution. Journal of African Earth Sciences, Vol. 42, pp. 41–60.

Reimold, W. U., and Koeberl, C. (2014). Impact structures in Africa – A review. Journal of African Earth Sciences, Vol. 93, pp. 57–175.

The Luizi impact crater on September 25, 2018 as a composite image based on TanDEM-X DEM and Copernicus Sentinel-2 (channel 2/3/4) data.

Luizi

Democratic Republic of the Kongo

latitude = 10°10'09" S / longitude = 28°00'21" E / diameter = 17 km / age = < 573 Myr

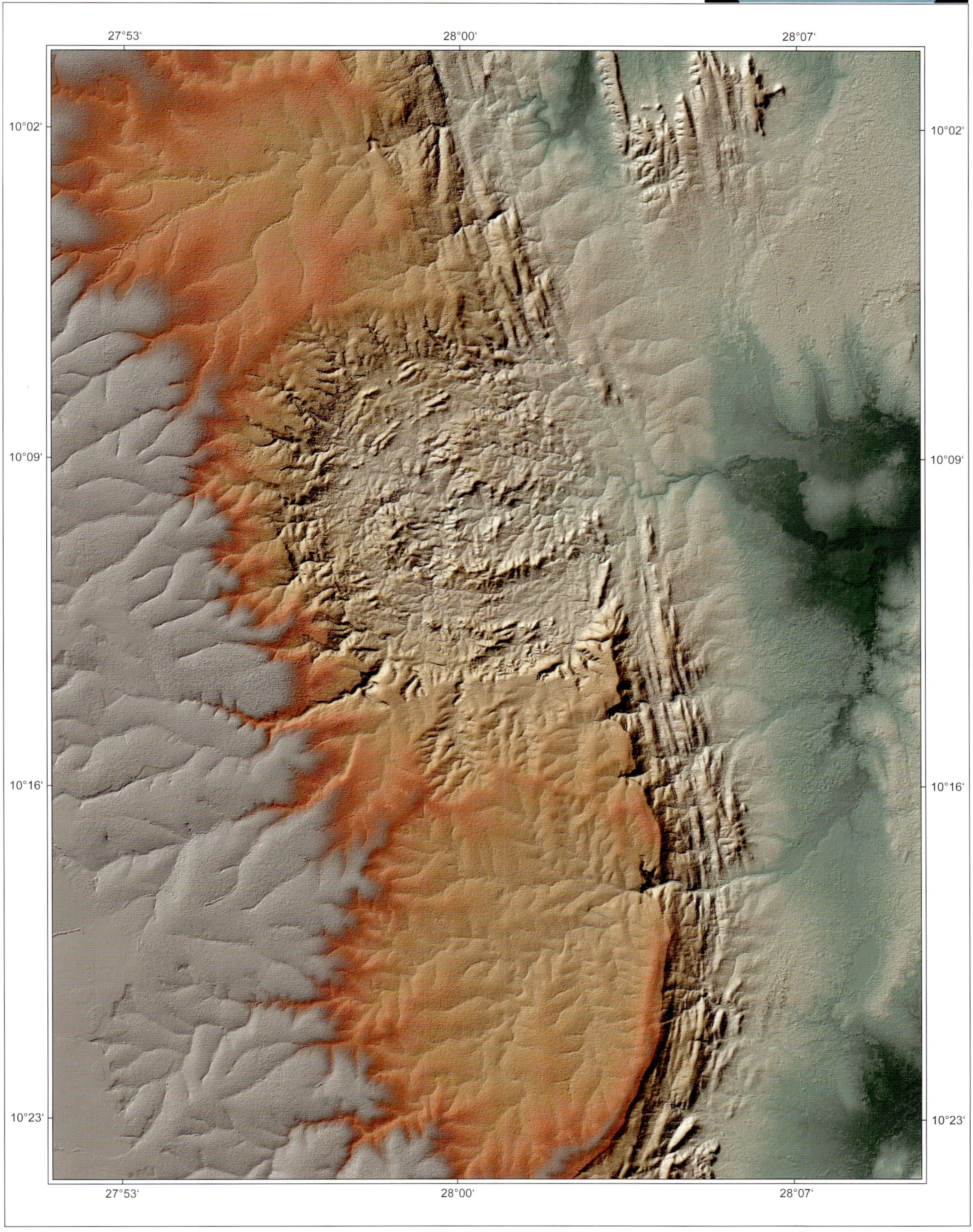

scale = 1 : 184 000 / scene = 33.1 × 44.2 km²

0 1.84 9.20 km

N

985 meter 1735

Morokweng
South Africa

Wolf Uwe Reimold

The second-largest confirmed impact structure in Africa is the 70 km wide Morokweng structure in the North West Province of South Africa. It is located in a semi-arid to arid terrane, which is part of the southern Kalahari Desert. The small town of Morokweng is located in the northern sector of the structure. The area of the impact structure is mainly covered by sand. Very rare rock exposures are essentially limited to river-beds, where Archean granitoids can be observed. In addition, there is a limited exposure – a low ridge – of horizontally disposed quartzite and banded iron formation near the hamlet of Heuningvlei – at the western margin of the structure. To the east of the Ganyesa Dome, Dwyka diamictite has been mapped in patches. Small inliers of amphibolitic schist occur in the central area and have been associated with the Archean Kraaipan Group that mainly occurs further to the east. To the west of the Dome, metasediments of the Proterozoic Transvaal Supergroup and Griqualand West Group occur, including quartzite, dolomite, jaspilite, and tillite, and extensive exposures of ironstones.

The area of the Morokweng structure is characterized by a sizable aeromagnetic anomaly over the Ganyesa Dome, a basement high made up of Archean granitoid. In the environs of the structure and beyond, the basement is overlain by Ventersdorp Supergroup volcanics and Transvaal Supergroup and Griqualand West Group carbonate and arenite. Based on the limited exposure in the actual area of the structure and the mapping of the wider region, the regional stratigraphy resembles that in the region around the **Vredefort** impact structure. A possible similarity of the target stratigraphy of Morokweng and Vredefort is also indicated by the nearly identical compositions of the Morokweng and Vredefort impact melt rocks.

The Morokweng impact structure is essentially not related to a gravity anomaly. The regional gravity field is determined by the regional pre-impact geology. The gravity anomaly over the impact structure was modelled as a broad positive anomaly of up to more than +20 mGal above regional background, however there is no indication that this could be related to impact lithologies. The Morokweng structure's main geophysical signature is found in the aeromagnetic anomaly map. Here a ca. 35–40 km wide central area interrupts a regional, northeast-southwest trending magnetic dike swarm thought to be related to the craton-wide, 2700 Myr Ventersdorp Supergroup magmatic event. The uplifted basement at Morokweng – in the region of the Ganyesa Dome – is rather magnetic and causes a weak but consistent long-wavelength magnetic high in the center of the impact structure. Over the width of this magnetic anomaly, the regional granitoid basement has been remagnetized, which was interpreted to indicate that there existed a coherent impact melt body extending over this entire area.

Morokweng impact melt rock, 5 × 4.5 cm in size, from drill core MWF-03 (photo courtesy: Ralf-Thomas Schmitt, Museum für Naturkunde Berlin, sample from W. U. Reimold).

Geophysical observations (gravity and aeromagnetics) fueled quite a bit of speculation regarding the size of this essentially buried impact structure. Values from more than 300 to 70 km were variably punted. The lower limit is, however, supported by three lines of evidence: The aeromagnetic data that have demonstrated the disruption of this dike swarm of pre-impact age at a distance of about 35–40 km from the center of the structure. Second, drill core from a deep exploration borehole located 35 km southwest of the center showed hardly any deformation that could be related to the impact event/structure. Finally, geophysical modelling of the Morokweng impact structure is also consistent with these other findings. Consequently, the diameter of Morokweng is generally accepted as 70–80 km.

The discovery and nearly everything that is known about this structure is based on analysis of drill cores. In 1995 Marco Andreoli of the University of the Witwatersrand and his team studied drill cores from the Ganyesa Dome and were the first to report the existence of impact melt rock, as well as shock deformation in underlying granitoids. Borehole stratigraphy involves – from top to bottom – about 7–110 m of calcrete and sandstone deposits of the Kalahari Group, followed by 90–870 m of impact melt rock. In two boreholes, the bottom contact between impact melt rock and underlying granitoid basement was intersected. The highly diverse thicknesses of impact melt rock intersections are suggestive of effects related to the collapse of the central uplift, as it is not possible to explain these different thicknesses based on the drilling locations with respect to the center of the structure. The very thick impact melt rock intersection also fueled some speculation about possible economic value of this impact melt body – in perceived analogy to the **Sudbury** Igneous Complex. However, these ideas remained without reward.

The first fully published report came from Branko Corner and co-workers in 1997, who investigated the regional geophysics and found – obviously erroneously – that an approximately 300 km wide, circular structure could be visualized in potential field data. As they also detected shock metamorphic deformation in the form of planar deformation features (PDFs) in quartz from a surface sample of a Transvaal Supergroup arenite, they could support the existence of an impact structure. Cornelis Bootsman and colleagues investigated the long-time evolution of the drainage system of the Molopo River, which in these parts of the northwest of South Africa forms the boundary with Botswana. From comparison of the pre-impact and post-impact drainage patterns, they could conclude that after the impact the drainage direction and the drainage pattern were both changed dramatically. Prior to the impact, at about 300 Myr ago, the regional drainage direction was towards the northwest, whereas at about 75 Myr ago, i. e., the time of filling of the Kalahari Basin, a shift to generally southward flow took place.

Much of the later work was dedicated to the impact melt rock that has been accessible through four drill cores. Two groups, one around Rodger Hart, and the second, Wolf Uwe Reimold with Christian Koeberl, respectively, found a significant enrichment of siderophile elements in comparison to the compositions of country rocks, and thus established the presence of a meteoritic component. Isotopic evidence showed that individual samples of melt rock may carry as much as 2–5 % of this component, which is high compared to samples from most other terrestrial impact structures. The impactor type became rather well constrained by chemical analysis to LL chondrite. This was confirmed when a small, 25 cm long chunk of the actual meteorite was detected in impact melt rock from one of the drill cores – quite an exceptional finding.

Morokweng

South Africa

latitude = 26°28‘ S / longitude = 23°32‘ E / diameter = 70 km / age = 145 ± 2 Myr

The age of the Morokweng impact melt rock was determined by two independent U-Pb single zircon dating studies to an aggregate age of 145 ± 2 Myr, which is within error limits the same age as that of the Jurassic-Cretaceous Boundary. This boundary is known for a significant mass extinction, that was, however, much less strong than that associated with the Cretaceous-Paleogene (K-Pg) boundary. An attempt to find evidence for impact in samples from the Jurassic-Cretaceous Boundary in southern England and northern France remained unsuccessful.

Location/Access

Of the two likely starting points to reach the Morokweng structure, Johannesburg is vastly preferable to Cape Town, as the travel distance of some 500 km, about 6 hours driving time, is significantly less than that for the 1150 km from Cape Town to the town of Morokweng. Heading west from Johannesburg on National Road N14, progress will be quite good until the town of Vryburg to the southeast of the structure, from where a secondary road will take you first to Ganyesa and then on to Morokweng. Basic supplies can be found in Morokweng, however the traveler wishing to stay longer, may want to find a base in Vryburg or Kuruman, the regional centers with excellent accommodation and provision possibilities.

Selected References

Andreoli, M. A. G., Ashwal, L. D., Hart, R. J., Smith, C. B., Webb, S. J., Tredoux, M., Gabrielli, F., Cox, R. M., and Hambleton-Jones, B. B. (1995). The impact origin of the Morokweng Ring Structure, Southern Kalahari, South Africa (abstract). Centennial Geocongress of the Geological Society of South Africa, Johannesburg, South Africa, pp. 541–554.

Bootsman, C. S., Reimold, W. U., and Brandt, D. (1999). Evolution of the Molopo drainage and its possible disruption by the Morokweng impact event at the Jurassic-Cretaceous boundary. Journal of African Earth Sciences, Vol. 29, pp. 669–678.

Corner, B., Reimold, W. U., Brandt, D., and Koeberl, C. (1997). Morokweng impact structure, Northwest Province, South Africa: geophysical imaging and shock petrographic studies. Earth and Planetary Science Letters, Vol. 146, pp. 351–364.

Hart, R. J., Andreoli, M. A. G., Tredoux, M., Moser, D., Ashwal, L. D., Eide, E. A., Webb, S. J., and Brandt, D. (1997). Late Jurassic age for the Morokweng impact structure, southern Africa. Earth and Planetary Science Letters, Vol. 147, pp. 25–35.

Henkel, H., Reimold, W. U., and Koeberl, C. (2002). Magnetic and gravity model of the Morokweng impact structure. Journal of Applied Geophysics, Vol. 49, pp. 129–147.

Maier, W.D., Andreoli, M. A. G., McDonald, I., Higgins, M. D., Boyce, A. J., Shukolyukov, A., Lugmair, G. W., Ashwal, L. D., Graser, P., Ripley, E. M., and Hart, R. J., (2006). Discovery of a 25-cm asteroid clast in the giant Morokweng impact crater, South Africa. Nature, Vol. 441, pp. 203–206.

Reimold, W. U., and Koeberl, C. (2003). Petrography and geochemistry of a deep drill core from the edge of the Morokweng impact structure, South Africa. In: Impact Markers in the Stratigraphic Record. Impact Studies Series, Koeberl, C., and Martinez-Ruiz, F. (eds.), Springer, Berlin Heidelberg, pp. 271–292.

Reimold, W. U., and Koeberl, C. (2014). Impact structures in Africa – A review. Journal of African Earth Sciences, Vol. 93, pp. 57–175.

Oasis
Libya

Wolf Uwe Reimold

The Oasis impact structure comprises an about 5.5 km wide inner ring of up to 100 m elevated hills, but the crater is actually much larger. In that, the structure does resemble the **Gosses Bluff** impact structure in Australia. Oasis is located about 120 km east-northeast of Al Kufrah oasis and about 80 km due south of the **B.P.** impact crater. Like the B.P. structure, Oasis is formed in Nubian Sandstone Formation thought to be 90–120 Myr old. Rocks of the inner ring predominantly dip outward. This ring surrounds a more or less flat, innermost plain with local dune sand accumulation and pavement style outcrops. The inner ring of strongly deformed, intensely folded and also faulted strata is surrounded by an also quite flat annular ring syncline. This morphology can be interpreted as suggesting rather significant erosion of the Oasis impact structure.

A view across the gravel and sand cover of the desert to the north of the Oasis impact structure, onto the prominent ring of hematized (limonitized), highly weathering-resistant chimney features (sand dikes?). They delimit the ring basin around the central uplift of Oasis on the northern side. Field of view about 3 km wide (photo: W. U. Reimold).

Findings of *Lepidodendron* and trace fossils in some strata of the inner ring clearly show that the Oasis structure was not only formed in Nubia Formation sandstone and conglomerate, but that the structure also involves sandstones, siltstones and claystones of Upper Carboniferous age. In the hills of the inner ring and in the innermost depression, some lithic breccia occurrences were noted, although no contact relationships to country rock could be observed due to very low outcrop and partial sand cover. The prominent hills of the inner ring are capped by erosion-resistant, siliceous sandstones of Lower Cretaceous age, whereas the surrounding flattish basin seems to

Oasis

Libya

latitude = 24°34‘33“ N / longitude = 24°24‘43“ E / diameter = 18 km / age = < 120 Myr

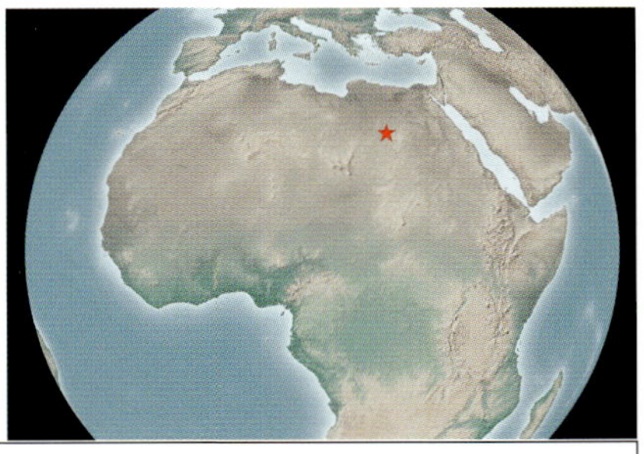

scale = 1 : 176 000 / scene = 31.7 × 42.3 km²

0 1.76 8.80 km

N

370 meter 625

An example of three of those possible sand dikes forming chimney features on the northern edge of the ring depression around the Oasis central uplift. Field of view about 300 m wide (photo: W. U. Reimold).

be underlain by less resistant Carboniferous strata. An easily notable quartz-pebble conglomerate that, based on the literature about the regional geology, could mark the base of the Carboniferous package occurs widespread throughout the structure. The rocks of the central parts of the structure are strongly kaolinitized, and some are rich in iron and manganese nodules. Locally, iron impregnation and chert formation were noted. This could be considered as evidence of a possible post-impact hydrothermal system, but because the iron and manganese impregnation is also observed in the outer reaches of the structure, it might as well be of pre- or post-impact age. Regionally outside the impact structure, the attitude of the strata is generally horizontal to gently subhorizontally warped. In contrast, at Oasis and especially within the inner ring, intense folding and faulting is observed, whereby the rock layers have locally up- to overturned orientations. In the annular basin, tangentially oriented fold structures extend to at least 12.5 km from the center of the impact structure, and small-scale (decimeter to centimeter) folding could be observed in sub-outcrop of the northeastern sector to a radial distance of 18 km from the center. No shatter cones have been detected and in the annular basin no obvious impact breccias either. Thus, it is thought that the current erosion level could expose a relatively deep section through the impact structure.

In the past, different diameters have been cited for Oasis, ranging from 5.1 to 11.5 and up to 18 km. Some observations made during a 2010 expedition by Roger Gibson of the University of the Witwatersrand, Johannesburg and Wolf Uwe Reimold might indicate that the impact structure could actually be significantly larger than previously thought. To date, however, the 18 km value is considered widely as a reliable value for the diameter of Oasis.

Multiple sets of "planar elements" (planar deformation features – PDFs) in quartz grains from orthoquartzite together with a glass-bearing micro-breccia were found in the inner depression early in the study of Oasis. Small interstices could be noted filled with brownish, partly devitrified glass with small sandstone fragments and shocked quartz. More recently, sets of PDFs, besides abundant planar fractures (PFs), were observed in quartz of several breccia samples from the innermost part of the structure. Samples collected during the 2010 expedition did not include glass-bearing material though.

No datable melt rocks, which might allow an attempt at dating the impact event with radiometric methods, have so far been found at Oasis. The only age constraint available so far, for a maximum age limit for this impact event, is the ill-defined age range of 90–120 Myr for the Nubia Formation. It can be speculated that the impact might have taken place prior to the formation of distinct, northwest-trending, normal faults thought to be of Tertiary, about 39 Myr, age. If this notion is correct, this would make the Oasis impact event significantly older than the age of 29 Myr for the **Libyan Desert Glass** in the Libyan Desert of southwestern Egypt, and remove the debated possibility that the Oasis impact could have anything to do with the formation of the Libyan Desert Glass.

Like B.P., Oasis was first mentioned by Truman Kohman and co-workers in 1967, who described the locations of these two structures based on remote sensing imagery. Bevan French and colleagues in 1972, and then 1974, provided evidence of shock metamorphism and, thus, for an origin by impact. Past visits to Oasis are reported as having been rather short. A large number of samples was, however, collected at Oasis when Gibson and Reimold spent nine days mapping parts of the Oasis structure in October 2010.

Location/Access

Like the B.P. impact structure, Oasis occurs in the remote southeastern part of Libya, within the triangle formed by the borders of Libya, Egypt, and Sudan. A traveler to Oasis will likely set out from the capital city of Tripoli, after having ensured that all travel permits are in hand and in order. The itinerary will follow a route on the main road along the Mediterranean Sea in direction of Benghazi, from Ahbadija taking the paved road south towards Al Kufrah Oasis. In this rather remote region, where smugglers roam abundantly, the traveller should have commissioned good guides and protection. The crater structure is located about 125 km to the east-northeast of Al Kufrah. Good four-wheel drive vehicles and equipment including multiple GPS and a satellite telephone are demanded. Above all, Libya has to return to good civil governance, before any attempt to reach the B.P. or Oasis structures can be considered again.

Selected References

French, B. M., Underwood, J R., and Fisk, E. P. (1974). Shock metamorphic features in two meteorite impact structures, south-eastern Libya. Geological Society of America Bulletin, Vol. 85, pp. 1425–1428.

Gibson, R. L., Reimold, W. U., Baegi, M., Crosta, A. P., Shbeli, E., Eshwedi, A., and Plescia, J. (2011). Remote sensing analysis and field mapping of the Oasis structure, Libya (abstract). 2nd Arab Impact Cratering and Astrogeology Conference (AICAC II), Casablanca, Morocco, pp. 47–48.

Koeberl, C., Reimold, W. U., and Plescia, J. (2005). BP and Oasis impact structures, Libya: remote sensing and field studies. In: Impact Tectonics, Impact Studies, Koeberl, C., and Henkel, H. (eds.), Springer, Berlin Heidelberg, pp. 161–190.

Kohman, T. P., Lowman, P. D. Jr, and Abdelkhalek, M. L. (1967). Space and aerial photography of the Libyan Desert glass area (abstract). 30th Annual Meeting of The Meteoritical Society Meeting, Moffett Field, California.

Reimold, W. U. and Koeberl, C. (2014). Impact structures in Africa – A review. Journal of African Earth Sciences, Vol. 93, pp. 57–175.

Roter Kamm
Namibia

Wolf Uwe Reimold

The Roter Kamm impact crater is a nearly circular, 2.5 km wide, structure in southern Namibia. The crater is located in the southern Namib Desert just west of the Aurusberge mountain range, about 80 km north of the town of Oranjemund. The impact crater was formed in 1200 Myr granitic gneiss of the Namaqua Metamorphic Complex (NMC). The interior of the crater and large parts of the crater rim are covered with eolian sands. Locally impact breccia has been sampled on the rim, and in the environs of the crater occur some exposures of Gariep Group marble.

The impact occurred into a three-layer target. Its topmost layer is formed by the two Namib ergs. The semi-consolidated Tsondab Formation erg occurs at the base of this layer, which is exposed in a prominent wind scoop north of the crater and is characterized by the presence of paleo-ostrich egg shells. Uppermost lies the unconsolidated Sossus Sand Formation, which represents the base onto which impact ejecta was deposited. This ejecta layer has been stabilized by calcrete, which has also stabilized the dune surfaces for many thousands of years. The underlying bedrock involves Gariep Group marble, schist, minor quartzite, and sandstone that overly the granitoids of the NMC basement. The majority of breccias found in outcrop along the rim crest are cataclasites. These rocks do not display any shock deformation. Many of these breccia occurrences have a radial orientation with respect to the center of the structure, which strongly indicates that these cataclasites are impact-related.

A number of conclusions can be drawn from the detailed geological work at the structure: (1) Rock foliation patterns along the rim were found to be quite variable and likely suggest that block rotation took place. (2) The chemical composition of the pre-impact target was highly heterogeneous. (3) Cataclastic gneisses occur widespread along the crater rim and are K-enriched compared to the host gneisses. (4) Significant remnants of the ejecta blanket still occur in the environs of the crater, more so in the north-northwest to west sector around the crater, where ejected blocks can still be found.

The locally prevailing wind directions must have changed periodically over the last 30 years, thereby revealing new geological information from time to time. In 1997 Wolf Uwe Reimold and his team found a patch of suevite on the northwestern crater rim. Only seven years later, in 2004, they discovered a previously unknown several meter-wide patch of impact melt rock. Thus, future visits to the crater may still allow to obtain new information.

A gravity model for the Roter Kamm crater by Bob Fudali indicated that the crater fill comprises some 800 m of sands overlying impact breccia. A similar conclusion was later reached by Dion Brandt and his team from newer gravity data that revealed a negative Bouguer anomaly of −8.8 mGal. A ground-penetrating radar study suggested local presence of ejecta below a relatively thin sand cover and that the crater has been degraded by not more than several tens of meters. Modeling of gravity and magnetic data, as well as the size of 2.5 km diameter strongly favor that the structure mainly formed in crystalline basement is of the simple, bowl-shape type.

The age of the Roter Kamm impact event was initially determined by ^{40}Ar-^{39}Ar step-heating analysis of a pebble-sized float of melt breccia to 3.7 ± 0.3 Myr. A more recent argon dating study of samples from the new locality of impact melt rock on the crater rim only yielded two rather poorly constrained ages of 4.69 ± 0.31 Myr and 5.06 ± 0.44 Myr. Thus, at present, the age of the Roter Kamm impact event is essentially constrained to younger than 5 Myr, which is in agreement with the stratigraphic ages for the aforementioned Sossus and Tsondab formations of the target.

The first to propose an impact origin for this structure were Robert Dietz in 1965 and Bob Fudali in 1973, on morphological and geophysical grounds. In 1989 Wolf Uwe Reimold and Roy Miller gave the first detailed geological account of the structure and reported first findings of shock metamorphic evidence in the form of planar deformation features (PDFs) in clasts of rare impact melt breccia samples. Lutz Hecht and co-workers in 2008 obtained a strong indication that the impact melt rock contained a very small meteoritic component of ordinary chondrite composition. About the same time, Miller published comprehensive mapping of the crater rim and surrounding areas, which included a record of remnants of impact ejecta.

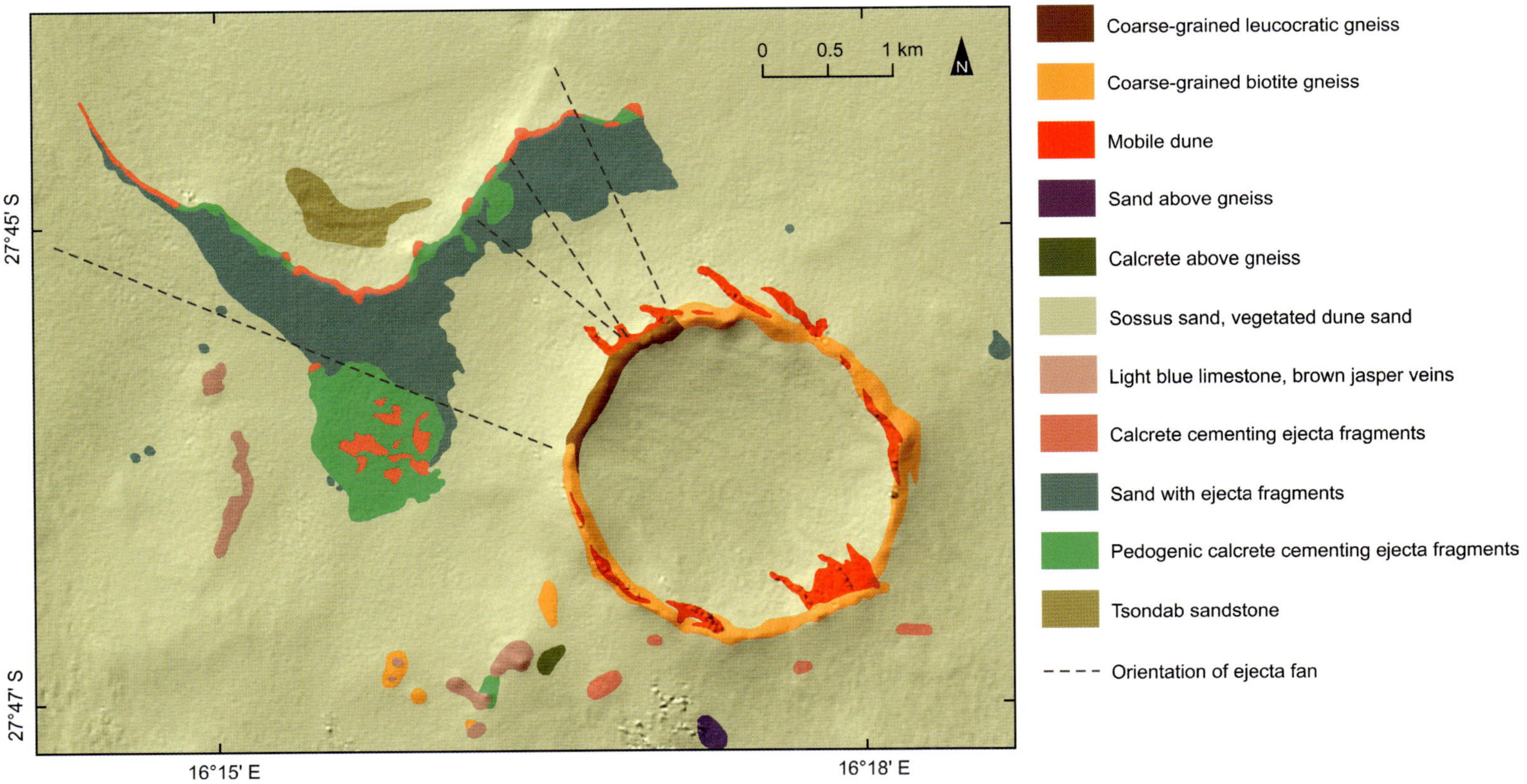

Geological map of the Roter Kamm crater with underlain TanDEM-X DEM (modified after Reimold and Koeberl, 2014 – based on Miller, 2010).

Cataclasite of brecciated quartz from a quartz vein; the matrix is blackened by tiny crystals of magnetite, rutile, and other dark minerals. Sample is 8.5 × 6 × 6 cm in size (photo courtesy: Ralf-Thomas Schmitt, Museum für Naturkunde Berlin, sample from W. U. Reimold).

Specimen of suevite from the Roter Kamm impact crater, 6 × 5 × 4 cm in size (photo courtesy: Ralf-Thomas Schmitt, Museum für Naturkunde Berlin, sample from W. U. Reimold).

Location/Access

Roter Kamm crater is located in the restricted area of the southern Namib-Naukluft Nature Reserve. Therefore, special permits are required to visit this environmentally protected area. Permits need to be issued by the relevant government agencies in Windhoek. The Geological Survey of Namibia in Windhoek may be able to advise.

Access to Roter Kamm should be from the small mining town of Rosh Pinah to the southeast of the crater. From there a four-wheel drive vehicle is required to travel past the Scorpion Pb-Zn mine towards the crater. It is not permitted to scale the crater rim by car; it is a must to leave vehicles beyond the rising slope of the crater rim. Those who get permission to enter the structure will find vehicle tracks in the desert sand that have already survived for 20 years. The desert environment is extremely fragile.

Selected References

Brandt, D., Reimold, W. U., Franzen, A. J., Koeberl, C., and Wendorff, L. (1998). The geophysical signature of the Roter Kamm impact crater, Namibia. Meteoritics and Planetary Science, Vol. 33, pp. 447–454.

Hecht, L., Reimold, W. U., Sherlock, S., Tagle, R., Koeberl, C., and Schmitt, R. T. (2008). New impact melt rock from the Roter Kamm impact structure, Namibia: further constraints on impact age, melt rock chemistry, and projectile composition. Meteoritics and Planetary Science, Vol. 43, pp. 1201–1218.

Miller, R. McG. (2008). Meteorite impact craters and meteorites. In: The Geology of Namibia, Miller, R.McG. (ed.), Ministry of Mines and Energy, Windhoek, Namibia, Chapter 26, pp. 1–8.

Miller, R. McG. (2010). Roter Kamm impact crater of Namibia: new data on rim structure, target rock geochemistry, ejecta, and meteorite trajectory. In: Large Meteorite Impacts and Planetary Evolution IV, Gibson, R. L., and Reimold, W. U. (eds.), Geological Society of America Special Paper, Vol. 465, pp. 489–508.

Reimold, W. U., and Miller, R. McG. (1989). The Roter Kamm impact crater, SWA/Namibia. Proceedings of the 19th Lunar and Planetary Science Conference, Houston, Cambridge University Press, pp. 711–732.

Reimold, W. U., and Koeberl, C. (2014). Impact structures in Africa – A review. Journal of African Earth Sciences, Vol. 93, pp. 57–175.

In 2004 the shifting sand on the crater rim revealed some of the hidden treasure – a previously not known patch of impact melt rock! Hammer, for scale, 35 cm long (photo: W. U. Reimold).

Roter Kamm

Namibia

latitude = 27°45'58" S / longitude = 16°17'21" E / diameter = 2.5 km / age = < 5 Myr

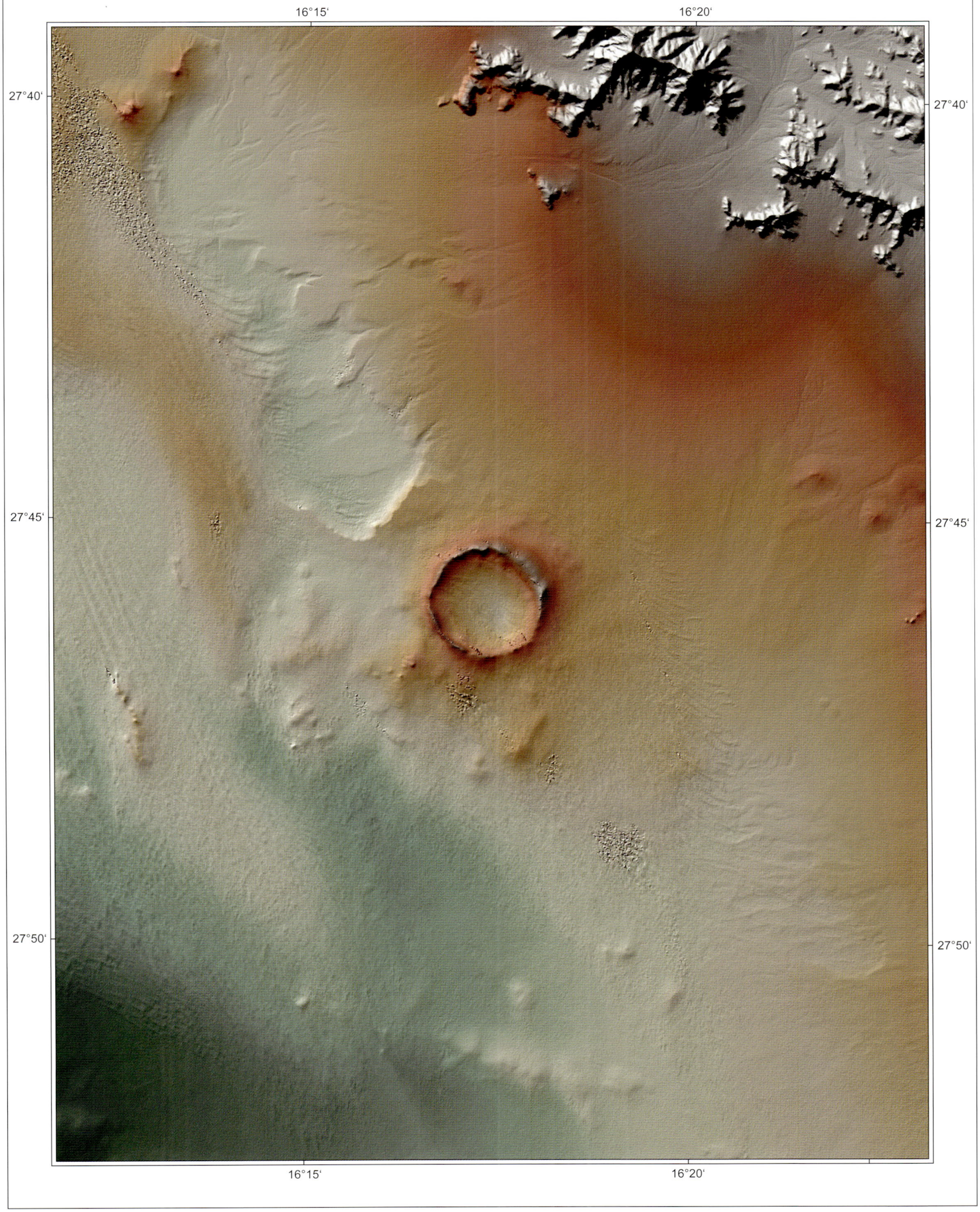

scale = 1 : 103 000 / scene = 18.6 × 24.8 km²

0 1.03 5.15 km

N

320 meter 1080

Roter Kamm crater on July 21, 2017 as a composite image based on TanDEM-X DEM and Copernicus Sentinel-2 (channel 2/3/4) data.

Talemzane

Algeria

Wolf Uwe Reimold

The Talemzane impact structure, which is also known variably as *Daiet El Maâdna* or simply *Maâdna*, is of the simple, bowl-shape type. It has a diameter of 1.75 km, and a relief of about 45 m, measured from the base of the central depression to the crest of the crater rim. It was formed in limestones of Senonian (Upper Cretaceous) or Eocene age. The limestones are strongly fractured, upturned, and – in the upper rim – overturned. Sizable blocks of limestone have been ejected outside of the crater. The crater wall intersects breccia dikes. Clastic or, possibly, reworked monomict breccia is found near the rim and on the crater floor. Different types of likely impact-generated breccias are present, both of monomict and polymict lithic character, in addition to several types of impact melt-bearing breccias. They show evidence for carbonate melting in the form of immiscible impact-generated $CaCO_3$-rich and silicate-rich melt phases. It appears that further detailed studies of the breccia lithologies might be rewarding. The limestones do contain some rare quartz, and some quartz grains showed the presence of impact-diagnostic planar deformation features (PDFs).

Ground magnetic and susceptibility studies have been conducted along radial profiles across Talemzane. The interior of the crater structure shows a complex pattern of positive magnetic anomalies, which could be related to local accumulations of magnetite. It has been mooted that this might be due to the local occurrence of bodies of impact melt. The most pronounced magnetic anomaly is superimposed by a distinct positive soil magnetic anomaly.

The age of the structure is estimated at 0.5 to 3 Myr, because of the limited degree of erosion observed. Other studies favored, in contrast, a Late Miocene to Early Pliocene age for the impact event, based on paleomagnetic and rock magnetic investigations.

This structure was first visited in 1951 and then investigated from the air. Roman Karpoff was the first to propose a meteorite impact origin for Talemzane in 1953, with which Talemzane became the first impact structure discovered in Algeria. Finally, Philippe Lambert and co-workers in 1980 obtained limited but definite evidence for the impact origin in the form of a few quartz clasts from an impact breccia from the flank of the crater, with poorly developed PDFs.

Talemzane became the first impact crater detected with spaceborne radar imagery when SIR-A, NASA's Shuttle radar mission, showed it as a prominent feature in 1981. The annular drainage system around the crater was interpreted by John McHone and Ronald Greeley as evidence for the presence of a shallow ring syncline in the environs of the crater in 1987. Modern high-resolution missions such as TanDEM-X may even indicate remnants of an ejecta blanket in the immediate environs of the crater structure.

Location/Access

Talemzane is located about 35 km south of the road between Messaad and Guettara. It is easily reached on a dirt track from the town of Hassi Delaa, which lies about 47 km to the northwest of the crater. Talemzane is the most frequently accessed impact site in Algeria.

Talemzane
Algeria

latitude = 33°18'52" N / longitude = 04°02'05" E / diameter = 1.75 km / age = 0.5 - 3.0 Myr

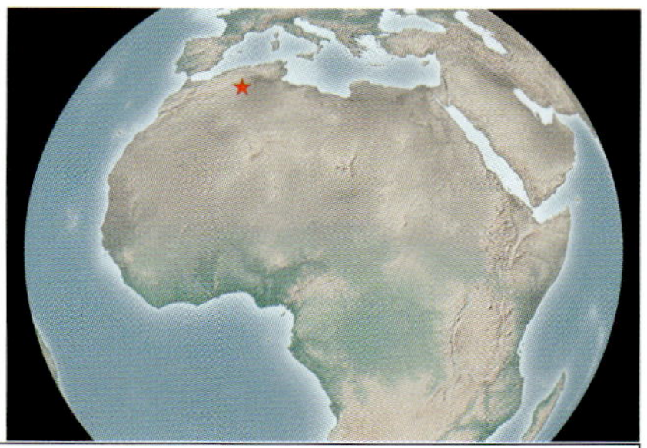

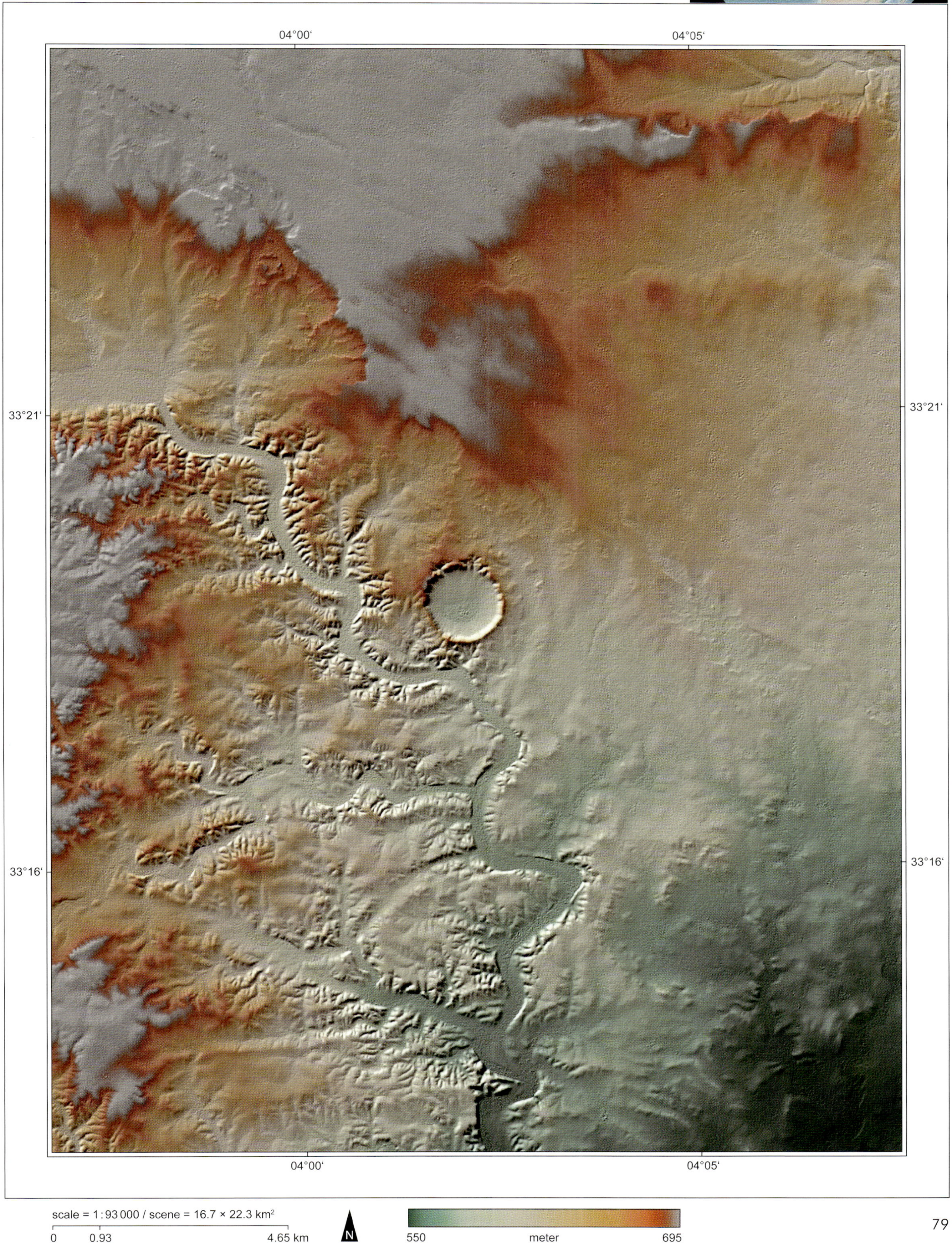

Talemzane monomict impact breccia in dolomite, cut surface, 3.5 × 3 × 1.5 cm in size (photo courtesy: Ralf-Thomas Schmitt, Museum für Naturkunde Berlin, sample from S. Romano, Bologna, Italy, 2015).

Selected References

Chabou, M. C. (2019). Meteorite impact structures in the Arab World: An overview. In: The Geology of the Arab World – An Overview, Bendaoud, A., Hamimi, Z., Hamoudi, M., Djemai, S., and Zoheir, B. (eds.), Springer Geology, pp. 455–506.

Karpoff, R. (1953). The meteorite crater of Talemzane in southern Algeria. Meteoritics, Vol. 1, pp. 31–38.

Lamali, A., Rochette, P., Merabet, N., Abtout, A., Maouche, S., Gattacceca, J., Ferrière, L., Hamoudi, M., ASTER Team, Meziane, E. H., and Ayache, M. (2016). Geophysical and magneto-structural study of the Maâdna structure (Talemzane, Algeria): Insights on its age and origin. Meteoritics and Planetary Science, Vol. 51, pp. 2249–2273.

Lambert, P., McHone, J. F., Dietz, R. S., and Houfani, M. (1980) Impact and impact-like structures in Algeria, Part 1. Four bowl-shaped depressions. Meteoritics, Vol. 15, pp. 157–179.

McHone, J. F., and Greeley, R. (1987). Talemzane: Algerian impact crater detected on SIR-A orbital imaging radar. Meteoritics, Vol. 22, pp. 253–264.

Sahoui, R., Belhai, D., and Jambon, A. (2013). Impact-generated carbonate melts from the Talemzane structure (Algeria). 44th Lunar and Planetary Science Conference, Lunar and Planetary Institute, Houston, Texas, Abstract #1184.

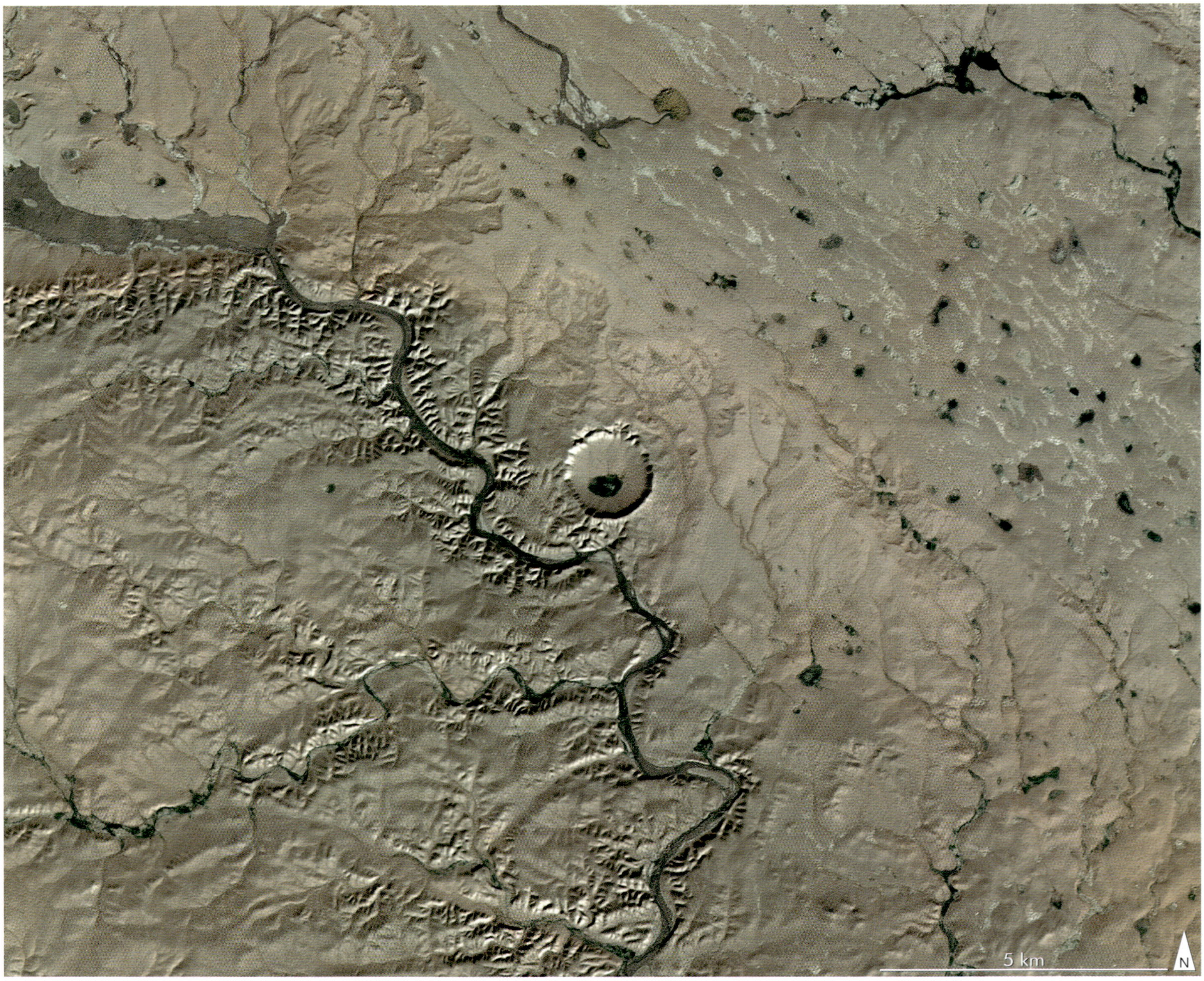

The Talemzane impact crater and environs on September 2, 2017 as a composite image based on TanDEM-X DEM and Copernicus Sentinel-2 (channel 2/3/4) data.

Tenoumer
Mauretania

Wolf Uwe Reimold

Tenoumer is a 1.9 km wide, nearly circular, well-preserved and consequently young, simple bowl-shaped crater. It is located in the Western Sahara of Mauretania, some 400 km northwest of **Aouelloul** impact crater.

The Tenoumer crater was formed on a peneplained surface of Precambrian terrane, which is constituted mostly of gneiss and granite, besides some mafic rocks such as amphibolite and an intrusion of olivine-gabbro. The basement is overlain by a thin cover of young, possibly Pliocene age sediment. From the top of the crater rim to the apparent crater floor the depth of the crater is about 100 m. The inner slopes of the crater are quite steep, whereas the outer rim slopes are only steep in their upper reaches and then flatten out into the surrounding plain. Based on geophysical data, unconsolidated post-impact sediments of the crater fill occur to a depth of 200–300 m, where impact breccias would be reached.

Earliest Tenoumer workers favored a volcanic origin. In 1948 Jacques Richard-Molard reported the presence of basalt lava and pumice, and suggested a volcanic explosion. Three years later, in 1951, André Allix was the first to propose an origin by impact but without presenting conclusive evidence. Theodore Monod and Charles Pomerol, in 1966, alleged the occurrence of a swarm of small "dikes" of "rhyodacitic lava" in concentric fractures around the crater and in outcrop just outside the crater rim and interpreted this as evidence for a possible volcanic origin of the crater. Finally, Bevan French and his team put the discussion to rest when they reported in 1970 up to 8 sets of planar deformation features (PDFs) in quartz grains from obviously strongly shocked inclusions of granite in the "lava", as well as inclusions of lechatelierite and diaplectic quartz glass and ballen quartz. This demonstrated that Tenoumer is of impact origin and that the previously alleged "lava" represents impact melt rock. Rb-Sr isotope data by this group showed that the melt was derived from the crystalline basement.

In 2005 Giovanni Pratesi and co-workers reported findings of an earlier Italian expedition to the crater. This included detailed petrography of the impact melt rocks, and extensive chemical analyses, including Platinum Group Element (PGE) data for melt rock samples and various regionally occurring lithologies. Their major element results showed that besides felsic granites and gneisses, a mafic component is required to make up the composition of the impact melt rock. In addition, some petrological detail, including evidence for liquid immiscibility between silicate melt and spherules and globules of calcite could be provided. No meteoritic component was detected in the impact melt rock.

Recent analyses of several Tenoumer bomb and lapilli sized melt rock samples provided further important details. Dina Schultze and co-workers derived whole-rock chemical compositions, which were interpreted as the result of considerable variation in relative proportions of melt from more siliceous rocks (mainly granitoids) and mafic rocks of the target. The impact melt rock samples are mainly of intermediate composition (andesite to basaltic andesite) but do contain significant amounts of olivine micro-phenocrysts in Mg- and Fe-rich samples. All textures described are typical for fast cooling – in fact indicating extreme disequilibrium conditions upon crystallization. It was suggested that the heterogeneity of Tenoumer melt rock samples had two main reasons: first, impact melting of various target lithologies resulted in mixing at different target rock proportions, on a local scale; and second, melt rock heterogeneity occurs at the thin section scale and is due to fast cooling under disequilibrium crystallization conditions. This took place during and after the dynamic ejection and emplacement of the melt bombs.

Two samples of impact melt rock from the Tenoumer impact crater (photo courtesy: Lutz Hecht, Museum für Naturkunde Berlin, samples from W. U. Reimold).

The age of Tenoumer crater was initially determined by French to 2.5 ± 0.5 Myr from K-Ar dating of melt rock. Later, contrasting values have been determined. In 2003, Dieter Storzer obtained a very young age of 21 400 ± 9700 years by the fission-track method. And still to date, the age controversy about Tenoumer is not finished: The recent work by Schultze and her team arrived at an age of 1.52 ± 0.14 Myr from ^{40}Ar-^{39}Ar step-heating analysis.

Location/Access

As with any travel these days in the Western Sahara, the traveler ought to be extremely careful because of civil strife. Apparently the Tenoumer crater can be accessed easily by four-wheel-drive transport.

Selected References

Allix, A. (1951). Note et correspondence à propos des cratères météoriques. Revue de Géographie de Lyon, Vol. 26, pp. 357–359.

French, B. M., Hartung, J. B., Short, N. M., and Dietz, R. S. (1970). Tenoumer crater, Mauritania: Age and petrologic evidence for origin by meteorite impact. Journal of Geophysical Research, Vol. 75, pp. 4396–4406.

Monod, Th., and Pomerol, C. (1966). Le cratère de Tenoumer et ses laves. Bulletin de la Société Géologique de France, Vol. S7–VIII, pp. 165–172.

Pratesi, G., Morelli, M., Rossi, A. P., and Ori, G. G. (2005). Chemical compositions of impact melt breccias and target rocks from the Tenoumer impact crater, Mauritania. Meteoritics and Planetary Science, Vol. 40, pp. 1653–1672.

Reimold, W. U., and Koeberl, C. (2014). Impact structures in Africa: A review. Journal of African Earth Sciences, Vol. 93, pp. 57–175.

Richard-Molard, J. (1948). Le cratère d'explosion de Tenoumer et l'existence probable d'une grande fracture rectiligne au Sahara occidental. Comptes Rendus de l'Académie des Sciences Paris, Vol. 227, pp. 213–214

Schultze, D. S., Jourdan, F., Hecht, L., Reimold, W. U., and Schmitt, R. T. (2016). Tenoumer impact crater, Mauritania: Impact melt genesis from a lithologically diverse target. Meteoritics and Planetary Science, Vol. 51, pp. 323–350.

Storzer, D., Sélo, M., Latouche, L., and Fabre, J. (2003). The age of Tenoumer crater, Mauritania, revisited. 34th Lunar and Planetary Science Conference, Lunar and Planetary Institute, Houston, Texas, Abstract #1183.

The Tenoumer impact crater on October 4, 2017 as a composite image based on TanDEM-X DEM and Copernicus Sentinel-2 (channel 2/3/4) data.

Tenoumer
Mauritania

latitude = 22°55'07" N / longitude = 10°24'20" W / diameter = 1.9 km / age = 1.52 ± 0.14 Myr

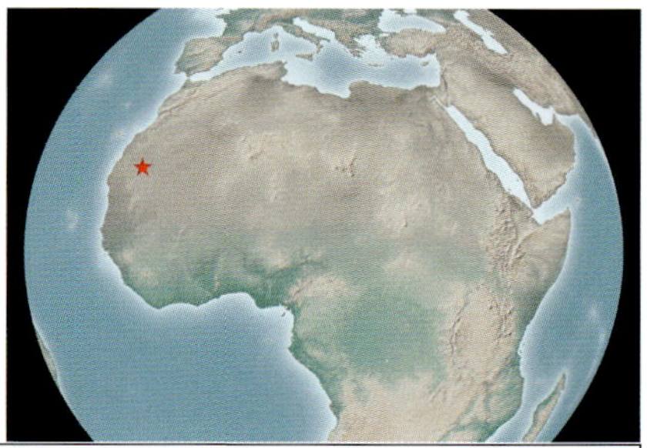

scale = 1 : 81 000 / scene = 14.5 × 19.4 km²

0 0.81 4.05 km

N

240 meter 360

Tin Bider
Algeria

Wolf Uwe Reimold

Tin Bider, also known as *Tademaït*, is a conspicuous structure with a diameter of about 6 km, with three prominent annular ridge features that are about 2, 3.5 and 6 km wide, respectively. Several other stratigraphic horizons also form semi-annular ridges. Tin Bider is situated on a mesa with escarpments surrounding the structure in the west, south and east.

Tin Bider was formed in Lower to Upper Cretaceous lithologies. Upper Cretaceous claystone and limestone formations form the various rings of the structure. The variable susceptibilities of these strata to erosion are the cause of the obvious multi-ring arrangement. Massive sandstones of Lower Cretaceous age occur in the innermost part of the structure. Apparently they have been uplifted by some 500 m from their original stratigraphic position. The strata forming the concentric ridges are complexly deformed and exhibit intense radial folding due to convergent movements during crater modification. This deformation was likely facilitated by the different rheological behaviors of the various target rocks.

The Tin Bider structure was first investigated in 1962 by J. Guillemot. He described a central core surrounded by two annular folds. Subsequently, in 1981 some field detail was provided by A. Attar, as well as by Philippe Lambert and co-workers. The team of Lambert reported on impact related deformation. Whilst neither macroscopic evidence in the form of shatter cones could be reported nor any breccia occurrences were observed, these teams observed quartz grains with up to 7 sets of decorated planar deformation features (PDFs) in samples from the innermost, apparently uplifted, massive sandstone. This provided definite evidence of impact deformation.

As the entire breccia fill that should have been associated with an impact structure has been removed by erosion, Tin Bider is thought to be quite deeply eroded. The age of the Tin Bider impact structure is only constrained stratigraphically, namely by the minimum age of the target rocks. Thus, the impact must have taken place at or since Lower Cretaceous time. Lambert and co-workers favored a Neogene origin though.

Obviously, Tin Bider is a prime target for further detailed geological analysis. Major aspects for future analysis should be macroscopic studies of shock deformation within a target made up of sedimentary rocks of very different rheological behavior. In addition, the well-exposed sandstones should be sampled in detail for a comprehensive micro-deformation study.

The Tin Bider impact structure on September 21, 2017 as a composite image based on TanDEM-X DEM and Copernicus Sentinel-2 (channel 2/3/4) data.

Location/Access

The structure lies on the Saharan Platform, some 265 km to the east-northeast of the city of In Salah. From there the structure can be reached with four-wheel drive vehicle transport on a rough desert track. Along the way to the structure, one traverses the Reg Acfer meteorite field, which has provided 400 meteorites to date.

Selected References

Attar, A. (1981). La structure de Tin Bider (plateau du Tinhert, Algérie). Un impact météoritique. Unpublished report, Sonatrach, 15 pp.

Chabou, M. C. (2019). Meteorite impact structures in the Arab World: An overview. In: The Geology of the Arab World – An Overview, Bendaoud, A., Hamimi, Z., Hamoudi, M., Djemai, S., and Zoheir, B. (eds.), Springer Geology, pp. 455–506.

Guillemot, J. (1962). Structures anticlinales circulaires d'origine mal connue. Photo interprétation. Technip Editions, fascicule 4.

Lambert, P., McHone, J. F., Dietz, R. S., Briedj, M., and Djender, M. (1981). Impact and impact-like structures in Algeria, Part 2. Multi-ringed structures. Meteoritics, Vol. 16, pp. 203–227.

Tin Bider

Algeria

latitude = 27°36'07" N / longitude = 05°06'44" E / diameter = 6 km / age = 50 - 66 Myr

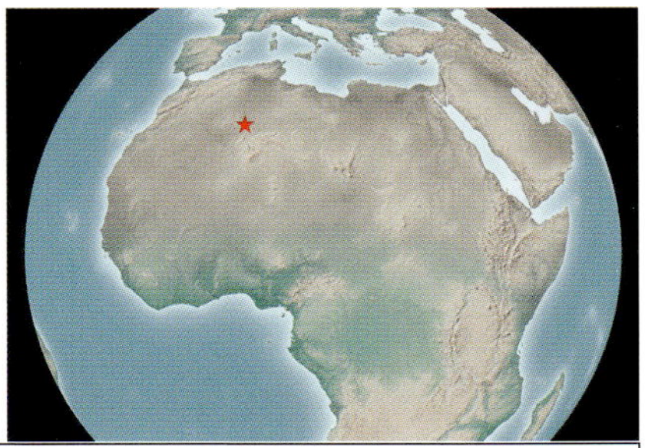

scale = 1 : 86 000 / scene = 15.5 × 20.7 km²

0 0.86 4.30 km

N

450 meter 675

Tswaing
South Africa

Wolf Uwe Reimold

The 1.13 km wide Tswaing impact crater is located at the northeastern border of Gauteng Province of South Africa, about 40 km north-northeast of the city of Pretoria (Tswane) and some 120 km north of Johannesburg. The crater structure has in the past been known as *Zoutpan*, *Salt Pan*, or *Pretoria Saltpan*; today's official name *Tswaing* means "Place of Salt" in the local Sotho language. Tswaing crater has the same diameter as **Meteor Crater** in Arizona, the probably best-known impact crater in the world, at 1.13 km. The very well-preserved crater has a prominent rim that rises up to 60 m above the surrounding terrain, and some 90 m above the apparent crater floor. Tswaing was formed in Nebo Granite of the Bushveld Complex, which was overlain at the time of impact by some Mesozoic sediments. In the interior of the crater occurs a lake, the width of which changes seasonally. This lake has been the subject of considerable environmental analytical work. Since the crater basin has been closed ever since its formation, the crater fill is regarded to have preserved a long paleoclimatic record.

Along the dirt-track from the shoreline of the crater lake up the southeastern crater rim, some outcrops illustrate the lithologies and structure of the rim section. From the bottom up, one sees steep, inward-dipping faults, which are followed by prominent anticlinal structures, and in turn, by a section with low-angle, inward-dipping faults. Finally, on top, occurs inverted stratigraphy with Nebo granite folded over younger Karoo age grit. The rim crest is formed by ejecta with less than 1 decimeter to meter size, frequently highly angular blocks of granite.

At a number of localities around the crater rim, the granite of the inner and top crater rim displays small exposures of various volcanic intrusives, including trachyte, phonolite, lamprophyre and carbonatite. Their presence was taken for a long time as evidence that the crater had something to do with the regional volcanism of the extensive field of the Pienaars River Volcanic Complex.

The origin of the Tswaing crater had been controversial ever since the beginning of the previous century. The famous South African geologist Percy Wagner set the scene in 1922, when he proposed a volcanic origin for the crater on the basis of the presence of the volcanic intrusives in the crater wall. He was opposed by Herbert Rohleder in 1933, who suggested a meteorite impact origin on the grounds of morphological considerations. His work was, unfortunately, largely ignored by the South African geological community. Much later, in 1995, a detailed geological study of the crater and its environs was carried out by Dion Brandt.

Drill cores helped to confirm the impact origin. Several short drill cores were already retrieved from the crater interior in the 1950s and 1970s, mostly to investigate the reserves of trona deposits in the crater lake that had been mined for several decades to provide a means for the neutralization of acid mine drainage in the gold mines of the Witwatersrand gold fields to the west and just south of Johannesburg. The material retrieved from these shallow drill cores did not provide conclusive evidence with respect to the origin of the crater. In 1988, Tim Partridge of the University of the Witwatersrand conducted a comparatively deeper drilling. A 200 m long drill core was recovered from the crater through the entire, 90 m thick crater fill, through 60 m of sand-like breccia, and the upper part of the crater floor (first fractured and brecciated basement formed by Nebo Granite, and then into undeformed granite). Lithic, mineral and even fossil fragments in the suevitic breccia allowed to reconstruct the likely composition of the pre-impact target. A thin cover of diamictite, a lacrustine deposit, likely overlay the Karoo grits that, in turn, covered the Nebo granite. The same year Wolf Uwe Reimold and co-workers found diagnostic evidence of shock metamorphism, in the form of planar deformation features (PDFs) in quartz and feldspar particles, diaplectic quartz glass, and numerous particles of glass and melt in the sandy breccia, which, thus, was revealed as an unconsolidated suevitic breccia. In addition, Christian Koeberl and his team showed a significant, up to 10 % meteoritic (chondritic) component in samples from this breccia. These findings could lay the 100 year long controversy about the origin of the crater to rest – Tswaing is an impact crater.

Dieter Storzer and co-workers obtained an age of 0.220 ± 0.052 Myr for the impact event from fission-track dating of impact glass particles from the suevitic breccia. In contrast, the volcanic lithologies occurring within the crater were shown by Brandt and Reimold to be ca. 1300 Myr old and, therefore, had to be part of the volcanic suite of the regional Pienaars River Alcaline Complex. The fission-track age for the impact event overlaps, within error limits, with the U-Th series

Panoramic view across the Tswaing meteorite crater, towards the northwestern crater rim. The crater is just about 1.13 km wide (photo: W. U. Reimold).

Tswaing

South Africa

latitude = 25°24‘31“ S / longitude = 28°04‘57“ E / diameter = 1.13 km / age = 0.220 ± 0.052 Myr

scale = 1 : 89 000 / scene = 15.9 × 21.3 km²

0 0.89 4.45 km

N

1040 meter 1255

age for the **Kalkkop** crater, which may suggest that these two impact events could be related.

Significant paleoenvironmental knowledge about vegetation and climate in this part of the southern hemisphere – for the period back to the impact 200 000 years ago – has been gleaned from the crater fill sedimentary record of Tswaing. Cyclical changes in climate and environment have been revealed through this long-term record. Further detail of this paleoenvironmental work can be found in the 2014 review of the African impact record by Reimold and Koeberl.

Location/Access

The Tswaing Crater is located due north of the city of Tswane (Pretoria). A straightforward access route would follow the N1 Freeway north to the Zambesiweg offramp north of Pretoria. Drive west to the end of Zambesiweg, then turn north and west again to the Onderstepoort Veterinary Research Station. From there follow the signs "Tswaing" or "Zoutpan" north for some 25 km through a densely populated area. The crater will become visible in the distance, to the front-left, several kilometers before the turn-off into the Tswaing Museum parking area is reached. Note that the Tswaing Crater Museum property is entirely surrounded by dense, and in part, informal settlements.

Selected References

Brandt, D., and Reimold, W. U. (1995a). The geology of the Pretoria Saltpan impact structure and the surrounding area. South African Journal of Geology, Vol. 98, pp. 287–303.

Brandt, D., and Reimold, W U. (1995b). The petrology of the volcanic rocks occurring in the Pretoria Saltpan impact crater and surrounding area. South African Journal of Geology, Vol. 98, pp. 304–318.

Koeberl, C., Reimold, W. U., and Shirey, S. B. (1994). Saltpan impact crater, South Africa: geochemistry of target rocks, breccias and impact glasses, and osmium isotope systematics. Geochimica et Cosmochimica Acta, Vol. 58, pp. 2893–2910.

Partridge, T. C. (ed.) (1999). Investigations into the Origin, Age, and Palaeoenvironments of the Pretoria Saltpan. Council for Geoscience, Memoir 85, 198 pp.

Reimold, W. U., Koeberl, C., and Brandt, D. (1999). The origin of the Pretoria Saltpan crater. In: Investigations into the Origin, Age and Paleoenvironments of the Pretoria Saltpan, Partridge, T. C. (ed.), Memoir 85, Council for Geoscience, Pretoria, pp. 35–54.

Reimold, W. U., and Koeberl, C. (2014). Impact structures in Africa – A review. Journal of African Earth Sciences, Vol. 93, pp. 57–175.

Storzer, D., Reimold, W. U., and Koeberl, C. (1999). Fission-track age of the Saltpan impact crater. In: Investigations into the Origin, Age and Paleoenvironments of the Pretoria Saltpan, Partridge, T. C. (ed.), Memoir 85, Council for Geoscience, Pretoria, pp. 64–71.

The Tswaing impact crater and surroundings on September 29, 2017 as a composite image based on TanDEM-X DEM and Copernicus Sentinel-2 (channel 2/3/4) data.

Vredefort
South Africa

Wolf Uwe Reimold

The center of the Vredefort impact structure is located about 130 km southwest of the city of Johannesburg. It straddles the borders between the Northwest, Gauteng, and Free State provinces of South Africa. With an original diameter estimated at 250–300 km and a well-defined age of 2023 ± 4 Myr, Vredefort is both the world's largest and one of the oldest confirmed impact structures.

The Vredefort structure is centered near the geographic focus of the structural remnant of the Witwatersrand basin. In fact, the Vredefort impact structure comprises the central Vredefort Dome and the entire surrounding basin. The Dome represents a unique, deep window into the geology of the upper and middle crust of the ancient Kaapvaal craton, and has regionally unique, spectacular rock deformation phenomena caused by the impact. The origin of the Vredefort Dome has long been controversial. Endogenic hypotheses included gas explosion or tectonic processes, in contrast to the exogenic impact cratering process. In the mid-1990s it was recognized that the entire Witwatersrand Basin with all its gold and uranium resources is the actual remnant of the Vredefort impact structure, and it has since become widely accepted that this gigantic impact lead to the preservation of the Witwatersrand ore resources.

The Witwatersrand Basin between Johannesburg in the northeast and the gold-mining town of Welkom in the southwest is completely underlain by the Archean crystalline basement of the Kaapvaal craton. Stratigraphically upward, this is overlain by Archean Dominion Group volcanics and metasediments in the western part of the basin around Klerksdorp town and in the area of the Vredefort Dome, and by the successive deposits of the metasedimentary Witwatersrand Supergroup, and the volcanics of the Ventersdorp Supergroup. This is followed by Proterozoic metasediments of the Transvaal Supergroup. Basement is exposed in a number of basement highs in the environs of the basin, e.g., in the Johannesburg Dome between Johannesburg and Pretoria, but none of these exposures shows the deformation phenomena exclusively related to the Vredefort Dome. The basement and Archean-Proterozoic supracrustals in the southern part of the Witwatersrand Basin, including the southern part of the Vredefort Dome, are obscured by a thick blanket of sediments and dolerites of the Mesozoic Karoo Supergroup.

This atlas presents two topographical maps for Vredefort: one showing the Vredefort Dome and surrounding parts of the Witwatersrand Basin and one zooming in on the Dome. This displays the terrain dichotomy between the rugged Vredefort Mountain Land in the north/northwest, along the so-called "collar" of the Dome, in the surroundings of the rather flat and largely agriculturally used "core" of the Dome. Excellent exposures of basement are provided by a host of former dimension stone quarries in the core, some of which offer three-dimensional views into the upper- and mid-crust of the craton. The 45–50 km wide core has numerous exposures of Archean granite-gneiss. The 20–25 km wide collar comprises up- and overturned supracrustal strata of the Witwatersrand and Ventersdorp supergroups. The inner part of the Archean gneiss complex (core) has rocks at granulite metamorphic grade, whereas the outer part is in amphibolite grade. The transition between these two terrains hosts a lot of pseudotachylitic breccia (see below). U-Pb single-zircon dating of various granitoids and ^{40}Ar-^{39}Ar age data for an amphibolite have indicated that the basement of the Dome is ca. 3100–3200 Myr old and underwent metamorphism around 3080 Myr ago. A few zircons indicated that the basement contains a component of 3300–3400 Myr age. Indeed, in the southeast of the dome is a large inlier of Archean greenstones (the Greenlands Formation), with komatiitic basalts and associated lithologies that resemble those of the Barberton Greenstone Belt. Limited dating of the Greenlands rocks suggests an age possibly older than 3400 Myr.

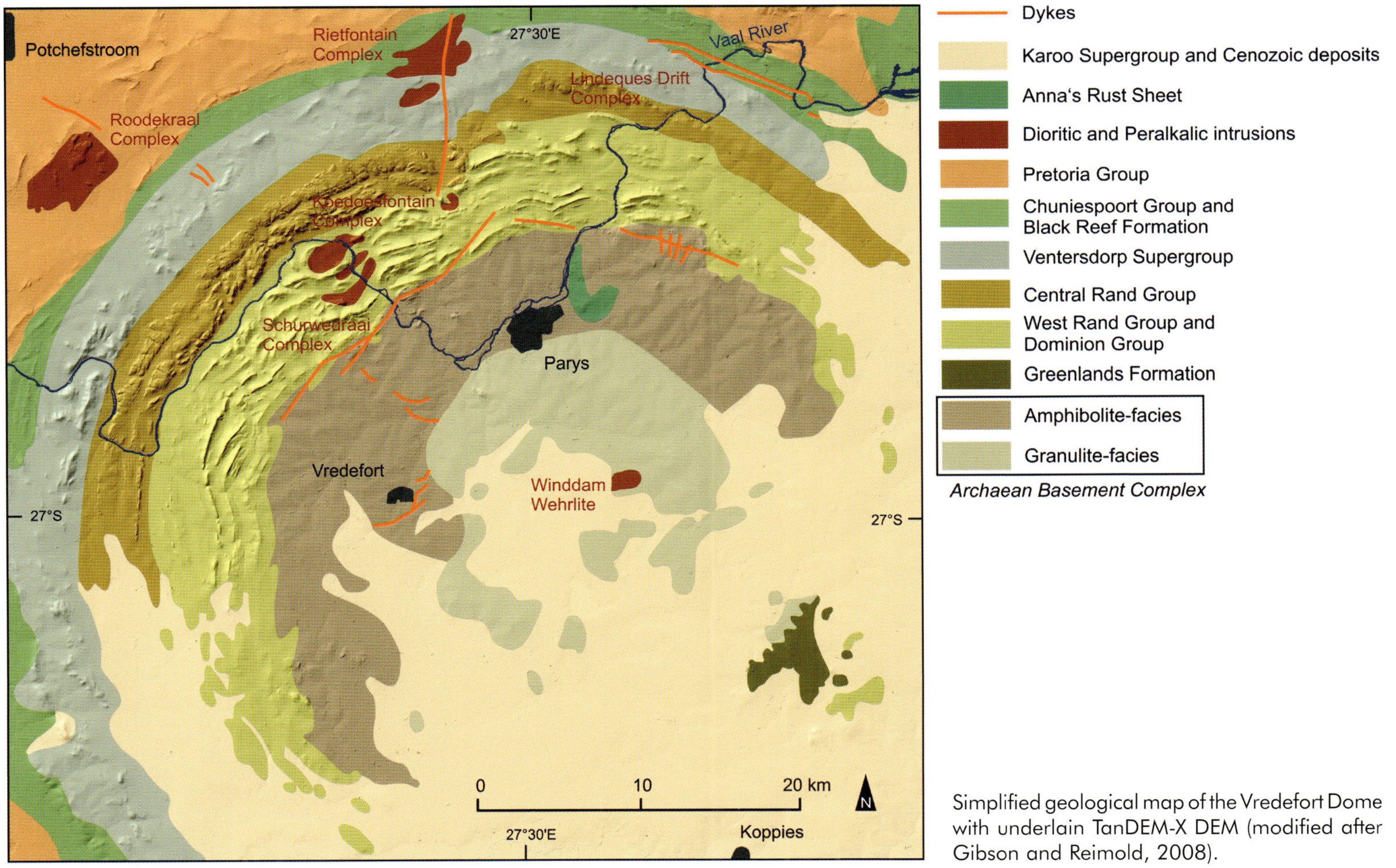

Simplified geological map of the Vredefort Dome with underlain TanDEM-X DEM (modified after Gibson and Reimold, 2008).

◁ A section of the here about 10 m wide dike of Vredefort Granophyre (impact melt rock) on the farm Daskop in the northern core of the Vredefort Dome (photo: W. U. Reimold).

A section of the Leeukop quarry, north of the town of Parys, in the granite-gneiss core of the Vredefort Dome. This approximately 10 m wide exposure shows a massive development of pseudotachylitic breccia (photo: W. U. Reimold).

The metasedimentary collar strata are intruded by numerous dioritic sills (epidiorite) thought to be of Ventersdorp age, about 2700 Myr, besides some Bushveld-age magmatic (alkali granitic and mafic) intrusions.

The Vredefort impact generated a vast amount of impact melt rock. While the melt that originally resided in the actual crater has been completely eroded, there are remnants of impact melt rock in the form of a series of 9 dikes that can be studied on the Vredefort Dome. The *Vredefort Granophyre* received its name from the distinct granophyric texture of its groundmass. The main Granophyre variety is regionally of homogeneous chemical composition. All Granophyre dikes carry a major clast component derived from the granitoid basement, besides significant proportions of clasts from Witwatersrand shale and quartzite. Granophyre has long been known to be related to the catastrophe that caused the doming event, as it is one of the rare lithologies on the dome that is not significantly affected by rock deformation. Most clasts are strongly annealed from the thermal overprint from an initially superheated melt, but in recent years both shock metamorphosed zircon and quartz have been identified. In addition, checkerboard feldspar and vermicular quartz texture testify to the strong thermal overprint caused by the impact.

The findings of shock metamorphosed clasts, and the recognition by Re-Os isotope analysis of a definite, albeit small, meteoritic component in the Granophyre in 1996 finally provided the proof that the Granophyre was impact melt rock – and laid the controversy about the origin of the Granophyre and, in fact, that about the genesis of the Vredefort Dome, to rest.

The second type of impact-generated melt rock is known as *pseudotachylitic breccia (PTB)*. Since the beginning of the 20th century the Vredefort Dome has been known as the type locality for "pseudotachylite" [historical spelling "pseudotachylyte"]. In structural geology, this term relates to "friction melt", numerous occurrences of which are known from tectonic fault and shear zones. In impact structures, however, breccias that are generally similar to pseudotachylite can be formed by several processes including friction melting, shock melting, decompression melting, and combinations of

◁ A flat exposure of pseudotachylitic breccia in Archean granite gneiss at Salvamento quarry north of the town of Parys, Vredefort Dome. Hammer, for scale, ca. 60 cm long (photo: W. U. Reimold).

Vredefort

South Africa

latitude = 27°00' S / longitude = 27°30' E / diameter = 250 - 300 km / age = 2023 ± 4 Myr

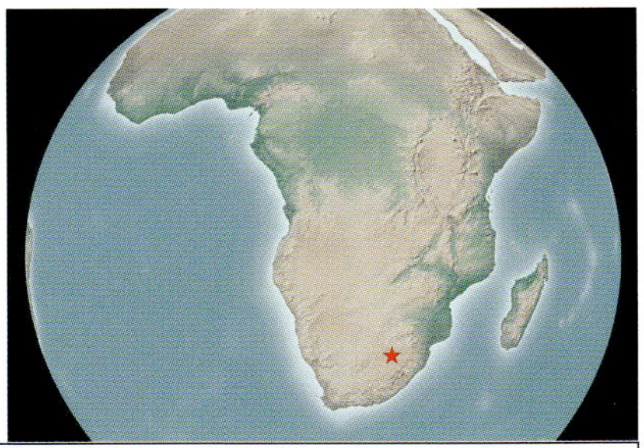

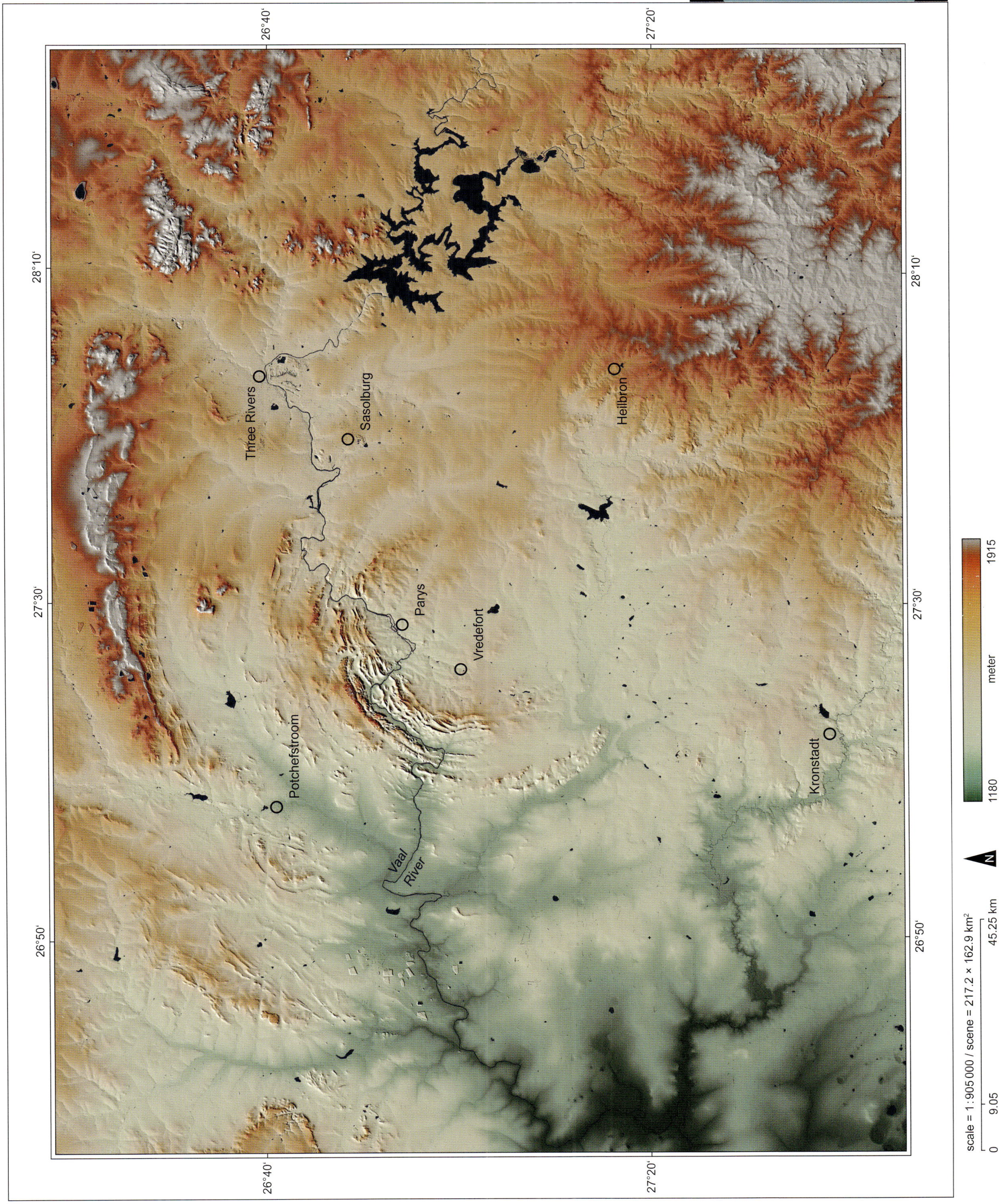

Many intricate shatter cones in a metapelite (shale of the Booysens Shale Formation), in the northwestern sector of the outer Vredefort Dome. Note that the apices of shatter cones point into opposite directions. Hammer, for scale, about 35 cm long (photo: W. U. Reimold).

these processes. The presence of coesite and stishovite in millimeter-wide Vredefort PTB veinlets strongly suggests that at least these thin occurrences represent shock compression melt, i.e., melting caused by decompression immediately after propagation of the shock front.

Shatter cones occur widespread in the metasedimentary collar rocks and occasionally in the granitoid core rocks of the Vredefort Dome. They frequently occur together with dense networks of narrow-spaced curviplanar joint sets that have been known as multipli-striated joint sets for 20 years.

Also the Archean mafic rocks of the Vredefort Dome show the effects of impact-related deformation. Here, a large (coin for scale 2 cm wide) boulder of pseudotachylitic breccia with clasts of 3074 million year old Dominion Group lava from Farm Koppieskraal, northwestern core of the Vredefort Dome (photo: W. U. Reimold).

Quartz in Vredefort rocks often contains single or – more rarely – multiple sets of micro-deformations in the form of planar fluid inclusion trails or "strings" of microscopic quartz crystals of different extinction behavior in the polarizing microscope. Detailed transmission electron microscopic study in 1994 revealed that these trails are indeed decorated planar deformation features (PDFs) – at the locations of now annealed PDFs. Since then shock deformed (planar fractures, twinning, granular shock texture) zircon from the Vredefort Dome, and shocked monazite derived from the Vredefort Dome in Vaal River sediment have been recorded. Roger Gibson and Wolf Uwe Reimold studied samples from a radial traverse across the core of the dome and revealed textures that can be related to shock metamorphism throughout the range of shock pressures from more than 45 GPa at the center of the dome to 10 GPa in the inner collar.

The combined evidence of impact-diagnostic shock metamorphic effects in quartz, zircon, and monazite, the occurrence of coesite and stishovite, shatter cones, the Vredefort Granophyre containing a meteoritic component, and the extraordinary volume of pseudotachylitic breccia that does not have a match amongst the known tectonic occurrences of pseudotachylite (friction melt) – all this leaves no doubt that the Vredefort structure represents indeed a very large impact structure. However, it took a very long time to resolve the controversy about the origin of this large and old structure.

Very early in the geological exploration of this region it was recognized that the Vredefort Dome is a local metamorphic "hot spot". In the Lower Witwatersrand and Dominion rocks of the collar the metamorphic grade is mid-amphibolite and significantly higher than the lower greenschist grade of the stratigraphically equivalent rocks further out in the Witwatersrand basin. In the core of the Vredefort Dome a metamorphic gradient from the center outward involves granulite facies and then amphibolite grade. The enhanced metamorphic grade of the collar rocks was, at first, considered the expression of a contact metamorphic aureole around a hidden pluton, but it was later revealed as having been caused already by high-grade Archean regional metamorphism. The metamorphism of the Witwatersrand collar strata, with an anti-clockwise pressure-temperature-time path – incompatible with contact metamorphism – was shown to have lasted from pre-impact through to post-impact time. The impact-generated lithologies are also affected by this second phase of metamorphic overprint. It is believed that this metamorphic event was related to the emplacement of the Bushveld complex, some 35 Myr prior to the Vredefort event. The Bushveld Complex is a huge magmatic intrusion located to the north of Johannesburg but extending with some outliers, e.g., the Losberg Complex, right into the Vredefort Structure. Following the onset of the second metamorphic event, the currently exposed rocks of the Vredefort Dome became subject to the impact event and related impact metamorphism. Detailed petrographic analysis established that shock and thermal metamorphism decrease progressively from the center of the dome outward.

Like the complex, multi-stage metamorphism of the Vredefort Dome, the deformation history of the region is complex and long. At least four deformation events with related ductile deformation features have been identified.

Vredefort Dome

South Africa

latitude = 27°00‘ S / longitude = 27°30‘ E / diameter = 250 - 300 km / age = 2023 ± 4 Myr

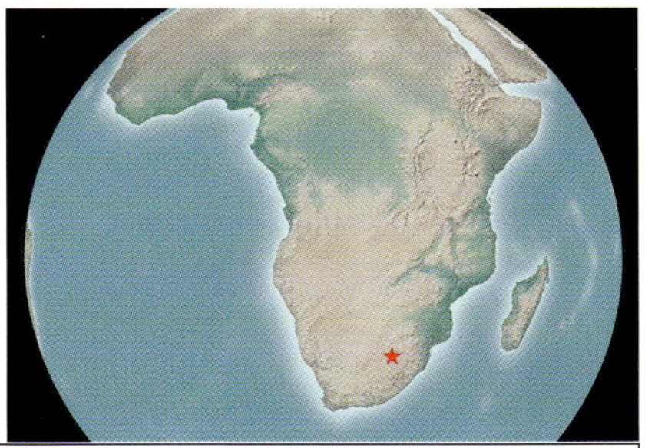

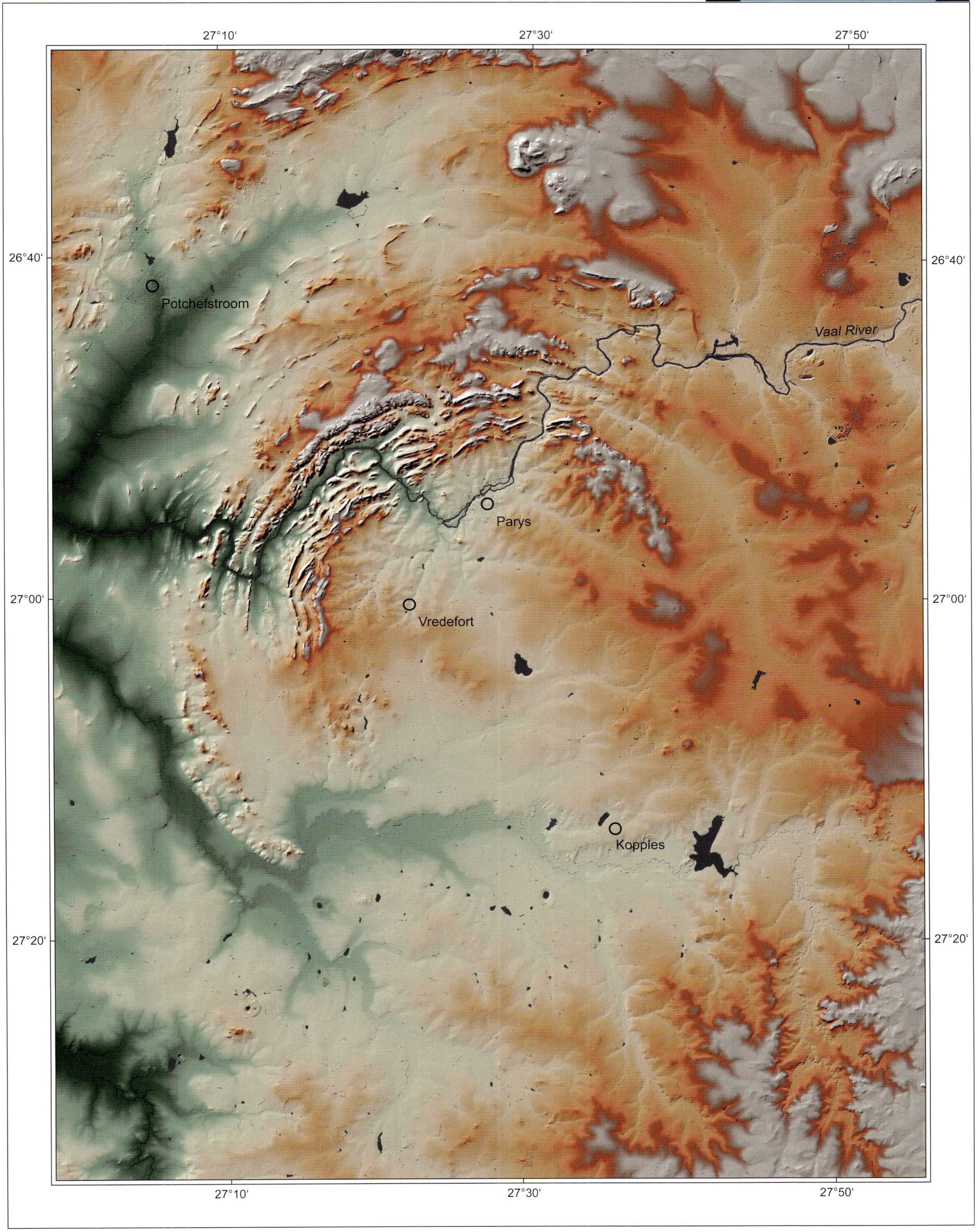

scale = 1 : 507 000 / scene = 91.3 × 121.7 km²

0 5.07 25.35 km

N

1240 meter 1755

The slightly overturned quartzite strata of the Hospital Hill Subgroup in the northwest sector of the Vredefort Dome. The flat terrain in the foreground represents Archean granite of the core of the dome. The length of the prominent outcrop in the background is about 1.5 km (photo: W. U. Reimold).

Following the confirmation of the impact origin of the Vredefort Structure, the tectonic setting of the wider region of the Witwatersrand Basin and its geophysical signatures were re-assessed. Thus, the Vredefort impact structure comprises a deeply eroded central uplift structure, the Vredefort Dome, of at present some 100 km width. This is surrounded by the – in the northern sector – up to 13 km deep Witwatersrand Basin, which forms the ring syncline around the Vredefort Dome. Its outer margin can be followed in the north along the basement high known as the *Rand Anticline*, at about 100 km distance from the center of the Vredefort Dome. The actual crater rim zone is completely eroded.

Steenkampsberg: An excellent view over the macro-folded and faulted Vredefort Mountain Land in the west-northwestern part of the collar of the Vredefort Dome is afforded from the Steenkampsberg hill, one of the most prominent elevations of the Vredefort Mountain Land. In the middle distance, the scene is about 2 km wide (photo: W. U. Reimold).

Vredefort Dome in August, 2018 as a composite image based on TanDEM-X DEM and Copernicus Sentinel-2 (channel 2/3/4) data.

The age of the Vredefort impact event has been derived from TIMS (thermal ionization mass spectrometry) U-Pb dating of authigenic, unshocked zircon grains from impact-generated pseudotachylitic breccia and the impact melt rock from Vredefort, the Vredefort Granophyre. This work yielded an age of 2023 ± 4 Myr. Since impact time, the Vredefort Dome has experienced between 5 and 10 km of erosion. The crustal cross-section of the Dome, from the outermost supracrustal rocks to the granitoids of the central core encompasses some 20–25 km thickness – a unique study possibility for anybody interested in understanding the makings of thick continental crust.

Seismic studies showed that the crust-mantle boundary to the west of the Vredefort structure occurs at a depth of about 40 km. The Vredefort Dome is characterized by a distinct, albeit weak central gravity anomaly that is explained by slightly elevated densities of the basement due to uplift of mid-crustal rocks as a consequence of impact. The area straddling the transition from amphibolite to granulite terrain exhibits a strong, negative aeromagnetic anomaly, which is interpreted as the result of remagnetization due to the impact event. Comprehensive geophysical modeling by integrating gravity, magnetic, and reflection and refraction seismic information allowed to completely reconstruct the impact structure and suggested an original diameter of 250 km. The fact that the original Vredefort impact structure extended over the entire width of the Witwatersrand basin means that the impact itself was responsible for the preservation of the long mined ore-bearing strata of the Witwatersrand Supergroup within the impact ring basin due to down-faulting along ring faults and subsequent cover of these large blocks with impact breccia. Hydrothermal activity followed immediately on the impact event, with fluids having circulated particularly in those strata that were strongly affected by impact-related mechanical damage. This model of impact-induced fluid flow provides a scenario in which the widespread authigenic deposition of gold observed in the Witwatersrand basin could be achieved.

The Vredefort impact structure is a very old and deeply eroded structure. Its original ejecta blanket has been entirely eroded. However, there are some ideas that maybe distal ejecta could have been preserved in the form of impact spherule deposits. A thick spherule layer was reported from the Ketilidian of southern Greenland. The age of this deposit is rather poorly constrained between ca. 1800 and 2100 Myr, but it has been suggested that this occurrence could represent a remnant of ejecta from either the Vredefort or the Sudbury impacts of ages within this interval. More recently, it was proposed that Vredefort ejecta might have been intersected in drill core from the Onega Basin of the Baltic Shield of Karelia, Russia. The drilled succession there contains spherule beds, the deposition of which is bracketed by two stratigraphic levels dated at 2050 ± 20 and 1980 ± 27 Myr – around the age of the Vredefort impact. Still, better chronological constraint is required, before a relationship to the Vredefort impact can be confirmed.

The uniqueness of the Vredefort Dome, now known as the central part of the impact structure, became already known in 1868 when coal exploration in the Karoo strata of the southern Witwatersrand basin was carried out by the government of the then Republic of the Orange Free State. The first detailed mapping of the Dome was done in the first decades of the 20th century. The regionally enhanced metamorphic pattern was long considered the result of contact metamorphism around a hidden pluton, and when the regional geophysical anomaly maps became available in the 1970s, the evidence for a small gravity high in the center of the dome and the strong magnetic anomaly seemingly related with an alkali granitic intrusion in the northwest collar seemed to favor this hypothesis strongly. Endogenic genetic processes were still considered by many as late as the 1990s. At the other end of the opinion spectrum, Reginald Daly in 1947, and Robert Dietz and Robert Hargraves, both in 1961, were the first to entertain ideas that the "Vredefort Ring" was the result of a large impact event. The latter two emphasized that shatter cones represented definitive evidence of impact. This interpretation rapidly became quite widely accepted overseas by the early 1970s – but not yet in South Africa. The controversy about the Vredefort origin was only concluded in the mid-1990s – due to the multidisciplinary and often collaborative studies between South Africa-based and overseas workers. This was largely spawned by a *Cryptoexplosions and Catastrophes in the Geological Record* conference that was held in July 1987 in the town of Parys in the core of the Vredefort Dome. Since then, the Vredefort impact structure has been subject of numerous studies and comprehensive publications. Amongst them are Reimold and Gibson's detailed historical and geological review of 2008 and their natural history of the Vredefort Structure in the Springer monograph from 2010.

In July 2005, a portion of the Vredefort Dome with spectacular exposures of regional geology and impact-related deformation was declared a World Heritage Site by UNESCO. However, from the side of the South African authorities a National Natural Heritage Site declaration is still required before this World Heritage status can become fully inscribed. Considering that this unique geological-planetological feature is located in a region inhabited by 15–20 million people, this structure represents a fantastic geo- and ecotouristic resource.

Location/Access

The Vredefort impact structure is transected by a net of good roads as well as numerous unpaved roads. Easy access from the city of Johannesburg is via the N1 South Freeway (in the direction of Bloemfontein) that actually tangentially crosses the Dome. Take the R13 offramp to Parys and continue in that direction. The town of Parys is located within the northern part of the core of the Dome. Travel-time from Johannesburg to Parys is just about 1 hour.

Selected References

Gibson, R. L., and Reimold, W. U. (2005). Shock pressure distribution in the Vredefort impact structure, South Africa. In: Large Meteorite Impacts III, Kenkmann, T., Hörz, F., and Deutsch, A. (eds.), Geological Society of America, Special Paper, Vol. 384, pp. 329–349.

Gibson, R. L., and Reimold, W. U. (2008). The Geology of the Vredefort impact structure. Memoir 97, Council for Geoscience, Pretoria, 181 pp.

Henkel, H., and Reimold, W. U. (1998). Integrated geophysical modeling of a giant, complex impact structure: anatomy of the Vredefort Structure, South Africa. Tectonophysics, Vol. 287, pp. 1–20.

Henkel, H., and Reimold, W. U. (2002). Magnetic model of the central uplift of the Vredefort impact structure, South Africa. Journal of Applied Geophysics, Vol. 51, pp. 43–62.

Ivanov, B. A. (2005). Numerical modeling of the largest terrestrial meteorite craters. Solar System Research, Vol. 39, pp. 381–409.

Johnson, M. R., Anhaeusser, C. R., and Thomas, R. J. (eds.) (2006). The Geology of South Africa. Geological Society of South Africa, Johannesburg and Council for Geoscience, Pretoria, 691 pp.

Reimold, W. U. (2006). Impact structures in South Africa. In: The Geology of South Africa, Johnson, M. R., Anhaeusser, C. R., and Thomas, R. J. (eds.), Geological Society of South Africa, Johannesburg and Council for Geoscience, Pretoria, pp. 629–649.

Reimold, W. U. and Gibson, R. L. (2010). Meteorite Impact! The Danger from Space and South Africa's Mega-Impact, the Vredefort Structure (3rd edition), Springer, Berlin Heidelberg, 337 pp.

Reimold, W. U., and Koeberl, C. (2014). Impact structures in Africa – A review. Journal of African Earth Sciences, Vol. 93, pp. 57–175.

Reimold, W. U., Hauser, N., Hansen, B. T., Thirlwall, M., and Hoffmann, M. (2017). The impact pseudotachylitic breccia controversy: Insights from first isotope analysis of Vredefort PTB and impact melt rock (Vredefort Granophyre). Geochimica et Cosmochimica Acta, Vol. 214, pp. 266–281.

Impact Structures – Further Confirmation Required

Ouarkziz

Algeria

Wolf Uwe Reimold

The eroded Ouarkziz structure in Algeria, also known as *Tindouf*, has a width of 3.5 km. The circular depression with an elevated rim is situated in a desert landscape with distinct northwest-southeast trending escarpments formed by gently southward dipping strata with competence contrast. The structure is superposed onto this regional-scale relief and is, thus, younger than the relief formation.

The structure was formed in Carboniferous limestones and shales– which represents an upper age limit for its formation. The orientation of the strata is upturned at the crater rim, where folding and faulting have also been described. The northern crater rim is elevated by about 130 m with respect to the crater depression. The inner parts of the crater structure are largely covered by alluvium.

Only few studies of Ouarkziz have been reported. In 1970 Jean Fabre and co-workers reported the presence of a breccia and that they observed "planar features" in quartz in samples from the inner crater rim. However, four decades later Philippe Lambert and Atmane Lamali did not accept this evidence as bona fide evidence for impact and requested further verification of these petrographic features as proper planar deformation features (PDFs). They distinguished three different, concentric zones in the structure: an outer zone with inward dipping concentric faults, followed towards the center by a zone with inward dipping strata, and in the central part a circular uplift feature with vertically dipping strata. Lambert and Lamali suggested that the impact occurred "after formation of the Mesozoic regional peneplain" and, thus, was of pre-Tertiary age.

In the absence of any new study of this structure, Ouarkziz would be a prime target for further geological exploration. It has been listed in the Earth Impact Database as a confirmed impact structure, and indeed all structural features are in favor of an impact structure. However, obviously the alleged evidence for shock metamorphism still needs to be verified. Also, the surrounding regional geology has not been investigated in detail – and it would be rewarding to investigate that further as well.

Location/Access

The Ouarkziz structure is relatively easy to visit by following the paved road between the town of Tindouf and the village of Hassi Mounir, about 26 km southwest of the former. When arriving at the military post of Foum Defili, the visitor has to turn west onto a desert track and follow it for 20 km to the structure. Military permission for a visit should be obtained, as the structure is located in a military zone close to the Algeria-Morocco border.

Selected References

Chabou, M. C. (2019). Meteorite impact structures in the Arab World: An overview. In: The Geology of the Arab World – An Overview, Bendaoud, A., Hamimi, Z., Hamoudi, M., Djemai, S., and Zoheir, B. (eds.), Springer Geology, pp. 455–506.

Fabre, J., Kazi-Tani, N., and Megartsi, M. (1970). Le rond de l'Ouarkziz (Sahara nordoccidental), un astroblème. Comptes Rendus de l'Académie des Sciences, Paris, Vol. D270, pp. 1212–1215.

Lambert, P., and Lamali, A. (2009). Impact structures in Algeria (Abstract). In: 1st Arab Impact Cratering and Astrogeology Conference, Amman, Jordan, pp. 50–53.

Panoramic view of the 3.5 km wide Ouarkziz structure (image courtesy: Charaf Chabou, Ferhat Abbas University, Sétif, Algeria).

The Ouarkziz structure on July 26, 2018 as a composite image based on TanDEM-X DEM and Copernicus Sentinel-2 (channel 2/3/4) data.

Ouarkziz

Algeria

latitude = 29°00‘38“ N / longitude = 07°33‘02“ W / diameter = 3.5 km / age = > 66 Myr

scale = 1 : 82 000 / scene = 14.8 × 19.8 km²

0 0.82 4.10 km

N

380 meter 690

7.2 North/Central America

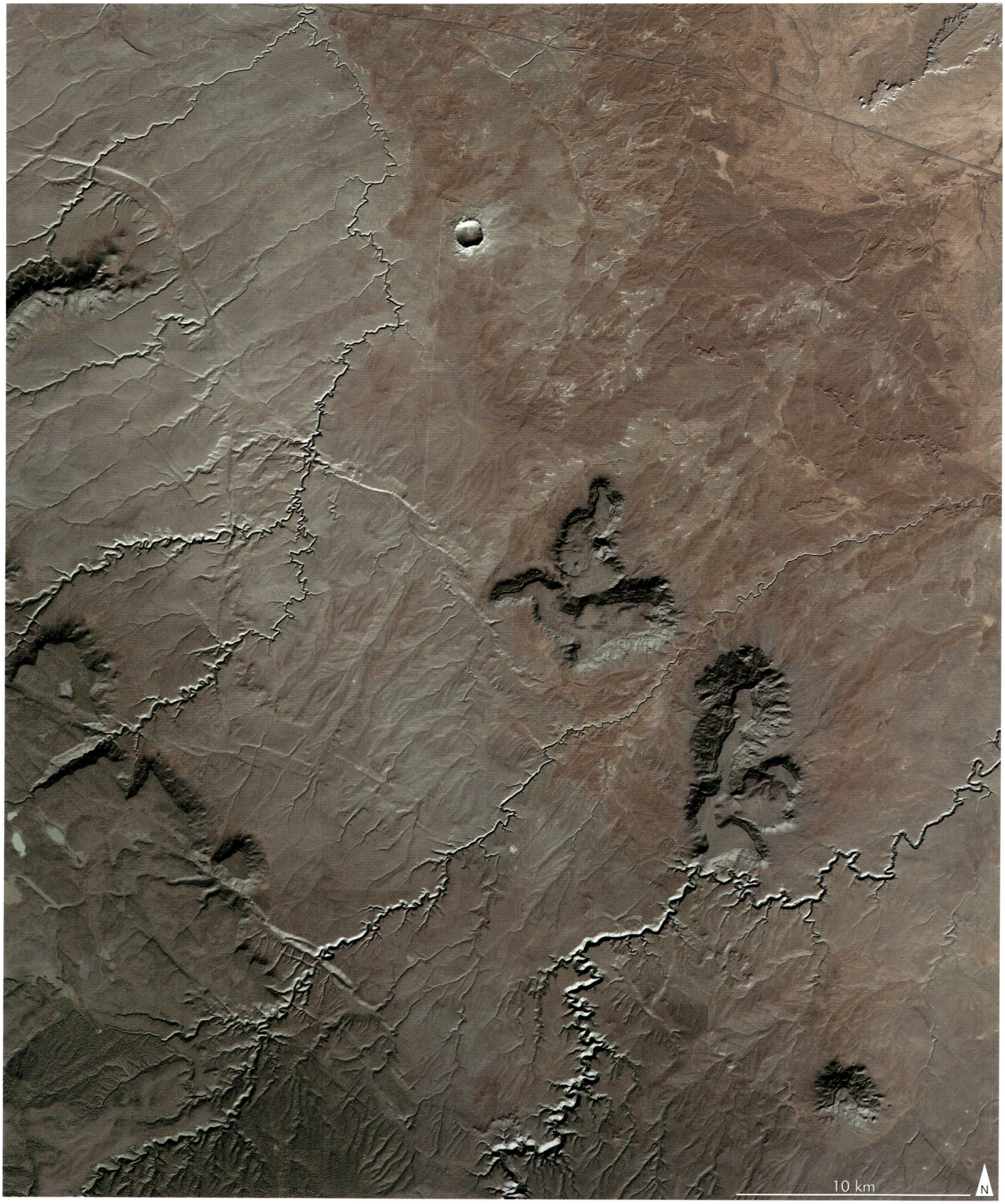

Meteor Crater in September 2018 - TanDEM-X / Copernicus Sentinel-2

North/Central American Impact Structures – Overview

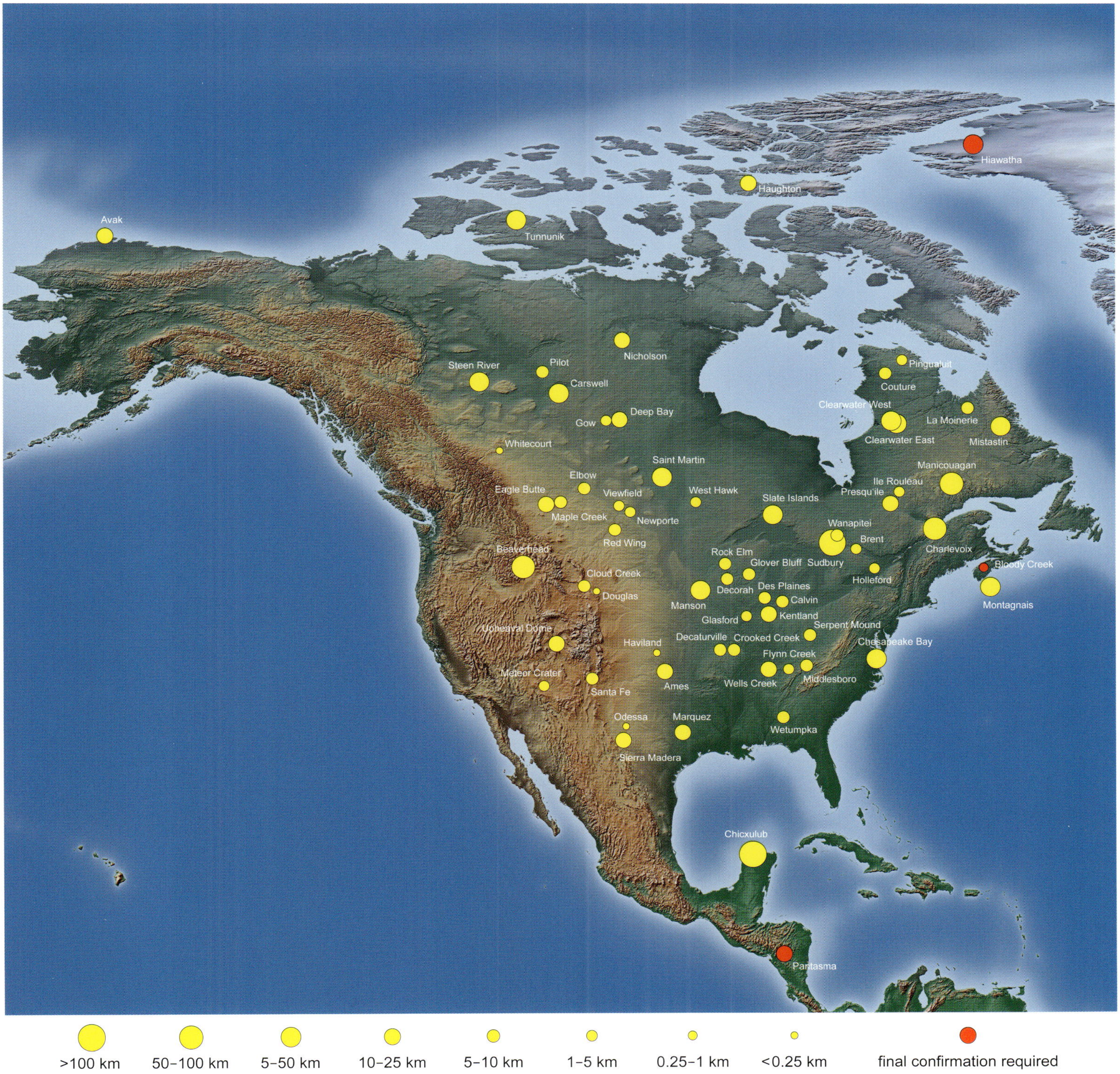

Table: North/Central American Impact Structures – Overview

ID	Name	Country/Region	Latitude	Longitude	Diameter (km)	Age (Myr)	Topographic Visibility
1	Ames[1]	Oklahoma, United States	36°15'N	98°12'W	15	470	no (buried)
2	Avak[2]	Alaska, United States	71°15'N	156°28'W	12	90–94	no (buried)
3	Beaverhead[3]	Montana, United States	44°36'N	112°58'W	75	< 900	no
4	Brent	Ontario, Canada	46°04'30"N	78°28'55"W	3	453	yes (buried)
5	Calvin[4]	Michigan, United States	41°49'48"N	85°57'00"W	8.5	450	no (buried)
6	Carswell	Saskatchewan, Canada	58°25'N	109°31'W	39	481.5 ± 0.8	yes
7	Charlevoix	Quebec, Canada	47°32'N	70°21'W	55	450 ± 20	yes
8	Chesapeake Bay[5]	Virginia, United States	37°14'N	76°01'W	85	35.2 ± 0.3	no (buried/submerged – Chesapeake Bay)
9	Chicxulub	Mexico	21°20'N	89°30'W	180	66	no (buried/submerged – Gulf of Mexico)
10	Clearwater East	Quebec, Canada	56°03'N	74°05'W	24	460–470	yes (Clearwater Lake)
11	Clearwater West	Quebec, Canada	56°12'N	74°30'W	32	290–300	yes (Clearwater Lake)
12	Cloud Creek[6]	Wyoming, United States	43°10'36"N	106°42'30"W	7	190 ± 20	no (buried)
13	Couture	Quebec, Canada	60°07'47"N	75°17'56"W	8	425 ± 25	no (submerged – Lake Couture)
14	Crooked Creek	Missouri, United States	37°50'05"N	91°23'44"W	7	323–348	yes
15	Decaturville	Missouri, United States	37°53'33"N	92°43'11"W	6	< 300	yes
16	Decorah	Iowa, United States	43°18'50"N	91°46'20"W	5.6	460–483	no (buried)
17	Deep Bay	Saskatchewan, Canada	56°24'N	102°59'W	13	95–102	no (submerged – Lake Reindeer)
18	Des Plaines[7]	Illinois, United States	42°02'42"N	87°52'24"W	8.8	< 280	no (buried)
19	Douglas Crater Field[8]	Wyoming, United States	42°40'N	105°28'W	max. 0.1	280	yes
20	Eagle Butte[2]	Alberta, Canada	49°42'N	110°30'W	15	< 66	no
21	Elbow[9]	Saskatchewan, Canada	50°58'N	106°45'W	8	201–358	no (buried)
22	Flynn Creek[10]	Tennessee, United States	36°17'N	85°40'W	3.8	382	no
23	Glasford[7]	Illinois, United States	40°36'06"N	89°47'06"W	10	455 ± 2	no (buried)
24	Glover Bluff	Wisconsin, United States	43°58'12"N	89°32'18"W	10	455–459	no
25	Gow	Saskatchewan, Canada	56°27'23"N	104°28'56"W	5	250	yes (Gow Lake)
26	Haughton	Nunavut, Canada	75°23'N	89°39'W	23	39	yes
27	Haviland[11]	Kansas, United States	37°34'57"N	99°09'49"W	0.017	0.020 ± 0.002	no (destroyed)
28	Holleford	Ontario, Canada	44°27'34"N	76°37'57"W	2.35	> 485	yes (buried)
29	Île Rouleau	Quebec, Canada	50°40'53"N	73°52'49"W	4	< 300	yes (Lac Mistassini)
30	Kentland[12]	Indiana, United States	40°45'N	87°24'W	12.5	< 107	no
31	La Moinerie	Quebec, Canada	57°26'10"N	66°35'09"W	8	400 ± 50	no (submerged – Lake La Moinerie)
32	Manicouagan	Quebec, Canada	51°23'N	68°41'W	100	214	yes
33	Manson[13]	Iowa, United States	42°35'N	94°33'W	35	74.1 ± 0.1	no (buried)
34	Maple Creek[2]	Saskatchewan, Canada	49°48'00"N	109°06'30"W	6	< 75	no (buried)
35	Marquez[3]	Texas, United States	31°17'00"N	96°17'30"W	15	58.3 ± 3.1	no (buried)
36	Meteor Crater	Arizona, United States	35°01'39"N	111°01'20"W	1.2	0.049	yes
37	Middlesboro	Kentucky, United States	36°37'03"N	83°43'39"W	6	290–300	yes
38	Mistastin	Labrador, Canada	55°53'N	63°19'W	28	36.6 ± 2	yes (Mistastin Lake)
39	Montagnais[14]	Nova Scotia, Canada	42°53'N	64°13'W	45	50.5 ± 0.8	no (buried/submerged – Atlantic Ocean)
40	Newporte[15]	North Dakota, United States	48°58'N	101°58'W	3	500	no (buried)
41	Nicholson Lake	Northwest Territories, Canada	62°40'N	102°42'W	12.5	389.0 ± 6.7	yes (submerged – Lake Nicholson)
42	Odessa Crater Field	Texas, United States	31°45'21"N	102°28'43"W	max. 0.17	0.0635 ± 0.0045	yes
43	Pilot	Northwest Territories, Canada	60°17'06"N	110°59'54"W	6	445 ± 2	no (submerged – Pilot Lake)
44	Pingualuit	Quebec, Canada	61°16'40"N	73°39'36"W	3.4	1.4 ± 0.1	yes

ID	Name	Country/Region	Latitude	Longitude	Diameter (km)	Age (Myr)	Topographic Visibility
45	Presqu'île	Quebec, Canada	49°43'N	74°50'W	22	< 500	no (Lac de la Presqu'île)
46	Red Wing[16]	North Dakota, United States	47°36'N	103°33'W	9	200–220	no (buried)
47	Rock Elm	Wisconsin, United States	44°43'25"N	92°13'41"W	6.5	410–460	yes
48	Saint Martin[17]	Manitoba, Canada	51°47'N	98°32'W	40	227.4 ± 0.8	no (buried)
49	Santa Fe[18]	New Mexico, United States	35°43'54"N	105°51'04"W	9.5	350–1200	no
50	Serpent Mound	Ohio, United States	39°01'56"N	83°24'10"W	8	256–290	yes
51	Sierra Madera	Texas, United States	30°35'44"N	102°54'42"W	12	< 100	yes
52	Slate Islands	Ontario, Canada	48°39'N	87°01'W	30	436 ± 3	yes (Lake Superior)
53	Steen River[19]	Alberta, Canada	59°30'N	117°38'W	22	108	no (buried)
54	Sudbury	Ontario, Canada	46°36'N	81°11'W	200	1849.3 ± 0.3	yes
55	Tunnunik	Northwest Territories, Canada	72°28'N	113°55'W	25	430–450	yes
56	Upheaval Dome	Utah, United States	38°26'09"N	109°55'40"W	8	66–100	yes
57	Viewfield[9]	Saskatchewan, Canada	49°35'N	103°04'W	2.5	190	no (buried)
58	Wanapitei	Ontario, Canada	46°44'44"N	80°44'43"W	7.5	37 ± 2	no (submerged – Lake Wanapitei)
59	Wells Creek	Tennessee, United States	36°22'34"N	87°39'33"W	12	70–325	yes
60	West Hawk	Manitoba, Canada	49°45'41"N	95°11'12"W	2.4	351 ± 20	no (submerged – West Hawk Lake)
61	Wetumpka	Alabama, United States	32°31'24"N	86°10'32"W	7	84.4 ± 1.4	yes
62	Whitecourt[20]	Alberta, Canada	53°59'57"N	115°35'50"W	0.036	< 0.0011	yes
Further confirmation required							
63	Bloody Creek[21]	Nova Scotia, Canada	44°45'00"N	65°14'35"W	0.4	unknown	no (submerged – Dalhousie Lake)
64	Hiawatha	Greenland	78°45'N	66°15'W	31	< 2.6	no (buried – Greenland ice sheet)
65	Pantasma	Nicaragua	13°21'49"S	85°57'14"W	14	0.815 ± 0.011	yes

Position references for selected impact structures:

1 Koeberl, C., Reimold, W. U., and Powell, R. A. (1994). Shocked quartz and impact melt rock at the Ames structure, Oklahoma (abstract). Meteoritics, Vol. 29, p. 483.
2 Westbroek, H. H. (1997). Seismic Interpretation of Two Possible Meteorite Impact Craters: White Valley, Saskatchewan and Purple Springs, Alberta. Thesis M.Sc., University of Calgary, 145 pp.
3 Dietz, R. S., and McHone, J. F. (1991). Astroblemes recently confirmed with shatter cones. Meteoritics, Vol. 26, pp. 332.
4 Milstein, R. L. (1994). The Calvin Impact Crater, Cass County, Michigan: Identification and Analysis of a Subsurface Ordovician Astrobleme. Thesis, Oregon State University, 114 pp.
5 Horton, J. W., Powars, D. S., and Gohn, D. S. (2005). Studies of the Chesapeake Bay impact structure – Introduction and discussion. In: Studies of the Chesapeake Bay Impact Structure – The USGS-NASA Langley Corehole, Hampton, Virginia and Related Coreholes and Geophysical Surveys, Horton, J. W., Powars, D. S., and Gohn, G. S. (eds.), USGS Professional Paper, Vol. 1688, A1–A24.
6 Stone, D. S., and Therriault, A. M. (2003). Cloud Creek structure, central Wyoming, USA: Impact origin confirmed. Meteoritics and Planetary Science, Vol. 38, pp. 445–455.
7 McHone, J. F., Sargent, M. L., and Nelson, W. J. (1986). Shatter Cones in Illinois: Evidence for meteoritic impacts at Glasford and Des Plaines. Meteoritics, Vol. 21, p. 446.
8 Kenkmann, T., Sundell, K. A., and Cook, D. (2018a). Exhumed Paleozoic impact crater strewn field near Douglas, Wyoming, USA: Evidence from micro-structural analysis, satellite, and drone Imagery. 49th Lunar and Planetary Science Conference, The Woodlands, Texas, Abstract #1469.
9 Grieve, R. A. F., Kreis, K., Therriault, R. M., and Robertson, P. B. (1998). Impact structures in the Williston Basin. Meteoritics and Planetary Science, Vol. 33, pp. A63–A64.
10 Evenick, J. C., Lee, P., and Deane, B. (2004). Flynn Creek impact structure: New insights from breccias, melt features, shatter cones, and remote sensing. 35th Lunar and Planetary Science Conference, League City, Texas, Abstract #1131.
11 Hodge, P. W. (1979a). The location of the site of the Haviland meteorite crater, Meteoritics, Vol. 14, pp. 233–234.
12 Koeberl, C., and Sharpton, V. L. (1993). Geochemical study of rocks from the Kentland, Indiana, impact structure: Progress report (abstract). Meteoritics, Vol. 28, p. 382.
13 Hartung, J. P., Kunk, M. J., and Anderson, R. R. (1990). Geology, geophysics, and geochronology of the Manson impact structure. Geological Society of America Special Paper, Vol. 247, pp. 207–221.
14 Jansa, L. F., Pe-Piper, G., Roberston, P. B., and Friedenreich, O. (1989). Montagnais: A submarine impact structure on the Scotian Shelf, eastern Canada. Geological Society of America Bulletin, Vol. 101, pp. 450–463.
15 Koeberl, C., and Reimold, W. U. (1995). The Newporte impact structure, North Dakota, USA. Geochimica et Cosmochimica Acta, Vol. 59, pp. 4747–4767.
16 Koeberl, C. Reimold, W. U., and Brandt, D. (1996). Red Wing Creek structure, North Dakota: Petrographical and geochemical studies, and confirmation of impact origin. Meteoritics and Planetary Science, Vol. 31, pp. 335–342.
17 Schmieder, M., Jourdan, F., Tohver, E., and Cloutis, E. A. (2014). ^{40}Ar-^{39}Ar age of the Lake Saint Martin impact structure (Canada) – Unchaining the Late Triassic terrestrial impact craters. Earth and Planetary Science Letters, Vol. 406, pp. 37–48.
18 Fackelman, S. P., Morrow J. R., Koeberl C., and McElvain, T. H. (2008). Shatter cone and microscopic shock-alteration evidence for a post-Paleoproterozoic terrestrial impact structure near Santa Fe, New Mexico, USA. Earth and Planetary Science Letters, Vol. 270, pp. 290–299.
19 Mazur, M., Hildebrand, A., and Stewart, R. (2003). Geophysical Studies of the Steen River Impact Structure, Alberta, Canada. 65th EAGE Conference and Exhibition, Stavanger, Norway, 4 pp.
20 Kofman, R. S., Herd, D. K., and Froese, D. G. (2010). The Whitecourt meteorite impact crater, Alberta, Canada. Meteoritics and Planetary Science, Vol. 45, pp. 1429–1445.
21 Spooner, I., Stevens, G., Morrow, J., Pufahl, P., Grieve, R., Raeside, R., Pilon, J., Stanley, C., Barr, S., and McMullin, D. (2009). Identification of the Bloody Creek structure, a possible impact crater in southwestern Nova Scotia, Canada. Meteoritics and Planetary Science, Vol. 44, pp. 1193–1202.

Ames
Oklahoma, United States

Thomas Kenkmann

Ames is a 15 km diameter, complex impact structure that is buried under a 2.75 km thick pile of Paleozoic sediments. The structure has been extensively drilled in the course of oil and gas exploration, and has been a strong hydrocarbon producer for decades. Morphology, gravity and magnetic characteristics are typical of complex impact structures.

The crater structure was formed in a sequence of Cambrian and Early Ordovician sediments of the Arbuckle Group. This sedimentary sequence overlies a crystalline basement of granites and metamorphosed igneous and sedimentary rocks of Proterozoic and Lower Cambrian age. Middle and Upper Ordovician shales were deposited after the impact and provide a biostratigraphic age for the impact event. The post-impact sedimentary sequence also includes Devonian and Carboniferous strata.

Geophysical surveying, and a dense array of boreholes that covers most parts of the crater enabled to map the buried crater geology. A structure map was compiled that traces a prominent sediment layer, the Upper Ordovician Sylvan shale. This post-impact layer visualizes the outline and morphology of the crater as it appeared shortly after its formation. It shows the presence of two concentric rings. The outer rim zone with a diameter of about 15 km is between 1.5 and 3 km wide. This zone is separated from an inner concentric high of about 5 km width by an annular trough. The relief between the ridges and the moat amounts to 60 m. The rocks of the outer rim consist mainly of fractured and brecciated Ordovician dolomite. The lower part of the moat is filled with a limestone that represents the prolific oil and gas producer. The inner ring structure probably results from the collapse of an over-heightened central uplift and is composed of brecciated Precambrian granites and rocks of the Cambro-Ordovician Arbuckle Group. The innermost part of the central domain forms a depression.

Some rocks recovered from this central region contain quartz and feldspar grains with planar deformation features (PDFs) and demonstrate the impact origin of the structure. Small fragments of a clast-rich impact melt rock were encountered in two drill cores. This impact melt rock is slightly enriched in siderophile elements, including iridium, which points to the presence of a meteoritic component. ^{40}Ar-^{39}Ar dating of melt rock gave scattered ages of around 300 Myr that are distinctly lower than the biostratigraphic age for the first post-impact sediments (Oil Creek Formation) of 470 Myr. The large scatter of radiometric ages is likely caused by resetting of the Ar isotopic system in a post-impact hydrothermal phase.

The crater anomaly was discovered in 1991 in the course of geophysical prospecting and was, at the time, named the "Ames hole". Debates on the origin of this anomaly focused on a tectonic graben structure or an idea of a collapsed sinkhole structure. Eventually the impact origin could be confirmed by Christian Koeberl and coworkers in 1994, who discovered evidence of shock metamorphism and possible traces of the impactor.

Location/Access

The Ames structure is located in Oklahoma's Major County, 25 km east of the county's major city, Fairview, and 100 km northwest of Oklahoma City. The site is situated in the area of Townships 20 and 21 N, Ranges 9 and 10 W, but there is no obvious topographic expression at the surface above the structure.

Selected References

Carpenter, B. N., and Carlson, R. (1992). The Ames impact crater. Oklahoma Geological Survey. Vol. 52, pp. 208–223.

Koeberl, C., Reimold, W. U., and Powell, R. A. (1994). Shocked quartz and impact melt rock at the Ames structure, Oklahoma (abstract). Meteoritics, Vol. 29, p. 483.

Koeberl, C., Reimold, W. U., Brandt, D., Dallmeyer, R. D., and Powell, R. A. (1997). Target rocks and breccias from the Ames impact structure, Oklahoma: Petrology, mineralogy, geochemistry and age. In: The Ames Structure and Similar Features, Johnson, K. S., and Campbell, J. A. (eds.), Oklahoma Geological Survey Circular, Vol. 100, pp. 169–198.

Koeberl, C., Reimold, W. U., and Kelley, S. P. (2001). Petrography, geochemistry, and ^{40}Ar-^{39}Ar ages of impact melt rocks and breccias from the Ames Impact Structure, Oklahoma: The Nicor Chestnut 18-4 drill core. Meteoritics and Planetary Science, Vol. 36, pp. 651–671.

Avak
Alaska, United States

Thomas Kenkmann

Avak is a well investigated, buried, 12 km wide, complex impact structure in Northern Alaska. The form of this structure is not perfectly circular. Three gas fields are being exploited within, and in the vicinity of, the impact structure. Due to the economic significance of the structure for gas exploration, a wealth of borehole and geophysical data has been gathered that constrain the three-dimensional shape of the structure.

Target rocks comprise low-grade metamorphic rocks of the basement and overlying sedimentary strata. Outside of the crater structure, the sedimentary rocks generally have subhorizontal attitude. The sedimentary sequence includes Ordovician and Silurian argillites, Upper Triassic to Lower Jurassic clastic shelf strata, and Lower Cretaceous shales. Pliocene-Pleistocene marine deposits, as well as Quaternary sands and gravels that are part of the permafrost ground form the cover of the structure.

The central uplift, in which Lower Cretaceous rocks have been uplifted by more than 500 m, measures about 4 km in diameter and is slightly elongated in west-southwest/east-northeast direction. The central uplift is surrounded by a 1.6 km wide ring syncline. Strata of the ring basin are situated up to 300 m lower than their regional position outside the crater. The ring syncline, in turn, is surrounded by a discontinuous concentric anticline that acted as a trap and reservoir for hydrocarbons. The Lower Jurassic Barrow Sand, lying at a depth of about 670 m, is the gas-producing layer. Outside of this rim anticline occurs a system of circumferential normal faults that define the perimeter of the Avak impact structure. The faults are inward-dipping, are partly of listric shape, and show some embayments. The anticline may be interpreted as a rollover anticline in the hanging-wall of the listric normal faults. It most likely began to form during the excavation stage of cratering by structural thickening along the rim of the transient cavity.

The central uplift consists of chaotically deformed Upper Triassic to Lower Cretaceous sedimentary strata. Planar deformation features (PDFs) in quartz were found in samples from a polymict breccia

Ames

Oklahoma / United States

latitude = 36°15' N / longitude = 98°12' W / diameter = 15 km / age = 470 Myr

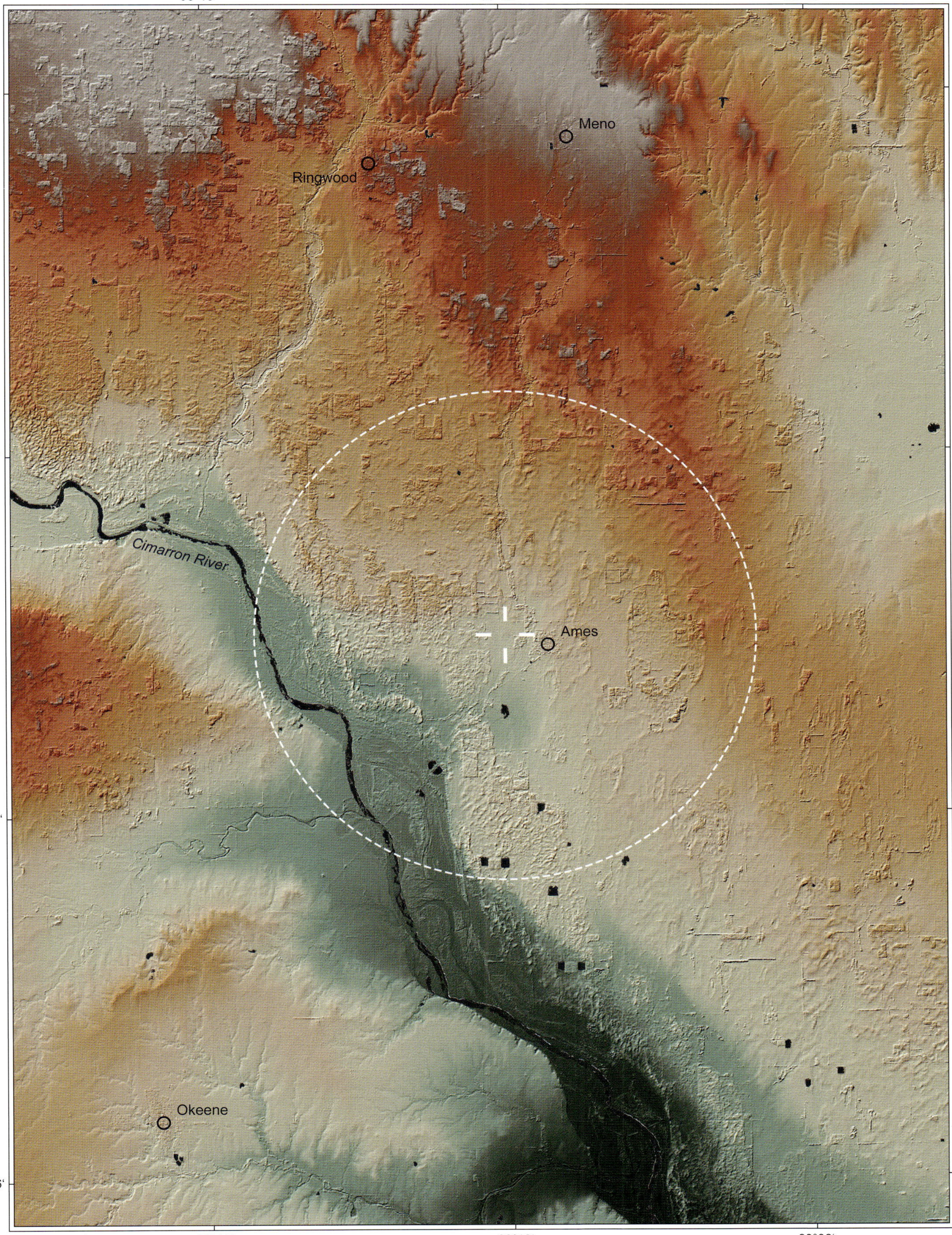

scale = 1:155 000 / scene = 27.9 × 37.1 km²

0 1.55 7.75 km

N

330 meter 435

lens. This breccia was encountered at a depth of about 700 m in a borehole drilled into the central uplift. Moreover, shatter cones occur in uplifted Jurassic sandstones of the central uplift.

The lack of a breccia infill of the annular moat of the crater and the observation that ejecta are missing in the surroundings of the structure indicate that this impact structure has been subject to severe erosion since its formation. Palynology of the ejecta matrix provides a middle to late Turonian age of 90–94 Myr for the Avak impact event.

The structure was investigated in the course of hydrocarbon exploration. The impact origin was confirmed by Charles Kirschner and co-workers in 1992.

Location/Access

Access to Avak can be an issue. The structure is situated some 12 km southeast of the village of Utqiagvik (Barrow) close to the shore of the Beaufort Sea / Arctic Ocean in Northern Alaska. This village has an airstrip, and the whole area is an industrial field for gas exploration and production. Three gas fields, the South and East Barrow Gas Fields and the Sikulik Gas Field, are in operation, which means that there could be restricted access.

Selected References

Banet, A. C., and Fenton, J. P. G. (2008). An examination of the Simpson core test wells suggests an age for the Avak impact feature near Barrow, Alaska. In: The Sedimentary Record of Meteorite Impacts, Evans, K., Horton, Jr., J. W., King, D. T., and Morrow, J. R. (eds.), Geological Society of America Special Paper, Vol. 437, pp. 139–145.

Collins, F. R., and Robinson, F. M. (1967). Exploration of Naval Petroleum Reserve No. 4 and adjacent areas, northern Alaska, 1944–53 – Part 5, Subsurface geology and engineering data. U.S. Geological Survey Open File Report, Geological Survey Professional Paper 305-K, pp. 140–145.

Kirschner, C. E., Grantz, A., and Mullen, M. W. (1992). Impact origin of the Avak Structure, Arctic Alaska, and genesis of the Barrow gas fields. The American Association of Petroleum Geologists Bulletin, Vol. 76, pp. 651–679.

Therriault, A. M., and Grantz, A. (1995). Planar deformation features in quartz grains from mixed breccia of the Avak structure, Alaska (abstract). 26th Lunar and Planetary Science Conference, Houston, Texas, Lunar and Planetary Institute, pp. 1403–1404.

◁ The area of the Avak impact structure on April 19, 2019 as a composite image based on TanDEM-X DEM and Copernicus Sentinel-2 (channel 2/3/4) data. Cross hairs and a dashed circle indicate position and extent of the buried structure. Note the numerous, partially frozen lakes in the structure and its environs.

Avak

Alaska / United States

latitude = 71°15' N / longitude = 156°28'W / diameter = 12 km / age = 90 - 94 Myr

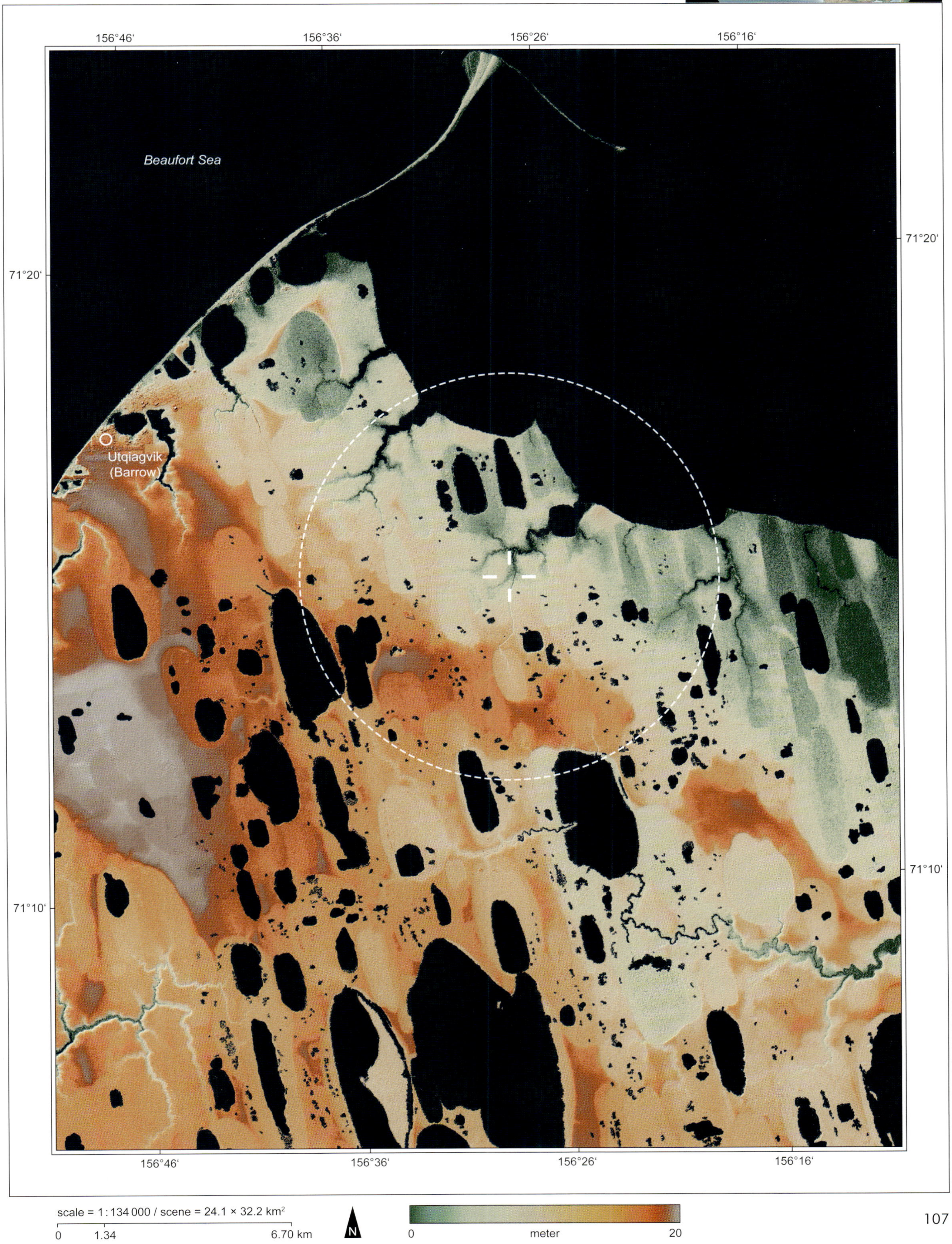

Beaverhead
Montana, United States

Thomas Kenkmann

Beaverhead is situated in the Idaho-Montana thrust belt of the Rocky Mountains, a terrane with a complex geological history. The structure is morphologically not identifiable but shatter cones are exposed in Late Proterozoic sandstones and Proterozoic gneisses over a large area of some 200 km². These occurrences suggest that the crater may have had an original size of at least 60 km, possibly even 75 km. Further indications of shock metamorphism are present in a large thrust sheet called the "Cabin Thrust" that involves Precambrian strata. This sheet was thrust eastward by several tens of kilometers onto Paleozoic limestones and younger sediments. Thus, the original location of the structure is assumed to be located further to the west. Thrusting occurred during the Sevier orogeny in the Late Cretaceous and postdates the impact event.

Units exposed in this region include Precambrian gneisses and sandstones of the Belt Supergroup. Paleozoic carbonate and clastic sedimentary rocks are most likely post-impact deposits. They include Cretaceous-Tertiary extrusive rocks, Paleocene sediments, Pleistocene glacial deposits, and Quaternary gravels and alluvium.

Shatter cones are primarily found in massive sandstones of probably Late Proterozoic age. Less well developed shatter cones also occur in Proterozoic gneisses. Most of the cones have cone axes of fairly steep dip. The area in which shatter cones are exposed measures 25 × 8 km. Shatter cone appearances are restricted to the Precambrian rocks of the nappe and are absent in Phanerozoic strata. Moreover, numerous breccia occurrences with flow-banded matrix of greenish to black color were observed in floatstones near Island Butte. These have been interpreted as pseudotachylite formed by frictional melting. Later, up to 1 m thick pseudotachylite dikes were found in an area of 50 km² within the limit of the shatter cone occurrences. More recent ^{40}Ar-^{39}Ar dating of muscovite and biotite from a breccia zone in high-grade gneiss and U-Pb ages obtained on zircon from the Beaverhead impact structure indicate that the structure formed at or after 900 Myr ago.

A large, 50 km wide, circular magnetic anomaly is centered on the town of Challis in Idaho. However, this anomaly is offset to the west by some 40 km with respect to the current area of impact evidence near Island Butte. This offset is interpreted to result from thin-skin eastward thrusting during the Laramide orogeny at 40–70 Myr ago. The deeper parts of the crater structure that would be responsible for the magnetic anomaly were not thrust eastward. Beaverhead was identified in 1989 by Robert Hargraves on the basis of shatter cone findings in sandstone.

Location/Access

Beaverhead is located in Beaverhead County in southwest Montana close to the border with Idaho. The nearest town is Challis, which has a small airpor-. Interstate 15 touches the Beaverhead structure at its eastern edge between Idaho Falls in the south and Butte in the north. Landmcrks of the area are Beaverhead Rock and the Beaverhead River, both within Beaverhead State Park. Beaverhead River forms, together with Red Rock River, the headwaters of the Missouri River.

The area of the buried Beaverhead ▷ impact structure on May 17, 2017 as a composite image based on TanDEM-X DEM and Copernicus Sentinel-2 (channel 2/3/4) data. The eastward thrust position near Island Butte and extent of Beaverhead are illustrated by cross hairs and a dashed circle.

Beaverhead

Montana / United States

latitude = 44°36' N / longitude = 112°58' W / diameter = 75 km / age = < 900 Myr

scale = 1 : 601 000 / scene = 108.2 × 144.3 km²

0 6.01 30.05 km

N

1325 meter 3715

Selected References

Fiske, P. S., Hargraves, R. B., Onstott, T. C., Koeberl, C., and Hougen, S. B. (1994). Pseudotachylites of the Beaverhead impact structure: Geochemical, geochronological, petrographic, and field investigations. Geological Society of America Special Paper, Vol. 293, pp. 163–176.

Hargraves, R. B., Christiansen, P. P., Cullicott, C. E., Deffeyes, K. S., Fiske, P. S., and Hougen, S. (1990). Shatter cones and shocked rocks in southwestern Montana: The Beaverhead impact structure. Geology, Vol. 18, pp. 832–834.

Hargraves, R. B., Kel ogg, K. S., Fiske, P. S., and Hougen, S. B. (1994). Allochthonous impact-shocked rocks and superimposed deformations at the Beaverhead site in southwest Montana. Geological Society of America Special Paper, Vol. 293, pp. 225–235.

Kellogg, K. S., Snee, L. W., and Unruh, D. M. (2003). The Mesoproterozoic Beaverhead impact structure and its tectonic setting, Montana-Idaho: ^{40}Ar-^{39}Ar and U-Pb isotopic constraints. Journal of Geology, Vol. 111, pp. 639–652.

Koeberl, C., and Fiske, P. S. (1991). Beaverhead impact structure, Montana: Geochemistry of impactites and country rock samples (abstract). Meteoritics, Vol. 26, pp. 358-359.

Brent
Ontario, Canada

Wolf Uwe Reimold

Brent impact crater is an approximately 3 km wide feature that is located near the northern limit of Algonquin Provincial Park of Canada's Ontario Province. It is named after the village of the same name, which is situated 4 km south of the structure. The structure stands out regionally as an up to 60 m deep circular depression, with a much smoother interior than the surroundings. The basin is more or less completely covered by glacial deposit. These deposits overlie post-impact sediments of Ordovician age. The original crater rim has been entirely eroded. In fact, it has been estimated that erosion has degraded the structure by some 400 m. Based on the drilling results and comparison against other, less eroded crater structures, the original crater diameter has been estimated at ca. 3.8 km. An extensive drilling record for Brent has allowed to reconstruct the structure in three dimensions. Its geometry is somewhat parabolic, with a depth to the crater floor at the center estimated at about 880 m. Drilling has also provided a complete cross-section through the crater fill, which has been used extensively in the literature as a type section through a final (modified) simple, bowl-shaped impact crater formed in a crystalline target.

The Brent crater was formed in granodioritic gneisses and minor amphibolite of the basement to the Grenville structural province of the Canadian Shield. The gneisses are thought to be 1480–1570 Myr old. They have been extensively overprinted by the Grenville metamorphism at ca. 900–1000 Myr ago. The damage zone of fracturing, local brecciation, and hydrothermal overprint around the crater is about 400 m wide. The crater fill is well-known from the drilling record across the entire structure – not less than 12 boreholes were drilled, with an aggregate core length of about 5 km core recovered. In the two central holes surficial Pleistocene glacial deposit was encountered, followed by some 263 m of undeformed Middle Ordovician post-impact sediments (an upper sequence of dolostones with interlayered gypsum and siltstone, and a lower sequence of limestone, sandstone, and siltstone). The lower sequence represents the well-laminated deposits of a saline crater lake. In contrast, the upper sequence is interpreted as the result of a marine transgression. The post-impact sediments overlie a lens-like deposit of allochthonous impact breccia of 630 m maximum thickness, below which the fractured and shock metamorphosed parautochthonous basement was reached.

The volume of the allochthonous impact breccia has been estimated at 2.1 km³. The main portion of the polymict allochthonous breccia is a lithic impact breccia with mainly unshocked and poorly shocked lithic and mineral clasts. In addition, there are breccia types with clasts of all levels of shock metamorphism, including altered glass fragments, i. e., impact breccia of suevite type. Even breccia with melted and vesiculated groundmass has been reported. Below about 605 m depth only very rare altered melt fragments have been encountered, but from about 825 m depth a 30 m thick lens of crystalline impact melt rock occurs. The parautochthonous basement rocks underneath the melt layer have been thermally metamorphosed over a further 15 m depth.

Shock metamorphic evidence, in particular planar deformation features (PDFs) in quartz, have been reported from the crater fill, especially from the impact melt rock. Shock overprint peters out with depth at about 60 m into the basement. Studies of siderophile elements, including the platinum group elements, have indicated that the impact-melt bearing breccias and the impact melt rock carry a component derived from a chondritic meteoritic projectile, possibly of L chondrite type.

The structure shows a well-defined, circular, negative Bouguer gravity anomaly of −6 mGal. Airborne as well as ground-based magnetic surveys exhibited a very low magnetic relief over the crater, whereas there is more contrast between different gneiss units outside of the structure. Reflection and refraction seismic data suggested that brecciation and fracturing extend to a maximum depth of 1.2 km, and fracturing may extend to 1 km around the structure.

The biostratigraphic age for the post-impact sediment fill is lower Middle Ordovician. More specifically, an Early Caradoc age of ca. 453 Myr has been cited. Only K-Ar radiometric ages have been reported, and what is worse, it seems that all dated material suffered significant alteration. The original data suggested a 450–500 Myr

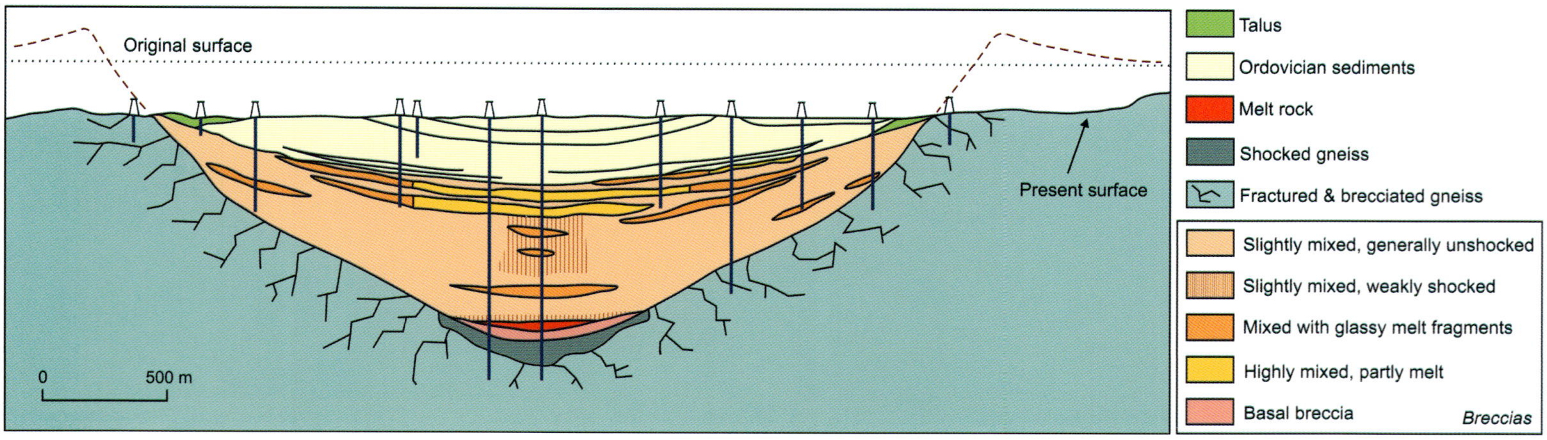

Schematic geological cross-section through the Brent impact structure, based on an extensive drilling program (modified after Grieve, 1978 and 2006).

Brent

Ontario / Canada

latitude = 46°04'30" N / longitude = 78°28'55" W / diameter = 3 km / age = 453 Myr

scale = 1:126 000 / scene = 22.6 × 30.2 km²

0 1.26 6.30 km

N

300 meter 555

age, in line with the biostratigraphic age, but when the required recalculation for revised decay constants is carried out, the K-Ar ages are much lower than the biostratigraphic age for the post-impact sediments. Therefore, the biostratigraphic age is currently the only good age constraint for the Brent impact.

The structure was first noted as an anomalous circular feature on aerial photographs by John A. Roberts in 1951. A first reconnaissance investigation was undertaken in 1951, followed by gravity, magnetic and seismic work two years later. The first two boreholes were sunk in 1955, and a first report on this drilling by Peter Millman and coworkers in 1960 already concluded that the Brent structure represented an impact structure. Prior to 1967, altogether 5 km of drill core were extracted, whereby all strata of this structure were recovered. Brent is arguably one of the best studied terrestrial impact structures.

Location/Access

The best starting-point for a visit to Brent crater is probably Toronto, also because of its international airport. From this metropolis a drive of about 470 km will take the visitor to the Algonquin Provincial Park and then the town of Brent at the shore of Cedar Lake. Leaving Toronto, take the Ontario Highway ON 400 North right up to the location Barrie. Here, take ON 11 for another 185 km. Take the highway exit Lake Nosbonsing Road/ON 654W, then Callander Bay Drive/ON94E, and finally Lansdowne Street/ON94E. After a further 9 km you will reach Trans-Canada Highway 17E. After 80 km eastward, turn south into Brent Road. After 34 km the crater structure is located ca. 400 m west of the road, just 4 km north of the town of Brent. On surface in the structure, there are only a few outcrops of cataclastic gneiss.

Selected References

Dence, M. R. (1968). Shock zoning at Canadian craters: Petrography and structural observations. In: Shock Metamorphism of Natural Materials, French, B. M., and Short, N. M. (eds.), Mono Book Corp., Baltimore, pp. 169–184.
Dence, M. R., Grieve, R. A. F., and Robertson, P. B. (1977). Terrestrial impact structures: Principal characteristics and energy considerations. In: Impact and Explosion Cratering, Roddy, D. J., Pepin, R. O., and Merrill, R. B. (eds.), Pergamon Press, New York, pp. 247–275.
Grieve, R. A. F. (1978). The melt rocks at Brent crater, Ontario, Canada. Proceedings of the 9th Lunar and Planetary Science Conference, Houston, Texas, Pergamon Press, New York, pp. 2579–2608.
Grieve, R. A. F. (2006). Impact Structures in Canada. Geological Association of Canada, GeoText No. 5, 210 pp.
Millman, P. M., Liberty, B. A., Clark, J. F., Willmore, P., and Innes, M. J. S. (1960). The Brent Crater. Ottawa Dominion Observatory Publications, Vol. 24, 43 pp.
Robertson, P. B., and Grieve, R. A. F. (1977). Shock attenuation at terrestrial impact structures. In: Impact and Explosion Cratering, Roddy, D. J., Pepin, R. O., and Merrill. R. B. (eds.), Pergamon Press, New York, pp. 687–702.

Calvin

Michigan, United States

Thomas Kenkmann

Calvin crater is a buried, 8.5 km wide, complex impact structure of Late Ordovician age in southwestern Michigan. The structure consists of a central dome, an annular depression, and an outer anticlinal rim zone.

The impact occurred into the Michigan Basin, where the Precambrian basement is situated at a depth of about 1200 m below surface. The sedimentary succession comprises about 530 m of Cambrian strata, predominantly sandstone and sandy dolomite, 330 m of Ordovician shales, limestone and dolomite, 210 m of Silurian dolomite and evaporites, and 224 m of Devonian shale, limestone and dolomite. The structure is entirely covered by material from a glacial ground moraine.

Exploration of three oil fields producing from Middle Devonian limestone has provided a wealth of subsurface data. The Calvin 20 and Juno Lake oil fields are along the western portion of the anticlinal rim, and the Calvin 28 field extends over the central uplift. In total 110 wells were drilled into the structure. In addition to the dense array of boreholes, the shape of the crater structure could also be constrained with seismic, gravity, magnetic, and resistivity data.

The shape of the crater was analyzed by the vertical position of a Devonian horizon, called the "Traverse Limestone", that was deposited after impact. The most striking feature is a gentle central uplift of 2 km diameter, with a stratigraphic uplift of 415 m that is partially surrounded by an annular depression of 1–2 km width. Here, the Traverse Limestone lies 28 m below its regional level and 41 m below equivalent strata in the central uplift. The encircling anticlinal rim is roughly 1.5 km wide. The target seems strongly disturbed and an intricate system of faults exists in the periphery of the crater structure.

Microbreccias occur in deep wells that penetrate the structure. The polymict breccia is composed of fractured and unfractured, subrounded to rounded, quartz grains embedded in a carbonate matrix, or floating shard-like carbonate fragments cemented by a carbonate matrix. A few single sets of planar deformation features (PDFs) in quartz could be identified. A collection of metallic microspherules of Ordovician age was also recovered from borehole samples. There are no radiometric age determinations available. From the stratigraphic record the event is older than the Early Silurian deposits but younger than the strata of the Late Ordovician Richmond Group – i.e., the impact occurred at about 450 Myr ago.

The structure was investigated by the Michigan Geological Survey in the course of hydrocarbon exploration. Initially, the name Calvin 28 was adopted from the oilfield Calvin 28 over the central uplift of Calvin.

Location/Access

The structure lies in a well-developed area and can easily be reached. South Bend in Indiana, about 30 km southwest of Calvin, provides an airport, and Edwardsburg Highway and Highway 62 lead in northeasterly direction to the village of Edwardsburg in the close vicinity of the Calvin structure. However, there is not much to see there, as material of a glacial ground moraine covers the surface. Glacial lakes form a chain in the northern part of the structure.

Selected References

Milstein, R. L. (1988). The Calvin 28 structure: Evidence for impact origin. Canadian Journal of Earth Sciences, Vol. 25, pp. 1524–1530.
Milstein, R. L. (1994). The Calvin Impact Crater, Cass County, Michigan: Identification and Analysis of a Subsurface Ordovician Astrobleme. Ph.D. Thesis, Oregon State University, 114 pp.

Calvin

Michigan / United States

latitude = 41°49'48" N / longitude = 85°57'00" W / diameter = 8.5 km / age = 450 Myr

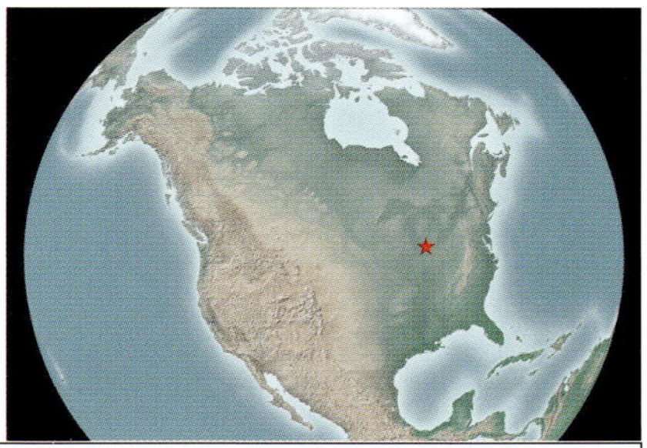

scale = 1 : 120 000 / scene = 21.7 × 28.9 km²

0 1.20 6.00 km

N

205 meter 350

Carswell
Saskatchewan, Canada

Thomas Kenkmann

The Carswell impact structure in northern Saskatchewan is a deeply eroded, 39 km wide circular structure. It is located in the western part of the Athabasca Basin. The original diameter of Carswell might have been as large as 50 km.

Carswell consists of an 18–20 km wide central uplift that is composed of shocked Archean basement rocks. This structure is surrounded by a downfaulted ring syncline, in which much younger deposits, such as the William River Subgroup and the Ordovician Carswell Formation, are preserved. The Carswell Formation forms morphological ridges and cliffs, and constrains the size of the crater structure. The Archean basement beneath the ring basin is composed of metamorphic rocks of the Churchill Province and includes 2600–3500 Myr old gneisses and komatiites, as well as metasedimentary rocks. The intracratonic Athabasca Basin is filled by up to two kilometer thick Precambrian sandstones, conglomerates, and fine-grained siliciclastic rocks. The Carswell Formation is the youngest formation in this basin and is composed of dolomites. Glacial till deposits of Quaternary age form the regional cover.

The structure is divided into three concentric ring features – an 18 to 20 km wide core of uplifted Archean to Paleoproterozoic basement rocks, a 5–8 km wide annular trough filled by sandstones of the William River Subgroup, and an outer, 3–6 km wide, ring where the Carswell Formation occurs. The latter is preserved because it was downfaulted during the modification stage of cratering.

The structure is deeply eroded, and allochthonous breccias are rare. Prominent is a breccia known as the *Cluff Breccia* that contains highly shocked basement-derived fragments with planar deformation features (PDFs) and planar fractures (PFs) in mineral fragments, partly melted fragments, and clasts with diaplectic quartz and feldspar. This breccia is altered to variable degree and occurs mainly as irregular veins and dikes of widths from several centimeters up to 100 m. A borehole drilled close to the southern margin of the central uplift intersected a 70 m thick sheet of impact melt rock and breccias. Moreover, pseudotachylitic breccia veins have been observed in the basement rocks, and there are also fragments of this material in the Cluff Breccia. The latter observation suggests that these pseudotachylitic breccias represent a formation of the early cratering process. Surface outcrops in the central uplift also expose shatter cones.

Uranium mineralization is concentrated along the southern margin of the central uplift along the contact between Archean basement rocks and sandstones of the William River Subgroup. The age of this mineralization has been estimated at 1000 Myr. This ore deposition took place prior to impact and was passively uplifted and tilted during later crater formation. The Carswell structure has been intensively explored for uranium.

The structure shows a central gravity high surrounded by an annular low of −10 mGal. Other concentric deviations from the regional field occur further outside. The magnetic signature is generally smoother than that of the surrounding terrain. In the crater center, however, a more complex magnetic relief exists. Recent ^{40}Ar-^{39}Ar dating of authigenic adularia that was separated from an impact melt rock from the Carswell impact structure yielded an age of 481.5 ± 0.8 Myr – interpreted as the age of the impact event.

The Carswell impact structure on ▷ May 17, 2017 as a composite image based on TanDEM-X DEM and Copernicus Sentinel-2 (channel 2/3/4) data.

Carswell

Saskatchewan / Canada

latitude = 58°25‘ N / longitude = 109°31‘ W / diameter = 39 km / age = 481.5 ± 0.8 Myr

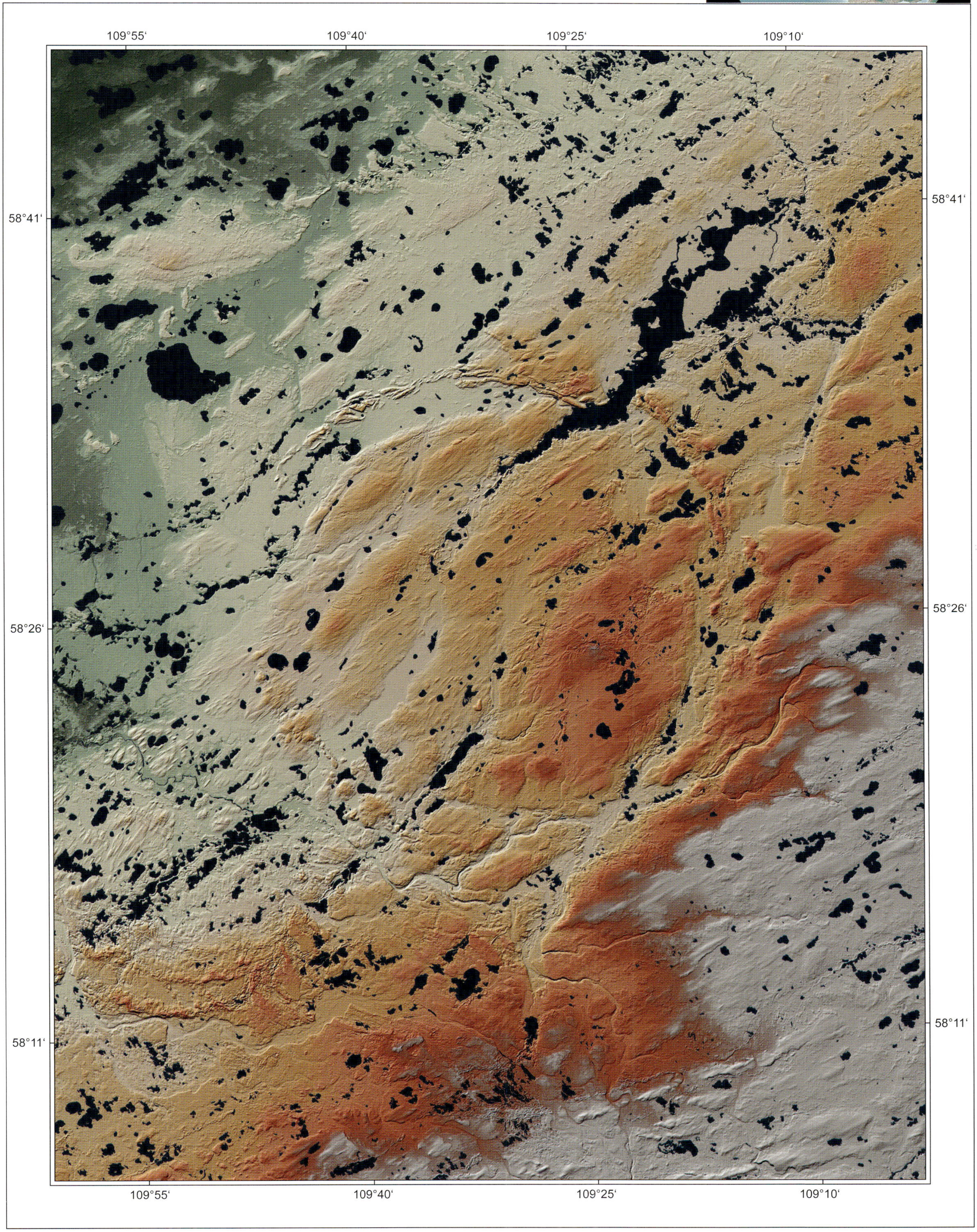

scale = 1 : 319 000 / scene = 57.4 × 76.5 km²

0 3.19 15.95 km

N

195 meter 575

The Dominion Observatory's meteorite crater research program led to the discovery of the Carswell structure in 1964 by Morris Innes. The structure was mapped in 1969 by K. Currie, who also noted shock metamorphic effects whilst still adhering to a "cryptovolcanic" origin.

Location/Access

The Carswell structure is located in a remote part of northern Saskatchewan, about 70 km from the southern shore of Lake Athabasca. Access is from the city of Prince Albert that sports a regional airport. From there, a drive of more than 700 km northwards is required, along the excellent highways 3 and 55. This is then followed from Green Lake by further riding along Northern Highway 155, before from La Roche onward unpaved Road 955 leads directly into the structure. Within the area of the structure, no services are provided for visitors.

Selected References

Alwmark, C., Bleeker, W., LeCheminant, A., Page, L., and Scherste'n, A. (2017). An early Ordovician ^{40}Ar-^{39}Ar age for the ~50 km Carswell impact structure, Canada. Geological Society of America Bulletin, Vol. 129, pp. 1442–1449.

Currie, K. L. (1969). Geological notes on the Carswell circular structure, Saskatchewan. Canadian Geological Survey of Canada Paper 67–32, 60 pp.

Genest, S., Robert, F., and Duhamel, I. (2010). The Carswell impact event, Saskatchewan, Canada: Evidence for a pre-Athabasca multiring basin? In: Large Meteorite Impacts and Planetary Evolution IV, Gibson, R. L., and Reimold, W. U. (eds.), Geological Society of America Special Paper, Vol. 465, pp. 543–570.

Grieve, R. A. F. (2006). Impact Structures in Canada. Geological Association of Canada, GeoText No. 5, 210 pp.

Harper, C. T. (1982). Geology of the Carswell Structure, Central Part. Saskatchewan Geological Survey, Report 214, 6 pp.

Innes, M. J. S. (1964). Recent advances in meteorite crater research at the Dominion Observatory, Ottawa, Canada. Meteoritics, Vol. 2, pp. 219–241.

Charlevoix

Quebec, Canada

Thomas Kenkmann

The Charlevoix impact structure is situated north of the St. Lawrence River about 100 km northeast of Quebec City, the capital of Quebec. Originally the structure was named after the town of La Malbaie. Two thirds of the structure are exposed. The southern part of the crater structure has been down-faulted by the St. Lawrence rift system and is submerged in the here 30 km wide St. Lawrence stream. Although deeply eroded, the semi-circular structure is clearly visible in the landscape and in remote sensing data. A hummocky central region, with the 770 m high Mont Eboulement as the highest peak, extends into the St. Lawrence River and is surrounded by an annular moat with a concentric drainage system. This, in turn, is framed by a ring of hills with peak heights over 900 m that mark the crater circumference. The structure, thus, has an apparent diameter of roughly 55 km.

Target rocks include gneisses, migmatites, anorthosites, and charnockitic rocks of the Grenville Province of the Canadian Shield. The sedimentary cover of these basement rocks is formed by Ordovician limestone and is locally preserved along the annular moat. In structural terms, this moat corresponds to a ring syncline, in which blocks and strata were down-sagged and down-faulted along listric faults. They now rest 1–2 km lower than their original stratigraphic position. The central uplift at Mont Eboulement, in contrast, shows a stratigraphic uplift in the order of 6 km.

Except for a single occurrence of impact melt rock, no allochthonous impact lithologies are preserved in the crater. Shatter cones are prominent features in its central region. They can be observed up to 12 km from the center and occur both in basement lithologies, as well as in the Ordovician limestone where they can reach sizes larger than a meter, e.g., at the shore of the St. Lawrence River. A systematic shock zoning study based on the crystallographic orientations of planar deformation features (PDFs) in quartz revealed maximum pressures of 23 GPa recorded in the center of the structure. The PDFs are generally decorated with fluid inclusions. Micro-thermometric analysis of the fluid inclusions indicated that the grains were annealed to about 500 °C due to post-shock heating. PDFs occur at distances of 2–10 km from the center.

The crater structure is associated with a discontinuous regional Bouguer gravity anomaly. For a long time the accepted age of Charlevoix determined by K-Ar dating of impact melt rock samples was 342 ± 15 Myr. However, recently laser ablation inductively coupled plasma-mass spectrometry U-Pb geochronology of shocked zircon grains in a vesicular-fluidal impact melt rock suggested an Ordovician to Silurian age of 450 ± 20 Myr for the impact. That means it happened during the phase of enhanced impact bombardment onto Earth. The impact structure was affected by the tectonics of the St. Lawrence

A large shatter cone in Ordovician limestone of the Charlevoix impact structure. The exposure is along the annular moat of the structure at the shore of the St. Lawrence River. Hammer for scale ca. 40 cm long (photo: T. Kenkmann).

Charlevoix

Quebec / Canada

latitude = 47°32' N / longitude = 70°21' W / diameter = 55 km / age = 450 ± 20 Myr

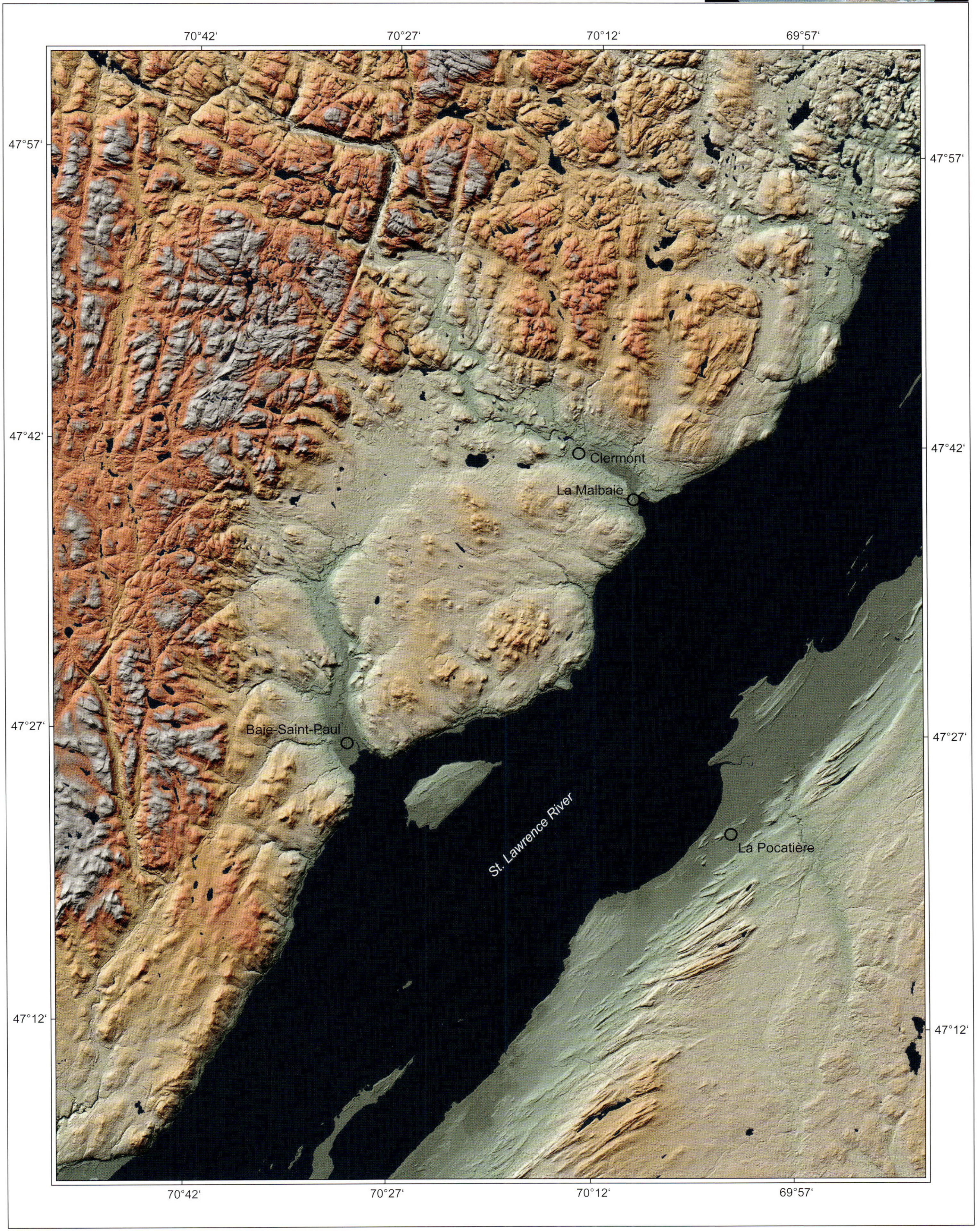

scale = 1 : 446 000 / scene = 80.2 × 106.9 km²

0 4.46 22.30 km

N

25 meter 1170

Rift in the Mesozoic, which lead to the down-faulting of the southern sector of the structure. Normal faults of this rift trend parallel to the shore of the St. Lawrence River and are still seismically active.

The structure was discovered in 1966 by Jehan Rondot in the course of a regional mapping survey. The impact origin was revealed based on the identification of shatter cones. Shortly after the macroscopic proof of shock metamorphism had been noted, abundant PDFs in quartz grains were detected as well.

Location/Access

The crater structure is easily accessible, for example from Quebec City on a highway to the north of the St. Lawrence River. The towns of Charlevoix and La Malbaie are situated along the semi-circular valley. The best shatter cones formed in Ordovician limestone can be found along the shore of the St. Lawrence River at 7–10 km distance from the center of the structure.

Selected References

Grieve, R. A. F. (2006). Impact Structures in Canada. Geological Association of Canada, GeoText No. 5, 210 pp.

Robertson, P. B. (1968). La Malbaie structure, Quebec. A Paleozoic impact site. Meteoritics, Vol. 4, pp. 89–112.

Robertson, P. B. (1975). Zones of shock metamorphism at the Charlevoix impact structure, Quebec. Geological Society of America Bulletin, Vol. 86, pp. 1630–1638.

Rondot, J. (1966). Geology of La Malbaie area, Charlevoix County. Department of Natural Resources, Quebec. Preliminary report, Vol. 544, 18 pp.

Schmieder, M., Shaulis, B. J., Lapen, T. J., Buchner, E., and Kring, D. A. (2019). In situ U–Pb analysis of shocked zircon from the Charlevoix impact structure, Quebec, Canada. Meteoritics and Planetary Science, Vol. 54, pp. 1808–1827.

Trepmann, C. A., and Spray, J. G. (2005). Planar microstructure and Dauphiné twins in shocked quartz of the Charlevoix impact structure, Canada. In: Large Meteorite Impacts III, Kenkmann, T., Hörz, F., and Deutsch, A. (eds.), Geological Society of America Special Paper, Vol. 384, pp. 315–328.

The Charlevoix impact structure on May 18, 2018 as a composite image based on TanDEM-X DEM and Copernicus Sentinel-2 (channel 2/3/4) data.

Chesapeake Bay
Virginia, United States

Thomas Kenkmann

The 85 km diameter Chesapeake Bay impact structure ranks among the best preserved impact structures on Earth, even though – or actually because – it lies buried under the southern end of the eponymous bay at the East Coast of the United States.

The target comprised three layers with very different strength properties. The basal layer is crystalline basement composed of meta-sedimentary and meta-igneous rocks of mainly Proterozoic age. This basement is covered by 400–1500 m of poorly consolidated, water saturated siliciclastic sands, silts and clays, as well as consolidated sediments, of Early Cretaceous to Eocene age, which form a seaward-thickening wedge. At the time of impact the sedimentary succession was submerged in the sea. The water depth ranged from 0 to 340 m. The Chesapeake Bay structure was formed in the Late Eocene, at 35.2 ± 0.3 Myr ago, in a shelf environment at the passive continental margin at the east coast of North America. The impact caused an enormous tsunami that strongly affected the crater structure and the deposition of impact lithologies.

The crater morphology and structure are constrained by correlation of lithostratigraphic and biostratigraphic data from various drill cores and well cuttings, borehole geophysical logs, and a number of seismic reflection traverses. Several core holes were drilled into the annular trough and one core hole was drilled into the inner crater basin. The most prominent feature of the Chesapeake Bay impact structure is its "inverted-sombrero" geometry with a deep inner basin surrounded by a shallower brim, which is presumably the result of the large competence contrast between the sedimentary and crystalline target materials. The structure's outer diameter of 85 km is defined by steep, 300–1200 m high fault scarps, which were dominantly formed in soft to partly consolidated siliciclastic sediments. The annular trough contains kilometer-scale, slumped mega-blocks of generally stratified Lower Cretaceous sediments. These mega-blocks rest on a relatively flat floor, which coincides with the transition from the sedimentary cover to the intact crystalline basement. An inner crater basin of about 35 km diameter was carved to a depth of one kilometer into the crystalline basement and contains an irregularly shaped central uplift about 10 × 12 km wide and 500–800 m high. The central uplift structure is surrounded by a moat. The inner crater basin has a collapsed margin and is surrounded by a raised rim.

A 1.76 km deep core hole was drilled by the International Continental Scientific Drilling Program (ICDP) together with the U.S. Geological Survey from the tip of the Delmarva Peninsula near the center of the crater structure. It revealed a representative section through Chesapeake Bay impact lithologies and the post-impact crater

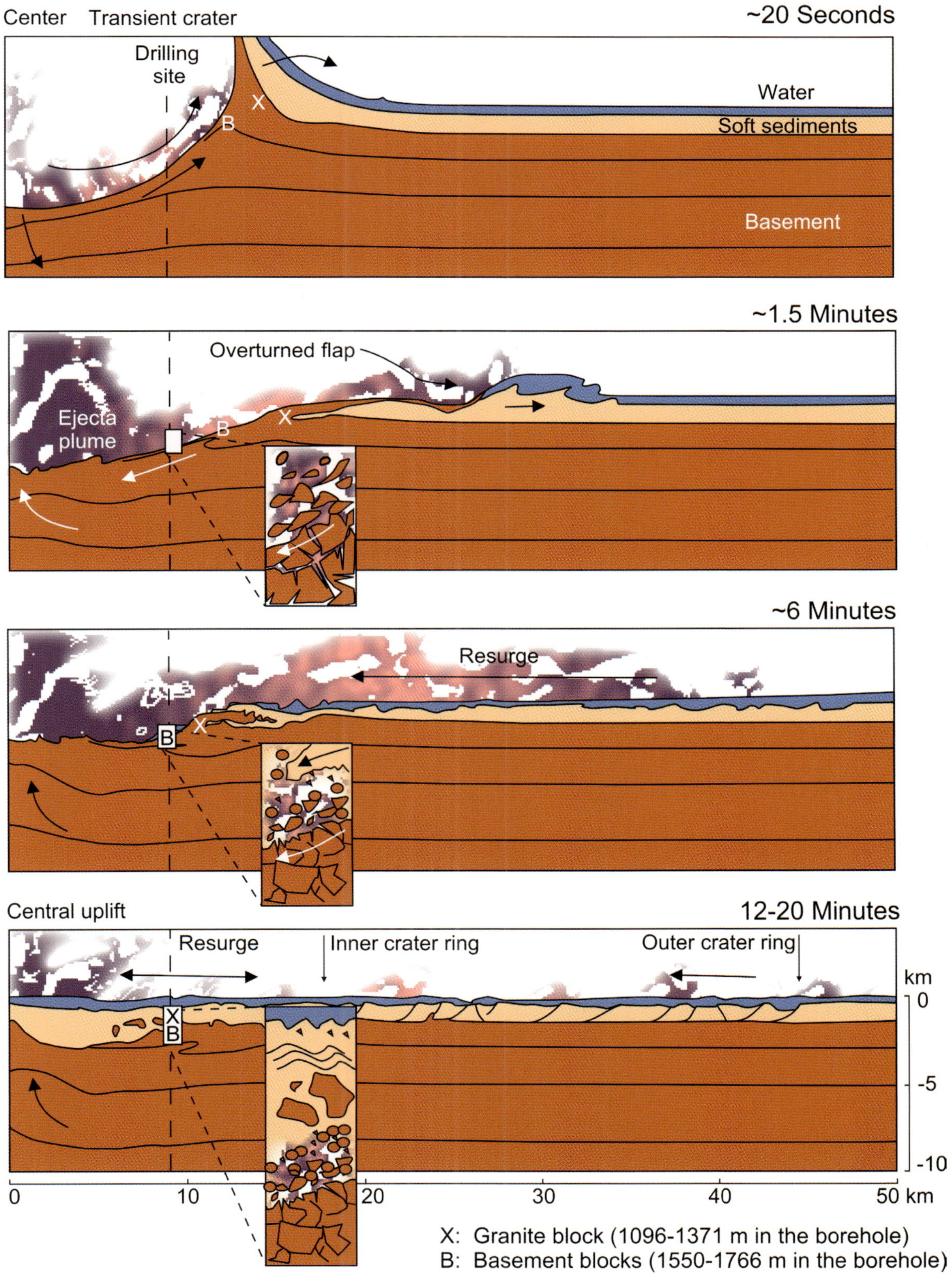

Schematic synopsis of crater formation for the marine Chesapeke Bay impact structure, as revealed by numerical modeling by Collins and Wünnemann (2005) and drill core analysis. Note the position of the ICDP borehole (modified after Kenkmann et al., 2009b).

Thin section micrograph of fractured and brecciated basement rock at 1.69 km depth, from the 1.76 km deep ICDP borehole drilled near the center of the Chesapeake Bay impact structure. The dark veins are altered pseudotachylitic breccia (photo: T. Kenkmann).

infill. The lowest unit consists of 215 m of fractured but only weakly shocked allochthonous basement material that contains a series of breccia dikes of variable thickness. Above this lowermost unit occur 154 m of suevitic breccia, with clasts that exhibit all different types of shock features. An overlying 26 m thick unit of gravelly quartz sand and large rock clasts indicates abrasive displacement of an overlying 275 m thick, massive, unshocked granitic megablock. It was washed from the edge of the inner crater into the crater moat by a forceful tsunami resurge. A minimum horizontal displacement of 5 km inward could be inferred. The uppermost and thickest crater-fill unit of 652 m comprises brecciated target sediments and polymict, sediment-dominated breccias. Glauconitic sediment-dominated breccias of this unit represent the main pulse of ocean-water resurge into the collapsing transient cavity. The post-impact crater fill has a thickness of 444 m at the drill site. Comparison of the post-impact sedimentary sequences inside and outside the structure indicates that compaction of the crater fill influenced the long-term sedimentation pattern of the region.

In the 1980s and 1990s the subsurface of Virginia's coastal plain was studied by Federal, State, and local authorities, oil companies, and universities. These investigations culminated in the discovery of the Chesapeake Bay impact structure in 1993 by David S. Powars, C. Wylie Poag, and co-workers. Final confirmation came from the identification of shock features in partially melted basement rocks of a borehole by Christian Koeberl and co-workers in 1996. In 2005 and 2006 three boreholes were drilled to a depth of 1766 m near the center of the structure at Eyreville Farm, in Northampton County in Virginia. This drilling initiated a great number of new geological, geochemical, geophysical and hydrological investigations.

The crater has environmental consequences still today, such as ground instability and land subsidence, or the disruption of coastal aquifers. The reason for the latter is that the brecciated ground of the crater area has allowed an inland saltwater wedge to form in these aquifers, upon which the population centers of Virginia, such as Hampton and Newport News that are in the immediate vicinity of the impact structure, rely heavily.

Wylie Poag found a number of seismic disturbances in the relatively undisturbed, flat-lying, coastal plain that can be explained as secondary craters. The closest possible secondary crater is located 20 km from the primary crater, and the farthest is 85 km away. Apparent diameters of the 23 identified secondary craters range from 0.4 to 4.7 km. Four of them have apparent diameters greater than 3 km.

The Chesapeake Bay impact structure is the source crater for the **North American tektite strewn field**, the oldest known on Earth. Of the large terrestrial tektite strewn fields, the North American is the one that was recognized last, in the late 1930s. First, the Bediasites were identified in Grimes County, Texas, followed by the Georgiaites in Dodge County, Georgia. The link between Chesapeake Bay and the North American strewn field relies on their similar ages. Furthermore, the isotopic parameters of target lithologies, crater fill breccias, and post-impact sediments conform to those of the North American tektites. Today the exact extent of this strewn field is a matter of debate and ongoing research. The tektite strewn field includes the Caribbean tektite occurrences but also extends into the Atlantic Ocean with discoveries at a drill site off the coast of New Jersey. The recognition of microtektites in cores from the eastern Pacific may even further expand the area of this strewn field. The shape of the strewn field and tektite distribution down-range from the source crater are in agreement with other tektite strewn fields and constrain the impact trajectory of the Chesapeake Bay impact event.

Location/Access

The Chesapeake Bay impact structure has no topographic expression. It extends over the southern Chesapeake Bay and Virginia coastline, the area of the southern Delmarva Peninsula, and off-shore regions. The center of the structure is close to Cape Charles on the Delmarva Peninsula. The deep core hole described above was drilled by the International Continental Scientific Drilling Program (ICDP) in conjunction with the U.S. Geological Survey into the inner crater basin, at about 9 km distance from the crater center, on Eyreville Farm in Northampton County. The area with the major cities of Hampton and Portsmouth lies about 230 km southwest of Washington DC and can be easily accessed by car from the US capital.

Selected References

Catchings, R. D., Powars, D. S., Gohn, G. S., Horton, J. W., Goldman, M. R., and Hole, J. A. (2008). Anatomy of the Chesapeake Bay impact structure revealed by seismic imaging, Delmarva Peninsula, Virginia, USA. Journal of Geophysical Research, Vol. 113, 23 pp.

Collins, G. S., and Wünnemann, K. (2005). How big was the Chesapeake Bay impact? Insights from numerical modeling. Geology, Vol. 33, pp. 925–928.

Deutsch, A., and Koeberl, C. (2006). Establishing the link between the Chesapeake Bay impact structure and the North American tektite strewn field: The Sr-Nd isotopic evidence. Meteoritics and Planetary Science, Vol. 41, pp. 689–703.

Gohn, G. S., Koeberl, C., Miller, K. G., Reimold, W. U., Browning, J. V., Cockell, C. S., Horton, J. W., Kenkmann, T., Kulpecz, A. A., Powars, D. S., Sanford, W. E., and Voytek, M. A. (2008). Deep drilling into the Chesapeake Bay impact structure. Science, Vol. 320, pp. 1740–1745.

Horton, J. W., and Izett, G. E. (2005). Crystalline-rock ejecta and shocked minerals of the Chesapeake Bay impact structure, USGS-NASA Langley core, Hampton, Virginia, with supplemental constraints on the age of impact. In: Studies of the Chesapeake Bay Impact Structure – The USGS-NASA Langley Corehole, Hampton, Virginia and Related Coreholes and Geophysical Surveys, Horton, J. W., Powars, D. S., and Gohn, G. S. (eds.), United States Geological Survey Professional Paper, Vol. 1688, pp. E1–E30.

Kenkmann, T., Collins, G. S., Wittmann, A., Wünnemann, K., Reimold, W. U., and Melosh, H. J. (2009b). A model for the formation of the Chesapeake Bay impact crater as revealed by drilling and numerical simulation. In: The ICDP-USGS Deep Drilling Project in the Chesapeake Bay Impact Structure: Results from the Eyreville Core Holes, Gohn, G. S., Koeberl, C., Miller, K. G., and Reimold, W. U. (eds.), Geological Society of America Special Paper, Vol. 458, pp. 571–585.

Koeberl, C., Poag, C. W., Reimold, W. U., and Brandt, D. (1996). Impact origin of Chesapeake Bay structure and the source for the North American tektites. Science, Vol. 271, pp. 1263–1266.

Poag, C. W., Koeberl, C., and Reimold, W. U. (2004). The Chesapeake Bay crater. Geology and Geophysics of a late Eocene Submarine Impact Structure. Impact Studies, Springer, Berlin Heidelberg, 522 pp.

Powars, D. S., and Bruce, T. S. (1999). The effects of the Chesapeake Bay impact crater on the geological framework and correlation of hydrogeologic units of the lower York-James peninsula, Virginia: US Geological Survey Professional Paper, Vol. 1612, 82 pp.

Chesapeake Bay
Virginia / United States

latitude = 37°14' N / longitude = 76°01' W / diameter = 85 km / age = 35.2 ± 0.3 Myr

scale = 1 : 569 000 / scene = 102.3 × 136.5 km²

0 5.69 28.45 km

N

0 meter 325

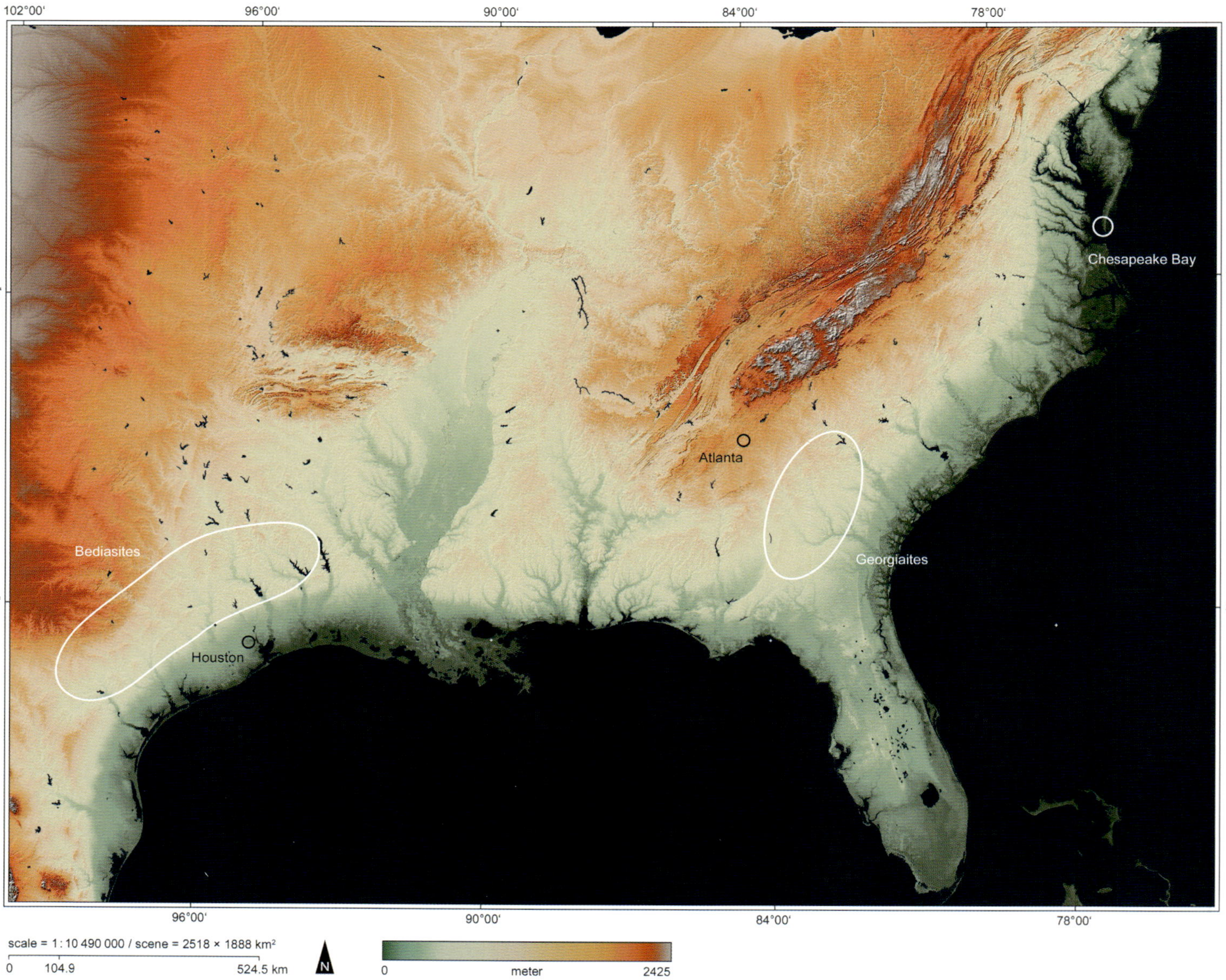

The location of the Bediasites and Georgiaites in the North American tektite strewn field with Chesapeake Bay, the source impact crater, in a physical map generated from the TanDEM-X 90m DEM.

Chicxulub

Mexico

Wolf Uwe Reimold

The Chicxulub impact structure is currently known as the world's third-largest of its kind, with a diameter of about 180 km. It straddles the northern coastline of the Yucatán peninsula of Mexico. The structure is famous for the fact that it is considered the smoking-gun for the tremendous impact event at the end of the Cretaceous period, through which life on our planet was decisively changed. This near-instantaneous event caused a global environmental catastrophe that meant the end of the era of the dinosaurs and the advent of the mammals.

The Chicxulub structure is the best preserved very large terrestrial impact crater yet discovered, because it was completely buried only a short time after the impact event. The structure is now preserved from erosion by more than 500 m of Tertiary platform carbonates. Chicxulub represents the largest known Phanerozoic impact structure. Based on evaluation of its strong gravity and magnetic signatures, its diameter was debated for a number of years. Values of 180–300 km were favored at times, with these upper and lower limits being important, as they would mean a difference of an order of magnitude of impact energy. This would be a major factor for any assessment of the impact-triggered environmental consequences. At variably 300, 250, or 180 km diameter one would also expect different morphologies for this structure – either a peak ring crater or a multi-ring basin structure. Today there is general acceptance of the 180 km size limit, and seismic reflection studies have confirmed this value and the existence of a peak ring geometry. At this size, Chicxulub ranks prominently in the triumvirate of terrestrial impact giants – together with the much older **Sudbury** and **Vredefort** structures that are, however, not completely preserved like Chicxulub.

The relatively flat topography of the area on the Yucatán peninsula sports less than 50 m of elevation gradient over 150 km. The structure's northern part is off-shore, the southern part is onshore. The only clear expression of the onshore portion of the impact structure is a semi-circular ring of "cenotes" (sinkholes) that traces a concentric fault at depth. Surface outcrops for the geological study of the impact structure are not accessible due to its deep burial. Consequently, investigating geology and morphology of the crater structure has had to rely entirely on geophysical data and drilling.

The Chicxulub impact occurred into a target geology comprising a crystalline basement composed of granite, gneiss, and amphibolite, as recognized in clasts of impact breccias. According to isotope

Chicxulub

Mexico

latitude = 21°20‘ N / longitude = 89°30‘ W / diameter = 180 km / age = 66 Myr

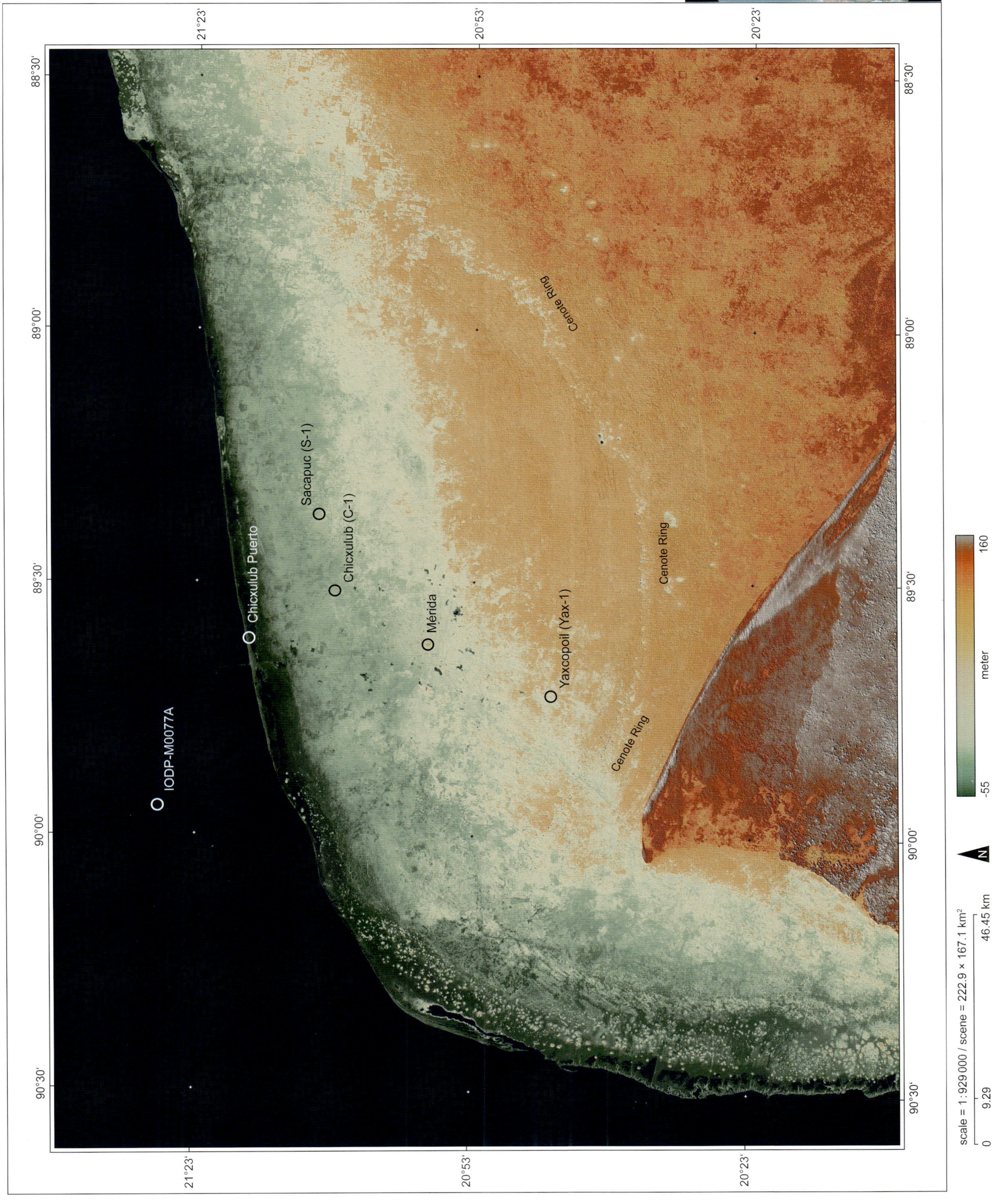

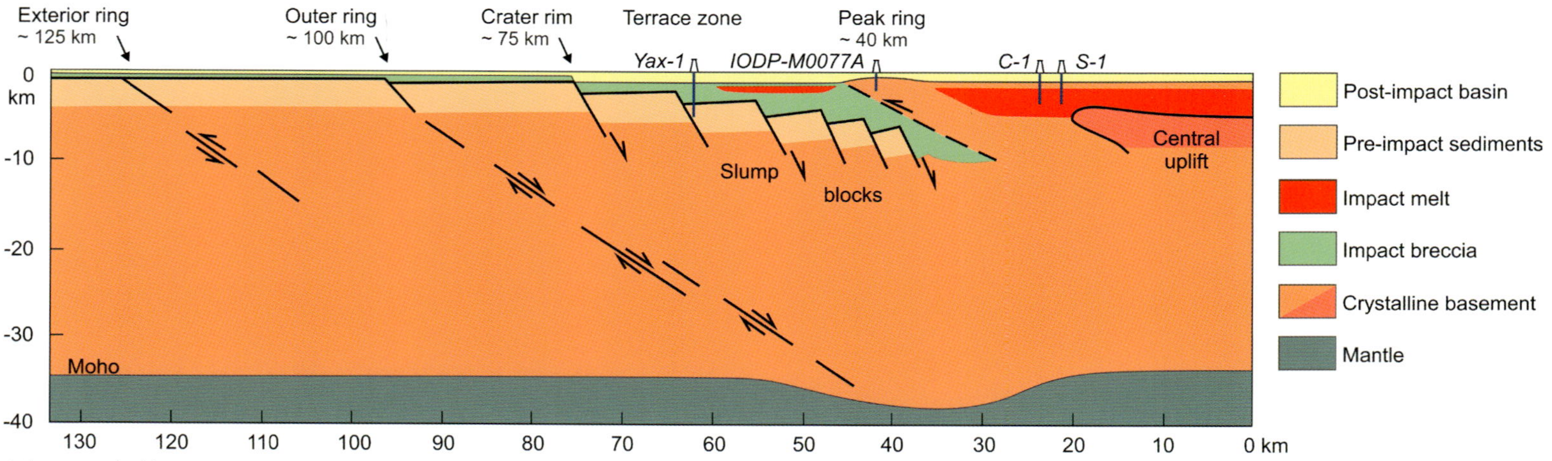

Schematic half cross-section of the Chicxulub impact structure (modified after J. V. Morgan et al., 2000).

geological studies, these lithologies probably belong to both Gondwanan, about 550 to 320 Myr ago, and Laurentian, 1500–1000 Myr, crust. This basement complex was overlain at the time of impact by a 3 km thick sequence of Cretaceous platform carbonates and other evaporites.

Chicxulub has distinct potential field anomalies, including a series of semi-continuous and roughly circular features in the gravity and magnetic data. Together with the results of a series of reflection seismic profiles, these geophysical data have been instrumental in determining the morphology of the buried impact structure. A distinct and continuous ring feature in the gravity data occurs at the outer edge of a gravity low at about 100 km from the center. This feature surrounds a central gravity high, which is open in the southwest sector. The outer edge of a gravity low occurs at about 180 km from the center and approximately coincides with a part of the onshore cenote ring in the eastern half of the structure. The main large-amplitude magnetic anomaly has a diameter of approximately 90 km and occurs just inside of the 100 km diameter gravity low.

While the structure was initially explored with geophysical methods, its geology was then also defined on the basis of a number of drilling campaigns. These not only provided the essential validation of geophysical models, but they also allowed to study the within-crater rock record. Three main drillings must be mentioned: first a series of boreholes was sunk by the National University of Mexico, and then two drilling campaigns were sponsored by the International Continental Scientific Drilling Program (ICDP) and by the International Scientific Ocean Discovery Program (IODP) jointly with ICDP, respectively. As a consequence of these results, the interior make-up of the structure is now reasonably well understood.

Chicxulub's schematic cross-section indicates that the peak ring structure of the central part of the collapsed crater structure is well defined. In its interior, impact lithologies occur, in the form of both massive impact melt rock and lithic and suevite impact breccias that were drilled in a number of wells. For example, boreholes Sacapuc-1 (S-1) and Chicxulub-1 (C-1) are the closest core holes about 10–15 km inside of the peak ring. They went to depths of 1530–1580 m, through some 360 m of suevitic impact breccia and about 110 m of impact melt rock, in the case of S-1, and some 175 m of suevitic breccia and 330 m of impact melt rock in C-1. Beyond the peak ring, a complex ring syncline and megablock zone ensue, with intercalations of melt and other impact breccias, between large blocks of target rock. The ICDP borehole Yaxcopoil-1 (Yax-1) was drilled in 2002 60 km south-southwest of the structure's center within the annular trough between the peak ring and the crater rim, down to a depth of 1511 m. Here a sequence of 615 m of deformed Cretaceous carbonates and sulfates was recovered below a 100 m thick unit of suevitic breccias and 795 m of post-impact Tertiary rocks. The outer parts of the impact structure are characterized by a series of ring faults related to the collapse of the initial crater cavity.

Ejecta from the Chicxulub structure have been discovered all over the world. At more than 100 locations worldwide, a distinct layer of ejecta has been identified directly at the stratigraphic boundary between the pre-impact Cretaceous era and the post-impact Paleogene. Close to (i.e., proximal to) the structure this boundary layer is thick and the deposits laid down directly after the impact event illustrate the turbulent marine environment after such a catastrophic event. Further away (distal) from the crater structure, the boundary layer thins out gradually, until in Europe it only measures a few centimeters in thickness. A very strong mass extinction of both marine and continental species is associated with this boundary. It includes the demise of the dinosaurs and was followed by the significant increase of mammalian species. Altogether it has been estimated that some 70 % of all species living on Earth prior to the impact event were killed off near-instantaneously.

Proximal ejecta are prominent in Mexico and further afield – for example in Belize and Haiti. This does include strongly disturbed beds around the Carribean, which have been interpreted as the result of impact-induced tsunami activity. Impact glass discovered in the 1980s on Haiti was dated by the ^{40}Ar-^{39}Ar method, and yielded the same ages as impact melt rock from drill core from the Chicxulub structure. All these data combine to a very precise age of just about 66 Myr, coincident with the biostratigraphic age of the K-Pg boundary.

In 2016, IODP and ICDP drilled a deep off-shore borehole (IODP-M0077A) into the crater's peak ring, whereby core was recovered from a depth of 505.7–1334.7 m below the sea floor. The main purpose for this technologically involved off-shore drilling project was to understand the formation of the peak ring of large impact structures and to carry out detailed analysis of the impact-related deformation. In addition, it was hoped to obtain further data that would allow to better constrain the catastrophic environmental consequences of this mega-impact event. Results have shown that the peak ring was formed from uplifted, fractured, and shocked, felsic basement rocks. The peak ring structure is a consequence of the gravitational collapse of an over-heightened central uplift that, while collapsing, overthrust the rocks of the surrounding downfaulted ring syncline. The peak ring is cross-cut by breccia dikes and shear zones and is characterized by unusually low density. Strong rock flow and shear deformation is indicated. In addition, studies of this new drill core have shown that once the crater cavity refilled with seawater, a new cosmos of life developed quite quickly, maybe as fast as hundreds to a few thousand years after this evolutionary hiatus.

The discovery history of Chicxulub is a showcase how deductive natural scientific research works. In the 1980s, the boundary clay at the interface between the Cretaceous (K) and Paleogene (Pg; then known as the Tertiary) periods was proposed – independently by the group around Luiz and Walter Alvarez of Stanford University and Jan Smit from the Vrije Universiteit in Amsterdam – to be a global layer of impact ejecta. For the first time ever large-scale impact was linked with the then already well-known K-Pg mass extinction. Shocked quartz quickly provided the proof that the boundary clay represented ejecta fallout. Soon shocked zircon grains were discovered as well. However, something vital remained missing: the so-called smoking-gun – in the form of an impact structure large enough to cause global fallout of ejecta material – and a mass extinction of global proportion. From the spatial distribution of thicknesses of the K-Pg event layer it became apparent in the 1980s that the missing crater should be located somewhere in North or Central America.

The subsurface structure of what was later known as the Chicxulub impact crater was initially detected through geophysical anomalies as early as the 1950s, which was then followed up by drilling of several hydrocarbon exploration wells by PEMEX, the Mexican oil and gas company. However, only some 30 years later, Antonio Camargo and Glen Penfield were looking into the same direction, also while searching for hydrocarbon deposits. Penfield was initially unable to obtain evidence that the geological feature was a crater and gave up his search, but in 1990, together with Alan Hildebrand, he – and then a host of others – obtained core samples that provided the evidence that the exploration target was an impact feature. Evidence for the impact origin of the crater came in the form of shocked quartz and shock-melted materials, including impact glasses from the wider region of the Carribean. Impact glasses and impact melt rock were dated to the age of the Cretaceous-Paleogene (K-Pg) boundary. The smoking-gun riddle had been solved – Chicxulub was the missing piece in the puzzle!

While the vast majority of impact workers and geoscientists have full-heartedly accepted the link between the Chicxulub impact and the global mass extinction, there are still other hypotheses that include the massive volcanic activity centered on the Deccan Trap region of India as a possible agent of catastrophe. One of these hypotheses relates to joint action between large-scale volcanic activity and impact. Mark Richards and co-workers presented a scenario in 2015 that involves two key stages: onset of Deccan volcanism significantly earlier than the Chicxulub impact, but subsequent to the impact catastrophe, the volcanism continued with renewed fervor. Both processes together would have been a lethal combination on the global scale.

The lift-boat Myrtle was used in 2016 for the IODP-ICDP expedition 364 to drill corehole M0077A 1334.7 m deep into Chicxulub's peak-ring (image courtesy: Michael Poelchau, University of Freiburg, Germany).

Drill core, 83 mm in diameter, recovered from borehole M0077A (Chicxulub impact structure). The granites of the peak ring are deformed to various degrees. Here, the lower half is crenulated by intense micro-cataclasis (image courtesy: Michael Poelchau, University of Freiburg).

Location/Access

The Chicxulub impact structure is centered off-shore from the northern coastline of the Yucatán peninsula in Mexico. To visit the onshore part, to investigate cenotes or the former drilling site on the farm Yaxcopoil, a potential tourist in all likelihood will start in the city of Mérida, which has an international airport. Road 176 will then lead to the town of Chicxulub and on to Chicxulub Puerto, with the latter located on the coast.

Selected References

Alvarez, L. W., Alvarez, W., Asaro, F., and Michel, H. V. (1980). Extraterrestrial cause for the Cretaceous-Tertiary Extinction. Science, Vol. 208, pp. 1095–1108.

Grieve, R. A. F., Reimold, W. U., Morgan, J., Riller, U., and Pilkington, M. (2008). Observations and interpretations at Vredefort, Sudbury and Chicxulub: Towards an empirical model of terrestrial impact basin formation. Meteoritics and Planetary Science, Vol. 43, pp. 855–882.

Kenkmann, T., and Schoenian, F. (2006): Ries and Chicxulub: craters on Earth provide insights for Martian impacts. Meteoritics and Planetary Science, Vol. 41, pp. 1587–1604.

Kettrup, B., and Deutsch, A. (2003). Geochemical variability of the Yucatàn basement: Constraints from crystalline clasts in Chicxulub impactites. Meteoritics and Planetary Science, Vol. 38, pp. 1079–1092.

Kring, D. A., Claeys, P., Gulick, S. P. S., Morgan, J. V., Collins, G. S., and the IODP-ICDP Expedition 364 Science Party. (2017). Chicxulub and the exploration of large peak-ring impact craters through scientific drilling. GSA Today, Vol. 27/10, pp. 4–8.

Morgan J., Warner M., and the Chixculub working group (1997). Size and morphology of the Chicxulub impact crater. Nature, Vol. 390, pp. 472–476.

Morgan, J. V., Warner, M. R., Collins, G. S., Melosh, H. J., and Christeson, G. L. (2000). Peak-ring formation in large impact craters: geophysical constraints from Chicxulub. Earth and Planetary Science Letters, Vol. 183, pp. 347–354.

Morgan, J. V., Gulick, S. P. S., Bralower, T., Chenot, E., Christeson, G., Claeys, P., Cockell, C., Collins, G. S., Coolen, M. J. L., Ferrière, L., Gebhardt, C., Goto, K., Jones, H., Kring, D. A., Le Ber, E., Lofi, J., Long, X., Lowery, C., Mellett, C., Ocampo-Torres, R., Osinski, G. R., Perez-Cruz, L., Pickersgill, A., Poelchau, M. H., Rae, A., Rasmussen, C., Rebolledo-Vieyra, M., Riller, U., Sato, H., Schmitt, D. R., Smit, J., Tikoo, S., Tomioka, N., Urrutia-Fucugauchi, J., Whalen, M., Wittmann, A., Yamaguchi, K. E., and Zylberman, W. (2016). The formation of peak rings in large impact craters. Science, Vol. 354, pp. 878–882.

Richards, M. A., Alvarez, W., Self, S., Karlstrom, L., Renne, P. R., Manga, M., Sprain, C. J., and Smit, J. (2015). Triggering of the largest Deccan eruptions by the Chicxulub impact. Geological Society of America Bulletin, Vol. 127, pp. 1507–1520.

Riller, U., Poelchau, M. H., Rae, A. S. P., Schulte, F. M., Collins, G. S., Melosh, H. J., Grieve, R. A. F., Morgan, J. V., Gulick, S. P. S., Lofi, J., Diaw, A., McCall, N., Kring, D. A., and IODP-ICDP Expedition 364 Science Party (2018). Rock fluidization during peak-ring formation of large impact structures. Nature, Vol. 562, pp. 511–518.

Schulte, P., Alegret, L., Arenillas, I., Arz, J. A., Barton, P. J., Bown, P. R., Bralower, T. J., Christeson, G. L., Claeys, P., Cockell, C. S., Collins, G. S., Deutsch, A., Goldin, T. J., Goto, K., Grajeles-Nishimura, J. M., Grieve, R. A. F., Gulick, S. P. S., Johnson, K. R., Kiessling, W., Koeberl, C., Kring, D. A., MacLeod, K. G., Matsui, T., Melosh, J., Montanari, A., Morgan, J. V., Neal, C. R., Nichols, D. J., Norris, R. D., Pierazzo, E., Ravizza, G., Rebolledo-Vieyra, M., Reimold, W. U., Robin, E., Salge, T., Speijer, R. P., Sweet, A. R., Urrutia-Fucugauchi, J., Vajda, V., Whalen, M. T., and Willumsen, P. (2010). The Chicxulub asteroid impact and mass extinction at the Cretaceous-Paleogene Boundary. Science, Vol. 327, pp. 1214–1218.

Clearwater East and Clearwater West

Quebec, Canada

Wolf Uwe Reimold

Clearwater Lake in northern Quebec comprises two somewhat circular parts that are divided by an array of islands. Each circular lobe does represent an impact structure. East Clearwater is the smaller one, with a currently accepted diameter of 24 km. The structure has been described as a possible central-peak crater. West Clearwater has long been considered to be 32 km wide and potentially of peak-ring geometry. This diameter value has been discussed in the literature as possibly only representing a lower limit, and recent modeling work has suggested an original crater diameter of 35–40 km. Both structures are quite deeply eroded, with the recent modeling having indicated a degradation by some 2 km.

West Clearwater lake is about 32 km wide. Onshore, the land rises from lake level to a poorly defined and discontinuous apparent rim that is elevated by about 50–100 m with respect to lake level. The lake features a ring of prominent islands at ca. 6–10 km from the center of the structure. This ring with elevations up to 100 m above lake level is interpreted as a possible erosional remnant of a peak ring. In addition, there are a few small islands around the center of the western lake – interpreted as a surface expression of a central uplift. Outcrops of parautochthonous basement, as well as impact melt rock and other impact breccias are accessible on these islands. The target was composed of amphibolite to granulite facies metagranites, -granodiorites, -tonalites, and metamorphosed felsic gneisses. On the central islands occur exposures of metagabbro as well. The target rocks belong to the regionally occurring 2600–2800 Myr Archean Bienville Sub-Province of the Superior Province of the Canadian Shield. This basement is transected by both metamorphosed and unmetamorphosed diabase dikes. It is thought that this crystalline target was overlain, at impact time, by a thin cover of Middle to Upper Ordovician limestone. Blocks of this lithology occur on the inner ring of islands, and limestone inclusions have been observed in impact melt breccia. Impactites include shocked rock of the crystalline basement, including cataclastic breccias with shatter cones and shocked quartz, and various types of impact breccia. Impact melt breccia occurs in the form of blocks of up to 100 m dimension. Drilling has shown a general stratigraphy for the crater (from bottom upward): fractured basement, allochthonous clast-rich and clast-poor breccia, and impact melt rock. The up to 20 m thick basal breccia has been described as suevitic, i.e., besides mineral and lithic clasts derived from the target, it does contain an impact melt component. This breccia is overlain by 15 m of clast-rich impact melt rock and, then, further 85 m (maximum preserved thickness) of clast-poor impact melt rock. The metagabbro has been shown to contain maskelynite, that is feldspar that shows a shock-induced transformation into a pseudo-amorphous glass – affected under solid state conditions. The granitic rocks of the island ring contain quartz with planar deformation features (PDFs).

No meteoritic component has been identified in West Clearwater impact melt rock. The drill cores from the two most central boreholes showed that shock pressure at surface is as high as 35 GPa, decreasing in magnitude to 7.5 GPa at the depth of 390 m. A recent, detailed shock barometric study by Auriol Rae and co-workers demonstrated that shocked quartz with PDFs is prominent up to about 8.5 km radial distance from the center of the structure, beyond which no further PDFs were observed. Shatter cones have been observed up to 11 km from the center. Shock pressure determined from PDF orientation analysis is reported as constant and at a level of 17.5 GPa for the inner part of the structure.

East Clearwater lake measures about 24 km in width. Inside of the lakeshore occur numerous islands regarded as part of the crater rim section. The impact structure is not exposed on surface. A crater diameter of 22 km is taken inside of the spray of near-shore islands. The interior of the structure has been investigated with drill core from two boreholes. A borehole into the center of the structure penetrated, below the water column, 12 m of glacial drift, some 25 m of post-impact sediments, and some 880 m of granitic and granodioritic gneisses. The upper 100 m of this basement section contain injections of impact melt and breccia dikes and veinlets. Shock pressure in these rocks reaches 30 GPa and decreases with depth to 15 GPa attained at the bottom of this borehole. Most prominent are developments

Clearwater East / West

Quebec / Canada

East: latitude = 56°03' N / longitude = 74°05' W / diameter = 24 km / age = 460 - 470 Myr
West: latitude = 56°12' N / longitude = 74°30' W / diameter = 32 km / age = 290 - 300 Myr

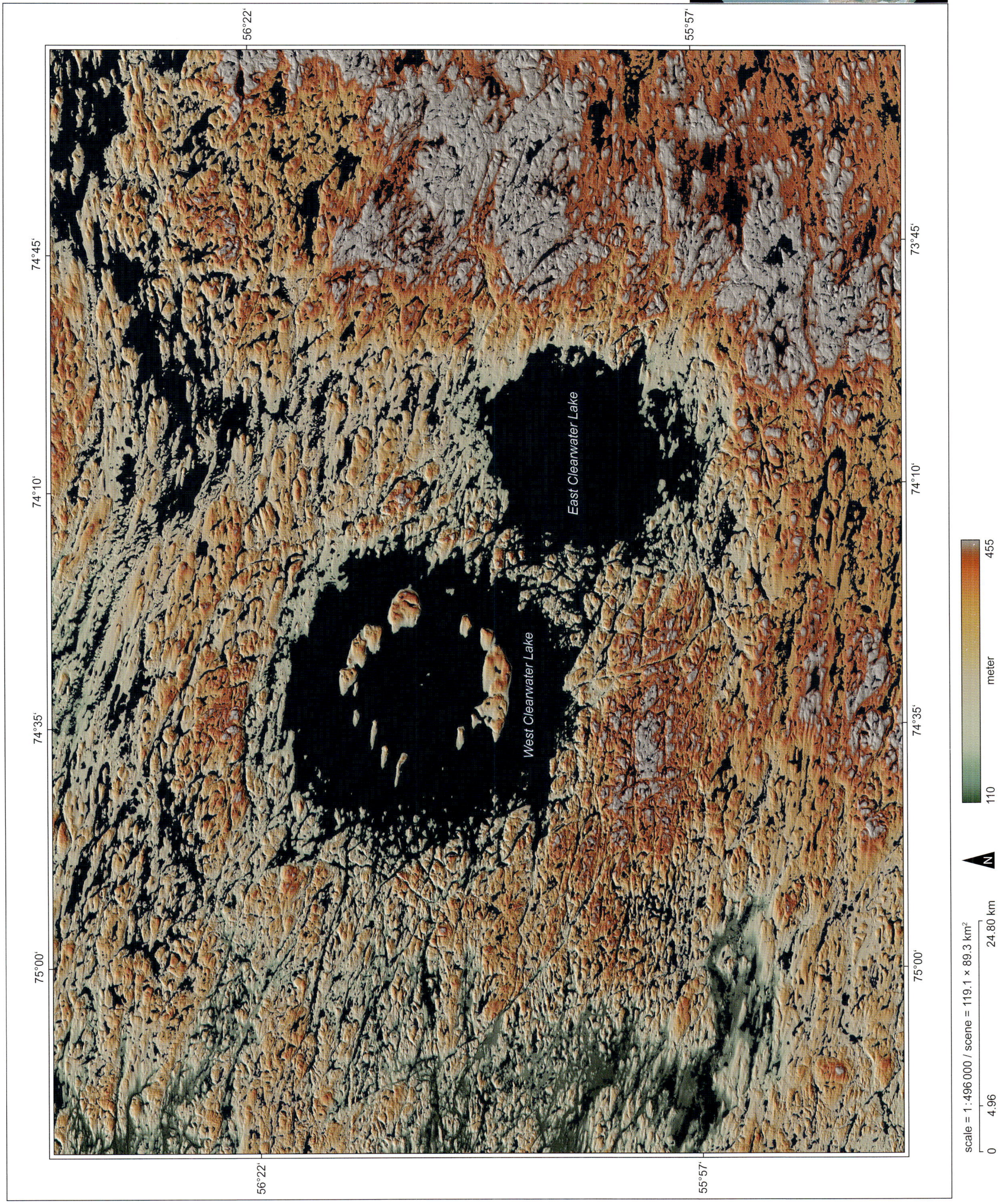

Impact breccia veins in drill core (20 × 3.2 cm in size) from borehole 1-63, at 332.3 m depth, from the Clearwater West impact structure. Base to the left, top to the right (photo courtesy: Ralf-Thomas Schmitt, Museum für Naturkunde Berlin, sample from Alexander Deutsch, previously Geological Survey of Canada, Ottawa).

Drill core (7.5 × 2.8 cm) of Clearwater East impact melt rock, from 300.2 m depth in borehole 2-63. Base to the left, top to the right (photo courtesy: Ralf-Thomas Schmitt, Museum für Naturkunde Berlin, sample from Alexander Deutsch, previously Geological Survey of Canada, Ottawa).

Drill core of impact melt rock (specimen 5.5 × 3.2 cm in size) from borehole 5-63, at 38.7 m depth, from the Clearwater West impact structure. Base to the left, top to the right (photo courtesy: Ralf-Thomas Schmitt, Museum für Naturkunde Berlin, sample from Alexander Deutsch, previously Geological Survey of Canada, Ottawa).

of PDFs in quartz. A second hole drilled off the central uplift, some 5 km towards the southeast, recorded about 46 m of glacial drift, some 112 m of post-impact sediment, a 36 m interval with poor core recovery, and finally 41 m of impact melt rock. The interval with poor core recovery actually represents a section of suevite, with mineral, lithic and melt clasts. The impact melt rocks are distinctly enriched in siderophile elements (Ni, Co, Ir) and Cr. These and related data have been interpreted to indicate a significant chondritic component in this melt rock, with exceptionally high enrichment in selected samples. Platinum group element (PGE) abundances initially indicated either a carbonaceous (C1) or L chondritic impactor. However, more recent data on PGE abundances in different meteorite classes, as well as Cr isotope ratios for a few East Clearwater melt rock samples, have been interpreted to favor a possibly L chondritic extraterrestrial component.

The regional Bouguer gravity anomaly map for the lake region depicts two gravity lows over the western and eastern parts of Clearwater Lake. The residual anomaly over the western structure measures −10 mGal, and that over the eastern structure −9 to −1 mGal. These anomalies have been modeled as depicting zones of reduced density due to fracturing. Aeromagnetic data over the two structures indicate two magnetic lows with reduction of magnetic gradients in contrast to the surrounding region. The rocks of the central uplift structures show susceptibilities and remanent magnetizations much lower than regional values.

These two Clearwater impact structures have long been thought to be the result of the impact of a binary asteroid. This was based on early dating results that suggested ages for samples from both impact structures at approximately 300–290 Myr. However, two recent investigations by the argon step-heating technique as well as (U-Th)/He dating yielded a contrasting age for Clearwater East, of 460–470 Myr, much older than for Clearwater West. Thus, the binary asteroid impact scenario has been rendered unlikely.

Apparently the first report of Clearwater Lake was made already in 1818 by the writer George Atkinson II, who came across the lake on a reconnaissance trip for the Hudson Bay Company. First suggestion that these two structures could represent impact crater structures was made by Carlyle Beals and co-workers in 1956. Over the subsequent 15 years, it was debated whether the structures could be of impact or endogenous (volcanic) origin. The Dominion Observatory of Ot-

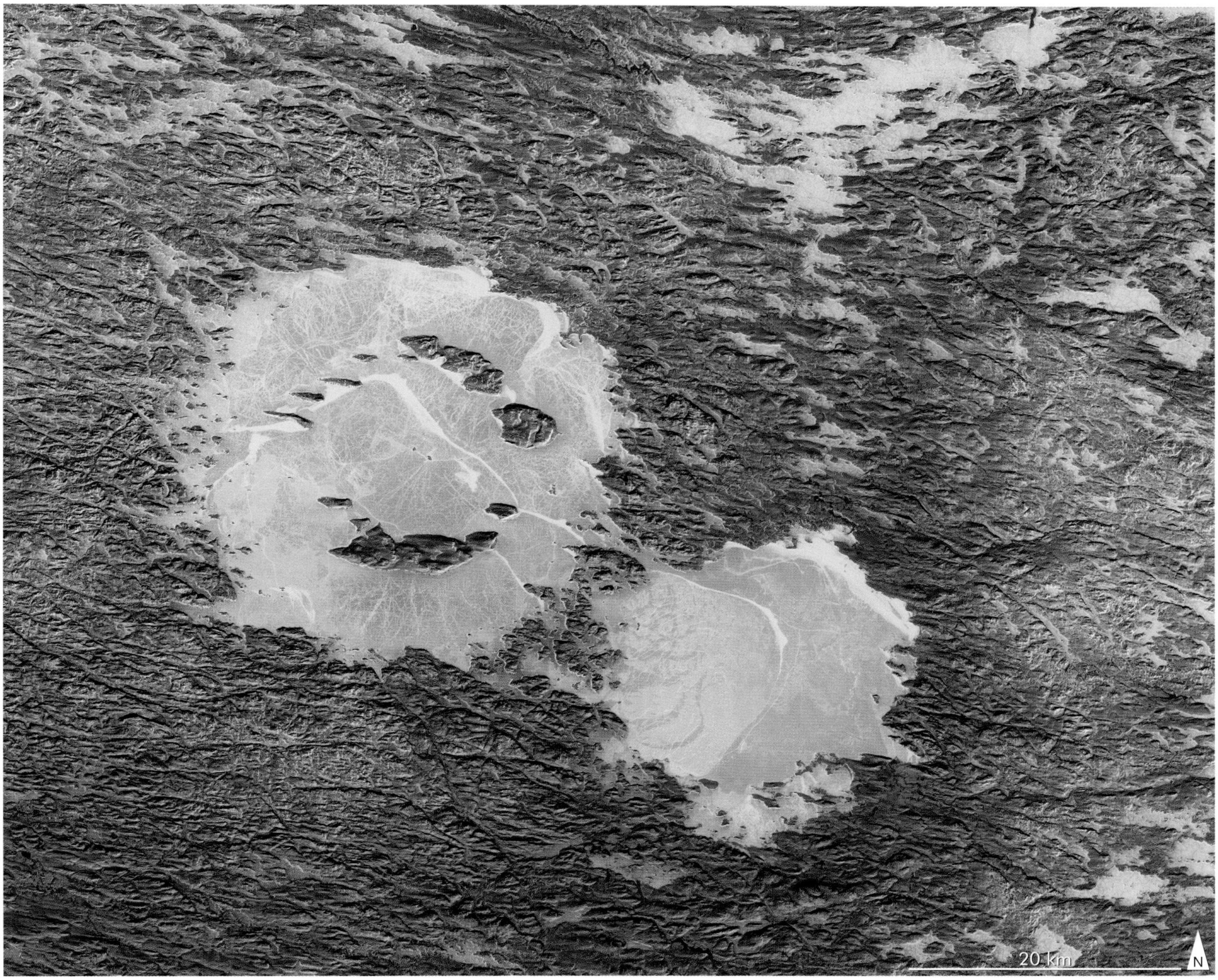

The Clearwater East/West impact structures on June 16, 2018 as a composite image based on TanDEM-X DEM and Copernicus Sentinel-2 (channel 2/3/4) data. Shortly before summer, both lakes are still ice-covered.

tawa recovered 5 drill cores from West Clearwater, and two from East Clearwater. Drilling was carried out from the ice cover during the winters of 1962/1963 and 1963/1964, together with a gravity survey. Recognition of shock deformation (PDFs in quartz) in the late 1960s then settled the debate in favor of impact. Further detailed investigations followed until the mid-1990s.

Location/Access

Clearwater Lake (in French *Lac à l'Eau Claire,* in native Cree *Lac Wiyâshâkimî*) is located in Arctic Canada, some 110 km due east of the southeastern lobe of Hudson Bay. It is also situated 200 km north of the village of Kuujjuaraapik at the mouth of the Great Whale River (*Poste-de-la-Baleine* and *Grande Rivière-de-la-Baleine*, in French). Access to the Clearwater impact structures requires planning of a full Arctic expedition. Approach to the lake would only be possible by sea-plane.

Selected References

Beals, C. S., Ferguson, G. M., and Landau, A. (1956). A search for analogies between lunar and terrestrial topography on photographs of the Canadian Shield. Journal of the Royal Astronomical Society of Canada, Vol. 50, pp. 203–250.

Grieve, R. A. F. (2006). Impact Structures in Canada. Geological Association of Canada, GeoText No. 5, 210 pp.

Hische, R. (1994). Geology of the Clearwater Impact Structures (English translation). Ph.D. Thesis, Westfälische Wilhelms-Universität Münster (Germany), 200 pp.

Phinney, W. C., Simonds, C. H., Cochran, A., and McGee, P. E. (1978). West Clearwater, Quebec, impact structure, Part II. Petrology. Proceedings of the 9th Lunar and Planetary Science Conference, Houston, Texas, Pergamon Press, New York, pp. 2659–2693.

Rae, A. S. P., Collins, G. S., Grieve, R. A. F., Osinski, G. R., and Morgan, J. V. (2017). Complex crater formation: Insights from combining observations of shock pressure distribution with numerical models at the West Clearwater Lake impact structure. Meteoritics and Planetary Science, Vol. 52, pp. 1330–1350.

Schmieder, M., Schwarz, W. H., Trieloff, M., Tohver, E., Buchner, E., Hopp, J., and Osinski, G. R. (2015). New ^{40}Ar-^{39}Ar dating of the Clearwater Lake impact structures (Québec, Canada) – Not the binary asteroid impact it seems? Geochimica et Cosmochimica Acta, Vol. 148, pp. 304–324.

Simmonds, C. H., Phinney, W., McGee, P., and Cochran, A. (1978). West Clearwater, Quebec, impact structure. Part I: Field geology, structure and bulk chemistry. Proceedings of the 10th Lunar and Planetary Science Conference, Houston, Texas, Pergamon Press, New York, pp. 2633–2658.

Von Engelhardt, W., Hörz, F., Stöffler, D., and Bertsch, W. (1968). Observations on quartz deformation in the breccias of West Clearwater Lake, Canada and the Ries Basin, Germany. In: Shock Metamorphism of Natural Materials, French, B. M., and Short, N. M. (eds.), Mono Book Corp., Baltimore, pp. 475–482.

Cloud Creek
Wyoming, United States

Thomas Kenkmann

The Cloud Creek structure is a complex impact crater of 7 km diameter in the Rocky Mountains of central Wyoming. The structure is buried below some 1200 m of Jurassic and Cretaceous sediments.

At the time of impact the target was composed of Precambrian crystalline basement overlain by Paleozoic and probably Triassic sedimentary formations, with an estimated original thickness of approximately 900 m. An unknown amount of rocks was removed by erosion prior to the deposition of the Middle Jurassic formations over the eroded Cloud Creek structure. Post-impact sediments are of Middle-Jurassic to Cretaceous age. The structure lies on a relatively stable block between two faults and folds. The Laramide compressional overprint is evident but did not cause strong tilting of the strata in the area of the crater structure.

There is a well-developed central peak composed predominantly of uplifted, brecciated, fractured, and faulted Mississippian and Pennsylvanian carbonate and clastic rocks. Internal stacking led to a thickening of strata. The central peak is bordered by faults, has a minimum diameter of about 1.4 km, and a stratigraphic uplift of about 520 m. The surrounding annular trough has a width of 1200 to 1600 m. A rim anticline encircles the moat. Several oil industry boreholes penetrated this rim anticline. Its limbs are bordered by inward-dipping fault zones that converge downward. Thin sections from drill cutting samples from a borehole on the southern perimeter of the central peak showed planar deformation features (PDFs) in a few quartz grains.

The age of the Cloud Creek structure can only be inferred from stratigraphy. The published age is between Late Triassic and Middle Jurassic, i.e., about 190 ± 20 Myr ago.

Reflection seismic data, gravity and magnetic mapping, along with data from ten exploration wells, have provided a sound knowledge of the crater morphology. The Cloud Creek structure is nearly circular in plane view, but with a slight north-northwest/south-southeast elongation. Interestingly Cloud Creek lies exactly on the trajectory inferred for the **Douglas** impact crater strewn field.

The Cloud Creek impact structure was detected by Donald Stone in 1999. Its confirmation dates to 2003.

Location/Access

The Cloud Creek impact structure is located near the center of the Casper Arch in central Wyoming, about 45 km north of the city of Casper and 115 km northwest of the Douglas impact crater strewn field. The structure is accessible from Natrona, where Road 126 departs northward from U.S. Highway 20 and leads, after about 20 km, to the impact structure.

Selected References

Stone, D. S. (1999). Cloud Creek: A possible impact structure on the Casper Arch, Wyoming. The Mountain Geologist, Vol. 36, Rocky Mountain Association of Geologists. pp. 212–234.

Stone, D. S., and Therriault, A. M. (2003). The Cloud Creek structure, central Wyoming, U.S.A. – Impact origin confirmed. Meteoritics and Planetary Science, Vol. 38, pp. 445–455.

Cloud Creek

Wyoming / United States

latitude = 43°10'36" N / longitude = 106°42'30" W / diameter = 7 km / age = 190 ± 20 Myr

scale = 1 : 103 000 / scene = 18.5 × 24.7 km²

0 1.03 5.15 km

N

1560 meter 1805

Couture
Quebec, Canada

Thomas Kenkmann

Like the **Pingualuit** crater, the 8 km wide Couture impact structure is situated on the arctic Ungava peninsula of Northern Quebec. The structure is entirely submerged in the lake of the same name. The bathymetry of the lake suggests that the diameter of the crater structure is somewhat smaller than that of the lake. The deepest point was measured to 120 m below lake level. In the center of the impact structure the lake is less deep, suggesting the possible presence of a central uplift.

Target rocks were Archean granites, granodiorites, and gneisses of the Superior Province that are cut by pegmatites and diabase dikes. Intense fracturing is observed in the rocks on the small islands close to the main shore. The presence of an impact-induced, narrow-spaced joint network caused the formation of the numerous small islands and the serrated appearance of the lake's shoreline. This halo of fractures can be traced for about 3 km onshore. A negative gravity anomaly of roughly −5 mGal that superposes the regional trend was recorded over the lake.

Impact lithologies occur as 1–5 m sized blocks in glacial till on the western shore. Different types of allochthonous polymict breccia have been reported, among them impact melt breccia. ^{40}Ar-^{39}Ar stepwise heating of such impact melt rock gave an age for the impact of 425 ± 25 Myr.

Carlyle Beals of the Dominion Observatory was the first to propose an impact origin for the Couture structure, already in 1955. As with other structures on the Canadian Shield this was based on the unusual circular outline of Lake Couture, in contrast to other shapes of the glacially formed lakes that are irregular or elongated and follow the regional joint system. The structure was named after Guillaume Couture (1617–1702), an early explorer who led an expedition to Hudson Bay in 1663. Until today our knowledge of the crater is limited and primarily based on the early work by Carlyle Beals.

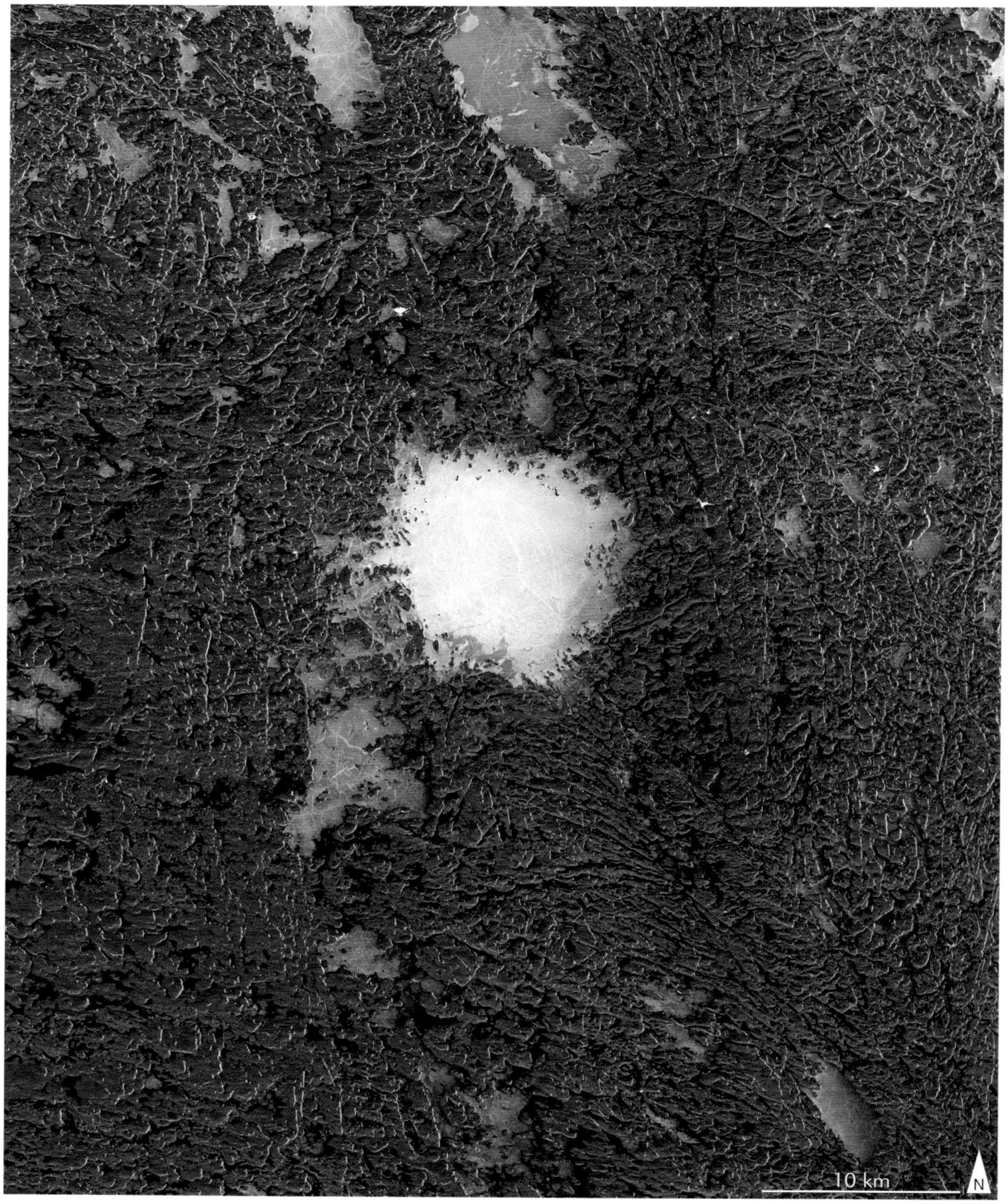

The Couture impact structure with the homonymous ice-covered lake on June 14, 2019 as a composite image based on TanDEM-X DEM and Copernicus Sentinel-2 (channel 2/3/4) data.

Location/Access

The impact structure is located on the arctic Ungava Peninsula about 160 km south of **Pingualuit** crater. There is no direct access to this structure. The closest Inuit villages are Povirnituq and Akulivik, 130 km and 170 km, respectively, to the west at Hudson Bay. Both settlements feature an airfield.

Selected References

Beals, C. S., Dence, M. R., and Cohen, A. J. (1967). Evidence for the impact origin of Lac Couture. Publications of the Dominion Observatory Ottawa, Vol. 31, pp. 409–426.

Grieve, R. A. F. (2006). Impact Structures in Canada. Geological Association of Canada, GeoText No. 5, 210 pp.

Couture

Quebec / Canada

latitude = 60°07'47" N / longitude = 75°17'56" W / diameter = 8 km / age = 425 ± 25 Myr

scale = 1:210 000 / scene = 37.9 × 50.5 km²

0 2.10 10.50 km

N

135 meter 285

Crooked Creek

Missouri, United States

Thomas Kenkmann

The Crooked Creek impact structure, a 7 km wide roughly circular disturbance, lies in Missouri's Crawford County about 130 km southwest of St. Louis between the Meramec and Crooked Creek rivers.

The crater is formed in Upper Cambrian to Ordovician sedimentary rocks, predominantly sandstones and dolomites that are intensely folded, faulted, and brecciated. Outside the structure these rocks are lying almost flat. Bounded by a series of normal faults, the structure consists of a peripheral ring syncline where Ordovician strata are downfaulted by some 80 m. The ring syncline encircles a heavily faulted central uplift. The oldest Cambrian beds are exposed along a ring anticline of the central uplift and have been uplifted by more than 300 m above their normal position. The presence of this anticline, along with younger bedrock in the center, suggests that the dome has been partly collapsed. Abundant hydrothermal mineralization (galena, sphalerite, marcasite, barite and vein quartz) can be observed in the brecciated core of the impact structure. Mining took place 100 years ago from a cluster of shallow shafts. The fractured area provided channel-ways for impact-induced hydrothermal circulation and precipitation.

Sedimentary rocks in the central uplifted core of the impact structure are faulted, intensely shattered, and brecciated. A network of narrow-spaced faults without large displacements characterizes these rocks. Shatter cones are ubiquitously present here, and injections of clastic dike occur. Folding is generally accommodated by brittle deformation and localized slip on complex shear-zone networks. The collective deformation on the fracture networks causes the apparent ductile behavior of impact-induced folds.

A study on reworked Conodont fossils in impact breccia suggests a Mississippian age of about 323–348 Myr.

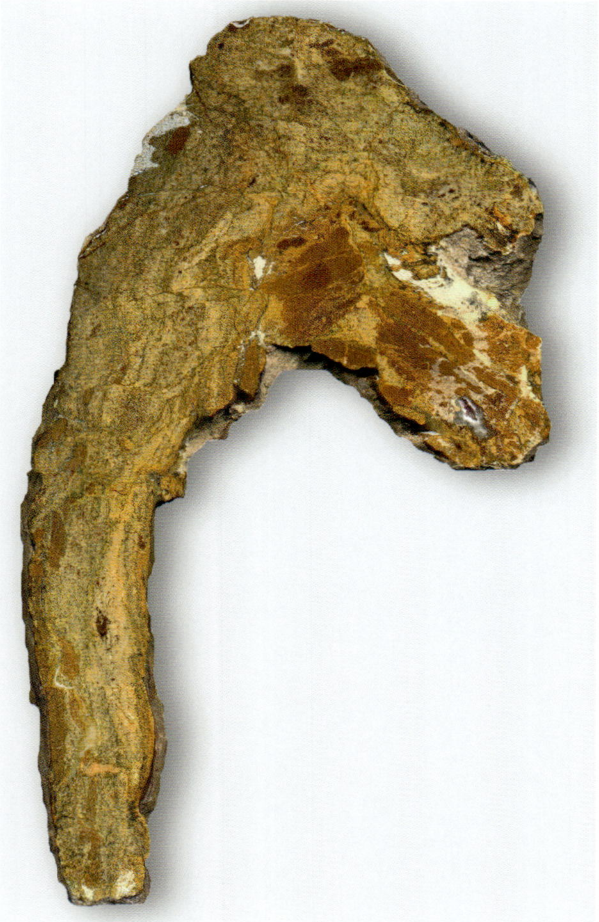

Impact-induced fold within calcareous dolomite from the central uplift of the Crooked Creek impact crater. Image width is about 15 cm (photo: T. Kenkmann).

Outcrop of brecciated bedrock, about 2 m wide, in the central uplift of the Crooked Creek impact structure (photo: T. Kenkmann).

Like a number of similar structures in the United States, Crooked Creek was originally described as a geologic disturbance, not an impact crater: essentially circular in outline with peripheral high-angle faults and highly deformed and brecciated rocks in the center that were uplifted to expose older rocks. A first report by V. Hughes dates back to 1911, while Herbert Hendricks provided the results of his intensive fieldwork in 1954. Although Hendricks already suggested that of the various possible origins, a meteoritic impact might be the best explanation, a subterranean explosion was still favored by many geologists for the Crooked Creek structure – and for all the other "disturbances" in the United States. Since no volcanic rocks were found in this and similar structures, they were usually termed "cryptovolcanic". Even in 1962, when shatter cones from Crooked Creek had become known, Thor Kiilsgaard and co-workers still favored "... *that the Crooked disturbance originated from a subterranean gaseous explosion.*" Only later studies drew the correct conclusions and confirmed an impact origin. The 1980 report by Robert Dietz and Philippe Lambert, where they documented shatter cones and planar deformation features (PDFs) of four different crystallographic orientations in quartz grains of the Upper Cambrian Lamotte sandstone, settled the debate once and for all.

Location/Access

Crooked Creek is easily accessible. It is located about 15 km south of Steelville, Missouri. The center of the structure can be reached via Wesco Road and Beers Road. However, the structure is hardly visible on the ground. The river giving Crooked Creek its name is winding from north to south through the eastern part of the structure.

Selected References

Bucher, W. H. (1933). Cryptovolcanic structures in the United States. 16th International Geological Congress, London. Report of the United States, Vol. 2, pp. 1055–1084.

Dietz, R. S., and Lambert, P. (1980). Shock metamorphism at Crooked Creek cryptoexplosion structure, MO (abstract). Meteoritics, Vol. 15, pp. 281–282.

Hendricks, H. E. (1954). The geology of the Steelville quadrangle, Missouri. Missouri Geological Survey and Water Resources [Report], 2nd Series, Vol. 36, 88 pp.

Hughes, V. H. (1911). Reconnaissance work: Missouri Geological Survey and Mines. Biennial Report for 1909–10, pp. 36–54.

Crooked Creek

Missouri / United States

latitude = 37°50'05" N / longitude = 91°23'44" W / diameter = 7 km / age = 323 - 348 Myr

scale = 1:100 000 / scene = 18.0 × 23.9 km^2

0 1.00 5.00 km

N

225 meter 425

Kenkmann, T. (2002). Folding within seconds. Geology, Vol. 3, pp. 231–234.
Kiilsgaard, T. H., Heyl, A. V., and Brock, M. R. (1962). The Crooked Creek disturbance, southeast Missouri. U.S. Geological Survey Professional Paper 450-E, pp. E14–E19.
Miller, J., Evans, K. R., Ethington, R. L., Repetski, J. E., Sandberg, C. A., and Thompson, T. L. (2007). Critical stratigraphic data from reworked Conodonts in impact breccias across Missouri's Ozark Dome (abstract). Geological Society of America South-Central and North-Central Joint Section Meeting, Lawrence, Kansas, Abstracts with Programs, Vol. 39, p. 62.
Mulvany, P. S. (2004). Field Trip II: Geology of the Crooked Creek Ring Structure, Crawford County, Missouri. Association of Missouri Geologists Field Trip Guide Book, 51st Annual Meeting, Rolla, Missouri, pp. 13–30.

Decaturville
Missouri, United States

Thomas Kenkmann

The Decaturville impact structure has a subdued topographic expression. It is marked by a ring of low, wooded hills inside a nearly complete circle of low-lying, unforested ground. Streams drain radially outward from the center of the area, a pattern in conspicuous contrast to the generally rectilinear stream network of the surrounding countryside. The structure, bounded by a fault, has a roughly elliptical outline, with its major axis trending northwest and measuring 6.3 km across; the minor axis is 5.5 km long.

The target rocks of the Decaturville structure range in age from Precambrian to Silurian. The Precambrian basement consists of schists intruded by economically interesting pegmatite dikes. Decaturville is located on the Ozark plateau, where the basement is overlain by a 540 m thick, flat-lying sequence of undisturbed sediments. They were deposited in the Upper Cambrian and Lower Ordovician, and in the Silurian. The dominant lithology is dolomite. All these formations are exposed on surface.

Inside the inward-dipping rim fault occurs a structurally depressed ring syncline of about 1.6 km width, which surrounds an about 2.4 km wide central uplift. Deformation in the ring syncline is highly complex. Strata here occur stratigraphically up to 128 m deeper relative to their normal position outside the structure. In contrast, the stratigraphic uplift in the central part of the structure amounts to 280–305 m. There exists even a surface outcrop of a single megablock of Precambrian basement rock that normally occurs about 540 m below surface. The block measures about 15 × 21 m in size and has a minimum thickness of 15 m as shown by drill core. Extensive drilling at the center of the structure has recovered more than 5500 m of core from 60 boreholes. Those from the center of the structure revealed a chaotic mixture of rocks, with all the formations from the top of the Precambrian basement through the Cambrian Potosi Dolomite having been recognized.

Monomict and polymict breccias, and megablocks occur along the annular moat and in the area of the central uplift. Shock features were found in quartz grains of the breccias and include planar deformation features (PDFs) and planar fractures (PFs). Shatter cones occur at the center of the structure in a circular area about 450 m across.

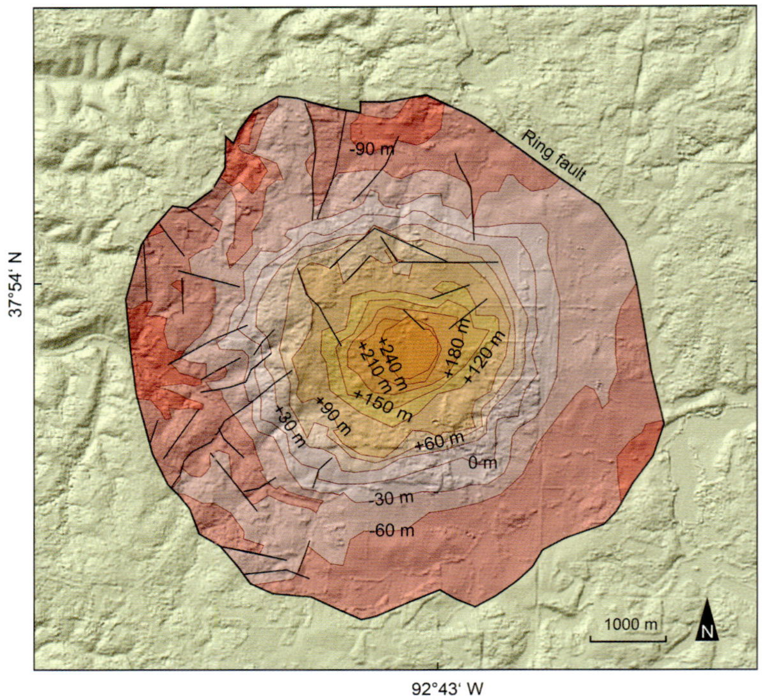

An isopleth map of the Decaturville impact structure, underlain by a TanDEM-X DEM. The isopleth pattern shows the vertical displacements of strata relative to the pre-impact configuration. Heavy solid lines indicate faults (modified after Offield and Pohn, 1979).

The Decaturville structure is centered at 4.8 km west of a conspicuous northwest-trending linear magnetic high. The gravity map for the structure indicates only a weak positive excursion of about +1 mGal. This is not surprising, as there is only weak density contrast between the rocks involved in the cratering process.

View towards the central uplift of the Decaturville impact crater (photo: T. Kenkmann).

Polymict lithic breccia of the Decaturville impact structure. The breccia is composed of different Paleozoic bedrock lithologies, dominantly dolomite. Width of the photo is 30 cm (photo: T. Kenkmann).

Decaturville

Missouri / United States

latitude = 37°53'33" N / longitude = 92°43'11" W / diameter = 6 km / age = < 300 Myr

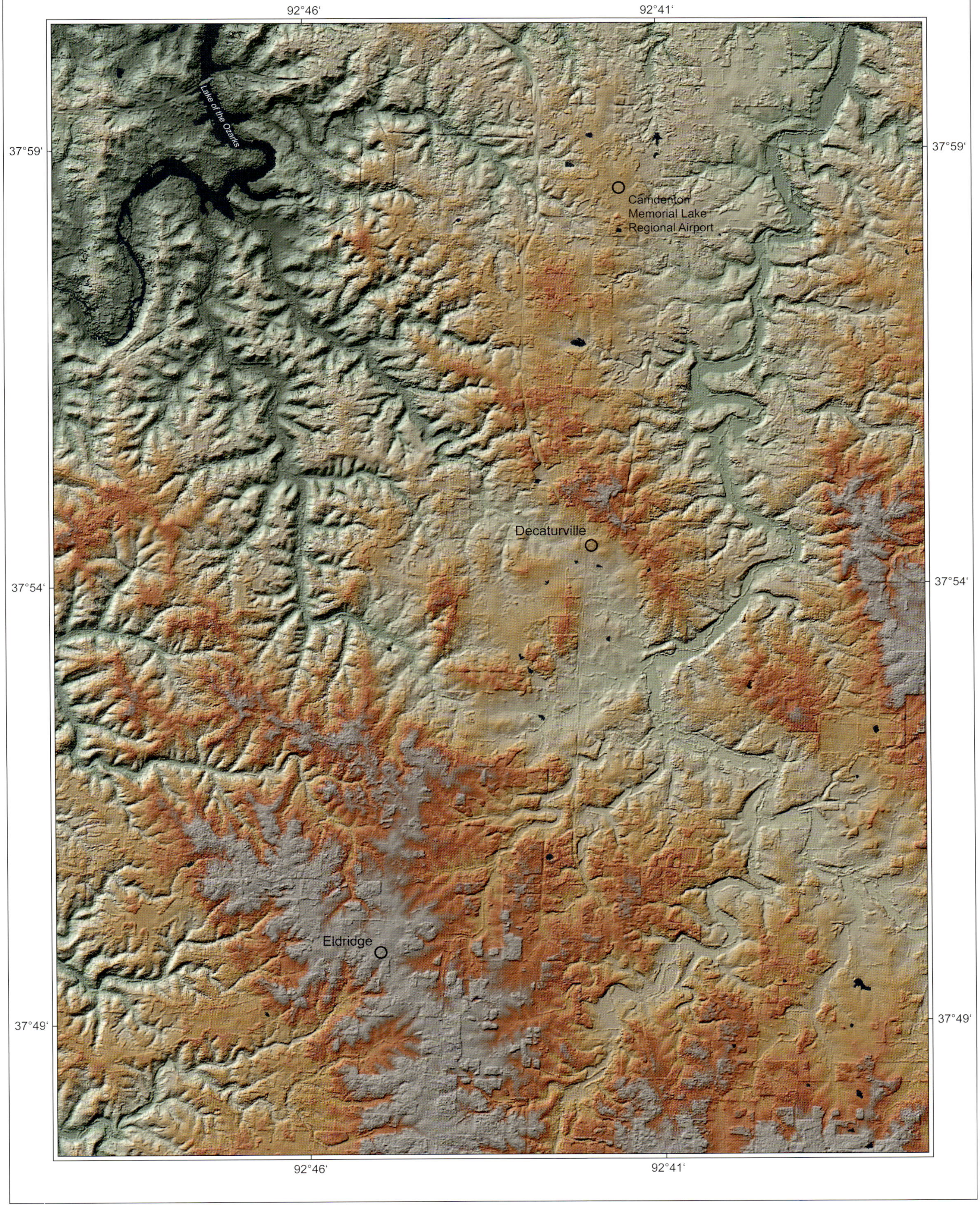

scale = 1 : 100 000 / scene = 18.0 × 23.9 km²

0 1.00 5.00 km

N

130 meter 380

The age of the Decaturville impact structure is not well constrained. The stratigraphic record suggests a post-Pennsylvanian age, but the structure may be even younger than Cretaceous.

In the first half of the 20th century Decaturville was among the structures in North America that were considered as being of "cryptovolcanic" origin. Another proposal suggested that the disturbance was located at an intersection of regional tectonic lines, where subterranean gaseous explosions ("cryptoexplosions") might have occurred. The riddle was solved in 1979 when Terry Offield and Howard Pohn of the U.S. Geological Survey investigated the structure in great detail and confirmed its impact origin.

Location/Access

The Decaturville structure is located in central Missouri. The closest large city is Springfield with a national airport. From Springfield, Interstate 44 runs to Lebanon (90 km), where State Highway 5 departs. After another 25 km this road intersects the impact structure. There are several outcrops to study the intense rock deformation.

Selected References

Nickerson, G. A. J., Spray, J. G., and Mayer, L. A. (2001). Investigation of integrated geologic and geophysical data using GIS; Crooked Creek and Decaturville impact structures, Missouri. Atlantic Geology, Vol. 36, pp. 67–68.

Offield, T. W., and Pohn, H. A. (1977). Deformation at the Decaturville impact structure, Missouri. In: Impact and Explosion Cratering, Roddy, D. J., Pepin, R. O., and Merrill, R. B. (eds.), Pergamon Press, New York, pp. 321–341.

Offield, T. W., and Pohn, H. A. (1979). Geology of the Decaturville impact structure, Missouri. U.S. Geological Survey Professional Paper, Vol. 1042, 48 pp.

Decorah

Iowa, United States

Thomas Kenkmann

Decorah is a recently confirmed impact structure of 5.6 km diameter. Except for just a few outcrops the structure is almost completely covered by Ordovician beds and Quaternary deposits. At depth, the circular structure forms a basin filled with polymict breccia and fossil-rich shales. A central uplift has not been noted so far.

The structure was formed on the Paleozoic Plateau region of Iowa, into which the Upper Iowa River and its tributaries have deeply incised their beds. Target rocks are undeformed but gently tilted sandstones, carbonates, and, to a lesser degree, shales of Upper Cambrian and Lower Ordovician age. The total thickness of the Paleozoic sequence is 620 m. Quaternary deposits, mainly colluvium, loess and glacial till, form a cover of up to 111 m thickness over much of the structure.

Surface exposures are limited to riverbank exposures of the post-impact so-called "Winneshiek Shale" and one deformed and brecciated outcrop of Lower Ordovician dolomite close to the rim of the structure. The outcrop shows closely spaced fracturing with multiple subparallel fracture sets, local monomict brecciation, and tight overturned folding.

Well cuttings from abundant water wells and samples from drill core indicated the existence of a circular basin with a maximum depth of about 210 m. It is filled completely by a series of uniquely local lithologies that are not found in the sedimentary succession of the surroundings of the structure. Polymict breccias and displaced Paleozoic strata are overlain by the Winneshiek Shale, an organic-rich shale rich in fossils. The shallow basin is discordant with the surrounding geology. Because of the crater infill, the Decorah impact structure displays virtually no relief. The structure is covered by the Ordovician St. Peter Sandstone, which indicates that the structure underwent severe post-impact erosion before the St. Peter Sandstone was deposited.

Two of the boreholes drilled into the crater fill breccias provided the critical samples for the discovery and analysis of shock-metamorphic features in quartz. Quartz grains in drill core samples of the breccia unit exhibit planar deformation features (PDFs) and planar fractures (PFs) in about 1 % of quartz grains. Until now, the crater floor below the breccias has not been studied for shock features.

The subsurface character of the Decorah structure remains vague and there is no indication as to whether or not the Decorah impact structure exhibits a central uplift. An airborne electromagnetic survey could successfully map the Winneshiek Shale as a conductor occurring in a circular area.

The current biostratigraphic age estimate for the Decorah impact is 460–483 Myr. Thus, this structure may belong to the group of Middle Ordovician impact craters formed as a consequence of collision in the asteroid belt at approximately 470 Myr ago.

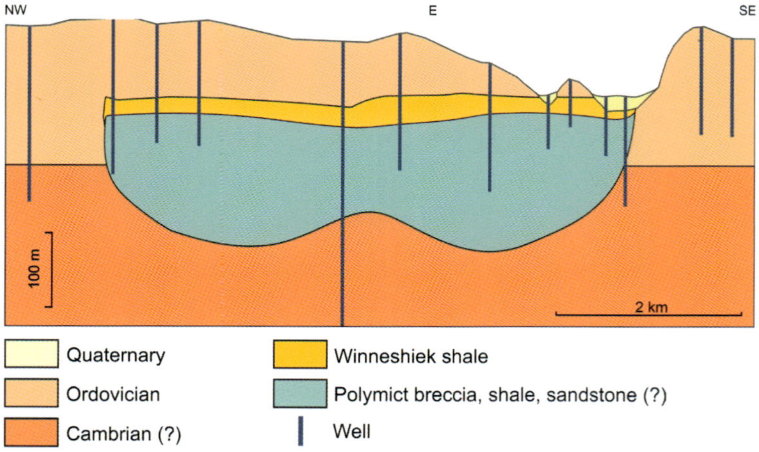

Schematic geological cross-section through the Decorah impact structure (modified after French et al., 2018).

The Decorah impact structure was recognized by H. Paul Liu and others in 2006 during the investigation of the fossil-rich Winneshiek Shale deposit that formed in this isolated basin. However, confirmation of the impact origin was only achieved in 2018 when Bevan French and co-workers found shock effects in quartz grains of the polymict crater fill breccias.

Location/Access

The Decorah structure is located in Winneshiek County, northeast Iowa. From Dubuque on the Mississippi River to the southeast of the structure it is a 140 km drive along U.S. Highway 52 to reach the city of Decorah, which extends over 40 % of the structure. The Upper Iowa River meanders through the southern part of the structure, where its incision has exposed the Winneshiek Shale and deformed strata. A single exposure of a strongly deformed and brecciated Ordovician dolomite has been described from the grounds of the Oneota Country Club.

Selected References

French, B. M., McKay, R. M., Liu, H. P., Briggs, D. E. G., and Witzke, B. J. (2018). The Decorah structure, northeastern Iowa: Geology and evidence for formation by meteorite impact. Geological Society of America Bulletin, Vol. 130, pp. 2062–2086.

Decorah

Iowa / United States

latitude = 43°18'50" N / longitude = 91°46'20" W / diameter = 5.6 km / age = 460 - 483 Myr

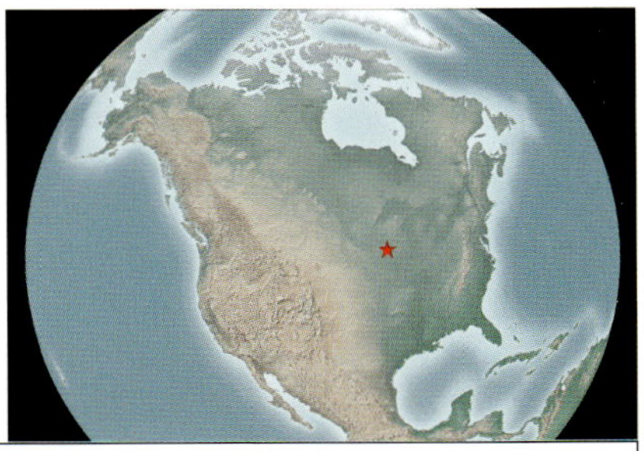

scale = 1 : 144 000 / scene = 25.9 × 34.5 km²

0 1.44 7.20 km

N

225 meter 405

Kass, M. A., Bedrosian, P. A., Drenth, B. J., Bloss, B. R., McKay, R. M., Liu, H., French, B. M., and Witzke, B. J. (2013). Modeling and inversion results from airborne geophysics over a buried impact structure in Decorah, Iowa, USA. Geological Society of America Abstracts with Programs, Vol. 45, p. 485.

Liu, H. P., McKay, R. M., Young, J. N., Witzke, B. J., McVey, K. J., and Liu, X. (2006). A new lagerstätte from the Middle Ordovician St. Peter Formation in northeast Iowa, USA. Geology, Vol. 34, pp. 969–972.

Deep Bay
Saskatchewan, Canada

Thomas Kenkmann

The Deep Bay impact structure of 13 km diameter is completely submerged in an isolated bay of Reindeer Lake in northeastern Saskatchewan. This almost circular bay of 11 km diameter is rather deep at 220 m depth, which is unusual for the otherwise relatively shallow lakes of glacial origin in the wider region. The deepest point is offset from the bay's center to the southeast and is part of an annular trough. The lake's depth over the center of Deep Bay amounts to 120 m. This shallower part corresponds to the structure's central uplift.

Target rocks are Archean and Proterozoic gneisses of the Churchill Province of the Canadian Shield. Rocks exposed along the shore of the bay do not show a shock metamorphic overprint but display brittle fracturing. A halo of enhanced fracture density is also recognizable in remote sensing data and extends up to a radial distance of 9–10 km from the submerged crater center. This led to a radial and concentric drainage pattern close to the shore of the bay. Boreholes were drilled off-shore near the eastern shore and in the western part of the bay. Two boreholes west of the center of the structure cored some 50 m of impact lithologies, each after penetrating through glacial drift and post-impact lacustrine sediments. The lithologies comprised suevitic breccia containing, amongst others, quartz with planar deformation features (PDFs) and ballen quartz.

Deep Bay is characterized by a negative gravity anomaly of about −20 mGal. Refraction seismic studies revealed a 25 % reduction in p-wave velocity over a region of 19 km diameter that corresponds to the fracture halo observed by means of remote sensing. The age of Deep Bay was narrowed down to 95–102 Myr, based on palynological studies of the shales deposited after the impact.

Interestingly an impact origin of Deep Bay was first proposed in 1951 by a biologist, D. Dawson, but his suggestion was opposed by the regional Geological Survey. The impact hypothesis was revived when a systematic search program for Canadian impact structures was initiated by Carlyle Beals and co-workers at the then Dominion Observatory in Ottawa, after the discoveries of the **Brent** and **Holleford** craters. An off-shore drilling program between 1956 and 1965 was key to prove the impact origin of Deep Bay crater by recovering critical shock effects.

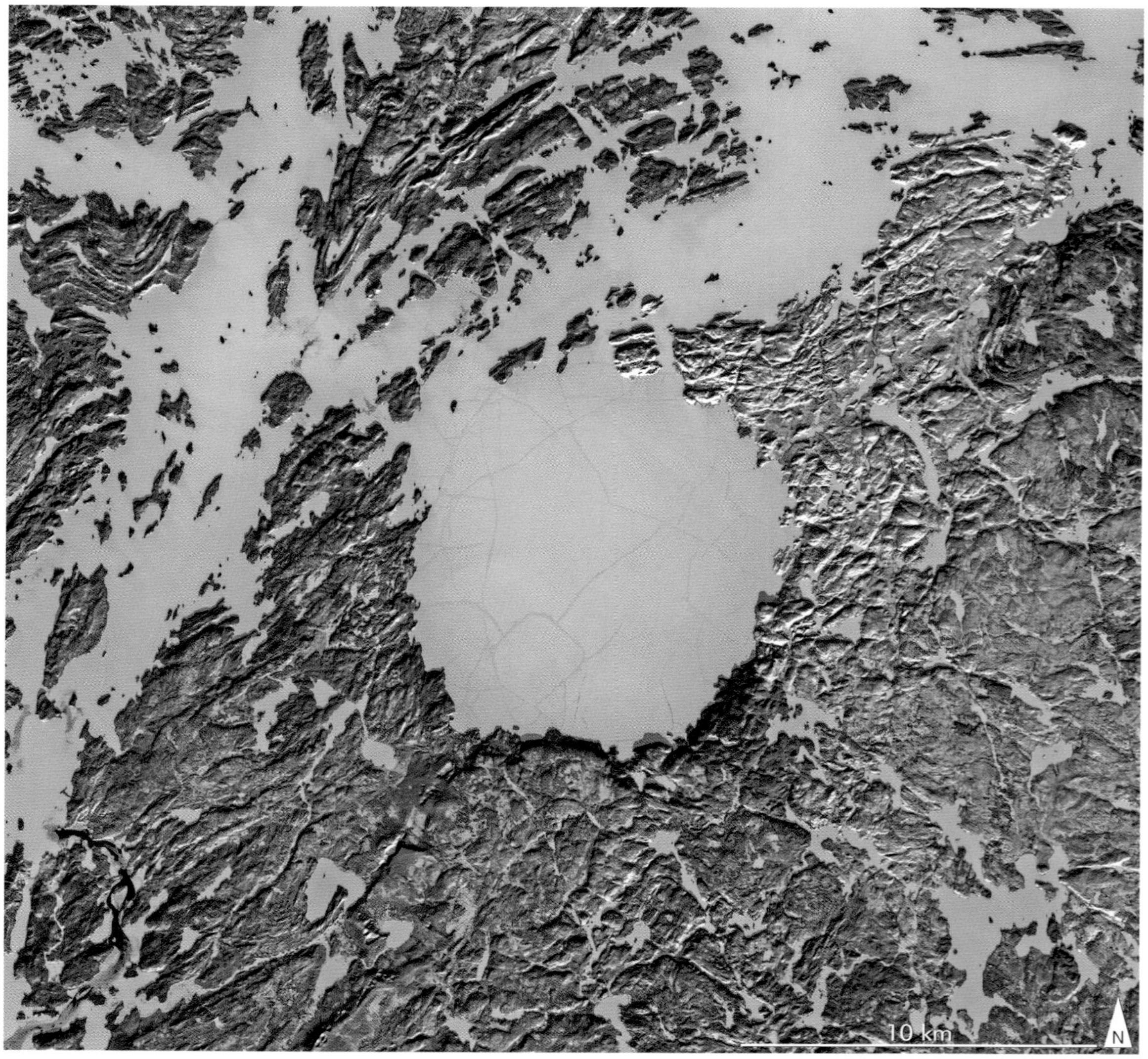

The ice-covered Deep Bay impact structure on January 5, 2019 as a composite image based on TanDEM-X DEM and Copernicus Sentinel-2 (channel 2/3/4) data.

Location/Access

Deep Bay is located in the eastern part of Saskatchewan, about 90 km east of the **Gow** impact structure. The closest village is Southend, about 12 km to the southwest of Deep Bay. Southend is the northern terminus of Saskatchewan Highway 102, which starts 220 km further south in La Ronge, where Saskatchewan Highway 2 comes to an end.

Selected References

Dence, M. R., Innes, M. J. S., and Robertson, P. B. (1968). Recent geological and geophysical studies of Canadian craters. In: Shock Metamorphism of Natural Materials, French, B. M. and Short, N. M. (eds.), Mono Book Corp., Baltimore, pp. 339–362

Innes, M. J. S. (1957). A possible impact crater at Deep Bay, Saskatchewan, Canada. Journal of the Royal Astronomical Society of Canada, Vol. 51, pp. 235–240.

Deep Bay

Saskatchewan / Canada

latitude = 56°24‘ N / longitude = 102°59‘ W / diameter = 13 km / age = 95 - 102 Myr

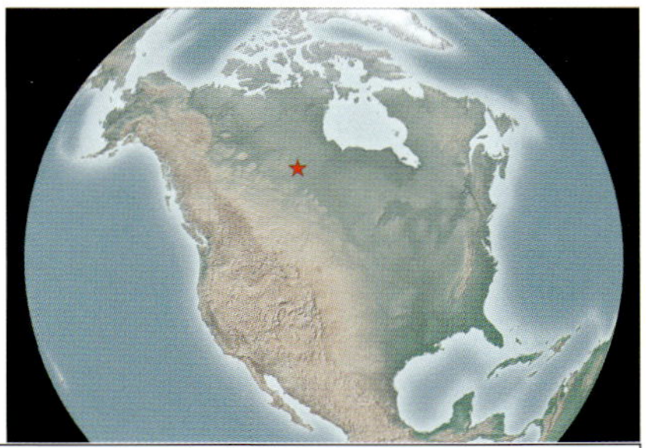

scale = 1 : 175 000 / scene = 31.4 × 41.9 km^2

0 1.75 8.75 km

N

330 meter 460

Des Plaines
Illinois, United States

Thomas Kenkmann

The Des Plaines structural disturbance represents a buried and eroded complex impact structure with a diameter of 8.8 km. The structure has been known from geophysical surveys and the investigation of 295 boreholes. The structure is buried beneath 25–60 m of glacial till and does not present any topographic indication on surface. The ages of the involved strata range from Early Ordovician to Pennsylvanian. The impact structure is surrounded by nearly horizontal and undisturbed dolomite beds of Silurian age.

The strata in the structure are strongly separated into fault-bounded blocks. Most of the faults appear essentially vertical, and the beds in the fault blocks are nearly horizontal. The vertical displacement on faults may be as large as 300 m. Des Plaines has a central uplift, which is also bounded by faults, with displacements of as much as 200 m. In the central part of the structure, Middle Ordovician St. Peter Sandstone has been uplifted closer to surface. However, the intensity of deformation in the central area appears to be only moderate. Because of this, Des Plaines is regarded to be eroded by possibly more than 300 m. The proof of the impact origin came from the discovery of shatter cone fragments in drill cores.

As Pennsylvanian rocks are present in the structure, it must be of Late Pennsylvanian or younger age, i.e., less than 280 Myr. A negative gravity anomaly of –8 mGal roughly coincides with the structure.

Like many other structures in the United States, Des Plaines was regarded as a "cryptovolcanic", and later as a "cryptoexplosion" structure. Grover Emrich and Robert Bergstrom published a detailed investigation of Des Plaines in 1962. Later, John McHone and others could prove the impact origin by detecting shatter cones in drill core.

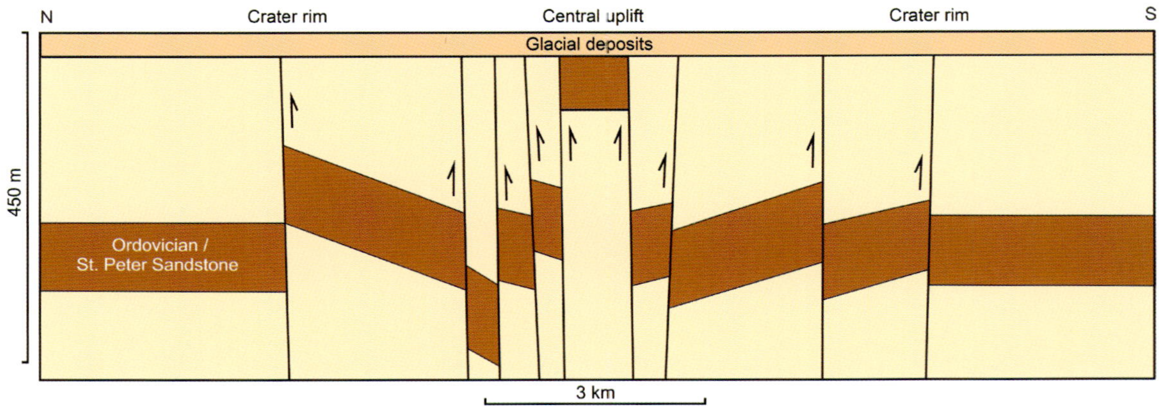

Schematic geological cross-section through the Des Plaines impact structure (modified after Emrich and Bergstrom, 1962).

Selected References

Emrich, G. H., and Bergstrom, R. E. (1962). Des Plaines disturbance, Northeastern Illinois. Geological Society of America Bulletin, Vol. 73, pp. 959–968.
McHone, J. F., Sargent, M. L., and Nelson, W. J. (1986). Shatter cones in Illinois: Evidence for meteoritic impacts at Glasford and Des Plaines (abstract). Meteoritics, Vol. 21, pp. 446.
Willman, H. B. (1971). Summary of the Geology of the Chicago Area: Illinois. Geological Survey Circular, Vol. 460, 77 pp.

Location/Access

The buried Des Plaines impact structure is located below the outskirts of the greater Chicago metropolitan area, some 8 km northeast of the International Airport Chicago O'Hare. The Des Plaines River crosses the area from north to south and some oxbow lakes occur along its alluvial plain.

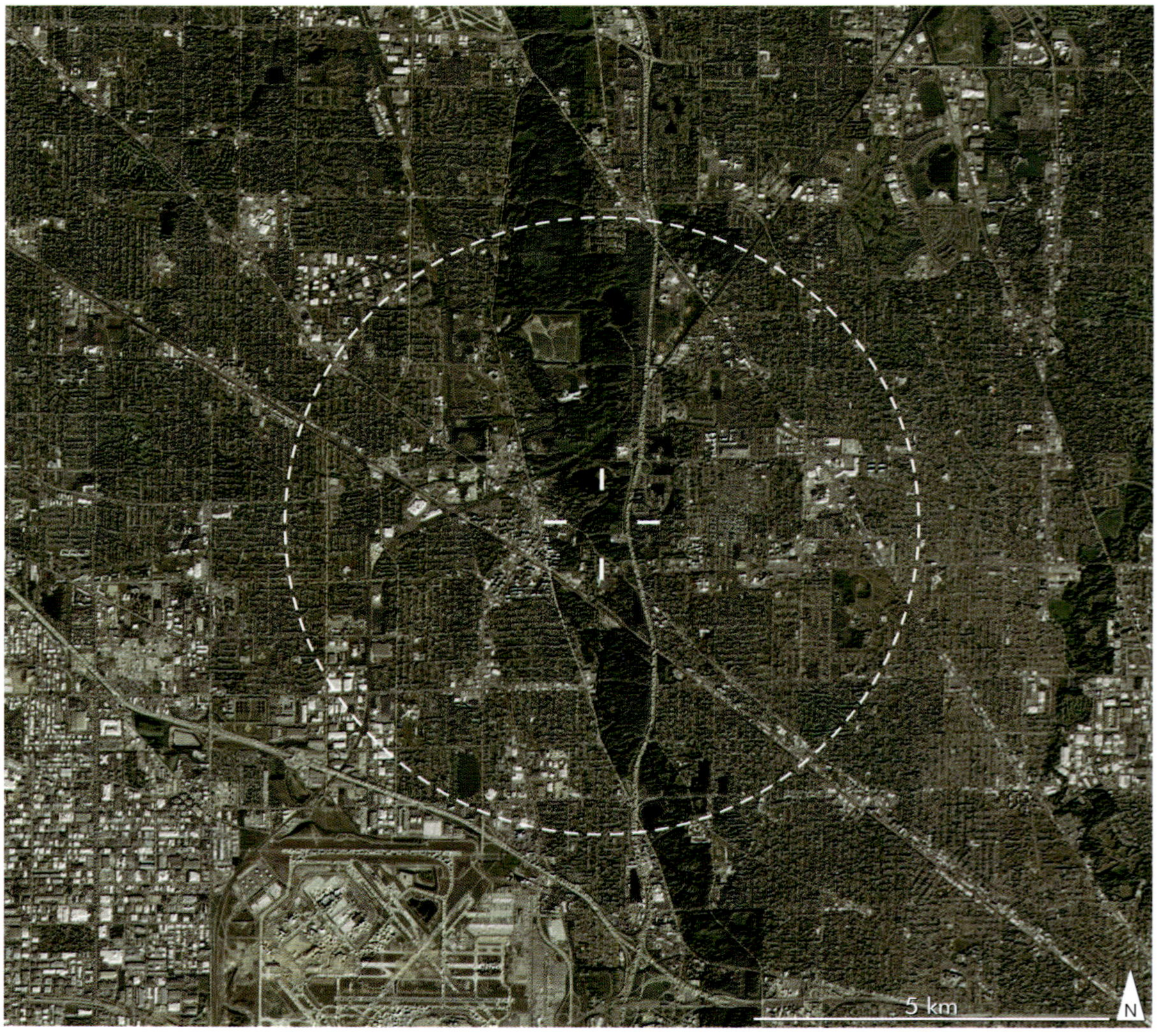

The urban area of the Des Plaines impact structure on July 15, 2019 as a composite image based on TanDEM-X DEM and Copernicus Sentinel-2 (channel 2/3/4) data. Cross hairs and a dashed circle indicate position and extent of the buried structure.

Des Plaines
Illinois / United States

latitude = 42°02'42" N / longitude = 87°52'24" W / diameter = 8.8 km / age = < 280 Myr

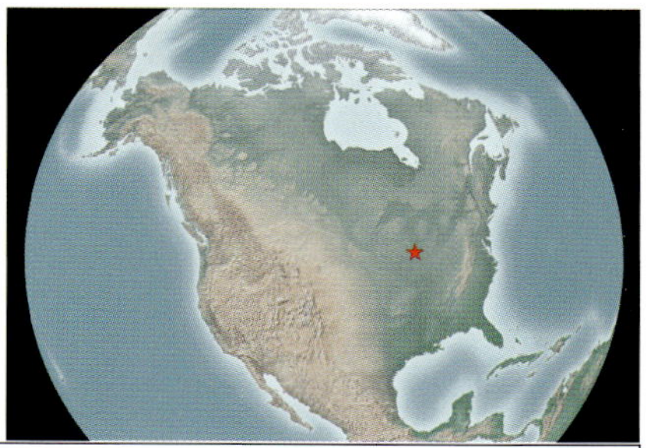

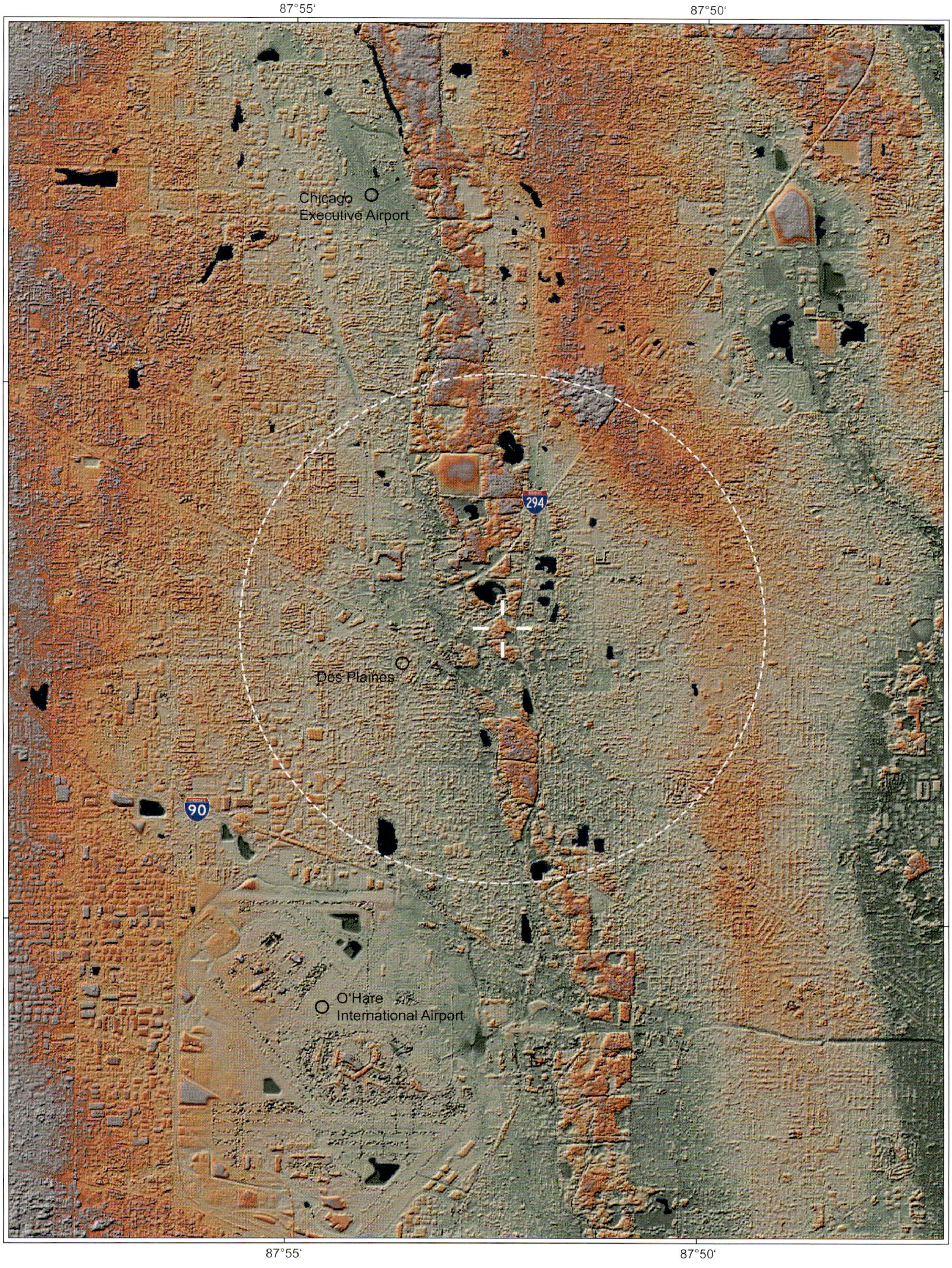

scale = 1 : 87 000 / scene = 15.7 × 21.0 km²

0 0.87 4.35 km

N

175 meter 230

Douglas Crater Field
Wyoming, United States

Thomas Kenkmann

Douglas is a collective term for a recently discovered strewn field of small impact craters. Currently more than fourty circular and ellipsoidal, in part confirmed or probable, but otherwise possible impact craters have been identified on the northeast flank of the Sheep Mountain anticline near Douglas, Wyoming. The crater structures are exposed in the uppermost quartz-cemented sandstone of the Permo-Carboniferous Casper Formation. The diameters of the crater pits range from 16 to 100 m. The exposed strewn field on Sheep Mountain has a minimum length of 7.5 km in southeast-northwest direction. Recently a 50 m crater could be confirmed that lies on an anticline further south within the same exposure of the uppermost Casper sandstone. This finding expands the crater strewn field size to roughly 13 × 7 km. Satellite and drone imagery has revealed crater shapes, orientations, and size. Some of the craters have crescent-shaped morphologies, with lowest spill points to the northeast, caused by strata tilting and erosional processes. Many of the craterlings possess the geomorphol-

The eastern slope of the Sheep Mountain anticline, where most of the craters of the Douglas crater strewn field were discovered. The craters occur stratiform in the uppermost Casper Sandstone in direct contact to the reddish Opeche Member of the Goose Egg Formation (photo: T. Kenkmann).

Variegated contact bed between the uppermost Casper Sandstone and the reddish Opeche Member of the Goose Egg Formation. The contact layer contains shocked quartz grains. Image is 1 m wide (photo: T. Kenkmann).

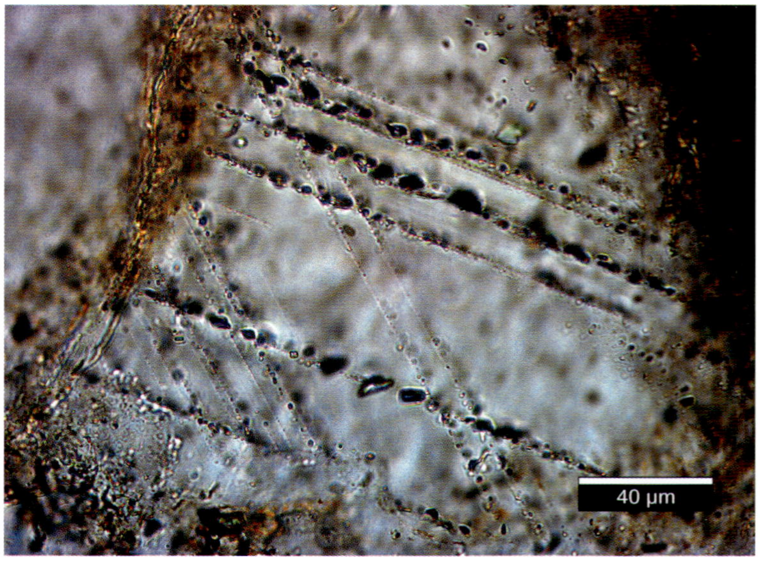

Cross-cutting sets of planar deformation features (PDFs) decorated with fluid inclusions, in quartz grain from crater 36 of the Douglas crater strewn field (photo: T. Kenkmann).

Douglas Crater Field

Wyoming / United States

latitude = 42°40‘ N / longitude = 105°28‘ W / diameter = max. 100 m / age = 280 Myr

scale = 1:71 000 / scene = 12.9 × 17.2 km²

0 0.71 3.55 km

N

1405 meter 1955

Panorama view of crater 1 of the Douglas crater strewn field, observed from its south-southwest crater rim. This elliptical crater measures 60 × 41 m. (photo: T. Kenkmann).

ogy of a simple, bowl-shape impact crater with a raised rim and overturned flap, apparently continuous ejecta blankets, and ovoid shapes oriented southeast to northwest coincident with the strike of the strewn field. Based on crater structure and ejecta distribution, an impact from the southeast towards the northwest has been inferred. Other craters exhibit lesser quality of the described features and are variably rated as "possible" or "probable".

Samples from the first discovered 60 m diameter crater display cross-cutting planar deformation features (PDFs) of various crystallographic orientations in quartz. PDFs and planar fractures (PFs) have so far been documented from 14 of the crater pits and additionally from a nearby stratigraphic profile that comprises a continuous section from the Casper Formation to the overlying Opeche Member of the Goose Egg Formation. Quartz grains exhibit extensive fracturing. The deformation affected rounded quartz grains but not quartzitic overgrowths on them. This implies that the impact occurred in unconsolidated sand prior to diagenesis. Fractures and PDFs are massively decorated with fluid inclusions suggesting that pore space was filled with water during impact. Shock lithification made craters more resistant to erosion, so that some became partial pedestal craters.

The craters have been exhumed from beneath the red-colored shales of the Permo-Triassic Goose Egg Formation and are formed in very resistant Casper sandstone. Strata were tilted by 15° east-northeast during the Laramide orogeny during the Upper Cretaceous and Paleogene. The craters occur only in a narrow stratigraphic band along strike at the top of the Casper Formation. It is assumed that there was originally an elliptical strewn field pattern but possible additional craters on the west side of the ellipse have been eroded away, and additional ones on the east side of the ellipse may lie below the Goose Egg Formation. Determination of the actual number of crater pits and their proofing as impact craters is currently underway. Additional craters may extend the length of the Douglas Strewn Field. The impact age is inferred from the age limits of regionally occurring strata, to 280 Myr.

Five circular depressions on Sheep Mountain were first recognized in a master thesis by A. Spelman in 1959, who did not interpret their origin. In an undergraduate fellowship report by H. Kastning and P. Huntoon of 1996 these depressions were interpreted as small impact craters – however, without providing proof for the validity of this hypothesis. This report stimulated a survey of these craters in 2017 by Thomas Kenkmann, Kent Sundell, and Douglas Cook, in the course of which shock metamorphism could be demonstrated. Ongoing research on this crater strewn field is aimed at constraining the size of the strewn field, the nature of the projectile, and the mechanism of bolide destruction. Future work will also investigate a potential link tc the 7 km diameter **Cloud Creek** impact crater that lies exactly down-range on the inferred impact trajectory some 120 km northwest of the Douglas crater strewn field.

Location/Access

The crater field is located on private land on the western slope of Sheep Mountain, the easternmost, north-south trending mountain range of the Rocky Mountains, some 13 km west of Douglas, Wyoming.

Selected References

Kastning, H., and Huntoon, P. W. (1996). Cluster of five small Pennsylvanian meteorite impact craters on Sheep Mountain near Douglas, Wyoming. In: NASA's Wyoming Space Grant Fellowship Program 1995-6 Space Science Research: Graduate and Undergraduate Fellowship reports, Ciardi, T. M., and Johnson, P. E. (eds.), Wyoming Space Grant Consortium, University of Wyoming, pp. 57–64.

Kenkmann, T., Sundell, K. A., and Cook, D. (2018a). Exhumed Paleozoic impact crater strewn field near Douglas, Wyoming, USA: Evidence from microstructural analysis, satellite, and drone imagery. 49th Lunar and Planetary Science Conference, Lunar and Planetary Institute, Houston, Texas, Abstract, #1469.

Kenkmann, T., Sundell, K. A., and Cook, D. (2018b). Evidence for a large Paleozoic impact crater strewn field in the Rocky Mountains. Scientific Reports, Vol. 8, Article #13246.

Schulze, T., Kenkmann, T., Poelchau, M. H., Sundell, K. A., and Cook, D. (2019). Douglas impact crater strewn field, WY, USA: a progress report. Large Meteorite Impact and Planetary Evolution VI, Brasilia, Abstract #5036.

Composite drone imagery of the southeastern Douglas crater strewn field. For details see Kenkmann et al., 2018b (drone imagery recorded by K.A. Sundell, Casper College, Wyoming).

Eagle Butte
Alberta, Canada

Thomas Kenkmann

The complex impact structure Eagle Butte is covered by glacial till, but still displays a vague morphological expression due to some hills arranged in a semi-circular fashion. The structure's diameter of 15 km is inferred from the extent of enhanced rock faulting.

Target rocks are Lower and Upper Cretaceous siliciclastic rocks of various grain sizes and coals resting on Mississippian beds. The Cretaceous rocks are strongly deformed. The stratigraphic uplift in the center is of the order of 300 m, but varies strongly depending on which strata are considered. The about 5.5 km wide central uplift is surrounded by a ring syncline, in which strata occur some 100 m below their original stratigraphic level outside the crater structure. The crater rim is intensely faulted. Shatter cones have been found at the surface near the peak of the central uplift. Planar deformation features (PDFs) were encountered in brecciated core samples.

The area of Eagle Butte is a target for the petroleum industry. The central uplift of the structure is marked by a +1.5 mGal positive Bouguer gravity anomaly. Many reflection seismic lines cross the structure, but only one has been published in the scientific literature. This profile and a number of boreholes constrain the widths of central uplift and annular trough, and the diameter of the crater rim. They indicate that the disturbance decays at a depth of about 1.3 km. Based on stratigraphy, the age of the impact is younger than Upper Cretaceous and older than the Pleistocene glaciation.

A structural anomaly at Eagle Butte has been known since the 1920s. Mike Dence suggested an impact origin in 1972 but the proof of this hypothesis via documentation of shock effects was only achieved in 1985.

Location/Access

Eagle Butte is located in southern Alberta. It is named after a nearby community. The Eagle Butte Road crosses the structure. It links Alberta Provincial Highway 41, the Buffalo Trail, in the south with the Trans-Canada Highway close to the town of Medicine Hat, some 40 km north of Eagle Butte.

Selected References

Grieve, R. A. F. (2006). Impact Structures in Canada. Geological Association of Canada, Geo¯ext No. 5, 210 pp.

Sawatzky, H. B. (1976). Two probable late Cretaceous astroblemes in Western Canada – Eagle Butte, Alberta, and Dumas, Saskatchewan. Geophysics, Vol. 41, pp. 1261–1271.

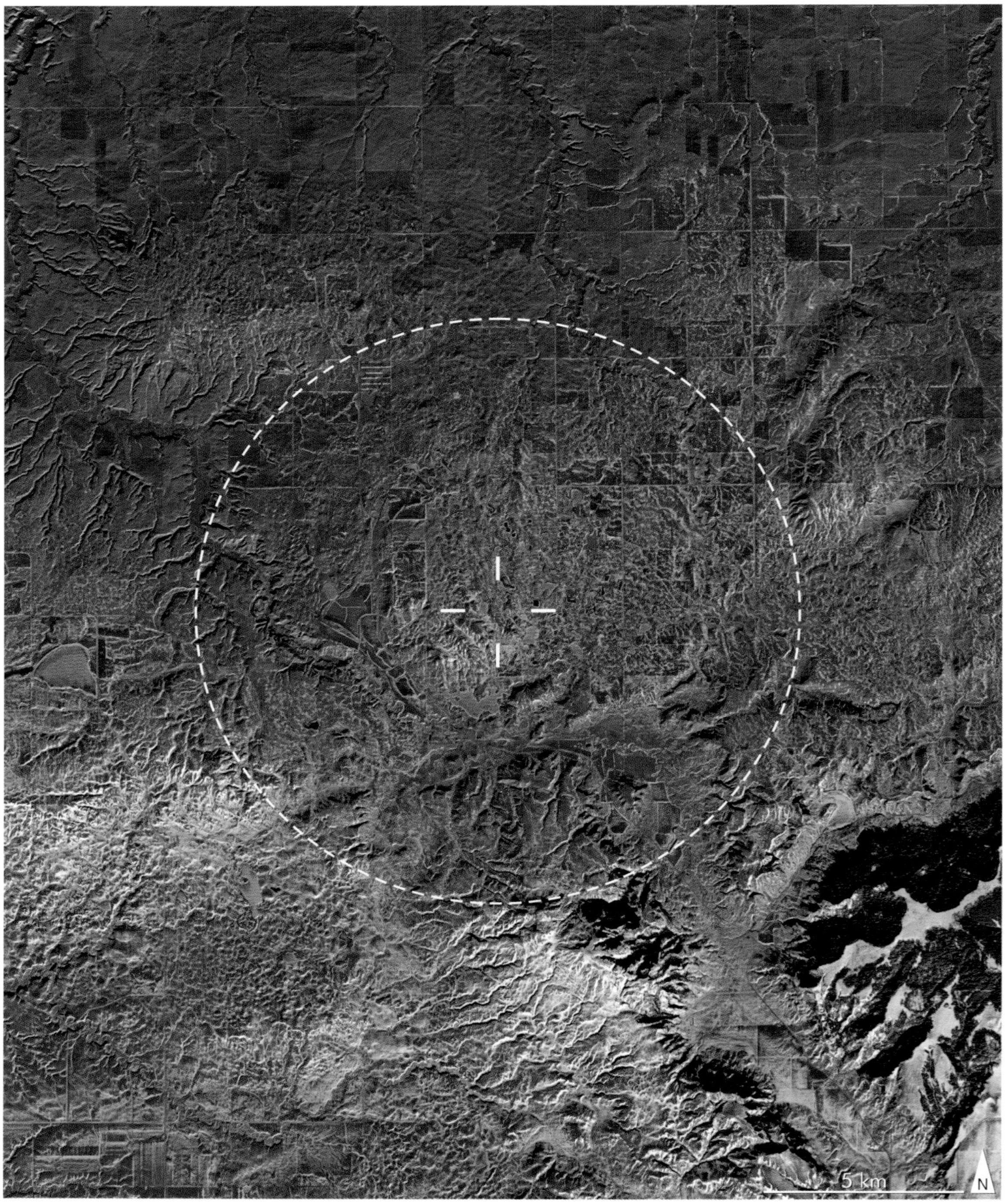

The Eagle Butte impact structure on January 11, 2019 as a composite image based on TanDEM-X DEM and Copernicus Sentinel-2 (channel 2/3/4) data. Cross hairs and a dashed circle indicate position and extent of the buried structure.

Eagle Butte

Alberta / Canada

latitude = 49°42‘ N / longitude = 110°30‘ W / diameter = 15 km / age = < 66 Myr

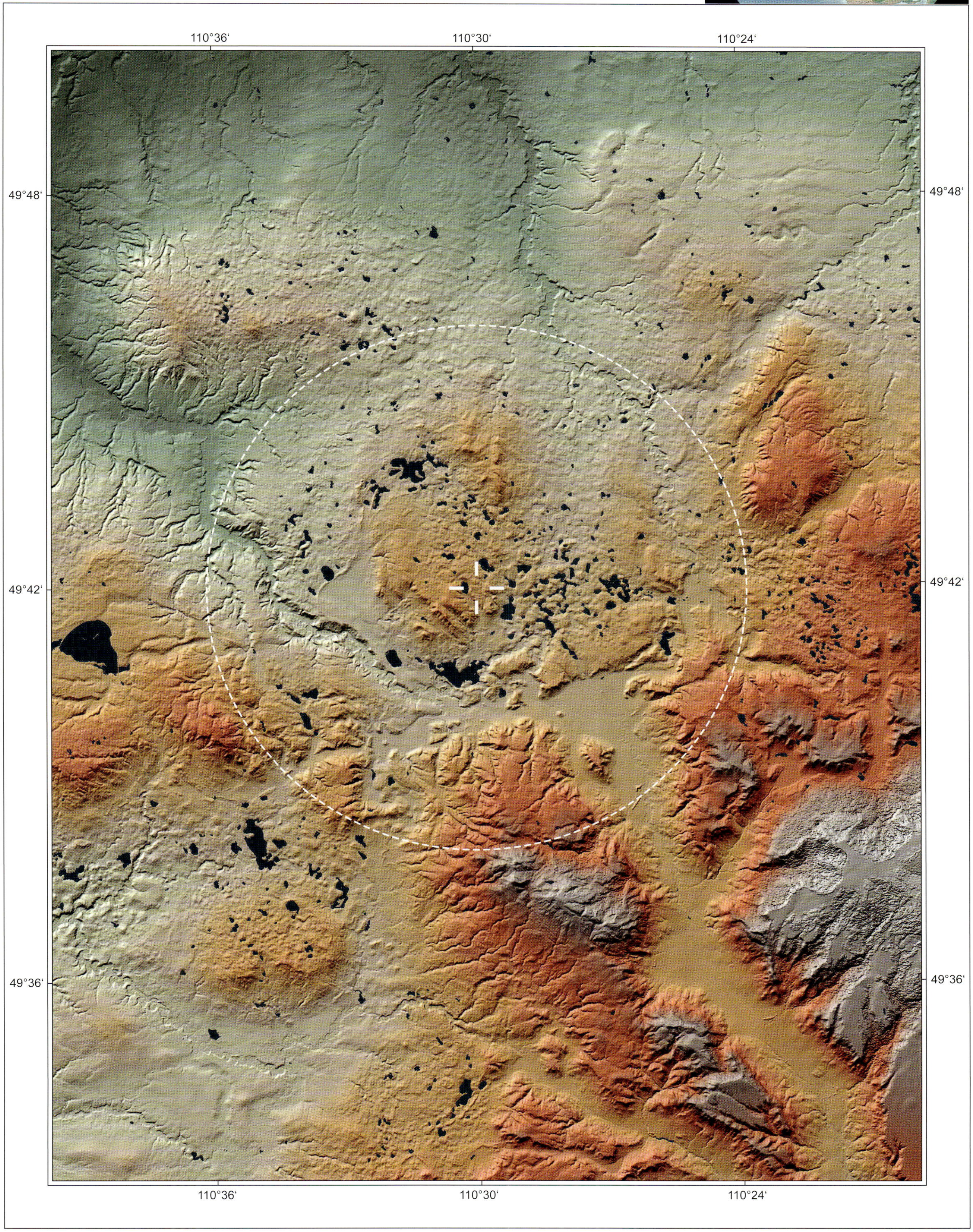

scale = 1 : 133 000 / scene = 23.9 × 31.9 km²

0 1.33 6.65 km

N

825 meter 1480

Elbow
Saskatchewan, Canada

Thomas Kenkmann

Although completely buried, the complex Elbow impact structure of 8 km diameter has a subdued morphological expression in the form of a 30 m high hill with a flat top that approximately corresponds to the position of the central part of the crater.

Target rocks affected by the impact comprise Silurian, Devonian, and Mississippian strata, dominated by carbonates. Quartz grains of a carbonate-dominated breccia were shown to contain planar deformation features (PDFs). A seismic cross-section and four boreholes revealed a central dome of uplifted and brecciated strata surrounded by a gentle ring syncline. The central uplift is about 3 km wide and associated with a stratigraphic uplift of circa 300 m. A positive +2 mGal Bouguer gravity anomaly corresponds to the extent of the central uplift.

The age of Elbow is not well constrained. Jurassic strata are unaffected by the impact. Devonian beds, however, are brecciated and, thus, an age between 201 and 358 Myr is favored for the impact.

The Elbow structure attracted the oil exploration industry in the 1940s, who drilled some boreholes and undertook a seismic survey. In the 1960s the structure was regarded as an anticlinal dome or a salt diapir. It was not until 1998 that Richard Grieve documented shock features that demonstrated the impact origin.

Location/Access

The Elbow impact structure is located south of the meandering South Saskatchewan River, which is dammed to form a winding lake. Access to the structure is via the Riverhurst village to the southwest on Sasketchewan Highway 42. Elbow has been given its name from the town of Elbow, about 20 km northeast, at the other side of the lake.

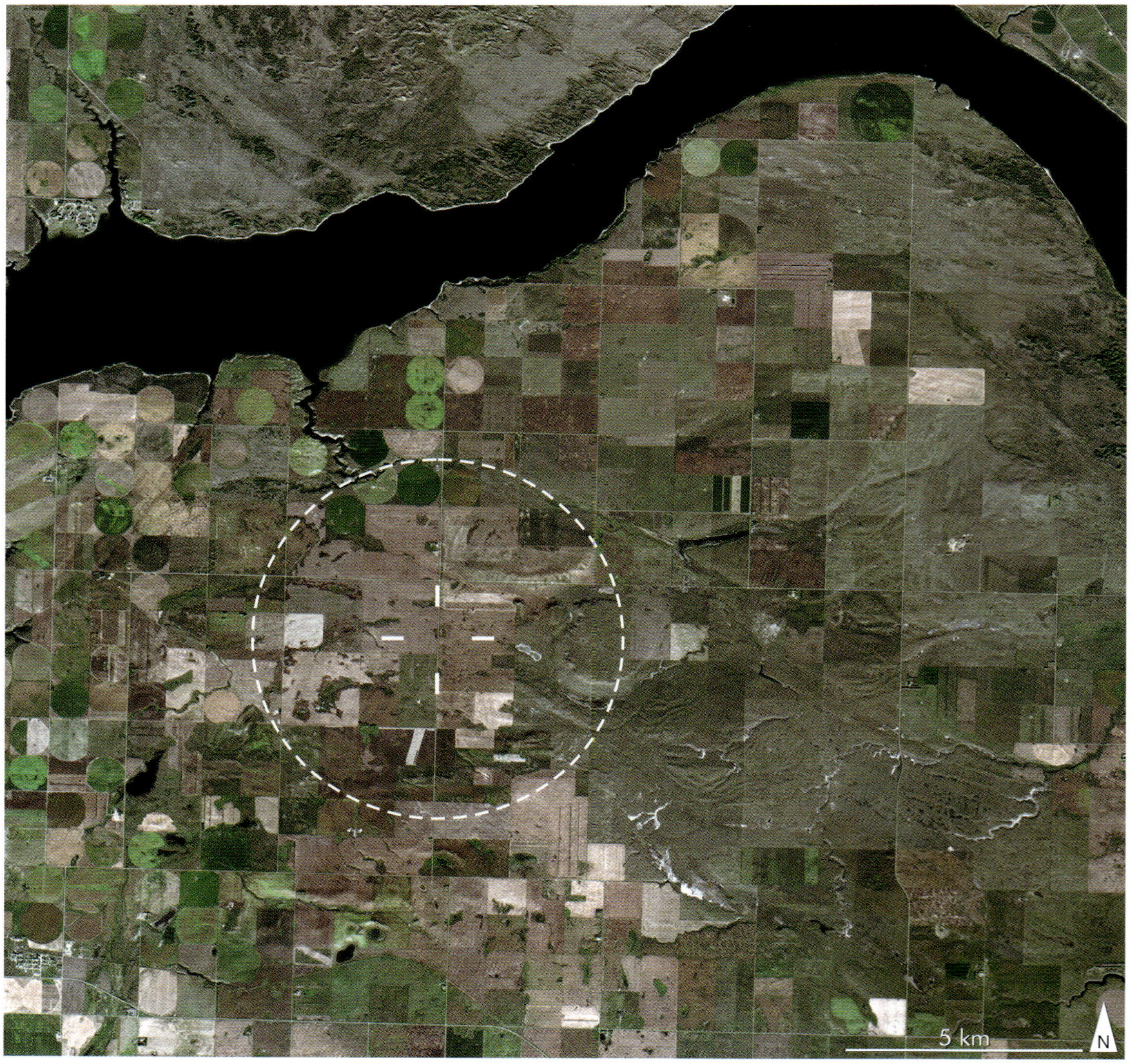

The Elbow impact structure on September 15, 2019 as a composite image based on TanDEM-X DEM and Copernicus Sentinel-2 (channel 2/3/4) data. Cross hairs and a dashed circle indicate position and extent of the buried structure.

Selected References

De Mille, G. (1960). Elbow structure of south-central Saskatchewan. Journal of the Alberta Society of Petroleum Geologists, Vol. 8, pp. 154–162.

Grieve, R. A. F. (2006). Impact Structures in Canada. Geological Association of Canada, GeoText No. 5, 210 pp.

Elbow

Saskatchewan / Canada

latitude = 50°58‘ N / longitude = 106°45‘ W / diameter = 8 km / age = 201 - 358 Myr

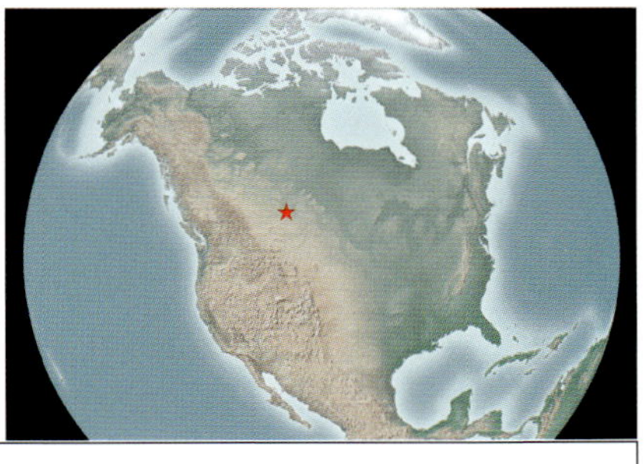

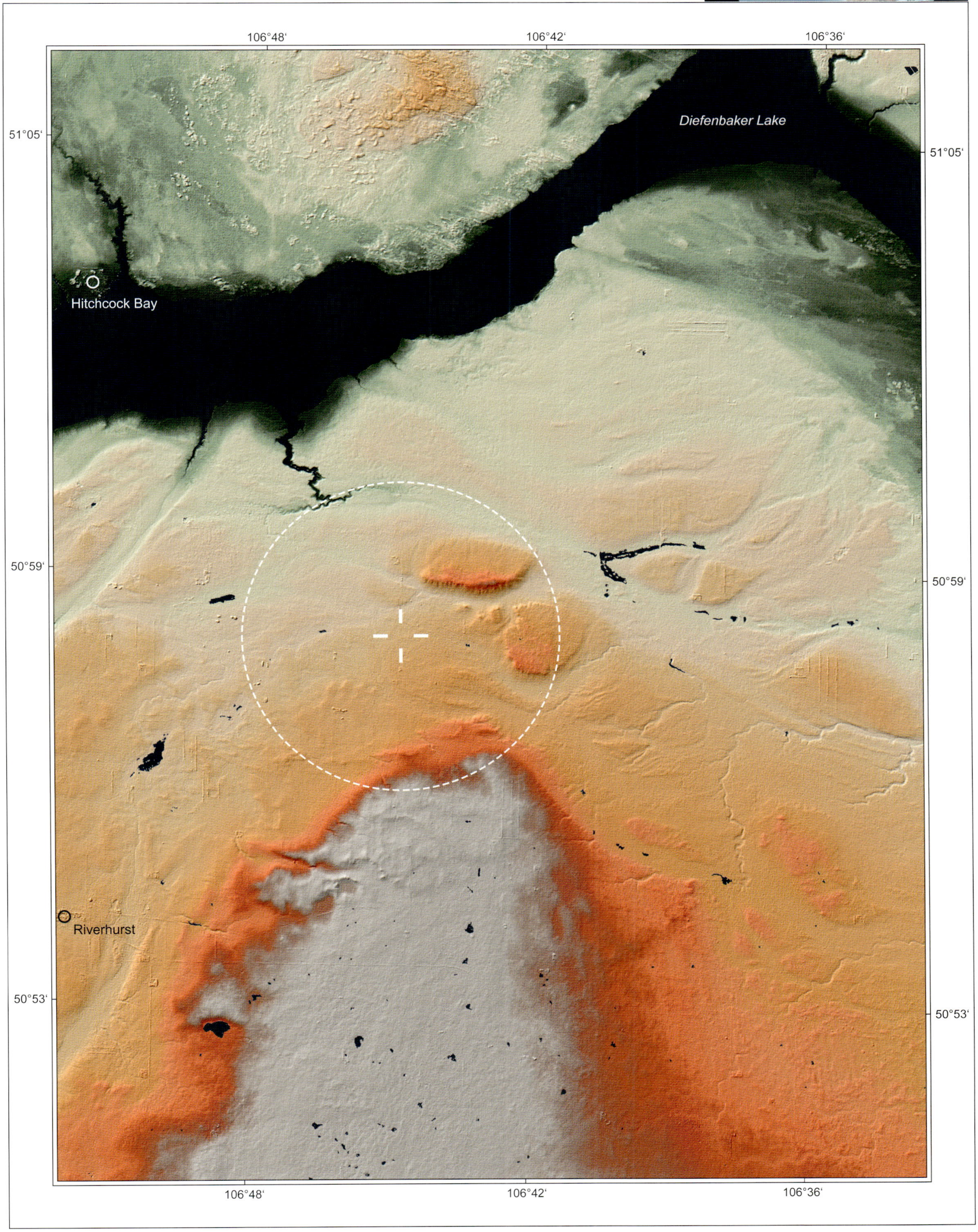

scale = 1:121 000 / scene = 21.8 × 29.1 km²

0 1.21 6.05 km

N

545 meter 660

Flynn Creek
Tennessee, United States

Thomas Kenkmann

The 3.8 km diameter Flynn Creek structure is among the smallest complex terrestrial impact craters. The crater is partly exposed and has a subdued morphology.

The target rocks for this impact were Ordovician sediments. The impact occurred in flat-lying limestones, dolostones, and shales. Target stratigraphy includes the Lower Ordovician Knox Group and the Upper Ordovician Leipers-Catheys Formation belonging to the Nashville Basin. In the course of the impact event, the crater cavity was filled by 40 m of polymict lithic breccia, composed predominantly of strata of the uppermost 150 m of the target (limestone and dolostone), which is followed by coarse-grained dolomitic sandstone and fine-grained dolomite. Dark marine mudstones of Upper Devonian age, known as the Chattanooga Shale, reach a thickness of 60 m above the crater infill, whereas elsewhere they are usually only about 6 m thick. This marine sedimentation commenced immediately after impact and suggests that the impact most likely occurred in very shallow water in a coastal environment. The shale is overlain by Mississippian strata that now cap parts of the Flynn Creek area.

The crater rim is partly exposed and indicates outward-directed thrust movement and vergent folding, with uplift of 50 m. Faults in the rim often grade into tight anticlinal folds or into monomict breccia occurrences. This deformation can be attributed to the excavation of the transient crater cavity. Further inside the structure occur fault-bounded terraces. Deformation in the ring syncline below the breccia lens is a thin-skin shallow feature. Boreholes show that the impact-induced deformation is nearly absent about 100 m beneath the crater floor below the 40 m thick breccia lens mentioned above. Strata in the ca. 0.75 km wide central uplift have been stratigraphically raised by about 450 m above their normal stratigraphic positions. The southern two-thirds of the central uplift are buried beneath post-impact strata. A large cave developed inside the central uplift provides insight into the megablock nature of the central uplift. While the periphery of the central uplift is built by large megablocks thrust onto each other, the interior of the central uplift is composed of smaller blocks. Rocks of the central uplift display shatter cones, and planar deformation features (PDFs) and planar fractures (PFs) in quartz, as well as very dense micro-twinning in calcite. Radiometric calibrations of Devonian Conodont zones suggest that the impact took place about 382 Myr ago.

A slightly brecciated, 11 × 11 × 7 cm sized piece of Flynn Creek limestone (photo courtesy: Ralf-Thomas Schmitt, Museum für Naturkunde Berlin).

The structure was first interpreted as a sinkhole. Later, a "cryptovolcanic" or "cryptoexplosion" origin was favored. Dave Roddy of the Astrogeology Division of the U.S. Geological Survey in Flagstaff investigated Flynn Creek in great detail and confirmed its impact origin by the documentation of shock deformation features in 1977. Based on drill core data he compiled a subsurface contour map that revealed the overall shape of the crater structure.

Location/Access

The Flynn Creek impact structure is located in Jackson County, Tennessee, about 100 km east of Nashville. Interstate 40 runs eastward from Nashville to Cookville, where local roads will bring the visitor into the structure. Although there is limited outcrop in the area, there are exposed faults, folds, and large outcrops of impact breccia within the crater.

Selected References

Milam, K. A., Deane, B., King, P. L., Lee, P. C., and Hawkins, M. (2006). From the inside of a central uplift: the view from Hawkins impact cave. 37th Lunar and Planetary Science Conference, Lunar and Planetary Institute, Houston, Texas, Abstract #1211.

Roddy, D. J. (1968). The Flynn Creek crater, Tennessee. In: Shock Metamorphism of Natural Materials, French, B. M. and Short, N. M., (eds.), Mono Book Corp., Baltimore, pp. 291–322.

Roddy, D. J. (1977a). Tabular comparisons of the Flynn Creek impact crater, United States, Steinheim impact crater, Germany, and Snowball explosion crater, Canada. In Impact and Explosion Cratering, Roddy, D. J., Pepin, R. A., and Merrill, R. B. (eds.), Pergamon Press, New York, pp. 125–161.

Roddy, D. J. (1977b). Pre-impact conditions and cratering processes at the Flynn Creek crater, Tennessee. In: Impact and Explosion Cratering, Roddy, D. J., Pepin, R. A., and Merrill, R. B. (eds.), Pergamon Press, New York, pp. 277–308.

Flynn Creek

Tennessee / United States

latitude = 36°17‘ N / longitude = 85°40‘ W / diameter = 3.8 km / age = 382 Myr

scale = 1: 90 000 / scene = 16.3 × 21.7 km²

0 0.90 4.50 km

N

110 meter 350

Glasford
Illinois, United States

Thomas Kenkmann

Glasford is a large buried and strongly deformed dome structure of impact origin. The crest of the apparent central uplift lies 350 m below the present surface, with the dome having a diameter of about 4 km. This could indicate that the crater diameter is possibly 2–3 times larger. The structure has been investigated by geophysics and drilling.

The recovered cores from a deep well drilled at the crest of the dome encountered at the base of the borehole deformed and tilted blocks of Cambrian formations that were uplifted about 300 m above their normal stratigraphic position. They comprise intensely contorted beds of red and green Ordovician shales, cataclastically deformed sandstones, and dolomites. Dips appear to be of random orientation and attain values up to 90° throughout much of the section. The disordered and brecciated section is overlain by about 500 m of younger Paleozoic strata that are completely undeformed. This suggests that the structure was formed around 450 Myr ago, an age which has recently been confirmed by the dating of graptolite fossils that yielded 455 ± 2 Myr. The post-impact succession overlying the breccia is thinned and arched over the dome and indicates that the central uplift affected the subsequent sedimentation as a local high by differential compaction of the lithologies of the impact crater.

The Glasford structure was originally mapped as a dome containing some coal formations. It had been regarded as a "cryptoexplosion" structure. John McHone and co-workers confirmed the impact origin by the documentation of shatter cones in 1986. Currently, the structure is used for storage of natural gas.

Location/Access

Glasford is situated 6 km north of the Illinois River and 10 km southwest of the city of Peoria with the General Wayne A. Downing international airport. The area is mostly in agricultural use.

Selected References

Buschbach, T. C., and Ryan, R. (1963). Ordovician explosion structure at Glasford, Illinois. Bulletin of the American Association of Petroleum Geologists, Vol. 47, pp. 2015–2022.

Dietz, R. S., and McHone, J. F. (1991). Astroblemes recently confirmed with shatter cones. Meteoritics, Vol. 26, pp. 332.

McHone, J. F., Sargent, M. L., and Nelson, W. J. (1986). Shatter cones in Illinois: Evidence for meteoritic impacts at Glasford and Des Plaines (abstract). Meteoritics, Vol. 21, p. 446.

Monson, C. C., Sweet, D., Segvic, B., Zanoni, G., Balling, K., Wittmer, J. M., Ganis, G. R., and Cheng, G. (2019). The Late Ordovician (Sandbian) Glasford structure: A marine target impact crater with a possible connection to the Ordovician meteorite event. Meteoritics and Planetary Science, Vol. 54, pp. 2927–2950.

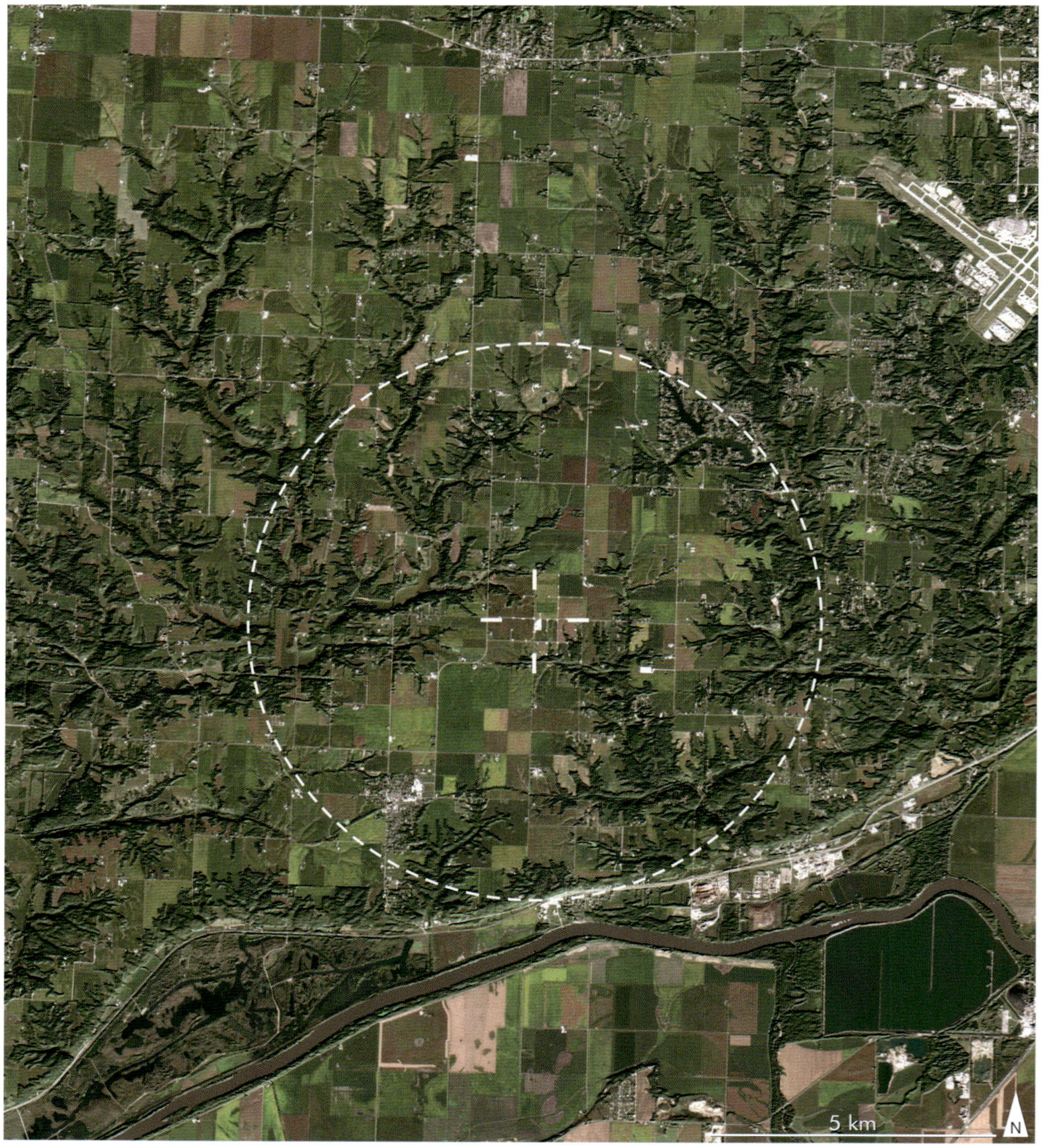

The Glasford impact structure on September 26, 2019 as a composite image based on TanDEM-X DEM and Copernicus Sentinel-2 (channel 2/3/4) data. Cross hairs and a dashed circle indicate position and extent of the buried structure.

Glasford

Illinois / United States

latitude = 40°36'06" N / longitude = 89°47'06" W / diameter = 10 km / age = 455 ± 2 Myr

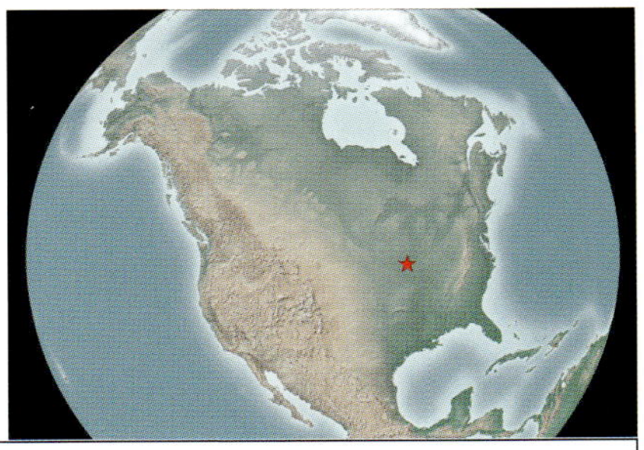

scale = 1:97 000 / scene = 17.4 × 23.3 km²

0 0.97 4.85 km

N

80 meter 240

Glover Bluff
Wisconsin, United States

Thomas Kenkmann

Glover Bluff is a partly exposed but poorly investigated crater of Ordovician age in Wisconsin, west of Lake Michigan. Glover Bluff comprises three hills, each of which rises a little more than 30 m above the surrounding area and exposes heavily fractured Lower Ordovician dolomite. Whether or not these outcrops can be regarded as the crest of a central uplift is still unclear. Exposure of sandstone strata in an arcuate arrangement south, southwest, and southeast of the bluff suggests a diameter of the impact structure of roughly 10 km, with the Bluff being situated in its approximate center. The geomorphology of the area was sculpted by glacial drift, and the region is therefore covered with glacial till. Dead ice kettle holes are abundant in the area.

The impact crater was formed in Cambrian sandstones and Lower Ordovician dolomites. The structure was most likely filled by Ordovician St. Peter Sandstone. This formation occurs from north to south from Minnesota to Arkansas and east-west from Illinois to Nebraska and South Dakota. It was deposited at 455–459 Myr ago. Glass-bearing bombs have been reported to occur stratiform in the St. Peter Sandstone. They were interpreted as ejecta with a possible link to the Glover Bluff impact. That would also suggest that the impact would have occurred into a basin under shallow water conditions. St. Peter Sandstone has been encountered in numerous water wells drilled into the Glover Bluff structure. Shatter cones occur in the dolomite of the possible central uplift. Some nice shatter cones reach a length of more than 7 cm.

An analysis of calcite grains from the impact structure and the regional environs provides an age estimate. It is suggested that the impact occurred earlier than the Alleghenian orogeny in the Late Paleozoic, 300 Myr ago. If the glass-bearing bombs within the St. Peter Sandstone originate from Glover Bluff, then the structure's age is 455–459 Myr.

The Bluff has been the site of quarry operations in Lower Ordovician dolomite. This dolomite was burned for mortar and crushed for road-building. Glover Bluff was initially considered as a tectonic graben structure until W. Read in 1983 discovered shatter cones in the dolomite of the central area.

Location/Access

Glover Bluff is located about 100 km north of the city of Madison, with which it is connected via Interstate 39. The central area of Glover Bluff is marked by the group of three hills that were quarried for dolomite.

Selected References

Ekern, G. L., and Thwaites, F. (1930). The Glover Bluff structure, a disturbed area in the Paleozoic of Wisconsin. Wisconsin Academy of Sciences, Transactions, Vol. 25, pp. 89–97.

Ernst, B., and Paulson, T. (2011). New constraints on the timing of deformation at the Glover Bluff structure near Coloma, WI, from a calcite twinning analysis (abstract). Geological Society of America Annual Meeting, Minneapolis, Minnesota, Abstracts with Programs, Vol. 43, p. 188.

Read, W. F. (1983). Shatter cones at Glover Bluff, Wisconsin. Meteoritics, Vol. 18, pp. 241–243.

Read, W. F. (1984). The circular structure at Glover Bluff: What and where it is (abstract). Meteoritics, Vol. 19, pp. 295–296.

Read, W. F. (1985). Impact bombs in Middle Ordovician marine sediments of southeastern Wisconsin (abstract). 16th Lunar and Planetary Science Conference, Lunar and Planetary Institute, Houston, Texas, pp. 687–688.

The Glover Bluff impact structure on April 19, 2019 as a composite image based on TanDEM-X DEM and Copernicus Sentinel-2 (channel 2/3/4) data. Cross hairs and a dashed circle indicate position and extent of the mostly buried structure.

Glover Bluff

Wisconsin / United States

latitude = 43°58'12" N / longitude = 89°32'18" W / diameter = 10 km / age = 455 - 459 Myr

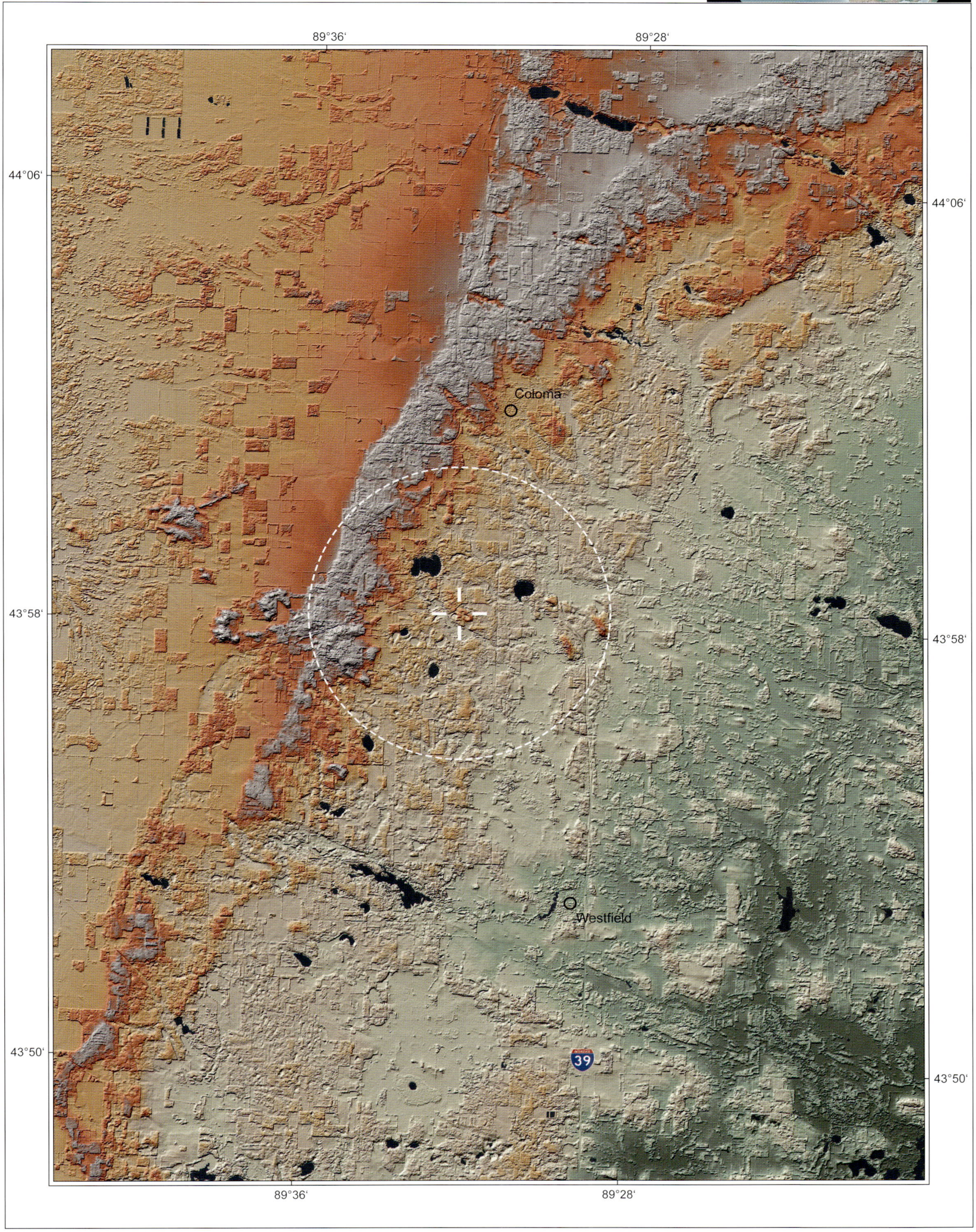

scale = 1 : 160 000 / scene = 28.7 × 38.3 km²

0 1.60 8.00 km

N

205 meter 410

Gow
Saskatchewan, Canada

Thomas Kenkmann

The deeply eroded Gow impact structure largely corresponds to the 4 km wide Gow Lake. The lake is roughly circular, except for the northeastern sector with bays that were formed by glaciation. The lake has a conspicuous island of 1 × 2 km size in its central area. The island reaches some 45 m above lake level, and water depth of the lake is about 40 m. The island is the remnant of the central uplift. With 5 km diameter, Gow is one of the smallest complex craters formed in crystalline target rocks known on Earth.

The target rocks belong to the Precambrian basement of the Churchill Province of the Canadian Shield. Fractured granites and quartzo-feldspathic gneisses are exposed around the perimeter of the lake. The island is formed mainly of lithic breccia composed of clasts of these lithologies, and by red and green impact melt rock varieties of granitic bulk rock composition. Trace element analyses indicated a small enrichment in siderophile elements and Cr, which is indicative of a possible meteoritic component. Breccias contain microscopic shock deformation features and strong thermal overprint, such as multiple sets of planar deformation features (PDFs) in quartz and ballen quartz.

A local negative gravity anomaly with an amplitude of −3 mGal is centered on the lake. It has been attributed mainly to highly fractured basement rocks occurring below the lake. Modelling of the gravity data suggested that this deformation may extend to a depth of 900 m.

A tentative minimum age of at least 100 Myr was initially proposed for this impact structure, based on its more heavily eroded nature in comparison to the **Deep Bay** impact structure, which is located only 90 km from Gow. In accordance with this, more recent ^{40}Ar-^{39}Ar dating of impact melt rock yielded an approximate age of 250 Myr.

In the early 1970s Gow Lake was studied during a regional gravity survey. Later M. Thomas and Morris Innes suggested the impact origin when they detected planar deformation features (PDFs) in quartz. Recently, Gordon Osinksi and his team revisited Gow Lake and reported a complete sequence of impactites forming the crater-fill of this impact structure.

Location/Access

The Gow crater structure is located some 90 km west of the **Deep Bay** impact structure in Canada's taiga environment. There is no infrastructure that would facilitate access. The closest settlement is the small village of Brabant Lake on Saskatchewan Highway 102, located at a distance of 60 km to the southeast of the structure.

Selected References

Grieve, R. A. F. (2006). Impact Structures in Canada. Geological Association of Canada, GeoText No. 5, 210 pp.

Osinski, G. R., Singleton, A. C., Ozaruk, A., and Hanson, J. R. (2012). New investigations of the Gow Lake impact structure, Saskatchewan, Canada: Impact melt rocks, astronaut training, and more. 43rd Lunar and Planetary Science Conference, The Woodlands, Texas, Abstract #2367.

Thomas, M. D., and Innes, M. J. S. (1977). The Gow Lake impact structure, northern Saskatchewan. Canadian Journal of Earth Sciences, Vol. 14, pp. 1788–1795.

The Gow impact structure with the homonymous lake on August 16, 2018 as a composite image based on TanDEM-X DEM and Copernicus Sentinel-2 (channel 2/3/4) data.

Gow

Saskatchewan / Canada

latitude = 56°27'23" N / longitude = 104°28'56" W / diameter = 5 km / age = 250 Myr

scale = 1:115 000 / scene = 20.8 × 27.7 km²

0 1.15 5.75 km

N

390 meter 520

Haughton
Nunavut, Canada

Thomas Kenkmann

Haughton crater, also known as *Haughton Dome*, on Devon Island in Canada's High Arctic is a well-preserved, complex impact structure with a present-day diameter of 23 km. The structure is only weakly eroded, with the amount of erosion estimated at less than 200 m. The structure can easily be recognized in air- and space-borne imagery due to the radial and concentric arrangement of faults and fractures, and the light grey color of impactites in the center of the structure that contrast with the surrounding sedimentary rocks.

The impact happened into the tectonically stable Arctic platform that comprises a 1880 m thick series of almost flat-lying, Lower Paleozoic sedimentary rocks. The sedimentary rocks overlie 2500 Myr old Archean metamorphic rocks of the Canadian Shield that crop out in the eastern part of Devon Island. They predominantly comprise granulite facies tonalitic and granitic gneisses, with intercalated metasedimentary rocks, into which felsic plutons were intruded at 1900 Myr ago. All pre-Paleozoic rocks are cut by 600–800 Myr old dolerite dikes. The sedimentary sequence ranges from the Middle Cambrian to the Ordovician-Silurian and is dominated by limestone and dolomite with subordinate amounts of evaporites, shale, and sandstone. Post-impact lacustrine sediments superpose the impactites in the northwestern and southeastern parts of the crater structure.

Haughton possesses a central uplift, but without any topographic expression. The core of the central uplift, composed of megablocks of Lower Ordovician age with vertical dips, has a width of about 2 km. A stratigraphic uplift of up to 1450 m is indicated. This core is surrounded by large fault-bounded and tilted blocks of Middle Ordovician to Upper Silurian age with moderate outward dips, and a structural ring of intensely faulted and uplifted strata that are either in vertical position or overturned. The majority of the innermost faults at Haughton dip away from the center. The extent of the central uplift is marked by the inner ring with a diameter of about 11 km. Further outside follows a depression, the axis of which has a diameter of 16 km. The depression is subdivided by blocks and radial and concentric faults. Many of the faults and folds can be explained by non-plane-strain-type deformation that is typical for impact cratering. The outer limit of this depression is defined by inward dipping, concentric normal faults, with 100 to 400 m vertical throws measured on slump blocks. These faults mark the present size of the crater structure. Structures of the northern and eastern part of the crater differ substantially from those of the southern and western parts. This asymmetry may reflect heterogeneities in the pre-crater geology or indicate a low-angle impact.

The Haughton impact structure on July 3, 2018 as a composite image based on TanDEM-X DEM and Copernicus Sentinel-2 (channel 2/3/4) data.

A variety of impact lithologies exists at Haughton. Monomict breccias represent moderately displaced target rocks. Polymict breccia with a lithic groundmass represents allochthonous ejecta and displaced material. This includes clasts of basement rocks and indicates that the transient cavity was excavated into the basement. The volumetrically dominant impact lithology that fills the central crater in an area of about 54 km^2 is a pale-grey unit that has been variably described as a polymict impact breccia with a clastic matrix or as a coherent carbonatite impact-melt rock layer, the groundmass of which contains calcite, silicate impact glass, and anhydrite. The latter interpretation is based on a high Si/Al content of calcite and liquid immiscibility microtextures between carbonate and silicate phases.

The type of projectile that created Haughton could not be unraveled from investigation of the impact melt rock lithologies. Shatter cones occur in abundance at Haughton. They can be found well developed within the area of the central uplift, within megablocks of the ejecta, and in clasts of allochthonous impact melt breccias. As carbonates form

Haughton
Nunavut / Canada

latitude = 75°23‘ N / longitude = 89°39‘ W / diameter = 23 km / age = 39 Myr

scale = 1 : 174 000 / scene = 31.3 × 41.7 km²

0 1.74 8.70 km

N

0 meter 415

A shatter cone in dolomite from the Haughton impact structure with a width of about 25 cm (Photo: T. Kenkmann).

the dominant target lithologies, classical shock indicators such as planar deformation features (PDFs), diaplectic glasses after quartz and feldspar, and high-pressure phases such as coesite are restricted to Precambrian metamorphic clasts within the breccias.

There are strong signs for impact-induced hydrothermal activity at Haughton. It resulted in the deposition of a series of alteration products such as quartz, calcite, marcasite, celestite, barite, and fluorite within cavities and fractures in the impact melt breccias, central uplift lithologies, and around the faulted crater rim.

A large negative gravity anomaly with a diameter of 24 km characterizes the Haughton impact structure. Within this overall negative anomaly occurs a local negative gravity anomaly of a few kilometers width, at the center of the structure, that corresponds to the core of the central uplift. This central negative gravity excursion is also associated with a strong positive magnetic anomaly.

Attempts to date the impact event with the fission-track method on apatite grains from shocked basement clasts yielded a Miocene age of 22.4 ± 1.4 Myr. A similar age of 23.4 ± 1.0 Myr was obtained by the argon-argon step-heating method on a granitic, impact melt-bearing basement clast. This agrees with a more recent zircon (U-Th)/He dating that provided an age of 23.5 ± 2.0 Myr. In contrast, laser-probe ^{40}Ar-^{39}Ar dating of impact melt material from shocked basement clasts yielded an older Eocene age of about 39 Myr for the Haughton impact event.

The Haughton impact structure was initially considered an evaporite dome. Michael Dence proposed an impact origin and in 1972 included Haughton in a list of possible impact structures. The impact origin for Haughton was confirmed in 1975 by Blyth Robertson and G. Mason, who discovered shatter cones and coesite-bearing gneiss clasts in impact breccia. Detailed petrographic and structural investigations have been carried out by a consortium of the University of Münster, Germany, and later by Gordon Osiniski and co-workers.

Location/Access

Haughton is located in the Canadian High Arctic on Devon Island, some 170 km east of the community of Resolute Bay located on neighboring Cornwallis Island. Visiting the structure requires comprehensive Arctic expedition planning and equipment.

Selected References

Bischoff, L., and Oskierski, W. (1988). The surface structure of the Haughton impact crater, Devon Island, Canada. Meteoritics, Vol. 23, pp. 209–220.

Grieve, R. A. F. (1988). The Haughton impact structure: Summary and synthesis of the results of the HISS project. Meteoritics, Vol. 23, pp. 249–254.

Grieve, R. A. F. (2006). Impact structures in Canada. Geological Association of Canada, GeoText No. 5, 210 pp.

Osinski, G. R., and Spray, J. G. (2001). Impact-generated carbonate melts: Evidence from the Haughton structure, Canada. Earth and Planetary Science Letters, Vol. 194, pp. 17–29.

Osinski, G. R., and Spray, J. G. (2005). Tectonics of the Haughton impact event, Devon Island, Canadian High Arctic. Meteoritics and Planetary Science, Vol. 40, pp. 1813–1834.

Osinski, G. R., Lee, P., Spray, J. G., Parnell, J., Lim, D., Bunch, T. E., Cockell, C. S., and Glass, B. (2005). Geological overview and cratering model of the Haughton impact structure, Devon Island, Canadian High Arctic. Meteoritics and Planetary Science, Vol. 40, pp. 1759–1776.

Redeker, H. J., and Stöffler, D. (1988). The allochthonous polymict breccia layer of the Haughton impact crater, Devon Island, Canada. Meteoritics, Vol. 23, pp. 185–196.

Robertson, P. B., and Mason, G. D. (1975). Shatter cones from Haughton Dome, Devon Island, Canada. Nature, Vol. 255, pp. 393–394.

Robertson, P. B. and Grieve, R. A. F. (1978). The Haughton impact structure. Meteoritics, Vol. 13, pp. 615–618.

Young, K. E., van Soest, M. C., Hodges, K. V., Watson, E. B., Adams, B. A., and Lee, P. (2013). Impact thermochronology and the age of Haughton impact structure, Canada. Geophysical Research Letters, Vol. 40, pp. 3836–3840.

Haviland

Kansas, United States

Thomas Kenkmann

The small Haviland meteorite crater had, upon formation, an elliptical outline and a size of only 17 × 10.7 m, with its long axis trending west-northwest to east-southeast. Haviland would rank among the smallest meteorite craters on Earth, if it still existed. However, the crater itself has been completely destroyed by excavation and many years of subsequent agricultural cultivation. Thus, the structure is no more identifiable, but meteorite fragments can still be found.

Haviland resulted from the fall of the Brenham meteorite. Brenham is a pallasite, a relatively rare type of stony-iron meteorite. Fragments with masses of 650 kg and 450 kg were found near Greensburg 10 km west-northwest of Haviland in 2005 and in 1949, respectively. In total, 4.3 tons of the Brenham meteorite were collected from the meteorite strewn field. The largest collection of Brenham meteorite specimens is housed at the Kansas Meteorite Museum and Nature Center in Haviland.

Haviland crater is the only known crater from the Brenham meteorite strewn field. Originally the crater was surrounded by an elevated rim. Small oxidized meteorite fragments could be recovered along the eastern and southern parts of the crater rim. During excavation of the crater, numerous meteorite pieces were found. The largest, weighing 39 kg, was recovered from a penetration funnel at a depth of 3 m, against the west wall of the crater. The terrestrial age of the Brenham meteorite, which also marks the age of crater formation, has been derived from cosmogenic ^{14}C in the silicate phase of a Brenham sample to amount to 20 000 ± 2000 years.

The Brenham meteorites were first recognized on Kimberly farm near the town of Haviland, Kiowa County, Kansas, in 1882. Even though searching for meteorites continued, the crater pit was overlooked for a long time. Only in 1933, Harvey Nininger and Jesse Figgins investigated it, at a time when the crater had already been strongly degraded.

Haviland

Kansas / United States

latitude = 37°34'57" N / longitude = 99°09'49" W / diameter = 17 m / age = 20 000 ± 2000 yr

scale = 1 : 64 000 / scene = 11.4 × 15.3 km²

0 0.64 3.20 km

N

595 meter 690

A 3.4 kg slice of the Brenham meteorite, on display at The University of Arizona Mineral Museum (image courtesy: The University of Arizona Mineral Museum).

Oxidized pieces of the Brenham meteorite, termed "meteorodes" by Harvey Nininger, from The University of Arizona Mineral Museum (image courtesy: The University of Arizona Mineral Museum).

Location/Access

The location of the former Haviland crater is situated on privately owned farmland 6 km southwest of the village of Haviland in Kiowa County. From Wichita, the next larger city, it is a 160 km drive on U.S. Highway 54 westwards to reach the site of the former crater pit.

Selected References

Hodge, P. W. (1979a). The location of the site of the Haviland meteorite crater. Meteoritics, Vol. 14, pp. 233–234.

Hodge, P. W. (1979b). Meteoritic material in the soil near two meteorite craters. Meteoritics, Vol. 14, pp. 422–423.

Honda, M., Caffee, M. W., Miura, Y. N., Nagai, H., Nagao, K., and Nishiizumi, K. (2002). Cosmogenic nuclides in the Brenham pallasite. Meteoritics and Planetary Science, Vol. 37, pp. 1711–1728.

Nininger, H. H., and Figgins, J. D. (1933). The excavation of a meteorite crater near Haviland, Kiowa County, Kansas. American Journal of Science, Vol. 28, pp. 312–313.

Holleford

Ontario, Canada

Thomas Kenkmann

Holleford, although it is a buried impact structure, has a well visible morphological expression. The structure is defined by a gentle, circular depression of 2.35 km width and 30 m depth. The structure is located within regionally occurring Ordovician limestone. These Ordovician limestones are younger than the crater and also fill the crater cavity. Due to variations in thickness, the compaction of the sedimentary cover was strongest in the central part of the crater and led to the observed down-sagging of the topography.

Target rocks are Proterozoic quartzites, marbles, and diabase dikes that were affected by the Grenville orogeny between about 1200 and 1000 Myr ago. The age of the impact is not well constrained, but it must have occurred prior to the deposition of the Ordovician limestones that filled the crater cavity.

About 930 m of drill core were recovered from three boreholes into the crater. One hole was drilled about 430 m east of the center of the structure through 228 m of post-impact sediments, before impact breccias were cored. This drilling did not reach the crater floor. Another borehole was drilled 760 m off the crater center and recovered the Ordovician sedimentary infill, impact breccias, and the shattered basement rocks below. Impact breccias contain shocked quartz grains with planar deformation features (PDFs) but a systematic study of the shock deformation has not been performed.

A geophysical survey revealed a negative residual Bouguer gravity anomaly of −2 mGal. This anomaly can be modeled with a sedimentary fill of, at maximum, 260 m thickness and an impact breccia lens of 240 m thickness – which correlates with the drilling results.

Holleford was recognized back in 1955 in the course of systematic remote sensing screening for circular structures of possible impact origin by the Dominion Observatory in Ottawa. Subsequent drilling in 1956 and 1957 of the crater exposed the breccias that provided the proof of the impact origin in the form of shock metamorphosed minerals.

Location/Access

The Holleford impact structure, named after the community of the same name, is situated in the agriculturally used and partly forested landscape of southeastern Ontario, about 27 km northwest of Lake Ontario and the town of Kingston. The crater structure lies on privately owned farmland. The local infrastructure provides easy access.

Selected References

Beals, C. S. (1960). A probable meteorite crater of Precambrian age at Holleford, Ontario. Publication of the Dominion Observatory Ottawa, Vol. 24, pp. 117–142.

Grieve, R. A. F. (2006). Impact Structures in Canada. Geological Association of Canada, GeoText No. 5, 210 pp.

Robertson, P. B., Dence, M. R., and Vos, M. A. (1968). Deformation in rock forming minerals from Canadian craters. In: Shock Metamorphism of Natural Materials, French, B. M., and Short, N. M. (eds.), Mono Book Corp., Baltimore, pp. 433–452.

Holleford

Ontario / Canada

latitude = 44°27'34" N / longitude = 76°37'57" W / diameter = 2.35 km / age = > 485 Myr

scale = 1 : 80 000 / scene = 14.4 × 19.2 km²

0 0.80 4.00 km

N

125 meter 215

Île Rouleau
Quebec, Canada

Thomas Kenkmann

Île Rouleau is a small island of 1 km size in the large, elongated Lac Mistassini in central Quebec. The island's shape is round, except for a southern tail, which is formed by glacial till deposits. Steep cliffs exist at the northern and western shores of the lake and expose heavily deformed and brecciated rocks that were eventually shown to be of impact origin. The rocks of the neighboring Manitounouc island, 1.5 km further east, and at the nearest shore of Lac Mistassini 5 km west of Île Rouleau do not show any type of deformation. From these dimensions a diameter of 4 km has been estimated for the impact structure.

The target rocks in the region are flat-lying, 1800 Myr old Proterozoic sedimentary rocks, predominantly dolomites. On Île Rouleau these rocks are strongly faulted, tightly folded, and often in upright position. Numerous breccia occurrences and breccia dikes of up to 6 m width, with dolomite and claystone fragments, occur here as well. Rare quartz grains exhibit shock effects in the form of planar deformation features (PDFs). Shatter cones are abundantly formed in the target dolomite and have been particularly found on the eastern and southwestern shores of the island. The island is interpreted as an expression of the central uplift of the crater that is otherwise not exposed. Neither geophysical nor geochemical data are available for the crater, and the impact age is only roughly estimated at less than 300 Myr based on the assumption that a crater of this small size would not be able to sustain erosion for a longer period.

The impact structure was discovered in the course of regional mapping in 1973 by J. Caty. Its name refers to the island where the impactites are exposed.

Location/Access

The nearest infrastructure is the town of Mistissini at the southern end of Lac Mistassini, 30 km south of Île Rouleau. The island is accessible by boat from this community. Mistassini can be reached via Quebec route 167, which starts at Lac Saint-Jean and continues to Lac Albanel, west of Lac Mistassini. From route 167 a 16 km drive leads to Mistissini town.

The Île Rouleau impact structure on May 29, 2018 as a composite image based on TanDEM-X DEM and Copernicus Sentinel-2 (channel 2/3/4) data.

Selected References

Caty, J. L., Chown, E. H., and Roy, D. W. (1976). A new astrobleme: Île Rouleau structure, Lake Mistassini, Quebec. Canadian Journal of Earth Sciences, Vol. 13, pp. 824–831.

Grieve, R. A. F. (2006). Impact Structures in Canada. Geological Association of Canada, GeoText No. 5, 210 pp.

Île Rouleau

Quebec / Canada

latitude = 50°40‘53“ N / longitude = 73°52‘49“ W / diameter = 4 km / age = < 300 Myr

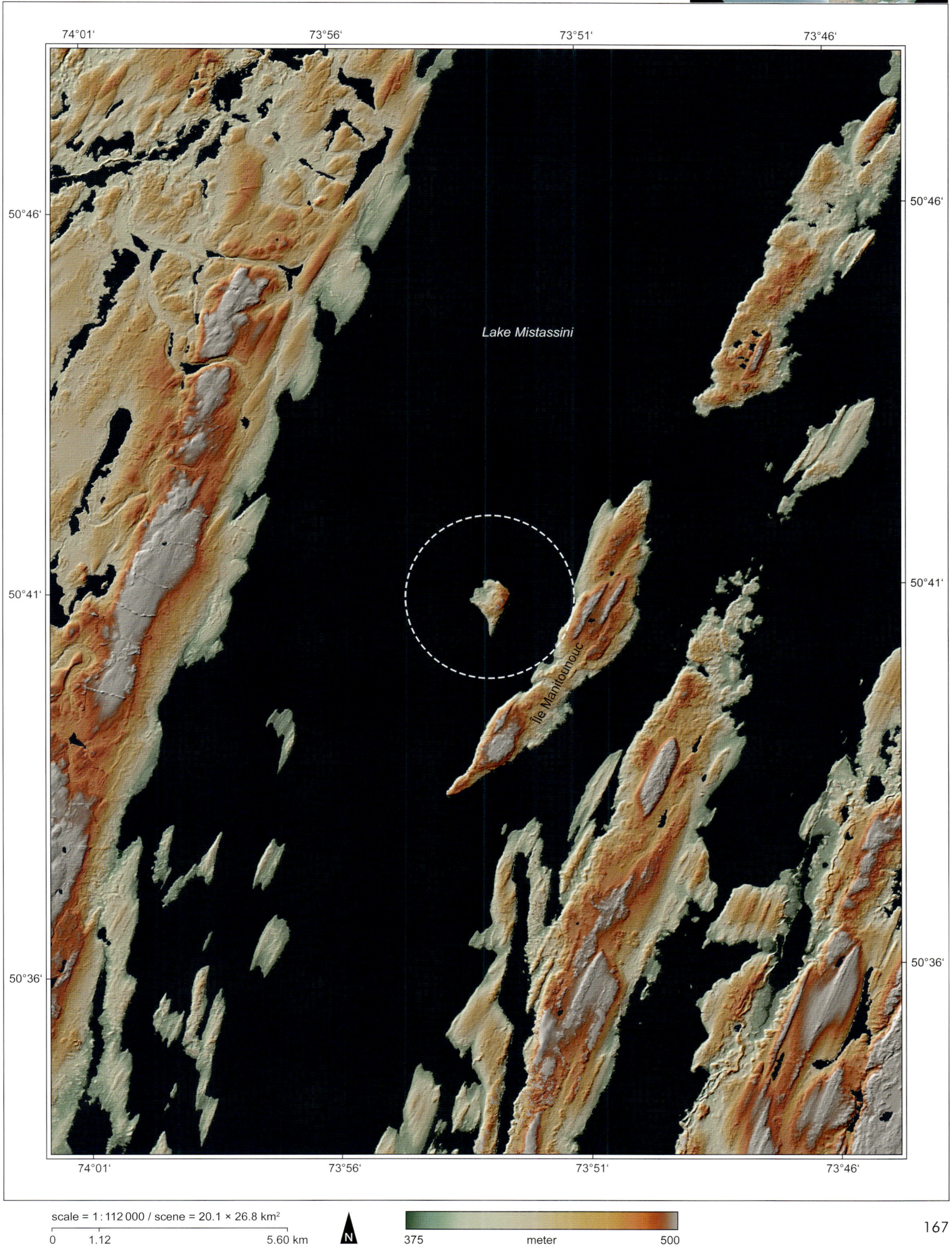

Kentland
Indiana, United States

Thomas Kenkmann

Kentland is a 12.5 km diameter complex impact structure. Morphologically the structure has only a subdued appearance. However, a large quarry in the core of the central uplift provides insights into the geological structure of this impact structure.

Kentland was formed in a geologically stable area. Precambrian basement rocks including granite and rhyolite lie 1680 m below the surface. The overlying sedimentary sequence comprises sandstone, carbonate, and shale formations of Cambrian and Ordovician age, and more than 700 m of Ordovician to Pennsylvanian strata are known to have been involved in the Kentland structure. Most of the impact structure is covered with glacial till that locally reaches a thickness of up to 40 m.

The central uplift has a diameter of about 4 km, and the stratigraphic uplift is more than 600 m, so that Lower Ordovician to Silurian rocks occur in outcrop. A ring depression of 1.5 to 2.0 km width, containing Pennsylvanian coals, surrounds the central uplift. A subdued structural high, elevated by about 15 m, occurs at a radial distance from the center of 6.2 km and is interpreted as the crater rim anticline. The central quarry exposes a large, steeply plunging to vertical synclinal fold and numerous smaller blocks dissected by high-angle normal and reverse faults.

The central uplift involves six, 300–1000 m large, and internally bent or folded megablocks of Ordovician strata in mostly upright position, and abundant smaller, fault-bounded blocks. In the innermost part of the central uplift the stratigraphic context could not be resolved, as the block size has decreased further. Sandstone and shale formations appear to be more ductilely deformed than more competent carbonates. The periphery of the central uplift consists of Silurian beds with attitudes of only 20–30°. Monomict and polymict breccias, as well as abundant shatter cones, are exposed in the quarry. Shatter cones occur in all lithologies but are most abundant in limestones. Shatter cone axes are at a high angle to bedding surfaces and point upwards after restoration into their presumed initial position. Petrographic studies of quartz grains from the breccias and other units show that less than one percent of quartz grains display basal planar deformation features (PDFs). From this a shock pressure of 5–10 GPa can be derived for the exposed rock of the central uplift, which indicates that Kentland must have been subjected to considerable erosion since the time of impact. Further shock microstructures include planar fractures (PFs) and feather features (FFs) in quartz.

Strongly tilted and multiply-faulted dolomite and limestone beds of Ordovician age in the central uplift of Kentland impact structure (photo: T. Kenkmann).

The Newton County stone quarry operates in carbonate rocks of the center of the intensely deformed central uplift of the Kentland impact structure (photo: T. Kenkmann).

Geological information on the moat and crater rim comes from drill cores. Generally, the dip of strata is much more gentle than in the central uplift. This central uplift correlates with a positive +3.5 to +4.0 mGal gravity anomaly, which, in turn, is surrounded by a ring-shaped -1 mGal negative anomaly at 3.2 km radial distance from the center. This correlates with the ring depression. A further +0.5 mGal positive anomaly at 6.2 km from the center coincides with the crater rim.

The age of the Kentland impact structure must be younger than the Early Pennsylvanian strata that are downfaulted in the

Kentland

Indiana / United States

latitude = 40°45‘ N / longitude = 87°24‘ W / diameter = 12.5 km / age = < 107 Myr

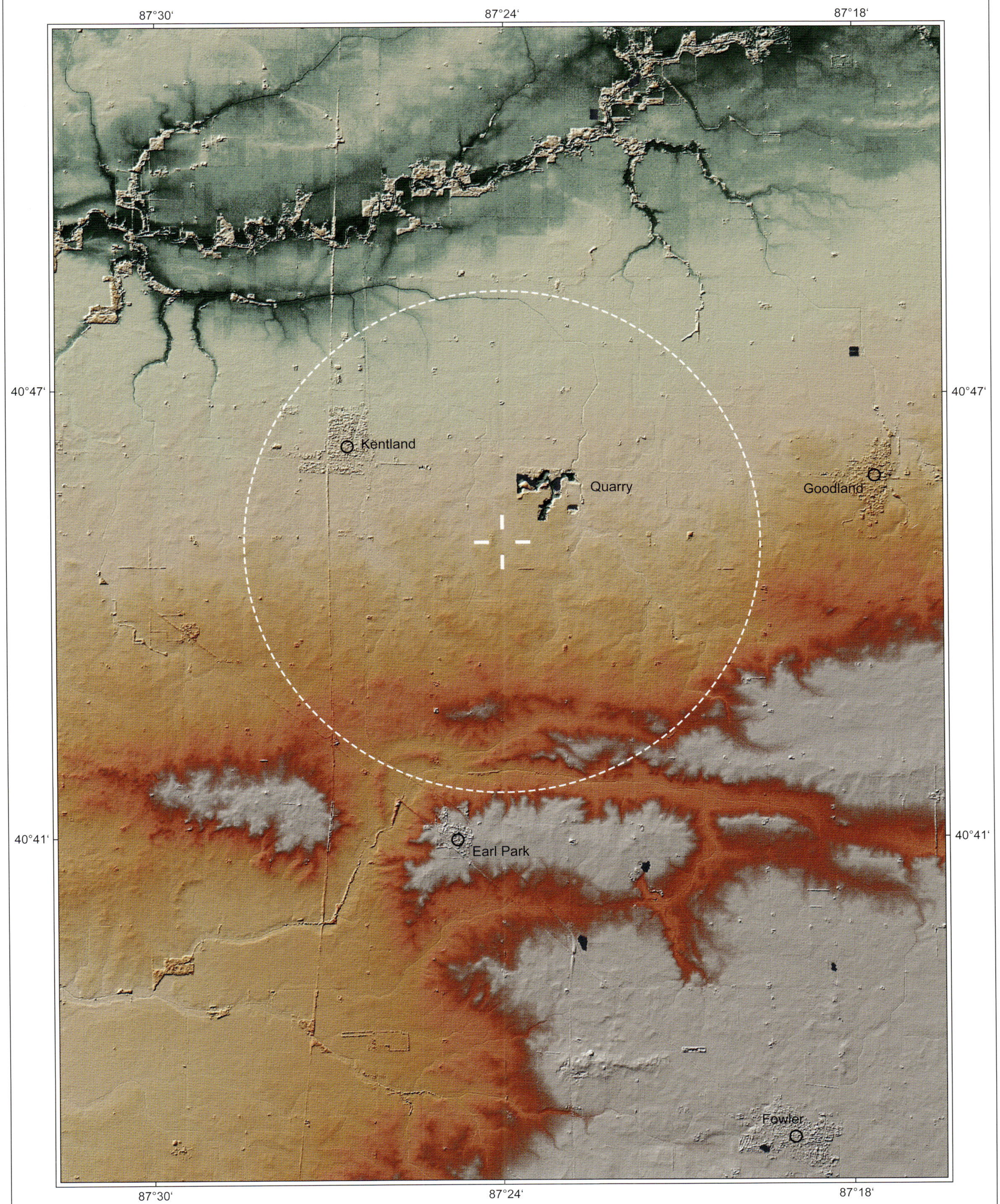

scale = 1 : 119 000 / scene = 21.4 × 28.5 km²

0 1.19 5.95 km

N

180 meter 270

ring depression, and older than the Pleistocene. Based on the remanent magnetic signature of the Kentland rocks, a paleomagnetic study suggested a 97 ± 10 Myr maximum age for the crater.

The disturbance of strata near Kentland has been known since, at least, the 1880's when a quarry started to excavate strongly deformed rocks of the central uplift. Robert Dietz provided proof for the impact origin of this structure already in 1947 through documentation of shatter cones. However, Ray Gutschick, who studied the structure and the progress in the quarry over three decades, declined to ascribe a particular cause to the structure's formation. Randy Laney and Randall Van Schmus in 1978 presented the structure in detail and documented further evidence of shock deformation.

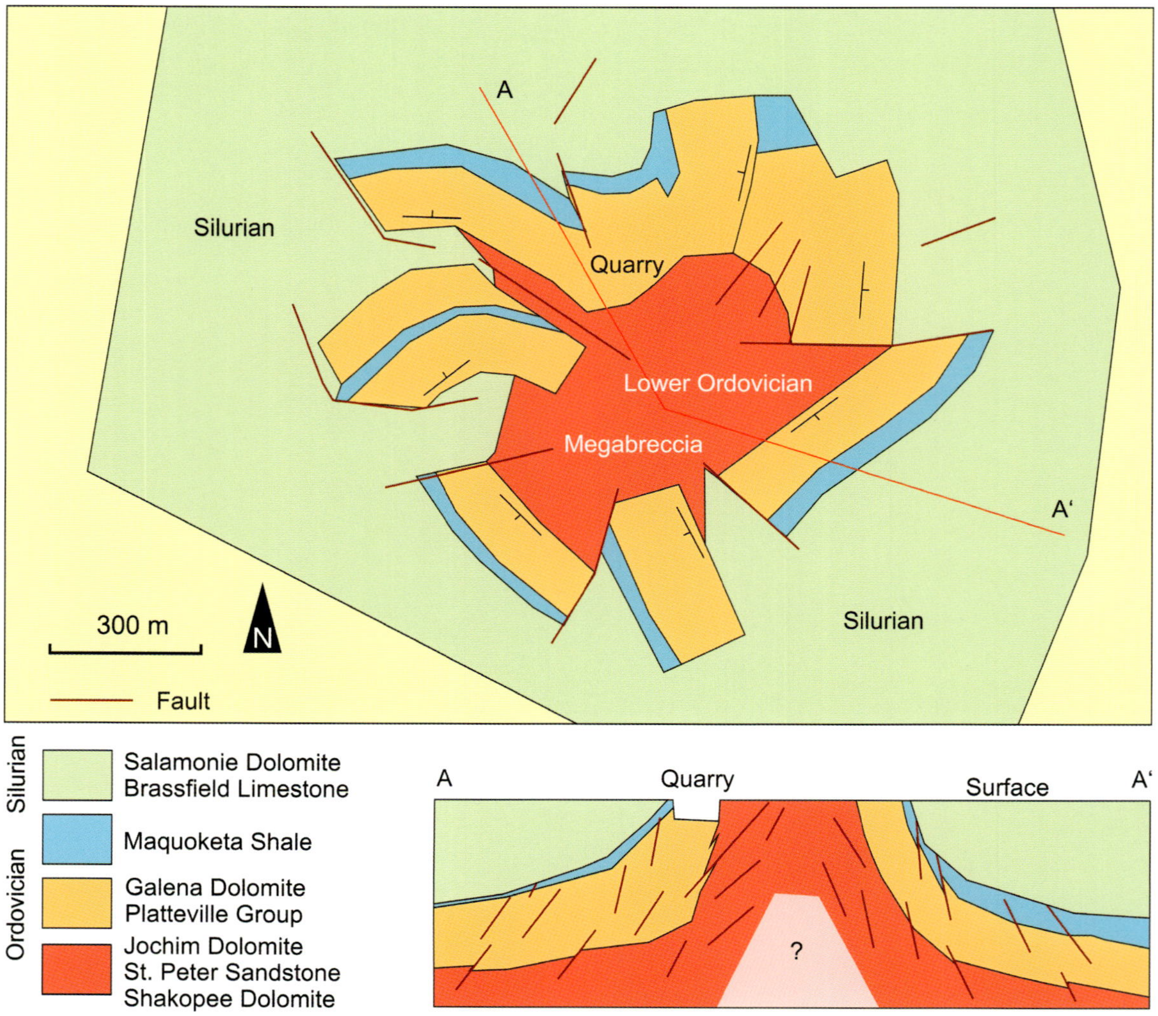

Generalized geological map and inferred cross-section of the central uplift of the Kentland impact structure (modified after Laney and Van Schmus, 1978).

Location/Access

The center of the Kentland impact structure lies approximately 4 km southeast of the town of Kentland in Newton County, Indiana. From there it is only a short drive on U.S. Highway 24 to where the Newton County Stone Company operates their large quarry in the northern part of the central uplift.

Selected References

Bjornerud, M. G. (1998). Superimposed deformation in seconds: Breccias from the impact structure at Kentland, Indiana (USA). Tectonophysics, Vol. 290, pp. 259–269.

Dietz, R. S. (1947). Meteorite impact suggested by the orientation of shatter cones at the Kentland, Indiana, disturbance. Science, Vol. 105, pp. 42–43.

Gutschick, R. C. (1982). Geology of the Kentland structural anomaly, northwestern Indiana – update. Purdue University, Department of Geosciences, and Geological Society of America, North-Central Section, Guidebook 4, 38 pp.

Laney, R. T., and Van Schmus, W. R. (1978). A structural study of the Kentland, Indiana, impact site. Proceedings of the 9th Lunar and Planetary Science Conference, Houston, Texas, Pergamon Press, New York, pp. 2609–2632.

Shrock, R. R., and Malott, C. A. (1933). The Kentland area of disturbed Ordovician rocks in northwestern Indiana. Journal of Geology, Vol. 41, pp. 337–370.

Weber, J. C. (2013). Impact geology: Central uplift, Kentland impact structure, Newton County Stone (Kentland) Quarry, Indiana, USA. In: Insights into the Michigan Basin, The Geological Society of America Field Guide, Vol. 31, pp. 1–7.

La Moinerie
Quebec, Canada

Thomas Kenkmann

The La Moinerie crater structure, the diameter of which has been estimated at 8 km, is largely submerged in Lake La Moinerie. The lake has an unusual circular to rectangular shape and is conspicuous in the Arctic region of northern Quebec where lakes are often elongated due to glacial overprint. There are several islands in the lake that expose impact lithologies. Loose glacial boulders from the northwestern shore have been investigated.

Target rocks are felsic gneisses of the Canadian Shield. Impactites include gneisses dissected by pseudotachylitic breccia, monomict and suevitic breccias with shocked quartz grains, as well as impact melt rock with a microcrystalline matrix of a somewhat more mafic composition than the felsic gneiss host rocks. ^{40}Ar-^{39}Ar stepwise heating of impact melt rock samples yielded a rough age for the impact of 400 ± 50 Myr. The regional Bouguer anomaly shows a local -4 mGal negative excursion over the lake.

It was the conspicuous shape of Lake La Moinerie that first attracted its investigation. The structure was visited in 1978 by David (Duff) Gold and co-workers. His report is the only documentation of this crater structure.

Location/Access

There is no nearby infrastructure in this remote region of the northern Quebec tundra, about 110 km south of Ugava Bay. Access requires comprehensive logistical planning.

La Moinerie

Quebec / Canada

latitude = 57°26'10" N / longitude = 66°35'09" W / diameter = 8 km / age = 400 ± 50 Myr

scale = 1: 194 000 / scene = 34.9 × 46.5 km²

0 1.94 9.70 km

N

200 meter 395

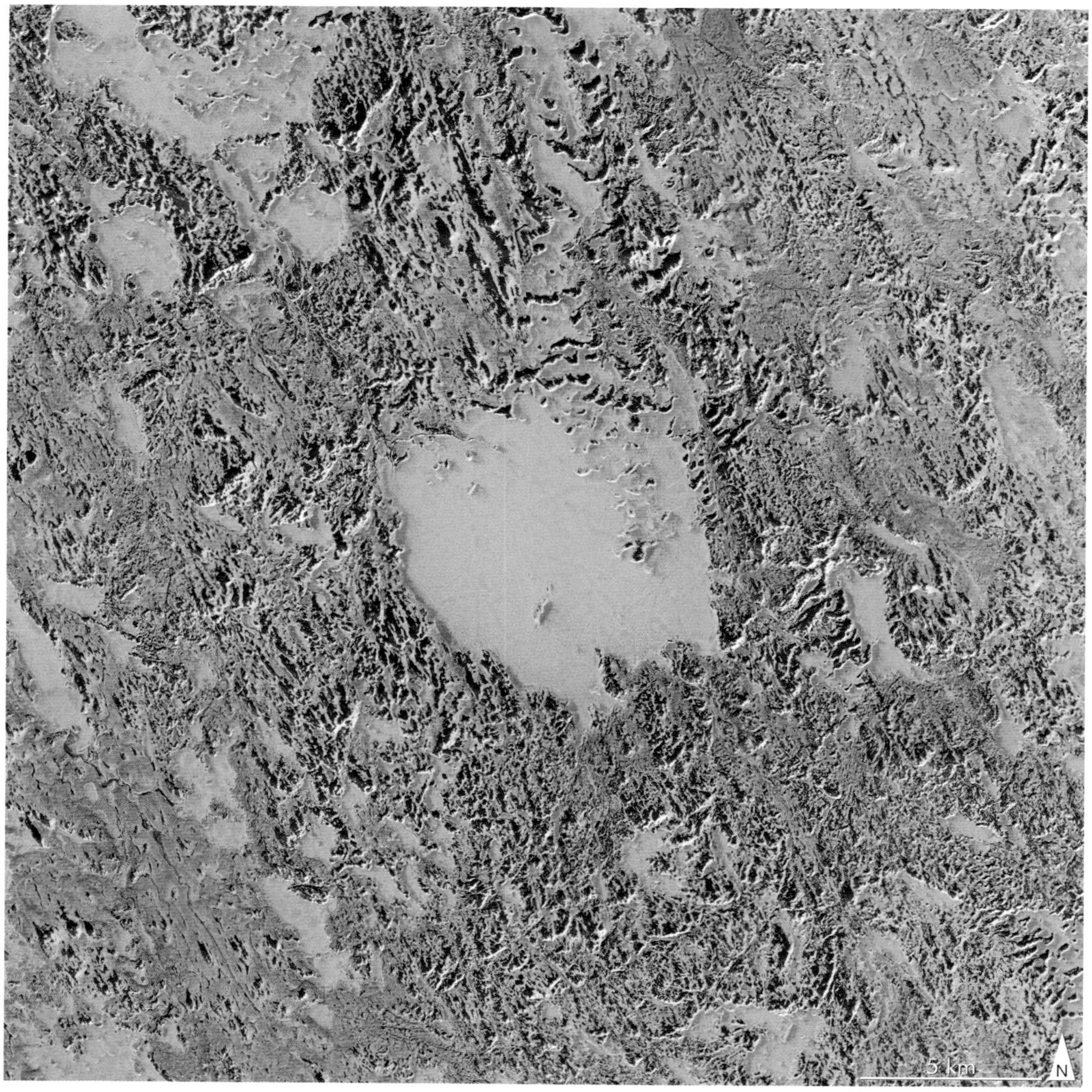

Selected References

Bottomley, R. J., York, D., and Grieve, R. A. F. (1990). ^{40}Argon-^{39}Argon dating of impact craters. Proceedings of the 20th Lunar and Planetary Science Conference, Houston, Texas, Cambridge University Press, pp. 421–431.

Gold, D. P., Tanner, J. G., and Halliday, D. W. (1978). The Lac La Moinerie crater: A probable impact site in New Quebec. Geological Society of America, Vol. 10, pp. 44.

Grieve, R. A. F. (2006). Impact Structures in Canada. Geological Association of Canada, GeoText No. 5, 210 pp.

The La Moinerie impact structure with the ice-covered homonymous lake on May 13, 2019 as a composite image based on TanDEM-X DEM and Copernicus Sentinel-2 (channel 2/3/4) data.

Manicouagan
Quebec, Canada

Wolf Uwe Reimold

The Manicouagan structure with a variably referred diameter of 90–100 km is one of the largest known impact structures on Earth. The structure is located in remote northern Quebec, 180 km northwest of the town of Sept-Îles. The impact occurred around 214 Myr ago, and it has been debated whether this large-scale catastrophic event could have been part of a Late Triassic multi-impact event – other structures of similar age are **Rochechouart** in France and **Saint Martin**, also in Canada – that might have caused or contributed to the minor mass extinction at the Triassic-Jurassic boundary.

The structure is named after the homonymous *Lac Manicouagan*, one of a two-lake assembly (the other lake is called *Lac Mouchalagne*) that partially surrounded the central part of the impact structure. The lake basins were flooded by the construction of the Daniel-Johnson hydroelectric dam in the 1960s. The Manicouagan Reservoir now forms an annular lake within the impact structure.

Manicouagan's prominent morphology is obvious on topographic maps. The outer diameter of the Manicouagan Reservoir measures 65 km and the reservoir has a width of about 10 km. Outside of the reservoir occurs a zone of fractured basement. Inside of the reservoir follows an inner plateau that is covered by a thick mass of coherent impact melt rock. And the centermost area is occupied by a 25 km wide remnant of the central uplift. The maximum extent of the fractured zone is generally taken as the diameter of the impact structure. The outer limit is accentuated on the western side of the structure by a series of annular-trending lakes. The inner plateau rises out of the annular reservoir in the form of steep cliffs or steep inclines. The innermost part is a topographically slightly depressed area with a 500 m high central elevation – formed by anorthosite – that is known as *Mont de Babel*. The somewhat more complex topography of the innermost zone has been tentatively discussed as perhaps indicating a peak-ring structure rather than a typical central peak; alternatively, erosional processes may be invoked for the formation of this complex morphology. The plateau is incised by a number of radial and annular trending valleys.

The target rocks at Manicouagan were 1170–1700 Myr old anorthosite, mangerite, charnockite, and granites, and other associated igneous rocks, as well as minor supracrustal rocks. In the western sector occur grey gneisses. All these lithologies were metamorphosed to amphibolite to granulite facies, and locally eclogite facies, as well as deformed during the 900–1200 Myr Grenville orogeny.

After peak metamorphic conditions at 1000–1100 Myr, crust was rapidly exhumed. Remnants of Middle Ordovician limestones and dolomites, as well as silt- and mudstones of the Trenton Group (also of this age) occur in the area of the reservoir. These supracrustals,

Manicouagan

Quebec / Canada

latitude = 51°23' N / longitude = 68°41' W / diameter = 100 km / age = 214 Myr

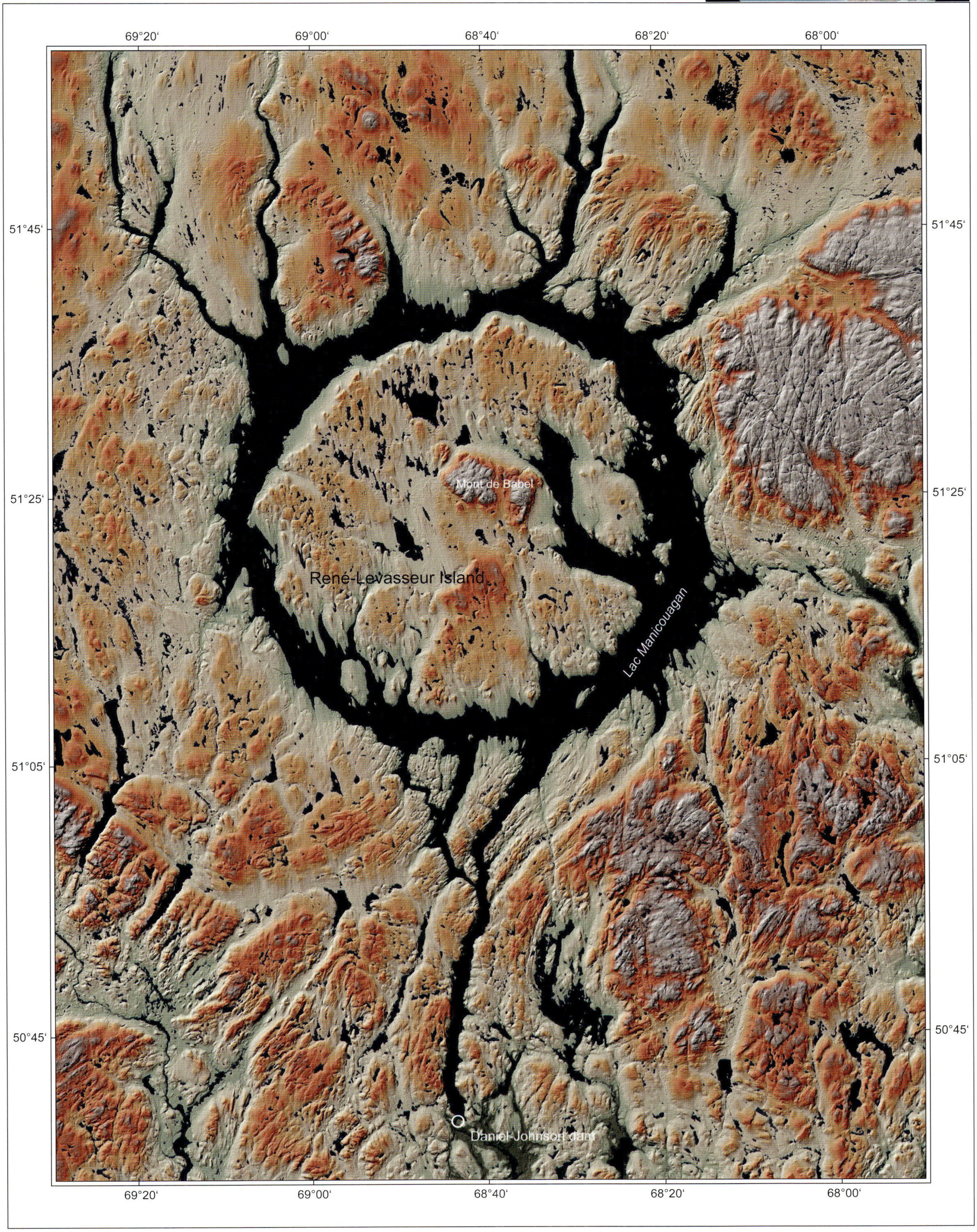

scale = 1 : 647 000 / scene = 116.6 × 155.5 km^2

0 6.47 32.35 km

N

145 meter 1105

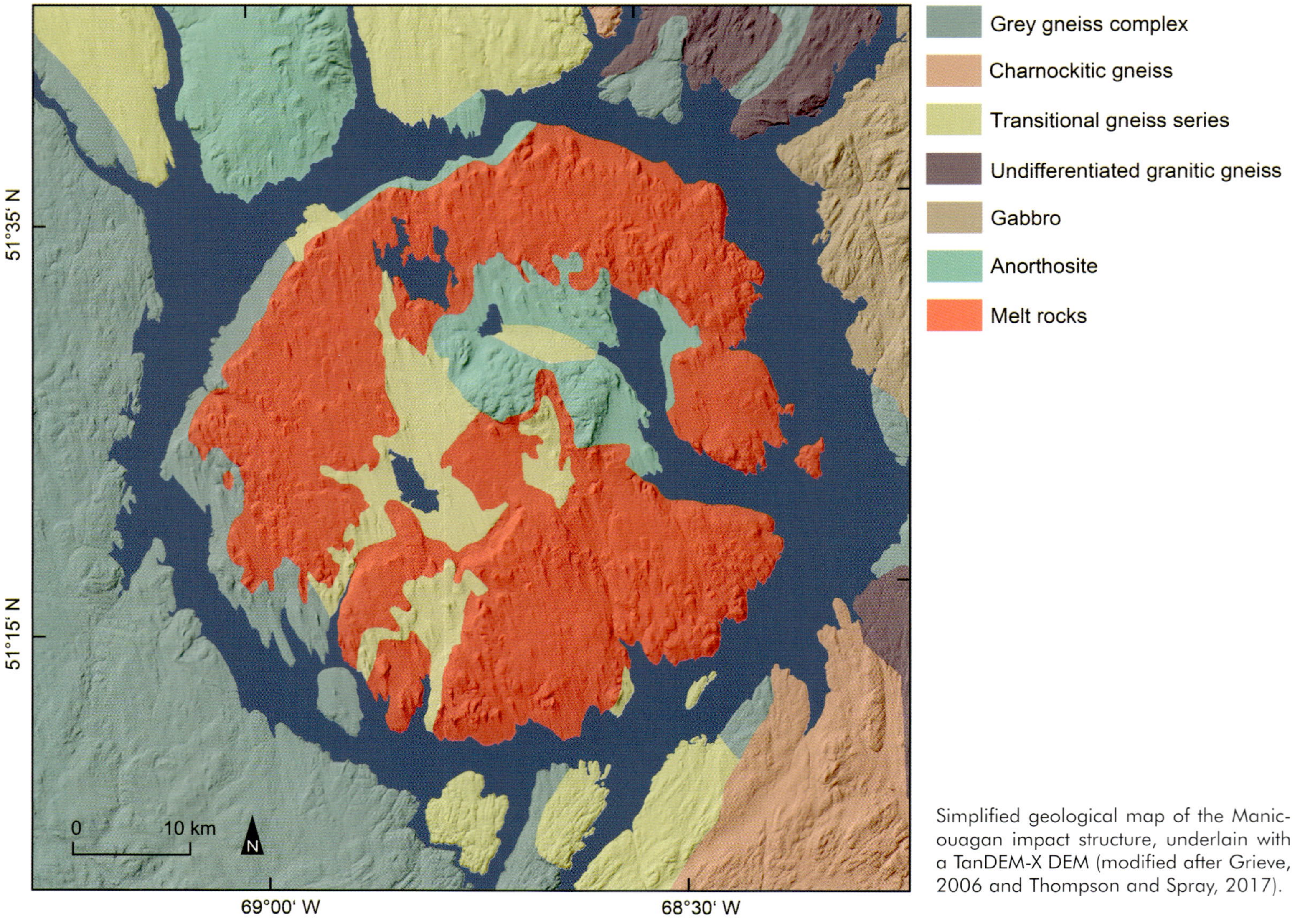

Simplified geological map of the Manicouagan impact structure, underlain with a TanDEM-X DEM (modified after Grieve, 2006 and Thompson and Spray, 2017).

at impact time, apparently formed a thin cover on top of the metamorphic basement. Blocks of such material have been observed in impact breccias.

Since impact in the Late Triassic, the Manicouagan region has not been significantly overprinted or buried. The structure has remained undeformed and is well exposed. The main post-impact modification was caused by recent glacial erosion. No ejecta have been located in the region because of this. However, it has been discussed whether Late Triassic sedimentary strata with shocked particles found near Bristol in southwest England could represent distal ejecta from this impact event.

A small piece of anorthosite, 14 × 6 × 4.5 cm in size, from the Manicouagan impact structure (photo courtesy: Ralf-Thomas Schmitt, Museum für Naturkunde Berlin, sample from Burkhard Dressler).

The basement rocks of the central peak and exposed in valleys incising the inner plateau are shock metamorphosed, and shock metamorphic grade has been utilized for analysis of the dimension of the impact structure. There is extensive development of diaplectic quartz and diaplectic plagioclase (also known as *maskelynite*, and actually there is a location called *Maskelynite Hill* in the central area at Manicouagan) glasses in the rocks from the center of the structure, planar deformation features (PDFs) in quartz and feldspar, and other minerals, up to more than 20 km from the center of the structure.

A nearly complete crater-fill sequence has been preserved. This includes shocked and unshocked, relatively little deformed basement, as well as fractured and brecciated basement. These autochthonous to parautochthonous units grade into allochthonous impact breccias, which include both lithic and suevitic types. Upward, this is followed by a basal clast-laden impact melt rock, and then a clast-poor to clast-free, well-crystallized impact melt complex. Impact melt and fault-related pseudotachylite can be found in coherent sections of basement, including the central uplift rocks.

Outcrop of impact melt rock occurs on the inner plateau, where a sheet-like body is exposed at up to 230 m thickness. It rests on a basal suevite layer up to 14 m thick or on fractured, shocked basement rocks. The melt sheet was subdivided into different packages based on the crystallization textures observed. Manicouagan impact melt has been for decades an example of well-homogenized impact melt. Felsic and mafic gneisses, as well as anorthosite, were likely mixed to generate this melt mixture.

In the last decades, the structure has been extensively evaluated by drilling for mineral exploration purposes. Some 18 km of drill core were extracted from 38 locations in the central parts of the structure. The sheet of impact melt rock, as well as the underlying basement, have, thus, been sampled, which has provided vast information on variation of the thickness of the melt body as well as basement geology and crater structure. Drilling includes three boreholes to depths in excess of 1.5 km. This has shown that the – on surface – flat appearing impact melt body has a highly variable thickness, i. e., footwall geometry, with melt rock thicknesses between 400 and 1500 m. The thickest melt section occurs in a deep central trough that is surrounded by a horseshoe-shaped zone of uplifted Precambrian basement. This uplifted unit is interpreted as a modified central peak structure, which partially breached the impact melt sheet.

Regional lithological variation is a strong factor in the interpretation of the gravity anomaly map. For example, a large positive anomaly in the northeast of the structure has been related to the Manicouagan gabbro complex. When such regional influences are removed, the resulting Bouguer gravity map shows a −4 to −10 mGal peripheral gravity low coincident with the annular moat (reservoir), which surrounds a central (relative) high of ± 0 mGal. The aeromagnetic map over the structure also displays prominent anomalies associated with local geology (such as the gabbro in the northeast), but there is also a strong positive (1000 to 2000 nT) central anomaly. It has been interpreted as a tabular, perhaps ring-like source that could be related to shock remanence of a mafic body. However, shock decomposition and alteration of mafic minerals, leading to magnetite formation, have been invoked as well. A refraction seismic project provided an indication of increasing seismic velocities with depth. Paleomagnetic work showed Triassic pole orientations for North America and allowed determination of a relatively short cooling interval of less than 10000 years for the impact melt sheet.

Age determinations have been carried out at Manicouagan since the late 1960s. With various geochronometers consistently ages around ca. 214 Myr were obtained. In the early 1990s a U-Pb zircon age of 214.1 ± 1 Myr was reported for Manicouagan impact melt rock. Recently, Marc Biren and co-workers analyzed 40 titanite grain fragments from 9 samples of metamorphosed anorthosite from the central uplift of the Manicouagan impact structure by the (U-Th)/He technique and obtained an age of 208.9 ± 5.1 Myr. They also reported an age for zircon of 213.2 + 2.3 Myr, obtained with the same method. Both these results are within error of the previous U-Pb zircon age of 214.1 + 1 Myr for the impact melt rock. Quartz with PDFs has been detected in southwest England at this level – and Manicouagan remains the strongest contender for the possible source.

View over the Manicouagan Reservoir towards Mont de Babel (photo: Martin Schmieder, University of Applied Sciences, Neu-Ulm, Germany).

The Daniel-Johnson dam at the southern end of the Manicouagan Reservoir (photo: Martin Schmieder, University of Applied Sciences, Neu-Ulm, Germany).

A cliff about 100 m high of impact melt rock at the eastern side of the Manicouagan impact structure (photo: Martin Schmieder, University of Applied Sciences, Neu-Ulm, Germany).

Since the late 19th century geological investigations have been carried out in the Manicouagan region. Right up to the mid-20th century, the apparently igneous rocks – now recognized as an impact melt sheet – were interpreted as being of volcanic origin. In 1956, Carlyle Beals and colleagues of the Dominion Observatory in Ottawa considered the annular shape of the lake basins as an indication for a possible impact structure. In 1963 a geophysical survey was carried out, as part of Beals' program to study impact structures on the Canadian Shield. Shortly thereafter, a report of maskelynite (diaplectic glass after plagioclase) in rocks from the center of the structure was published – first evidence for impact. A considerable amount of work has since been carried out on Manicouagan, making it one of the best investigated terrestrial impact structures. This includes detailed mapping in the 1970s, shock metamorphic studies in the 1980s, and detailed analysis of the impact melt sheet ever since. This culminated in the already mentioned drilling campaign and subsequent studies.

The Manicouagan impact structure is a conspicuous landmark from low earth orbit. This image with the ice- and snow-covered Manicouagan Reservoir at its center was captured by the astronauts aboard the International Space Station on February 23, 2003 (photo: NASA-ISS006-E-34153).

Location/Access

The Manicouagan impact structure is located in the Côte-Nord region of Quebec, about 250 km north of the St. Lawrence stream. The best time to visit the structure seems to be in March, when the ice on the reservoir is really thick and one can skidoo around the reservoir and see the impact melt rock cliffs. Alternatively, viewing is recommended immediately after the ice has melted and the water level is at its lowest because of periodic water release through the dam ports. Outcrops can then be observed along the lower shoreline.

To reach the structure, a good starting point might be Quebec City with its large airport, and then advancing on road QC-362 along the northern shore of the St. Lawrence stream via Malbaie to Baie-Comeau. From there take road QC-389 to Relais-Gabriel to the east of the Manicouagan reservoir. The distance along this road is somewhere around 300 km. Accommodation at the structure is scarce. On the eastern shore of the reservoir is a Refuge du Prospecteur that in the past has been operated by a company called "Station Uapishka". From there it is possible to reach excellent exposures along the inner shoreline and on islands, by boat. Exploration of the interior of the structure requires helicopter support, and the nearest helicopter station seems to be in the town of Sept-Îles.

Selected References

Biren, M. B., and Spray, J. G. (2011). Shock veins in the central uplift of the Manicouagan impact structure: Context and genesis. Earth and Planetary Science Letters, Vol. 303, pp. 310–322.

Biren, M. B., van Soest, M., Wartho, J.-A., and Spray, J. G. (2014). Dating the cooling of exhumed central uplifts of impact structures by the (U-Th)/He method: A case study at Manicouagan. Chemical Geology, Vol. 377, pp. 56–71.

Dressler, B. (1990). Shock metamorphic features and their zoning and orientation in the Precambrian rocks of the Manicouagan Structure, Quebec. Tectonophysics, Vol. 171, pp. 229–245.

Eitel, M., Gilder, S. A., Spray, J. G., Thompson, L. M., and Pohl, J. (2016). A paleomagnetic study of the Manicouagan impact structure: Implications for crater formation and geodynamo effects. Journal of Geophysical Research – Solid Earth, Vol. 121, pp. 436–454.

Floran, R. J., Grieve, R. A. F., Dence, M. R., Phinney, W. C., Warner, J. L., Blanchard, D. P., and Simonds, C. H. (1978). Manicouagan impact melt, Quebec. I. Stratigraphy, petrology, and chemistry. Journal of Geophysical Research, Vol. 83, pp. 2737–2759.

Grieve, R. A. F. (2006). Impact structures in Canada. Geological Association of Canada, GeoText No. 5, 210 pp.

Grieve, R. A. F., and Floran, R. J. (1978). Manicouagan impact melt, Quebec. 2. Chemical interactions with basement and formational processes. Journal of Geophysical Research, Vol. 83, pp. 2761–2771.

Spray, J. G., Thompson, L. M., Biren, M. B., and O'Connell-Cooper, C.D. (2010). The Manicouagan impact structure as a terrestrial analogue site for lunar and martian planetary science. Planetary and Space Science, Vol. 58, pp. 538–551.

Thompson, L. M. and Spray, J. G. (2017). Dynamic interaction between impact melt and fragmented basement at Manicouagan: The suevite connection. Meteoritics and Planetary Science, Vol. 52, pp. 1300–1329.

Young, K. E., Mercer, C. M., van Soest, M. C., Hodges, K. V., Wartho, J. A., and Biren, M. B. (2014). An examination of noble gas geochronology and thermochronology in the context of dating impact events. 45th Lunar and Planetary Science Conference, Lunar and Planetary Institute, Houston, Texas, Abstract #2670.

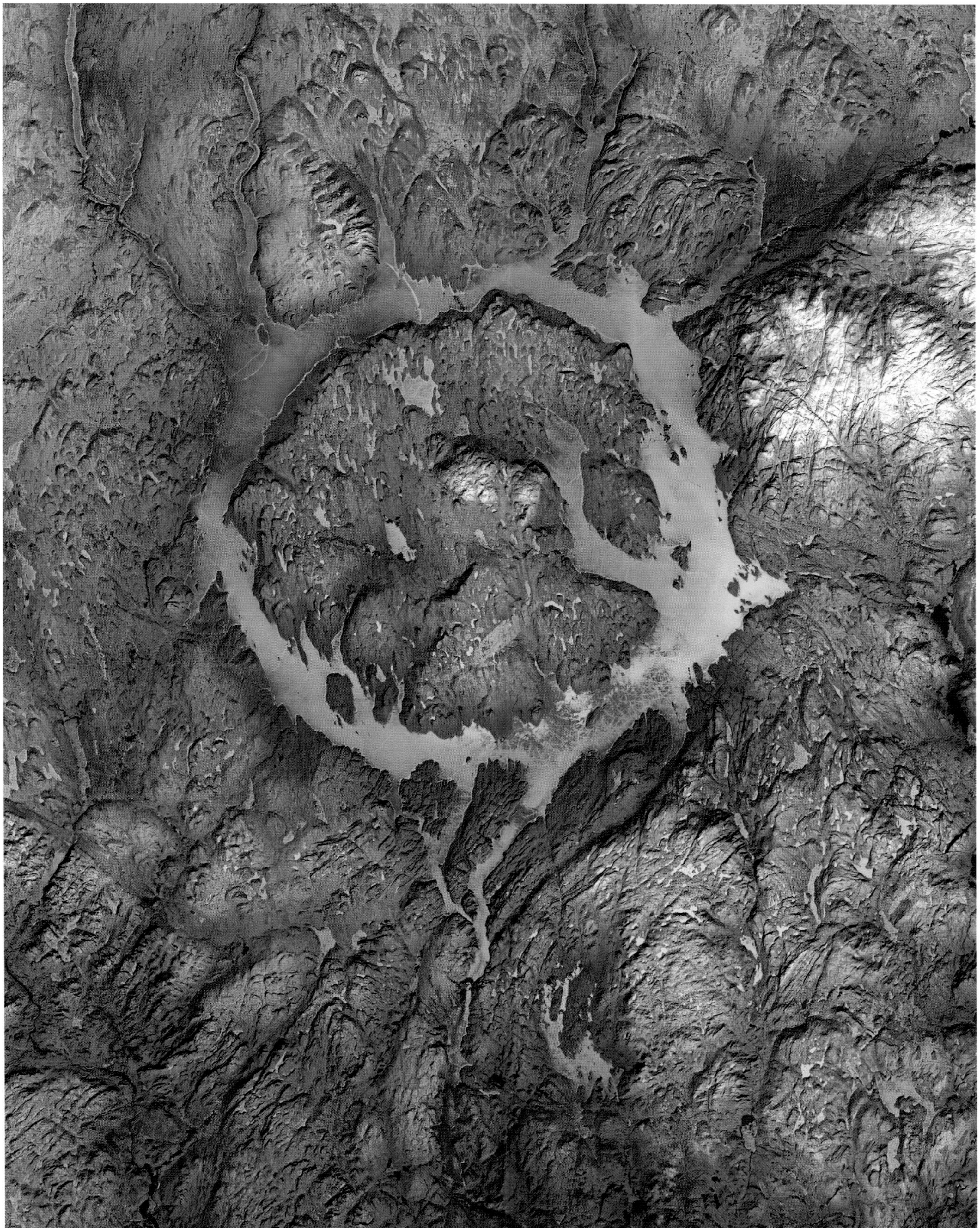

The Manicouagan impact structure on May 23, 2018 as a composite image based on TanDEM-X DEM and Copernicus Sentinel-2 (channel 2/3/4) data.

Manson

Iowa, United States

Thomas Kenkmann

Manson is a well-preserved, large, complex impact structure in northwest central Iowa. It lies buried below 30–100 m of glacial till and does not exhibit a surface expression. The target at Manson, as gauged from regional information, was composed of an Early to Middle Proterozoic granitic and metamorphic basement, on which Precambrian siliciclastics are deposited in a southeastward thickening sequence. Some 800 m of Paleozoic carbonates and 100 m of Cretaceous shales form the upper parts of the crust near Manson. The structure is covered by Quaternary glacial deposits.

This standard stratigraphic sequence is disturbed in a region of 35 km width that represents the apparent crater diameter of Manson. Near its center, granitic basement is uplifted. The central uplift has a minimum diameter of 7.5 km, and the structural uplift has been estimated at about 2.8 km. An annular trough with a diameter of about 21 km surrounds the central uplift, and a terrace zone of downfaulted Paleozoic and Cretaceous strata follows further outside. A sub-surface escarpment formed by a circumferentially striking normal fault defines the structural rim of the Manson structure. A gravity survey identified locations of low-density brecciated rocks and high-density uplifted crystalline rocks. Further geophysical analysis included aeromagnetic and ground magnetic surveys and reflection seismic studies.

The structure was perforated by 12 scientific drillings, in which a shock metamorphic overprint could be demonstrated. Planar deformation features (PDFs) occur in quartz, feldspar, and apatite grains of breccia samples. One of the holes drilled on the flank of the central uplift penetrated through an impact melt breccia layer of 40 m thickness. Dating this melt layer by total-fusion ^{40}Ar-^{39}Ar analysis of sanidine yielded an age of 74.1 ± 0.1 Myr. Hence, the Manson impact has no causal connection to the Cretaceous-Paleogene (K-Pg) mass extinction at 66 Myr, as had been previously assumed.

At various places in eastern South Dakota and Nebraska, a layer of shocked quartz and feldspar grains exists in the lower part of the Upper Cretaceous Pierre Shale. Due to a systematic increase in grain sizes of shocked particles towards the Manson structure, this impact structure has been considered the source of this distal ejecta layer, which was confirmed when dating of this layer also revealed an age of 74 Myr.

Attention was drawn to Manson at the beginning of the 20th century by the unusual quality of the groundwater that was caused by the uplift of granitic rocks. Soon the anomalous geology of the area was recognized. Like many other North American impact craters the structure was first considered as a "cryptovolcanic", later as a "cryptoexplosive" feature. In 1966 Nicholas Short identified shock deformation in quartz of granitic rocks of the central uplift. Attempts to date the Manson impact in the 1980s initially revealed ages indistinguishable from that of the Cretaceous-Tertiary boundary. This stimulated a geophysical survey and drilling program in 1991–1992 by the Iowa Geological Survey Bureau and the U.S. Geological Survey. Some 1200 m of drill core were extracted and analyzed in an international consortium effort. This work also yielded Manson's actual age.

Location/Access

The Manson impact structure of northwest central Iowa lies in an area of flat, cultivated farm land. The town of Manson is close to the center of the circular structure. The city of Fort Dodge, with a regional airport, is located about 30 km to the east. From there, U.S. Highway 20 intersects the Manson structure at its southern end.

Selected References

Anderson, R. R., and Hartung, R. B. (1992). The Manson impact structure; its contribution to impact materials at the Cretaceous/Tertiary Boundary. Proceedings of the 22nd Lunar and Planetary Science Conference, Houston, Texas, pp. 101–110.

Anderson, R. R., Witzke, B. J., and Roddy, D. J. (1996). The drilling of the 1991–1992 Geological Survey Bureau and U.S. Geological Survey Manson impact structure research cores. Geological Society of America Special Paper, Vol. 302, pp. 45–88.

Grieve, R. A. F. (1989). Manson structure implicated. Nature, Vol. 340, pp. 428–429.

Hartung, J. B., and Anderson, R. R. (1988). A summary of data on the Manson impact structure (abstract). 19th Lunar and Planetary Science Conference, Lunar and Planetary Institute, Houston, Texas, pp. 455–456.

Hartung J. B., and Anderson, R. R. (1996). A brief history on investigations of the Manson impact structure. Geological Society of America Special Paper, Vol. 302, pp. 31–43.

Izett, G. A., Cobban, W. A., Obradovich, J. D., and Kunk, M. J. (1993). The Manson impact structure: ^{40}Ar-^{39}Ar age and its distal impact ejecta in the Pierre shale in southeastern South Dakota. Science, Vol. 262, pp. 729–732.

Keiswetter, D., Black, R., and Steeples, D. (1996). Seismic reflection analysis of the Manson Impact Structure, Iowa. Journal of Geophysical Research, Vol. 101, pp. 5823–5834.

Koeberl, C., and Anderson, R. R. (eds.) (1996). The Manson impact structure: Anatomy of an impact crater. Geological Society of America Special Paper, Vol. 302, 457 pp.

Plescia, J. B. (1996). Gravity investigation of the Manson impact structure, Iowa. Geological Society of America Special Paper, Vol. 302, pp. 89–104.

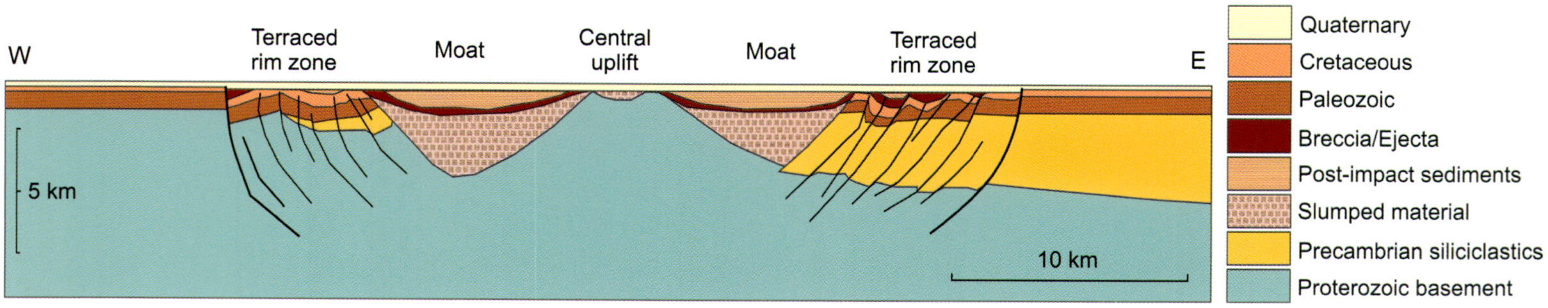

Geological cross-section for the Manson impact structure (modified after Anderson and Hartung, 1992).

Manson

Iowa / United States

latitude = 42°35‘ N / longitude = 94°33‘ W / diameter = 35 km / age = 74.1 ± 0.1 Myr

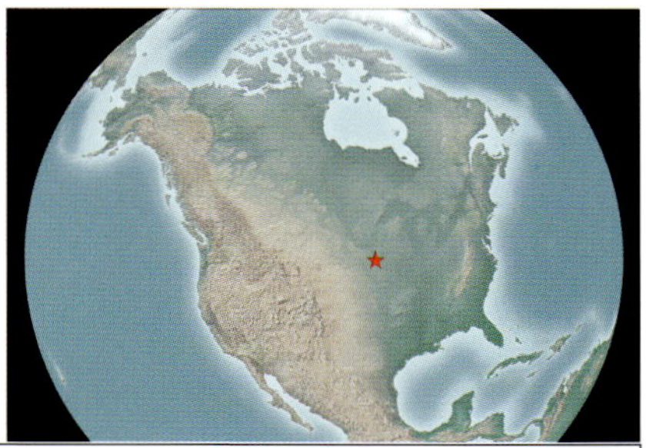

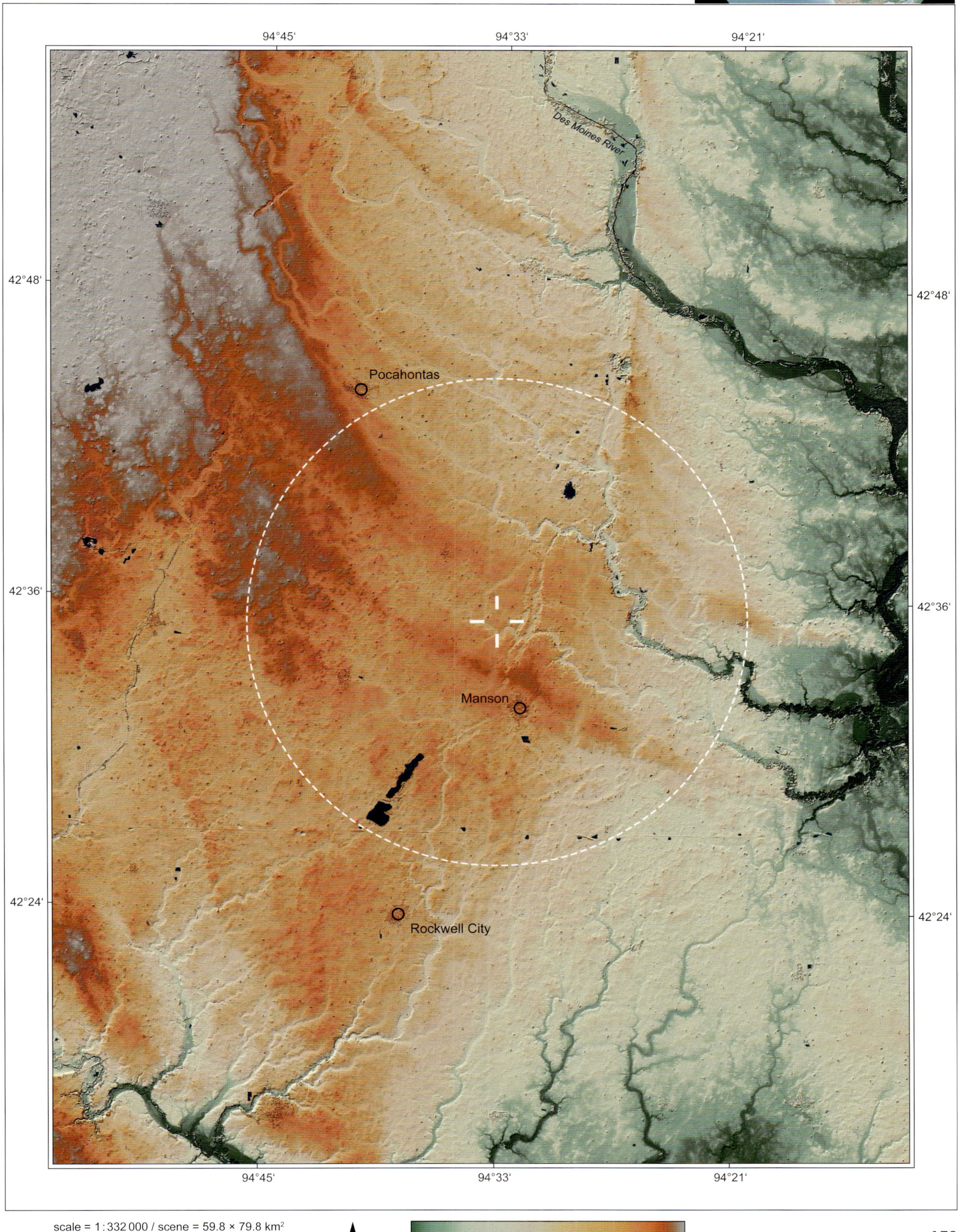

scale = 1:332 000 / scene = 59.8 × 79.8 km²

0 3.32 16.60 km

N

300 meter 445

Maple Creek
Saskatchewan, Canada

Thomas Kenkmann

Maple Creek in southwestern Saskatchewan is a circular impact structure of about 6 km diameter with a central uplift. It has a subdued topographic surface expression. The structure is also known as the *White Valley Structure*. It is located in the Western Canadian Sedimentary Basin. Target lithologies involved in the cratering process are exclusively sedimentary rocks of Carboniferous to Upper Cretaceous age.

The structure has been explored by four reflection seismic profiles and 12 boreholes. The central uplift has a diameter of about 3 km, with a stratigraphic uplift of roughly 600 m. A ring syncline around the uplift has a width of about 1 km. Here, Upper Cretaceous strata were down faulted. The central uplift is composed of shale and siltstones, with some sandstone units. Strata repetition, fault zones, and breccia occurrence were encountered during drilling of the central uplift. Shock features in the form of decorated planar deformation features (PDFs) were found in quartz grains of Upper Cretaceous sandstone of the central uplift.

Seismic disturbance decreases with depth and vanishes at a depth of 1.3 km, corresponding to the position of Mississippian strata. A positive gravity anomaly of +2.5 mGal characterizes the central uplift of Maple Creek.

The youngest strata affected by the impact event are of Upper Cretaceous age. The impact event must be somewhat younger than 75 Myr.

The anomalous character of Maple Creek was already recognized in the 1950s. Results of investigations for hydrocarbons by the petroleum industry remained classified. A first reflection seismic study and gravity survey accessible to the public was published by Malcolm Gent and co-workers in 1992. The structure gained attention as a possible kimberlite diatreme. Hans-Henrik Westbroek and Robert Stewart proposed an impact origin in 1995 but failed to find convincing evidence. Only in 1998 Richard Grieve and colleagues could document unequivocal evidence for shock deformation.

Location/Access

Maple Creek is situated some 25 km southeast of the town of Maple Creek at the northern flank of the forested Cypress Hills. The structure is accessible via road 724 – leaving the town of Maple Creek eastwards, and, after about 30 km, following a dirt road heading off towards the south.

Selected References

Gent, M. R., Kreis, L. K., and Gendzwill, D. (1992). The Maple Creek structure, southwestern Saskatchewan. Summary of Investigations 1992, Saskatchewan Geological Survey, Rep. 92-4, pp. 204–208.

Grieve, R. A. F. (2006). Impact Structures in Canada. Geological Association of Canada, GeoText No. 5, 210 pp.

Grieve, R. A. F., Kreis, K., Therriault, A. M., and Robertson, P. B. (1998). Impact structures in the Williston Basin. Meteoritics and Planetary Science, Vol. 33, pp. A63–A64.

Mazur, M. J., Stewart, R. R., and Hildebrand, A. R. (1999). Seismic characterization of buried possible meteorite impact structures. 30th Lunar and Planetary Science Conference, Houston, Texas, Abstract #1393.

Westbroek, H. H., and Stewart, R. R. (1995). The White Valley structure, Saskatchewan: a possible meteorite impact crater. CREWES research report, Chapter 19, 18 pp.

Maple Creek
Saskatchewan / Canada

latitude = 49°48'00" N / longitude = 109°06'30" W / diameter = 6 km / age = < 75 Myr

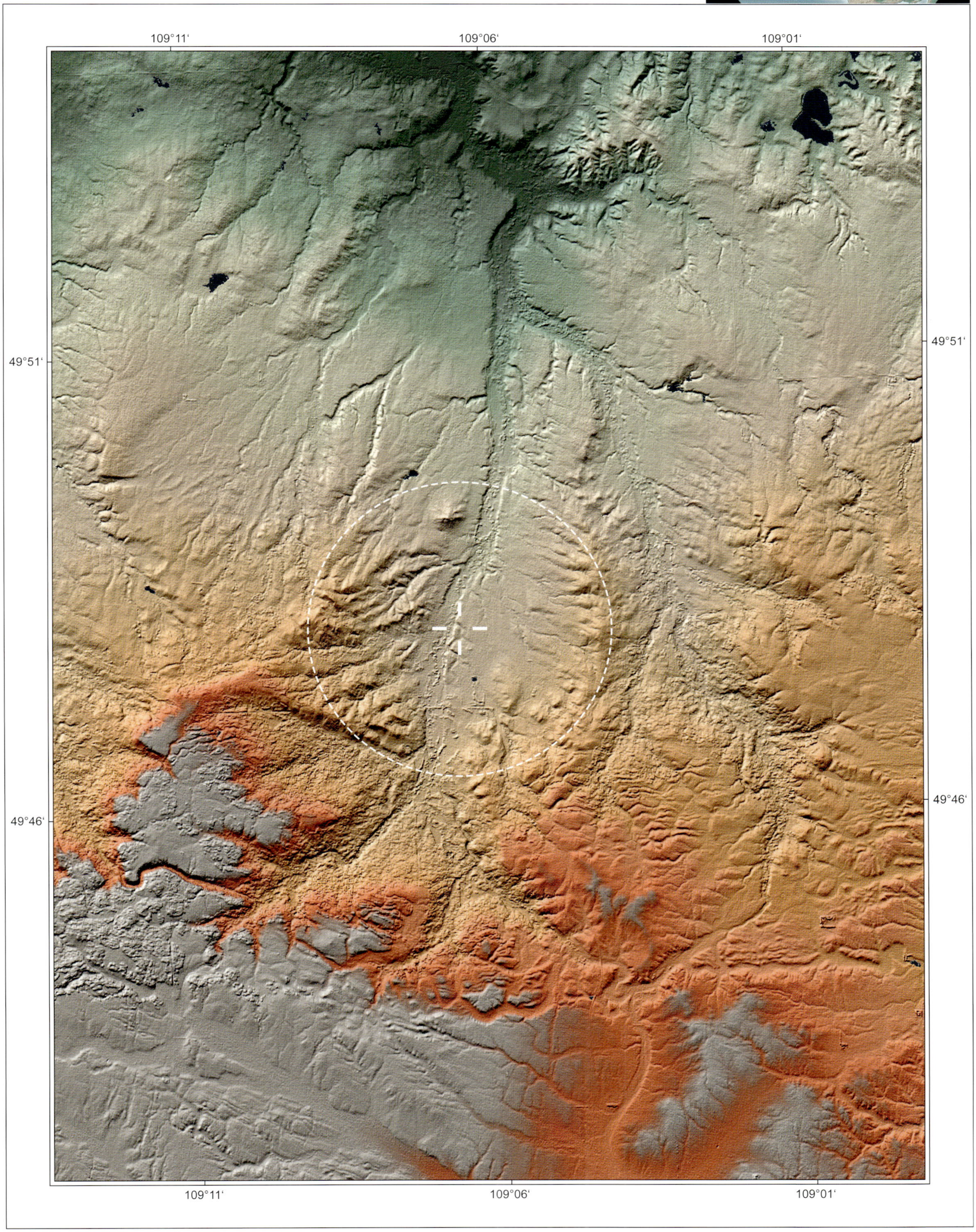

scale = 1 : 95 000 / scene = 17.1 × 22.8 km²
0 0.95 4.75 km
N
775 meter 1220

Marquez
Texas, United States

Thomas Kenkmann

Marquez, also known as *Marquez Dome*, is a mostly buried complex impact crater of about 15 km diameter. Its main feature is a steep, dome-like central rise, associated with a stratigraphic uplift of 1120 m that peaks through the overlying strata and is accessible in a few outcrops.

The Marquez impact structure is situated in the East Texas Basin, which was part of the Gulf of Mexico throughout large parts of the Mesozoic and Cenozoic era and was filled with as much as 5500 to 7000 m of sediments. At the time of impact, the bolide hit a near-shore environment with water depths of probably less than 50 m. The target comprised flat-lying Cretaceous to Eocene sediments, mainly marly limestone, clays, and silts. The younger sedimentary strata were unconsolidated and water-saturated.

The prominent feature of Marquez is the central dome of about 2.5 km diameter, which is a central uplift with almost vertical flanks as indicated by surface outcrop. The zone of deformation associated with the cratering event is limited to a depth of 1630 m. Boreholes were drilled into the central uplift at two locations, 1.1 and 2 km from the center of the structure. They revealed a shock metamorphic overprint of strata. No impact breccias were recovered but the rocks collected from the central zone show shatter cones and planar deformation features (PDFs) in quartz.

There are no surface exposures of the crater rim but seismic reflection data reveal a peripheral fault zone of about 15 km diameter. Further concentrically striking and shallowly inward dipping normal faults were mapped, but the strata are not strongly down-faulted and the dip of the strata is gentle to moderate.

The structure was intensely explored by the petroleum industry, who acquired reflection seismic, gravity, and magnetic data. Twenty-nine industry wells were sunk within the area of the crater structure. Currently gas wells are in operation. Geophysical well logs could be correlated from borehole to borehole and enabled the reconstruction of the geological structure in three dimensions. The magnetic survey did not yield a circular pattern consistent with the central uplift and the crater outline. The residual gravity data show a positive anomaly of +2.2 mGal with a diameter of 1700 m. Over the inner part of this area, the positive anomaly decreases to +1.8 mGal. The positive anomaly of the central uplift is surrounded by a subdued gravity low of −0.4 mGal. Apatite fission track dating on samples of the central uplift yielded an age of 58.3 ± 3.1 Myr that is thought to correspond to the stratigraphic age of the structure.

Marquez was originally interpreted as the surface expression of a salt dome. J. Gibson and Virgil Sharpton re-interpreted the dome in 1989 and 1990 as a buried complex impact structure with a central uplift, and presented evidence for shock metamorphic overprint in the form of shatter cones and PDFs in quartz grains.

Shatter cone in limestone, 7 × 5 × 3 cm in size, from the Marquez impact structure (photo courtesy: Ralf-Thomas Schmitt, Museum für Naturkunde Berlin, sample from T. Kenkmann).

Location/Access

The Marquez Dome is located in Leon County, Texas, about halfway between Houston and Dallas, near the Interstate 45. Use the exit marked Centerville and drive 25 km westward to the village of Marquez. The crater center lies northwest of this town. Most areas have restricted access.

Selected References

Buchanan, P. C., Koeberl, C., and Reid, A. M. (1998). Impact into unconsolidated, water-rich sediments at the Marquez Dome, Texas. Meteoritics and Planetary Science, Vol. 33, pp. 1053–1064.

Gibson, J. W. Jr., and Sharpton, V. L. (1989). Marquez Dome: A newly recognized complex impact structure in east Texas (abstract). AGU Spring Meeting, Baltimore, Maryland, EOS, Transactions of the American Geophysical Union, Vol. 70, p. 383

Herrick, R. R., and Sharpton, V. L. (1997). Shallow seismic test at Marquez Impact Structure. 28th Lunar and Planetary Science Conference, Lunar and Planetary Institute, Houston, Texas, Part 2, pp. 557–558.

Sharpton, V. L., and Gibson, J. W., Jr. (1990). The Marquez Dome impact structure, Leon County, Texas (abstract). 21st Lunar and Planetary Science Conference, Lunar and Planetary Institute, Houston, Texas, pp. 1136–1137.

Wong, A. M., Reid, A. M., Hall, S. A., and Sharpton, V. L. (2001). Reconstruction of the subsurface structure of the Marquez impact crater in Leon County, Texas, based on well-log and gravity data. Meteoritics and Planetary Science, Vol. 36, pp. 1443–1455.

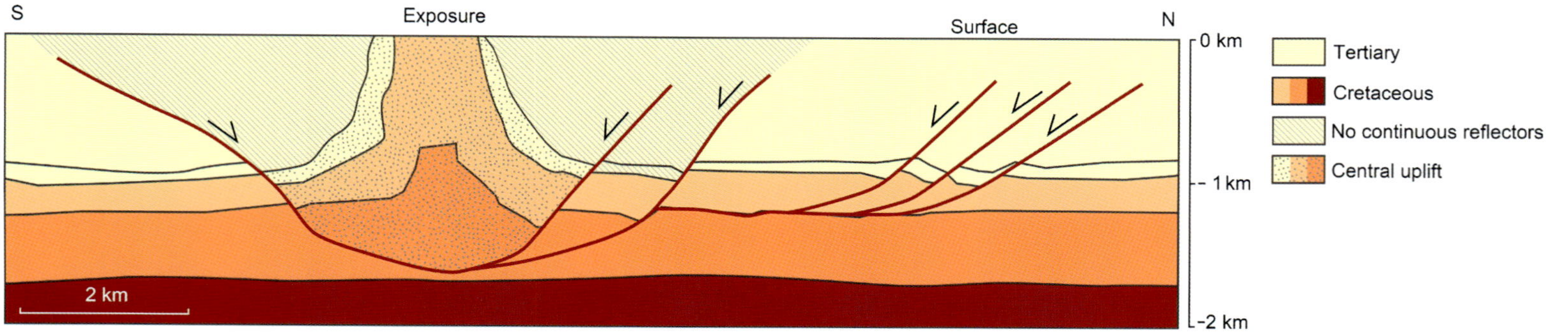

The modeled subsurface structure of the Marquez Dome impact structure, based on seismic reflection data (modified after Wong et al., 2001).

Marquez

Texas / United States

latitude = 31°17‘00“ N / longitude = 96°17‘30“ W / diameter = 15 km / age = 58.3 ± 3.1 Myr

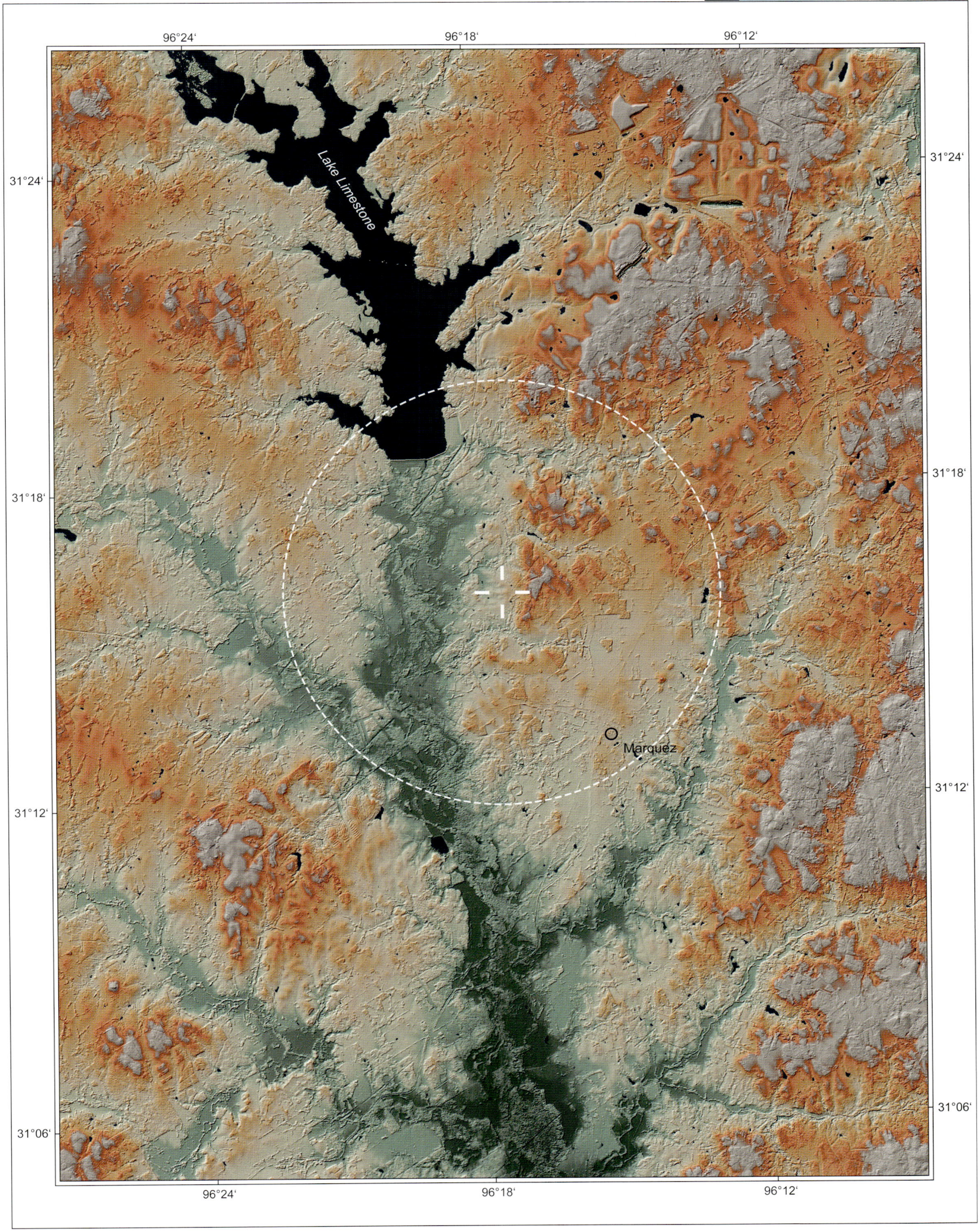

scale = 1 : 165 000 / scene = 29.7 × 39.5 km²
0 1.65 8.25 km
N
85 meter 215

Meteor Crater
Arizona, United States

Thomas Kenkmann

Meteor Crater, also known as *Barringer Crater* or *Coon Butte*, is the prime example of a young, well-preserved, simple impact crater. It was formed on the flat plains of the Southern Colorado Plateau in Arizona, when a circa 50 m sized bolide collided with Earth. We know the asteroid's composition from the multitude of small iron meteorites and shrapnel-like fragments of the so-called "Canyon Diablo" meteorite that have been collected around the crater. While passing the atmosphere, the aerodynamic drag caused the disintegration of the small asteroid shortly before impact.

With a diameter of about 1.2 km, Meteor Crater is an impact crater of the simple, bowl-shape geometry. It is surprisingly well preserved for terrestrial standards. It has been estimated that only 15 to 20 m of erosion off the rim crest has taken place. The rim still rises some 30–60 m above the surrounding flat Colorado Plateau of some 1700 m altitude. The apparent crater floor shows accumulation of roughly 30 m of sediment deposited since the impact. It lies on a thick lens of brecciated and melted rock. A part of these rocks fell back from the ejecta cloud above the crater, another part never left the crater and slumped down from the originally steep transient cavity wall. Nowadays talus covers the lower parts of the inner crater wall but leaves the upper 80–100 m of the stratigraphic sequence exposed. The upper part of the slope is rugged and steep, with angles larger than 45°, and the strata dip away from the inner rim. The lowest and oldest stratum observable in the crater wall is the Permian Coconino sandstone. The main unit exposed in the crater wall is the 80 m thick Kaibab Formation. The top unit is the Triassic Moenkopi Formation that is easily traced along the crater wall because of its reddish-brown, massive sandstone bank. The Moenkopi Formation marks the land surface at the time of impact. This formation and the strata underneath have been uplifted. The formations bend upward and form a recumbent fold along the crater rim, with the Moenkopi Formation occupying the fold hinge. The upper limb is in overturned position and forms the proximal part of the ejecta blanket. The ejecta blanket further out comprises blocks of various sizes, which cause the hummocky surface appearance of the crater environs. The thickness of the ejecta blanket decreases with distance.

A section of the crater rim at Meteor Crater, with structurally uplifted dolomite beds of the Kaibab Formation in the foreground (photo: T. Kenkmann).

While the outline of many simple craters is circular, the shape of Meteor Crater strongly deviates from a circle and resembles more a square. The quadrangular shape is attributed to the occurrence of joint sets oriented parallel to the diagonals of the crater. Joint surfaces in rocks are formed by tectonic stresses, and these joints were already present prior to the impact event on the Colorado Plateau. Such joints extend over kilometers as smooth vertical fractures. Along the joint planes, excavation during the impact event was more effective, because the rocks here were less resistant. Consequently, the crater grew wider along the joint trends, which led to the quadrangular crater outline. Large faults can be observed in the corners of the crater, with a scissor-type displacement. The rim uplift correlates with the crater shape and is highest in the corners of the crater. The differential uplift occurs via the emplacement of wedge-shape thrust slices in the steep walls of the cavity during crater growth.

The age of Meteor Crater has been measured by various techniques, including thermoluminescene and cosmogenic nuclides. They arrived at similar values of about 49 000 or 50 000 years. Recently, however, recalibration of production rates of cosmogenic nuclides have yielded slightly different values that could shift the impact which formed Meteor Crater by about 10 000 years into the past.

Simple craters like Meteor Crater are among the most frequent morphological features on solar system bodies with solid surfaces. Therefore, this structure has provided a training camp for Apollo astronauts to familiarize them before their visit to the moon, with hands-on knowledge about impact craters and deformation caused by the impact

Meteor Crater has straight rim segments, separated by corners in which radial faults occur. Here, a view towards the southwest corner of the structure (photo: T. Kenkmann).

Meteor Crater

Arizona / United States

latitude = 35°01‘39“ N / longitude = 111°01‘20“ W / diameter = 1.2 km / age = 49 000 yr

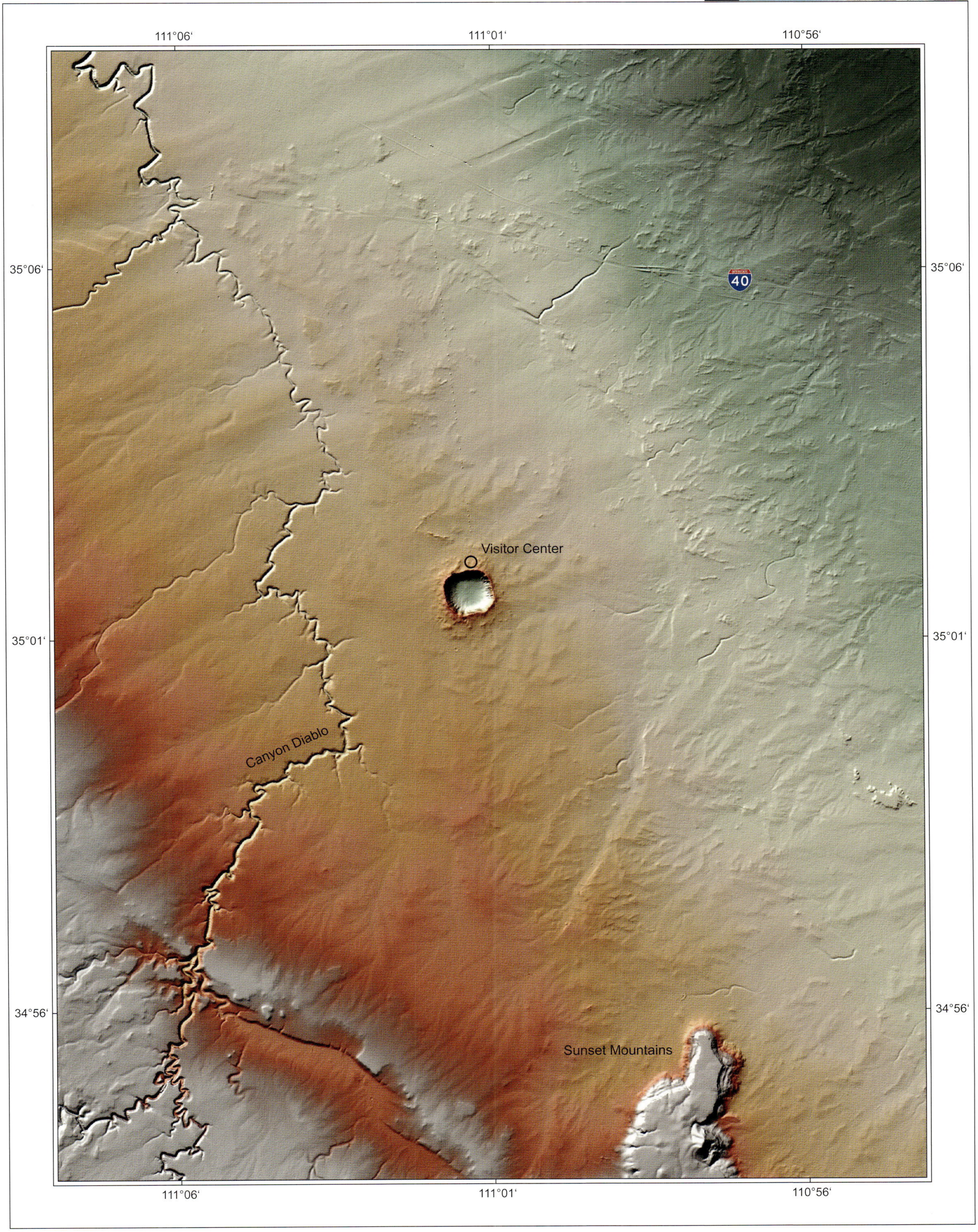

scale = 1 : 117 000 / scene = 21.1 × 28.1 km^2

0 1.17 5.85 km

N

1515 meter 1825

Panoramic view from the eastern crater rim over the rim and crater depression, the latter of which sports fresh green vegetation after a period of rain fall (Photo: T. Kenkmann).

process. Meteor Crater still provides a first-rate training locality of astronauts and the next generation of impact cratering workers.

Meteor Crater was the first geological structure on Earth where the impact origin was confirmed. However, despite the fact that a huge number of iron meteorites of undisputable extraterrestrial origin had been collected in the environs of the crater already by the 1890s, it took several decades until its impact origin was finally accepted by most geologists.

The principal protagonists in the scientific debate about the origin of the structure were Grove Gilbert, one of the most renowned geoscientists at the end of the 19th century, who then was chief geologist of the US Geological Survey and president of the Geological Society of Washington (GSW), and Daniel Moreau Barringer, a young mining entrepreneur. Gilbert was convinced that the crater had been produced by a violent volcanic steam explosion similar to maar volcanoes known, for example, from the Eifel region of West Germany. This was apparently substantiated by the fact that the crater was situated close to a volcanically active district, the San Francisco Volcanic Field. Also, Gilbert's team did not find a magnetic anomaly in the crater, so that he concluded that no iron meteorite could be buried underneath the crater floor. Consequently, he rejected the impact hypothesis and published this as a GSW presidential address in the prestigious journal *Science*. He claimed that iron meteorites were coincidentally concentrated in this region.

Initially unaware of Gilbert's work but convinced of the formation of the crater by impact of an extraterrestrial projectile, Barringer obtained the crater property in 1903 and formed the Standard Iron Company to extract the iron and nickel metal from the crater that he expected to be provided by the extraterrestrial bolide and that were buried beneath the crater floor. Barringer started to investigate the crater in detail. He found that iron meteorites are concentrically distributed around the crater and increased in abundance towards the rim. He also observed iron meteorites buried beneath fractured rocks of the ejecta and correctly concluded that both iron meteorites and fractures must have been formed at the same time in a forceful event. In addition, he became aware that close to the rim of the crater the succession of strata had been overturned and occurred now upside-down, a phenomenon that is now well-known from the rims of many impact craters as the "inverted stratigraphy of the overturned flap". Finally, he noticed that some of the rocks were heavily deformed or even pulverized. Barringer and his colleague Benjamin Tilghman drilled several boreholes into the structure down to a depth of more than 400 m and could not find any volcanic material below the crater floor. He published his work in 1905 and bluntly criticized Gilbert and his conclusion of a volcanic steam explosion.

The geological community, however, would not be ready for a long time to accept Barringer's work, because Gilbert was such an influential adversary. This only changed when Eugene Shoemaker, the well-known pioneer of impact cratering and planetology, re-examined the crater. He compared the structure with nuclear

View towards the northwest from the crater rim of Meteor Crater. The boulders in the foreground, a few meters in size, belong to the ejecta blanket. At the horizon, at a distance of about 50 km, is the San Francisco Volcanic Field (photo: T. Kenkmann).

Meteor Crater observed from the access road to the museum and visitor center. The raised rim and hummocky relief of the ejecta blanket surrounding it are clearly visible (photo: T. Kenkmann).

explosion craters and found many analogies. In 1960 he and his colleague Edward Chao discovered coesite and stishovite at Meteor Crater. Such high-pressure minerals of quartz form close to the Earth's surface exclusively from the interaction of upper crustal rocks with impact-generated shock waves. This was the ultimate proof for the origin of the crater from the hypervelocity impact of an iron meteorite.

Shock physics also explains the absence of a large iron meteorite beneath the crater floor. During the impact event the immense kinetic energy of the extraterrestrial projectile approaching at a speed of more than 11 km/sec is converted instantaneously, and to a large degree, into heat that causes melting and vaporization of the target rock as well as the bulk of the bolide.

Barringer could not commercially use his property owing to the lack of any large ore body, and yet, he and his descendants have been prime supporters and sponsors of impact crater research ever since. The family's Barringer Crater Company has successfully run the visitor center situated on the crater rim. The Barringer Crater Company annually sponsors the prestigious Barringer Medal and Award for a leading researcher in the field of impact cratering and has provided countless students of meteoritics and impact cratering with travel grants to participate in the Annual Meeting of The Meteoritical Society, one of the most important planetary science conferences, and other conferences with a major impact cratering focus. An extensive field guidebook on Meteor Crater was published by David Kring, a scientific advisor of the Barringer Crater Company.

Location/Access

Barringer Crater is easily accessible. It is located east of the city of Flagstaff in Arizona just off Interstate 40 exit 233. The visitor center offers plenty of information, an interactive discovery center, Astronaut Memorial Park, and more. There are three different lookout points to view the crater from the rim.

Selected References

Barringer, D. M. (1910). Meteor Crater in northern central Arizona. Paper presented at the National Academy of Sciences, Princeton University, 16 November 1909, 24 pp. (plus 18 plates and 3 maps).

Barrows, T. T., Magee, J., Miller G., and Fifield, L. K. (2019). The age of the Wolfe Creek meteorite crater (*Kandimalal*), Western Australia. Meteoritics and Planetary Science, Vol. 54, pp. 2686–2697.

Grant, J. A., and Schultz, P. H. (1993). Erosion of ejecta at Meteor Crater, Arizona. Journal of Geophysical Research, Vol. 98, pp. 15033–15047.

Hörz, F., Mittlefehldt, D. W., See, T. H., and Galindo, C. (2002). Petrographic studies of the impact melts from Meteor Crater, Arizona, USA. Meteoritics and Planetary Science, Vol. 37, pp. 501–531.

Kring, D. A. (1997). Air blast produced by the Meteor Crater impact event and a reconstruction of the affected environment. Meteoritics and Planetary Science, Vol. 32, pp. 517–530.

Kring, D. A. (2017) Guidebook to the Geology of Barringer Meteorite Crater, Arizona (a.k.a. Meteor Crater), 2nd edition. Lunar and Planetary Institute Contribution No. 2040: https://www.lpi.usra.edu/publications/books/barringer_crater_guidebook

Meteor Crater Company: http://www.barringercrater.com

Meteor Crater Visitor Center: http://www.meteorcrater.com

Poelchau, M. H., Kenkmann, T., and Kring, D. A. (2009): Rim uplift in simple craters: the effects of target heterogeneities and trajectory obliquity. Journal of Geophysical Research, Vol. 114. doi: 10.1029/2008JE003235.

Roddy, D. J. (1978). Pre-impact geologic conditions, physical properties, energy calculations, meteorite and initial crater dimensions and orientations of joints, faults and walls at Meteor Crater, Arizona. Proceedings of the 9th Lunar and Planetary Science Conference, Houston, Texas, Pergamon Press, New York, pp. 3891–3930.

Shoemaker, E. M., and Kieffer, S. W. (1974). Guidebook to the Geology of Meteor Crater, Arizona. 37th Annual Meeting of The Meteoritical Society, Arizona State University Center for Meteorite Studies, Tempe, Arizona, 66 pp.

Middlesboro

Kentucky, United States

Thomas Kenkmann

Middlesboro is a 6 km diameter, complex impact structure that forms a peculiar depression in the hilly Appalachian mountain land. The circular structure has a small central uplift, in which strata are uplifted by 245 m with respect to their original stratigraphic position.

Target rocks are siliciclastic sediments of Pennsylvanian age, ranging from conglomerates over sandstones to shales and siltstones. Some coal seams occur as well. Most likely rocks of the underlying Mississippian were also affected by the impact. This rock sequence is relatively undisturbed outside the structure. However, the area belongs to an overthrust sheet and is bracketed by several thrust faults. Thrusting was to the northwest and occurred at the end of the Late Pennsylvanian / Early Permian in response to collisional tectonics at the end of the Paleozoic. From coal rank an original burial depth of roughly 4 km has been estimated for the strata exposed today.

Middlesboro is delineated by steeply dipping, circumferential normal faults. In this outer area the strata dip away from the crater and are upturned. Brittle deformation increases inward and the strata show some folding and brecciation. Tight folding occurs particularly in the shale units. The core composed of steeply dipping conglomeratic sandstones is bounded by steep fault zones. Rocks are shattered and brecciated here, down to the grain scale. Pebbles and matrices of the conglomerates display quartz mosaicism and contain planar fractures (PFs) and planar deformation features (PDFs). Faults contain brecciated, silt-sized quartz derived from adjacent rocks, set within a semi-glassy matrix. These materials have been identified as cataclasites and possibly pseudotachylites that were produced during central uplift formation.

The impact event occurred after the regional tectonic thrusting, which happened in the Carboniferous during the Variscan/Appalachian orogeny. Thus, the impact may have occurred at the end of the Pennsylvanian or in Early Permian time.

Initially the circular depression associated with strong deformation of rocks was attributed to regional tectonic overthrusting. The

The Middlesboro impact structure on October 17, 2018 as a composite image based on TanDEM-X DEM and Copernicus Sentinel-2 (channel 2/3/4) data.

Middlesboro

Kentucky / United States

latitude = 36°37‘03“ N / longitude = 83°43‘39“ W / diameter = 6 km / age = 290 - 300 Myr

scale = 1: 119 000 / scene = 21.4 × 28.5 km²

0 1.19 5.95 km

N

285 meter 980

structure was first described as a probable ancient meteorite impact crater by Kenneth Englund and John Roen – as early as 1962. Englund subsequently mapped the structure in detail. While it was also considered as a "cryptoexplosion" structure, Robert Dietz could finally confirm the impact hypothesis in 1966 when he found shatter cones at Middlesboro.

Location/Access

The Middlesboro structure is located in the southeast corner of Kentucky, in the Appalachian Plateau country at the border to Tennessee. The city of Middlesboro completely occupies the crater depression. The central uplift crops out at a golf course and shows the most intense deformation including shatter cone occurrences. The largest city nearby is Knoxville, Tennessee – about 75 km to the south. From Knoxville, Tennessee State Route 1, followed by State Route 32 from Tazewell, leads directly to Middlesboro.

Selected References

Black, D. F. B. (1964). Cryptoexplosive structure near Versailles, Kentucky. U.S. Geological Survey Professional Paper, Vol. 501-B, pp. B9–12.
Dietz, R. S. (1966). Shatter cones at the Middlesboro structure, Kentucky. Meteoritics, Vol. 3, pp. 27–29.
Englund, K. J., and Roen, J. B. (1963). Origin of the Middlesboro Basin, Kentucky. U.S. Geological Survey Professional Paper, Vol. 450-E, pp. E20–E22.
Hower, J. C., Greb, S. F., Kuehn, K. W., and Eble, C. F. (2009). Did the Middlesboro, Kentucky, bolide impact event influence coal rank? International Journal of Coal Geology, Vol. 79, pp. 92–96.

Mistastin

Labrador, Canada

Wolf Uwe Reimold

The Mistastin, or *Mistastin Lake*, impact structure is, at about 28 km apparent crater diameter, a complex impact structure. This diameter value represents a mean of the distances from the center to the outermost prominent hills that are thought to be aligned along the largely eroded crater rim of this apparently oval structure. The impact structure was formed about 36 Myr ago. It has been eroded, but indications of the terraced rim and remnants of the central uplift are still observed. The inner parts of the structure are covered by Lake Mistastin, while the environs of the lake and, in fact, of the structure are locally covered by 0.5–5 m thick glacial deposits and soil/vegetation.

The crystalline basement of the Mistastin Lake region, which formed the target for this impact event, is dominated by anorthosite, mangerite, granodiorite, and quartz monzonite. Mistastin Lake is located in the upper northeast quadrant of the granodioritic Mistastin Lake batholith. Within the confines of the structure the rocks are fractured, which is distinct from the appearance of the rocks of the surrounding terrain. The lack of fracturing is also obvious in spaceborne imagery.

The rather good exposure at Mistastin allows the investigation of a near complete section through a mid-size complex impact crater

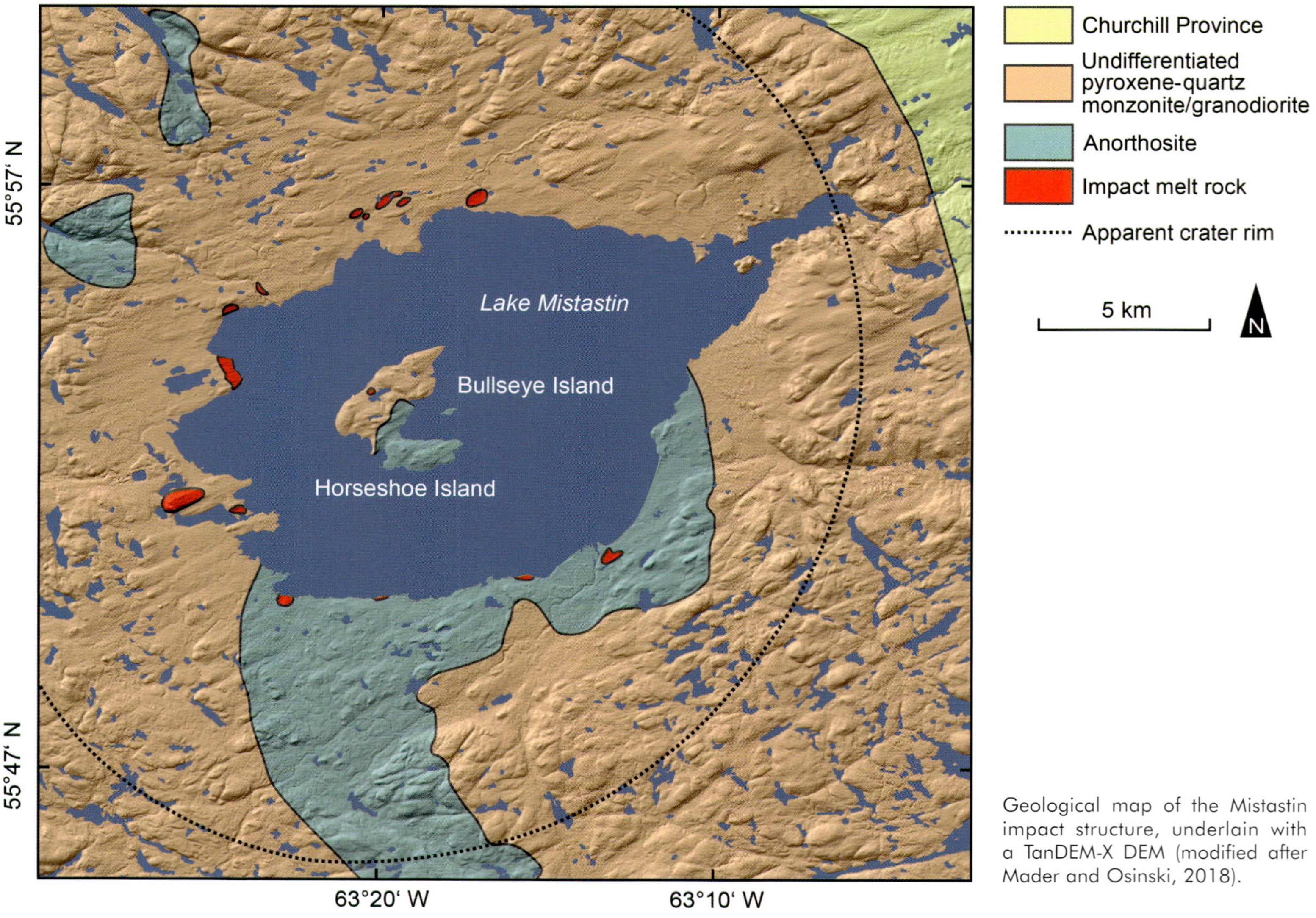

Geological map of the Mistastin impact structure, underlain with a TanDEM-X DEM (modified after Mader and Osinski, 2018).

Mistastin

Labrador / Canada

latitude = 55°53' N / longitude = 63°19' W / diameter = 28 km / age = 36.6 ± 2 Myr

scale = 1:306 000 / scene = 55.1 × 73.5 km²

0 3.06 15.30 km

N

40 meter 685

Exposure of massive impact melt rock on Horseshoe Island in Lake Mistastin. Note the prominent, vertical cooling fractures. Hammer for scale about 40 cm long. (image courtesy: G. Osinski).

formed in crystalline target rock. The fact that anorthosite represents one of the major target lithologies at this structure renders it comparable with impact craters formed in anorthositic lunar crust.

The structure has been extensively studied, but mainly with a focus on the impact melt rocks. Much new information has become available in recent years in the course of a dedicated investigation by the group around Marianne Mader and Gordon Osinski. They reported extensively on a suite of impactites preserved in parts of the Mistastin Lake impact structure. This includes shocked basement rocks, monomict impact breccias, polymict lithic and suevitic impact breccias, and impact melt rocks. While the structure has been significantly eroded – in some parts down to the crater floor, there are still ejecta deposits, and the uplifted rim and the central uplift are still partially preserved.

The topography from the edge of Mistastin Lake to the rim section is somewhat stepped. The highest and furthest hills have been interpreted as the apparent expression of the crater rim. This yields a 30 × 26 km area for this impact structure. Two distinct zones have been defined between the lakeshore and apparent crater rim; they are marked by steep changes in elevation, changes in fracture and fault patterns of the rocks, extent of rock exposure, and in changes of drainage patterns. Horseshoe and Bullseye islands near the center of the lake are thought to represent remnants of the central uplift of this complex impact structure. Horseshoe Island is 3 km wide and rises up to some 130 m above lake level.

Structural mapping revealed two zones that define a collapsed rim region and a central uplift region. The zones are delimited by raised rings marked by hills and short, discontinuous faults. Impactites have been reported to be unevenly distributed around the lake and to occur on Horseshoe Island. These rocks primarily occur within the innermost part of the structure. They are best exposed along the shore of the lake and on steep banks of streams. Along the rim section, barren hill tops are favored for exposure. It has been reported that Mistastin Lake shows evidence of impact melt ejecta that flowed over ballistically emplaced, polymict lithic and melt-poor impact breccias within the outer zone of the impact structure.

Around the lakeshore and on Horseshoe Island isolated occurrences of impact melt rock are found. In the southwest, a location known as "Discovery Hill" exhibits a 80 m high cliff face of impact melt rock with excellent columnar jointing. The impact melt rocks occur above discontinuous allochthonous impact breccia, i.e., suevite and lithic impact breccia, which, in turn, overlies parautochthonous, monomict brecciated basement. The latter also contains dikes of impact glass, suevite, and pseudotachylitic breccia.

The impact melt rocks at Mistastin have been studied in great detail. They seem to form an essentially flat-lying, shallowly inward-dipping (towards the center of the structure), sheet-like body. Basement-derived clasts occur up to block-size of several meters. Geochemical analysis has shown that the impact melt rock is essentially formed from anorthosite and mangerite, which are the basement lithologies exposed in the central part of the structure. Very limited chemical evidence for a possible enrichment of Ir and Ni above basement rock values has been obtained to date. It is still unclear whether the Mistastin impact melt rock contains an impactor component.

Evidence for impact has been observed in the form of shatter cones, as well as occurrence of shock effects including planar deformation features (PDFs) in quartz, and diaplectic quartz and feldspar glasses in basement rocks from the central island.

A regional gravity survey indicated a Bouguer gravity low of about −15 mGal extending over the southeastern sector of the impact structure. This anomaly has been related to the fact that this area is underlain by comparatively low-density anorthosite. The magnetic data over the structure have been tentatively argued to suggest a slight smoothing of gradients over the central part of the structure. However, whether this magnetic pattern can be related to impact overprint is uncertain. A more recent detailed paleomagnetic and magnetic investigation showed that only basement rock residing less than 1 m from impact melt rock was thermally overprinted by the impact event. And basement rocks up to 6–7 km from the center of the structure did not experience a thermal effect from the impact in excess of 200 °C.

Since the 1970s, an ^{40}Ar-^{39}Ar step-heating age of 36 ± 4 Myr has been given for the Mistastin Lake impact, calculated from plateau ages for six impact melt rock samples. More recently dating of zircon from impact melt rock by the low-temperature (U-Th)/He technique and of matrix of impact melt rock by the laser argon microprobe method yielded an age of 36.6 ± 2 Myr.

First geological studies of the area around Mistastin Lake were carried out in the late 1960s by the Geological Survey of Canada, after they had been alerted to the occurrence of prominent igneous rocks in the area by geologists of British Newfoundland Exploration. Subsequent work resulted in recognition of the structure's origin from a large impact event. This was based on findings of shatter cones and microscopic evidence of shock metamorphism. As with numerous other impact structures, the origin of Mistastin Lake continued to be debated in terms of a possible volcanic or "cryptoexplosion" origin, into the 1970s. However, all subsequent workers on this structure have agreed that the evidence for impact is overwhelming and diagnostic.

Location/Access

The Mistastin impact structure is located in a remote part of central Labrador, about 120 km north of Schefferville in Quebec. It seems that access is limited to air transport, i.e., helicopter or seaplane. Clearly elaborate preparations including permitting are required for an expedition to this remote impact structure.

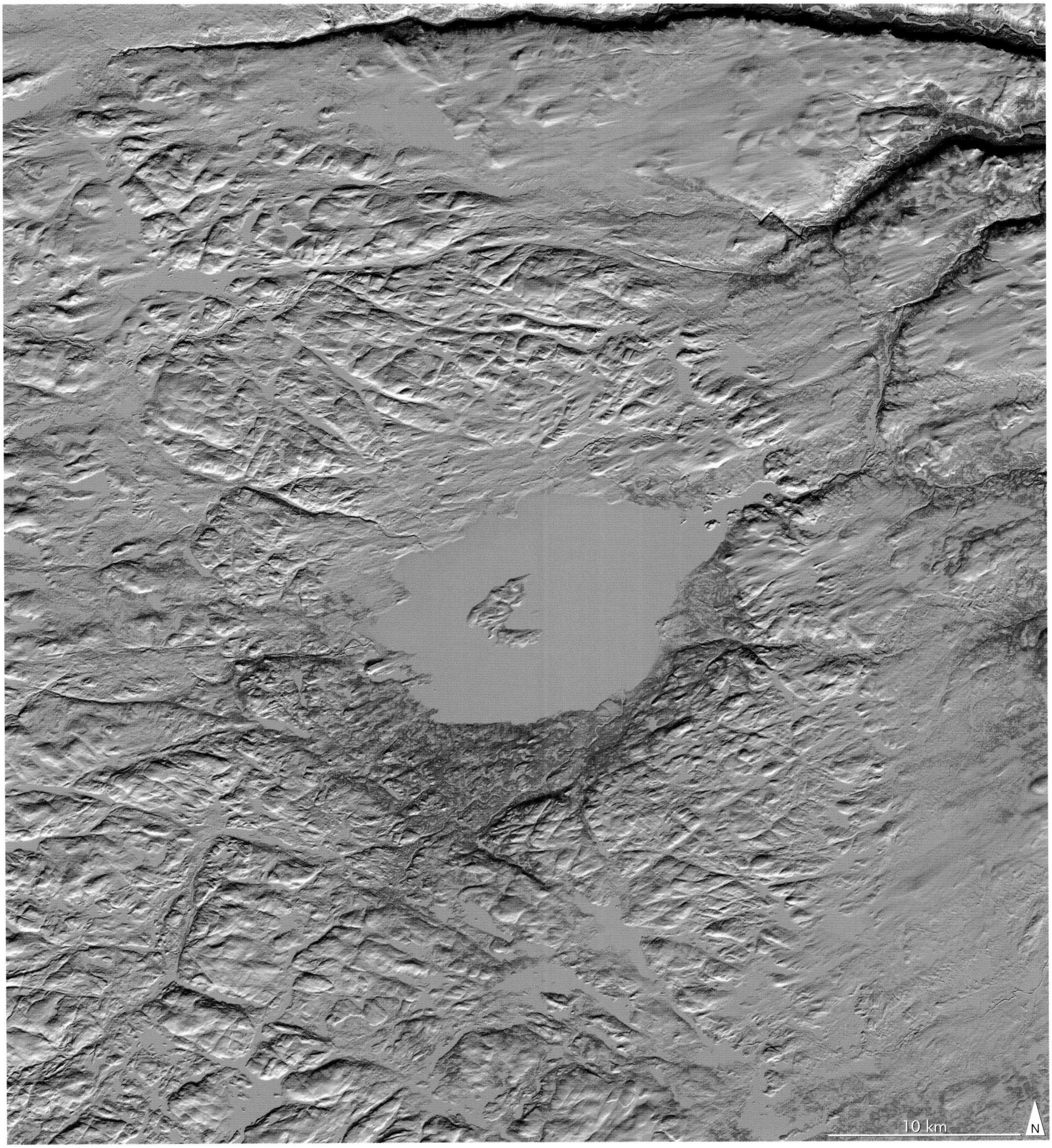

The Mistastin impact structure with the homonymous ice-covered lake on February 14, 2019 as a composite image based on TanDEM-X DEM and Copernicus Sentinel-2 (channel 2/3/4) data.

Selected References

Currie, K. L. (1971). Geology of the resurgent cryptoexplosion crater at Mistastin Lake, Labrador. Geological Survey of Canada, Bulletin 207, 62 pp.

Grieve, R. A. F. (2006). Impact Structures in Canada. Geological Association of Canada, GeoText No. 5, 210 pp.

Hervé, G., Gilder, S. A., Marion, C. L., Osinski, G. R., Pohl, J., Petersen, N., and Sylvester, P. J. (2015). Paleomagnetic and rock magnetic study of the Mistastin Lake impact structure (Labrador, Canada): Implications for geomagnetic perturbation and shock effects. Earth and Planetary Science Letters, Vol. 417, pp. 151–163.

Mader, M. M., and Osinski, G. R. (2018). Impactites of the Mistastin Lake impact structure: Insights into impact ejecta emplacement. Meteoritics and Planetary Science, Vol. 53, pp. 2492–2518.

Marion, C. L., and Sylvester, P. J. (2010). Composition and heterogeneity of anorthositic impact melt at Mistastin Lake crater, Labrador. Planetary and Space Science, Vol. 58, pp. 552–573.

Young, K. E., Mercer, C. M., van Soest, M. C., Hodges, K. V., Wartho, J. A., and Biren, M. B. (2014). An examination of noble gas geochronology and thermochronology in the context of dating impact events. 45th Lunar and Planetary Science Conference, Lunar and Planetary Institute, Houston, Texas, Abstract #2670.

Montagnais
Nova Scotia, Canada

Thomas Kenkmann

Montagnais is located in a shelf environment, off-shore Nova Scotia, at a water depth of 115 m. The buried structure is 45 km in diameter and is mainly known from reflection seismic studies.

Target rocks in this shelf area are low-grade meta-greywackes and quartzite of Cambrian to Ordovician age, on which sedimentary rocks of Mid Jurassic to Eocene age rest unconformably. They were water-saturated and partly unconsolidated at the time of impact. The structure is covered by more than 500 m of Cenozoic sedimentary rocks.

The central uplift of Montagnais is quite large. It has a diameter of 11.5 km with a relatively flat top. At the summit exists a 3.2 km wide and 200 m deep central pit. The stratigraphic uplift is at least 1.8 km, and the topographic height amounts to 500 m with respect to the surrounding crater fill breccia. An annular moat with a width of some 10 km, downfaulted by up to 1000 m, surrounds the central uplift. It is filled with allochthonous breccia of approximately 500 m thickness. A borehole was drilled into the central pit on top of the central uplift and revealed under 510 m of post-impact sedimentary rocks more than 550 m of polymict breccia, before reaching the strongly tilted Paleozoic strata of the central uplift. The suevitic breccia contains a variety of lithologies, including granites from the basement, that indicate that the excavation depth of the transient cavity exceeded 2 km depth. Breccia fragments display varied shock metamorphic overprint. Planar deformation features (PDFs) and planar fractures (PFs) have been described, as well as impact melt fragments, in these suevitic breccias. Suevite interfingers with impact melt rock layers.

Most of the information on the structure of Montagnais was gathered from reflection seismic profiles. A gravity survey showed a +8 mGal positive Bouguer gravity anomaly. This anomaly corresponds to the central rise, in which dense basement rocks were uplifted. A weak negative gravity excursion of −1 to −2 mGal surrounds the central uplift and correlates with the annular moat that is filled with comparatively less dense brecciated material. The magnetic anomaly pattern is dominated by the signatures of basement lithologies.

The age of the Montagnais impact is Early Eocene. Radiometric age determinations using the K-Ar technique on impact melt rock yielded ages between 49.9 and 55.8 Myr. ^{40}Ar-^{39}Ar step-heating experiments on melt rock samples by Richard J. Bottomley and Derek York gave an age of 50.5 ± 0.8 Myr, which is the currently accepted age for the event. The marine impact did not have a detectable effect on the regional biosphere.

The structure was first identified in reflection seismic sections that were gathered in the course of exploration activities by an oil company. In 1974 a borehole was drilled in the center of the structure in the course of oil exploration. Revisiting of the borehole samples in 1987 by Lubomir Jansa and Georgia Pe-Piper led to the discovery of shocked minerals in breccia samples and the confirmation of the marine impact event.

Location/Access

The submerged and buried impact crater is located off-shore, 200 km south of Halifax.

Selected References

Bottomley, R. J., and York, D. (1988). Age measurement of the submarine Montagnais impact crater. Geophysical Research Letters, Vol. 15, pp. 1409–1412.

Deptuck, M. A., and Campbell, D. C. (2012). Widespread erosion and mass failure from the ~51 Ma Montagnais marine bolide impact off southwestern Nova Scotia, Canada. Canadian Journal of Earth Sciences, Vol. 49, pp. 1576–1594.

Grieve, R. A. F. (2006). Impact Structures in Canada. Geological Association of Canada, GeoText No. 5, 210 pp.

Jansa, L. F., and Pe-Piper, G. (1987). Identification of an underwater extraterrestrial impact crater. Nature, Vol. 327, pp. 612–614.

Jansa, L. F., Pe-Piper, G., Robertson, P. B., and Friedenreich, O. (1989). Montagnais: A submarine impact structure on the Scotian shelf, eastern Canada. Geological Society of America Bulletin, Vol. 101, pp. 450–463.

Pilkington, M., Jansa, L. F., and Grieve, R. A. F. (1995). Geophysical studies of the Montagnais impact crater, Canada. Meteoritics, Vol. 30, pp. 446–450.

Poag, C. W., Plescia, J. B., and Molzer, P. C. (2002). Ancient impact structures on modern continental shelves: The Chesapeake Bay, Montagnais, and Toms Canyon craters, Atlantic margin of North America. Deep-Sea Research Part II: Topical Studies in Oceanography, Vol. 49, pp. 1081–1102.

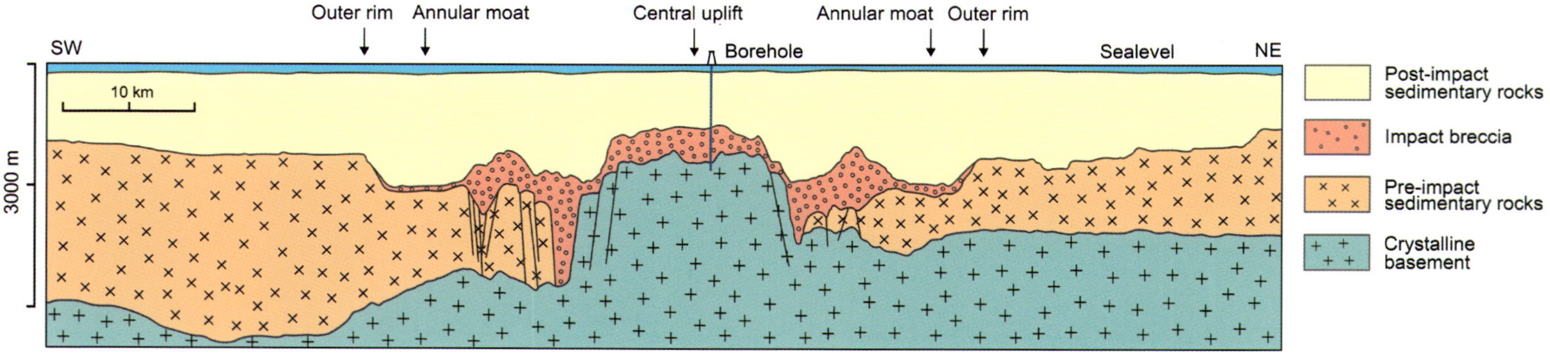

Schematic geological cross-section through the Montagnais impact structure (modified after Poag et al., 2002).

Montagnais

Nova Scotia / Canada

latitude = 42°53‘ N / longitude = 64°13‘ W / diameter = 45 km / age = 50.5 ± 0.8 Myr

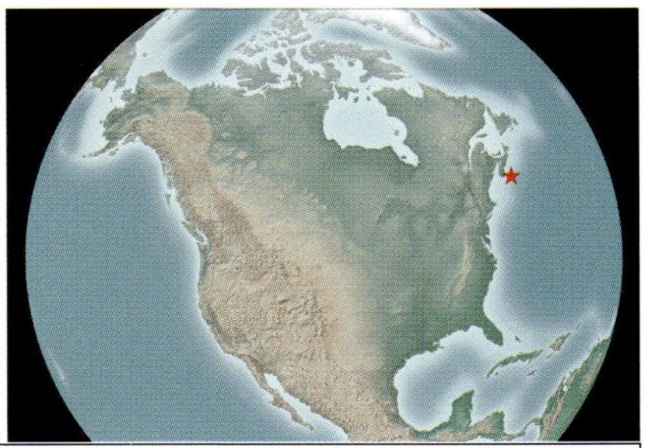

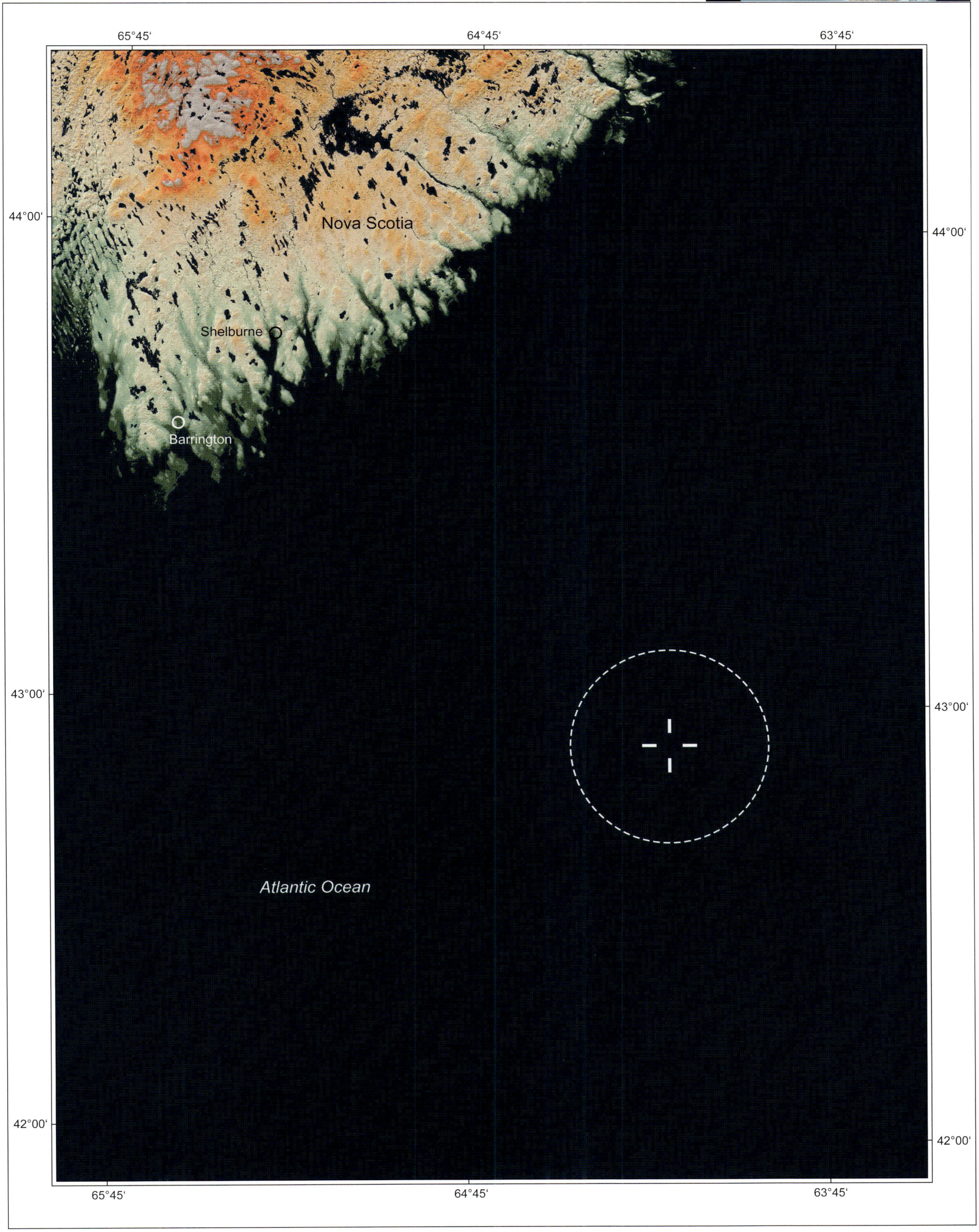

scale = 1 : 1 114 444 / scene = 201.0 × 267.0 km²

0 11.14 55.70 km

N

0 meter 205

Newporte
North Dakota, United States

Thomas Kenkmann

Newporte is a 3 km diameter impact structure located just south of the border between the United States and Canada. The entire crater is buried to a depth of almost 3 km. The structure lacks a central uplift and is considered a simple, bowl-shape crater. Seismic, well-log, and drill core data support a bowl-shaped crater geometry with a raised rim. The Newporte impact structure was formed in the Williston Basin that is well known for its oil resources.

The granitic Precambrian basement to the structure is unconformably overlain by sediments of Cambrian age, the so-called "Deadwood Formation". The Cambrian sandstones are oil producing. Ordovician, Mississippian, and Devonian carbonates are undeformed and younger than the impact, which occurred around 500 Myr ago.

Most of the drillings penetrated the rim of the structure that is uplifted by up to 60 m in the northeastern sector of the crater. The inner depression is 100 m deep. One of the cores contains a 10 m thick brecciated layer within the Deadwood Formation, at a depth of 2880 m. The Precambrian basement rocks and the Lower Cambrian Deadwood Formation are highly fractured and brecciated. Three types of breccia were distinguished: Granite-derived breccia with localized melting, sediment-derived breccia, and mixed types. The latter shows the strongest shock metamorphic overprint, with many planar deformation features (PDFs) having been found in quartz and K-feldspar. The impact most likely occurred during the deposition of the Deadwood Formation. The petroleum at Newporte has a different geochemical signature than that of other producing fields in the Williston Basin. This suggests that the oil producing layers were probably formed in isolation in the bowl-shaped crater depression.

Exploration for hydrocarbons led to the discovery of the Newporte structure in 1977. Richard Donofrio interpreted Newporte as a possible impact structure in 1981, without, however, giving any proof. This came in the 1990s when several research teams presented petrographic evidence for shock metamorphic overprint related to the crater breccias.

Location/Access

The structure lies at the border between the Unites States and Canada in Renville County, North Dakota. The nearest town is Sherwood, 25 km east of the structure. From there State Highway 107 NW leads directly to Newporte.

Selected References

Clement, J. H., and Mayhew, T. E. (1979). Newporte discovery opens new pay. Oil and Gas Journal, Vol. 77, pp. 165–172.

Donofrio, R. R. (1997). Survey of hydrocarbon-producing impact structures in North America: Exploration results to date and potential for discovery in Precambrian basement rock. In: Ames Structure in Northwest Oklahoma and Similar Features. Origin and Petroleum Production, Johnson, K. S. and Campbell, J.A. (eds.), Oklahoma Geological Survey, Norman, Oklahoma, Circular 100, pp. 17–29.

Forsman, N. F., Gerlach, T. R., and Anderson, N. L. (1996). Impact origin of the Newporte structure, Williston Basin, North Dakota. American Association of Petroleum Geologists Bulletin, Vol. 80, pp. 721–730.

Koeberl, C., and Reimold, W. U. (1995). The Newporte impact structure, North Dakota, USA. Geochimica et Cosmochimica Acta, Vol. 59, pp. 4747–4767.

Reimold, W. U., Koeberl, C., Gibson, R. L., and Dressler, B. O. (2005). Economic Mineral Deposits in Impact Structures: A Review. In: Impact Tectonics, Koeberl, C., and Henkel, H. (eds.), Springer, Berlin Heidelberg, pp. 479–552.

The Newporte impact structure on June 6, 2019 as a composite image based on TanDEM-X DEM and Copernicus Sentinel-2 (channel 2/3/4) data. Cross hairs and a dashed circle indicate position and extent of the buried structure.

Newporte

North Dakota / United States

latitude = 48°58‘ N / longitude = 101°58‘ W / diameter = 3 km / age = 500 Myr

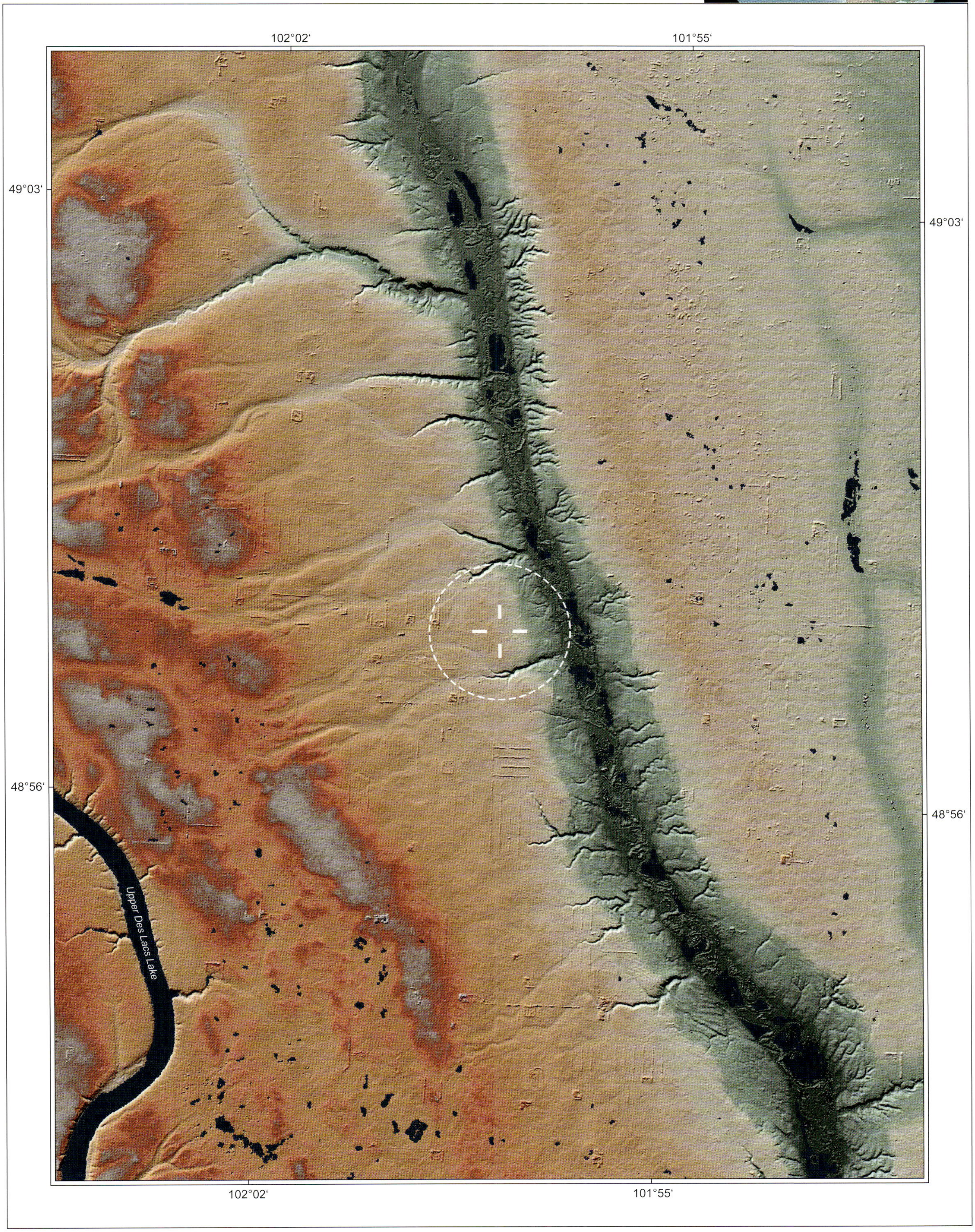

scale = 1: 103 000 / scene = 18.5 × 24.6 km²

0 1.03 5.15 km

N

485 meter 575

Nicholson Lake
Northwest Territories, Canada

Thomas Kenkmann

The Nicholson Lake impact structure is an eroded, complex crater of 12.5 km diameter. Its morphology lacks a characteristic circular surface expression due to deep glacial scour. A number of islands and a large peninsula on the western lakeshore have revealed impact lithologies. The central uplift rises approximately 50 m above the lake level, whereas the lake level itself lies 65 m below the mean regional topographic level. No elevated crater rim is notable.

The Nicholson structure was formed in the western part of Churchill Province of Canada. This craton area with rocks that range in age from 3400 to 1800 Myr has undergone a complex geological history. The basement includes gneisses and greenstones that have been intruded by various granitoids. At the time of impact, the Precambrian basement was overlain by Middle Ordovician limestones. These strata have largely been removed on a regional scale, but sedimentary rocks are documented as fragments in impact breccias. Two outliers of down-faulted Ordovician limestone and dolomites occur in the western part of the moat of the structure.

The present level of erosion roughly coincides with the ancient crater floor, below which fractured basement rocks can be found. Crude shatter cones are locally developed in these rocks and the shocked gneiss and granites show other shock effects such as planar deformation features (PDFs) and planar fractures (PFs) in quartz. Monomict and polymict clastic breccias are regarded as crater fill breccia. They are locally overlain by inclusion-rich impact melt rock. The latter is exposed on a number of small islands and the promontory on the western side of the lake. Also suevites are present.

The presence of melt-bearing breccias allowed to radiometrically date the structure: ^{40}Ar-^{39}Ar geochronology indicated an age for the formation of the Nicholson Lake structure of less than 400 Myr. An U-Pb isotopic study of apatites from basement rocks yielded a lower intercept age of 389 ± 6.7 Myr, which places the impact event into the Middle Devonian. Impact melt rock contains a meteoritic signature. The elements Ni, Cr and Co are distinctly more enriched than Ir and Au, which tentatively has been interpreted to suggest an olivine-rich achondritic projectile type.

The Nicholson Lake impact structure is associated with a negative Bouguer gravity anomaly of −7.5 mGal, which can be attributed to the intense fracturing of the basement rocks down to a depth of about 2 km below the surface. Aeromagnetic data show a circular zone with a low magnetic relief that corresponds to the extent of the impact structure.

A first reference to the unusual occurrence of Ordovician limestone and shattered rocks at Nicholson Lake dates back to 1896 and is accredited to the pioneering Canadian geologist Joseph Tyrell. Michael Dence and Morris Innes conducted a field reconnaissance survey in 1965 and discovered evidence of shock metamorphism.

Location/Access

The Nicholson impact structure is located in the remote north of Canada in the Northwest Territories. There is no infrastructure nearer than Fort Reliance, which is located 400 km west of Nicholson. Seaplanes are the means of transport to reach the impact structure.

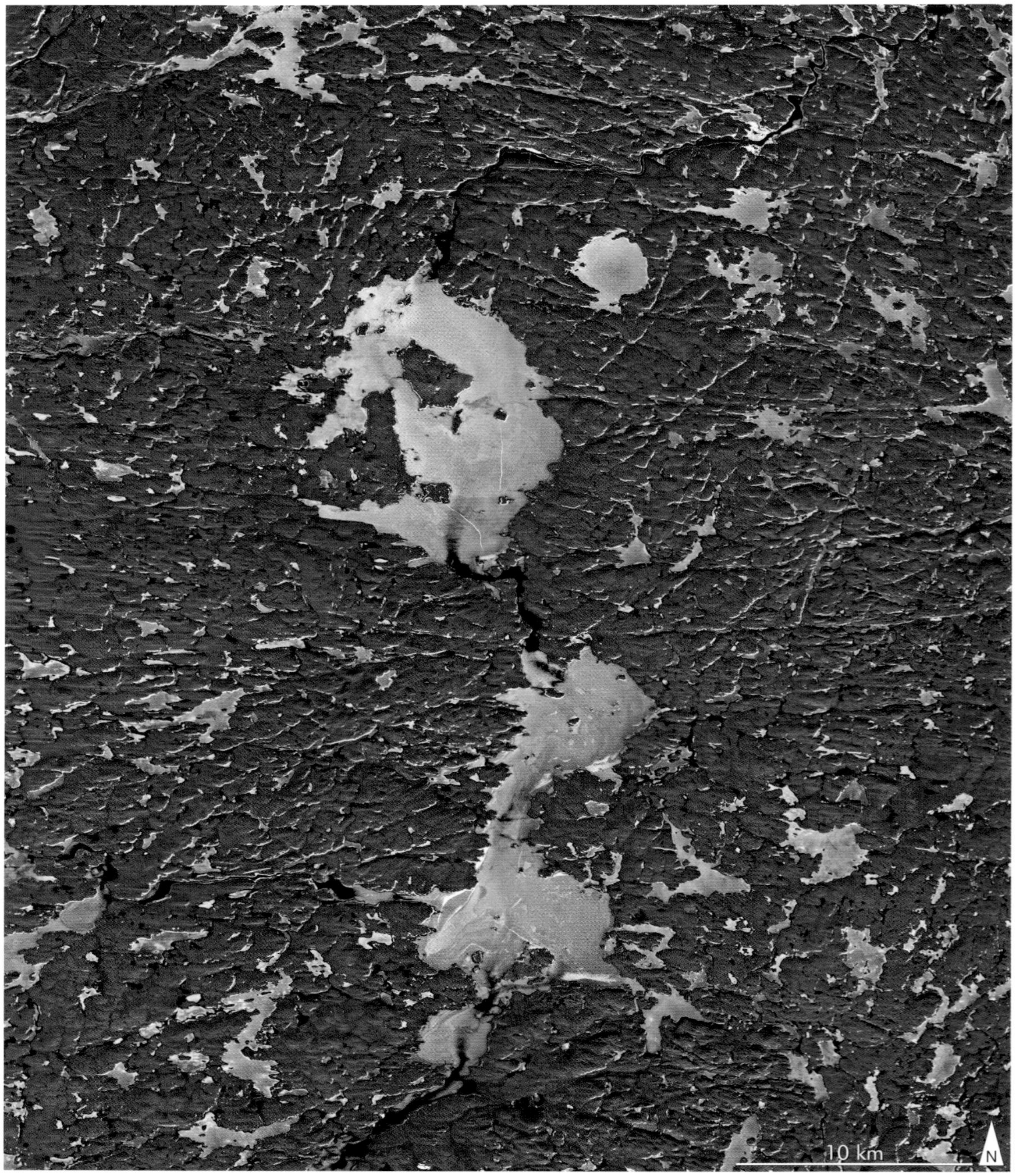

The Nicholson impact structure with the homonymous ice-covered lake on June 13, 2018 as a composite image based on TanDEM-X DEM and Copernicus Sentinel-2 (channel 2/3/4) data.

Nicholson Lake

Northwest Territories / Canada

latitude = 62°40‘ N / longitude = 102°42‘ W / diameter = 12.5 km / age = 389.0 ± 6.7 Myr

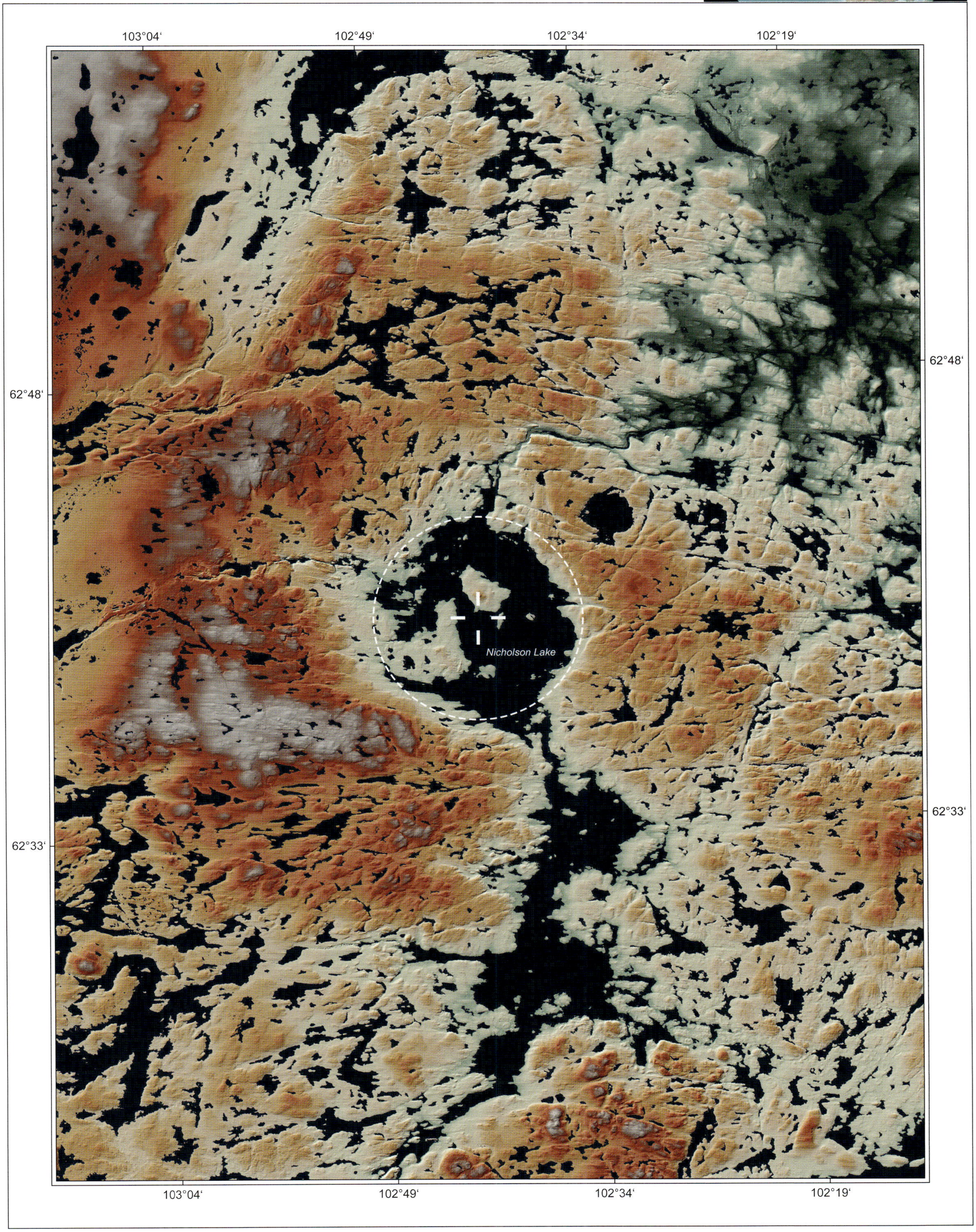

scale = 1 : 288 000 / scene = 51.9 × 69.2 km^2

0 2.88 14.40 km

N

240 meter 405

Selected References

Dence, M. R., Innes, M. J. S., and Robertson, P. B. (1968). Recent geological and geophysical studies of Canadian craters. In: Shock Metamorphism of Natural Materials, French, B. M., and Short, N. M. (eds.), Mono Book Corp., Baltimore, pp. 339–362.

Grieve, R. A. F. (2006). Impact Structures in Canada. Geological Association of Canada, GeoText No. 5, 210 pp.

Innes, M. J. S., and Dence, M. R. (1965). Nicholson Lake and Pilot Lake craters, N.W.T., Canada. Meteoritical Society Meeting, Odessa, Texas.

McGregor, M., McFarlane, C. R. M., and Spray, J. G. (2017). The Nicholson Lake impact structure, Canada: shock features and age of formation. 48th Lunar and Planetary Science Conference, The Woodlands, Texas, Abstract #2151.

Wolf, R., Woodrow, A. B., and Grieve, R. A. F. (1980). Meteoritic material at four Canadian impact craters. Geochimica et Cosmochimica Acta, Vol. 44, pp. 1015–1022.

Odessa Crater Field

Texas, United States

Thomas Kenkmann

The Odessa crater strewn field in Texas consists of five craters, the largest measuring 152 × 183 m with a somewhat irregular shape. The other four craters, located in close vicinity, are considerably smaller with diameters ranging from 3 to 25 m. The main crater is defined by an uplifted rim that is elevated by about 1 m with respect to the surrounding area; however, its height is also irregular. Limestones of the rim zone are in upright position and locally even overturned, and they form a crater rim anticline. The morphometry of the main crater with a low depth-to-diameter ratio suggests an oblique impact from either northeast or southwest.

In the Odessa area the bedrock is limestone and shale, which is underlain by sandstone and shale. All these strata are of Lower Cretaceous age. The surface is formed by altered eolian sediments of Pleistocene age, and the regional surface soil. The craters are essentially filled with post-impact debris, so that the basin center of the main crater is only 6 m below the rim. The filling of the main crater includes a three meter thick basal zone of fall-back ejecta material, a 4 m zone of interbedded gravel, silt, and clay, and 20 m of sand, silt, clay, and mud. Erosion has not been severe, as indicated by the fact that the surrounding area is still covered with an ejecta deposit of about 1 m thickness, as measured immediately adjacent to the crater.

A 9 × 6.5 × 5 cm sized limestone sample from the southeastern part of the Odessa impact crater (photo courtesy: Ralf-Thomas Schmitt, Museum für Naturkunde Berlin, sample from Michael Poelchau).

A large number of meteorites was found at the rim of the crater and in the surroundings. The largest piece had a mass of 150 kg. A total mass of approximately 5500 kg of meteoritic material was collected from the crater cavity. The projectile was classified as a coarse octahedrite, a subgroup of iron meteorites. The crater was used for the validation of numerical hydrocodes that are used to simulate impact processes in geologic materials.

The finding of an elephant tooth and equus fossils between 6 and 3 m below the surface indicated that the upper crater fill was of late Pleistocene age. Dating by optically stimulated luminescence revealed that the impact took place immediately prior to about 63 500 ± 4500 years ago.

The Odessa meteorite crater was first described by Arthur Bibbins in 1926, and by Daniel Moreau Barringer in 1928. Initially Odessa was considered to be formed in one event together with **Meteor (Barringer) Crater**. More precise dating, however, discarded this scenario. Later, four smaller craters were found in the environs of the main crater, and thousands of iron meteorites were discovered in the course of intensive field study from 1939 to 1941, and from 1958 to 1960.

Location/Access

Odessa can easily be reached from the Texas Interstate 20 connecting Dallas with the town of Odessa. In Odessa, the Meteorite Crater Road branches off southward from Interstate 20 to the crater and the Meteor Crater Museum.

Selected References

Barringer, B. (1967). Historical notes on the Odessa meteorite crater. Meteoritics, Vol. 3, pp. 161–168.

Barringer, D. M. Jr. (1928). A new meteor crater. Proceedings of the Academy of Natural Sciences of Philadelphia, Vol. 80, pp. 307–311.

Bibbins, A. B. (1926). A small meteor crater in Texas. Engineering and Mining Journal Press, Vol. 121, p. 932.

Holliday, V. T., Kring, D. A., Mayer, J. H., and Goble, R. J. (2005). Age and effects of the Odessa meteorite impact, western Texas, USA. Geology, Vol. 33, pp. 945–948.

Odessa Crater Field

Texas / United States

latitude = 31°45‘21“ N / longitude = 102°28‘43“ W / diameter = max. 170 m / age = 63 500 ± 4500 yr

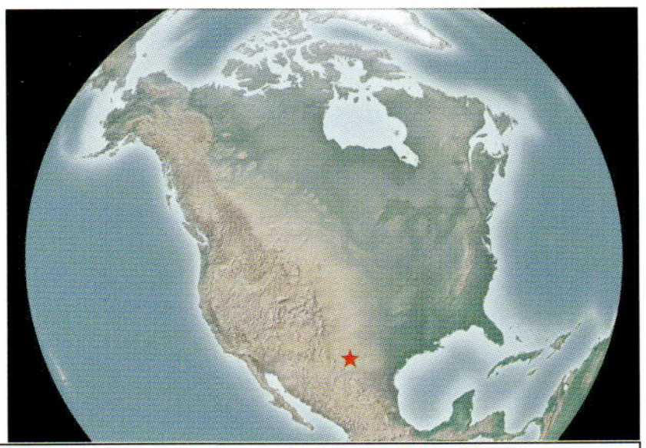

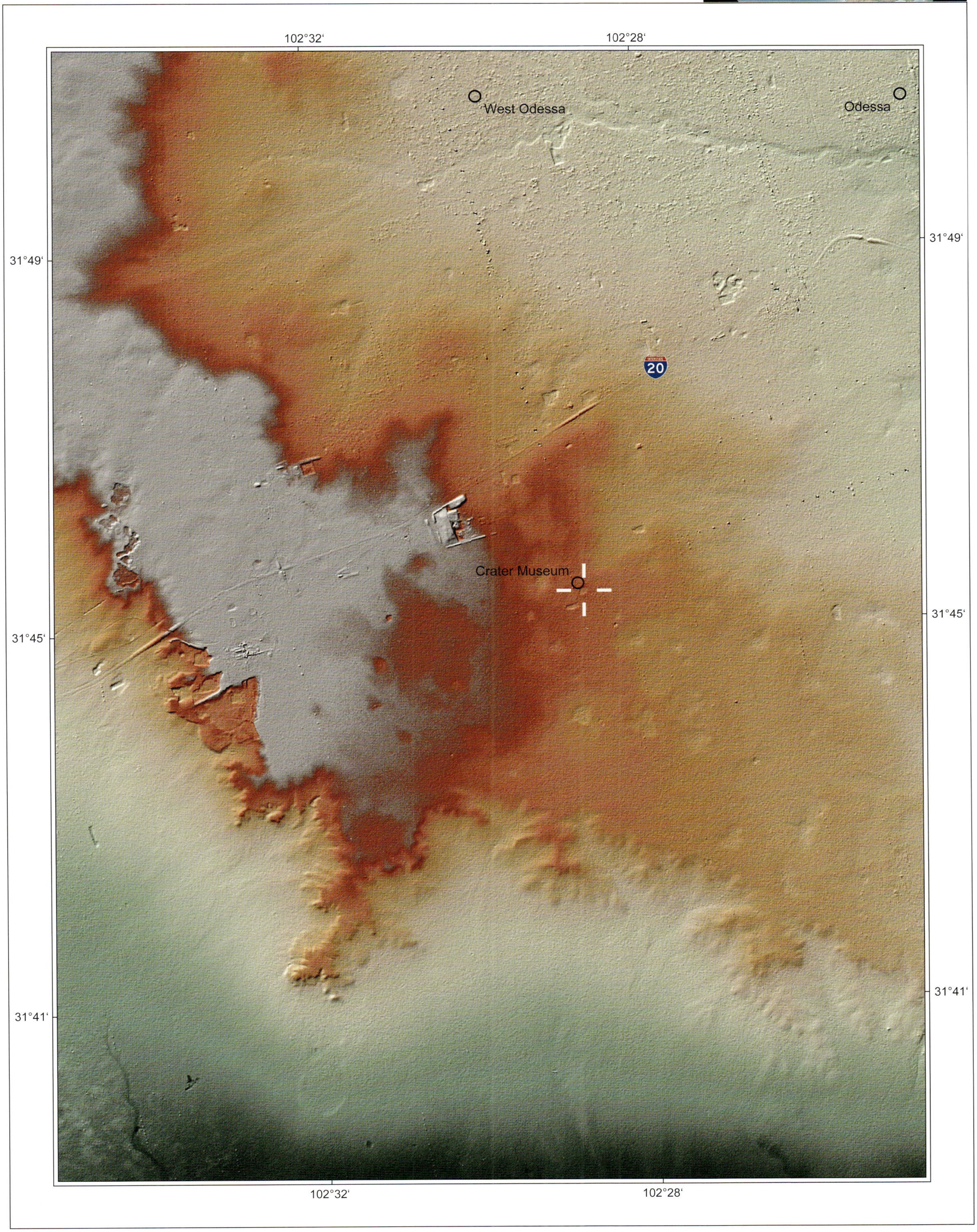

scale = 1 : 92 000 / scene = 16.6 × 22.1 km^2

0 0.92 4.60 km

N

850 meter 960

Pilot

Northwest Territories, Canada

Thomas Kenkmann

Pilot is a conspicuous circular to square shaped lake of roughly 5–6 km diameter in an area that is otherwise characterized by glacial morphology with narrow, elongated lakes following the orientation of joints. Except for the lake, the topography is non-specific. The surrounding area rises 60 m above the lake level, whilst the lake has a depth of 50–70 m. The crater is deeply eroded. Its size is given at 6 km, which roughly corresponds to the length of the diagonals across the lake. The structure was named "Pilot", as the lake's shape served as a navigation aid for pilots.

Target rocks are Proterozoic gneisses of granitic to granodioritic composition of the Churchill Province, which is part of the Canadian Shield. Shock effects were discovered in specimens from glacial till collected at the western shore of the lake. The boulders are fractured and brecciated gneisses. A small cobble includes some fine-grained, clast-rich impact melt rock that has the same composition as the granitic host rock. Planar deformation features (PDFs) were found in quartz and K-feldspar grains.

The Bouguer gravity map across the region displays a negative anomaly of −6 mGal, roughly coincident with Pilot Lake. The center of this anomaly is at the northern shore of the lake. Aeromagnetics does not exhibit any characteristic deviation from the regional anomaly pattern. Based on the single impact melt rock sample, an absolute radiometric age could be obtained using ^{40}Ar-^{39}Ar dating. According to this, the impact event occurred 445 ± 2 Myr ago. Pilot Lake was visited in 1965 by Morris Innes and Michael Dence, who found evidence for shock metamorphism in quartz-bearing rocks. This has remained the only inspection of the site until today.

Location/Access

Pilot is situated 60 km northeast of Fort Smith. This settlement is at the border between Alberta and the Northwest Territories. Seaplanes are the means of transport for reaching Pilot Lake, with the Taltston River Airport located just 20 km northwest of the crater.

Selected References

Bottomley, R. J., York, D., and Grieve, R. A. F. (1990). ^{40}Argon-^{39}Argon dating of impact craters. Proceedings of the 20th Lunar and Planetary Science Conference, Houston, Texas, Cambridge University Press, pp. 421–431.

Dence, M. R., Innes, M. J. S., and Robertson, P. B. (1968). Recent geological and geophysical studies of Canadian craters. In: Shock Metamorphism of Natural Materials, French, B. M., and Short, N. M. (eds.), Mono Book Corp., Baltimore, pp. 339–362.

Grieve, R. A. F. (2006). Impact Structures in Canada. Geological Association of Canada, GeoText No. 5, 210 pp.

The area of the Pilot impact structure with the homonymous lake on September 17, 2018 as a composite image based on TanDEM-X DEM and Copernicus Sentinel-2 (channel 2/3/4) data.

Pilot

Northwest Territories / Canada

latitude = 60°17‘06“ N / longitude = 110°59‘54“ W / diameter = 6 km / age = 445 ± 2 Myr

scale = 1 : 199 000 / scene = 35.8 × 47.7 km²

0 1.99 9.95 km

N

210 meter 350

Pingualuit
Quebec, Canada

Thomas Kenkmann

The Pingualuit crater, which was formerly known as *Chubb Crater*, *Ungava Crater*, and later *New Quebec Crater*, is the northernmost impact structure on mainland Canada. The crater has an almost circular shape with a diameter of 3.4 km. The infilling lake is 2.8 km wide and 250 m deep. The lake neither has an inlet nor an outlet. It is known as one of the world's clearest lakes and represents an important archive for paleo-climate studies. The lake's circular outline is very conspicuous in that region of the arctic tundra that is dominated by glacial landforms with elongated lakes that follow the regional joint systems in the rocks. The inner slope of the crater rim dips inward at about 30° and is covered with glacial debris and talus. The crater rim stands out some 120 m above the surrounding terrain. But for all that, the crater is not completely pristine. An ejecta blanket surrounding the crater was removed by advancing and retreating glaciers in the late Pleistocene. Pingualuit is a relatively young impact crater, by geological standards. The crater age was estimated applying the ^{40}Ar-^{39}Ar method to impact melt rock samples,

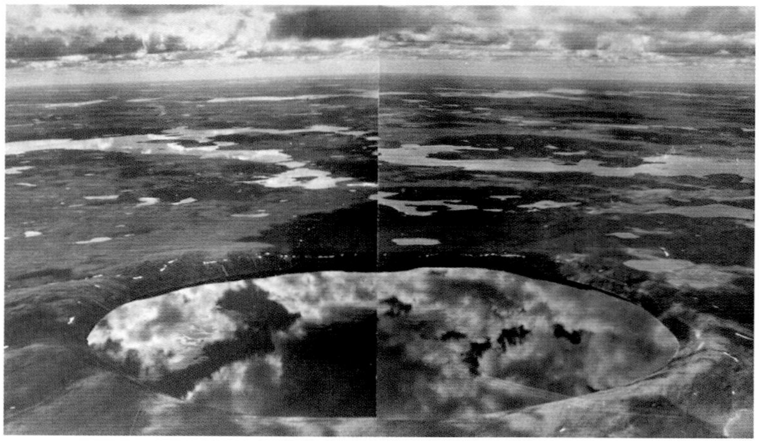

An early aerial view of the Pingualuit impact crater from 1960 (photo: National Archives of Canada).

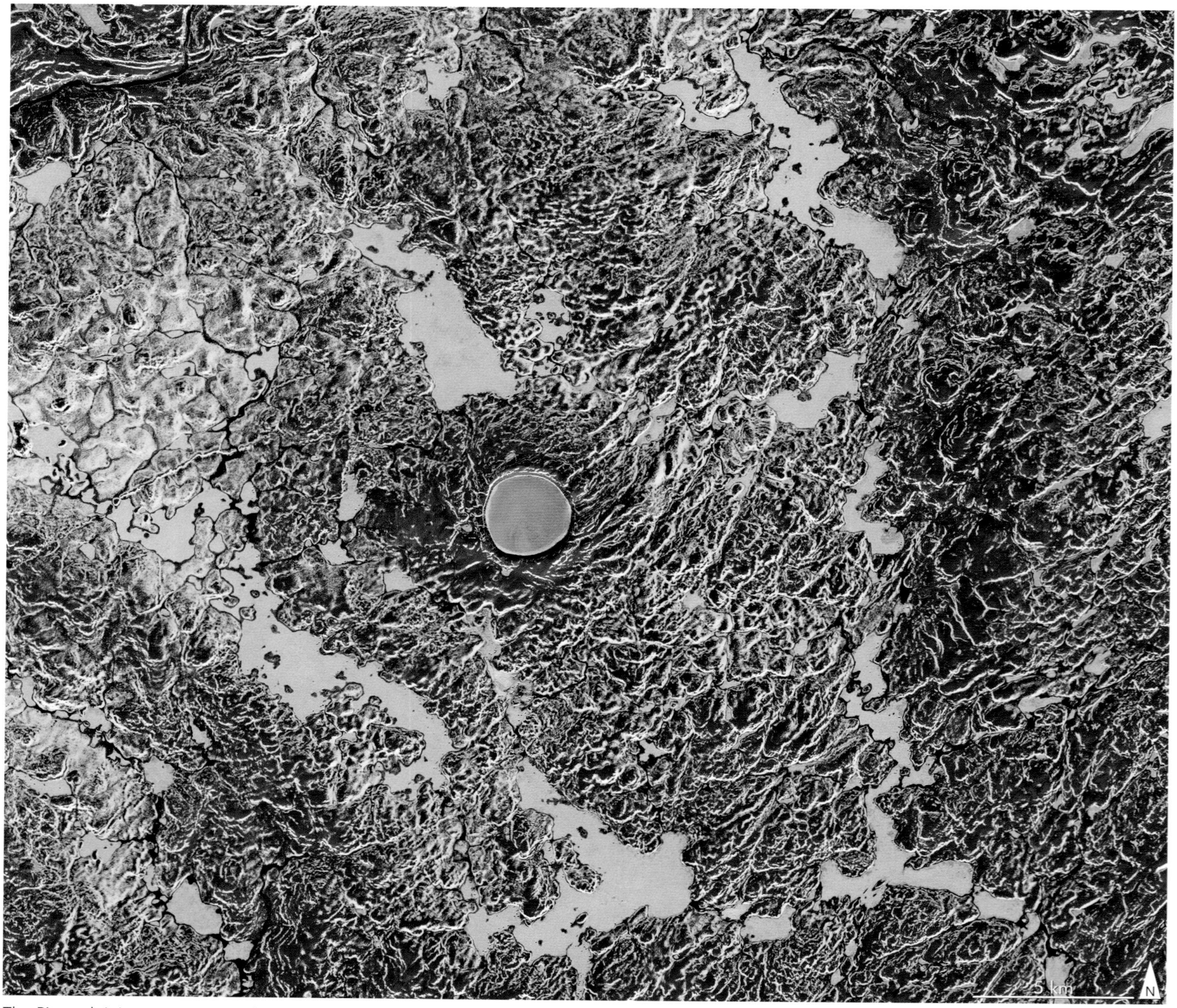

The Pingualuit impact crater and surroundings on June 4, 2017 as a composite image based on TanDEM-X DEM and Copernicus Sentinel-2 (channel 2/3/4) data.

Pingualuit
Quebec / Canada

latitude = 61°16'40" N / longitude = 73°39'36" W / diameter = 3.4 km / age = 1.4 ± 0.1 Myr

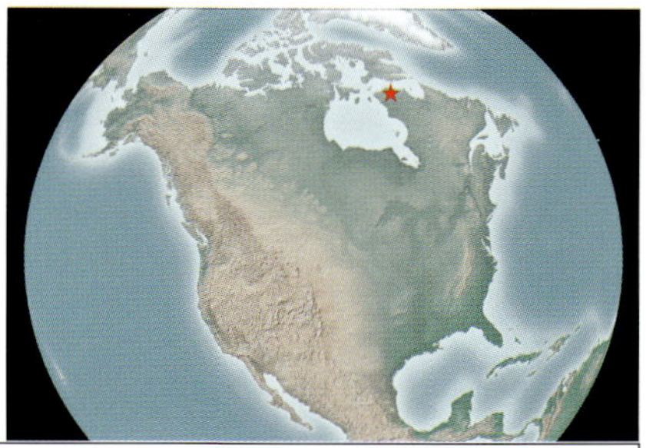

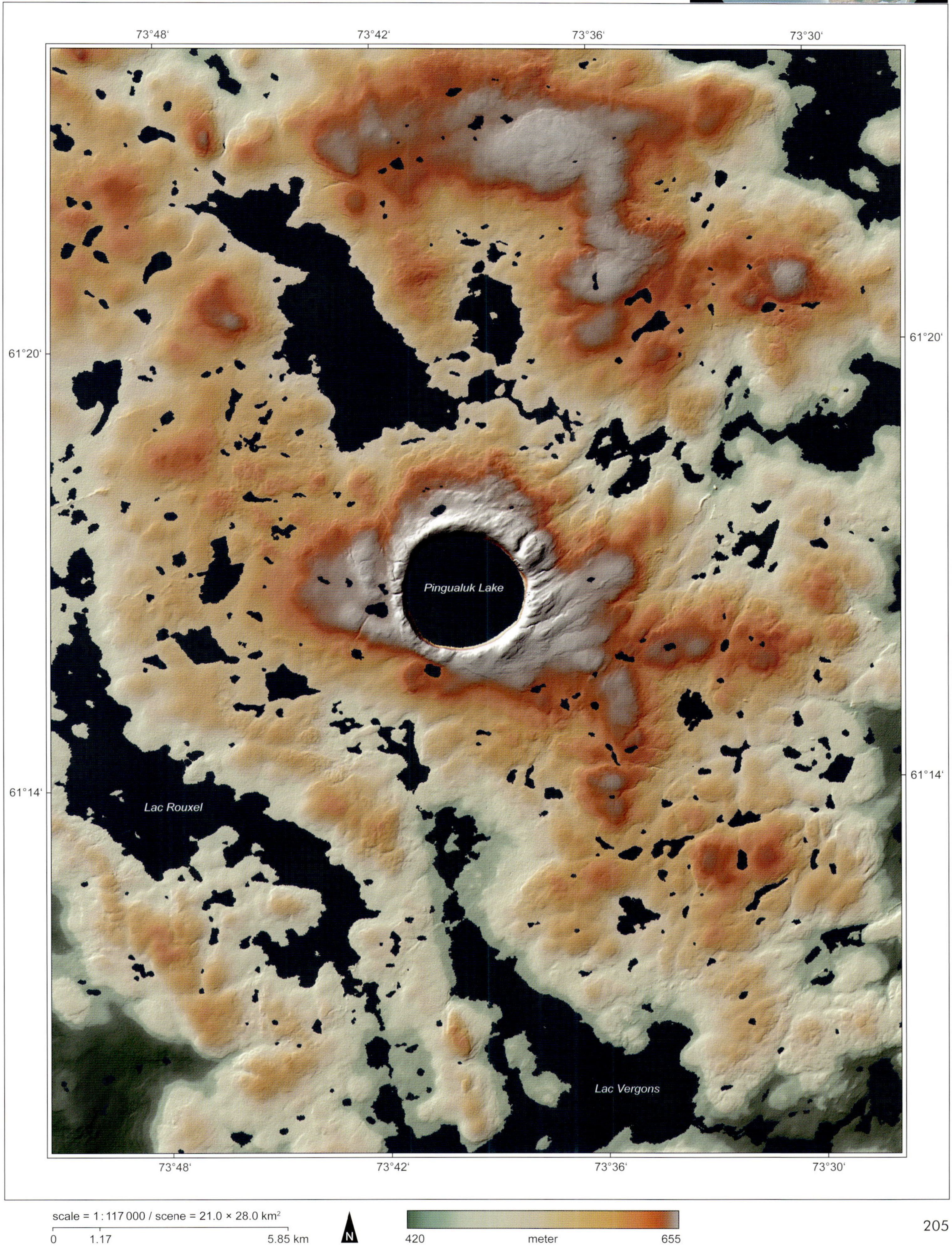

but partially only poorly developed age plateau features were obtained. Those measurements with uncertainties of less than 10 % yielded an average age of 1.4 ± 0.1 Myr. Target rocks are Archean gneisses and granodiorites of the Superior Province.

Pingualuit is among the largest simple craters on Earth. In cross-section the crater forms a bowl-shaped cavity with a raised crater rim. However, the thicknesses of the breccia lens and of the sedimentary fill within the crater are poorly constrained. A seismic survey indicated a sedimentary infill of possibly 150 m with a thickness of 75 m beneath the talus deposits. An east-west trending gravity profile showed a negative anomaly of -6 mGal when corrected for elevation, water, post-impact sediments, and the regional gravity gradient. Such a negative anomaly is characteristic for simple craters and indicates a loss in density by intense impact-induced rock fragmentation and presence of low-density breccias.

Geochemical studies of some of the melt-bearing rocks showed that the melt is enriched in Co, Ni, and Cr, compared to the country rock. Although the Ir concentration of these rocks is anomalously low, a chondritic meteorite component was suggested.

Pingualuit was the first impact crater discovered in Canada based on an aerial survey in 1946 by pilots of the Royal Canadian Airforce. However, it took another 15 years to prove the impact origin through analysis of a melt rock sample with shock metamorphosed clasts that had been recovered from the inner slope of the crater rim. The number of available samples is still very limited.

Location/Access

The crater is located on the arctic Ungava Peninsula, in the administrative region of Nord-du-Quebec, in Quebec, Canada. The crater is part of Pingualuit National Park. Closest Inuit settlements are Papiit 190 km to the west and Salluit with an airfield 150 km to the northwest. Full Arctic expedition planning would be required for a visit to the structure.

Selected References

Currie, K. L. (1965) The geology of the New Quebec crater. Canadian Journal of Earth Sciences, Vol. 2, pp. 141–160.
Currie, K. L., and Dence, M. R. (1963). Rock deformation in the rim of the New Quebec crater, Canada. Nature, Vol. 198, p. 153.
Grieve, R. A. F. (2006). Impact Structures in Canada. Geological Association of Canada, GeoText No. 5, 210 pp.

Presqu'île
Quebec, Canada

Thomas Kenkmann

Lac de la Presqu'île is roughly rhombic to circular in shape and contains a large central peninsula that is regarded as the center of this deeply eroded impact structure. The diameter of 20–24 km is inferred solely from the occurrences of shatter cones. Apart from the lake outline, no other specific morphological features exist. The area is almost completely covered by several meters of glacial till, and outcrops are sparse.

The crater is located in the Archean Abitibi greenstone belt of the Superior Province of the Canadian Shield. In the area of the crater structure, the belt comprises a series of basalt flows intercalated with andesitic and more felsic igneous rocks. All of them are metamorphosed to greenschist facies. Their schistosity strikes east-west and dips vertically. These rocks were intruded by a large tonalite body that is only weakly foliated. It occupies the central and northwestern parts of the crater area.

Shatter cones are exposed at three sites at the lake shore. The best preserved shatter cones, however, occur in outcrops along the Obatogamau River and an unnamed tributary five kilometers east of the lake. They are formed in fine-grained basalt and rhyodacite. Cone segments up to 60 cm long have been found. The presence of shatter cones provides definitive evidence that there has been an impact event, although no other shock effects and impact lithologies have been documented. If the center of Lac de la Presqu'île is taken as the center of the impact structure, then the distribution of shatter cones suggests a radius of about 12 km – permitting to infer a crater diameter of about 20–24 km.

The gravity and aeromagnetic data in this region are unspecific and dominated by the tonalite intrusion. They give no further evidence to constrain the size of this deeply eroded crater structure. The age of the impact event cannot be reliably constrained either. Given the high degree of erosion, an estimated upper age limit of 500 Myr is based on the ages of affected lithologies.

The structure was discovered by shatter cone findings during a regional mapping campaign of the Ministry of Energy and Resources of the Quebec Government in 1990.

Location/Access

This impact structure is located roughly halfway between Quebec City and the southeastern tip of Hudson Bay. The town Chapais is located only 8 km northwest of the proposed center of the crater structure and lies within its limits. There are abundant forest roads across the crater. Among them, Forest Road 1009 crosses the Obatogamau River, where shatter cones have been found.

Selected References

Grieve, R. A. F. (2006). Impact Structures in Canada. Geological Association of Canada, GeoText No. 5, 210 pp.
Higgins, M., and Tait, L. (1990). A possible new impact structure near Lac de la Presqu'île, Quebec, Canada. Meteoritics, Vol. 25, pp. 235–236.

Presqu‘île

Quebec / Canada

latitude = 49°43‘ N / longitude = 74°50‘ W / diameter = 22 km / age = < 500 Myr

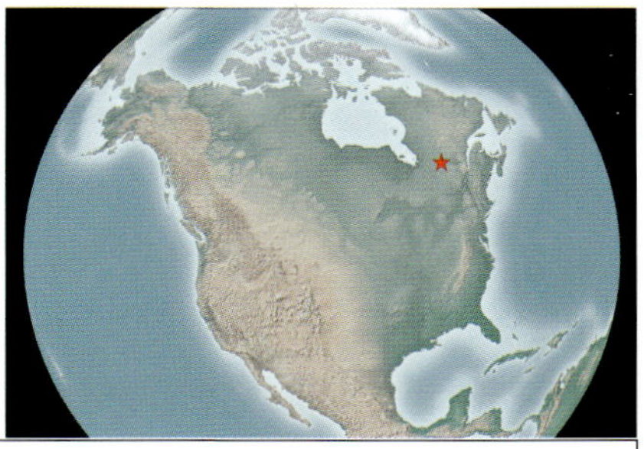

scale = 1 : 248 000 / scene = 44.6 × 59.4 km²

0 2.48 12.40 km

N

335 meter 565

Red Wing
North Dakota, United States

Thomas Kenkmann

Red Wing, or *Red Wing Creek*, is a complex impact structure in the western part of North Dakota. The structure has a diameter of 9 km. It lies hidden beneath a 2 km pile of sedimentary rock. The crater structure is situated close to the center of the intracratonic and oil-rich Williston Basin. The Red Wing structure itself is rich in oil resources, which have been exploited since the 1970s.

The Williston Basin contains sedimentary rocks ranging in age from Cambrian to Tertiary. Their maximum thickness is 4500 m. These sedimentary rocks are underlain by a Precambrian basement composed of igneous and metamorphic rocks. Because of the interest by the hydrocarbon industry, the structure has been extensively explored by seismic reflection profiling and drilling. In seismic cross-sections the Cambrian and Ordovician formations appear intact, whilst the overlying Devonian and Mississippian (Carboniferous) formations are strongly disturbed. Rocks of Triassic and Jurassic formations fill the crater depression. The structure has a central uplift with a maximum diameter of about 6 km. Mississippian units, which are composed predominantly of carbonates that are intercalated with sandstone, siltstone, shales, and evaporites. They are uplifted in the central rise by about 1 km above their normal stratigraphic position in the region. These rocks are steeply inclined and intensely faulted. Breccias were encountered during drilling at 2000 to 2800 m depth. The central uplift is surrounded by a ring syncline, where rock formations occur roughly 50 m lower than normally in the region. The ring depression is surrounded by an outer ring of slightly uplifted deformed rocks.

Shatter cone fragments were found in drill cuttings of these boreholes. The discovery borehole drilled close to the center of the structure also showed quartz with planar deformation features (PDFs) in up to three sets of different crystallographic orientation per host grain.

Red Wing has been intensely studied by the petroleum industry. The discovery well contained 850 m of producing carbonates. Reflection seismic profiles enabled to define the crater size and shape. The magnetic survey did not reveal a circular signature characteristic for impact craters. The central uplift shows a +1.5 mGal gravity anomaly. The oldest crater infill is of Triassic age and indicates an impact age of 200–220 Myr.

Red Wing's impact origin was brought to attention when, in 1975, Richard Brenan and co-workers reported the discovery of shatter cone fragments in drill core samples from the center of Red Wing. Later, in 1996, Christian Koeberl and colleagues found unambiguous shock metamorphic effects in quartz grains from drill cuttings.

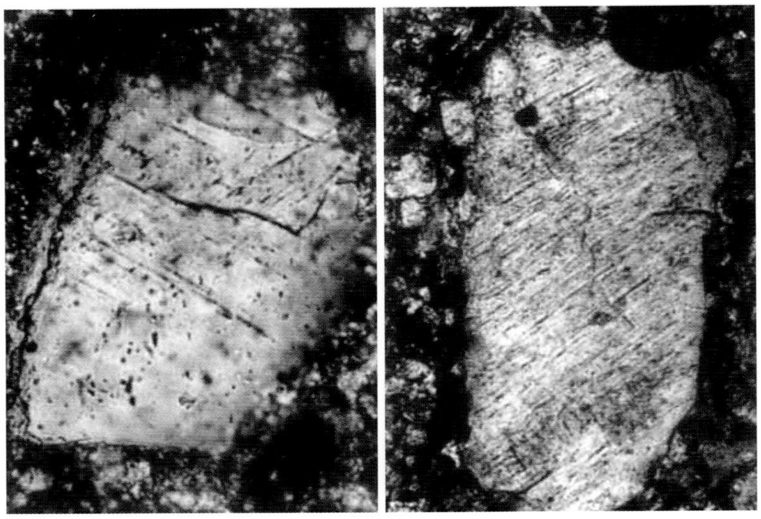

Photomicrographs of quartz grains with PDFs from the Red Wing impact structure. A quartz grain with two sets of PDFs (left), and a quartz grain with PDFs, some of which are decorated with oxide blebs (on the right). The width of the left image is 365 mm and that of the right 565 mm (images courtesy: Christian Koeberl, Naturhistorisches Museum Wien, Austria).

Location/Access

The buried Red Wing structure is located in North Dakota's McKenzie County, about 30 km southwest of Watford City, and 60 km south of Williston in Montana. The northern unit of the Theodor Roosevelt National Park lies in the immediate environs of the structure, to the east, and the Little Missouri River runs to the south of the structure.

Selected References

Brenan, R. L., Peterson, B. L., and Smith, H. J. (1975). The origin of Red Wing Creek structure: McKenzie county, North Dakota. Wyoming Geological Association Earth Science Bulletin, Vol. 8, 41 pp.

Koeberl, C., Reimold, W. U., and Brandt, D. (1996). Red Wing Creek structure, North Dakota: Petrographical and geochemical studies, and confirmation of impact origin. Meteoritics and Planetary Science, Vol. 31, pp. 335-342.

Sawatzky, H. B. (1977). Buried impact craters in the Williston basin and adjacent area. In: Impact and Explosion Cratering, Roddy, D. J., Pepin, R. O., and Merrill, R. B. (eds.), Pergamon Press, New York, pp. 461–480.

Red Wing

North Dakota / United States

latitude = 47°36‘ N / longitude = 103°33‘ W / diameter = 9 km / age = 200 - 220 Myr

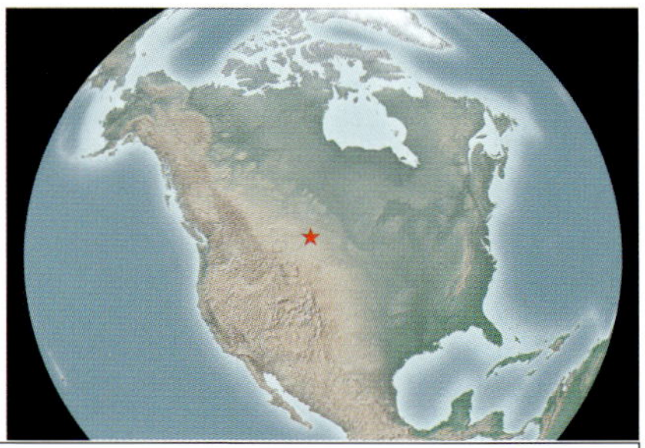

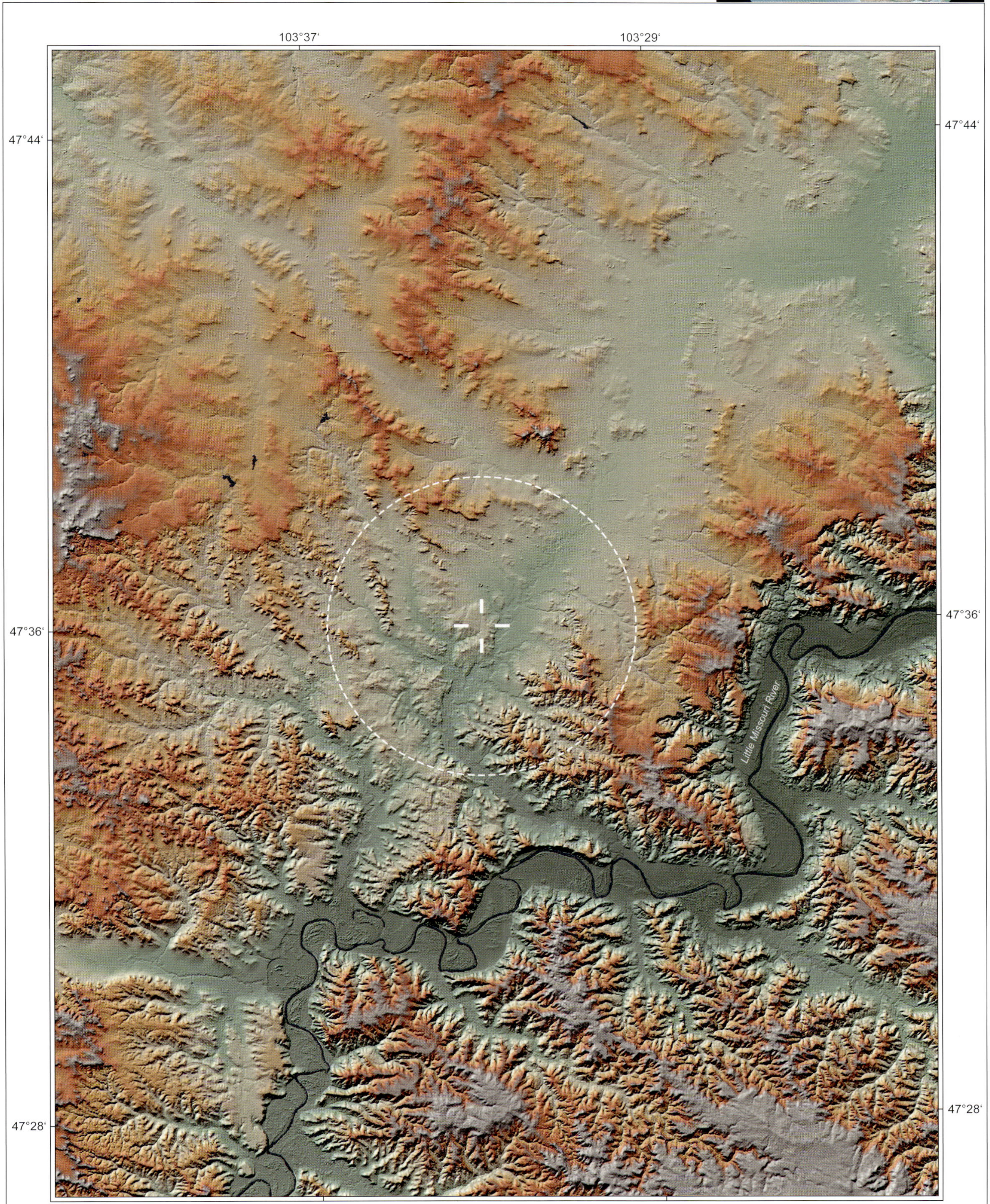

scale = 1 : 143 000 / scene = 25.8 × 34.4 km²

0 1.43 7.15 km

N

585 meter 825

Rock Elm
Wisconsin, United States

Thomas Kenkmann

Rock Elm, located in western Wisconsin, is a 6.5 km diameter impact crater with a complex structure. The structure has a subdued morphological expression involving a slightly depressed moat and a central uplift that forms a central hill that stands 80 m above the surrounding stream valleys.

Rock Elm was formed on a stable continental platform of Paleozoic strata, overlying a basement of Precambrian volcanic, sedimentary, and metasedimentary rocks that lies at a depth of 250–300 m beneath the surface. The Paleozoic units were deposited from Late Cambrian to Middle Ordovician time in a shallow-water marine environment. The Cambrian strata mainly comprise silty to sandy dolostone, with local chert, sandstone, siltstone, and shale. St. Peter Sandstone and carbonates are important lithologies in the Lower and Middle Ordovician strata. The bedrock geology is partly obscured by younger glacial till and alluvium, which may reach 15 m thickness.

Rock Elm has a well-defined crater rim fault that separates the deformed rocks of the structure from the normal stratigraphic sequence outside. This fault is clearly defined in the southern and eastern parts of the structure where it is marked by an arcuate ridge and a vertical throw of 45 m. Several fault-bounded blocks of kilometer size occur along the crater rim fault. The largest of these is internally deformed into a series of tight anticlines and synclines. Further inside of the structure lies an annular basin 4–5 km in diameter that surrounds the central uplift. This basin is filled with a generally flat-lying, 48 m thick, sedimentary sequence of post-impact age. These rocks are known as the Rock Elm shale and the Washington Road sandstone, and do not occur in the area outside the crater. The central uplift consists of an oval 1 × 2 km wide area of poorly exposed and deformed Upper Cambrian sandstone that dips at 20–30° towards the outside. The stratigraphic uplift amounts to about 250–300 m.

Breccia from the central uplift is reported as float. It is composed of lithic clasts of variable size and composition within a matrix of sand-sized particles. Shatter cones, one of the diagnostics for an impact structure, have not been found at Rock Elm. However, the deformed sandstones and breccias of the central uplift show a shock metamorphic overprint. Quartz grains exhibit planar deformation features (PDFs) in various crystallographic orientations, planar fractures (PFs), and feather features (FFs). A few years ago reidite, a high-pressure $ZrSiO_4$ polymorph diagnostic of impact, was found in these rocks – indicating a shock pressure of 20 GPa or more.

The residual gravity map displays no detectable anomaly specifically associated with the crater structure. Likewise, magnetic anomalies observed in the area of Rock Elm reflect variations in structure and composition of the underlying Proterozoic basement, rather than any effect created by the impact.

The age of the impact is constrained by unconformably overlying sedimentary strata of the basin-fill. However, the exact age of the basin-filling sedimentary rocks remains unclear. It is roughly estimated that Rock Elm formed 460–410 Myr ago. Thus, the impact may be linked to the Late Ordovician asteroid shower that was caused by a major collision in the asteroid belt. Post-impact erosion reduced the structure by 250–350 m.

The Rock Elm impact structure on November 18, 2018 as a composite image based on TanDEM-X DEM and Copernicus Sentinel-2 (channel 2/3/4) data.

Rock Elm

Wisconsin / United States

latitude = 44°43‘25“ N / longitude = 92°13‘41“ W / diameter = 6.5 km / age = 410 - 460 Myr

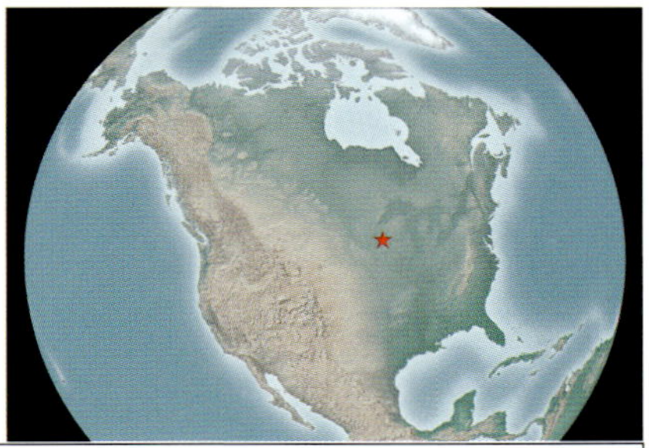

scale = 1 : 120 000 / scene = 21.7 × 28.9 km^2

0 1.20 6.00 km

N

205 meter 395

The Rock Elm structure has been known as a circular area of anomalous deformation for several decades. William Cordua stated 1985 that the structure has the basic characteristics of a "cryptoexplosion" structure but lacks definite evidence of meteorite impact. Other workers proposed that the structure was formed due to volcanism and hydrothermal activity associated with deep-seated basement faults. Bevan French and co-workers in 2004 reported shocked quartz grains with PDFs and PFs, and their diligent work confirmed the impact origin of the Rock Elm breccia.

Location/Access

Rock Elm is located about 90 km southeast of Minneapolis in Pierce County and can easily be accessed by car. From Minneapolis follow Interstate Highway 10 in easterly direction till Eastworth. From there take Interstate 72 that skirts the southern edge of the impact structure. The structure's name is derived from the small village of Rock Elm that lies within the moat of the crater structure. The landscape is characterized by rolling hills, farmland and forests; outcrop conditions are poor.

Selected References

Cavosie, A. J., Erickson, T. M., and Timms, N. E. (2015). Nanoscale records of ancient shock deformation: Reidite ($ZrSiO_4$) in sandstone at the Ordovician Rock Elm impact crater. Geology, Vol. 43, pp. 315–318.

Cordua, W. S. (1985). The Rock Elm structure, Pierce County, Wisconsin, a possible cryptoexplosion structure. Geology, Vol. 13, pp. 372–374.

French, B. M., Cordua, W. S., and Plescia, J. B. (2004). The Rock Elm meteorite impact structure, Wisconsin; geology and shock-metamorphic effects in quartz. Geological Society of America Bulletin, Vol. 116, pp. 200–218.

Saint Martin

Manitoba, Canada

Thomas Kenkmann

Saint Martin, also known as *Lake Saint Martin*, is an eroded complex crater structure of about 40 km diameter. The structure has only a subdued surface expression, and outcrops are limited. Target rocks are 2387 Myr old Archean granites of the Superior Province craton of the Canadian Shield. They are unconformably overlain by 200–400 m thick Ordovician to Silurian carbonates. However, a large part of the Paleozoic strata has been eroded since the impact and is only preserved as fragments in impact breccias.

The central uplift structure is built up by Precambrian and Lower Paleozoic rocks. An annular crater moat is filled with a sequence of impactites including carbonate-dominated and granitic lithic impact breccias, impact melt rocks, and suevitic breccias. The structure was partially filled by Jurassic sediments – mainly red-colored siliciclastics, anhydrite and gypsum. Finally, the crater fill is overlain by 30 m of Quaternary till.

Impact lithologies became mainly accessible in drill cores. Among them are monomict limestone breccia, crystalline rock breccia, suevites, and impact melt rocks. The limestone breccia is very pure in composition and must have originated from strata – most likely of Devonian age – that were eroded away before deposition of the Jurassic post-impact beds. The uplifted Precambrian basement of the central uplift shows a shock metamorphic overprint with planar deformation features (PDFs) in quartz and feldspar. The granitic basement is cut by pseudotachylitic breccia. Suevites contain target rock fragments of all shock stages. In addition to samples from drill core, impact melt rocks occur in some outcrops in an onlapping situation at the edge of the central uplift. One borehole situated close to the center of the structure encountered – beneath 52 m of Jurassic post-impact sediments – 265 m of suevitic breccia that grade into coherent impact melt rock at the bottom of the borehole. The impact melt rock has a bulk rock composition similar to the Precambrian basement lithologies with a small admixture of carbonate from the Paleozoic cover.

Saint Martin is associated with a negative Bouguer gravity anomaly of −14 mGal. A relative gravity high in the center of the structure correlates with the uplift of comparatively denser rocks. Saint Martin is also characterized by a magnetic low that interrupts the regional aeromagnetic pattern of the region. The center of the crater structure has a very strong negative magnetic anomaly of −1000 nT that is associated with the formation of hematite from the alteration of impact lithologies.

Several radiometric dating approaches by the K-Ar and Rb-Sr methods suggested a Permo-Triassic age of 200–250 Myr for the impact. John Spray and co-workers proposed that Saint Martin was part of a multiple impact event, together with the **Manicouagan**, **Rochechouart** and **Obolon** craters, at that time. However, recent and more precise age determination using ^{40}Ar-^{39}Ar dating of recrystallized K-feldspar from impact melt rock and partially molten granite of the central uplift yielded an age of 227.4 ± 0.8 Myr for Saint Martin, and in comparison with the ages of these other three structures this excludes the possibility of a multiple impact scenario.

The structure was of economic interest. The petroleum industry drilled boreholes in 1966 and 1968, which recovered 760 m of core. Gypsum and anhydrite were also mined for some time in an area centered on the village of Gypsumville. Hugh McCabe and Barry Bannatyne presented a first comprehensive study of the crater structure and provided convincing evidence for its impact origin in 1970. However, they did not commit themselves to the impact hypothesis and, instead, argued if shock metamorphic features could be formed by volcanic eruption. But as it is now certain that shock effects can only be formed in impact scenarios, Saint Martin's acceptance as an impact crater can be dated back to 1970.

Location/Access

Saint Martin is located in south-central Manitoba. The structure can be reached from Winnipeg via Highway 6. Large parts of Lake Saint Martin and the small town of Gypsumville lie within the impact structure. There are three outcrops of impact melt rock in the periphery of the central uplift.

Selected References

Currie, K. L. (1970). New Canadian cryptoexplosion crater at Lake St Martin, Manitoba. Nature, Vol. 226, pp. 839–841.

Grieve, R. A. F. (2006). Impact Structures in Canada. Geological Association of Canada, GeoText No. 5, 210 pp.

McCabe, H. R., and Bannatyne, B. B. (1970). Lake St. Martin crypto-explosion crater and geology of the surrounding area. Manitoba Department of Mines and Natural Resources, Mines Branch, Geological Paper, Vol. 3/70, 79 pp.

Reimold, W. U., Barr, J. M., Grieve, R. A. F., and Durrheim, R. J. (1990). Geochemistry of the melt and country rocks of the Lake St. Martin impact structure, Manitoba, Canada. Geochimica et Cosmochimica Acta, Vol. 54, pp. 2093–2111.

Schmieder, M., Jourdan, F., Tohver, E., and Cloutis, E. A. (2014). ^{40}Ar-^{39}Ar age of the Lake Saint Martin impact structure (Canada); unchaining the late Triassic terrestrial impact craters. Earth and Planetary Science Letters, Vol. 406. pp. 37–48.

Spray, J. G., Kelley, S. P., and Rowley, D. B. (1998). Evidence for a late Triassic multiple impact event on Earth. Nature, Vol. 392, pp. 171–173.

Saint Martin

Manitoba / Canada

latitude = 51°47‘ N / longitude = 98°32‘ W / diameter = 40 km / age = 227.4 ± 0.8 Myr

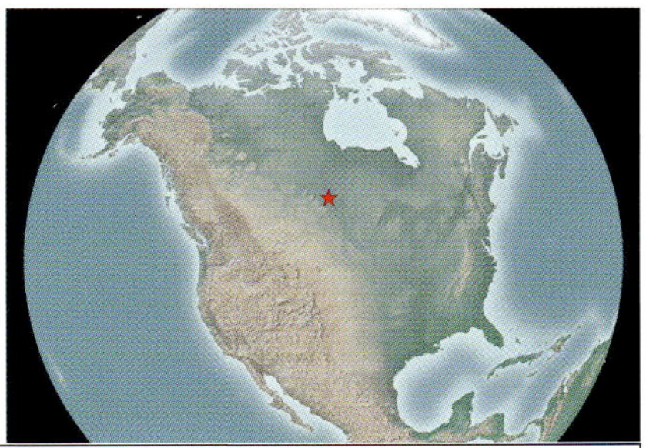

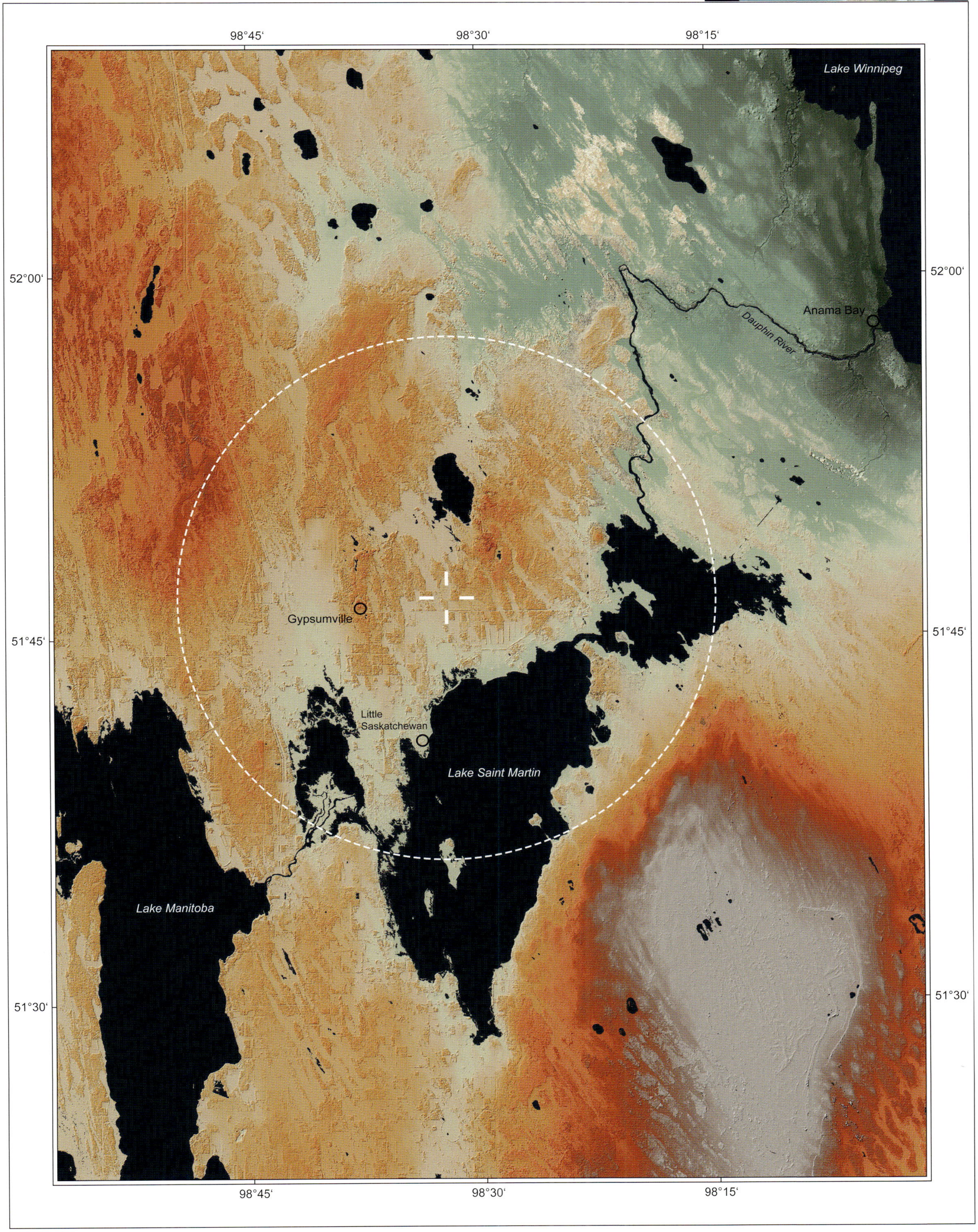

scale = 1 : 358 000 / scene = 64.5 × 86.0 km²

0 3.58 17.90 km

N

200 meter 315

Santa Fe
New Mexico, United States

Thomas Kenkmann

Santa Fe is an apparently deeply eroded impact structure with neither typical morphological nor structural features. Just a section of the original impact structure has been preserved near the city of Santa Fe. The structure, or its remnant, was recognized only by the occurrence of large, in situ shatter cones, in an area of about 5 km^2, with an east-west extent of about 3 km. Due to the lack of a crater morphology, the reconstruction of the original crater size is difficult. The currently given size estimates of 6–13 km diameter for the Santa Fe impact structure are exclusively based on the extent of shatter cone occurrence and the application of scaling relationships for these shock features.

The oldest target rocks in this area are about 1008–1700 Myr old, metamorphosed intrusive and extrusive rocks such as granites, rhyolites, and basalts, as well as metamorphosed siliciclastic sediments. The latter include various types of schists and quartzites. Proterozoic rocks are unconformably overlain by relatively undeformed quartzite and limestone of Mississippian age, and limestone, siltstone, shale, and minor sandstone of Pennsylvanian age.

A high degree of tectonic overprint and a deep level of erosion have hampered a detailed reconstruction of the impact structure. Shatter cones, some with lengths up to 2 m, are the dominant feature of the Santa Fe impact site. They occur in both intrusive igneous and metamorphic supracrustal rocks. The largest cones are developed in homogeneous and massive granitoids. The cones show all characteristics of shatter cones including sub-cone ridges, diverging, and bifurcating striations. In most cases, the apices of the cones point upward at relatively consistent orientations.

Along with the shatter cones occur other shock metamorphic features: Mineral grains close to shatter cone surfaces show rare decorated planar fractures (PFs) and planar deformation features (PDFs) in quartz. From the crystallographic orientation of the PFs and PDFs a peak shock pressure of about 5–10 GPa has been estimated. This range is well within the prime pressure regime for shatter cone formation. Shatter cones may have thin melt films on their surfaces. Recently, detrital zircon and apatite, as well as xenotime from a shatter-cone in granite, all with planar shock deformation have been described. Approximately 2 km west of the shatter cone area, a 75 m thick breccia is preserved that overlies the crystalline basement and is intercalated with Mississippian rocks. Whether this represents impact breccia, needs to be confirmed.

Within 1 km of the shatter cone area, the crystalline rocks are broken by numerous faults with meter scale displacements and at random orientations. Within 2.5 km, the rocks are still deformed by numerous faults. The fracturing appears to have occurred in-situ and does not involve significant displacements. The age of the Santa Fe impact is poorly constrained between about 350 and 1200 Myr, based on chronostratigraphic information only, namely the age of the deformed granitoids and the age of undeformed supracrustal rocks. Shatter cones were first observed during field work in the area by Tim McElvain in 2005. Subsequent mapping and petrographic investigations by Siobhan Fackelman and McElvain led to the detection of diagnostic shock evidence and confirmation of impact.

Location/Access

The Santa Fe impact structure is located in the western foothills of the southern Sangre De Cristo Mountains. This mountainous region is located approximately 8 km northeast of Santa Fe, New Mexico. Impressive shatter cones can be observed in excellent road cuts along the New Mexico State Highway 475 that leads from Santa Fe up to Ski Santa Fe, a resort at an elevation of about 3150 m.

One of the many large shatter cones along a roadcut several kilometers from the city of Santa Fe. Conference badge for scale 8 cm long (photo: W. U. Reimold).

Selected References

Cavosie, A. J., Montalvo, P. E., Timms, N. E., and Reddy, S. M. (2016). Nanoscale deformation twinning in xenotime, a new shocked mineral, from the Santa Fe impact structure (New Mexico, USA). Geology, Vol. 44, pp. 803–806.

Fackelman, S. P., McElvain, T. H., Morrow, J. R., and Koeberl, C. (2007). Shatter cone exposures indicate a new bolide impact structure near Santa Fe, New Mexico. 38th Lunar and Planetary Science Conference, Lunar and Planetary Institute, Houston, Texas, Abstract #1207.

Fackelman, S. P., Morrow, J. R., Koeberl, C., and McElvain, T. H. (2008). Shatter cone and microscopic shock-alteration evidence for a post-Paleoproterozoic terrestrial impact structure near Santa Fe, New Mexico, USA. Earth and Planetary Science Letters, Vol. 270, pp. 290–299.

Lugo, C. M., and Cavosie, A. J. (2014). First report of shocked zircon at the Santa Fe impact structure (USA). 45th Lunar and Planetary Science Conference, Lunar and Planetary Institute, Houston, Texas, Abstract #1839.

McElvain, T. H., Read, A. S., Petersen, M. T., Elston, W. E., Newsom, H. E., and Cohen, B. A. (2006). Possible impact structure in the Sangre de Cristo Mountains near Santa Fe, New Mexico: a preliminary report. Geological Society of America Abstracts with Programs, Vol. 38, p. 298.

Santa Fe

New Mexico / United States

latitude = 35°43'54" N / longitude = 105°51'04" W / diameter = 9.5 km / age = 350 - 1200 Myr

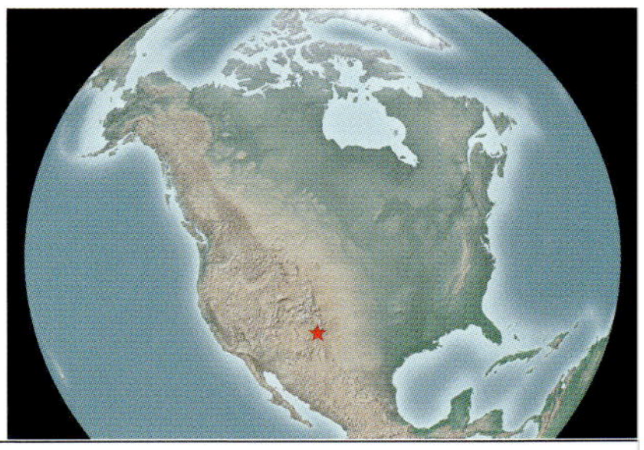

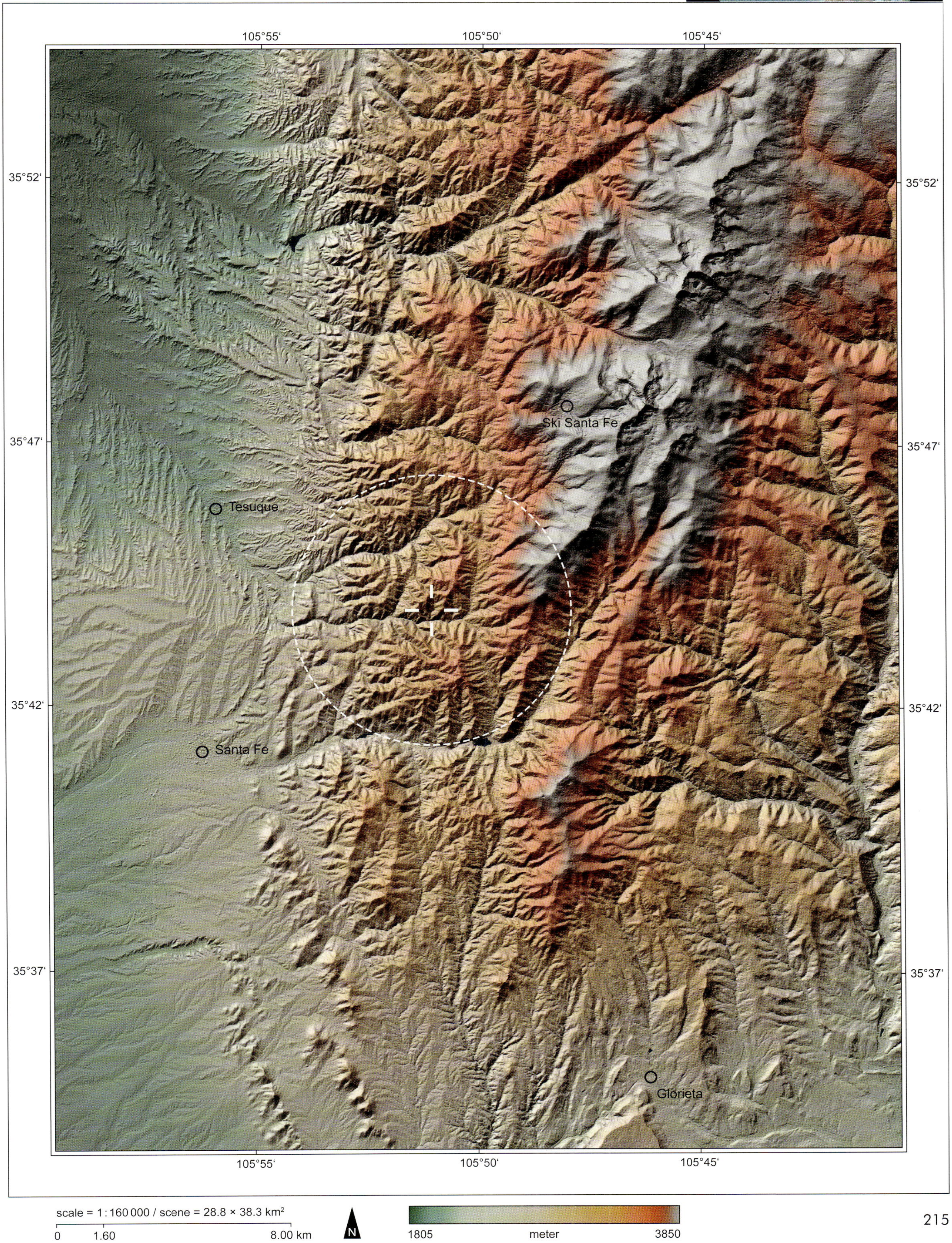

Serpent Mound
Ohio, United States

Thomas Kenkmann

Serpent Mound in south-central Ohio is an eroded complex impact structure with a diameter of about 8 km and a distinct morphological expression. The structure is slightly elliptical with an elongation in north-south direction. The origin of this geological disturbance was discussed controversially for a long time. Serpent Mound is named after an effigy mound, i.e., a mound in the shape of a snake, built by native Americans about a thousand years ago, which is located within the structure.

Serpent Mound is situated at the edge of the Appalachian Basin. Target rocks consist of Precambrian basement overlain by roughly 1100 m of Paleozoic sedimentary rocks that range in age from Late Cambrian to Early Mississippian. Due to flexural loading by the Appalachian mountain belt, the strata in the western foreland dip gently towards the east. The occurrence of a more than 300 m thick sequence of Late Mississippian and Pennsylvanian sedimentary strata in the proximity to the impact structure suggests that these rocks may also have covered the area when the impact occurred. The Paleozoic sedimentary rocks comprise dolomite, limestone, shale, sandstone, and minor siltstone and chert.

Structural units of Serpent Mound are a central uplift, a transition zone, and an outer annular ring graben. The central uplift is built by strongly disturbed Ordovician and Silurian rocks. Radial anticlines are important structural features. Due to their resistance to erosion, the strata of the central uplift form a topographic high in the center of the structure. The stratigraphic uplift amounts to at least 122 m. The transition area is characterized by a topographic low, in which mostly Silurian rocks crop out that are dissected by radial and concentric faults and fractures. In the annular ring graben further outward fault-bounded strata are down-faulted by as much as 245 m with respect to their original positions. Fault breccias are sealed with sphalerite, which indicates hydrothermal activity after impact. The ring graben shows a relief inversion, as the down-faulted rocks form a topographic high of resistant Mississippian and Devonian strata.

Two boreholes were drilled into the central part of the structure, with total core lengths of 903 m and 629 m, respectively. They document the disturbed nature of the rocks of the central uplift. The drilled rocks provide evidence of shock metamorphism in the form of multiple sets of planar deformation features (PDFs) in quartz grains, as well as the presence of clasts of altered impact melt rock. Geochemical analysis of impact breccias showed minor enrichments in the abundances of the siderophile elements Cr, Co, Ni, and Ir over average target composition, suggesting the presence of a minor meteoritic component. The level of erosion and other data suggest that Serpent Mound formed between 256 and 290 Myr ago.

The structure lies on a major regional positive gravity anomaly that extends from Michigan over Indiana into Ohio and Kentucky. The gravity pattern of the structure itself is complex, with local negative anomalies in the northwestern and southeastern sectors – including large parts of the central uplift.

The geological disturbance at Serpent Mound was known for a long time, and a "cryptoexplosion" origin was favored by many while the origin of the structure was discussed controversially. Robert Dietz reported shatter cones associated with Serpent Mound in 1960 and suggested a meteorite impact origin. This was substantiated in 1961 by Alvin Cohen and co-workers, who documented coesite, a high pressure polymorph of SiO_2. Stephen Reidel and his team, in turn, discounted the discovery of coesite in 1982 as a misidentification and favored an endogenic origin for the tectonic disturbance in 1982. Final proof of the impact origin was, however, provided by Richard Carlton and co-workers in 1998, who described PDFs and impact melt rocks from the Serpent Mound impact structure.

Location/Access

The hilly area of Serpent Mound in Ohio is farmland and forest with little outcrop. The crater is easily accessible, e.g., via the Appalachian Highway from Cincinnati that is located 100 km west of Serpent Mound.

The Serpent Mound impact structure on December 21, 2019 as a composite image based on TanDEM-X DEM and Copernicus Sentinel-2 (channel 2/3/4) data.

Serpent Mound

Ohio / United States

latitude = 39°01'56" N / longitude = 83°24'10" W / diameter = 8 km / age = 256 - 290 Myr

scale = 1 : 104 000 / scene = 18.7 × 24.9 km²

0 1.04 5.20 km

N

170 meter 420

Selected References

Carlton, R. W., Koeberl, C., Baranoski, M. T., and Schumacher, G. A. (1998). Discovery of microscopic evidence for shock metamorphism at the Serpent Mound structure, south-central Ohio: confirmation of an origin by impact. Earth and Planetary Science Letters, Vol. 162, pp. 177–185.

Cohen, A. J., Bunch, T. E., and Reid, A. M. (1961). Coesite discoveries establish cryptovolcanics as fossil meteorite craters. Science, Vol. 134, pp. 1624–1625.

Dietz, R. (1960). Meteorite impact suggested by shatter cone in rock. Science, Vol. 131, pp. 1781–1784.

Reidel, S. P., Koucky, F. L., and Stryker, R. J. (1982). The Serpent Mound disturbance, southwestern Ohio. American Journal of Science, Vol. 282, pp. 1343–1375.

Schedl, A. (2006). Applications of twin analysis to studying meteorite impact structures. Earth and Planetary Science Letters, Vol. 244, pp. 530–540.

Sierra Madera
Texas, United States

Thomas Kenkmann

The about 12 km diameter Sierra Madera complex impact structure is morphologically well visible. It includes central hills, an annular depression, and an outlying ring of low hills. These features correspond to a central uplift, a ring syncline and the crater rim, respectively.

The target at Sierra Madera is composed of a 6000 m thick sequence of sedimentary rocks that range in age from Cambrian to Lower Cretaceous. This sequence rests on a Precambrian granitic basement of 900 Myr age. Paleozoic strata up to the base of the Permian are about 1,600 m thick. The thickest sedimentary pile is from the Permian period that involves, from bottom to top, 3000 m of monotonous black shales interbedded with fine-grained sandstone and some limestone and dolomite of Lower Permian age (Wolfcamp Series). The next 800 m, the Hess Formation, also comprises interbedded shale, fine-grained sandstone, dolomite, limestone, and marl, but the carbonate content becomes increasingly dominant towards the top. Limestone and dolomite form the upper part of the Permian. There is a strong gradient in strata thickness from south to north. The youngest Permian strata vary in thickness from 500 m in the southern to 1300 m in the northern part of the structure. The Permian sequence is covered by a veneer of Triassic conglomerates that is, in turn, overlain by Lower Cretaceous limestones. Most of these have been eroded away. Quaternary deposits reach a thickness of 50 m in the annular moat surrounding the central uplift.

The exposed central uplift is about 8 km in diameter and corresponds to the central hills of Sierra Madera. The oldest outcropping strata from the Lower Permian indicate a stratigraphic uplift of 1200 m. Strata of the central uplift are folded and faulted. Cataclastically deformed rocks include both monomict and polymict breccias that occur widespread throughout the central uplift. Typical features of Sierra Madera's central uplift are outward plunging radial folds, which accommodated the convergent flow of inward moving rocks during crater modification. Sandstones behaved extremely incompetent during deformation and were injected into neighboring strata in the form of dikes. Deformation culminated in the center of the structure. The surrounding ring syncline is roughly 1–1.5 km wide and is made up of Lower Cretaceous rocks that are overlain by Quaternary deposits. Further outward follows a chain of outer hills that span about 0.8 km in width. Here, strata are uplifted and expose Upper Permian and Lower Cretaceous rocks. Concentric normal faults dip toward the crater, and folds with concentric axes are present here as well. This constitutes the typical deformation of an impact crater rim.

Shatter cones occur particularly abundant in the inner part of the central uplift and occasionally can be found out to 6.5 km from the center of the impact structure. Shock features in quartz such as planar deformation features (PDFs) and planar fractures (PFs) are ubiquitous in polymict breccias of the central uplift but rare to absent in monomict breccias and shattered rocks.

Boreholes drilled into the structure in search for oil and gas showed that the impact deformation extends to a depth of 1.8–2.4 km beneath the present surface. A geophysical traverse indicates a negative gravity anomaly of −1.5 mGal over the zone of brecciation in the center and a residual positive anomaly of +0.5 mGal associated with a positive magnetic anomaly to the southeast of the zone of brecciation. A precise age for the impact could not be determined so far. The impact must have occurred, however, after the deposition of the exposed Lower Cretaceous rocks and, therefore, would have to be younger than circa 100 Myr.

The impact event was modelled by numerical simulation. While the principal structural features of the crater could be reproduced, some mismatches remained. They could be minimized, if an initially somewhat larger crater of more than 12 km diameter would be assumed. Scaling the crater to the dimensions of other craters suggested that about 600 m of erosion could have taken place since the time of impact.

The unusual geological character of Sierra Madera was recognized as early as 1930 by Philip King in the course of a mapping campaign of the nearby Glass Mountains. He suggested that an igneous intrusion may have played a role in formation of

Monomict limestone breccia, 11.5 × 6.5 × 3 cm in size, from the Sierra Madera impact structure (photo courtesy: Ralf-Thomas Schmitt, Museum für Naturkunde Berlin; sample from M. C. Pope, Tamu, Texas).

Sierra Madera

Texas / United States

latitude = 30°35'44" N / longitude = 102°54'42" W / diameter = 12 km / age = < 100 Myr

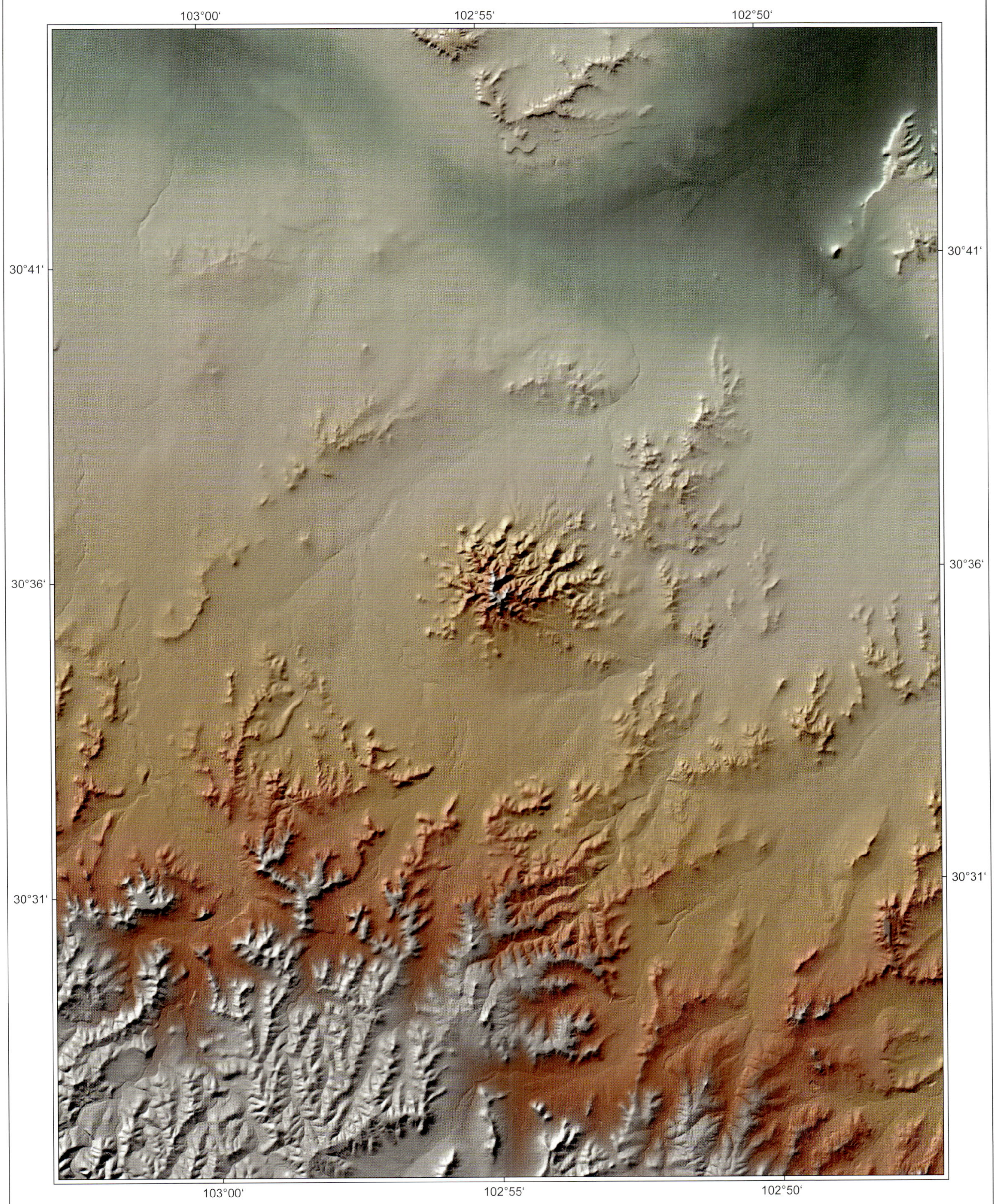

scale = 1 : 140 000 / scene = 25.3 × 33.7 km^2

0 1.40 7.00 km

N

965 meter 1520

the structure. Later, Sierra Madera was also considered as being of "cryptovolcanic" or "cryptoexplosive" origin. An origin by impact could be confirmed in 1961 by Richard Eggleton and Eugene Shoemaker, who described shatter cones and breccias in the structure. Detailed geological surveying and mapping followed in the late 1960s and early 1970s by Howard Wilshire and co-workers of the U.S. Geological Survey.

Location/Access

Sierra Madera ("wooden hills" in Spanish) is the name for a group of hills occurring over about 5 km. They rise 200 m above the desert plains between the Glass Mountains in the south and Fort Stockton in the north. Fort Stockton in Pecos County in west Texas is located 33 km north of the crater structure. Highway 385, leading from Fort Stockton southward to Big Bend National Park, traverses the structure.

Selected References

Eggleton, R. E., and Shoemaker, E. M. (1961). Breccia at Sierra Madera, Texas. U.S. Geological Survey Professional Paper, Vol. 424-D, pp. D151–D153.

Goldin, T. J., Wünnemann, K., Melosh, H. J., and Collins, G. S. (2006). Hydrocode modeling of the Sierra Madera impact structure. Meteoritics and Planetary Science, Vol. 41, pp. 1947–1958

Howard, K. A., Offield, T. W., and Wilshire, H. G. (1972). Structure of Sierra Madera, Texas, as a guide to central peaks of lunar craters. Geological Society of America Bulletin, Vol. 83, pp. 2795–2808.

Van Lopik, J. R., and Geyer, R. A. (1963). Gravity and magnetic anomalies of the Sierra Madera, Texas, "Dome". Science, Vol. 142, pp. 45–47.

Wilshire, H. G., and Howard, K. A. (1968). Structural patterns in central uplifts of cryptoexplosion structures as typified by Sierra Madera. Science, Vol. 162, pp. 258–261.

Wilshire, H. G., Offield, T. W., Howard, K. A., and Cummings, D. (1972). Geology of the Sierra Madera cryptoexplosion structure, Pecos County, Texas. U.S. Geological Survey Professional Paper, Vol. 599-H, 42 pp.

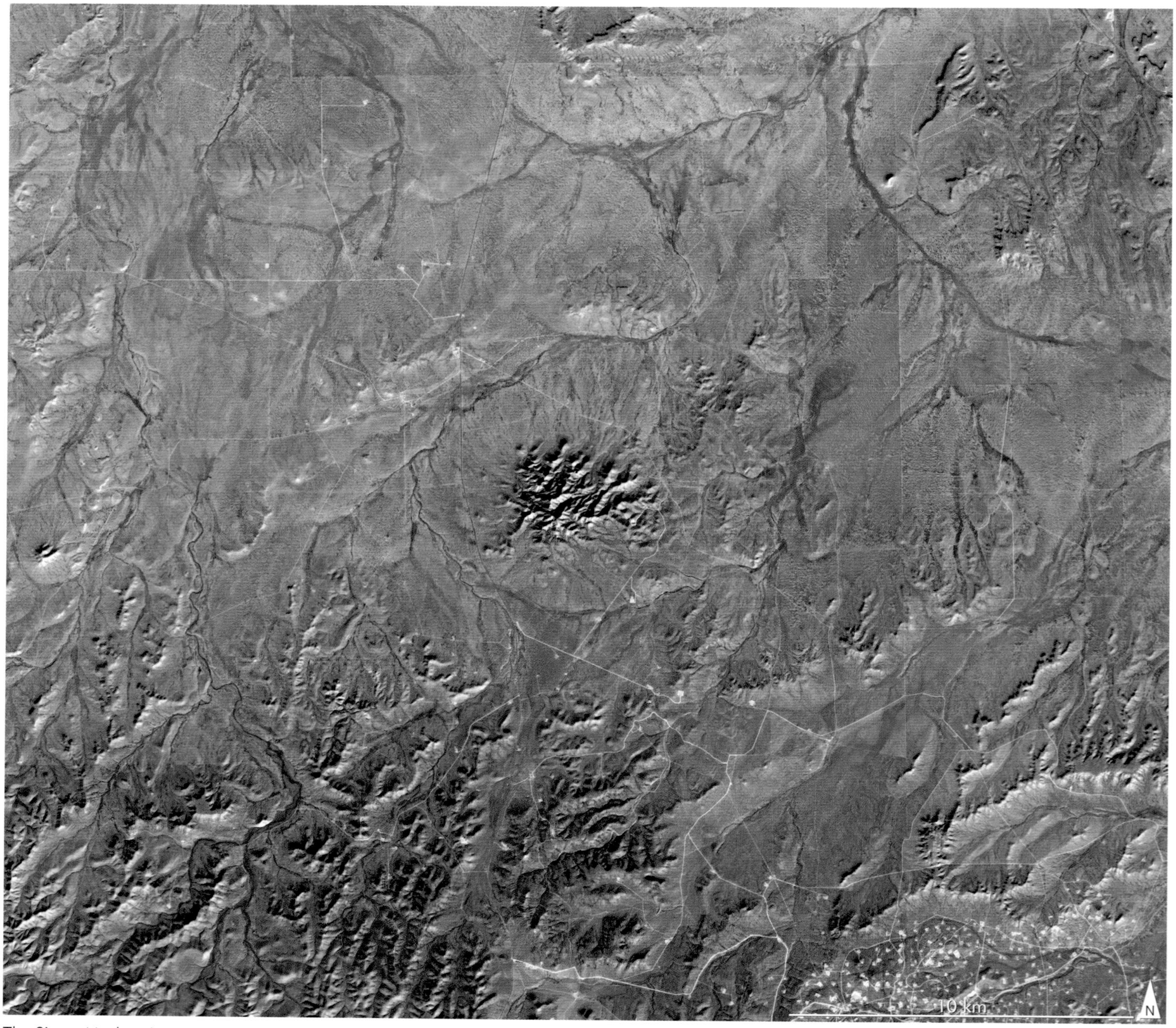

The Sierra Madera impact structure on March 7, 2018 as a composite image based on TanDEM-X DEM and Copernicus Sentinel-2 (channel 2/3/4) data.

Slate Islands
Ontario, Canada

Thomas Kenkmann

The Slate Islands impact structure is, to a large degree, submerged in Lake Superior. However, the central uplift of this roughly 30 km diameter crater is exposed in a near-circular cluster of islands, called the Slate Islands, that are elevated by up to 120 m above the lake level. These islands are surrounded by a deep but discontinuous annular trough where the lake becomes more than 250 m deep. The trough axis has a diameter of 15 to 16 km and is surrounded by an irregular annular ridge that rises to 100 m water depth. The ridge is interpreted as the crater rim. In the northern sector the crater rim trends inward of the shoreline.

The impact structure was formed on the Superior Province of the Canadian Shield. Exposed target rocks are a complex suite of Archean volcanic and subvolcanic rocks that range in composition from rhyolite to basalt. Intrusive rocks are diorites and granites. All these rocks have been regionally metamorphosed to greenschist facies and are cut by diabase dikes. They are unconformably overlain by basalts and some sandstones.

The Slate Islands impact structure is famous for its shatter cones that occur ubiquitously on the islands. Some are described as fracture surfaces more than ten meters high. Moreover, polymict breccia dikes injected into the target rocks are frequently exposed along the eastern shore of Patterson Island, the largest island in the group. The width of these exposures varies from a few centimeters to almost 60 m. The dikes have a clastic matrix and contain abundant angular fragments, some of which are shocked. Among the clasts are exotic lithologies that are not otherwise exposed in the region. Clasts within the dikes also occasionally show shatter cones and demonstrate that shatter cones form early in the crater forming process during the passage of the shock wave, whereas the emplacement of the dikes takes place somewhat later during the growth and collapse of the crater cavity. The mean composition of some of the dikes is mafic. Some have been interpreted to be pseudotachylites – but further work is required to investigate the true nature of these pseudotachylitic breccias.

Reports of local suevite and impact melt rock occurrences at the eastern shore of Patterson Island may require further proof. Such occurrences would indicate that the crater is not very deeply eroded and that locally the crater floor is exposed. The investigation of the orientations of planar deformation features (PDFs) in shocked quartz from the central uplift showed that the maximum shock pressure recorded in the rocks decays from about 20–25 GPa at the center to about 5 GPa at a distance of 5–7 km.

Geophysical studies conducted at Slate Islands include gravimetry, seismics, and magnetics. A negative gravity anomaly is associated with the structure. Its center is offset to the west from the centrally located island group. This could be, however, a result of the limited number of gravity stations. Seismic studies suggested that the ring depression surrounding the central uplift is about 50 % deeper than the bathymetry due to partial infill with sediment. The trough is bounded by peripheral faults at 24 km from the center. According to the seismic studies, the diameter of the central rise is about 12 km, and the structural uplift amounts to 1100 m. The extent of magnetic resetting in the structure decreases with increasing distance from the center. From paleomagnetic studies a rough age between 500 and 900 Myr has been estimated for the impact event. A ^{40}Ar-^{39}Ar apparent age spectrum from what has been reported as a pseudotachylite sample suggested an age of 436 ± 3 Myr.

The Slate Islands were mapped in 1974 by R. Sage of the Ontario Geological Survey. He recognized the complexities of the geological situation and described, inter alia, numerous breccia occurrences and shatter cones. Shortly after this, in 1975 and 1976, Henry Halls, Richard Grieve, and Blythe Robertson documented deformation lamellae in quartz and feldspar in these rocks, in addition to the already known shatter cones, and further supported the impact origin for this structure.

Giant shatter cone, about 10 m long, on Slate Island in McGreevy Harbour. Note the owner of the photo at the base of the shatter cone for scale (image courtesy: Chuck O'Dale, www.craterexplorer.ca).

Location/Access

The Slate Islands structure is situated mostly in Lake Superior. The central part of the structure is exposed on the unpopulated group of islands of homonymous name. Patterson and Mortimer islands are the largest islands of the archipelago. These and a dozen nearby islands are located about 13 km south of the northern shore of Lake Superior's Terrace Bay, which corresponds to the rim of the crater. The town of Terrace Bay is connected to Sault Ste. Marie and Thunder Bay via the Trans-Canada Highway.

Selected References

Dressler, B. O., Sharpton, V. L., and Schuraytz, B. C. (1998). Shock metamorphism and shock barometry at a complex impact structure: Slate Islands, Canada. Contributions to Mineralogy and Petrology, Vol. 130, pp. 275–287.

Grieve, R. A. F. (2006). Impact Structures in Canada. Geological Association of Canada, GeoText No. 5, 210 pp.

Grieve, R. A. F., and Robertson, P. B. (1976). Variations in shock deformation at the Slate Islands impact structure, Lake Superior, Canada. Contributions to Mineralogy and Petrology, Vol. 58, pp. 37–49.

Halls, H. C. (1975). Shock-induced remanent magnetization in Late Precambrian rocks from Lake Superior. Nature, Vol. 255, pp. 692–695.

Halls, H. C., and Grieve, R. A. F. (1976). The Slate Islands: a probable complex meteorite impact structure in Lake Superior. Canadian Journal of Earth Sciences, Vol. 13, pp. 1301–1309.

Sage, R. P. (1978). Diatremes and shock features in Precambrian rocks of the Slate Islands, northeastern Lake Superior. Geological Society of America, Bulletin 89, pp. 1529–1540.

Sharpton, V. L., Dressler, B. O., Herrick, R. R., Schnieders, B., and Scott, J. (1996). New constraints on the Slate Islands impact structure, Ontario, Canada. Geology, Vol. 24, pp. 851–854.

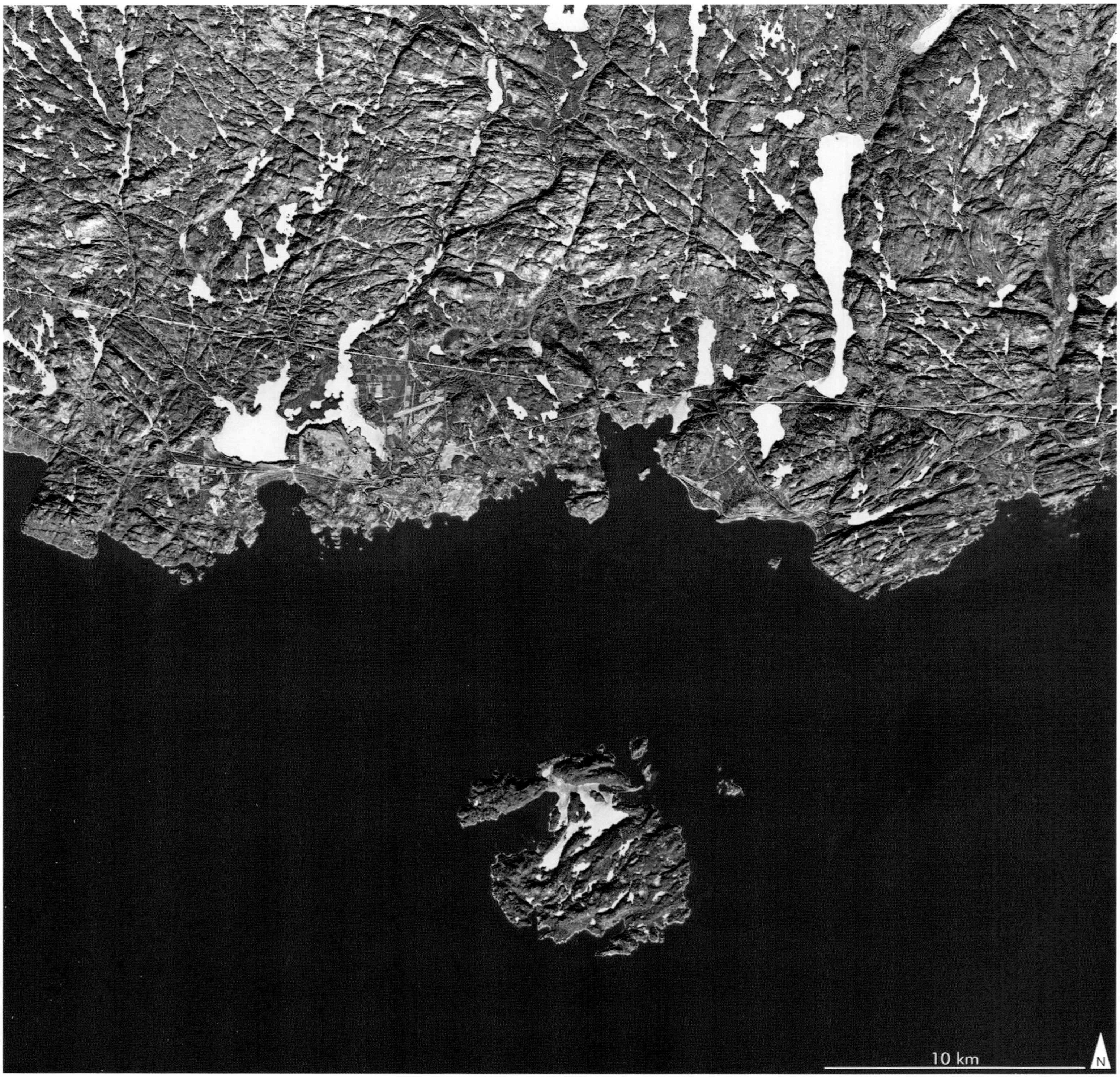

Bottom: The Slate Islands impact structure on April 19, 2018 as a composite image based on TanDEM-X DEM and Copernicus Sentinel-2 (channel 2/3/4) data.

Slate Islands

Ontario / Canada

latitude = 48°39‘ N / longitude = 87°01‘ W / diameter = 30 km / age = 436 ± 3 Myr

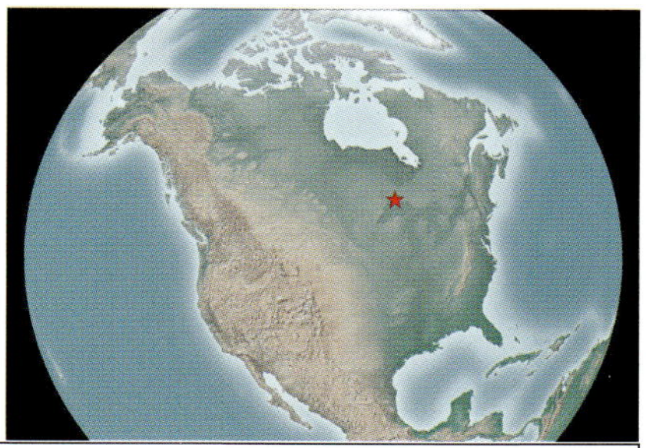

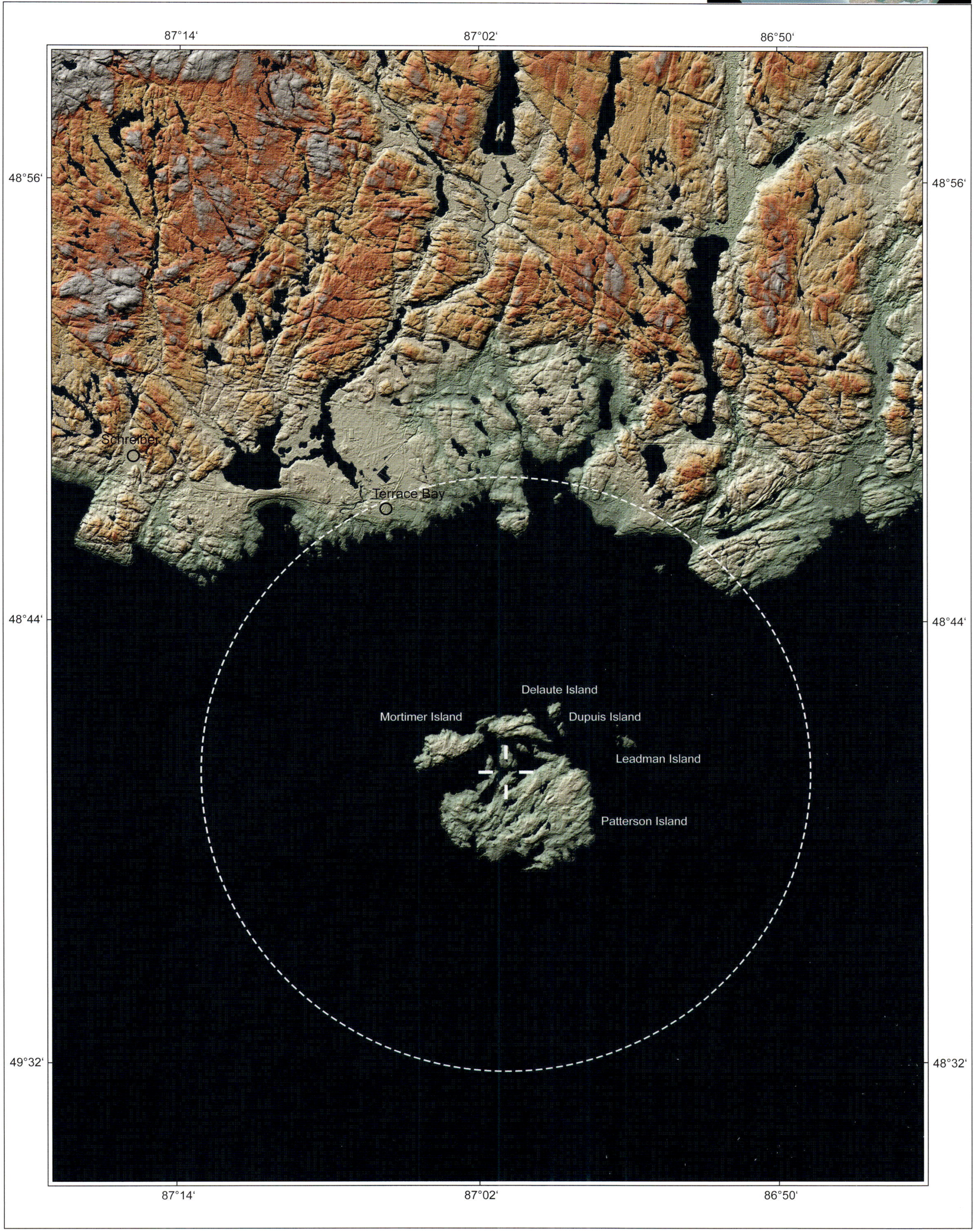

scale = 1 : 237 000 / scene = 42.6 × 56.8 km^2

0 2.37 11.85 km

N

180 meter 590

Steen River
Alberta, Canada

Thomas Kenkmann

Steen River is a buried complex impact structure without surface expression. The diameter is not well constrained but listed as about 22 km. Steen River is of great economic interest and is regarded as the largest known impact structure within the oil and gas producing Western Canadian Sedimentary Basin.

Target rocks are Precambrian crystalline basement rocks, primarily granitic gneiss, that are overlain by Devonian conglomerates, sandstones, and mudstones. There is a major unconformity between these Devonian strata and the overlying post-impact shales of Lower Cretaceous age that, together with Quaternary deposits, amount to 200 m thickness. Extensive erosion must have occurred before the Cretaceous sediments buried the crater. In this period of erosion the crater was peneplained.

The Precambrian basement shows a stratigraphic uplift of about 1000 m. The central uplift has a diameter of 4 km at its top but widens to about 8 km at depth. A thick breccia drapes the central uplift and surrounding disturbed sedimentary units. This polymict impact breccia contains lithic and mineral clasts, as well as melt rock, and thus deserves the classification as suevite. Impact melt in suevite forms rims enveloping granitic clasts (mantled clasts), and occurs as layers of agglomerated impact melt rock. The annular trough surrounding the central uplift is deeper in the northwestern sector of the crater where the basement drops 500 m below the regional level and contains brecciated target rocks. In contrast, in the south the basement is downfaulted by only 180 m. The rim strata are uplifted by 20–50 m with respect to the surrounding geology. Rocks of the central uplift and the disturbed zone show zeolite mineralization in fractures and indicate a hydrothermal overprint (likely hydrothermal activity at just postimpact time).

The structure is characterized by a negative Bouguer gravity anomaly of −5 mGal that is superposed onto the regional trend. However, the signature is complex, as the deepest part of the annular trough also corresponds to a local minimum. The aeromagnetic signature is inconclusive. Most of the seismic data gathered by hydrocarbon exploration companies are classified.

Recent work on biostratigraphy, as well as U-Pb and ^{40}Ar-^{39}Ar ages for bentonite layers, has confirmed a Lower Cretaceous age of about 108 Myr for the basal portion of the post-impact shale formation, which lies close to the contact to the crater-fill breccias.

The structure was first recognized during seismic surveying. It was drilled in 1963 by a hydrocarbon exploration company; shock features could be identified in drill core samples by Maurice Carrigy and Nicholas Short in 1968. Numerous exploration wells have been drilled near the crater and along the crater rim since then.

Location/Access

Steen River in northern Alberta is located about 750 km north of Edmonton and 170 km from the southern shore of the Great Slave Lake. A visitor has to travel from Edmonton towards the northwest on several of Alberta's highways until, at Grimshaw, Highway 35 branches off northward. Then it is still a 370 km drive to Steen River. The town High Level is the closest large settlement at about 120 km south of the impact structure.

The area of the Steen River impact structure on May 26, 2019 as a composite image based on TanDEM-X DEM and Copernicus Sentinel-2 (channel 2/3/4) data. Cross hairs and a dashed circle indicate position and extent of the buried structure. Smoke from wildfires covers the southeastern part of the scene.

Steen River

Alberta / Canada

latitude = 59°30' N / longitude = 117°38' W / diameter = 22 km / age = 108 Myr

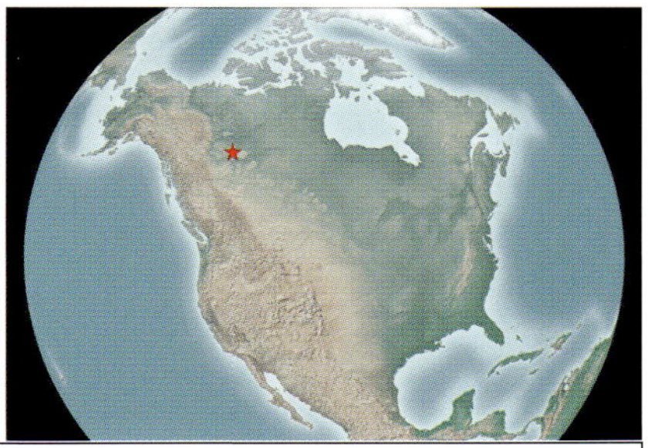

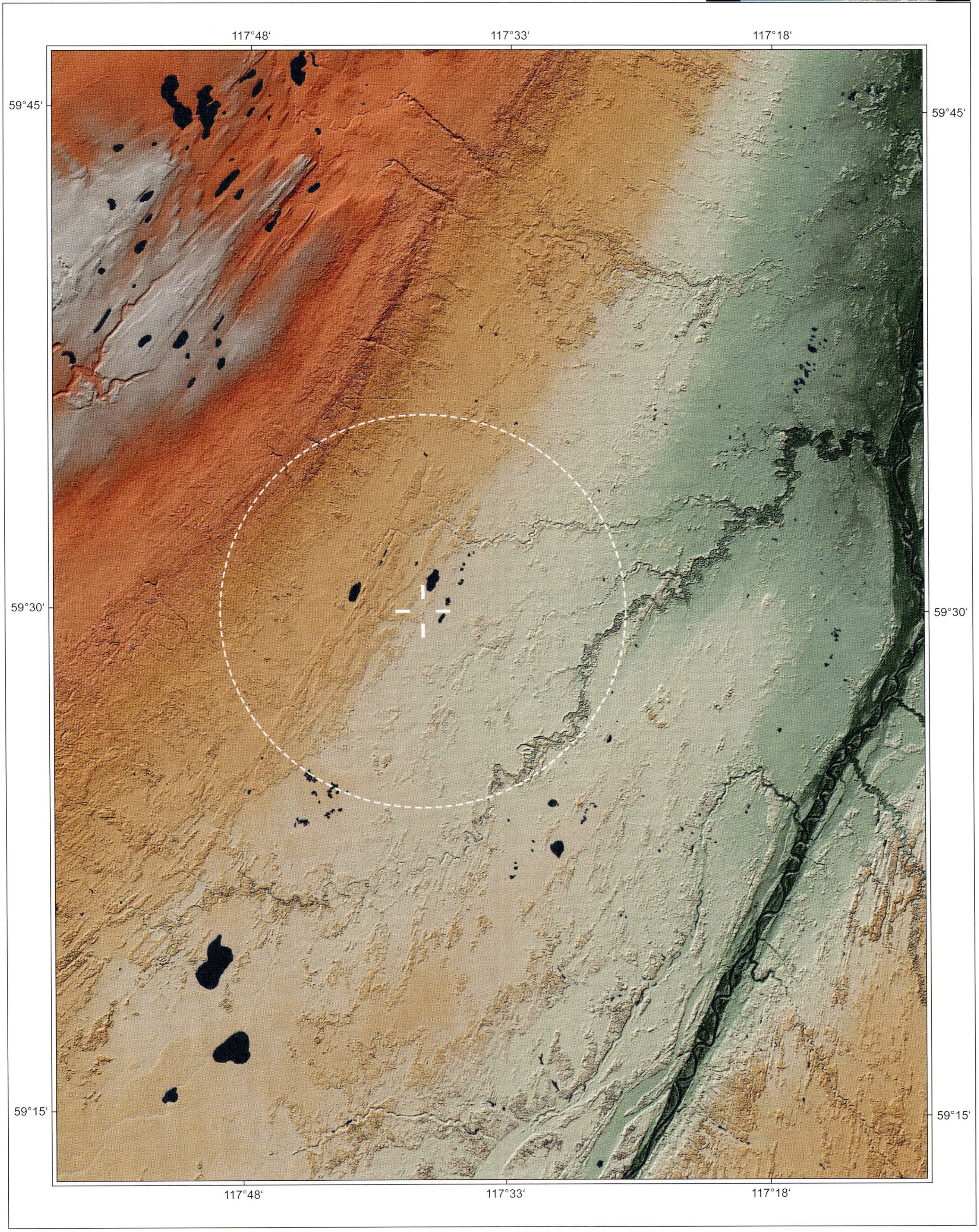

scale = 1 : 261 000 / scene = 47.1 × 62.8 km²

0 2.61 13.05 km

N

290 meter 760

Selected References

Carrigy, M. A., and Short, M. N. (1968). Evidence of shock metamorphism in rocks from the Steen River structure, Alberta. In: Shock Metamorphism of Natural Materials, French, B. M. and Short, N. M. (eds.), Mono Book Corp., Baltimore, pp. 367–378.

Grieve, R. A. F. (2006). Impact Structures in Canada. Geological Association of Canada, GeoText No. 5, 210 pp.

Maclagan, E. A., Walton E. L., Herd, C. D. K., and Dence, M. (2018). Investigation of impact melt in allochthonous crater-fill deposits of the Steen River impact structure, Alberta, Canada. Meteoritics and Planetary Science, Vol. 53, pp. 2285–2305.

Short, N. M. (1968). Petrographic study of shocked rocks from the Steen River structure, Alberta. In: Shock Metamorphism of Natural Materials, French, B. M. and Short, N. M. (eds.), Mono Book Corp., Baltimore, pp. 374–378.

Walton, E. L., Sharp, T. G., Hu, J., and Tschauner, O. (2017). Investigating the response of biotite to impact metamorphism: Examples from the Steen River impact structure, Canada. Meteoritics and Planetary Science, Vol. 53, pp. 75–92.

Sudbury
Ontario, Canada

Wolf Uwe Reimold

The Sudbury structure is located in Canada's Ontario Province some 400 km north of the city of Toronto. Sudbury's origin has been debated ever since the discovery of this exceptional geological feature around the end of the 19th century.

In the central part of the structure occurs the so-called "Sudbury Igneous Complex (SIC)" that forms an elliptical body ca. 27 × 60 km in extent and of ca. 2.5 km thickness. The SIC represents a huge pool of differentiated impact melt rock. The entire Sudbury structure covers a much larger area, of some 15 000 km². To date the original diameter of the Sudbury impact structure, prior to tectonic overprint, is generally estimated at 200–250 km. This makes the Sudbury structure the second largest impact structure currently known on Earth, after the **Vredefort** structure in South Africa. Around the SIC, fractured and brecciated rocks up to 250 m thick occur and represent part of a strongly brecciated crater floor formed by rocks of the Archean basement of the Superior and Southern provinces of the Canadian Shield. These rocks are mostly exposed to the north and east of the SIC. Furthermore, there are the supracrustal rocks of the Proterozoic Huronian Supergroup adjacent to the SIC, to the south. Overlying the SIC occur impact breccias, mudstones and graywackes of the Whitewater Group.

Sudbury is synonymous with one of the world's foremost Ni-Cu and Platinum Group Element deposits. These ore bodies are mainly concentrated in the lowermost part of the SIC, the so-called "Sublayer", and along a number of the so-called "Offset Dikes" that trend radially-concentrically to the SIC into both the Proterozoic and Archean target rocks. The total ore reserve has been estimated at 1.65×10^9 tons of ore at 1.2 % Ni and 1.05 % Cu. The cumulative value of ore from more than 100 years of production from the Sudbury structure has been estimated at far more than US$ 100 billion. Isotopic and geochemical data were most important in demonstrating that at least 50 % of a crustal component are required to model the formation of the main mass of the SIC adequately. Isotopic studies of ores, including platinum group elemental and Re-Os isotopic data, have shown that the ores could be obtained from mixtures of Superior Province and Huronian metasedimentary rocks, with some admixture of Archean to Proterozoic mafic rocks.

The SIC was dated by various methods, but principally by the U-Pb isotope method on zircon to 1849.3 ± 0.3 Myr and the Offset Dikes to 1852 ± 4 Myr. Similar ages were also obtained for the Sublayer and inclusions in the Sublayer, and also for some Sudbury Breccia.

The Sudbury structure is located at the present boundary of the metavolcanic and metasedimentary sequence belonging to the Proterozoic Huronian Supergroup in the south and the Archean basement rocks in the north. Impact evidence includes widespread occurrence of shatter cones all around the SIC, and up to a distance of 17 km from it. Pseudotachylitic breccias, the so-called "Sudbury Breccias", form numerous, often – in the vicinity of the SIC – very large, bodies. In fact, they form the largest dikes of such material known anywhere in the world. Sudbury Breccia occurrences are concentrated in a

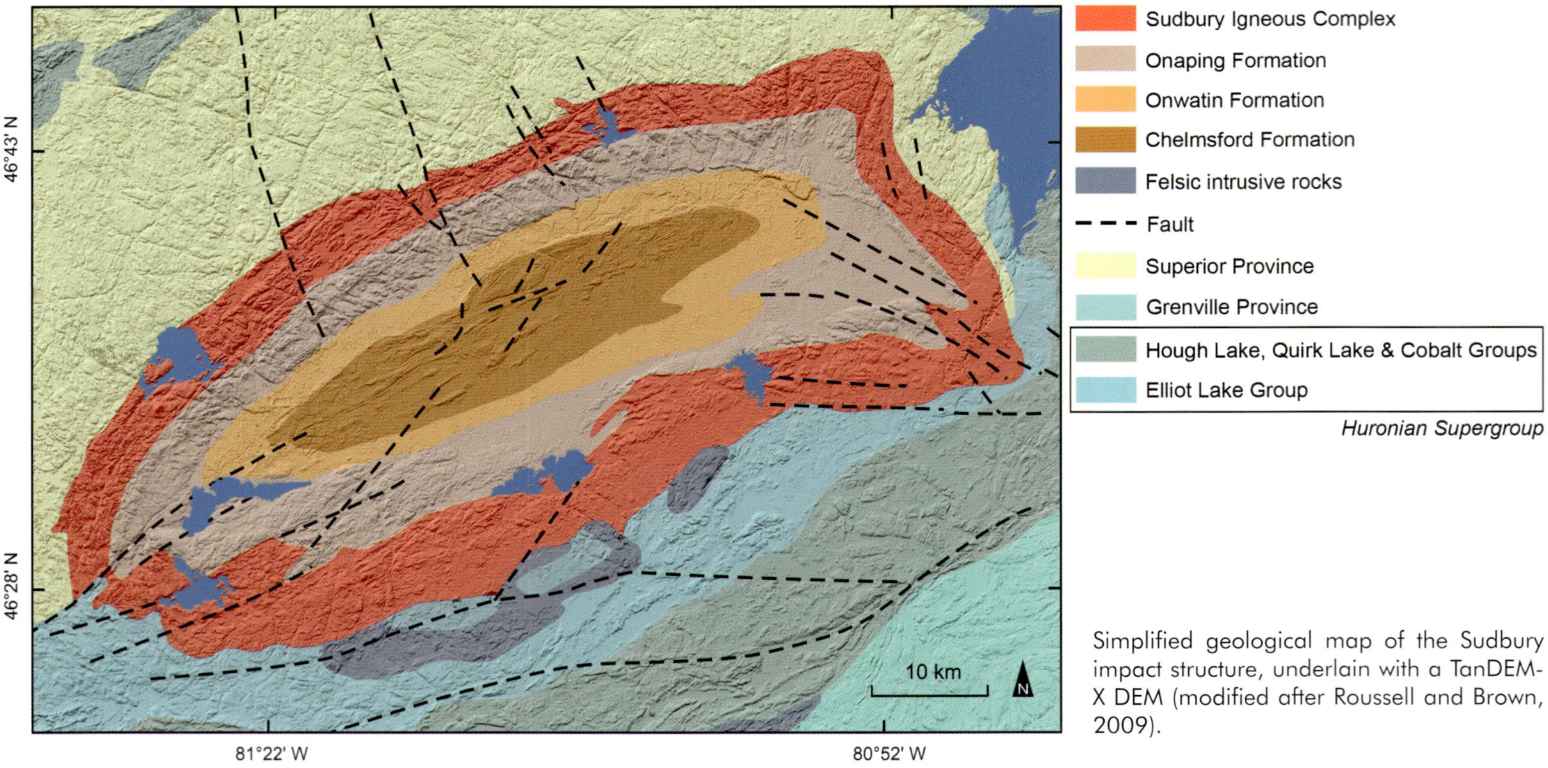

Simplified geological map of the Sudbury impact structure, underlain with a TanDEM-X DEM (modified after Roussell and Brown, 2009).

Sudbury

Ontario / Canada

latitude = 46°36‘ N / longitude = 81°11‘ W / diameter = 200 km / age = 1849.3 ± 0.3 Myr

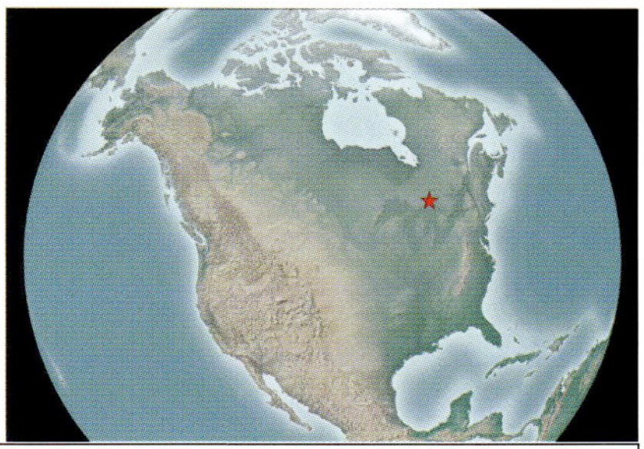

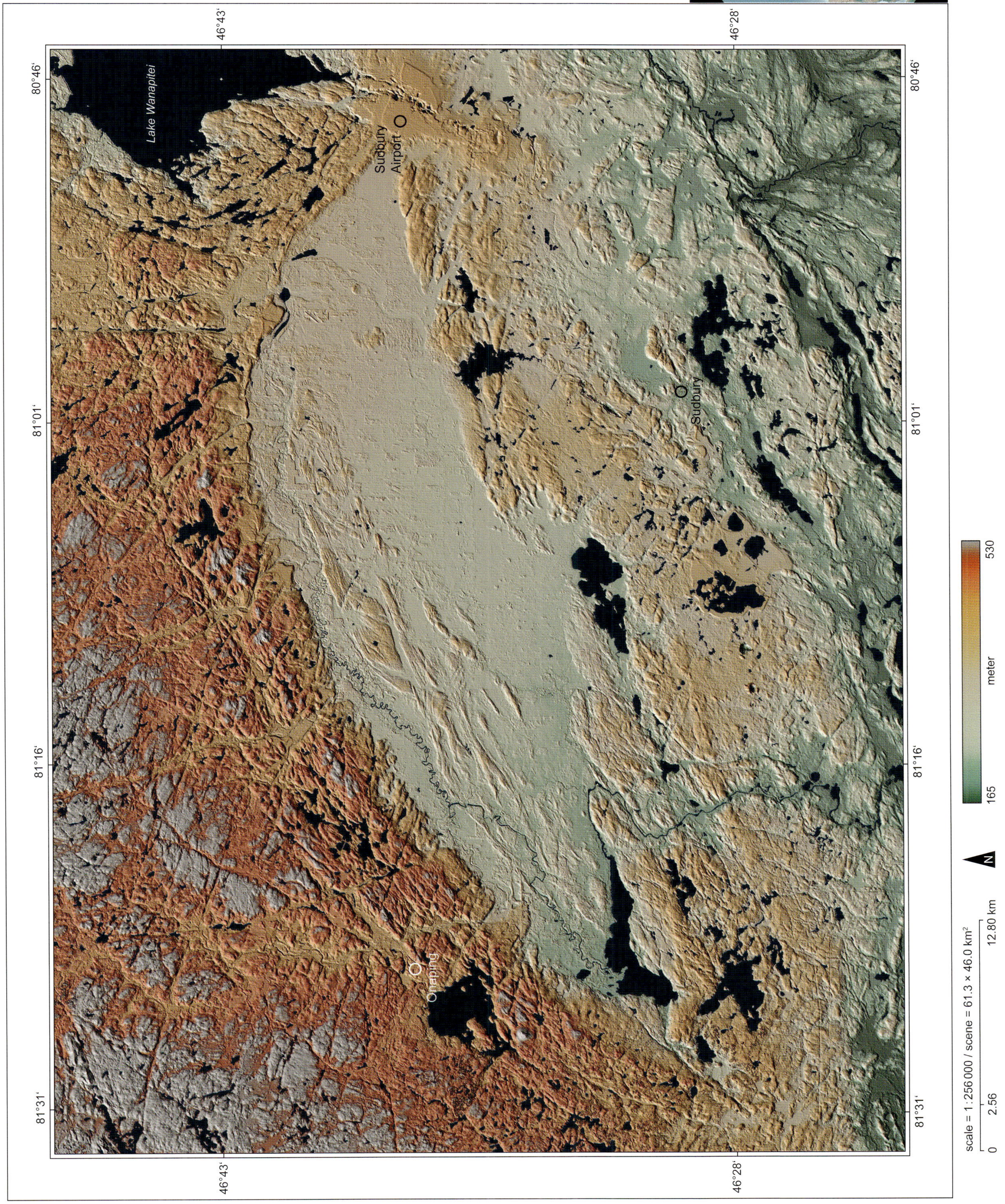

The Sudbury impact structure is a world-class copper and nickel deposit. Sudbury's skyline is dominated by the towering Copper Cliff smelter "Superstack" (photo: T. Kenkmann).

5–15 km inner zone around the SIC but have been found as far as 80 km northeast from the SIC. These breccias that actually involve different types of brecciated material with and without apparent melt groundmass, have been reported to occur in distinct zones of breccia concentration at various distances around the SIC, predominantly to the north. The zones have been interpreted as possible expressions of ring structures of a possible multi-ring impact structure. The Sudbury Breccia comprises a variety of breccia types, including purely clastic as well as melt-bearing breccias, and some that have been recrystallized due to the thermal metamorphic overprint from the SIC.

The vertical stratigraphy of the SIC comprises the Sublayer at the bottom, then above, the Norite, Quartz-Gabbro and Granophyre sequences. The Sublayer generally consists of a noritic-quartz dioritic rock containing mafic and ultramafic xenoliths, as well as inclusions derived from the various target rocks. The SIC has been interpreted as a differentiated impact melt sheet. The Offset Dikes host as much as 50 % of the ore resources of the Sudbury structure, and are also regarded as impact melt from the Sudbury impact that was intruded into radial and concentric fractures.

The Whitewater Group representing the post-impact fill of the Sudbury impact basin consists – from bottom to top – of impact melt breccias and suevitic breccias of the Onaping Formation, mudstones of the Onwatin Formation, and wackes of the Chelmsford Formation. Near the bottom of the Onwatin Formation is the stratigraphic position of the Vermilion Member that consists of a distinct carbonate-chert sequence with interbedded argillitic limestone and dolostone.

A huge amount of geophysical work has been carried out on the Sudbury structure, much of it in the course of exploration of the phenomenal ore resources. Seismic reflection studies have shown that the structure is asymmetric at depth: the North Range dips at ca. 30° towards the south, and in the South Range a series of southward-dipping reflectors indicate thrust faulting. This is the result of overprint from the Penokean orogeny that roughly coincided with the time of formation of the structure. In addition, 800 million years after the Sudbury impact event, the Grenvillian Thrust Front affected the region of the Sudbury structure from the southeast. It has been alleged that a thermal overprint related to the Penokean (about 1860–1830 Myr) and Grenvillian (about 1250–1000 Myr) orogenies occurs up to 50 km to the north of the North Range around the SIC. Extensive hydrothermal alteration exists widespread in the Sudbury Structure. There is a thermal aureole in the immediate environs of the SIC, within which quartz and feldspar and associated planar deformation features (PDFs) are more or less annealed. In the North Range this zone of recrystallization is about 500 m wide, but the thermal aureole extends beyond 1 km from the complex. Between distances from about 500 m to 8 km from the SIC PDFs can be recognized again.

A several centimeter wide vein of pseudotachylitic breccia in a foliated gneiss of the basement of the Sudbury Structure (North Range). Lens cap for scale about 6 cm wide (photo: W. U. Reimold).

It was not easy to convince the geological community that the exceptional Sudbury structure was not the product of a massive magmatic emplacement event but resulted from the impact of a large extraterrestrial projectile. Several first-order aspects of the structure and its exploration did not facilitate the acceptance of this: the elliptical shape of the SIC, the long-term focus on the SIC only, and as the name "Sudbury Igneous Complex" implies, on magmatic processes for the formation of the complex and magma differentiation as a means of generating the ore resources. While it was long ago recognized that impact played an integral part in the formation of the structure, the enormous size of the igneous complex and its differentiation were strong hindrances for the formulation of a simple impact model. Only from about 1985 has the impact origin of the Sudbury structure and of the SIC been generally accepted.

The impact origin for Sudbury was first mooted by Robert Dietz in 1962. This pioneer of both plate tectonics and impact studies predicted that shock metamorphic evidence would be found at Sudbury, and soon – only 2 years later – identified shatter cones there himself. PDFs were reported from quartz in clasts of the Onaping Formation by Bevan French in 1967 and Mike Dence in 1972. Further shock metamorphic studies in the footwall rocks of the northern circumference of the SIC by Burkhard Dressler followed, and finally PDFs were recognized in quartz of some inclusions in the SIC by Ann Therriault.

Still today, this huge structure remains a sought-after study object. For instance, the development of the Onaping Formation has remained enigmatic for decades and the source of metals in the target lithologies has not been resolved completely. Furthermore, uncertainties still exist as to whether the Sudbury impact structure was a peak-ring or a multi-ring basin and concerning the deformation mechanisms of rock flow during transient cavity formation and crater modification. And the authigenic zircon growth in SIC lithologies has triggered the interesting idea that the numerous much older, entirely unshocked Archean zircon grains (older than 4000 Myr in age) known from the Jack Hills region of Western Australia might not represent early crustal growth but rather crystals that crystallized from giant impact melt bodies formed in those early Earth bombardment times.

Location/Access

The Sudbury Basin is located a few kilometers to the northwest of the city of Sudbury, about 400 km north-northwest of Toronto. From this metropolis, Sudbury is reached by car on excellent Highway 400. Greater Sudbury Airport also receives several daily flights from Toronto. Many prominent outcrops are accessible with a normal car. Geological information can be obtained from the Ministry of Northern Development and Mines, on Ramsay Lake Road, and interesting displays on the regional geology and Sudbury impact event are set up in the Science North Center at the intersection of Highway 80 and Ramsay Lake Road.

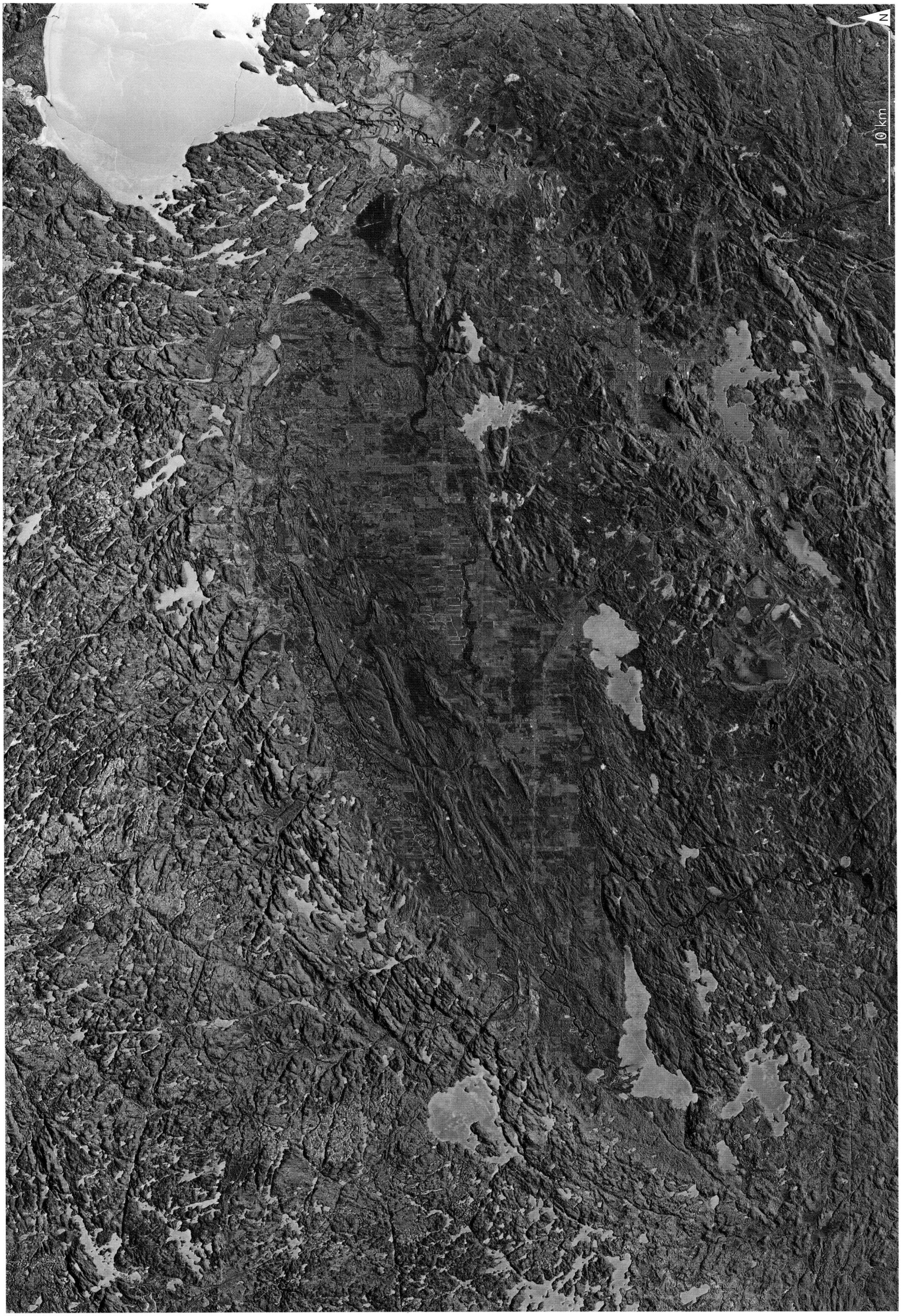

The Sudbury impact structure on April 30, 2018 as a composite image based on TanDEM-X DEM and Copernicus Sentinel-2 (channel 2/3/4) data.

Selected References

Dietz, R. S., and Butler, L. W. (1964). Shatter-cone orientation at Sudbury, Canada. Nature, Vol. 204, pp. 280–281.

Dressler, B. O. (1984). The effects of the Sudbury event and the intrusion of the Sudbury Igneous Complex on the footwall rocks of the Sudbury structure. In: The Geology and Ore Deposits of the Sudbury Structure, Pye, E. G., Naldrett, A. J., and Giblin, P. E. (eds.), Ontario Geological Survey, Special Volume 1, pp. 97–136.

Grieve, R. A. F. (2006). Impact Structures in Canada. Geological Association of Canada, GeoText No. 5, 210 p.

Grieve, R. A. F., and Stöffler, D. (1991). The Sudbury Structure: Controversial or Misunderstood? Journal of Geophysical Research, Vol. 96, pp. 22,753–22,764.

Grieve, R. A. F., Reimold, W. U., Morgan, J., Riller, U., and Pilkington, M. (2008). Observations and interpretations at Vredefort, Sudbury, and Chicxulub: Towards an empirical model of terrestrial impact basin formation. Meteoritics and Planetary Science, Vol. 43, pp. 855–882.

Kenny, G. G., Whitehouse, M. J., and Kamber, B. S. (2016). Differentiated impact melt sheets may be a potential source of Hadean detrital zircon. Geology, Vol. 44, pp. 435–438.

Lightfoot, P. C. (2016). Nickel Sulfide Ores and Impact Melts: Origin of the Sudbury Igneous Complex. Elsevier Publishers, Amsterdam, 662 pp.

Naldrett, A. J. (2003). From impact to riches: Evolution of geological understanding as seen at Sudbury, Ontario. Geological Society of America Today, Vol. 13, pp. 4–9.

Pye, E. G., Naldrett, A. J., and Giblin, P. E. (eds.) (1984). The Geology and Ore Deposits of the Sudbury Structure. Ontario Geological Survey, Special Volume 1, 603 pp.

Reimold, W. U., Koeberl, C., Gibson, R. L., and Dressler, B. O. (2005). Economic mineral deposits in impact structures: A review. In: Impact Tectonics, Koeberl, C. and Henkel, H. (eds.), Springer, Berlin Heidelberg, pp. 479–552.

Riller, U. (2005). Structural characteristics of the Sudbury impact structure, Canada: Impact-induced versus orogenic deformation – A review. Meteoritics and Planetary Science, Vol. 40, 1723–1740.

Roussell, D. H., and Brown, G. H. (2009). A Field Guide to the Geology of Sudbury, Ontario. Open File Report 6243, Ontario Geological Survey, Sudbury, 222 pp.

Spray, J. G. (1997). Superfaults. Geology, Vol. 25, pp. 627–630.

Therriault, A. M., Fowler, A. D., and Grieve, R. A. F. 2002. The Sudbury Igneous Complex: A differentiated impact melt sheet. Economic Geology, Vol. 97, pp. 1521–1540.

Ubide, T. P., Guyett, C., Kenny, G. G., O'Sullivan, E. M., Ames, D. E., Petrus, J. A., Riggs, N. and Kamber, B. S. (2017). Protracted volcanism after large impacts: Evidence from the Sudbury impact basin. Journal of Geophysical Research, Planets, Vol. 122, pp. 701–728.

Tunnunik

Northwest Territories, Canada

Thomas Kenkmann

Tunnunik is a large impact structure on Victoria Island in Canada's High Arctic region that was only discovered a few years ago. The structure is deeply eroded but is still recognizable as an anomalous circular feature formed by a series of concentric ridges that curve around the central area. The circular pattern has a diameter of 25 km; it seems to mainly represent the remnant of a central uplift. The dimension of the entire crater structure should, thus, be larger than this. The size has been variably estimated at 28 to 50 km.

The oldest sedimentary rocks in the area are of Precambrian age of roughly 1000 to 720 Myr. These rocks are cut by diabase dikes with an age of 723 Myr that reach thicknesses of up to 3 m in the structure. At the base of the overlying Paleozoic succession are Cambrian sandstones, siltstones, shales, and dolostones, which are followed upward by Ordovician and Silurian dolostones, and Silurian carbonates and shales. The Paleozoic strata on northwestern Victoria Island generally dip a few degrees to the northwest. Consequently, thick sedimentary sequences of Devonian strata

The Tunnunik impact structure on July 14, 2018 as a composite image based on TanDEM-X DEM and Copernicus Sentinel-2 (channel 2/3/4) data.

Tunnunik
Northwest Territories / Canada

latitude = 72°28‘ N / longitude = 113°55‘ W / diameter = 25 km / age = 430 - 450 Myr

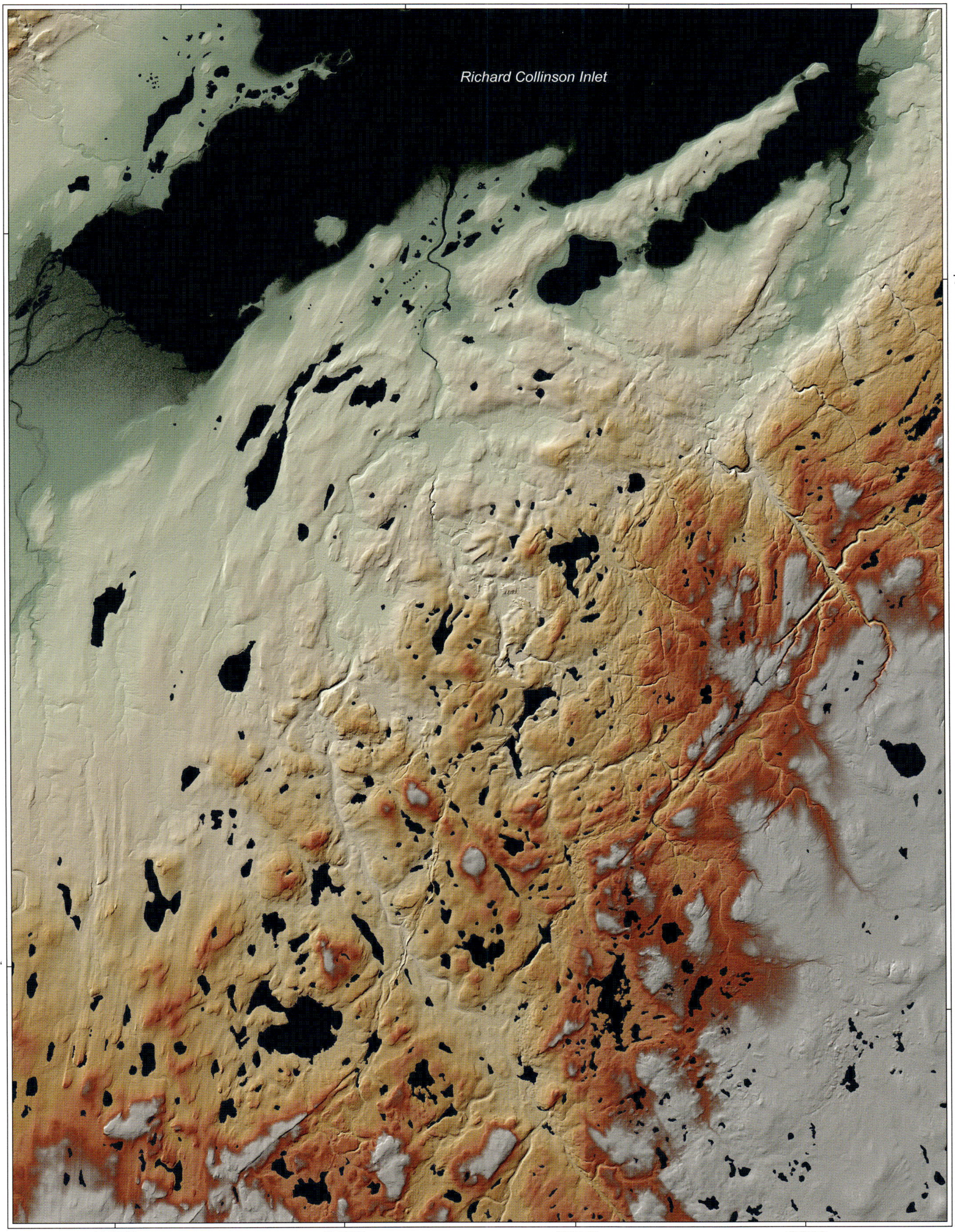

scale = 1:258 000 / scene = 46.4 × 61.8 km^2

0 2.58 12.90 km

N

0 meter 270

reaching almost 3000 m in thickness are preserved further to the northwest of the impact structure on Banks Island. It is very likely that several kilometers of these Devonian strata once covered the target area at the time of impact as well. Erosion estimates for the structure amount to at least 700 m, but values of possibly 5000 m have been suggested as well.

The sedimentary succession is covered by a layer of Quaternary deposits, which largely obscure the underlying geology. Only a few thousand years ago the glaciers retreated from this region, so glacially derived landforms such as polished and striated rocks, till, glacial erratic boulders, gravel deposits, drumlins, and eskers are omnipresent.

The core of Tunnunik's central uplift consists of Precambrian rocks that are surrounded by Cambrian and Ordovician strata. The Precambrian strata are cut by diabase dikes. Bedding dips quaquaversally outward from the core and becomes shallower with distance. Dips in the central portion are steep, locally even sub-vertical. Radial faults, as well as concentric thrust faults associated with drag folds, can be observed. Polymict carbonate breccias exist in places, of which some seem to be connected to mapped fault zones.

Shatter cones occur widespread in all units of the central uplift. The best developed cones with diverging grooves and branching subcone ridges are formed in fine-grained lithologies. The abundance of shatter cones decreases towards the edge of the uplift. Most of the shatter cones have stratigraphically upward-pointing apices, but downward-facing apices are also common.

Regional faults trending west-southwest/east-northeast can be followed across the impact structure and, thus, appear to be younger than the impact event. These faults are attributed to the opening of the Canada Basin prior to the Early Cretaceous. Thus, the impact age can be bracketed between 450 Myr, the uppermost age of impact-deformed strata, and approximately 130 Myr, the maximum age for the regional tectonic faulting. A further age constraint comes from paleomagnetism. The sedimentary rocks were re-magnetized after the impact, for which an age of 430–450 Myr could be determined.

Tunnunik does not show geophysical anomalies, neither with respect to gravity nor to aeromagnetics.

Early reports about Tunnunik date back to the 1970s, when Canadian petroleum exploration companies noted a dome-like uplift of Precambrian strata. The discovery of the Tunnunik impact structure was then made in 2010 in the course of a regional mapping campaign of northwestern Victoria. A first publication about this feature by Keith Dewing and colleagues was issued in 2013.

Location/Access

The Tunnunik impact structure is located south of the Richard Collinson inlet of northwestern Victoria Island, Northwest Territories, Canada. The structure is named after the traditional Inuvialuit name for this inlet. There is no infrastructure available in this Arctic region, so that access requires extensive expedition planning and preparation.

Selected References

Choe, B.-H., Tornabene, L. L., Osinski, G. R., and Newman, J. D. (2019). Remote predictive mapping of the Tunnunik impact structure in the Canadian Arctic using multispectral and polarimetric SAR data fusion. Canadian Journal of Remote Sensing, Vol. 44, https://doi.org/10.1080/07038992.2018.1544846

Dewing, K., Pratt, B. R., Hadlari, T., Brent, T., Bédard, J., and Rainbird, R. H. (2013). Newly identified "Tunnunik" impact structure, Prince Albert Peninsula, northwestern Victoria Island, Arctic Canada. Meteoritics and Planetary Science, Vol. 48, pp. 211–223.

Lepaulard, C., Gattacceca, J., Swanson-Hysell, N., Quesnel, Y., Demory, F., and Osinski, G. R. (2019). A Paleozoic age for the Tunnunik impact structure. Meteoritics and Planetary Science, Vol. 54, pp. 740–751.

Newman, J. D., and Osinski, G. R. (2016). Geological mapping of the Tunnunik impact structure, Victoria Island, Canadian High Arctic. 47th Lunar and Planetary Science Conference, Lunar and Planetary Institute, Houston, Texas, Abstract #1591.

Upheaval Dome
Utah, United States

Thomas Kenkmann

The Upheaval Dome structure in southeastern Utah is a spectacular, deeply eroded crater carved into the high mesa of the Colorado Plateau. The present topography at Upheaval Dome is entirely erosional in nature and does not reflect the original crater form. The structure has been eroded by 1.5–2 km since the time of impact and provides 3D insight into the subsurface of the crater; hence, typical impact lithologies such as allochthonous breccia or suevite are not preserved. Upheaval Dome consists of a central uplift of about 2.5 km diameter, which is surrounded by a ring syncline, and a crater rim monocline of approximately 5 km width. The latter defines the apparent extent of the structure. Initially the diameter of the Upheaval Dome impact crater would have been about 7–8 km. The inner part of the central uplift is topographically the opposite of an uplift and forms a morphological depression of approximately 1.4 km diameter.

The strata exposed in the center comprise rocks of uppermost Permian (White Rim Sandstone), Triassic (Moenkopi Formation, Chinle Formation), and Jurassic (Wingate Sandstone, Kayenta Formation, Navajo Sandstone) age, which are located at progressively increasing distance from the center. The Navajo Sandstone is preserved along the axis of the ring syncline and represents the youngest formation. At the perimeter of the structure a system of centrally-dipping listric faults can be observed that flatten within the ring syncline to detachments and then rise towards the central uplift. Faulting on these shear zones is found to evolve from normal to bedding-parallel, and eventually reverse in the central uplift. This is thought to be the reason for thinning of the Wingate Sandstone in several parts near the perimeter. While the layered Kayenta Formation displays complex internal thrusting on branches of these faults, the massive Wingate and Navajo sandstones show radial folds, plunging away from the center. These structures are interpreted as signs for convergent material flow. Towards the crater center the radial folds become tighter, and fold limbs are sheared off, so that a stack of imbricated thrust slices is developed. A 3D analysis of the deformation pattern showed that the arrangement of structural elements in the inner part of the Upheaval Dome structure roughly displays a bilateral symmetry, trending northwest. This indicates a material transport of "top to the southeast", which is interpreted to be the result of an oblique impact from the northwest. Strata in the center are usually in upright position, in contrast to those of the wider Colorado Plateau where the same strata are flat-lying. During crater collapse, the pervasively crushed White Rim Sandstone flowed locally as a granular medium, resulting in the formation of clastic dikes. Acting as a lubricant, this mobilized layer accommodated the complex folding and faulting of the overlying strata.

Due to the deep level of erosion, the vast majority of shocked rocks has been removed over time. A shock metamorphic overprint was only locally found in rocks of the Kayenta Formation in the form of multiple sets of cross-cutting planar deformation features (PDFs) in quartz. The features are annealed, exhibit dense tangles of dislocations, and are decorated with fluid inclusions.

Upheaval Dome

Utah / United States

latitude = 38°26'09" N / longitude = 109°55'40" W / diameter = 8 km / age = 66 - 100 Myr

Panoramic view of the central depression of the Upheaval Dome structure – seen from the south. The central depression represents the remnant of the deeply eroded central uplift. This situation represents a classical example of relief inversion (photo: T. Kenkmann).

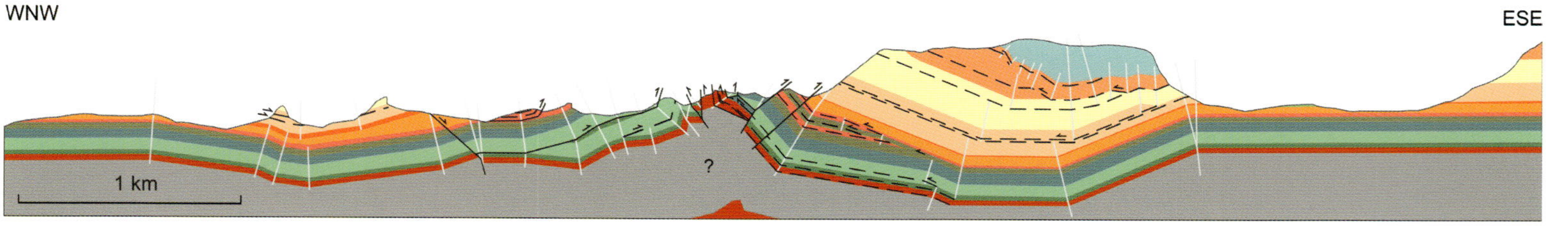

Schematic west-northwest/east-southeast cross-section of the Upheaval Dome impact structure. Modified after Kenkmann et al. (2005).

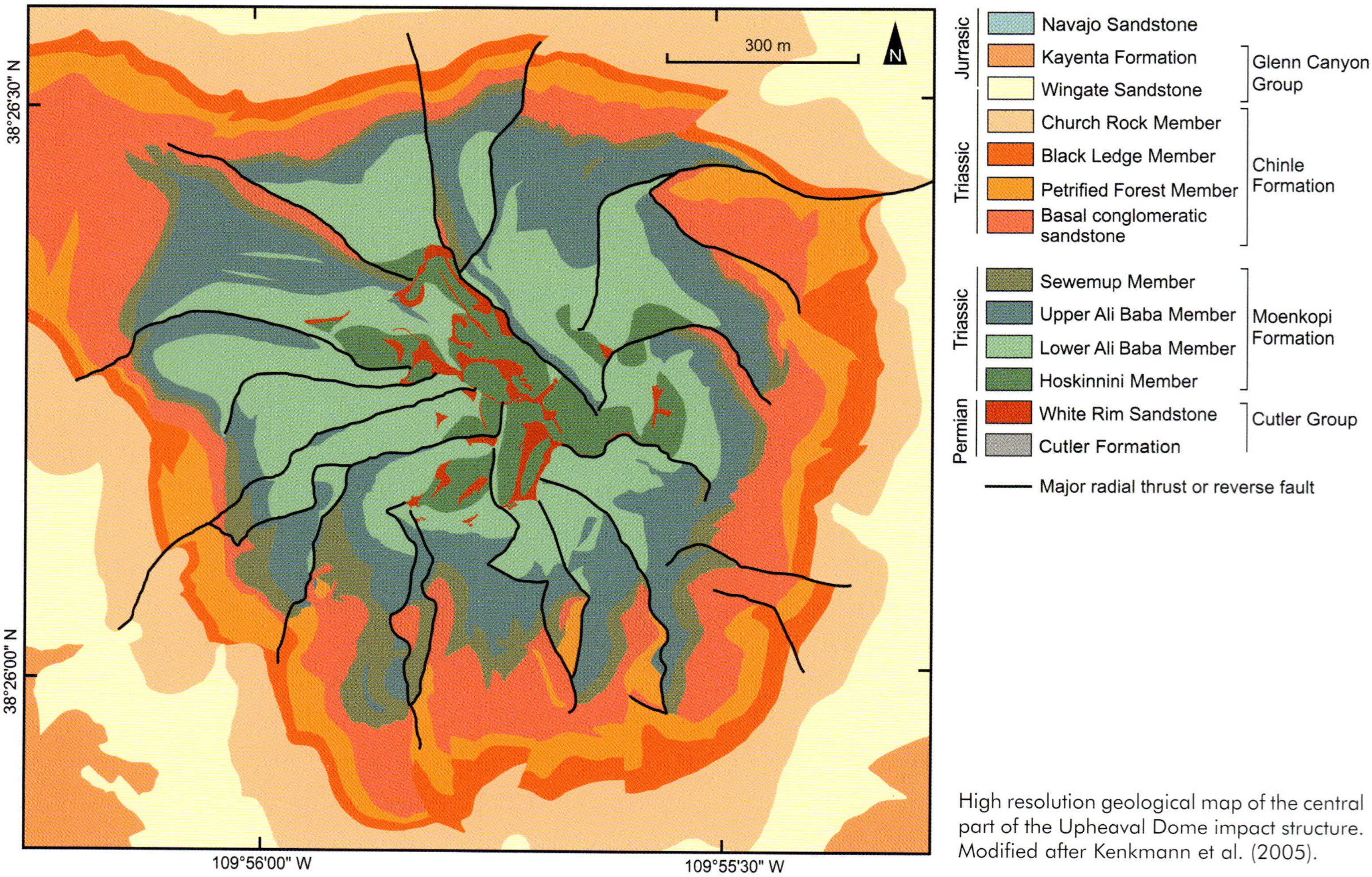

High resolution geological map of the central part of the Upheaval Dome impact structure. Modified after Kenkmann et al. (2005).

Seismic studies revealed that the deformation of the central uplift ceases with increasing depth and did not reach deeper than the Paradox Formation, thereby excluding the previous suggestion that the structure could represent a salt dome or a pinched-off salt diapir. Based on lithostratigraphy and the reconstructed amount of erosion since the time of impact, the most likely age for the impact is Upper Cretaceous.

There was a long-lasting debate about the origin of the Upheaval Dome structure. It started in 1936 when Walter Bucher postulated a "cryptovolcanic" origin. In 1968 Richard Mattox published the hypothesis that the structure was the result of salt flow in the underlying Paradox Formation. An impact origin for Upheaval Dome was first proposed by Eugene Shoemaker and co-workers in 1983, based on structural arguments. Still, Martin Jackson and his team interpreted the structure in 1998 as a salt dome, although no salt is present at Upheaval Dome. They suggested that the structure was formed as a pinched-off salt diapir, whereby the salt flowed through a thin throat and the diapir itself would have been eroded away by now. This was disputed by Thomas Kenkmann and co-workers who showed that the deformation inventory is compatible with the formation and collapse

Upheaval Dome provides the best three-dimensional view into the interior of a central uplift formed in sedimentary strata. The Permian White Rim sandstone and the Triassic Moenkopi Formation are the oldest strata in the center of the structure (photo: T. Kenkmann). ▷

of an impact crater. They mapped the crater and derived a three-dimensional model of its structure. Eventually a shock metamorphic overprint of sandstone samples could be documented in 2008 by Elmar Buchner and Thomas Kenkmann, which finally solved the long-standing debate about the enigmatic formation of Upheaval Dome.

Location/Access

Upheaval Dome is situated in the northern part of the Canyonlands National Park on the *Island in the Sky*, a high mesa north of the confluence of the Colorado and Green rivers. *Island in the Sky* can be reached from Moab on a 50 km drive via US 191 and UT 313. Visitors can easily walk to a spectacular vantage point overlooking the central depression of the crater. More insight is provided by the syncline loop, a 14 km long challenging trail that follows the canyons around Upheaval Dome and requires route-finding, climbing and scrambling through boulder fields, with considerable elevation change. When crossing Upheaval Canyon on this hike, a turn-off into the central depression along the Upheaval Canyon is possible. This interior part uncovers the extremely complex and wonderful geology of the central uplift.

The central part of the Upheaval Dome crater structure. The tower is composed of sandstone of the Hoskinini Member of the Moenkopi Formation, which is interspersed with dikes of fluidized White Rim sandstone (photo: T. Kenkmann).

Selected References

Buchner, E., and Kenkmann, T. (2008). Upheaval Dome, Utah: Impact origin confirmed. Geology, Vol. 36, pp. 227–230.

Kenkmann, T. (2003). Dike formation, cataclastic flow, and rock fluidization during impact cratering: an example from the Upheaval Dome structure, Utah. Earth and Planetary Science Letters, Vol. 214, pp. 43–58.

The ring syncline of the Upheaval ▷ Dome impact structure, as seen from Upheaval Canyon to the northwest of the crater structure. From bottom to top, the Chinle, Wingate, Kajenta and Navajo formations are visible (photo: T. Kenkmann).

The Upheaval Dome impact structure in the Canyonlands National Park on September 22, 2018 as based on Copernicus Sentinel-2 (channel 2/3/4) data.

Kenkmann, T., Jahn, A., Scherler, D., and Ivanov, B. A. (2005). Structure and formation of a central uplift: a case study at the Upheaval Dome impact crater, Utah. In: Large Meteorite Impacts III, Kenkmann, T., Hörz, F., and Deutsch, A. (eds.), Geological Society of America Special Paper, Vol. 384, pp. 85–115.

Kriens, B. J., Shoemaker, E. M., and Herkenhoff, K. E. (1999). Geology of the Upheaval Dome impact structure, southeast Utah. Journal of Geophysical Research, Vol. 104, pp. 18867–18887.

Scherler, D., Kenkmann, T., and Jahn, A. (2006). Structural record of an oblique impact. Earth and Planetary Science Letters, Vol. 248, pp. 28–38.

Shoemaker, E. M., and Herkenhoff, K. E. (1983). Impact origin of Upheaval Dome, Utah. Eos, Transactions American Geophysical Union, Vol. 44, p. 747.

Viewfield

Saskatchewan, Canada

Thomas Kenkmann

Viewfield is a buried, oil and gas producing impact crater. The crater has no surface expression. Intensive seismic surveying and drilling constrained the size and shape of the crater. The morphology of this structure is simple bowl-shaped, corresponding to a small diameter of 2.5 km. The apparent depth measures 180 m, and a wide raised rim of about 60 m occurs in the northeastern sector. The cavity has a somewhat polygonal outline and is slightly elongated in north-south direction.

Viewfield is located in Canada's intracratonic Williston Basin. The basal Cambrian to Ordovician target seems to be unaffected by the impact event. The main target rocks are of Devonian and Mississippian age. Fragments of these strata are the main constituents of the allochthonous breccia lens that fills the crater cavity. The elevated crater rim consists of Mississippian limestone and Triassic-Jurassic strata. Post-impact sediments are of Jurassic and Cretaceous age and bury the crater by more than 1300 m. They are undeformed but the Jurassic beds have sagged into the crater depression.

Shock features, namely planar deformation features (PDFs) in quartz grains, were found in Triassic to Jurassic debris from the crater rim. These grains are rounded and indicate re-deposition immediately after the impact, possibly by water resurge.

There are no gravity and magnetic anomalies associated with the small crater structure. Reserves of 10 million barrels of oil and 4.5 billion cubic feet of gas were estimated in 1997 for the structure. Gas and oil are produced from the Mississippian target rocks as well as from the allochthonous breccia lens that fills the crater.

The impact age is constrained by the lithologies affected by the impact. Currently, the accepted age is 190 Myr.

The structure was drilled by an oil company already in 1957 but production started only in 1969. H. Sawatzky proposed an impact origin in 1972 based on borehole data and seismic profiles. In 1998 Richard Grieve and co-workers documented shock metamorphic evidence for the impact.

Viewfield

Saskatchewan / Canada

latitude = 49°35‘ N / longitude = 103°04‘ W / diameter = 2.5 km / age = 190 Myr

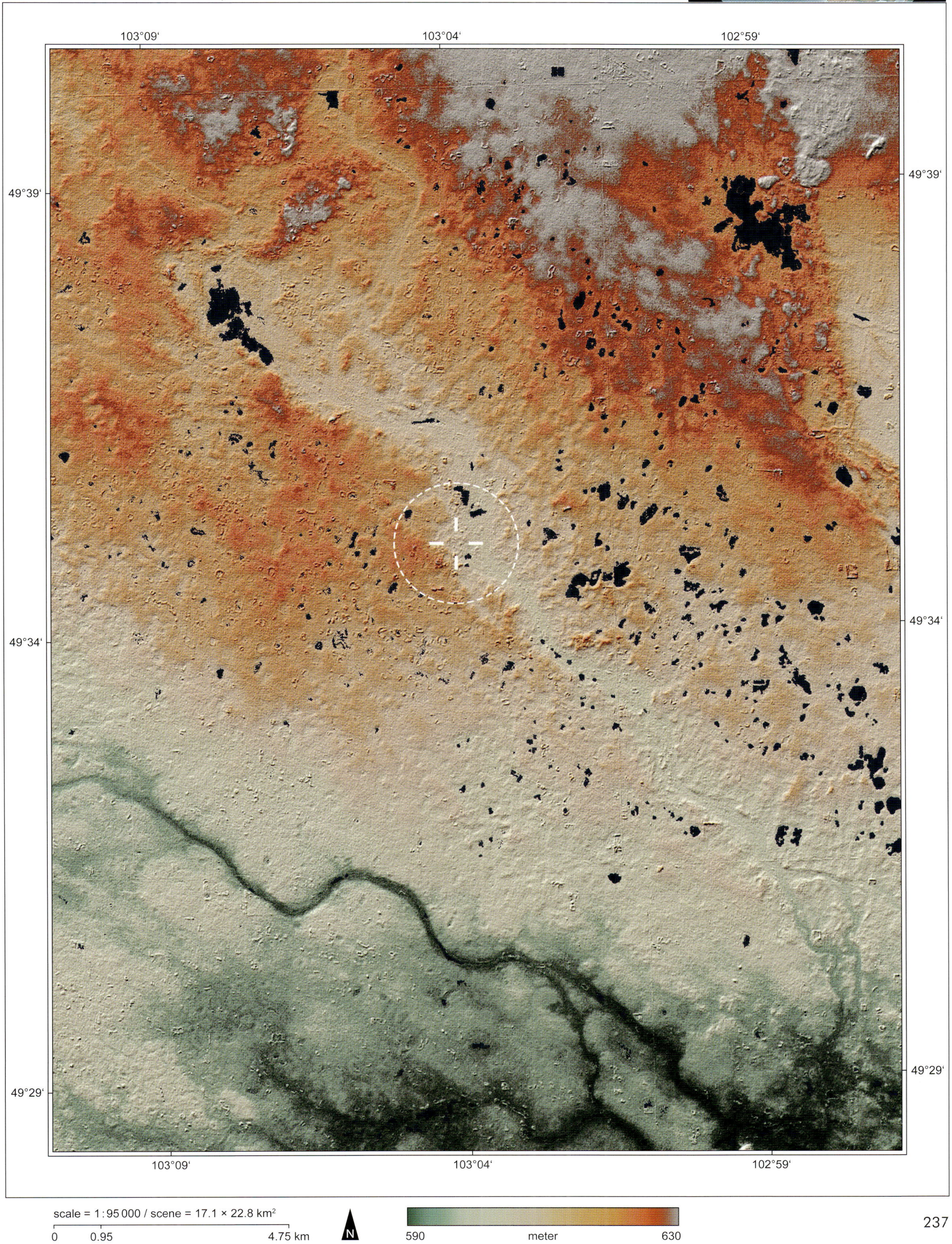

Location/Access

Viewfield is located in southeast Saskatchewan. There the flat surface topography is characterized by shallow glacial ground moraine lakes. Viewfield lies between the town of Estevan 50 km to the south and the Trans-Canadian Highway 90 km to the north, both of which are linked by Highway 47. From Winnipeg in Manitoba it is a 450 km drive on the Trans-Canada Highway to the west, until Highway 47 leads southwards between Grenfell and Summerberry.

Selected References

Grieve, R. A. F. (2006). Impact Structures in Canada. Geological Association of Canada, GeoText No. 5, 210 pp.

Grieve, R. A. F., Kreis, K., Therriault, A. M., and Robertson P. B. (1998). Impact structures in the Williston Basin. Meteoritics and Planetary Science, Vol. 33, pp. A63–A64.

Sawatzky, H. B. (1972). Viewfield – a producing crater? Canadian Society of Exploration Geophysicists Journal, Vol. 8, pp. 22–40.

Sawatzky, H. B. (1975). Astroblemes in the Williston Basin. Journal of the Canadian Society of Exploration Geophysicists, Vol. 10, pp. 23–38.

Wanapitei

Ontario, Canada

Thomas Kenkmann

The 7.5 km diameter Wanapitei impact structure is located within the rim zone of the originally 200–250 km diameter, 1850 Myr old **Sudbury** impact structure. The Wanapitei structure is not exposed and completely submerged within Lake Wanapitei. This lake has a circular shoreline in its northern part, but shows some glacially formed channels and embayments in the southern part. However, the bathymetry of the lake depicts the crater outline with a circular basin form. The deepest point was measured at −105 m and is located where the circular depression and a glacial trough meet. The crater corresponds to that part of the lake that is devoid of islands. The bathymetry neither indicates the presence of a central uplift nor of an elevated rim. A concentric joint pattern around Lake Wanapitei is related to the impact and caused by the formation of some arcuate lakes around the crater.

A slice of suevite, 19 × 15 × 4.5 cm in size, from the Wanapitei impact structure (photo courtesy: Ralf-Thomas Schmitt, Museum für Naturkunde Berlin, sample from Dieter Stöffler).

Target rocks are Archean granites, gneisses, and greenstones of the Superior Province, which were overlain by metasedimentary rocks of the Early Proterozoic Huronian Supergroup. These include metamorphosed arenites, greywackes, and conglomerates, with a small contribution from carbonate rocks. Intrusions of gabbro and diabase associated with minor granophyre and granodiorite crosscut the Archean and metasedimentary units.

Shatter cones occur in a few places at the shore and on islands of Lake Wanapitei. Also planar deformation features (PDFs) in quartz have been observed in bedrock from three locations: on a tiny island in southwestern Lake Wanapitei, and in two places near the shoreline of the southern part of the lake. However, these shock features are from outside the Wanapitei structure and are most likely related to the Sudbury impact structure, within which the Wanapitei structure is located.

Shock features related to the Wanapitei impact crater exclusively stem from glacial floatstones derived from the Lake Wanapitei area. The rounded boulders were transported in southerly direction by the last glaciation and were scoured either from the bottom of the lake and/or from an ejecta blanket around the impact structure.

Up to 10 % of the glacial drift contains evidence of shock metamorphism. A wide variety of shock effects has been found, including PDFs in quartz, diaplectic quartz and feldspar glasses, impact glass, ballen quartz, and coesite. Impact lithologies are shocked basement lithologies, glass-rich suevitic breccia, and vesiculated impact melt

Lake Wanapitei from the town of Skead (photo: M. Gottwald).

Wanapitei

Ontario / Canada

latitude = 46°44‘44“ N / longitude = 80°44‘43“ W / diameter = 7.5 km / age = 37 ± 2 Myr

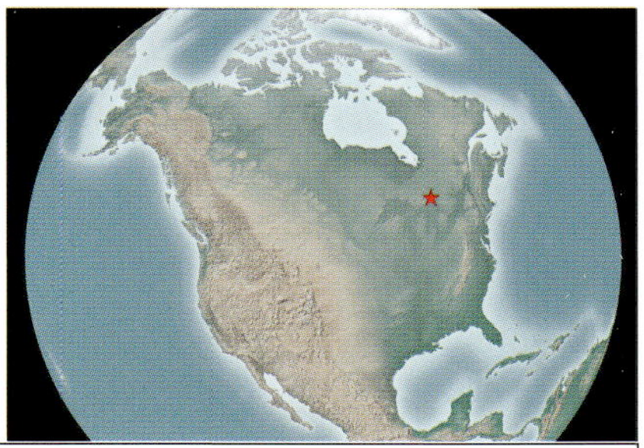

scale = 1 : 183 000 / scene = 32.9 × 43.9 km^2

0 1.83 9.15 km

N

35 meter 505

rock. The bulk composition of highly shocked and vesiculated impact melt rock boulders is close to that of the country rocks, but that of the suevite is highly variable. Impact melt rock has a high silica content of 75 wt%. Early studies of siderophile elements in the impact melt rocks indicated 1–2 % of a meteoritic component. Interelement ratios suggest that the Wanapitei projectile was a chondrite, possibly an L or LL chondrite. Analysis of platinum group elements (PGE) confirmed a chondrite as the most likely type of impacting body. ^{40}Ar-^{39}Ar step-heating dating of impact melt rock samples gave an age of 37 ± 2 Myr.

There is a distinct circular Bouguer gravity anomaly of –15 mGal over the lake. The aeromagnetic signature is relatively featureless. A gravity survey in 1969 generated the interest in the structure, and evidence for impact was provided shortly thereafter in 1972 by Michael Dence and Joe Popelar through the documentation of coesite and other shock effects.

Location/Access

Access to Wanapitei, located in the central part of Canada's Ontario Province, is much the same as for the Sudbury impact structure. The city of Sudbury is reached by car on excellent Highway 400 from Toronto. A 30 km drive from downtown Sudbury to the northeast brings the visitor to the Greater Sudbury Airport, which lies close to the southern tip of Lake Wanapitei.

Selected References

Dence, M. R., and Popelar, J. (1972). Evidence for an impact origin for Lake Wanapitei, Ontario. In: New Developments in Sudbury Geology, Guy-Bray, J. V. (ed.), Geological Association of Canada Special Paper, No. 10, 117 pp.

Dressler, B. O. (1982). Geology of the Wanapitei Lake area, District of Sudbury. Ontario Geological Survey Report, Vol. 213, 131 pp.

Dressler, B. O., Crabtree, D., and Schuraytz, B. C. (1997). Incipient melt formation and devitrification at the Wanapitei impact structure, Ontario, Canada. Meteoritics and Planetary Science, Vol. 32, pp. 249–258.

Grieve, R. A. F. (2006). Impact Structures in Canada. Geological Association of Canada, GeoText No. 5, 210 pp.

Grieve, R. A. F., and Ber, T. (1994). Shocked lithologies at the Wanapitei impact structure, Ontario, Canada. Meteoritics, Vol. 29, pp. 621–631.

Wells Creek
Tennessee, United States

Thomas Kenkmann

Wells Creek is a conspicuous, circular structure of 12 km diameter that is located close to the small city of Erin at the banks of the Cumberland River. It is surrounded by flat-lying strata that characterize the plateau topography of northern Tennessee.

The most striking feature about Wells Creek is the core of the structure that is composed of intensely deformed Ordovician Knox Dolomite. This central dome is surrounded by four roughly circular faults that are, in turn, offset by radial faults. These faults define an about 800 m wide outer annular graben, where strata are downfaulted by some 50 m. This graben also constrains the size of the structure. It is predominantly filled with rocks of Mississippian age. Further inward occur an annular horst, likewise about 800 m wide, and an inner graben that has been downthrown by, on average, some 80 m. This circular inner graben (or syncline) is 2–3 km wide and composed of Mississippian St. Louis Limestone. This ring syncline surrounds the central uplift. The involvement of Lower Ordovician Knox Dolomite indicates a stratigraphic uplift of about 1000 m above the regional level of this stratum. Rocks of the central uplift are heavily deformed and brecciated. Delicate shatter cones of various sizes with fully developed apices occur widespread in the central uplift. The strong uplift and the orientation of shatter cone axes with respect to tilted bedding surfaces have been interpreted to suggest that the crater was formed by a bolide striking the Earth at a steep to almost vertical angle. The structure is also well recognized by the associated gravity signature. Wells Creek shows a circular negative gravity anomaly with a small gravity high in its center. The latter corresponds to the uplift of relatively denser Knox Dolomite.

Delicate shatter cones formed in Ordovician dolomite of the central uplift of the Wells Creek impact structure. Sample size: 5 × 7 × 8 cm (photo: T. Kenkmann).

According to stratigraphic considerations the impact responsible for the formation of the Wells Creek structure must have occurred between the end of the Mississippian at 325 Myr ago and prior to the deposition of the Cretaceous Tuscaloosa Formation at 75 Myr ago.

Wells Creek attracted the attention of geologists quite early. James Safford compiled the first detailed map of the structure already between 1889 and 1893. Later, in 1936, Walter Bucher mapped the structure, unaware of the previous mapping, and suggested a "cryptoexplosive" origin. However, in the same year, a meteorite impact origin was already favored by John Boon and Claude Albritton. Based on his fundamental work on shatter cones and their connection to the impact process, Bob Dietz could substantiate the hypothesis of an impact here, which was followed a few years later by a detailed structural analysis by Richard Stearns and co-workers.

Wells Creek

Tennessee / United States

latitude = 36°22'34" N / longitude = 87°39'33" W / diameter = 12 km / age = 70 - 325 Myr

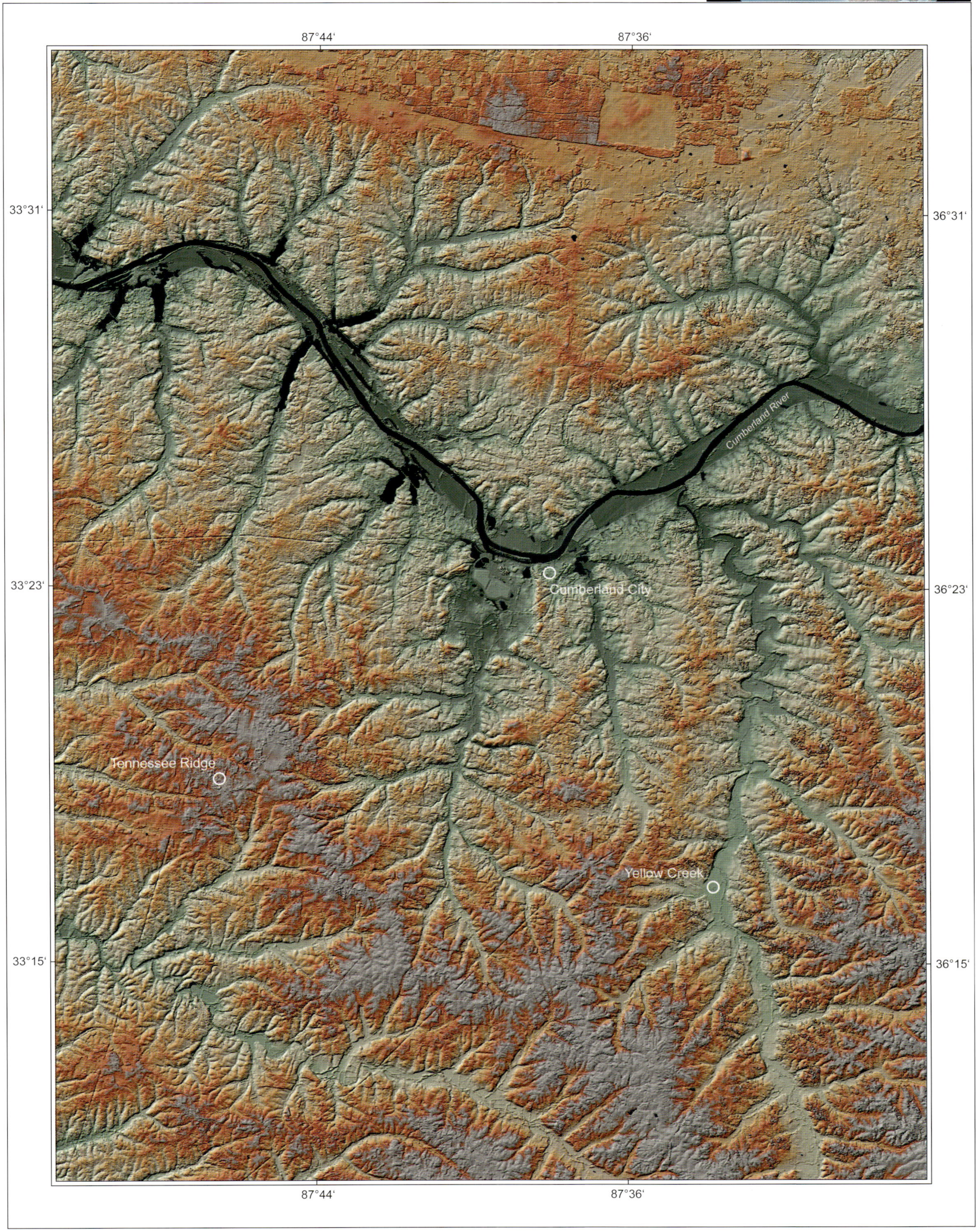

scale = 1 : 186 000 / scene = 33.5 × 44.7 km²

0 1.86 9.30 km

N

75 meter 280

Location/Access

The structure lies about 110 km northwest of Nashville and 50 km southwest of Clarksville. Wells Creek can be reached via State Highways 13 and 149 from Clarksville or via State Highway 149 from Erin. Cumberland City is situated close to the center of the structure. The Cumberland River flows through the northern part of the crater.

Selected References

Boon, J. D., and Albritton, C. C. (1936). Meteorite craters and their possible relationship to "cryptovolcanic structures". Field and Laboratory, Vol. 5, pp. 7–9.

Bucher, W. H. (1936). Cryptoexplosion structures caused from without or from within the Earth? American Journal of Science, Vol. 261, pp. 597–649.

Dietz, R. S. (1959). Shatter cones in cryptoexplosion structures (Meteorite Impact?). Journal of Geology, Vol. 67, pp. 496–505.

Ford, J. R. H., Orchiston, W., and Clendening, R. (2012). The Wells Creek meteorite impact site and changing views on impact cratering. Journal of Astronomical History and Heritage, Vol. 15, pp. 159–178.

Stearns, R. G., Wilson, C. W., Jr., Tiedemann, H. A., Wilcox, J. T., and Marsh, P.S. (1968). The Wells Creek structure, Tennessee. In: Shock Metamorphism of Natural Materials, French, B. M., and Short, N. M. (eds.), Mono Book Corp., Baltimore, pp. 323–338.

West Hawk
Manitoba, Canada

Thomas Kenkmann

The West Hawk impact crater is associated with an unusually deep lake of 110 m water depth in southeast Manitoba that is roughly 4 km wide and slightly elongated in northeast-southwest direction. The crater is completely submerged in this lake and has an estimated diameter of 2.4 km. It has a simple, bowl-shape morphology.

Target rocks belong to an Archean granite-greenstone belt of the Superior Province of the Canadian Shield. Lithologies include metamorphosed andesites, basalts, tuffs, and banded gneisses. A borehole was drilled into the center of the lake and passed through more than 90 m of glacial sediments. Beneath that, nearly 330 m of slump and fallback breccia were encountered. The top and base of the allochthonous breccia lens, which has a volume of 0.45 km^3, have been estimated to occur at 340 and 670 m below the estimated crater rim, which is not preserved. The breccia lens rests on more than 300 m of fractured and brecciated bedrock of the crater floor. Cataclasis and injection of shocked breccia into open fractures are typical for this rupture zone.

Numerous shock features demonstrate an impact origin for this structure. Among them are abundant planar deformation features (PDFs) and planar fractures (PFs) in quartz, as well as diaplectic glasses and devitrified melt clasts in the breccia. Shock-induced twinning and granulation is observed in carbonates. Several meter-thick layers of impact melt rock have also been drilled. Highest shock levels occur at the top and at the base of the breccia lens. The breccias are hydrothermally altered. Zoisite and secondary calcite are developed extensively within the breccia lens as a result of solution and neo-mineralization; serpentine formed within the breccia matrix, in a network of veinlets cutting across fragments, and in the damage zone; sodium-calcium zeolites form in what is likely replaced melt material at the crater base.

A circular Bouguer gravity anomaly of -6 mGal was found over the crater. The ground magnetic pattern has relatively smooth relief over the area of the crater, as observed for many other impact structures as well. The age for the West Hawk impact is not well constrained. Radiometric ages based on the K-Ar whole-rock method gave mixed ages between the impact age and the age of the target rocks that were obviously not completely reset. An age obtained by fission track dating yielded 351 ± 20 Myr.

West Hawk Lake was identified as a possible impact site in the course of the systematic impact crater survey of Canada's Dominion Observatory by Carlyle Beals and co-workers in 1960. Mapping of the gravity signal over the lake by Ian Halliday in 1963 provided further support for the impact hypothesis. The final proof by documentation of unequivocal evidence of shock metamorphism came from Halliday in 1967 after the drilling into the structure had been accomplished.

Location/Access

West Hawk Lake can easily be reached on Canada's Highway 1, the Trans-Canadian Highway. The lake is situated close to the border between Ontario and Manitoba about 160 km east of Winnipeg.

Selected References

Beals, C. S., Innes, M. J. S., and Rottenberg, J. A. (1960). The search for fossil meteorite craters. Ottawa Dominion Observatory Contributions, Vol. 4, pp. 1–31.

Grieve, R. A. F. (2006). Impact Structures in Canada. Geological Association of Canada, GeoText No. 5, 210 pp.

Halliday, I., and Griffin, A. A. (1963). Evidence in support of a meteoritic origin for West Hawk Lake, Manitoba, Canada. Journal of Geophysical Research, Vol. 68, pp. 5297–5305.

Halliday, I., and Griffin, A. A. (1967). Summary of drilling at the West Hawk Lake crater. Journal of the Royal Astronomical Society of Canada, Vol. 61, pp. 1–8.

Short, N. M. (1967). Petrographic evidence for an impact origin for the West Hawk Lake Structure, Manitoba, Canada. Eos, Transactions of the American Geophysical Union, Vol. 48, p. 147.

Short, N. M. (1970). Anatomy of a meteorite impact crater: West Hawk Lake, Manitoba, Canada. Geological Society of America Bulletin, Vol. 81, pp. 609–648.

West Hawk

Manitoba / Canada

latitude = 49°45'41" N / longitude = 95°11'12" W / diameter = 2.4 km / age = 351 ± 20 Myr

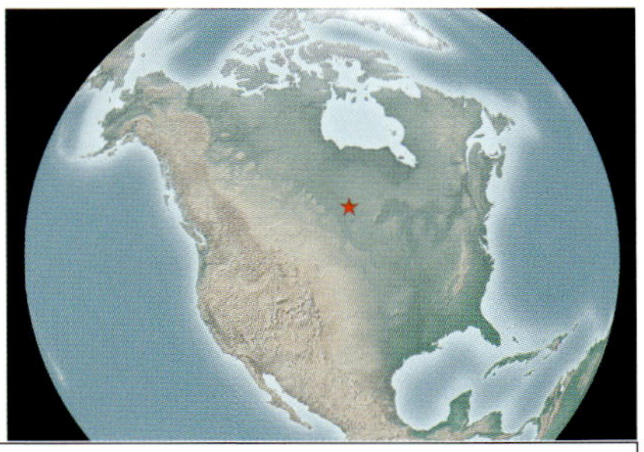

Wetumpka
Alabama, United States

Thomas Kenkmann

Wetumpka is an impact structure that was formed in a marine target in, what is now, the inner Gulf coastal plain of Alabama. The diameter has been variably referred as 6.5 to 7.3 km. Devoid of a distinct central uplift, the crater structure seems to be somewhat transitional between a simple, bowl-shaped crater and a complex impact structure. The crater structure appears topographically as a prominent, semi-circular, rimmed feature.

The target rocks at Wetumpka at the time of impact were submerged beneath 30–100 m of water. They consisted of poorly consolidated, Upper Cretaceous, shallow-marine sediments and comprised, from top to bottom, approximately 30 m of chalky ooze, 30 m of paralic marine sand, and 60 m of clayey sand and gravels. These strata rested on crystalline basement of gneisses and schists.

The present exposure of the crater shows a 6.1 km semi-circular crystalline rim in the western, northern and eastern sectors, which has up to 87 m of positive relief. Encircled by this is a terrain with low relief composed of undeformed to highly disturbed Upper Cretaceous rocks that are partly chaotically mixed to a tectonic mélange. This mélange comprises a mixture of lithic blocks of 10–18 m diameter and diamictite. In the central part of the structure occurs an area of 0.75 km^2 of breccia with sub-rounded to sub-angular clasts of schists and gneisses from the crystalline basement, and Upper Cretaceous sandstone clasts. Blocks reach 25 m size. The impact-crater mélange was likely formed by mass movements induced by a forceful backwash from the transient crater rim, as the impact occurred in a shallow marine environment.

The Wetumpka impact structure on January 5, 2019 as a composite image based on TanDEM-X DEM and Copernicus Sentinel-2 (channel 2/3/4) data.

Wetumpka

Alabama / United States

latitude = 32°31'24" N / longitude = 86°10'32" W / diameter = 7 km / age = 84.4 ± 1.4 Myr

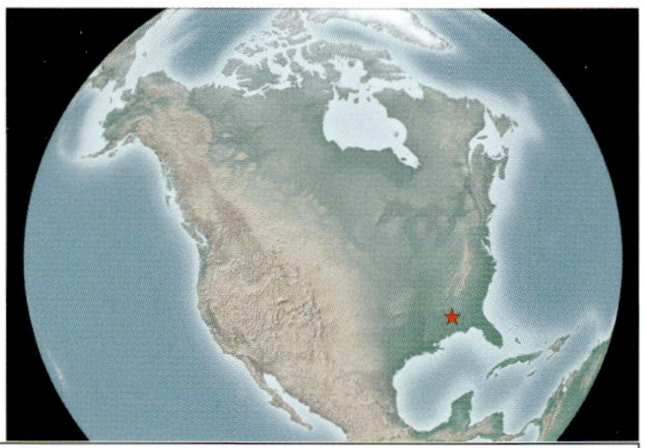

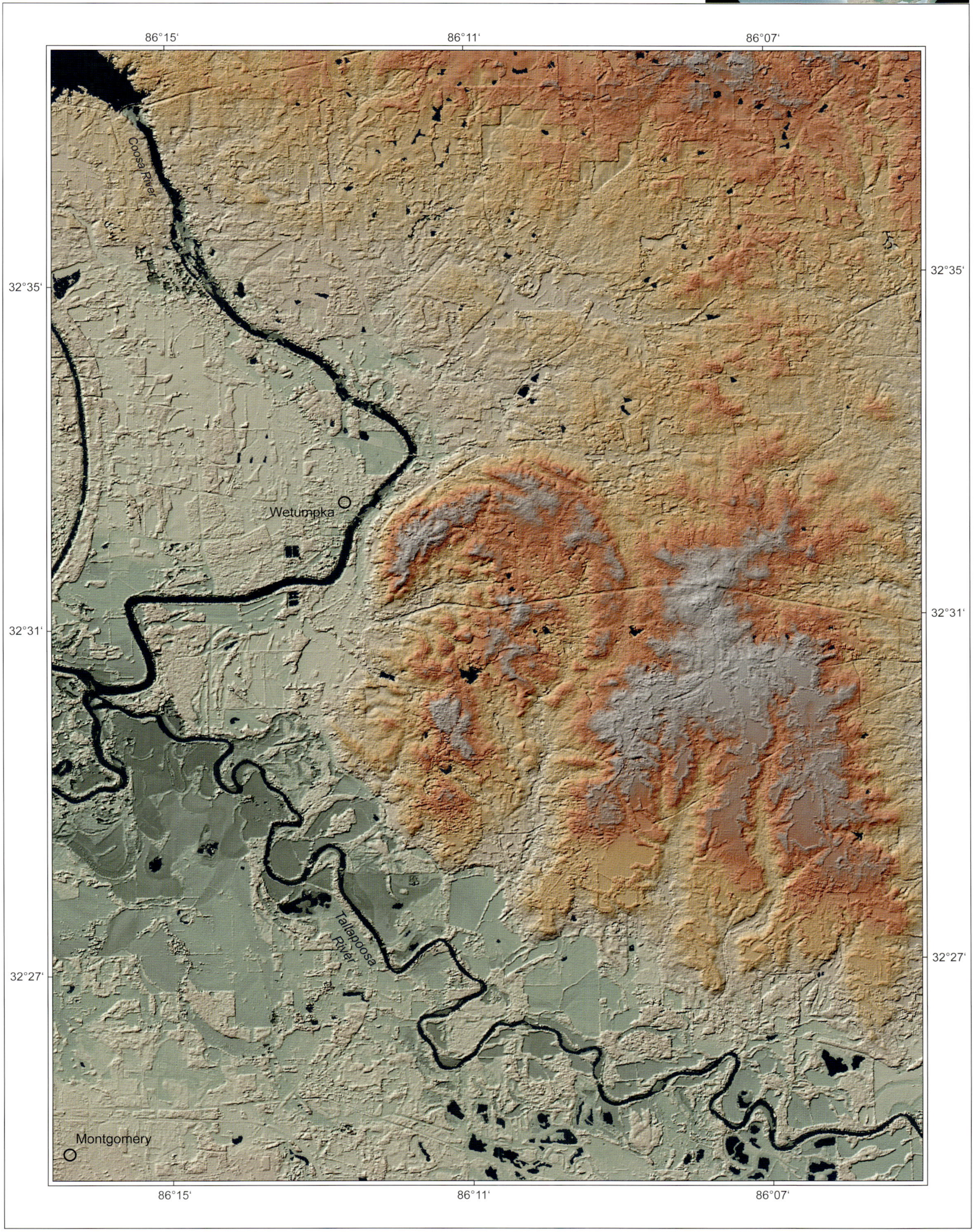

scale = 1 : 101 000 / scene = 18.2 × 24.2 km²

0 1.01 5.05 km

N

35 meter 200

The horseshoe-shaped crater rim is indicative for a low-angle oblique impact from the northeast. Two boreholes were drilled near the center and revealed that the crater cavity was filled by fall-back material plus resurge, which was then followed by a second seawater resurge event. The lower crater fill unit comprises polymict impact breccias, which contain quartz grains displaying shock-characteristic multiple sets of planar deformation features (PDFs). Breccia also exhibits slight concentrations of the siderophile elements Ir, Co, Ni, as well as of Cr, which may be indicative of a small extraterrestrial component.

Wetumpka has been dated by both biostratigraphic and radiometric methods. Using planktonic foraminiferal biozones, an age of 83.5 Myr was determined. Results of a single crystal apatite and zircon (U-Th)/He dating study on five impact breccia samples from drill core from the central uplift area yielded a mean age of 84.4 ± 1.4 Myr, in good agreement with the biostratigraphic age.

The Wetumpka structure has been the subject of geological study since the late 19th century. Since then, speculations about a possible impact origin have also been voiced. Like with many other impact craters in North America, Wetumpka was initially classified as a "cryptoexplosion" structure and was still interpreted as such by Thornton Neathery and co-workers in 1976. David King Jr. and colleagues confirmed the impact origin in 1999 and proceeded to identify Wetumpka as an impact structure formed in a shallow-marine environment.

Location/Access

The Wetumpka impact structure is located in Elmore County, Alabama and is easily accessible. Driving north on US Highway 231 from the city of Montgomery one can see the elevated northwestern rim of the impact structure to the east of the highway as one enters the city limits of Wetumpka.

Selected References

King Jr., D. T., Neathery, T. L., Petruny, L. W., Koeberl, C., and Hames, W. E. (1999). Evidence confirming meteoritic impact at Wetumpka crater, Alabama, USA. Meteoritics and Planetary Science, Vol. 34, pp. 63–64.

King Jr., D. T., Neathery, T. L., Petruny, L. W., Koeberl, C., and Hames, W. E. (2002). Shallow-marine impact origin of the Wetumpka structure (Alabama, USA). Earth and Planetary Science Letters, Vol. 202, pp. 541–549.

King, D. T., and Ormö, J. (2011). The marine-target Wetumpka impact structure examined in the field and by shallow core-hole drilling. Geological Society of America Special Paper, Vol. 483, pp. 287–300.

King Jr., D. T., and Petruny, L. W. (1999). Wetumpka impact crater guidebook, Auburn, Alabama, 29 pp.

Neathery, T. L., Bentley, R. D., and Lines, G. C. (1976). Cryptoexplosion structure near Wetumpka, Alabama. Geological Society of America Bulletin, Vol. 87, pp. 567–573. [Reprinted in Alabama Geological Society Guidebook No. 20 (1983) and No. 34c (1997)].

Wartho, J. A., van Soest, M. C., King, D. T., and Petruny, L. W. (2012). An (U-Th)/He age for the shallow-marine Wetumpka impact structure, Alabama, USA. Meteoritics and Planetary Science, Vol. 47, pp. 1243–1255.

Whitecourt
Alberta, Canada

Thomas Kenkmann

Whitecourt is a small, bowl-shaped meteorite impact crater of 36 m diameter. It was formed in an unconsolidated sedimentary target. The crater has a raised rim of about 4 m elevation above the environs, whereby the southwestern part is higher than the northeastern sector. The crater cavity is 6 m deep; the crater walls have relatively steep slopes of 28° angle, on average.

The Whitecourt impact occurred into loose, fine-grained Quaternary sand, silt, and mud. These materials were deposited as till when the Laurentide Ice Sheet advanced and retreated in this region. This glacial till unconformably overlies heterogeneous fluvial mudstone and sandstone of Paleocene and Upper Cretaceous age. Some 3 m of allochthonous material fills the crater cavity; hence, the true depth of the crater is 8.9 m, measured from the unaffected target surface. This infill can be described as a diamictite, with clasts up to 5 cm in size. Shocked mineral grains or melted material are not present.

An overturned ejecta flap, commonly associated with raised rims of simple impact craters, is missing. The ejecta blanket surrounding the rim consists of diamictite and fine sand. A few quartz grains containing single sets of unindexed parallel and planar deformation lamellae could be identified within the ejecta. Melt is present neither in proximal nor in distal ejecta. However, thousands of meteorites have been collected since the discovery of the crater with the aid of metal detectors. They include shrapnel-shaped fragments and regmaglypted meteorites with individual masses up to more than 10 kg. There is a clear concentration of ejecta and meteorites towards the east-northeast. Consequently, this direction is thought to represent

The Whitecourt impact crater in fall after the leaves are off the trees, with a person about 1.8 m tall for scale (image courtesy: Chris Herd, University of Alberta, Edmonton, Canada).

Whitecourt

Alberta / Canada

latitude = 53°59'57" N / longitude = 115°35'50" W / diameter = 36 m / age = < 1100 yr

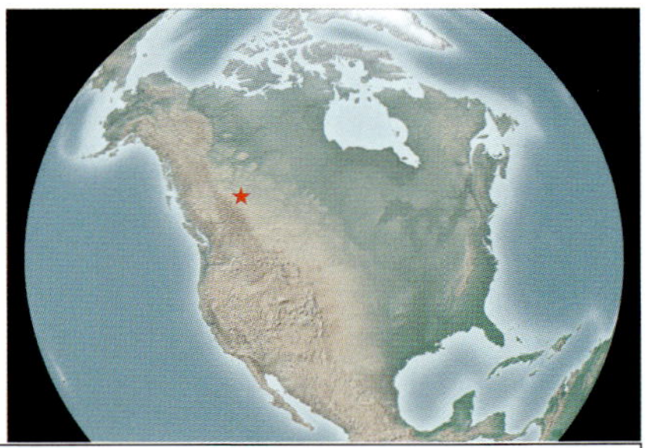

scale = 1:60 000 / scene = 10.7 × 14.3 km²

0 0.60 3.00 km

N

755 meter 1035

A 3D image of the Whitecourt impact crater looking northwest, based on LiDAR bare earth model data, with 2× vertical exaggeration (image courtesy: Chris Herd, University of Alberta, Edmonton, Canada).

the down-range direction of projectile trajectory. The most distally recovered meteorite could be located over 520 m to the east of the crater rim. As of 2015 the total known mass of all meteoritic material collected amounted to 229.5 kg.

The superposition of the crater onto a geomorphologically young landscape indicates that the crater is of Late Holocene age. Radiocarbon dating of charcoal samples from the ejecta provided an age of less than 1100 years. It has been reconstructed that the meteoroid, an iron meteorite of group IIIAB, struck the surface at a moderately oblique angle while traveling toward the northeast. A suggestion has been made that the meteoroid broke up in the atmosphere at a rather low altitude. No other crater pits have been discovered in the surrounding area though.

Whitecourt remained hidden for a long time, although forest roads exist nearby. The structure remains undetectable in areal or space-borne imagery due to the patchy vegetation in this area and the rather small size of the crater. The crater discovery in 2008 by Christopher Herd and colleagues was possible when LiDAR data from the area had been processed by Airborne Imaging, Inc. to produce a bare-earth digital elevation map.

Location/Access

The small Whitecourt crater is located in central Alberta, some 17 km south-southeast of the town of Whitecourt, and about 150 km northwest of Edmonton. The crater is accessible from Highway 43 connecting these towns. After branching southward off the highway some 15 km driving on forest roads takes one close to the crater. However, aspen forest and shrubs obscure the presence of the crater – in particular during the growth season. A 200 × 200 m area centered on the crater has been declared a *Protected State Site*, in which searching for meteorites and impactites is prohibited.

Selected References

Herd, C. D. K., Froese, D. G., Walton, E. L., Kofman, R. S., Herd, E. P. K., and Duke, M. J. M. (2008). Anatomy of a young impact event in central Alberta, Canada: Prospects for the missing Holocene impact record. Geology, Vol. 36, pp. 955–958.

Kofman, R. S., Herd, C. D., and Froese, D. G. (2010). The Whitecourt meteorite impact crater, Alberta, Canada. Meteoritics and Planetary Science, Vol. 45, pp. 1429–1445.

Newman, J. D., and Herd, C. D. K. (2015). Mineralogy, petrology, and distribution of meteorites at the Whitecourt Crater, Alberta, Canada. Meteoritics and Planetary Science, Vol. 50, pp. 305–317.

Impact Structures – Further Confirmation Required

Bloody Creek

Nova Scotia, Canada

Wolf Uwe Reimold

Bloody Creek is a small, somewhat elliptical structure in southwestern Nova Scotia, with a mean diameter of only 400 m. Its northwest-trending major axis amounts to 420 m, whereas its southwest-trending minor axis measures 350 m in length. The small structure was described as a possible impact crater with a near-continuous geomorphic rim around the 10 m deep interior that has an infill of lacustrine sediments and peat. The area of the structure was flooded in the 1990s in the course of the development of the Bloody Creek hydroelectric facility.

The structure was formed in Devonian granite of the 388–372 Myr old South Mountain Batholith. Ground penetrating radar observations indicated that the crater-form has a depressed inner floor, delimited on the outside by a 1 m high, buried scarp. Furthermore, heterogeneous material was described below the crater floor, which was interpreted as deposits from the collapse of the transient cavity walls of a small impact crater. This material is overlain by stratified and faulted lacustrine and wetland-related sediments. Rim rocks were described to commonly show grain comminution, and there was presence of a polymict lithic microbreccia. Feldspar and biotite displayed kink-bands, and quartz and feldspar contained single or multiple sets of closely-spaced, so-called "planar microstructures (PMs)". In addition, some quartz displayed mosaicism or reduced birefringence. From chlorite plastic deformation and flow microtextures were reported. The "PMs" were related to planar deformation features (PDFs), based on their form and crystallographic orientations. It was concluded that these alleged PDF orientations were consistent with up to 25 GPa shock pressures.

The structure was first noted in 1987 in the course of an aerial survey, and recorded in print by George Stevens in 1995. A detailed report was finally presented in 2009 by Ian Spooner and co-workers. The structure may be significantly eroded as implied by its low depth/diameter ratio and indications of rather strong shock pressures in samples from the top of the crater rim. However, an impact onto glacier ice during the end stages of the Wisconsinian glaciation around 12 000 years before present (BP) could have resulted in dissipation of some impact energy into the ice, which then would explain the current subdued morphology of the Bloody Creek structure. No age has been determined so far for the Bloody Creek structure.

The optical microscopic evidence presented by Spooner and his team appears quite promising, so that we include this structure in our atlas. However, the nature of the planar microdeformation features in quartz and feldspar ought to be clarified further by state-of-the-art methodology.

Bloody Creek

Nova Scotia / Canada

latitude = 44°45‘00“ N / longitude = 65°14‘35“ W / diameter = 400 m / age = unknown

scale = 1 : 59 000 / scene = 10.6 × 14.1 km²

0 0.59 2.95 km

N

20 meter 285

An aerial view with Dalhousie Lake and Corbett Lake in the foreground. The Bloody Creek structure is submerged in Dalhousie Lake just above the center of this photo (image courtesy: Chuck O'Dale, www.craterexplorer.ca).

Location/Access

The crater site is located along the Bloody Creek stream in Annapolis County, Nova Scotia, approximately 10 km south of Bridgetown. Because of the flooding of the area, the structure has been submerged in Lake Dalhousie. This lake lies immediately west of Morse Road that links Bridgetown with the community of West Dalhousie in the south.

Selected References

Spooner, I., Stevens, G., Morrow, J., Pufahl, P., Grieve, R., Raeside, R., Pilon, J., Stanley, C., Barr, S., and McMullin, D. (2009). Identification of the Bloody Creek structure, a possible impact crater in southwestern Nova Scotia, Canada. Meteoritics and Planetary Science, Vol. 44, pp. 1193–1202.

Stevens, G. R. (1995). Meteor Crater, Nova Scotia. The Journal of the Royal Astronomical Society of Canada, Vol. 89, pp. 111–113.

Stevens, G. R., Spooner, I. S., Morrow, J. R., Pufahl, P., Raeside, R., Grieve, R. A. F., Stanley, C., Barr, S., and McMullin, D. (2007). Physical evidence of a late-glacial (Younger Dryas?) impact event in southwestern Nova Scotia (abstract). Eos Transactions, American Geophysical Union, Fall Meeting Supplement, Vol. 88.

Hiawatha

Greenland

Thomas Kenkmann

Hiawatha is a recently discovered circular structure in the High Arctic of northwest Greenland. The structure is entirely buried under the ice sheet. A recently published paper suggests an impact origin for this proposed structure, based on the observation of shock deformation effects in quartz grains from sediment sampled outside of the structure. However, the evidence for the suggested impact origin leaves room for criticism so that further work is desired to finally resolve the origin of this structure.

Hiawatha is named after the homonymous glacier which emerges from a semicircular lobe at Greenland's ice sheet margin and forms a narrow tongue of 10 km length extending onto the ice-free Inglefield Land. There, exposed country rocks consist of highly metamorphosed Precambrian rocks of the Inglefield Mobile Belt and range in age between 1980 and 1740 Myr. The semicircular ice sheet lobe covers the proposed Hiawatha structure, whose subglacial topography was detected in the framework of NASA's IceBridge campaign to study changes of the polar ice caps. The structure could be detected by using an advanced airborne radar sensor by Kurt Kjær and co-workers at the end of 2018.

A brittle deformation overprint with local brecciation of the bedrock was described in close proximity to the ice cover and the supposed rim. The structure comprises a flat, circular to slightly asymmetric depression of about 31 km diameter, with an elevated rim and a rim-to-floor elevation difference of some 300 m. A subdued raised center is interpreted as a relic of a central uplift. Two winding subglacial channels converge immediately before they intersect the southeastern flank of the circular structure and enter the depression. A downstream valley crosses the northwestern rim. Ice flows through this valley and forms the tongue-shaped terminus of Hiawatha Glacier.

For several angular quartz grains in a sample taken from the glaciofluvial sediment of the floodplain at this terminus crosscutting deformation effects were reported and considered planar deformation features (PDFs). Moreover, toasted quartz, ballen quartz, as well as glasses and variably devitrified glasses were described as product of an impact event. For micro-breccias with a glassy matrix it was claimed that they displayed elevated concentrations of Ni, Co, Cr, platinum group elements, and Au compared to local bedrock. This was interpreted as a 0.01 to 0.05 % admixture of a type IVA iron meteorite to crustal material.

Hiawatha
Greenland

latitude = 78°45‘ N / longitude = 66°15‘ W / diameter = 31 km / age = < 2.6 Myr

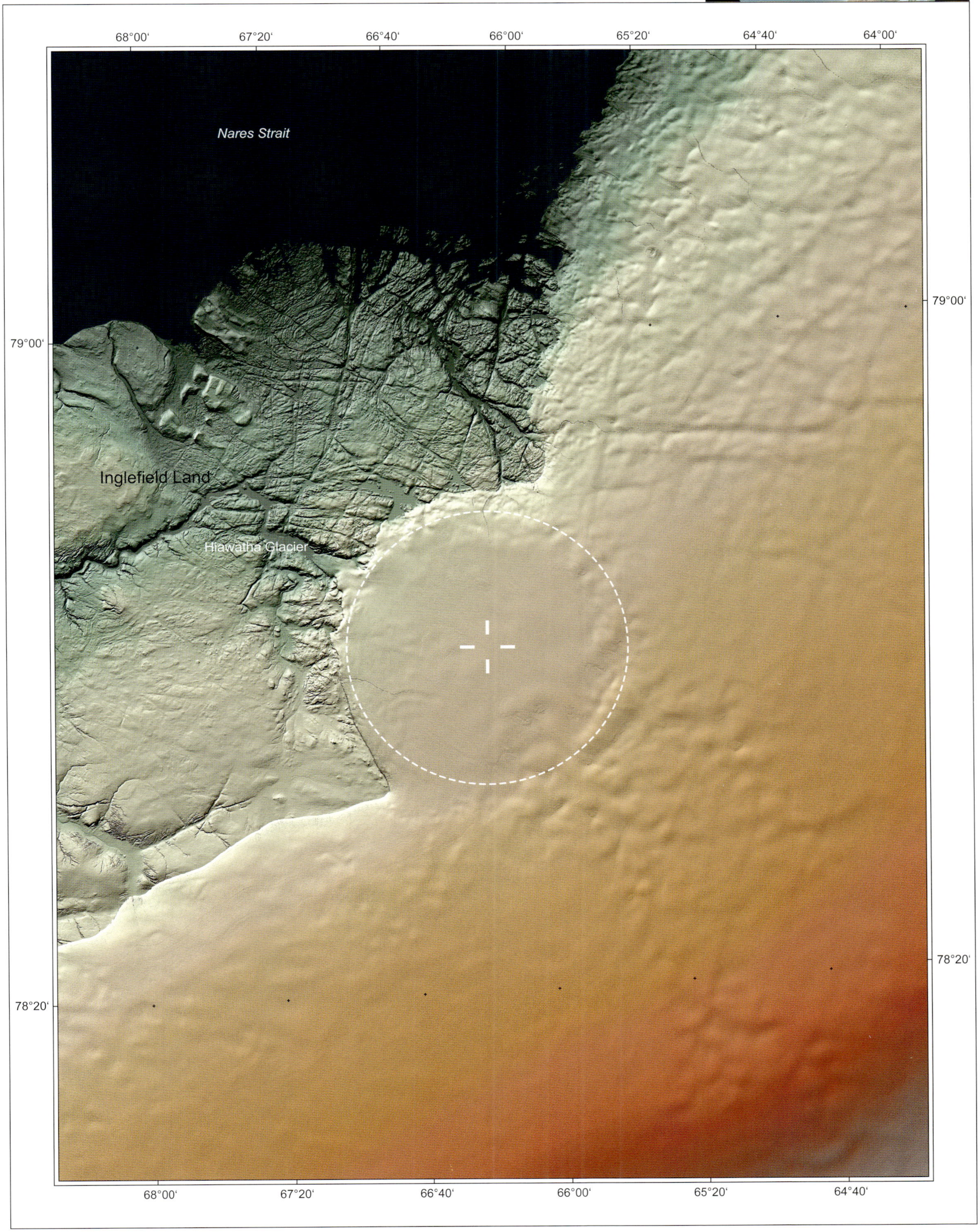

scale = 1 : 532 000 / scene = 95.9 × 127.8 km²

0 5.32 26.60 km

N

0 meter 1690

From the anomalous radiostratigraphy of the overburden ice sheet a Pleistocene age for the potential impact was proposed. The degraded morphology of the Hiawatha structure, that deviates from the normal geometry of a complex crater of this size, was explained by a combination of glaciofluvial and subglacial erosion of the rim and central uplift, sediment deposition within the crater, and post-impact rim collapse.

Further investigations are desired to confirm an impact origin of the Hiawatha structure. They may include geophysical surveying, radiometric dating of melt rock samples, and, most important, drilling to obtain in situ samples. This work should aim at verifying unequivocal shock effects and to explain Hiawatha's shallow depth in combination with a lack of a clear central uplift, which is unexpected for a rather fresh complex crater of this size with a proposed Pleistocene age between only 2.6 Myr and 12 000 years. Furthermore, the lack of signatures for a recent impact in Greenland's ice and rock records is puzzling. No signs of an ejecta layer exist, although the adjacent ice-free Inglefield Land should offer good opportunities to search for that.

The Hiawatha structure on August 31, 2018 in an image based on Copernicus Sentinel-2 (channel 2/3/4) data. The Hiawatha glacier is the ice tongue emanating from the circular margin of Greenland's ice sheet.

Location/Access

The possible impact structure is completely covered by the northwestern margin of Greenland's ice sheet in Inglefield Land. Largest infrastructure nearby is the Thule Air Base about 250 km to the southwest. Access to Hiawatha glacier requires planning for a full Arctic expedition.

Selected References

Kjær, K. H., Larsen, N. K., Binder, T., Bjørk, A. A., Eisen, O., Fahnestock, M. A., Funder, S., Garde, A. A., Haack, H., Helm, V., Houmark-Nielsen, M., Kjeldsen, K. K., Khan, S. A., Machguth, H., McDonald, I. Morlighem, M., Mouginot, J., Paden, J. D., Waight, T. E., Weikusat, C., Willerslev, E., and MacGregor, J. A. (2018). A large impact crater beneath Hiawatha Glacier in northwest Greenland. Science Advances, Vol. 4, eaar8173.

Pantasma

Nicaragua

Thomas Kenkmann

Pantasma is a recently proposed impact structure of about 14 km diameter, located in the volcanic highlands of Nicaragua. The structure forms an almost circular depression that is surrounded by steep hills rising 100–600 m above the interior depression. Pantasma is locally deeply dissected by the Pantasma River that discharges out of the basin to the north. The circular depression was previously regarded as an Oligocene-Miocene volcanic caldera, which was based on its location within a belt of Oligocene volcanic and volcanoclastic rocks. Country rocks range in composition from andesite to rhyolite. They include lava flows, ignimbrites and volcanoclastic layers, which are interbedded with fluvial sediments. The volcanic flows have total thicknesses between 0.3 and more than 1.5 km. They have been dated to 25–32 Myr. The underlying basement consists of Paleozoic schist, but it is not exposed in the vicinity of the structure. Owing to tropical weathering, outcrops are rare. Some outcrops occur at the edge of the depression and further outside the structure. They essentially show only volcanic formations. Further inside of the depression, a few outcrops with breccia have been described, as well as up to meter-sized breccia blocks.

Pantasma

Nicaragua

latitude = 13°21‘49“ N / longitude = 85°57‘14“ W / diameter = 14 km / age = 0.815 ± 0.011 Myr

scale = 1 : 245 000 / scene = 44.2 × 58.9 km²

0 2.45 12.25 km

N

275 meter 1640

View of the circular Pantasma depression in Nicaragua, from the south-southwest side that is the highest part of the rim. This image provides an excellent impression of the intense agricultural use of this area and the locally very dense, tropical vegetation (image courtesy: Pierre Rochette, Aix-Marseille University, Aix-en-Provence, France).

Pierre Rochette and co-workers studied a few samples collected near the center. One breccia sample is polymict and carried subangular, millimeter- to decimeter-size clasts of predominantly volcanic origin. However, there are also clasts of some Paleozoic metamorphic rocks, and the breccia broadly resembles altered suevite. They also found glass in such breccia, in which they allege to have documented high-pressure phases. The high pressure mineral coesite was identified by Raman spectroscopy. Reidite, a high pressure mineral of zircon, was identified indirectly by orientation mapping using electron backscatter diffraction (EBSD). This mineral has been documented to occur in shocked rocks from several impact structures. In addition, the presence of an extraterrestrial Cr isotopic signal in a polymict breccia sample from Pantasma, seemingly compatible with an admixture of an ordinary chondrite impactor, has been claimed.

Pantasma – a boulder of polymict breccia (Image courtesy: P. Rochette, Aix-Marseille University, Aix-en-Provence, France).

The cooling age of Pantasma glass has been obtained by ^{40}Ar-^{39}Ar geochronology and has been interpreted as the age of the alleged impact, at 0.815 ± 0.011 Myr.

While the data presented by Rochette and his team are sophisticated and solid, further proof appears to be required before Pantasma can be accepted as an impact structure. All studied rocks were sampled from river gravels or individual boulders, rather than from in situ outcrop. The glass samples could even represent tektites completely unrelated to the depression. Generally, it is particularly difficult to prove impact in a volcanic province where melt rocks and glasses can be of volcanic origin. Even high pressure phases might occur in such melt rocks in xenoliths from deep crustal or mantle sources. It is also notable that the interior part of the structure does not indicate anything that might hint at the occurrence of a central stratigraphic uplift.

Three specimens of glass from Pantasma. For detail see Rochette et al., 2019. (Image courtesy: P. Rochette, Aix-Marseille University, Aix-en-Provence, France).

The Pantasma structure on January 4, 2019 as a composite image based on TanDEM-X DEM and Copernicus Sentinel-2 (channel 2/3/4) data.

Pantasma was first recognized by Leo Kowald, a mathematician from Germany in 2006. While preparing for a journey through Nicaragua, he spotted Pantasma in a Google Earth scene. In a letter to the editor of the German astronomical magazine *Sterne und Weltraum* Kowald announced his finding and speculated about a possible impact origin. The recent detailed study by Pierre Rochette and co-workers, however, is the first extensive investigation of the structure.

Location/Access

The Pantasma depression is easily accessible via Nicaragua's Highway 43. The structure is host to the municipality of Las Praderas. The basin is extensively covered by agricultural fields. Outcrops are found mostly along road cuts, in a few small quarries, and in riverbeds.

Selected Reference

Rochette, P., Alac, R., Beck, P., Brocard, G., Cavosie A. J., Debaille, V., Devouard, B., Jourdan, F., Mougel, B., Moustard, F., Moynier, F., Nomade, S., Osinski, G. R., Reynard, B., and Cornec, J. (2019). Pantasma: Evidence for a Pleistocene circa 14 km diameter impact crater in Nicaragua. Meteoritics and Planetary Science, Vol. 54, pp. 880–901.

7.3 South America

Araguainha in August 2018 – TanDEM-X/Copernicus Sentinel-2.

South American Impact Structures – Overview

Table: South American Impact Structures – Overview

ID	Name	Country/Region	Latitude	Longitude	Diameter (km)	Age (Myr)	Topographic Visibility
1	Araguainha	Brazil	16°47'S	52°59'W	40	252–259	yes
2	Campo del Cielo Crater Field[1]	Argentina	27°38'S	61°42'W	max. 0.08	0.004	yes
3	Carancas[2]	Peru	16°39'52"S	69°02'38"W	0.014	0.000012	yes
4	Cerro do Jarau	Brazil	30°12'28"S	56°32'21"W	13	< 137	yes
5	Monturaqui	Chile	23°55'40"S	68°15'42"W	0.36	0.663 ± 0.090	yes
6	Riachão	Brazil	07°43'00"S	46°38'36"W	4.1	< 300	yes
7	Santa Marta	Brazil	10°10'19"S	45°13'58"W	10	< 93	yes
8	São Miguel do Tapuio	Brazil	05°37'36"S	41°23'18"W	21	< 300	yes
9	Serra da Cangalha	Brazil	08°04'45"S	46°51'24"W	13.7	< 300	yes
10	Vargeão Dome	Brazil	26°48'53"S	52°09'49"W	12.4	< 137	yes
11	Vista Alegre	Brazil	25°57'12"S	52°41'23"W	9.5	111–134	yes
Further confirmation required							
12	Colônia Basin	Brazil	23°52'10"S	46°42'22"W	3.6	5–36	yes
13	Nova Colinas	Brazil	07°09'33"S	46°06'30"W	5	< 300	yes
14	Rio Cuarto Crater Field	Argentina	32°52'S	64°11'W	max. 4.5	< 0.114	yes

Position references for selected impact structures:

1 Cassidy, W. A., and Renard, M. L. (1996). Discovering research value in the Campo del Cielo, Argentina, meteorite craters. Meteoritics and Planetary Science, Vol. 31, pp. 433–448.

2 Kenkmann, T., Artemieva, N. A., Wünnemann, K., Poelchau, M. H., Elbeshausen, D., and Nuñéz del Prado, H. (2009). The Carancas meteorite impact crater, Peru: geologic surveying and modeling of crater formation and atmospheric passage. Meteoritics and Planetary Science, Vol. 44, pp. 985–1000.

Araguainha
Brazil

Wolf Uwe Reimold

The Araguainha impact structure, also known as the *Araguainha Dome*, is – with a diameter of 40 km – the largest confirmed impact structure in South America. The structure straddles the border between Mato Grosso and Goiás states of central Brazil. Araguainha is a large, complex impact structure, with a 10–12 km wide central uplift, which is surrounded by the ring basin, followed by the crater rim section.

Araguainha Dome was formed in Paleozoic sedimentary rocks of the northeasternmost part of the Paraná Basin, and the crystalline basement underneath. The stratigraphic units that were affected by the impact are – from the oldest to the youngest – the crystalline basement that is composed of a porphyritic alkali granite that intruded into phyllites and meta-sandstones possibly belonging to the Cuiabá Group; the Devonian Furnas and Ponta Grossa formations, the Carboniferous Aquidauana Formation, and the Permian Passa Dois Group. The Devonian is represented by 800 m of fluvial white sandstones and conglomerates (Furnas Formation), and marine siltstones and minor sandstones (Ponta Grossa Formation). The Aquidauana Formation involves a 700–800 m thick section of Carboniferous sandstones and conglomerates that are overlain by stratified siltstones and sandstones. The Passa Dois Group comprises 20–40 m of Permian siltstone, chert and carbonate of the Iratí Formation and 80–100 m thick Permian-Triassic siltstones and subsidiary sandstones and chert of the Corumbataí Formation.

The metamorphosed rocks of what is thought to be the Cuiabá Group form the oldest stratigraphic unit of the central uplift of Araguainha. The rocks have experienced three metamorphic phases: a regional metamorphism linked to the Neoproterozoic Brasiliano-Pan African Cycle, then a thermal metamorphism related to the intrusion of the Araguainha alkali granite, and finally, the shock metamorphism related to the impact. The central peak involves a 4–5 km wide inner plug of porphyritic alkali granite. It is surrounded by a belt of occurrences of polymict impact breccia and, in some parts, by a 1–2 km wide collar of highly deformed sandstones and quartzites of the Furnas Formation.

The Furnas Formation is very prominent at the central uplift. It represents a 400–800 m thick pile of clastic quartzose strata (arenites and conglomerates) that have been segmented into fault-bounded blocks of varied rotation, resulting in a megablock zone around the granite core. The most prominent exposures occur in the northeastern Serra da Arnica, and to the southeast in the Morro Pelado hill. The Furnas Formation megablocks are strongly folded, at the meter to hundred meter scale. To achieve this in brittle quartzitic rock demands the enormous strain associated with impact.

The annular basin between the central uplift and the outer rim comprises strata of the Ponta Grossa Formation and the lower part of the Aquidauana Formation. Along the Araguaia River meandering around the eastern and southern sectors of the core of the central uplift, the Ponta Grossa sandstone is tightly folded. There are two ring features within the basin, at 10–12 km and 14–18 km from the center of the structure. The inner ring marks the extent of deformation caused by the outward movement of strata due to the collapse of the central uplift. A 700–800 m thick section of Carboniferous sandstones and conglomerates occurs in the main fill of the annular basin. The Permian and Permian-Triassic strata are well preserved as normal fault-bounded lenses in the outer basin.

Impactites from the Araguainha impact event occur in the central uplift. They include impact breccia with melt matrix (impact melt rock), as well as red and gray dikes of melt rock, and polymict and monomict breccias. The dikes are characterized by angular to subrounded clasts of granite and schlieren of mobilized feldspar that are

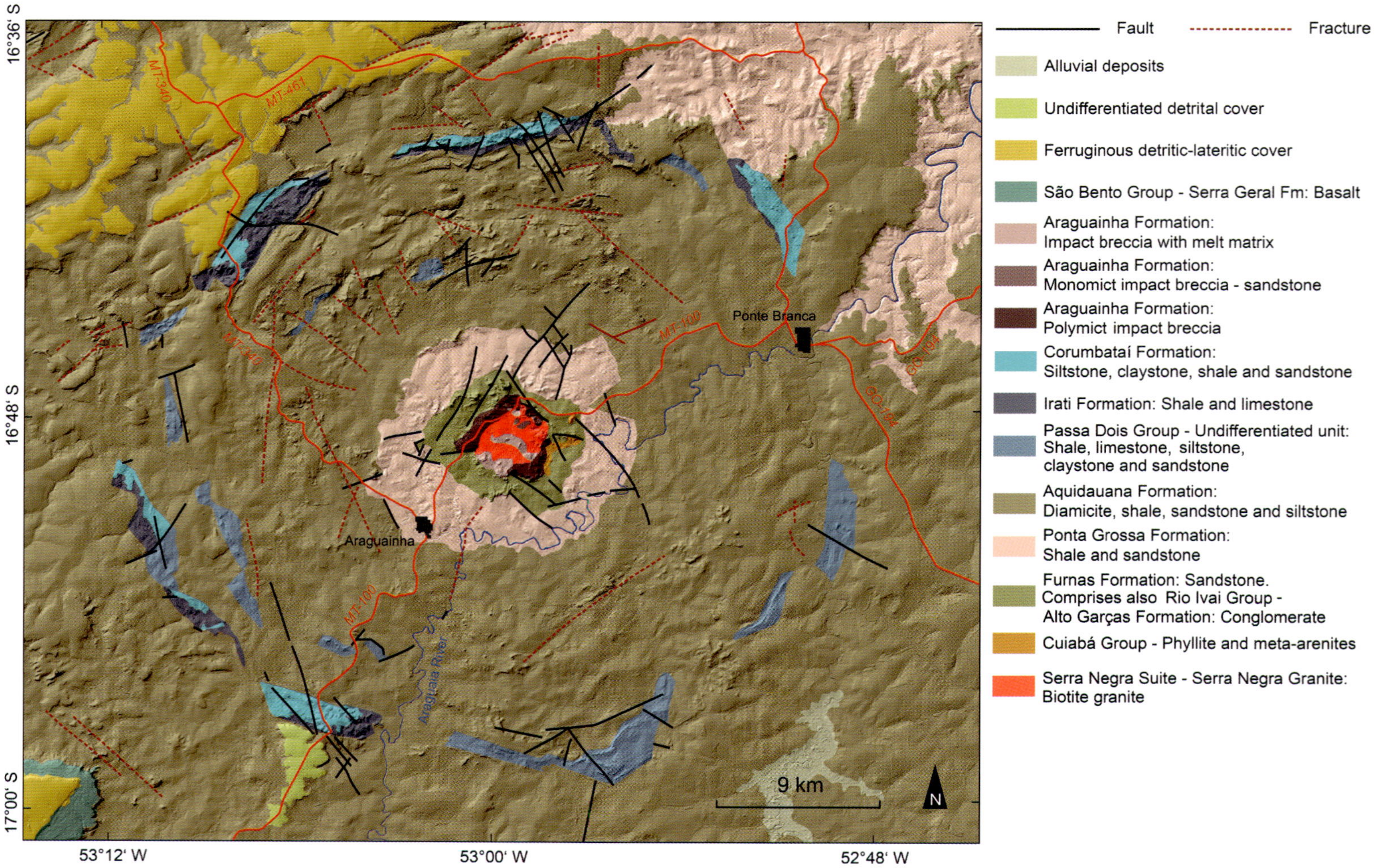

Geological map of the Araguainha impact structure with underlain TanDEM-X DEM (modified after Crósta et al., 2019a).

Araguainha

Brazil

latitude = 16°47‘ S / longitude = 52°59‘ W / diameter = 40 km / age = 252 - 259 Myr

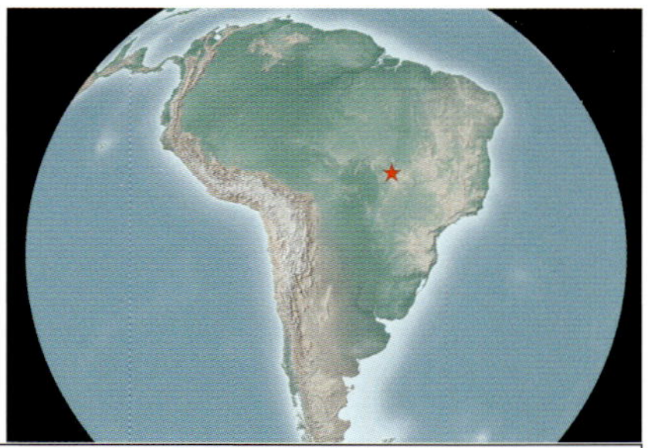

scale = 1:373 000 / scene = 67.1 × 89.4 km²

0 3.73 18.65 km

N

325 meter 890

Araguinha Dome, central uplift. A section of the Serra de Arnica, strongly deformed (including folded) strata of the Furnas Formation. Width of field of view about 200 m (photo: W. U. Reimold).

and Re-Os (rhenium-osmium) isotope analyses.

Araguainha Dome has been extensively studied with geophysical methods. Already in the 1970s, the region was covered by an aerial geophysical survey. The structure was then considered an alkaline igneous intrusion with possible potential for uranium mineralization. In terms of radiometrics, Araguainha Dome stands out conspicuously in the region. A circular potassium anomaly covers the granitic core. This is surrounded by an annular area of low gamma-ray signal, which likely corresponds to the occurrence of Furnas sandstone. This area is, in turn, surrounded by another circular anomaly of high potassium, thorium and uranium, which reflects the extent of occurrence of the Ponta Grossa Formation. Continuing towards the border of the structure, there is an ample area of very low gamma signal with sparse, small areas of low potassium. At approximately 20 km from the center, and surrounding almost the entire structure, a striking circular anomaly of high potassium marks the rim of the structure. This is interpreted to correspond to the occurrence of the deformed strata of the Passa Dois Group.

The magnetic data from this aerial survey indicate a depth to basement underneath the structure, which progressively decreases from a maximum of 1500 to 1700 m from the rim towards the center where basement is exposed. The magnetic anomaly pattern over the entire structure strongly contrasts with the regional field. Aeromagnetic data allow to determine the magnetic signature of the granite core. The orientation of magnetic lineation and magnetic foliation shows a regular pattern at the edge of the central peak, consistent with the fabric of sedimentary strata at the structure's inner collar. A complex anisotropy pattern emerged for the center of the structure.

The Araguainha structure was the first large impact structure discovered in South America. First reference to an anomalous circular structure in the northeasternmost part of the Paraná Basin came in 1969 from A. Northfleet and co-workers. They interpreted the structure as the result of a Cretaceous alkaline igneous intrusion, which had uplifted and deformed the Paleozoic sedimentary rocks. Because of this interpretation, and the economic potential of similar Cretaceous alkaline intrusions in the Paraná Basin, a geological

immersed in a fine-grained groundmass. Large outcrops of polymict impact breccia occur in the northern and northeastern sectors of the central uplift, and to a lesser degree in the southern part. All stratigraphic formations are represented as clasts in these breccias. In addition, one occasionally finds melt clasts. Overall, however, the distribution of lithic breccia without, and suevite with, melt clasts has not been fully established yet.

Melt rock occurs in the central granite in the form of blobs up to several meters in width, or as veins and dikes cross-cutting the granite and ranging in thickness from less than 10 to some 100 cm, and in length up to tens of meters. These blobs and veins/dikes contain cataclastic mixtures of shocked and unshocked granitic material, as well as melt. These melt rocks have been variably called pseudotachylite, pseudotachylitic breccias, or impact melt rock. The melt rocks in the core have essentially similar composition to that of the granite host rock. How they were formed and emplaced has remained unclear and a matter of debate.

Some of the melt rock clasts in suevitic polymict impact breccia were derived from granite, and others from quartz-rich sedimentary precursor strata. To date, no chemical traces of a meteoritic component have been discovered either, despite some platinum group element

A large boulder of metasedimentary rock likely derived from the Cuiabá Group, with a large shatter cone. Found just northeast of the small town of Araguainha, in the southwestern sector of the central uplift of the Araguainha impact structure. Coin, for scale, with a diameter of 2.5 cm (photo: W. U. Reimold).

A nice but small shatter cone (both the male and female parts are displayed) from phyllite assigned to the Cuiabá Group, which occurs around the granite core of the Araguainha Dome. Scale in centimeters (image courtesy: A. Crósta).

A several meter wide outcrop of impact melt rock in the northwestern part of the core of the Araguainha Dome (Fazenda Santa Maria). Alvaro Crósta, University of Campinas, and a regularly sized backpack, for scale (photo: W. U. Reimold).

A 7-cm-wide vein of impact melt rock in alkali granite host rock of the core of the Araguainha impact structure (photo: W. U. Reimold).

survey of the Araguainha Dome was conducted, with the outcome that Araguainha was considered a "cryptovolcanic" structure. Then in 1973, Robert Dietz and Bevan French reported the identification of two probable, eroded impact structures on images of NASA's new Landsat Program: Araguainha Dome and **Serra da Cangalha**. Their – for the time – novel interpretation was based on the typical morphology of eroded impact structures. A brief visit to Araguainha by Bob Dietz provided some first evidence of shock deformation – the documentation of shatter cones, and distinct sets of planar deformation features (PDFs) in several crystallographic orientations in quartz grains from Araguainha breccia.

Only at the beginning of the 1980s and 1990s came comprehensive evidence for impact at Araguainha. Detailed geological and geomorphological mapping was carried out by Alvaro Crósta, and then by Wolf von Engelhardt and co-workers, accompanied by thorough petrographic and microstructural analysis of the various rock types. All lithologies were shown to exhibit abundant evidence of shock deformation. There are excellent outcrops with shatter cones, especially in the rocks thought to belong to the Cuiabá Group. PDFs occur abundantly in quartz of all lithologies. Crósta and his team, in 1981, presented PDFs in quartz, selective melting of feldspar, and evidence for the occurrence of diaplectic quartz and feldspar glass in the granite.

The best ages for the Araguainha impact are the result of recent SIMS (Secondary Iron Mass Spectroscopy) U-Pb dating efforts with monazite from an impact melt rock sample; ages fall into the range from 252 to 259 Myr. This is, within error limits, coincident with the age of the global mass extinction at the Permian-Triassic Boundary. Considering that the much more powerful Chicxulub impact at the Cretaceous-Paleogene Boundary 66 Myr ago, at a time when Earth's environment was also severely stressed by massive volcanic eruptions, "only" killed off some 70 % of Earth's life, one should expect that an even more deadly impact event would be required to wipe out some 90 % of lifeforms at the Permian-Triassic Boundary. And yet, the 40 km diameter Araguainha structure is in the focus, as it has been estimated that an enormous amount of the powerful greenhouse gas methane might have been released as a consequence of the Araguainha impact from the wider Paraná Basin. Eric Tohver and team proposed in 2013 that methane could have been released by impact-induced seismic activity from the carbon-rich rocks of the Permian Iratí Formation from a major part of the Paraná basin. In a later study these authors showed that the products of seismic shaking (seismites) occur indeed widely in the basin. In any case, the coincidence of the Araguainha impact and the mass extinction has not been exactly confirmed with respective, coeval ages.

Location/Access

Araguainha Dome can be, except during the rainy season, readily reached with a reliable car. It is recommended to start at either Brasília or Goiânia, which can be reached easily by plane. From Brasília the 750 km journey to the small town of Ponte Branca close to the eastern rim of the structure will take about a day on the interstate BR-060 from Brasília to Anápolis, then the route 153 to Goiânia, again the BR-060 to Rio Verde, Jataí and Mineiros, and on to the turnoff to the small town of Portelândia. From there an initially paved, later unpaved and eventually very sandy road will lead after 50 km into the small town of Ponte Branca. Ponte Branca sports a small hotel with rather basic facilities. If a more comfortable hotel is desired, one should continue on the main thoroughfare BR-060 to Alto Araguaia, from where a reasonable road leads into the impact structure from the southwestern side.

Araguainha suevite, 9 × 6.5 × 5 cm in size, from the southern core of the central uplift (photo courtesy: Ralf-Thomas Schmitt, Museum für Naturkunde Berlin, sample from Stefanie Fischer).

Selected References

Crósta, A. P., Gaspar, J. C., and Candia, M. A. F. (1981). Feições de metamorfismo de impacto no Domo de Araguainha. Revista Brasileira de Geociências, Vol. 11, pp. 139–146.

Crósta, A. P., Reimold, W. U., Vasconcelos, M. A. R., Hauser, N., Oliveira, G. J. G., Mazivieiro M. V., and Góes, A. M. (2019a). Impact cratering: The South American record – Part 1. Geochemistry – Chemie der Erde, Vol. 79, pp. 1–61.

Dietz, R. S., and French, B. M. (1973). Two probable astroblemes in Brazil. Nature, Vol. 244, pp. 561–562.

Hauser, N., Reimold, W. U., Cavosie, A. J., Crósta, A. P., Schwarz, W. H., Trieloff, M., Da Silva Maia de Souza, C., Pereira, L. C., Rodrigues, E. N., Brown, M. (2019). Linking shock textures revealed by BSE, CL and EBSD with U-Pb data (LA-ICP-MS and SIMS) in zircon from the Araguainha impact structure, Brazil. Meteoritics and Planetary Science, 54, pp. 2386–2311.

Northfleet, A. A., Medeiros, R. A., and Muhlmann, H. (1969). Reavaliação dos dados geológicos da Bacia do Paraná. Boletim Técnico da Petrobrás, Vol. 12, pp. 291–346.

Tohver, E., Cawood, P. A., Riccomini, C., and Trindade, R. I. F. (2013). Shaking a methane fizz: Seismicity from the Araguainha impact event and the Permian–Triassic global carbon isotope record. Palaeogeography, Palaeoclimatology, Palaeoecology, Vol. 387, pp. 66–75.

Von Engelhardt, W., Matthai, S. K., and Walzebuck, J. (1992). Araguainha impact crater, Brazil. I. The interior part of the uplift. Meteoritics, Vol. 27, pp. 442–457.

Yokoyama, E., Trindade, R. I. F., Lana, C. R., Souza-Filho, C. R., Baratoux, D., Marangoni, Y. R., and Tohver, E. (2012). Magnetic fabric of Araguainha complex impact structure (Central Brazil): Implications for deformation mechanisms and central uplift formation. Earth and Planetary Science Letters, Vol. 331–332, pp. 347–359.

Campo del Cielo Crater Field

Argentina

Wolf Uwe Reimold

The Campo del Cielo meteorite/crater field occurs across the stateline between Argentina's provinces of Santiago del Estero and Chaco, in a vast plain that once was covered by dense forest but today is mainly farmland. There are at least 20–22 small meteorite craters and penetration funnels close to the town of Gancedo, some 1000 km northwest of Buenos Aires, in an area between 27°30'S and 27°40'S latitude and 61°30'W and 61°50'W longitude.

A large number of, in part huge, iron meteorites classified as type IA iron meteorite, has been recovered from an 18.5 × 4 km area. The largest meteorite is said to have weighed 37 tons, and the second largest, only recovered in 2016 within the Gancedo town limits, weighs 30.8 tons. It has been estimated that the total mass of fallen meteoritic matter could have been as much as 300 tons. The strongly extended strewn field infers that the angle of the fall was very low. The largest meteorites were recovered in the central part of the strewn field.

The craters range in size up to 65–115 m. The four largest have been classified as "explosion-analog craters"; they lack sizable projectile masses in the interior and have a lot of shattered meteorite pieces, *shrapnel*, in their immediate environs. These craters have reasonably symmetrical morphologies with raised rims. In contrast, the other depressions that have been investigated in some detail have a complex, elongated form that is wider in the uppermost section and then constricts to a rather straight funnel in the lower section. These forms are strongly inclined and have been interpreted as "penetration funnels". In some of them sizable projectile masses have been identified. Magnetic anomalies over other funnels suggest that further large masses of meteorite are still awaiting recovery. The impact azimuth has been variably estimated at 75° north to 40° northeast, based on the apparent eccentricity of crater forms, and the angle of impact was suggested as very low, possibly 9–16° from the horizontal.

The crater field was formed in a homogeneous, thick loess target. The crater depressions have been substantially filled in since their formation – in the course of prehistoric human activity, periodic rainfall, and cattle grazing in this area. Results of ^{14}C dating of three charcoal samples collected in a crater delimit the age of the meteorite fall to about 4000 years before present (BP).

The Campo del Cielo site has been investigated for more than four centuries, since Spanish conquistadores searching for precious metal came across a huge piece of iron, estimated to have weighed about 14 tons, which was then known as the *Meson de Fierro* (Large Table of Iron). This piece disappeared at some time prior to an expedition into the Chaco in 1783. Only since the early 20th century have some detailed investigations of this crater field been made. In 1926 J. Nagera described the four largest depressions near the town

The 37 ton "El Chaco" meteorite. This is the second-largest meteorite recovered on Earth, after the 60-ton Hoba meteorite of Namibia (photo: W. U. Reimold).

One of the largest, multi-ton "specimens" on display in the Parque Natural Cultural Pigüen N'Onaxa around the Campo del Cielo Visitors Center near the town of Gancedo, Chaco, Argentina (photo: W. U. Reimold).

Campo del Cielo Crater Field

Argentina

latitude = 27°38' S / longitude = 61°42' W / diameter = max. 80 m / age = 4000 yr

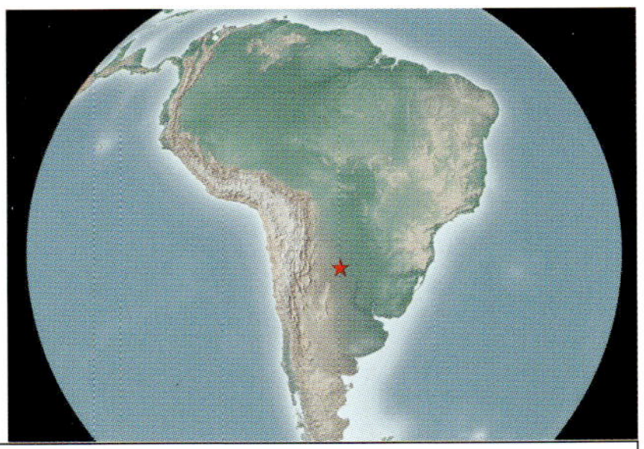

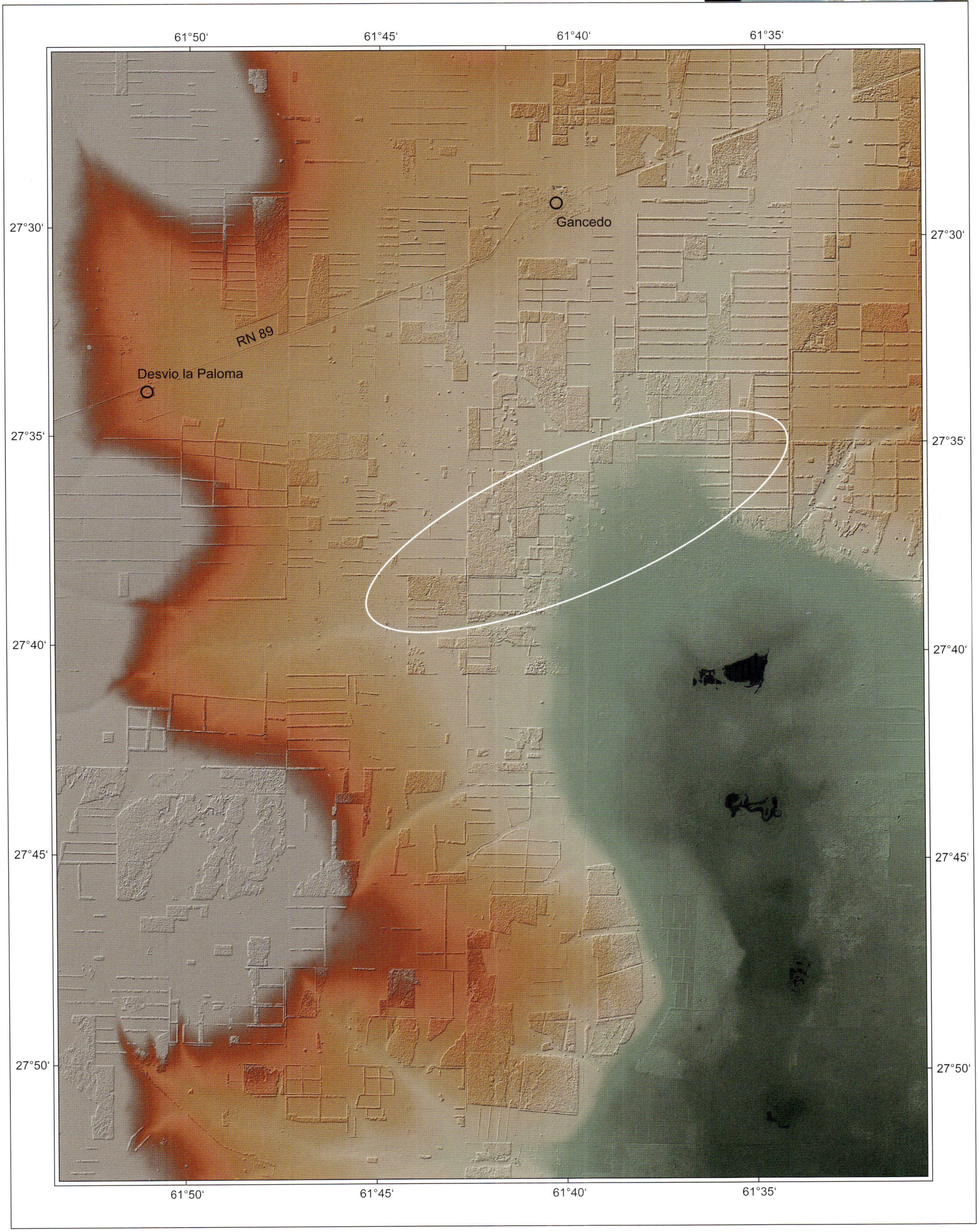

scale = 1 : 208 000 / scene = 37.4 × 49.8 km²

0 2.08 10.39 km

N

45 meter 155

of Gancedo as having small meteorite fragments associated with them, whilst seven years later L. Spencer was the first to suggest that these depressions might represent meteorite craters. Modern studies commenced in the late 1960s and early 1970s and led to the identification of sixteen craters in addition to the four originally described by Nagera. The most recent, detailed assessments of the crater field were provided by William Cassidy and Marc Renard in 1996 and Mario Vesconi and his team in 2011.

The wall-sized display cabinet with meteorite finds in the William Cassidy Hall of the Visitors Center at Campo del Cielo (photo: W. U. Reimold).

Location/Access

The crater field is readily accessible, travelling on reasonable National Roads (RN) towards the city of Gancedo. There are some interesting and educational exhibits in a very nice exhibition center in the Reserva Natural Piguem N'Onaxa ***"Campo del Cielo"***, where a strong effort is being made to protect the finds. In the grounds around the center, some of the largest meteorites are on display. However, there are also reports about theft and smuggling of iron meteorites from this area.

Selected References

Cassidy, W. A., and Renard, M. L. (1996). Discovering research value in the Campo del Cielo, Argentina, meteorite craters. Meteoritics and Planetary Science, Vol. 46, pp. 935–949.

Crósta, A. P., Reimold, W. U., Vasconcelos, M. A. R., Hauser, N., Oliveira, G. J. G., Maziviero, M. V., and Góes, A. M. (2019b). Impact cratering: The South American record – Part 2. Geochemistry – Chemie der Erde, online first, Vol. 79, pp. 191–220.

Nagera, J. J. (1926). Los hoyos del Campo del Cielo y el meteorito. Direccion General de Minas, Geologia y Hidrologia, Buenos Aires, Argentina.

Spencer, L. J. (1933). Meteorite craters as topographic features of the earth's surface. Geographic Journal, Vol. 81, pp. 227–248.

Vesconi, M. A., Wright, S. P., Spagnuolo, M., Jacob, R., Cerrutti, C., Garcia, L., Fernandez, E., and Cassidy, W. A. (2011). Comparison of four meteorite penetration funnels in the Campo del Cielo crater field, Argentina. Meteoritics and Planetary Science, Vol. 46, pp. 935–949.

Carancas

Peru

Thomas Kenkmann

The 14.2 m wide Carancas impact crater was formed on September 15, 2007, at noon (exactly at 11.45 h local time, 16.45 UTC), some 12 km south of Lake Titicaca. This fall/impact is among the rare impact events that could be witnessed. Furthermore, Carancas represents the smallest individual impact crater on Earth. Before the impact occurred, a bright fireball was observed in the sky from several places along the southern shore of Lake Titicaca in Bolivia and Peru.

After its formation the Carancas crater immediately filled with groundwater. Note the ejecta blocks on the elevated crater rim and slump blocks at the inner crater wall (photo: T. Kenkmann).

The meteorite hit the ground of the Puna region south of the village of Carancas, at an altitude of 3826 m. The fireball and its trajectory suggested an atmospheric passage from east-northeast towards west-southwest. Surprisingly, the meteoroid did not break apart upon passage through the atmosphere and, consequently, did not produce a strewn field. The meteorite impacted into unconsolidated soil and sand that were water-saturated at a depth of 1–1.5 m. Residents of the community of Carancas, situated about 3 km north-northwest of the point of impact, heard a loud explosion, which was followed by the formation of a mushroom cloud composed of disseminated water droplets, dust, and fine-grained particles. The crater cavity immediately filled with water, as the ground-water table was breached. Ascending water bubbles were observed after the impact. They were most likely formed by the release of air that was trapped and compressed into the target in front of the impacting meteoroid.

The Carancas crater is roughly circular in shape and has an average diameter of 14.2 m, measured from rim crest to rim crest. The crater formed at the edge of an about 1 m high creek bank. The embankment influenced the crater morphology. The crater rim on the terrace is higher and the cavity has higher and steeper slopes at the terrace, with the steep slopes having induced slumping of meter-sized soil blocks. The average height of the crater rim amounts to 54 cm above target level. The calculated crater volume is about 160 m^3. The ejecta blanket has a maximum thickness along the rim crest. The ejecta consist of blocks of dark soil and reddish sand that form a continuous blanket to about 4 m from the crater rim. The continu-

Carancas

Peru

latitude = 16°39'52" S / longitude = 69°02'38" W / diameter = 14 m / age = 12 yr

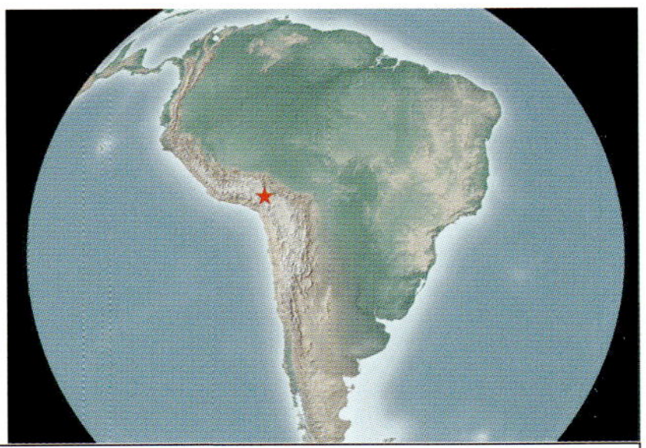

scale = 1:77 000 / scene = 13.9 × 18.5 km²

0 0.77 3.85 km

N

3800 meter 3860

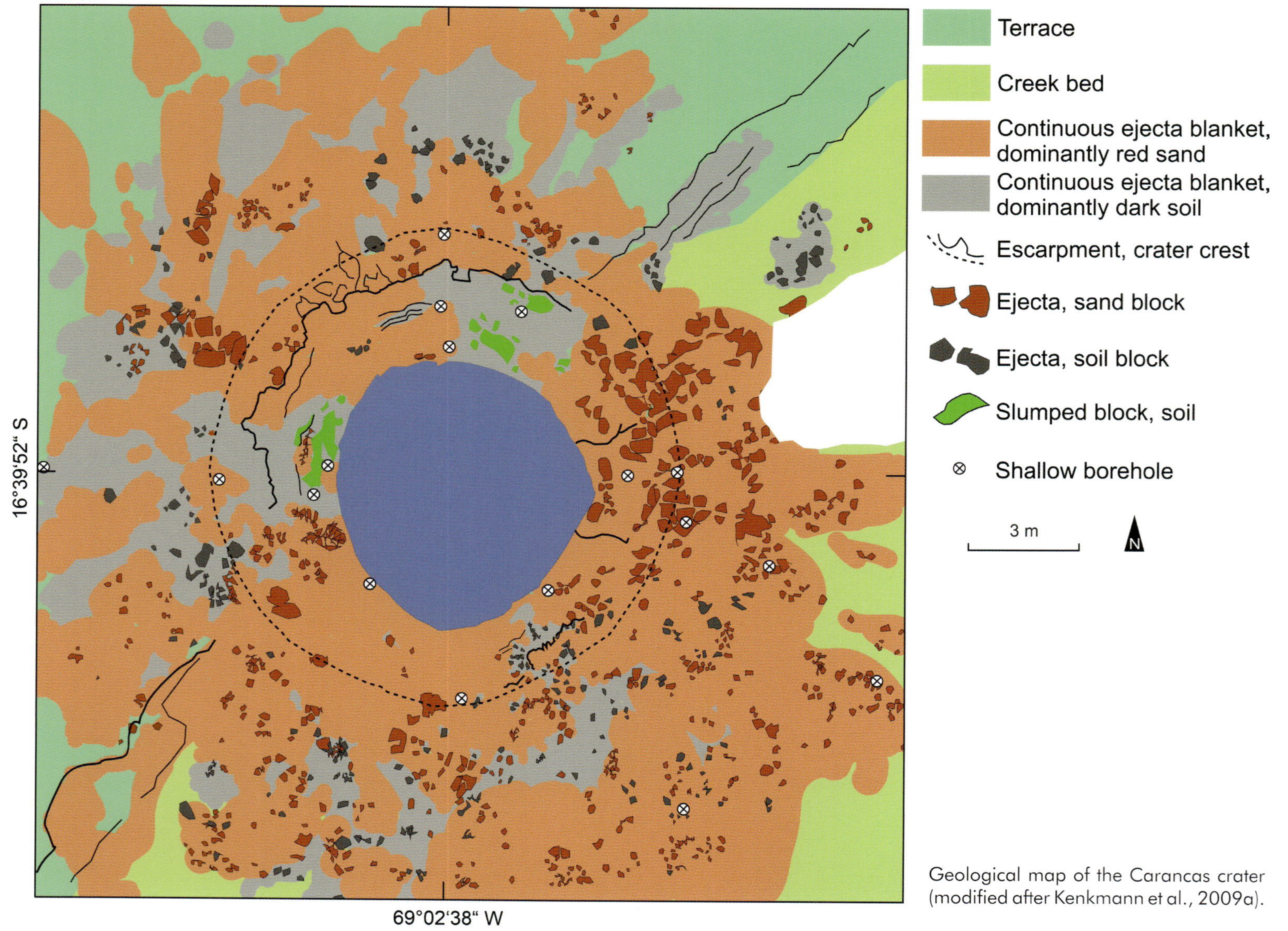

Geological map of the Carancas crater (modified after Kenkmann et al., 2009a).

Aerial photograph of the Carancas meteorite crater taken from the east from a height of about 7 m. The crater is 14.2 m across and formed at the edge of a creek bank. The photo was taken 3.5 months after the impact (photo: T. Kenkmann).

ous ejecta blanket in the southeast creek bed has a wider extent than on the higher terrace and is dominated by sand and soil blocks that locally are 0.5 m in size. A maximum radial extent of ejecta deposition of 234 m from the crater was determined.

Samples were studied from 18 shallow drillings and from surface probing. A few grains display transgranular fractures, which have been linked to the impact event. Shock features in quartz or feldspar could not be identified and even the fracture densities in the sample constituents are not particularly high. Microscopic examination suggests that stresses were locally sufficient for tensile fracturing and shear fracturing but not for shock deformation to be initiated.

What makes the Carancas event unique is the fact that a stony meteoroid was capable of penetrating the atmosphere without fragmentation. It could be shown that a low cosmic velocity, a very shallow entry angle into the upper atmosphere, and a low projectile mass would be necessary to minimize aerodynamic stresses and increase the probability of a survival of stony meteoroids with a standard dynamic strength of a few MPa. Alternatively, a meteoroid with a very high strength could be prevented from atmospheric disruption. However, the Carancas meteorite has been classified as a rather fragile H4-5 ordinary chondrite. Abundant fragments of the meteorite could be collected in the crater and in its close proximity. Numerical modeling of the meteoroid's passage through the atmosphere revealed strong meteoroid deceleration and a deflection from its initial shallow entry angle of 20° towards a steeper trajectory with a final impact at an angle of 40–60° and an impact velocity of 350–600 m/s, all of which would have prohibited the production of a shock wave and shock effects in target minerals.

The impact caused a worldwide sensation among scientists and a number of teams started to investigate the event with partly conflicting results. Crater structure and morphology of the crater were mapped in detail by Thomas Kenkmann and Michael Poelchau. As part of this study the atmospheric passage and the impact were modelled by Natalia Artemieva and Kai Wünnemann. The entry of the Carancas meteoroid produced infrasound waves that were recorded by two infrasound arrays of the International Monitoring System for the enforcement of the Comprehensive Nuclear-Test-Ban Treaty (CTBT) 80 km northeast and 1617 km southeast of the crater. These data were analyzed by Peter Brown and co-workers as well as by A. Le Pichon. While Le Pichon expected an initial energy release during atmospheric entry in the range of 1–8 kt TNT equivalent, the Brown team estimated 0.06–0.32 kt TNT equivalent. Both studies inferred a rather low atmospheric entry speed of the meteoroid in the order of 12 km/s.

Slumped soil blocks indicating synthetic as well as antithetic block rotations (photo: T. Kenkmann).

Location/Access

The crater can be reached from the town of Desaguadero, which is situated at the southern shore and outlet of Lake Titicaca at the Peruvian/Bolivian border about 12 km north of the point of impact. The closest village accessible on a dirt road is Carancas. The crater has undergone severe destruction and erosion since its formation, so that a wooden hut was constructed over the crater to prevent further degradation.

Selected References

Borovicka, J., and Spurný, P. (2008). The Carancas meteorite impact – Encounter with a monolithic meteoroid. Astronomy and Astrophysics, Vol. 485, pp. L1–L4.

Brown, P., ReVelle, D. O., Silber, E. A., Edwards, W. N., Arrowsmith, S., Jackson, L. E., Tancredi, G., and Eaton, D. (2008). Analysis of a crater-forming meteorite impact in Peru. Journal of Geophysical Research, Vol. 113, 13 pp.

Crósta, A. P., Reimold, W. U., Vasconcelos, M. A. R., Hauser, N., Oliveira, G. J. G., Maziviero, M. V., and Góes, A. M. (2019b). Impact cratering: The South American record – Part 2. Geochemistry – Chemie der Erde, Vol. 79, pp. 191–220.

Kenkmann, T., Artemieva, N. A., Wünnemann, K., Poelchau, M. H., Elbeshausen, D., and Nuñéz del Prado, H. (2009a). The Carancas meteorite impact crater, Peru: geologic surveying and modeling of crater formation and atmospheric passage. Meteoritics and Planetary Science, Vol. 44, pp. 985–1000.

Le Pichon, A., Antier, K., Cansi, Y., Hernandez, B., Minaya, E., Burgoa, B., Drob, D., Evers, L. G., and Vaubaillon, J. (2008). Evidence for a meteoritic origin of the September 15, 2007, Carancas crater. Meteoritics and Planetary Science, Vol. 43, pp. 1797–1809.

Cerro do Jarau

Brazil

Wolf Uwe Reimold

The about 13 km wide Cerro do Jarau structure is located very close to the Uruguay-Brazil border in Quaraí county in the southwestern portion of the state of Rio Grande do Sul. In the interior of the structure occurs a complex terrain about 6 km wide, which comprises a series of semi-circular ridges that rise some 200 m above the surrounding terrain. The environs around these Cerro do Jarau hills feature rather flat 'pampa' plains, as they are typical for this region in southern Brazil, Uruguay, and Argentina. In the surroundings of the elevated terrain, the drainage pattern is somewhat annular. The high ridges are composed of folded and silicified sandstone and thought to belong to the Botucatu Formation, whereas the exposures in the surrounding areas correspond to basaltic lava flows of the Serra Geral Formation.

Over a 13 km wide area, the two rock types that occur within the structure, namely basalt and sandstone, exhibit deformation in the form of brecciation, whereas outside of this limit the basalt is undeformed. The core of the structure that is composed of silicified sandstones has an elliptical shape of 6.5 km diameter in north-south direction, and 5.2 km in east-west direction. Following the most recent geological and morphological analysis, the outer limit of the structure, which is not obvious in topography, likely coincides with the semi-circular drainage patterns of the Guarupá stream in the north and northwest, and the Quaraí-Mirim creek to the south and southwest. These drainage channels join to form the Quaraí river, which delimits the structure to the west.

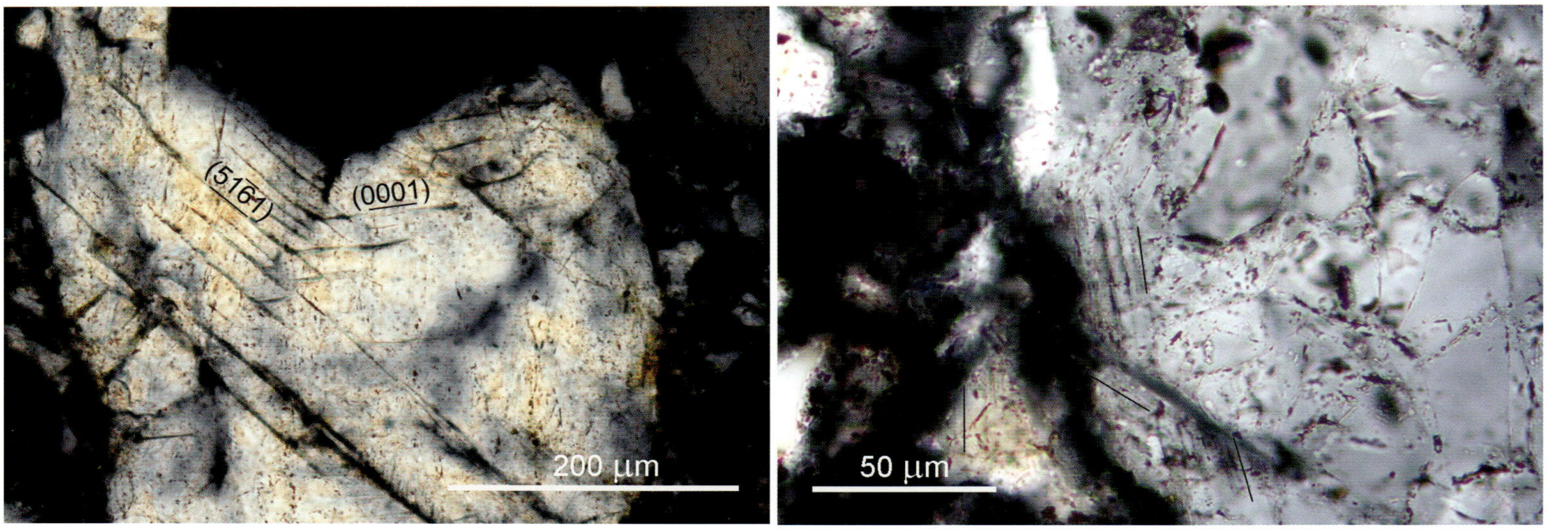

Two examples of shock deformation features in quartz from sandstone samples from the Cerro do Jarau impact structure. Relatively widely spaced features are PFs (planar fractures), whereas the planar deformation features (PDFs) are much more closely spaced and here represented by rather short, mostly marginal sets emphasized on the right side by single, thin black lines. Cross-polarized light images (photos: W. U. Reimold).

The area of the Cerro do Jarau impact structure on December 2, 2018 as a composite image based on TanDEM-X DEM and Copernicus Sentinel-2 (channel 2/3/4) data.

The geology of the inner structure is characterized by two different types of sandstone. In the innermost part, with outcrops at comparatively lower topographic levels, occur sandstones showing characteristics of fluvial deposition. They have been attributed to the Middle Jurassic Guará Formation of the São Bento Group. Around these fluvial sandstones but still within the limits of the inner part of Cerro do Jarau, there are exposures of eolian sandstones with cross-stratification at the meter scale. These sandstones are believed to belong to the Upper Jurassic Botucatu Formation.

Based on a ground gravimetric survey of Cerro do Jarau, a positive, semi-circular Bouguer gravity anomaly larger than +6 mGal occurs in the environs of the central part of the structure. There is also a positive anomaly exceeding +7 mGal in the northeastern portion of the structure, extending towards the center. The inner core of the structure exhibits a negative anomaly surrounding a small positive anomaly.

The age of the structure is only constrained by the age of the impact-affected Serra Geral basalts, which is of the order of 127–137 Myr, and which provides a maximum age for the impact event.

The structure has been known to geologists since the late 1960s from aerial photography. Its origin has been the subject of debate ever since. The first reference to a possible impact origin was made in 1987 – based purely on the circular appearance of the structure. Tell-tale impact (shock) indicators have been repeatedly invoked

Cerro do Jarau

Brazil

latitude = 30°12‘28“ S / longitude = 56°32‘21“ W / diameter = 13 km / age = < 137 Myr

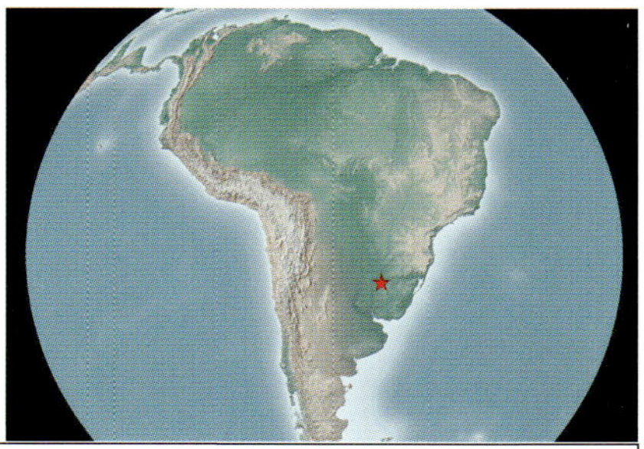

scale = 1 : 153 000 / scene = 27.6 × 36.8 km²

0 1.53 7.65 km

N

65 meter 305

since, but that evidence has remained uncertain. The first evidence to strongly support an impact origin for Cerro do Jarau included a report of brecciation of sandstone and basalt by Alvaro Crósta and his team in 2010, as well as deformation features in these two rock types: striated fracture surfaces that resembled multiply-striated joint surfaces related to the shatter cone phenomenon, as well as micro-deformation features in quartz. The conclusive evidence, in the form of multiple occurrences of planar deformation features (PDFs) in quartz, together with abundant planar fractures (PFs) and feather features (FFs), has only just been obtained by Wolf Uwe Reimold and co-workers in samples from the central part of the structure. Geology, geophysics, and shock petrography are all consistent with Cerro do Jarau representing a mid-size, complex impact structure with a central uplift where lower target strata were uplifted from their pre-impact stratigraphic position.

Location/Access

Cerro do Jarau is located in a remote part of Brazil, but it is readily accessible via good roads. Options could be to travel overland from, for example, Sao Paulo via Curitiba (1600 km), or from Porto Alegre (600 km). There is an airport in the town of Uruguaiana, 125 km to the northwest of Cerro do Jarau – and again with good road access from there. The airport of Uruguaiana is served by flights from both Sao Paulo and Porto Alegre.

Selected References

Crósta, A. P., Lourenço, F. S., and Priebe, G. H. (2010b). Cerro Jarau, Rio Grande do Sul: a possible new impact structure in Southern Brazil. In: Large Meteorite Impacts and Planetary Evolution IV, Gibson, R. L. and Reimold, W. U. (eds.), Geological Society of America Special Paper, Vol. 465, pp. 173–190.

Crósta, A. P., Reimold, W. U., Vasconcelos, M. A. R., Hauser, N., Oliveira, G. J. G., Maziviero, M. V., and Góes, A. M. (2019b). Impact cratering: The South American record – Part 2. Geochemistry – Chemie der Erde, Vol. 79, pp. 191–220.

Giacomini, B. B., Leite, E. P., and Crósta, A. P. (2017). 3D gravimetric investigation of the Cerro do Jarau structure, Rio Grande do Sul, Brazil. Meteoritics and Planetary Science, Vol. 52, pp. 565–583.

Reimold, W. U., Crósta, A. P., Hasch, M., Kowitz, A., Hauser, N., Sanchez, J. P., Amarante Simões, L. S., de Oliveira, G. J., and Zaag, P. T. (2018). Shock deformation confirms the impact origin for the Cerro do Jarau, Rio Grande do Sul, Brazil, structure. Meteoritics and Planetary Science, Vol. 54, pp. 2384–2397.

Monturaqui

Chile

Wolf Uwe Reimold

Monturaqui is a small, simple bowl-shape impact crater of approximately 350 × 370 m size, with a maximum depth of 34 m compared to the level of the environs. It is located in a remote part of the southern Atacama Desert in the Andean Precordillera of northern Chile, in the rugged desert hills of the Monturaqui mountain range. The crater is crossed by an old llama trail that connects the coast of the Pacific with San Pedro de Atacama and northwestern Argentina. The morphology of the crater is nearly circular, with a slight northwest-southeast elongation. The target rocks comprise Ordovician granite of the basement that is cut by 1–2 m wide mafic dikes. The granite is overlain by an up to 5 m thick layer of Pliocene ignimbrite. Both major target rocks, granite and ignimbrite, crop out along the entire crater rim. Detailed geological mapping showed that there is a continuous ejecta zone. Although granite is seemingly the dominant target rock, much of the granitic ejecta seems to be missing. Mixed ejecta composed of granite and ignimbrite are the most common type of ejecta. The crater rim has been eroded by about 5 m since crater formation.

No meteorite fragments have been found but glassy aggregates containing shocked and unshocked rock fragments, as well as metallic inclusions, were recovered. The ejecta deposits around the crater are described as cindery, highly vesicular, glassy aggregates containing ignimbrite, granite and dark impact melt. Some of these ejecta resemble volcanic bombs, with their twisted morphologies. Shock micro-deformation includes kink bands in biotite and apatite, abundant planar fractures (PFs) in quartz and feldspar, as well as planar deformation features (PDFs) in quartz, minerals with phase transformations, and complete mineral melting. Quartz, plagioclase and K-feldspar grains are partially to completely converted to diaplectic glass. Coesite was detected as rare, fine-grained crystals in diaplectic quartz glass. The matrix of the impactites consists of green and brown glass that carries rounded inclusions of nickel-iron, up to 2 mm in size. Analysis of metal spherules from ejecta allowed to determine the nature of the projectile as a coarse group I octahedrite. A geophysical survey yielded a negative Bouguer gravity anomaly of about −1 mGal at the center of the crater, which has been explained as a consequence of intense fracturing/brecciation of the target rocks in this area.

In 1975, the crater age was estimated at older than 0.1 Myr from thermoluminescence analysis. Meanwhile an absolute age for the formation of the crater has been obtained at 0.5–0.6 Myr. It is based on the terrestrial age of the impactor through measuring the residual activities of some nuclides, such as, e.g., ^{10}Be, ^{26}Al, ^{36}Cl, in iron-nickel spherules. In addition, a surface exposure age was determined by measuring in-situ produced ^{10}Be in granite outcrops that were exposed, after impact, to cosmic radiation. This method yielded an age of 0.2–0.25 Myr, but it is argued that this age was likely an underestimation because of the ongoing erosion of the crater rim. A similar result with a mean age of 0.663 ± 0.090 Myr was obtained with the low temperature (U-Th)/He dating technique from analysis of 22 apatite and 10 zircon grains.

The Monturaqui crater was discovered in 1962 from the study of aerial photographs and then confirmed by ground truthing three years later by Joaquin Sánchez and William Cassidy.

Location/Access

The crater is located in the high Atacama Desert close to the Chile-Argentina-Bolivia border, at 120 km directly south of San Pedro de Atacama and 30 km south of the Salar de Atacama. To reach the crater is a 308 km drive of about 6 hr duration from the city of Antofagasta. Follow the Panamericana Norte/Ruta 5 towards Lomas Bayas, then onto the B-385, B-241, and B-55. After 40 km on this latter road, Monturaqui will be close. However, access then will be very difficult in the Cordón de Lila range to the south of the locality of Peine.

Selected References

Buchwald V. F. (1975). Handbook of Iron Meteorites: Their History, Distribution, Composition, and Structure. University of California Press, Berkeley, California, 1426 pp.

Bunch, T. E., and Cassidy, W. A. (1972). Petrographic and electron microprobe study of the Monturaqui impactite. Contributions to Mineralogy and Petrology, Vol. 36, pp. 95–112.

Crósta, A. P., Reimold, W. U., Vasconcelos, M. A. R., Hauser, N., Oliveira, G. J. G., Maziviero, M. V., and Góes, A. M. (2019b). Impact cratering: The South American record – Part 2. Geochemistry – Chemie der Erde, Vol. 79, pp. 191–220.

Peate, I., Van Soest, M. C., Wartho, J. A., Cabrol, N., Grin, E., Piatek, J., and Chong, G. (2010). A novel application of (U-Th)/He geochronology to constrain the age of small, young meteorite impact craters: A case study of Monturaqui Crater, Chile. 31st Lunar and Planetary Science Conference, Lunar and Planetary Institute, Houston, Texas, Abstract #2161.

Monturaqui

Chile

latitude = 23°55‘40“ S / longitude = 68°15‘42“ W / diameter = 360 m / age = 0.663 ± 0.090 Myr

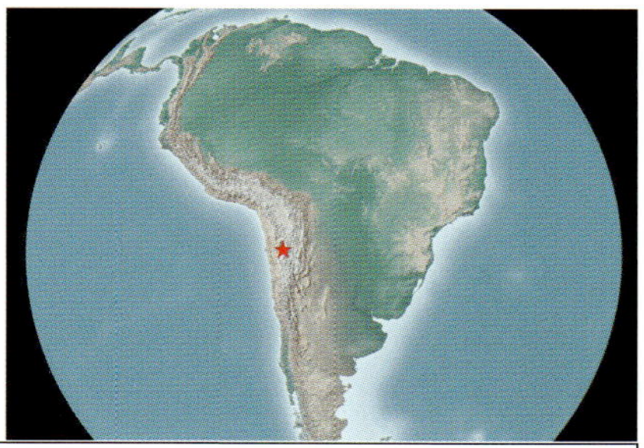

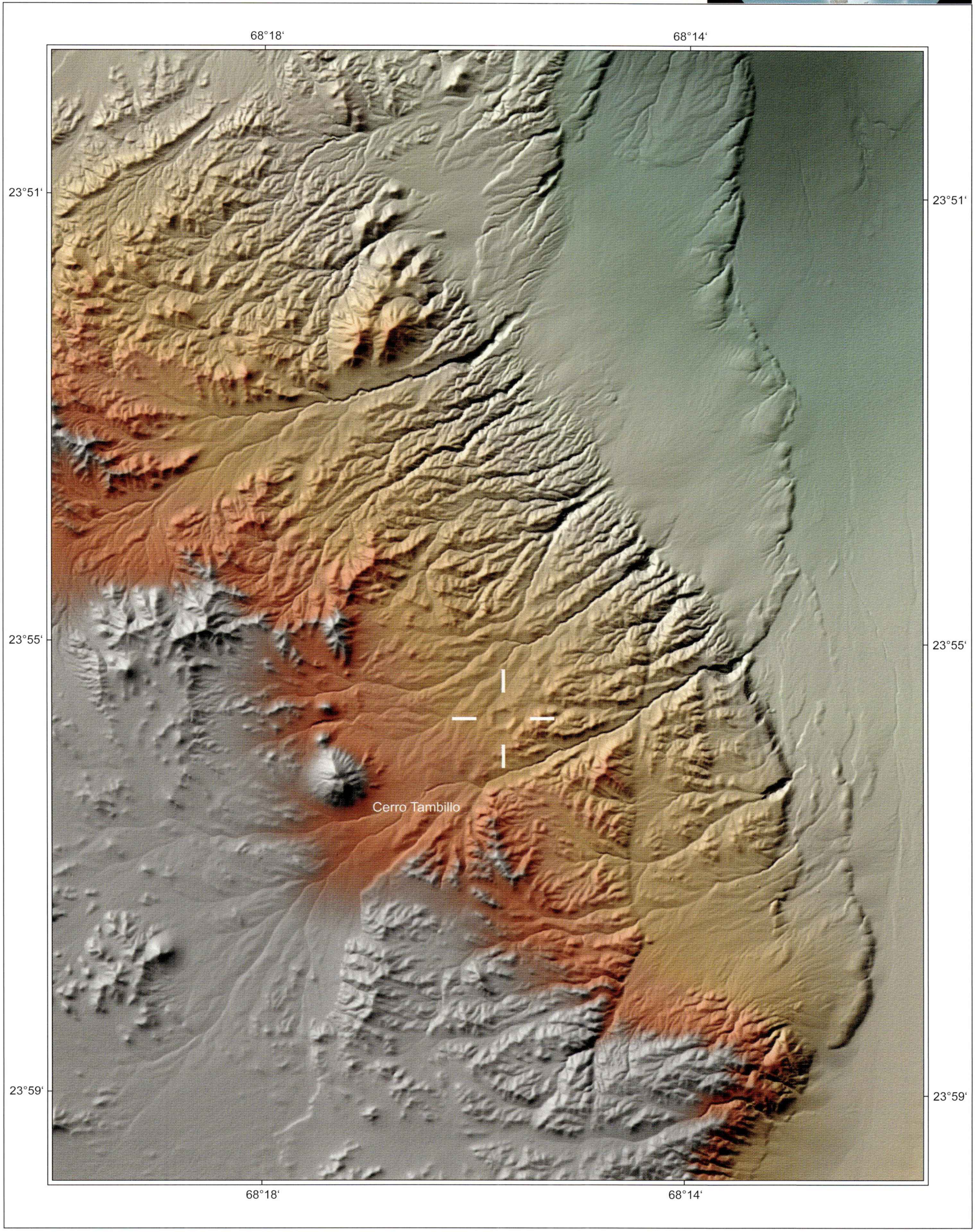

scale = 1:78 000 / scene = 14.0 × 18.6 km²

0 0.78 3.90 km

N

2355 meter 3415

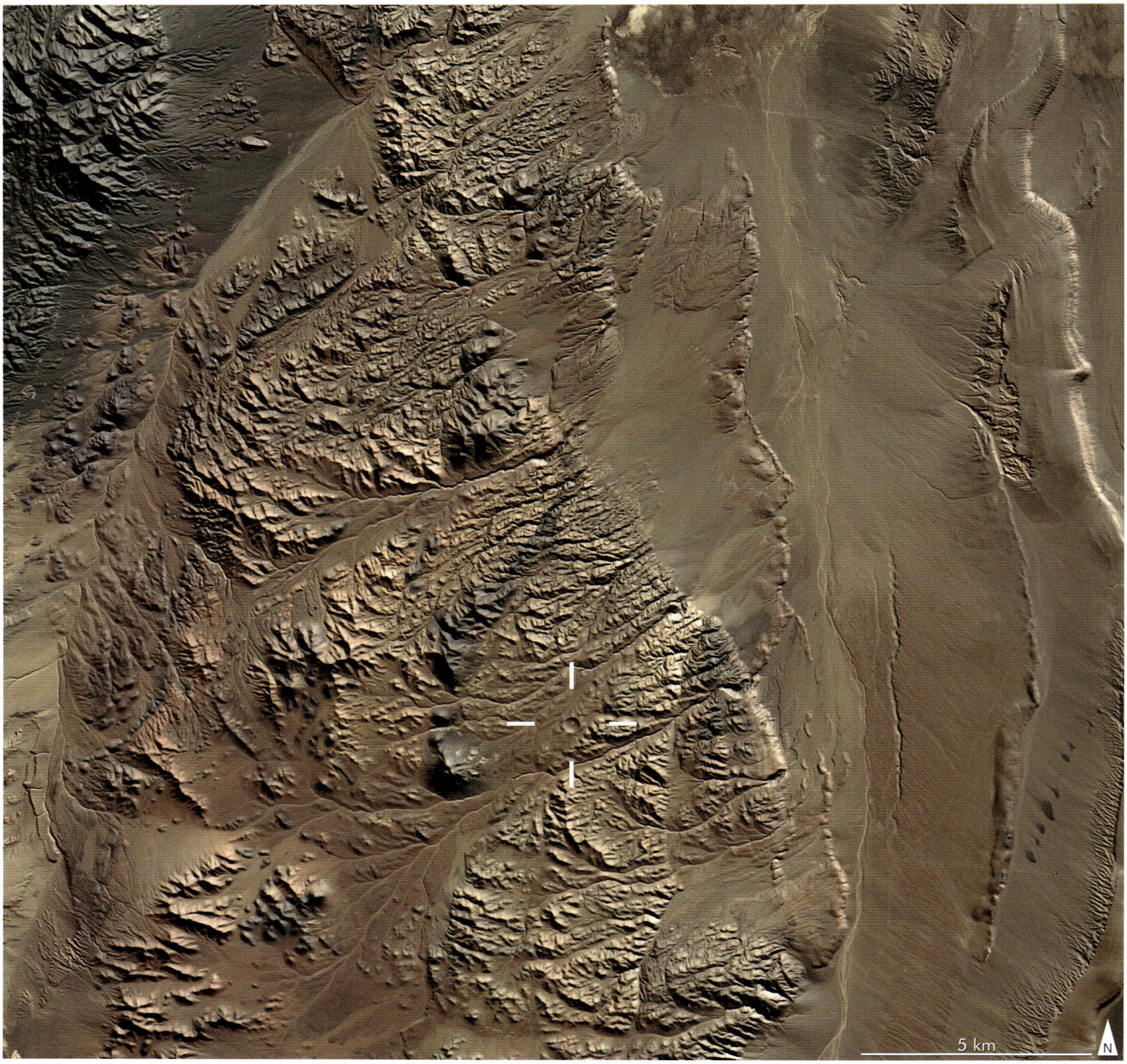

The Monturaqui crater and environs on June 23, 2018 as a composite image based on TanDEM-X DEM and Copernicus Sentinel-2 (channel 2/3/4) data. The location of the impact crater is illustrated by cross hairs.

Rathbun, K., Ukstins-Peate, I., Drop, S., and Gutierrez, F. (2016). A new geologic map of Monturaqui meteorite impact crater, Chile: Implications for satellite-based geologic mapping of small craters. 47th Lunar and Planetary Science Conference, Lunar and Planetary Institute, Houston, Texas, Abstract #2583.

Sanchez, J., and Cassidy, W. (1966). A previously undescribed meteorite crater in Chile. Journal of Geophysical Research, Vol. 71, pp. 4891–4895.

Ugalde, H., Valenzuela, M., and Milkereit, B. (2007). An integrated geophysical and geological study of the Monturaqui impact crater, Chile. Meteoritics and Planetary Science, Vol. 42, pp. 2153–2163.

Valenzuela, M., Rochette, P., Bourlès, D. L., Braucher, R., Faestermann, T., Finkel, R. C., Gattacceca, J., Korschinek, G., Merchel, S., Morata, D., Poutivtsev, J., Rugel, G., and Suavet, C. (2009). The age of Monturaqui impact crater. 72nd Annual Meeting of the Meteoritical Society, Nancy, France, Abstract #5185.

Riachão
Brazil

Wolf Uwe Reimold

The 4.1 km wide Riachão impact structure is located in Maranhão State of northeastern Brazil. The structure was formed on flat-lying Upper Carboniferous to Permian sedimentary rocks of the Balsas Group of the Parnaíba Basin. A generally flat to gently undulating relief, with a dendritic, radial fluvial drainage pattern, characterizes the structure. It can be subdivided into the rim, an inner ring basin of ca. 1.5 km maximum width, and a small central uplift structure that is located off-center, to the southeast, with respect to the surrounding ring basin. The diameter of the central uplift as estimated from geophysical data amounts to about 1.4 km. The overall horseshoe-like morphology is open in the northwest. The rim is elevated about 50 m above the environs of the structure, and the maximum elevation of the central uplift is only 30 m above the inner ring basin. Most of the area is covered by soil, vegetation, and alluvial sediment, and only a few scattered outcrops of sandstone and a sedimentary breccia of the Pedra de Fogo Formation occur in the structure and its environs. The scattered outcrops on the elevated rim represent mainly lithic sandstone with sandstone intraclasts. The rock dips at angles larger than 40° toward the northwest. In some places on the rim a clastic breccia with sandstone clasts of 0.2–10 cm sizes occurs, which is now interpreted as a sedimentary breccia of the Pedra de Fogo Formation. In the annular depression occur a few outcrops of laminated siltstone and calcilutite; these strata are dipping gently at 10°–20° toward east-northeast–west-southwest and northwest–southeast. At the transition to the central uplift, brecciated chert is observed. These lithologies also belong to the Pedra de Fogo Formation.

The central uplift is composed of a fine- to medium-grained, seemingly massive sandstone, with steep dips into various directions. This sandstone is related to the lowermost stratigraphic unit in the Riachão structure, the Piauí Formation. The Riachão structure seems to be deeply eroded, as strongly suggested by its subdued topography.

No shatter cones have been found at Riachão. Microscopic deformation in quartz of thin sections of sandstone from the central uplift mostly comprises features indicative for low shock deformation. Only one quartz grain from a sample of Piauí sandstone exhibited basal planar deformation features (PDFs). Otherwise feather features (FFs) and planar fractures (PFs) have been observed.

Geophysical studies at Riachão comprise aerial gamma-spectrometry, gravity, and magnetic data. A well-defined gamma-ray signature characterizes the rim and the central uplift. Whilst the rim displays low concentrations of radioactive elements, the core shows a high level of potassium (K), which has been related to the phyllosilicate cement of Piauí sandstones that were stratigraphically uplifted in the central uplift. A positive large-wavelength gravity anomaly of about +1 mGal covers the entire structure, whereas negative anomalies occur in the environs of it. The magnetic data indicate an elongated anomaly roughly trending northwest-southeast across the central uplift.

Direct radiometric age determination for the formation of the Riachão structure has not been possible so far, due to the absence of suitable melt-bearing material. Therefore, the age of the Permian Pedra de Fogo Formation of 271 to 299 Myr, apparently the youngest formation affected by the impact event, provides an unconstrained upper age limit for the impact event.

It has also been suggested that the formation of Riachão could perhaps be related to the formation of the **Serra da Cangalha** impact structure, which is located only about 40 km to the south-southwest, at the border to the neighboring state of Tocantins. However, to date there is no supporting evidence to validate this suggestion.

Riachão was first noted in 1966 by H. Ojeda and F. Bembom, who described the anomalous radial drainage pattern in this region but interpreted the structure as of cryptovolcanic origin. A possible impact origin was then suggested in 1978 by John McHone and Robert Dietz, based on photographs taken by the astronauts of the 1975 Apollo-Soyuz Test Project. McHone provided also the first geological map of Riachão, who referred to "impact breccia with chert clasts" on the rim and quartz grains with sets of PFs. However, bona fide evidence for an impact origin was only presented recently when Mariana Maziviero and co-workers published a detailed remote sensing investigation and assessment of the geometry and morphology of the structure in 2013.

Location/Access

The starting point for a journey to Riachão, and also to Serra da Cangalha, could be the capital of Tocantins State, the large city of Palmas. It has an airport with connections to many other Brazilian cities. It is also a place for car rental, with a four-wheel drive car required for access to both structures. To get as fast as possible to Riachão, state road TO-226 is recommended for some 500 km until the town of Araguaína, from where the journey should go east to Carolina, with a ferry transfer across the Tocantins River into Maranhão State, and on to the village of Riachão, for a total distance of some 250 km from Araguaína. From there, an unpaved road heading south will lead to the structure. Alternatively, coming from the community of Campos Lindos located a few kilometers northeast of the Serra da Cangalha impact structure, a rather poor unpaved road will take the intrepid traveller to Riachão. GPS navigation is highly recommended for both routes, as there are many minor roads leading to farms. It is now also possible to fly to Araguaína, for example from the cities of Brasília or Goiânia.

Selected References

Abrantes Jr., F. R. (2016). O Permo-Triássico da Bacia do Parnaíba, norte do Brasil: implicações paleoambientais, paleoclimáticas e paleogeográficas para o Pangea Ocidental. Ph.D. thesis, Universidade Federal do Pará, 130 pp.

Crósta, A. P., Reimold, W. U., Vasconcelos, M. A. R., Hauser, N., Oliveira, G. J. G., Maziviero, M. V., and Góes, A. M. (2019a). Impact cratering: The South American record – Part 1. Geochemistry – Chemie der Erde, Vol. 79, pp. 191–220.

Maziviero, M. V., Vasconcelos, M. A. R., Crósta, A. P., Góes, A. M., Reimold, W. U., and Carneiro, C. C. (2013). Geology and impact features of Riachão structure, northern Brazil. Meteoritics and Planetary Science, Vol. 48, pp. 2044–2058.

McHone, J. F. (1986). Terrestrial Impact Structures: Their Detection and Verification with Two New Examples from Brasil. Ph.D. thesis, University of Illinois at Urbana-Champaign, Urbana, Illinois, 210 pp.

McHone, J. F., and Dietz, R. S. (1978). Astroblemes in Brazil. Proceedings, 74th Geological Society of America Cordilleran Section Meeting, 116 pp.

Vasconcelos, M. A. R., Crósta, A. P., and Molina, E. C. (2010). Geophysical characteristics of four possible impact structures in the Parnaíba Basin, Brazil: Comparison and Implications. In: Large Meteorite Impacts and Planetary Evolution IV, Gibson, R. L., and Reimold, W. U. (eds.), Geological Society of America Special Paper, Vol. 465, pp. 201–217.

The Serra da Cangalha (bottom left) and Riachão (indicated by cross hairs) impact structures on August 31, 2018 as a composite image based on TanDEM-X DEM and Copernicus Sentinel-2 (channel 2/3/4) data.

Riachão

Brazil

latitude = 07°43‘00“ S / longitude = 46°38‘36“ W / diameter = 4.1 km / age = < 300 Myr

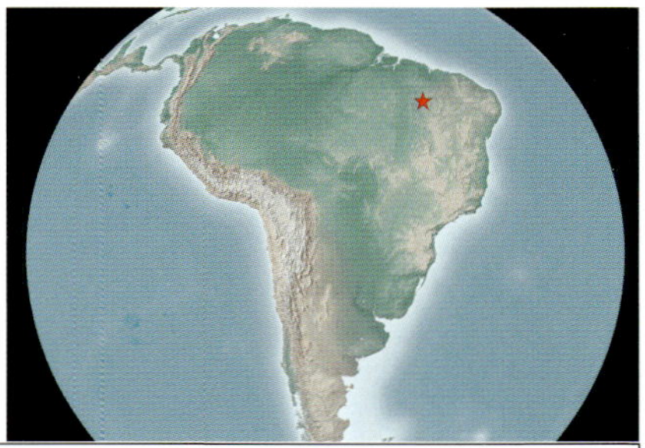

scale = 1:129 000 / scene = 23.2 × 31.0 km²

0 1.29 6.45 km

N

245 meter 380

Santa Marta
Brazil

Wolf Uwe Reimold

The Santa Marta impact structure is located in the southwest part of Piauí State in northeastern Brazil. It is a reasonably well preserved, complex impact structure of 10 km diameter. Santa Marta has a prominent, raised rim around the eastern and southern part of the structure, as well as a 2 km wide central uplift.

The Santa Marta structure was formed in sedimentary strata of the Parnaíba Basin as well as of the Sanfranciscan Basin. In this region the sequences of these two different sedimentary basins overlap. The structure is composed of rocks belonging to the Ordovician-Silurian Serra Grande and Devonian Canindé groups – both related to the Parnaíba Basin, and the Cretaceous Areado (Abaeté, Quiricó and Três Barras formations) and Urucuia (Posse Formation) groups related to the Sanfranciscan Basin. An undeformed detrital cover unconformably overlies these sequences mainly in the western and south-central part of the structure, as well as the basin strata to the west of it. This cover has been correlated with the regionally occurring Cenozoic Chapadão Formation.

Besides the effect of differential impact-induced stress over the interior of the structure, the occurrence of different rock types has been held responsible for the different styles of deformation from the center of the structure to the crater rim. The inner rings within the Santa Marta structure are interpreted as a direct consequence of the variation in lithologies. Four distinct morpho-structural domains can be recognized: Outer rim, annular basin, and a central elevation that is further divided into two distinct sub-domains: elevated plateau and central uplift. The central elevated plateau comprises the detrital cover that partially overlies the deformed, e. g., shatter-cone bearing, Paleozoic units of the central uplift.

From the edge of the central elevated plateau outwards, there are Devonian sandstones and pelites of the Cabeças and Pimenteiras formations, respectively. These strata occur up to the concentric but discontinuous inner rings formed by Silurian conglomeratic sandstones and conglomerate of the Serra Grande Group. In the inner part of the annular basin, laterally juxtaposed, intensely deformed Silurian-Devonian pelitic and Cretaceous rudaceous strata can be observed. Juxtaposition took place along shear zones. Along the outer edge of the structure deformed sandstones of the Posse Formation and the Três Barras Formation are found.

There are two types of lithic impact breccia in the Santa Marta structure: polymict breccias and, comparatively less abundant and more limited in distribution, monomict breccias. Only the monomict sandstone breccia and one of the numerous samples of polymict breccia from the top of Sinésio Hill, which is where the cover rocks of the central elevated plateau are best exposed, contain shocked mineral grains. It is thought that these shocked particles in the polymict breccia sample are detritus that was incorporated into fluvial

Part of the plateau in the western part of the Santa Marta structure, formed by polymict lithic breccia. Field of view about 500 m (photo: W. U. Reimold).

The uppermost part of the plateau in the Santa Marta impact structure (Sinesio's Hill), which is formed by Cenozoic, post-impact sediments. Width of field of view about 200 m (image courtesy: A. Crósta).

Santa Marta

Brazil

latitude = 10°10‘19“ S / longitude = 45°13‘58“ W / diameter = 10 km / age = < 93 Myr

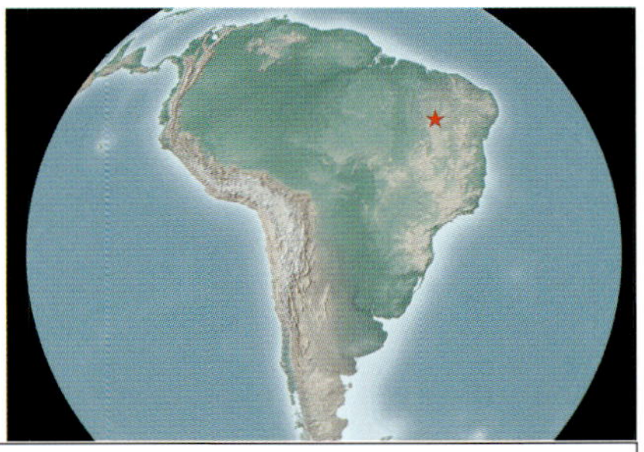

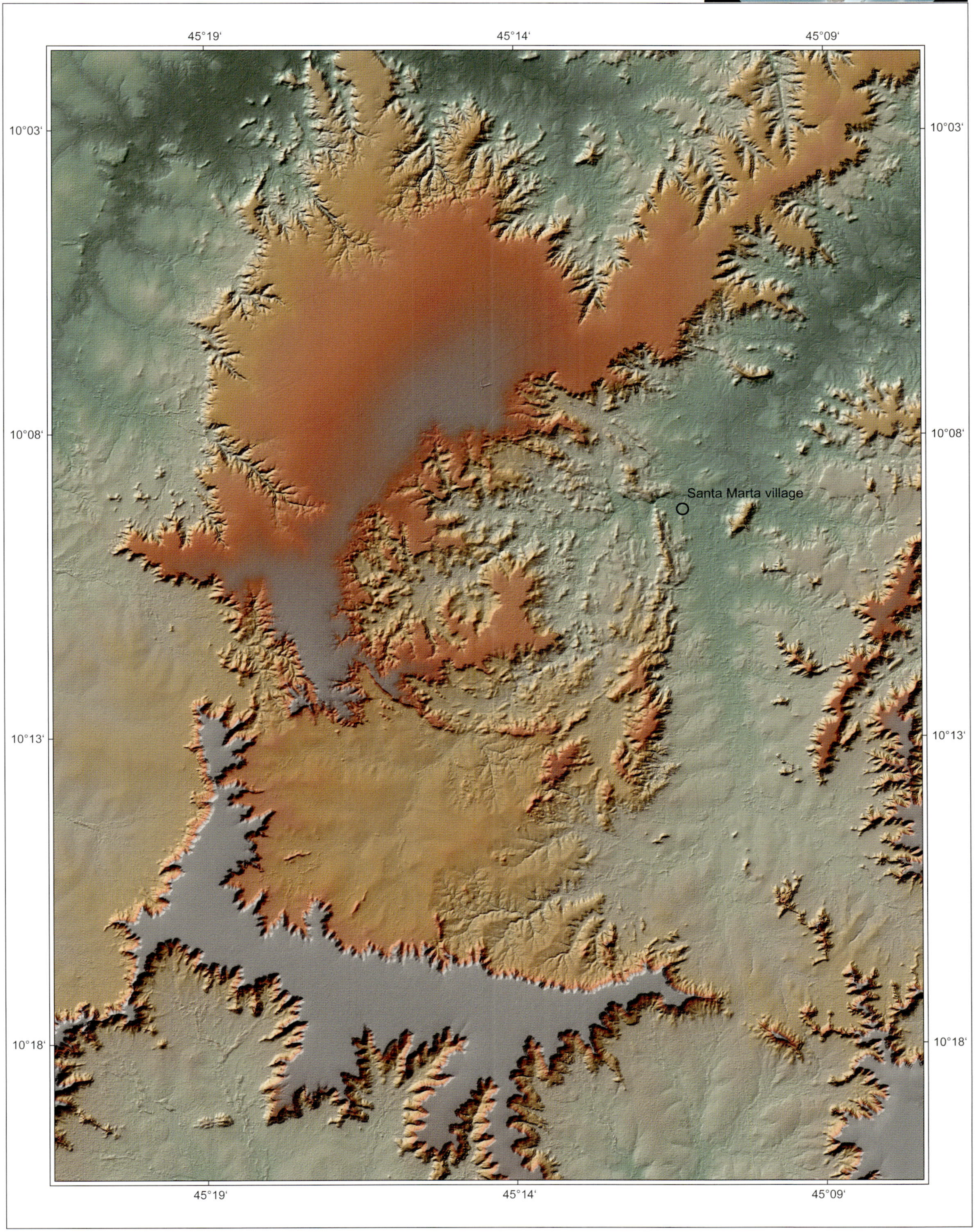

scale = 1 : 142 000 / scene = 25.6 × 34.2 km²

0 1.42 7.10 km

N

425 meter 795

At the base of Sinesio's Hill: contact between the underlying polymict impact breccia of the base of the central uplift with overlying, likely fluviatile, Cenozoic cover rock. This sediment is also polymict with respect to its clast content but essentially does not carry many shocked particles (image courtesy: A. Crósta).

Shatter cones in ferruginous sandstone from the base of Sinesio's Hill in the western part of the Santa Marta impact structure. The specimen is about 15 cm wide (image courtesy: A. Crósta).

cover deposits. Shatter cones can be found in sandstones and clasts of monomict sandstone breccia that have been tentatively related to the Cretaceous Areado Group at several locations in the central part of the structure. Shatter cones have also been detected in mostly well-rounded quartzite pebbles and blocks that occur as large clasts in conglomerates – possibly related to the Cretaceous Abaeté Formation – in the annular basin near the outer edge of the structure.

Planar deformation features (PDFs) were observed in quartz grains of polymict impact breccia samples from the central part of the structure. Feather features (FFs) appear commonly associated with PDFs in intensely deformed sandstone, with these occurrences increasing in frequency towards the center.

A ground gravimetric survey of Santa Marta revealed a northeast-southwest oriented negative anomaly partially coinciding with the Cenozoic cover of the central elevated plateau, and a positive anomaly of about +2 mGal in the central-eastern region. The eastern rim of Santa Marta is characterized by a distinct positive anomaly. Considering a diameter of 10 km for Santa Marta, an uplift in the center of about 1000 m is suggested, in accordance with the estimated depth to crystalline basement on the basis of modelling of the regional magnetic and gravity data.

Santa Marta: shatter cone in quartzite, a 16 × 11.5 × 8 cm specimen (photo courtesy: Ralf-Thomas Schmitt, Museum für Naturkunde Berlin, sample from W. U. Reimold).

The upper age limit for the Santa Marta impact event is estimated at 93–100 Myr. It is based on the stratigraphic age of the sandstones of the Posse Formation of the Urucuia Group, which occur along the rim of the structure and represent the youngest formation affected by the impact event.

Santa Marta was mentioned as a potential impact structure in 2000 by Sharad Master and J. Heymann, based on the circular geomorphology observed by remote sensing. They suggested the name "Gilbués" after a nearby town of that name. This name was revised in 2010 to "Santa Marta structure" by Marcos Vasconcelos and his team in accordance with the name of the village located directly at the eastern rim of the structure. However, the confirmation of the impact origin for Santa Marta had to await four more years until Grace de Oliveira and co-workers recorded abundant evidence of shock deformation in the form of shatter cones and shock micro-deformation effects in quartz.

Location/Access

The Santa Marta structure is located just west of Piauí state road 135, about 30 km north of the town of Corrente, and some 50 km south of the town of Gilbués, in the southern part of Piauí state of Brazil. To reach Corrente, it is recommended to fly into the town of Barreiras, which has a national airport, and follow the R135 for some 240 km in northerly direction. Barreiras is served by two airlines from Brasília and Salvador. Once the village of Santa Marta has been reached, a four-wheel drive car can travel directly into the structure, but roads there are poor and few. To explore the structure, a guide with horses or burros may be found at a local farm.

Selected References

Crósta, A. P., Reimold, W. U., Vasconcelos, M. A. R., Hauser, N., Oliveira, G. J. G., Maziviero, M. V., and Góes, A. M. (2019a). Impact cratering: The South American record – Part 1. Geochemistry – Chemie der Erde, Vol. 79, pp. 191–220.

Master, S., and Heymann, J. (2000). A possible new impact structure near Gilbués in Piauí Province, Northeastern Brazil. Meteoritics and Planetary Science, Vol. 35, Supplement, p. A105.

Oliveira, G. J. G. de, Vasconcelos, M. A. R., Crósta, A. P., Reimold, W. U., Góes, A. M., and Kowitz, A. (2014). Shatter cones and planar deformation features confirm Santa Marta in Piauí State, Brazil, as an impact structure. Meteoritics and Planetary Science, Vol. 49, pp. 1915–1928.

The Santa Marta impact structure on May 25, 2018 as a composite image based on TanDEM-X DEM and Copernicus Sentinel-2 (channel 2/3/4) data.

Oliveira, G. J. G. de, Chamani, M. A. C., Góes, A. M., Crósta, A. P., Vasconcelos, M. A. R., and Reimold, W. U. (2017). Geological investigation of the central portion of the Santa Marta impact structure, Piauí State, Brazil. Brazilian Journal of Geology, Vol. 47, pp. 673–692.

Vasconcelos, M. A. R., Crósta, A. P., and Molina, E. C. (2010). Geophysical characteristics of four possible impact structures in the Parnaíba Basin, Brazil: Comparison and Implications. In: Large Meteorite Impacts and Planetary Evolution IV, Gibson, R. L., and Reimold, W. U. (eds.), Geological Society of America Special Paper, Vol. 465, pp. 201–217.

São Miguel do Tapuio
Brazil

Wolf Uwe Reimold

The 21 km wide São Miguel do Tapuio structure is located in northern Piauí State, northeastern Brazil. Like the **Serra da Cangalha**, **Riachão**, and **Santa Marta** impact structures, São Miguel do Tapuio is hosted by sedimentary strata, mostly sandstone, of the Parnaíba Basin. The structure has been named after the nearest town of that name. On satellite imagery the structure exhibits a rugged morphology over its entire interior, in contrast to the surrounding terrain. There is also a distinctly annular and radial drainage pattern. The structure has a raised rim and several concentric inner rings. This rim is elevated by some 120 m over the surrounding terrain; the width of the central ring amounts to about 5 km. The strata there rise up to 150 m above the environs of the structure.

São Miguel do Tapuio has been known since the 1980s. Its potential origin as an impact structure has been based on observations regarding a morphology comprising an outer rim, inner ring structures, and a central peak area. It was also noted that some of the sandstones of the interior seemed structurally deformed. These strata belong mainly to the Cabeças Formation, but some strata of the Pimenteiras Formation (both are of Devonian age) also occur in the southeastern sector of the structure. In contrast to these rocks in the interior, none of these sedimentary strata are deformed or recrystallized outside the structure.

There are a number of publications that claimed to report shock micro-deformation, but none of this was substantiated. Quartzose sandstones and quartz-pebble conglomeratic sandstones with abundant fractures and micro-fracturing were described. Planar fractures (PFs) so far have been observed very rarely. Neither breccias nor shatter cones, or melt rock, were found. However, most samples analyzed so far were not derived from the innermost part of the structure, with exception of some samples by Jackson A. Martins and co-workers of 2010. However, their claim to have identified (PFs) and feather features (FFs) did not withstand careful inspection of their imagery.

During a recent field campaign by Alvaro Crósta and Marcos Vasconcelos in October 2017, deformed sandstones were sampled at several locations at the central peak. Breccia-like material was identified for the first time, and a possible shatter cone was collected. First petrographic analysis of thin sections of monomict sandstone breccia by Wolf Uwe Reimold unveiled PFs and FFs in numerous micro-clasts in two samples, as well as very rare planar deformation features (PDFs). This confirms São Miguel do Tapuio as the second-largest impact structure known from South America.

Geophysical work was done over the structure in order to investigate whether there could be a hidden magmatic body. Neither did these results essentially support an impact origin for this structure nor has an igneous intrusion been found. Low-resolution regional airborne geophysical data showed patterns not exactly in line with expectations for a large complex impact structure, but also not impossible depending on the local geological scenarios. More detailed gravimetry with higher resolution gave a Bouguer anomaly map with significant resemblance to a corresponding digital elevation model, both of which portray well the horseshoe-shape of the structure.

The age for this impact event is not known yet. Datable impact melt rocks have not been observed yet, so that the only constraint on the maximum age of the structure is stratigraphic. The youngest strata affected by the impact at São Miguel do Tapuio are likely Carboniferous. Thus, an upper limit of 300 Myr can be assigned to this impact event.

Location/Access

The town nearest to the structure is the homonymous São Miguel do Tapuio, where simple lodging can be found. The structure is accessible via the RN 407 road that passes along its eastern side. There are subsidiary roads that permit access to the crater rim, and a north-south trending track crosses the structure just east of the central elevated area. A four-wheel drive vehicle is highly recommended. Access to the central elevated area is difficult, as there are no roads or trails. The rugged topography, combined with high temperatures and very dry weather, also make hiking difficult. Anyone venturing to hike to the central area should look for a local guide.

Selected References

Crósta, A. P., Reimold, W. U., and Vasconcelos, M. A. R. (2019). Cerro Do Jarau and São Miguel do Tapuio: Two newly confirmed large impact structures in Brazil. 50th Lunar and Planetary Science Conference, Lunar and Planetary Institute, Houston, Texas, Abstract #3042.

Crósta, A. P., Reimold, W. U., Vasconcelos, M. A. R., Hauser, N., Oliveira, G. J. G., Mazivieiro M. V., and Góes, A. M. (2019a). Impact cratering: The South American record – Part 1. Geochemistry – Chemie der Erde, Vol. 79, pp. 1–61.

Martins, J. A., Castelo Branco, R. M. G., Castro, N. A., Peaulvast, J-P., and Lima, Jr. S. B. (2016). Morphological analysis of the São Miguel do Tapuio circular structure, Piauí – Brazil. Revista Brasileira Geomorfologia, Vol. 17, pp. 763–781.

Vasconcelos, M. A. R., Crósta, A. P., and Molina, E. C. (2010). Geophysical characteristics of four possible impact structures in the Parnaíba Basin, Brazil: Comparison and Implications. In: Large Meteorite Impacts and Planetary Evolution IV, Gibson, R. L. and Reimold, W. U. (eds.), Geological Society of America Special Paper, Vol. 465, pp. 201–217.

One of the rare grains with shock deformation so far observed in sandstone samples from the São Miguel de Tapuio impact structure. Cross polarized light (photo: W. U. Reimold).

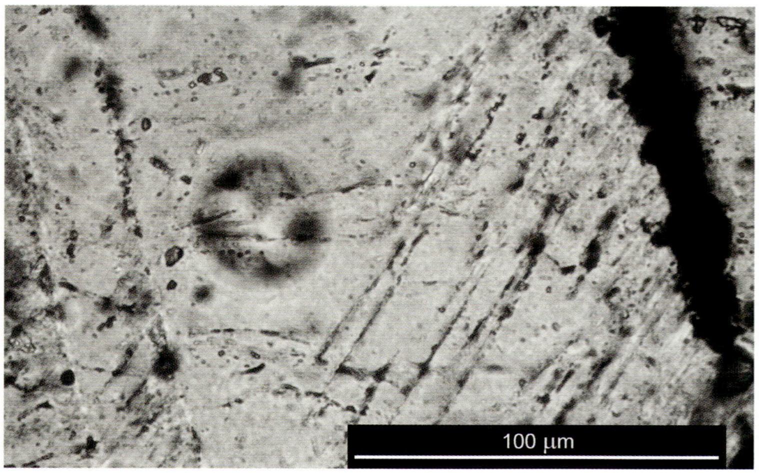

Short PDFs in another quartz grain from a São Miguel de Tapuio sandstone specimen. Cross polarized light (photo: W. U. Reimold).

São Miguel do Tapuio

Brazil

latitude = 05°37‘36“ S / longitude = 41°23‘18“ W / diameter = 21 km / age = < 300 Myr

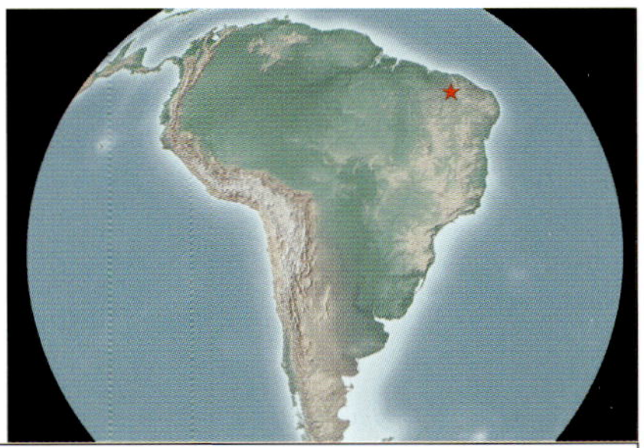

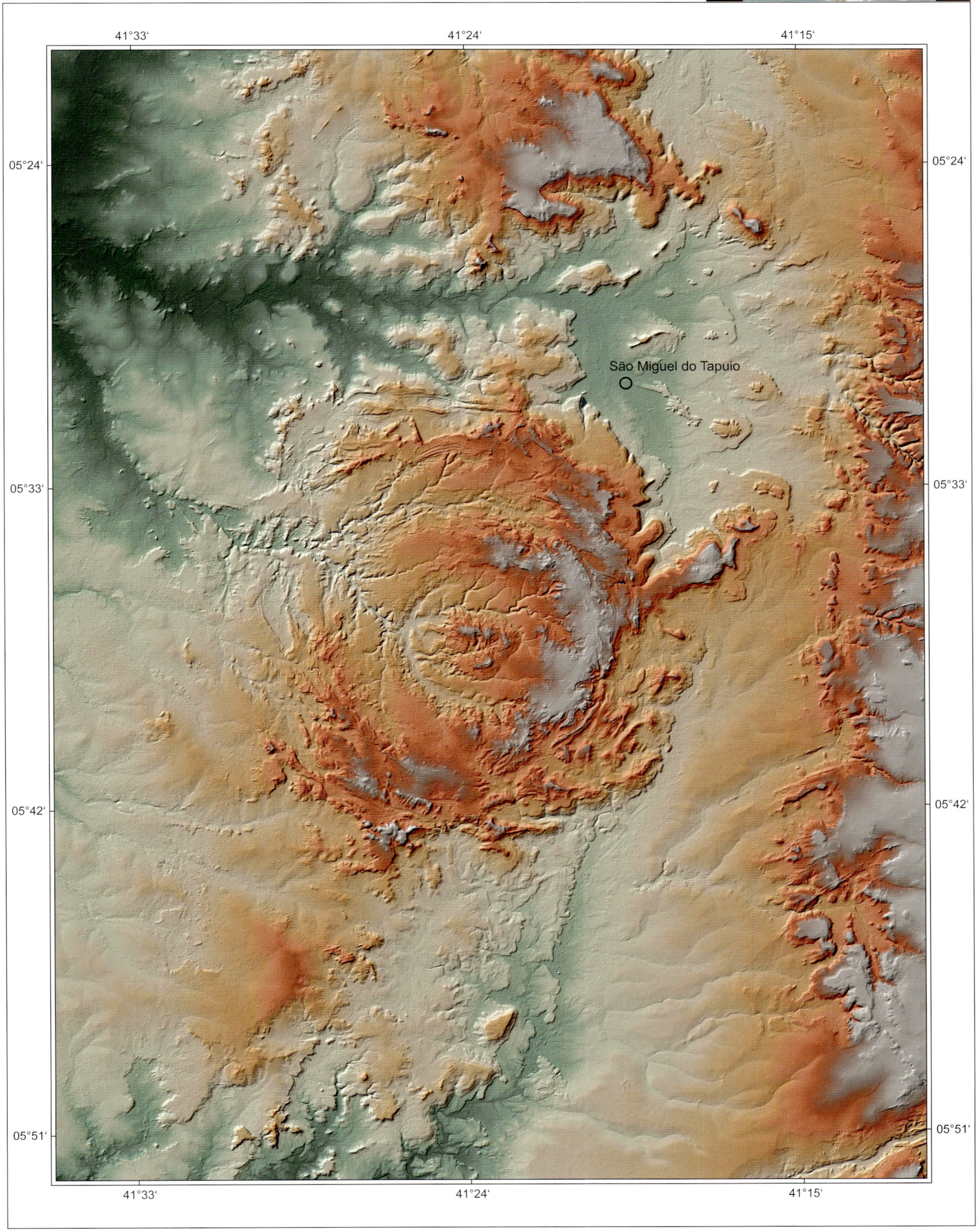

scale = 1 : 242 000 / scene = 43.6 × 58.1 km²

0 2.42 12.10 km

N

185 meter 600

The São Miguel do Tapuio impact structure on September 9, 2018 as a composite image based on TanDEM-X DEM and Copernicus Sentinel-2 (channel 2/3/4) data.

Serra da Cangalha
Brazil

Wolf Uwe Reimold

The Serra da Cangalha ("Horse Collar Range" in Portuguese) impact structure has an intriguing morphology. This 13.7 km wide, complex impact structure is located in the northeasternmost part of Tocantins State of east-central Brazil. The structure was formed in Paleozoic to Mesozoic sedimentary rocks of the Parnaíba Basin; it is considerably eroded but still shows a remarkable morphology. From the center outward several morpho-structural features are distinguished: a prominent inner ring of 3.2 km diameter around a central depression 2.2 km wide; a first intermediate ring feature of 5.8 km diameter; an incomplete second intermediate ring of 11 km diameter; and the outer ring which corresponds to the crater rim of 13.7 km diameter. The ring-like features are caused by selective erosion of tilted strata. A prominent collar of silicified sandstone surrounds the innermost depression. This collar is up to 350 m high and forms the most prominent morphological feature of the central uplift. The sandstone strata of the collar are steeply dipping and partly overturned in the northern and western sectors and strongly folded at the meter- to hundred-meter scale. This intensely silicified sandstone collar has been significantly more resistant to erosion than the surrounding rocks, thus forming a buttress standing out in the regional landscape. It is responsible for the name of the structure.

The periphery of the central uplift of Serra da Cangalha corresponds to the morphological ring feature at 5.8 km. This periphery is formed by the Carboniferous Piauí Formation, followed inward by the prominent collar of silicified sandstone strata of the equally Carboniferous Poti Formation, and the central depression is underlain by mostly shale and some siltstone of the Devonian Longá Formation. The Longá rocks are much more susceptible to weathering and erosion than the other siliciclastic rocks, which has caused the formation of the central depression. Information about the strata of the Longá Formation comes mostly from drill cores obtained in the course of diamond exploration in the 1970's.

The annular basin around the central uplift has ample occurrences of strata of the Pedra de Fogo Formation. These are mainly brecciated chert layers, as well as silicified sandstone and minor mudstone. In some areas, such as in the northeast and west, the Pedra de Fogo rocks exhibit centimeter- to meter-scale open folds. The rim of the structure is formed by flat-topped, topographically high, plateaus. Here, mostly rocks, i.e., laminated mudstone with abundant chert and subordinate coarse- to medium-grained and well sorted pink sandstone, of the Pedra de Fogo Formation occur, with some occurrences of underlying sandstones of the Piauí Formation. The Piauí Formation has frequent intercalations of purple/gray mudstone and fine- to medium-grained, red sandstone, usually with parallel bedding or some incipient sigmoidal cross-bedding. Beyond the rim the attitude of the strata is nearly horizontal, and rocks do not display significant deformation besides some gentle warping. In the northwestern sector of the structure, at a road cut that intersects the rim, a prominent concentrically trending monocline in sandstone of the Piauí Formation can be observed, whose inner limb dips gently toward the crater interior.

The structure has been strongly investigated with geophysical methods. Magnetotelluric data suggest that the deformation at Serra da Cangalha was restricted to a depth of 2 km. Airborne gamma-ray data show a well-defined K-Th-U (potassium-thorium-uranium)

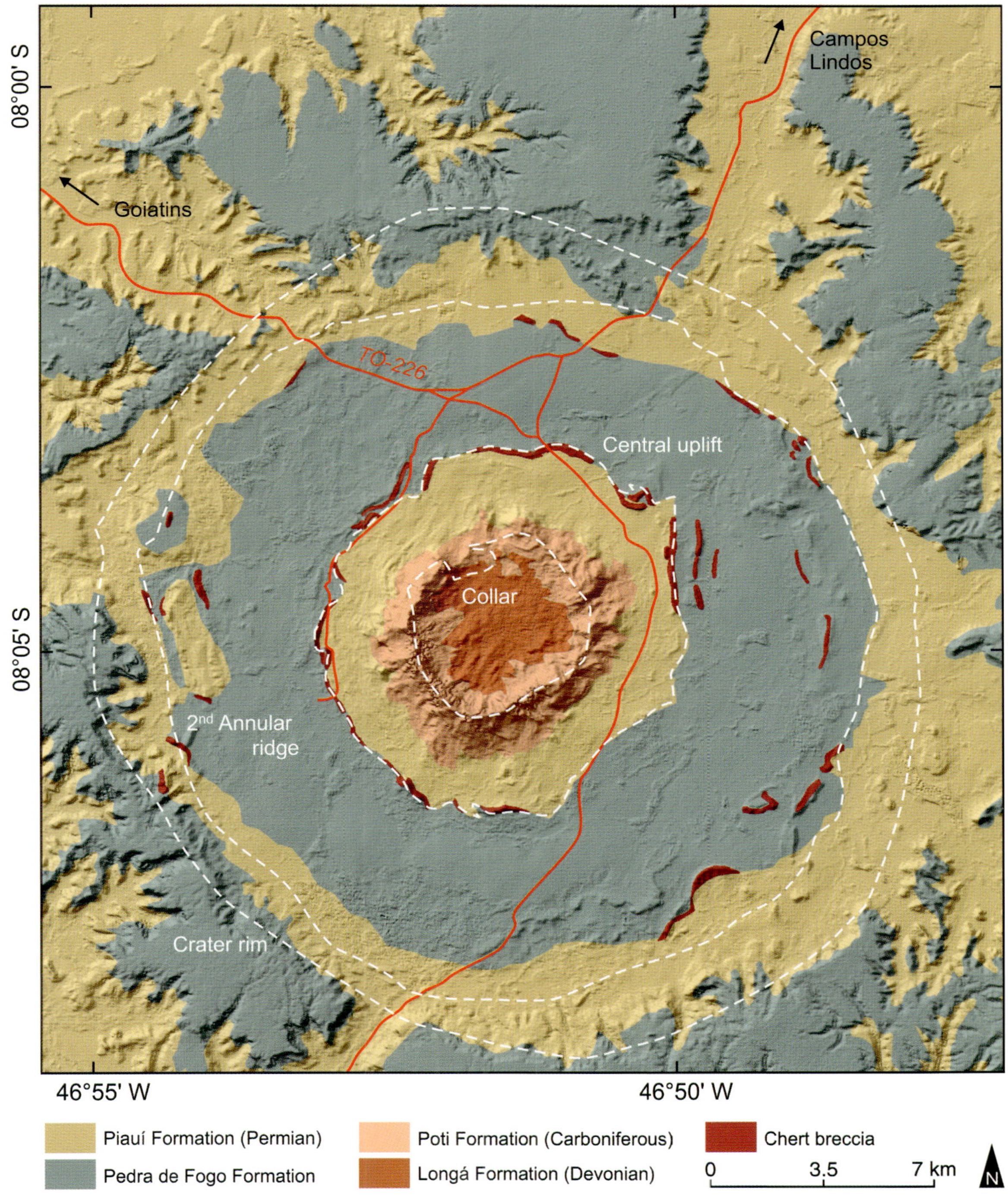

Geological map of the Serra da Cangalha impact structure underlain by a TanDEM-X DEM (modified after Kenkmann et al., 2011 and Vasconcelos et al., 2013).

Aerial photograph of the so-called "collar" of the central uplift of the Serra da Cangalha impact structure. View towards the south (image courtesy: A. Bartorelli, provided by A. Crósta).

anomaly over the annular basin around the central uplift. An estimation of the average depth to basement, based on regional aeromagnetic data, yielded a depth of 2.4 km. High resolution magnetic data indicated magnetic sources at 200–500 m depth below the center of the structure, and even deeper magnetic sources below the entire structure, at around 1.9 km depth. The Bouguer gravity anomaly map is characterized by a weakly positive anomaly of +1 mGal over the central area, and a low gravity signature of −2 mGal at the northwestern rim. Gravity modeling suggested disturbed basement in the innermost part of the structure to a depth of 2–3 km.

Shock deformation features include shatter cones in coarse-grained sandstone from two locations in the central depression. The shatter cones are smaller than 1 decimeter in size. The sandstone likely belongs to the Longá Formation. Also at a few locations in the central depression and in the periphery of the central uplift rare monomict and polymict breccias have been found. Feather features (FFs) were reported in quartz grains of sandstones from the Longá and Poti formations, including some quartz grains from shatter cones. Planar fractures (PFs) were described in quartz grains from the same stratigraphic units, in polymict breccias and in shatter cones; whilst planar deformation features (PDFs) occur in coarse sandstone from the drill cores through Longá Formation in the central depression, and in shatter cones. Based on the preserved shock features in the central uplift that suggest peak shock pressures of not more than 10 GPa, and the apparent lack of allochthonous breccia fill in the crater moat, Thomas Kenkmann and co-workers estimated about 500 m of erosion since the time of impact.

The age of Serra da Cangalha has been delimited to less than 270 Myr, the age of the youngest strata from the mid-Permian affected by the formation of the structure.

The discovery of the Serra da Cangalha structure dates back to the early 1970's. As part of a regional exploration program for industrial diamonds, three boreholes were drilled at the center of this circular structure in order to investigate its possible igneous intrusive origin – the structure was suspected to represent a possible kimberlite. First reference to a pos-

Strata of the collar of Serra da Cangalha show prominent radial folding. The fold axis plunges steeply towards the outside in its lower part and bends to an overturned orientation in its upper part (photo: T. Kenkmann).

Gently inward dipping beds of the Piauí Formation of the Serra da Cangalha impact structure. The strata are gently folded at the meter scale. Viewing direction is from the northwestern crater syncline towards the central uplift (photo: T. Kenkmann).

Serra da Cangalha

Brazil

latitude = 08°04'45" S / longitude = 46°51'25" W / diameter = 13.7 km / age = < 300 Myr

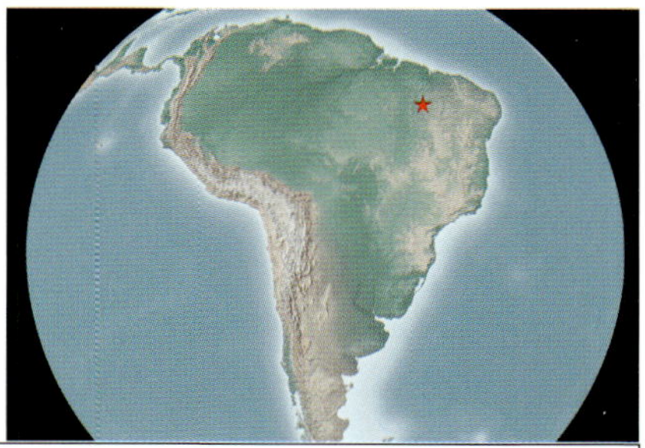

scale = 1 : 190 000 / scene = 34.2 × 45.5 km²

0 1.90 9.50 km

N

195 meter 660

sible impact origin was made by Robert Dietz and Bevan French in 1973 based on the circular morphology of the structure documented by some of the first images of the Landsat satellite. Reconnaissance fieldwork followed. Later, in 1986, John McHone reported the anomalous local deformation of the rocks and low-shock pressure deformation features such as intense grain fracturing and curvilinear features in quartz. However, only recently in the course of a geological mapping campaign of the structure, bona fide evidence for the impact origin of Serra da Cangalha was presented by the teams of Thomas Kenkmann in 2011 and Marcos Vasconcelos in 2013, with the discovery of unambiguous shock deformation features.

Location/Access

The Serra da Cangalha structure is located in the Brazilian state of Tocantins, directly at the northeastern border to the neighboring state of Maranhão. A likely starting point for exploration of this structure will be the capital of Tocantins, the large city of Palmas, which boasts excellent air travel connections. Taking the best roads, the journey should go from Palmas to Guaraí on the TO-226, and from there on the TO-431 and TO-10 roads to Goiatins, and then on the TO-226 to Serra da Cangalha and the small town of Campos Lindos. The total distance is roughly 500 km. The TO-226 skirts the impact structure on its northern side, and Campos Lindos is located about 10 km towards the northeast. Here, simple hotel accommodation can be found. Alternatively, flights can be taken, for example from Brasília or Goiânia, to the town of Araguaína, from where good road access is available to Campos Lindos.

Selected References

Crósta, A. P., Reimold, W. U., Vasconcelos, M. A. R., Hauser, N., Oliveira, G. J. G., Maziviero, M. V., and Góes, A. M. (2019a). Impact cratering: The South American record – Part 1. Geochemistry – Chemie der Erde, Vol. 79, pp. 191–220.

Dietz, R. S., and French, B. M. (1973). Two probable astroblemes in Brazil. Nature, Vol. 244, pp. 561–562.

Kenkmann, T., Vasconcelos, M. A. R., Crósta, A. P., and Reimold, W. U. (2011). The complex impact structure Serra da Cangalha, Tocantins State, Brazil. Meteoritics and Planetary Science, Vol. 46, pp. 875–889.

McHone, J. F. (1986). Terrestrial Impact Structures: Their Detection and Verification with Two New Examples from Brasil. Ph.D. thesis, University of Illinois at Urbana-Champaign, Urbana, Illinois, 210 pp.

Vasconcelos, M. A. R., Leite, E. P., and Crósta, A. P. (2012). Contributions of gamma-ray spectrometry to terrestrial impact crater studies: the example of Serra da Cangalha, northeastern Brazil. Geophysical Research Letters, Vol. 39, L04306.

Vasconcelos, M. A. R., Crósta, A. P., Reimold, W. U., Goes, A. M., Kenkmann, T., and Poelchau, M. H. (2013). The Serra da Cangalha impact structure, Brazil: Geological, stratigraphic and petrographic aspects of a recently confirmed impact structure. Journal of South American Earth Sciences, Vol. 45, pp. 316–330.

Vargeão Dome

Brazil

Wolf Uwe Reimold

The Vargeão Dome impact structure, located in the western part of Santa Catarina State of southern Brazil, has a diameter of 12.4 km. It is named after the town of Vargeão that is located inside of the southern rim of the structure. Vargeão Dome is a complex impact structure with a central uplift of about 3 km diameter. Like the other two impact structures of the southern Paraná Basin, **Vista Alegre** and **Cerro do Jarau**, the structure was formed in igneous and sedimentary rocks of the São Bento Group (Pirambóia, Botucatu and Serra Geral formations). The structure exhibits distinct annular and radial fractures in the basalts of the Jurassic-Cretaceous Serra Geral Formation, as well as in the sandstone outcrops in the central part of the structure. There are exposures of four flows of Serra Geral basalts in the structure, with the lower three having basaltic composition, and the fourth and uppermost flow having a more felsic composition. On the escarpments of the southern rim, these volcanics are well exposed. Faulting along the rim resulted in large blocks

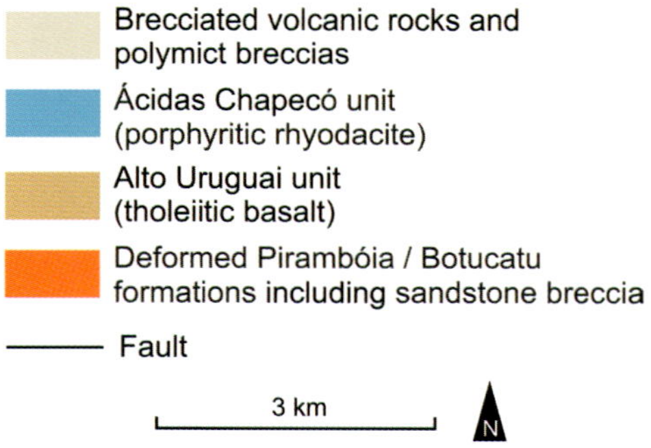

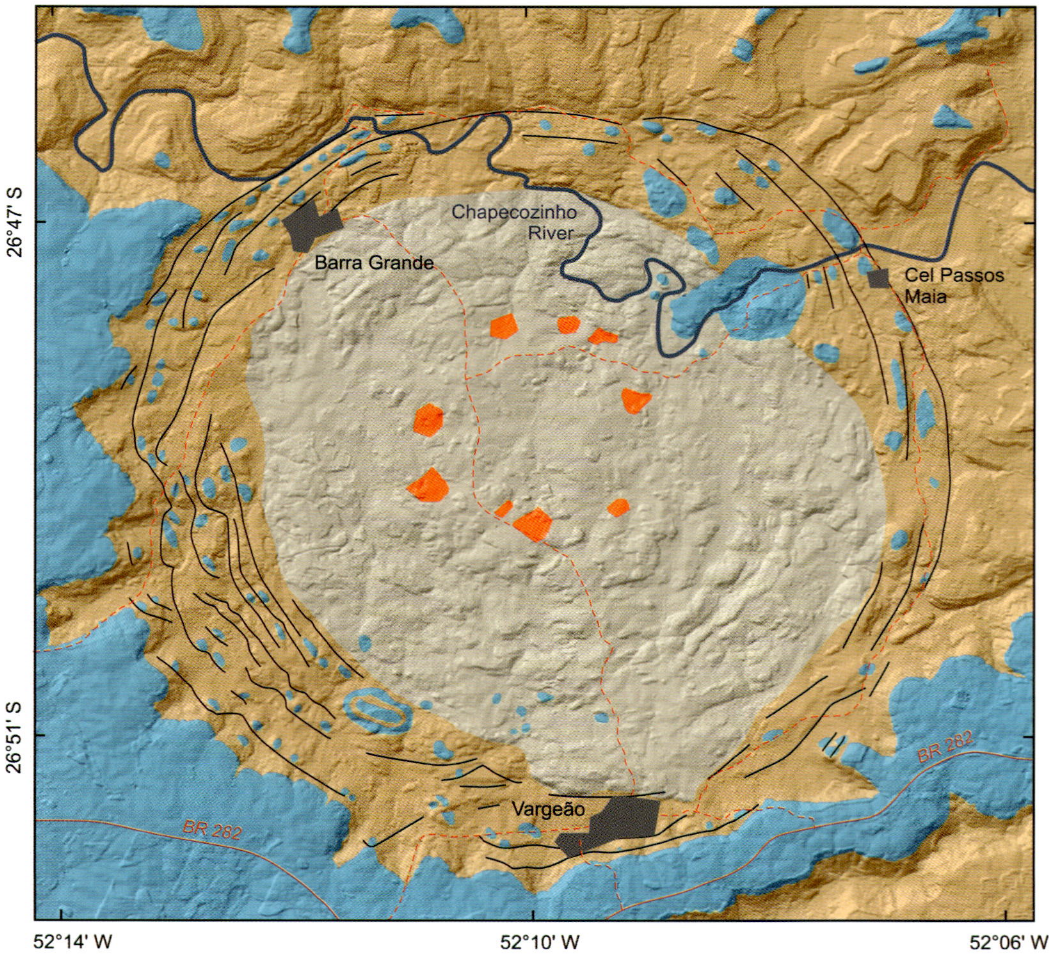

Schematic geological map of the Vargeão Dome impact structure, underlain by a TanDEM-X DEM (modified after Crósta et al., 2012b).

Vargeão Dome

Brazil

latitude = 26°48‘53“ S / longitude = 52°09‘49“ W / diameter = 12.4 km / age = < 137 Myr

scale = 1 : 143 000 / scene = 25.8 × 34.4 km²

0 1.43 7.15 km

N

520 meter 1130

Monomict impact breccia formed in tholeiitic basalt of the Vargeão impact structure. Hammer head for scale ca. 20 cm wide (image courtesy: A. Crósta).

Monomict impact breccia after basalt from Vargeão (image courtesy: A. Crósta).

of undeformed volcanic rocks slumping from the upper portions of the rim towards the interior of the structure, where they are currently found with tilted orientations.

Outcrop of polymict impact breccia formed from basalt and sandstone target rocks, Vargeão impact structure. Hammer, for scale, ca 35 cm long (image courtesy: A. Crósta).

Shatter cones formed on basalt from the Vargeão impact structure (image courtesy: A. Crósta).

The structure has the circular and multi-annular morphology due to extensive sub-vertical faulting that has been related to the modification stage of crater formation: uplift of the sandstone in the central area and rim collapse, followed by the gravitational collapse of the central uplift. The topography of Vargeão involves up to 150 m high escarpments between the outer rim and the inner portion of the structure.

Highly deformed sandstones occur in the central uplift – in structural contact with brecciated volcanics. The sandstones are related to the Pirambóia-Botucatu succession of the Late Permian to Early Cretaceous São Bento Group. The Botucatu eolian deposits rest unconformably on the fluvio-eolian deposits of the Pirambóia Formation, and the strata of both these formations are overlain by the Serra Geral volcanic flows. Cataclasis of the sandstones is such that it is not possible in the field or in thin section to differentiate between the two sandstone types.

Based on seismic data from oil exploration in the 1980's, Vargeão Dome appears as a relatively shallow crustal structure, as one would expect for an impact structure. The rim is marked by concentrically striking listric faults, and the central uplift is obvious. The sedimentary strata may have been uplifted by 1.1 km. Most of the rocks within the structure show some degree of deformation ranging from fracturing and brecciation to melting. The rock types in the interior of the structure are unshocked or slightly-to-moderately fractured basalt and rhyodacite in the outer parts of the central uplift and along the rim of the structure, and brecciated, fractured, or melted volcanic rocks and sandstones. There are also monomict breccias of all three major target rocks, and polymict breccias containing fragments of all three lithologies. Exposures of these breccias are, however, patchy.

The degree of brecciation varies between locations. The comprehensive alteration of matrices of breccia veinlets has precluded the identification of the original breccia matrix type. Locally, such breccia veinlets display fluidal texture and may comprise some dark glassy material. Geochemical analysis of such breccias, in comparison with compositions of target rocks outside the structure, shows that their compositions are intermediate to the basalt and rhyodacite endmember compositions. The Vargeão polymict breccias are composed of volcanic rocks plus sandstone.

Sandstones form up to several hundred meter sized, fault-bounded blocks that are arranged in a semi-annular pattern around the center of the structure. An oil exploration well drilled 22 km to the northeast and outside of the rim of Vargeão Dome puts the Botucatu Formation at a normal depth of 980 m and the Pirambóia Formation at a normal depth of 1100 m. The sandstones within the structure are not only extensively deformed and recrystallized but they also form abundant monomict breccias. Shatter cones have been found at a few locations within sandstones and also in tholeiitic basalt from a location in the central part of Vargeão Dome near exposures of polymict breccia. These small cones may contain volcanic glass with incipient crystal-

The Vargeão Dome impact structure on October 9, 2018 as a composite image based on TanDEM-X DEM and Copernicus Sentinel-2 (channel 2/3/4) data.

lization and devitrification. A red coating occurs on striated surfaces, which has been interpreted as alteration product or, alternatively, to represent an altered melt film. Further shock microstructures include abundant planar fractures (PFs) and rare feather features (FFs). This may be indicative for significant degradation of this structure.

The maximum age for Vargeão Dome, as well as for Vista Alegre and Cerro do Jarau, is determined by the age of the regional Serra Geral volcanism, estimated at 127 to 137 Myr. A single laser-ablation inductively coupled plasma mass spectrometry age of 123.0 ± 1.4 Myr for zircon from a Vargeão Dome sample from the outside of the southern rim of the structure has been reported. As it is unlikely that material from this remote setting with respect to the center of the structure would have been completely reset by the impact, this age may not be representative of the impact event.

A. Paiva Filho and co-workers were the first to note a circular geomorphological feature from interpretation of airborne radar images in 1978. They gave the structure its name. As they noted that the sandstones in the central part of the structure belonged stratigraphically below the Serra Geral lava flows, they suggested that this might be caused by a buried igneous intrusion that had uplifted these strata. Three years later, E. Barbour Jr. and W. Corrêa investigated the structure for its potential for oil and gas resources and produced the first geological map of Vargeão Dome. They emphasized a tectonic nature of the sandstone outcrops in the center and detected the sandstone monomict breccias. In 1987 Alvaro Crósta compared the deformation at Vargeão Dome with that at other impact structures, including Araguainha Dome, which, at that time, was the only confirmed impact structure in Brazil. Eventually the findings of shatter cones and the admittedly limited shock deformation confirmed the impact structure status of this intriguing structure. Since then detailed geological and geophysical studies have been carried out at Vargeão by various teams.

Vargeão Dome, Vista Alegre, and Cerro do Jarau represent a unique resource for the study of impact effects on mafic and intermediate volcanic rocks. Already a number of studies are being performed with a view to better understand such settings, as analogs to impact structures in basalt on other planetary bodies with extensive volcanic deposits. This also involves studies of hydrothermal alteration of impact-affected volcanics, which on Mars or other volcanics-covered bodies might present sites of evolution of early life.

Location/Access

Vargeão Dome is located in the western part of the state of Santa Catarina. The village of Vargeão within the structure is reached via state roads over about 80 km from the city of Chapecó that has good airline service from a number of major cities in Brazil. The interior of the structure has a number of unpaved roads, usually in good conditions, particularly in the dry season from May to September.

Selected References

Barbour, Jr. E. and Corrêa, W. A. G. (1981). Geologia da Estrutura de Vargeão, SC. São Paulo, Consórcio CESP/IPT, 33 pp. (unpublished report, Paulipetro).

Crósta, A. P. (1987). Impact Structures in Brazil. In: Research in Terrestrial Impact Structures, Pohl, J. (ed.), Springer Vieweg, Wiesbaden. pp. 30–38.

Crósta, A. P., Kazzuo-Vieira, C., Pitarello, L., Koeberl, C., and Kenkmann, T. (2012b). Geology and impact features of Vargeão Dome, southern Brazil. Meteoritics and Planetary Science, Vol. 47, pp. 51–71.

Crósta, A. P., Reimold, W. U., Vasconcelos, M. A. R., Hauser, N., Oliveira, G. J. G., Mazivieiro M. V., and Góes, A. M. (2019a). Impact cratering: The South American record – Part 1. Geochemistry – Chemie der Erde, Vol. 79, pp. 1–61.

Paiva Filho, A., Andrade, C. A. V. de, and Scheibe, L. F. (1978). Uma janela estratigráfica no oeste de Santa Catarina: o Domo de Vargeão. In: 30° Congresso Brasileiro de Geologia, Recife, Brazil, Proceedings, Vol. 1, pp. 408–412

Vista Alegre

Brazil

Wolf Uwe Reimold

The Vista Alegre impact structure is located in the western part of Paraná State of southern Brazil. The structure measures 9.5 km in diameter. It is formed in tholeiitic basalt and sandstones of the Late Permian to Early Cretaceous São Bento Group – especially the Pirambóia, Botucatu, and Serra Geral formations – of the Paraná Basin. In this, the geology of Vista Alegre is similar to that of **Vargeão Dome** and **Cerro do Jarau**. The Vista Alegre structure was named after the rural village of that name in Coronel Vivida County.

Vista Alegre is a complex impact structure that is surrounded by a pronounced, up to 150 m high, circular rim along much of its

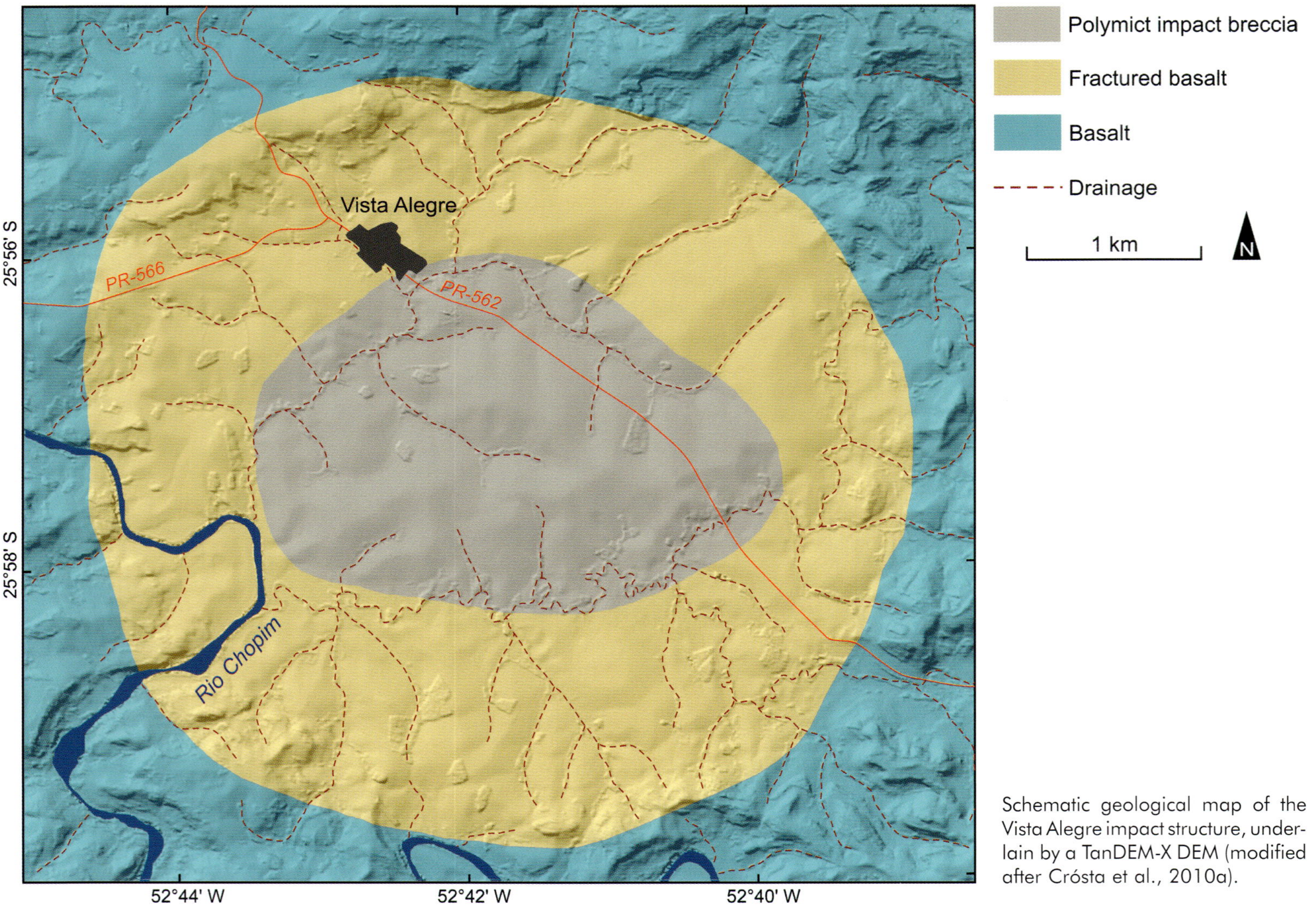

Schematic geological map of the Vista Alegre impact structure, underlain by a TanDEM-X DEM (modified after Crósta et al., 2010a).

Vista Alegre

Brazil

latitude = 25°57‘12“ S / longitude = 52°41‘23“ W / diameter = 9.5 km / age = 111 - 134 Myr

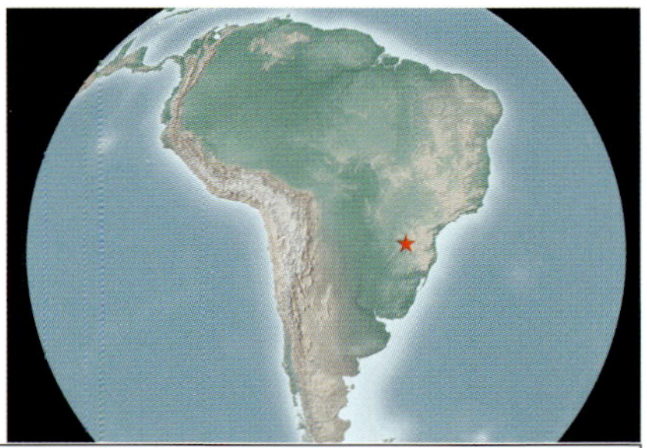

scale = 1: 139 000 / scene = 25.0 × 33.4 km²

0 1.39 6.95 km

N

425 meter 885

Panoramic view of the rim and interior of the Vista Alegre impact structure taken from the northeast rim. The width of this view is approximately 6 km (image courtesy: A. Crósta).

Shatter cone on a basalt clast in polymict impact breccia from Vargeão Dome (image courtesy: A. Crósta).

An assemblage of well-developed, small shatter cones, as a whole about 15 cm wide, formed in basalt. Found in a quarry near the town of Vista Alegre (image courtesy: A. Crósta).

perimeter. There are steep escarpments along the inner slopes of the rim. The interior of the structure does not show much topography, in comparison with the surrounding terrain. The morphology is determined by gently rolling hills, with a relief of not more than 50 m. In the area outside of the structure, there are elevation differences of more than 200 m between hilltops and river valleys.

A central uplift structure in the central part is not evident, but it has been inferred from the occurrence of blocks of deformed sandstone and is supported by gravimetric data. The sandstones are related to the Pirambóia-Botucatu succession that normally occurs stratigraphically some 700–800 m below the present surface in the southern part of the Paraná Basin. The sandstones should be covered by the multiple basalt flows of the Serra Geral Formation. Much of the interior of the structure is covered by meter thick soil, on which intense crop-farming takes place. The escarpments along the rim of the structure exhibit terrace-like features, formed by differential erosion along the contacts between different lava flows. Four to five distinct flows are recognized along the edge of the Vista Alegre structure.

The geological map of the Vista Alegre structure could only be compiled from a limited number of outcrops. Along the inside of the rim, and in an annular area around 3 km inward, only fractured basalt occurs. This fracturing seems to be more intense within the Vista Alegre structure than outside. No diagnostic shock deformation has been found in basalt from this structure.

Polymict impact breccia occurs in rare outcrops in the central part of the structure. It occurs mainly along the Quieto and Surubim streams and in a small quarry near the village of Vista Alegre. This quarry is located at the side of the paved road between the village and the town of Coronel Vivida. The quarry contains a geosite information board with introductory information about the formation of the Vista Alegre impact structure.

The parautochthonous Vista Alegre impact breccia is composed of a matrix of pelitic material mostly derived from Serra Geral basalt. Clasts ranging in size from millimeters to decimeters are composed of basalt, sandstone, and siltstone. Of the three impact structures occurring in the southern Paraná Basin, only Vista Alegre exhibits such polymict impact breccia, which could indicate that this structure is relatively less eroded in comparison with Vargeão Dome and Cerro do Jarau. Some of the basalt clasts, especially from the quarry breccia, contain shatter cone aggregates. Furthermore, shocked quartz with one or two sets of planar deformation features (PDFs) has been reported from sandstone clasts from the breccia.

A gravimetric survey produced a Bouguer anomaly map that shows a remarkable zonation into circular and concentric gravimetric lows and highs. A semi-circular gravity high along the rim of the structure completes the Bouguer map. This gravimetric bulls-eye pattern is very similar to that for Vargeão Dome.

An attempt to determine the age of the Vista Alegre impact by ^{40}Ar-^{39}Ar step-heating experiments on different aliquots of a melt fragment from polymict impact breccia yielded an apparent weighted mean age of 115 ± 4 Myr. This is considered a minimum age for the impact because of inherent alteration of this melt specimen. Conservatively, the age of Vista Alegre is estimated between about 111 Myr (the apparent minimum age obtained for the melt) and 134 Myr, the current age constraint for the Serra Geral volcanism.

The Vista Alegre impact structure was first proposed in 2004 by Alvaro Crósta and co-workers, after ground truthing a circular feature interpreted from a satellite image. In the following years Crósta, in cooperation with different teams, unveiled the details of the Vista

The Vista Alegre impact structure on October 9, 2018 as a composite image based on TanDEM-X DEM and Copernicus Sentinel-2 (channel 2/3/4) data.

Alegre impact structure. This resulted in a publication in 2010 that convincingly confirmed the impact.

Location/Access

The Vista Alegre structure is located in southern Paraná State, some 250 km east of Foz do Iguaçu. Some 200 km due south lies the city of Chapecó that has an airport served from several major cities in Brazil. From Chapecó to Coronel Vivida, and then on to Vista Alegre, one travels on various reasonably maintained interstate and state roads. From Coronel Vivida, state road PR-562 leads west, after 10 km, to Vista Alegre, cutting across the structure.

Selected References

Crósta, A. P., Kazzuo-Vieira, C., and Schrank, A. (2004). Vista Alegre: a newly discovered impact crater in Southern Brazil. Meteoritics and Planetary Science, Vol. 39 (Supplement): A-28.

Crósta, A. P., Koeberl, C., Furuie, R. A., and Kazzuo-Vieira, C. (2010a). Vista Alegre, southern Brazil: a new impact structure in the Paraná flood basalts. Meteoritics and Planetary Science, Vol. 45, pp. 181–194.

Crósta, A. P., Jourdan, F., and Koeberl, C. (2012a). ^{40}Ar-^{39}Ar dating of the Vista Alegre crater, Brazil. 34th International Geological Congress, Brisbane, Australia, 1 p.

Crósta, A. P., Reimold, W. U., Vasconcelos, M. A. R., Hauser, N., Oliveira, G. J. G., Mazivieiro M. V., and Góes, A. M. (2019a). Impact cratering: The South American record – Part 1. Geochemistry – Chemie der Erde, Vol. 79, pp. 1–61.

Impact Structures – Further Confirmation Required

Colônia Basin
Brazil

Wolf Uwe Reimold

The Colônia Basin structure is found in São Paulo state of southeastern Brazil. The northern part of the structure is occupied by a part of the Vargem Grande community, as well as a large penitentiary. The location of this likely simple, bowl-shape basin structure with an apparent diameter of 3.6 km is close to the Atlantic Mountain Range. There is a prominent ring of hills around the basin that are elevated by up to 125 m above the inner depression. The interior is largely a swampy, alluvial plain that is drained by the Vargem Grande stream with a radial, centripetal drainage pattern, with an outlet through the eastern rim.

The structure was formed in Neoproterozoic crystalline basement of the Ribeira Fold Belt. Schist, quartzite, gneiss, migmatite, diorite, quartzdiorite, and mylonites of these various rock types comprise the regional basement. The structure is located near the southeastern edge of the Paulistano Plateau, a geomorphological feature of likely Neogene age that is characterized by relatively flat relief at 700 to 900 m altitude. Paleogene mudstone, related to the Resende Formation of the São Paulo Basin, occurs in the southern parts of the structure. Neogene colluvial and alluvial deposits also occur mainly in the inner part of the basin.

Geophysical studies revealed a distinct circular gravity low over the structure. The depth to basement at the center of the structure has been variably estimated in the past between 200 and 450 m, with the upper limit seemingly being more accurate. An initial seismic survey indicated a low seismic velocity zone in the upper part of the basement below the basin, which suggests the presence of less dense rocks. Sediment fragments recovered from groundwater drilling comprise Precambrian rock fragments in a matrix-supported conglomerate at about 260 m depth, which can be interpreted as a fanglomeratic deposit derived from erosion of the rim. The deposition of this layer preceded the initial infill of the depression with organic-rich and fine-grained sediments.

Colônia is located in the Brazilian Atlantic Forest. It is one of the rare basin features that have accumulated sediment without interruption over perhaps several million years – thus preserving a possibly long-term paleo-environmental, and for the wider region important, paleo-climatic record. Therefore efforts have been made for nearly a decade to achieve a deep drilling project. As part of these preparations, a second seismic survey was conducted by geophysicists of the University of Sao Paulo between 2010 and 2016. The 2010 seismic profiles showed three zones with distinct seismic signatures. The top zone, with a maximum thickness of 270–280 m near the basin center, is interpreted to represent organic-rich clayey sediments and clastic material. Then follows, with depth, a second zone with a maximum thickness of 65 m near the center that could be a transition from sediment fill to the lowermost seismic unit, which is thought to represent fragmented basement rock. The lowermost, about 50 m thick zone is attributed to possibly deformed basement rocks in its upper part, grading into less deformed basement below. The 2016 seismic reflection data suggest a sediment fill thickness between 15 m at the edge of the basin and 400 m at the center of the basin.

A poorly-constrained age range for the Colônia Basin structure exists, from 5 to 36 Myr. The lower limit is derived from extrapolation of the sedimentation rate based on a shallow drilling of laminated sediment to a depth of 8 m only. The upper age limit is provided by the age of the youngest geological formation in the structure, the Resende Formation mudstone.

The Colônia structure has been discussed for 50 years by many groups as a possible impact structure. While finding conclusive evidence in favor of any other mode of formation failed, the conclusive evidence that would confirm impact is still outstanding. The first detailed discussion of the Colônia Basin structure was made in 1961 by R. Kollert

◁ The Colônia Basin on July 16, 2018 as a composite image based on TanDEM-X DEM and Copernicus Sentinel-2 (channel 2/3/4) data.

Colônia Basin

Brazil

latitude = 23°52‘10“ S / longitude = 46°42‘22“ W / diameter = 3.6 km / age = 5 - 36 Myr

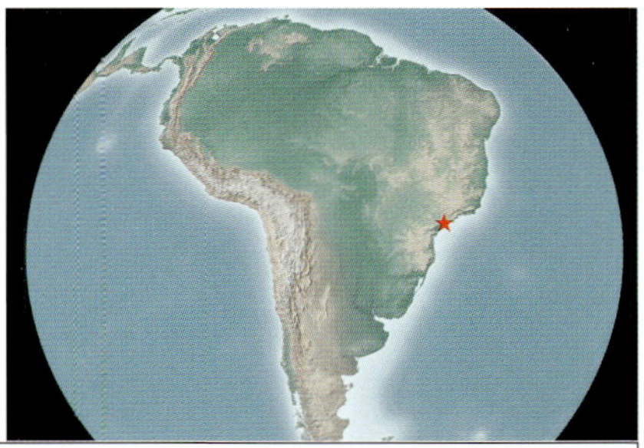

scale = 1:73 000 / scene = 13.1 × 17.4 km²

0 0.73 3.65 km

N

680 meter 915

and co-workers, who initiated geophysical investigation and were the first to refer to a possible impact origin. Claudio Riccomini and co-workers discussed alternative processes to explain the origin of the structure in 2011, but sinkhole formation, presence of a basin formed above a structural interference pattern, presence of a magmatic intrusion, or even "cryptoexplosive" activity are not supported by facts. Consequently, for many impact cratering has remained the preferred agent – despite lack of unambiguous evidence for it. While some authors have meanwhile claimed to have identified shock metamorphic micro-deformation features in thin sections of rock chips from groundwater drilling, this evidence has remained controversial and unconfirmed. The impact hypothesis is supported by the overall geometry of the structure, because the geology does not provide any other plausible explanation for the formation of such a distinct circular shape by endogenic processes.

Recently a 50 m deep drilling was accomplished. The full evaluation of this sedimentary record is in progress and should provide a much improved extrapolated estimate for the age of the sediment pile. However, foremost will be the gain from this core regarding the paleo-climatic history of the Atlantic Rainforest Domain.

Location/Access

Colônia, being located in the southern outskirts of the sprawling city of São Paulo, is easily accessible by road. From downtown São Paulo, the distance is about 40 km, but due to usual heavy traffic, the travel time by car may be two hours.

Selected References

Crósta, A. P., Reimold, W. U., Vasconcelos, M. A. R., Hauser, N., Oliveira, G. J. G., Maziviero, M. V., and Góes, A. M. (2019b). Impact cratering: The South American record – Part 2. Geochemistry – Chemie der Erde, Vol. 79, pp. 191–220.

Prado, R., Espin Fenoll, I., Miura, G., Ullah, I., Crósta, A. P., Zanon dos Santos, R., Elis, W., Machado, E., Reimold, W. U., Diogo, L., and Riccomini, C. (2018). Geophysical investigation of the Colônia structure, Brazil. Meteoritics and Planetary Science, Vol. 54, pp. 2357–2372.

Riccomini, C., Crósta, A P., Prado, R. L., Ledru, M. P., Turq, B. J., Sant'Anna, L. G., Ferrari, J. A., and Reimold, W. U. (2011). The Colônia structure, São Paulo, Brazil. Meteoritics and Planetary Science, Vol. 46, pp. 1630–1639.

Nova Colinas

Brazil

Wolf Uwe Reimold

Recently, a new impact structure, roughly 5 km across, became known from Maranhão State in northeastern Brazil. This structure, termed Nova Colinas, is located about 41 km north-northwest of the town of Balsas and 17 km east-southeast of the village with the homonymous name Nova Colinas. Interestingly, this new impact structure can be found 90 km east-northeast of the **Riachão** (Maranhão State) and 130 km east-northeast of the **Serra da Cangalha** (Tocantins State) impact structures. The morphology of the structure is expressed by several annular ridge segments that surround a more flattish interior terrain. However, within this interior, there is a small, about 0.8 km wide, structurally complex central elevation. The outer ridge system is disrupted by a stream on the western side, and it is also open towards the northeast.

The Nova Colinas structure occurs in sedimentary strata of the Silurian-to-Triassic Parnaíba Basin, mainly arenites and subordinate argillites of the Balsas Group, i. e., mainly the Sambaíba and Motuca formations, possibly in the southernmost part some Pedra de Fogo Formation. At the northernmost limit of the structure, volcanics of the Mosquito Formation may occur, and some associated diabase/microgabbro dikes were mapped by the Brazilian Geological Survey (CPRM) in the outer southwestern sector. Based on this regional geology, the structure could be as old as about 300 Myr, or – if Mosquito volcanics covered the area prior to formation of the Nova Colinas structure – as old as 200 Myr.

Nova Colinas had actually been indicated by the CPRM as a ring feature on their 1 : 750 000 map of Maranhão state. However, they failed to investigate the possible significance of this feature as an impact structure. In early 2019, Mark Jessell of the University of Western Australia drew attention to the ring-shaped magnetic anomaly and suggested that the radiometrics over Nova Colinas resembled those of the Riachão impact structure. Since then, Wolf Uwe Reimold and Marcos Vasconcelos have undertaken a first field visit and sampled a large number of arenite samples in the southern and eastern parts of the structure. Detailed micropetrographic analysis of these samples revealed significant evidence for shock metamorphism in four samples. Planar deformation features (PDFs), planar fractures (PFs), and feather features (FFs) were found in abundance and provide unequivocal proof for the impact origin of the Nova Colinas structure. Many other samples showed evidence of brecciation (monomict impact breccia) but without diagnostic shock metamorphic effects.

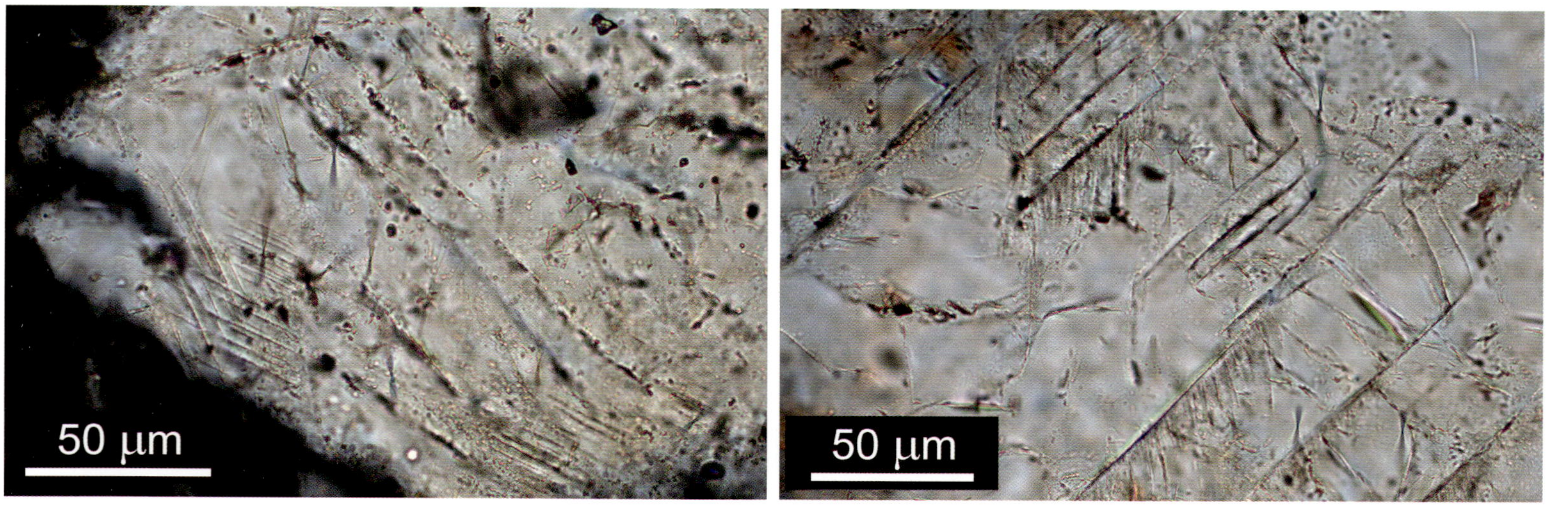

Two examples of impact-diagnostic shock deformation in quartz from sandstone samples from the interior of the Nova Colinas impact structure. Left: Planar deformation features at two crystallographic orientations – typically close to the edge of the host quartz grain, as abundantly observed in quartz from fine- to medium-grained shocked sandstone. Right: A quartz grain with planar fractures, some of which are actually feather features (image courtesy: L. Ferrière, Natural History Museum, Vienna).

Nova Colinas

Brazil

latitude = 07°09'33" S / longitude = 46°06'30" W / diameter = 5 km / age = < 300 Myr

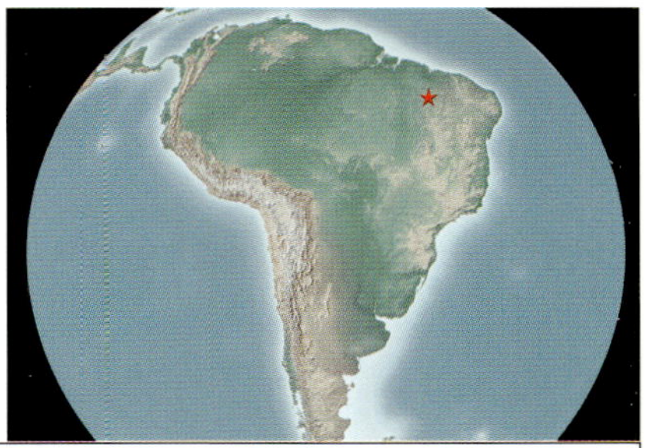

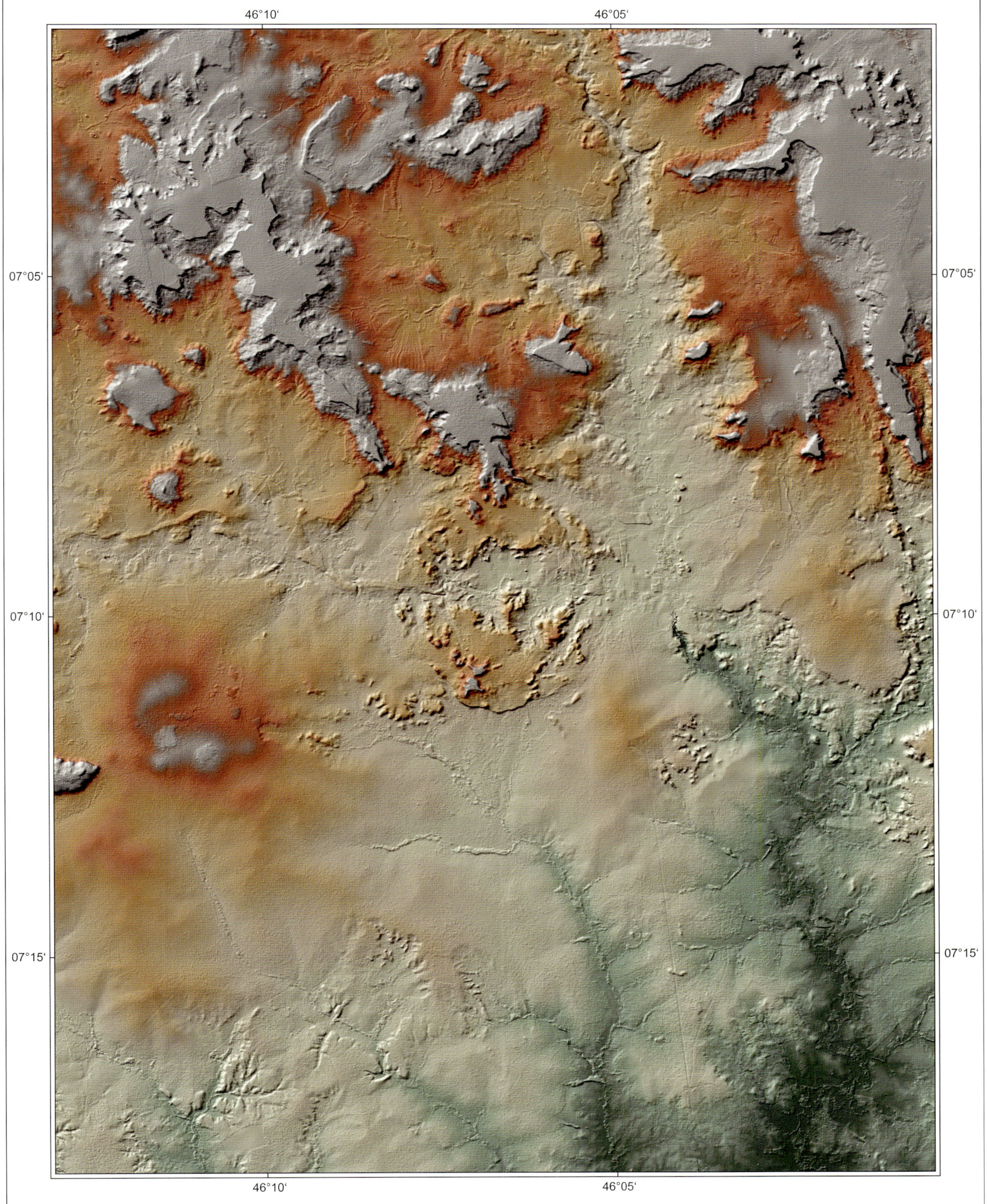

scale = 1 : 129 000 / scene = 23.2 × 31.0 km²

0 1.29 6.45 km

N

245 meter 590

The most strongly shocked samples originate from the very interior of the structure. It appears that the structure is significantly eroded, and the shock metamorphic degree in exposed rock is definitely less than 20 GPa.

Given the fact that samples from the interior of the structure provided definitive evidence for shock metamorphism in quartz, we think it is viable to include this new impact structure in our atlas. However, since this discovery has occurred only very recently and the findings are presently only reported in a conference contribution to be read at the 50th Congress of the Geological Society of Brazil (SBG), we still place Nova Colinas in this section at the end of the South American chapter.

Location/Access

The Nova Colinas impact structure is located in Maranhão State in northeastern Brazil about 41 km north-northwest of the town of Balsas and 17 km east-southeast of the village with the homonymous name Nova Colinas. There is a small airport in the city of Araguaína (Tocantíns State) with occasional flights from Brasília. From Araguaína a good road takes the traveler over 280 km, via the towns of Carolina and Riachão, to the city of Balsas in the state of Maranhão. Balsas features some reasonable hotels and, thus, can serve as a base for an expedition to Nova Colinas. There are two ways to reach the impact structure: (1) back towards the west in direction of Riachão and then towards the north on an unpaved road to the village of Nova Colinas. From there, the structure can be reached on farm roads in easterly direction. (2) The alternative route is directly from Balsas on a well-paved road in northerly direction. Eventually, a farm-road leads towards the west and into the area of the structure.

Selected References

Reimold, W. U., Vasconcelos, M. A. R., Jessell, M., Crósta, A. P., Ferrière, L., Hauser, N., Gottwald, M., and Baratoux, D. (2020). Nova Colinas – A new complex impact structure in Maranhão State, Brazil. Contribution to 50th Congress of the Sociedade Brasileira de Geologia (SBG).

The Nova Colinas impact structure on August 11, 2019 as a composite image based on TanDEM-X DEM and Copernicus Sentinel-2 (channel 2/3/4) data.

Rio Cuarto Crater Field
Argentina

Wolf Uwe Reimold

Numerous elongated, morphological depressions occur in the generally flat "Pampas" just north of the Rio Cuarto River and the city of Rio Cuarto in north-central Argentina. This area is within the eolian-loessic subhumid region of the Chacopampean Plains Geomorphological Province, very close to the Pampean Sierras Geomorphological Province. The originally described structures occur within a roughly 50 × 15 km wide area along a general southwest-northeast trend. They are formed in the regional, up to 50 m thick loess cover.

After their discovery in 1992 by Peter Schultz, and Ruben Lianza, who reported an anomalous parallel alignment of 11 oblong and clearly rimmed depressions of up to 4.5 × 1.1 km in size, the origin of these structures has been controversial. Some groups have favored oblique, highly low-angle impact of an extraterrestrial projectile, whereas others have discussed that these depressions were scooped out by wind deflation.

Supportive indications for an impact origin were reported by Peter Schultz and his Argentine, US and Austrian collaborators. This includes findings of frothy glass, interpreted as impact glass that also contains baddeleyite, the product of thermal decomposition of zircon to silica + ZrO_2, as well as rare shocked quartz grains. Elevated abundances of Cr, Ni and Ir in such glass samples were related to the presence of a meteoritic component. The geometries of the structures closely resembled the outcomes of laboratory impact experiments at very low-angle projectile trajectories. Two H4/H5 chondrite fragments found in the area were initially thought to be relics of a Rio Cuarto meteorite projectile.

Further work on the Rio Cuarto glasses showed that there were vesicular, layered or folded, vesicle-poor splash form, as well as hollow microsphere types. These glasses contained abundant mineral clasts, mainly quartz and feldspar, derived from the regionally occurring loess. It was put forward that these glasses did not resemble fulgurite – melt material produced by lightning strike. The mass of the largest glass bomb was given as ca. 0.5 kg. However, there were also 1–2 cm sized glass spherules with swirled glass structure, and much smaller (33–100 µm) hollow spherules. Occurrences of lechatelierite, i. e., silica glass with a glass transformation temperature in excess of 1600 °C, and diaplectic quartz glass were added to the list of pro-impact evidence. The water content of Rio Cuarto glasses was shown to be low, which is typical for impact glasses.

Based on ^{10}Be studies, carbon dating and argon chronology an estimated age for this alleged impact event was determined at 10 000 years or less. This would open the possibility that early inhabitants of the pampas plains might have witnessed this impact event. ^{40}Ar-^{39}Ar and fission track ages of two different glass types from Rio Cuarto yielded – surprisingly – two age populations. One type of glass gave ages of less than 10 000 years but more weathered glasses yielded a significantly older age of 114 000 ± 26 000 years. Consequently, it was concluded that the Rio Cuarto glass deposits resulted from two separate Quaternary impacts.

A second hypothesis for Rio Cuarto supports the impact origin for the Rio Cuarto glasses but rejects it for the Rio Cuarto "craters". Instead, wind "blowout" was put forward as a viable explanation for the formation of the depressions in such an extended region. Phil Bland and co-workers carried out a detailed satellite remote sensing study of the Pampean plains. It revealed not less than several hundred elongate depressions with high length-to-width ratios, even at a site 400 km to the south of the original study region of Schultz and Lianza. When Bland and his colleagues visited 52 of these structures in the field, it became obvious that their geometries resembled those of the 11 structures originally reported by Schultz and Lianza. They found basement material at the bottom of several structures, discovered two more meteorite fragments, as well as glass within two of the new structures, and at a site called Santa Rosa 400 km to the south of the original study region. Having seemingly extended the region with such structures significantly, they mooted that a bolide would have had to have a ca. 0.5 km original size prior to breaking up in the atmosphere. The new meteorites were classified as two further different types from the earlier finds – a chondrite and a basaltic achondrite, and yielded different exposure ages of 36 000 ± 4000 years and 52 000 years. They rejected that these meteorites were related to the formation of the Rio Cuarto structures and instead related their own, and the earlier findings, to local meteorite accumulation.

Another remote sensing mapping by Alberto Cione and coworkers also found hundreds of elongated structures in a wide area of the Pampean region, in the Argentinian provinces of Córdoba, San Luis, La Pampa, and Buenos Aires. This team provided evidence for the formation of these elongated depressions as wind blowout structures. The region is well-known for strong winds and a constant wind direction that corresponds to the long axes of the wind deflation depressions.

The argon-argon age for Rio Cuarto glass of 0.57 Myr reported by Bland and his team turned out to be similar to the age of glasses found 800 km to the south. Therefore these widely distributed occurrences were discussed as possibly representing a tektite strewn field. If this would be the case, a recent, larger than 5 km sized complex impact crater remains to be discovered below the loess cover of the Pampean Plains of Argentina.

In conclusion, there is significant evidence for the formation of the Rio Cuarto glasses by impact. The origin of the elongated Rio Cuarto depressions has, however, remained controversial. Intriguing evidence suggests that the glasses may not be of local origin but could be derived from a sizable impact event distal to Rio Cuarto that still remains to be discovered.

Location/Access

The city of Rio Cuarto is located some 200 km south of Córdoba city, with which it is linked by the RN36 road. Both Córdoba and Rio Cuarto have airports. The most prominent elongated depression can be reached from Rio Cuarto by travelling on road RN36 northward to Alcira. There turn eastward towards General Cabrera on road RP11. After about 15 km one comes close to the southern tip of a depression strewn field. A slightly smaller twin structure can be found 10 km to the southwest.

Selected References

Bland, P. A., de Souza Filho, C. R., Jull, A. J. T., Kelley, S. P., Hough, R. M., Artemieva, N. A., Pierazzo, E., Coniglio, J., Pinotti, L., Evers, V., and Kearsley, A.T. (2002). A possible tektite strewn field in the Argentinian pampa. Science, Vol. 296, pp. 1109–1111.

Cione, A. L., Tonni, E. P., San Cristóbal, J., Hernández, P. J., Benítez, A., Bordignon, F., and Perí, J. A. (2002). Putative meteoritic craters in Río Cuarto (Central Argentina) interpreted as eolian structures. Earth Moon Planets, Vol. 91, pp. 9–24.

Crósta, A. P., Reimold, W. U., Vasconcelos, M. A. R., Hauser, N., Oliveira, G. J. G., Maziviero, M. V., and Góes, A. M. (2019b). Impact cratering: The South American record – Part 2. Geochemistry – Chemie der Erde, Vol. 79, pp. 191–220.

Schultz, P. H., and Lianza, R. E. (1992). Recent grazing impacts on the Earth recorded in the Rio Cuarto crater field, Argentina. Nature, Vol. 355, pp. 234–237.

Schultz, P. H., Koeberl, C., Bunch, T., Grant, J., and Collins, W. (1994). Ground truth for oblique impact processes: new insight from the Rio Cuarto, Argentina, crater field. Geology, Vol. 22, pp. 889–892.

Schultz, P. H., Zarate M. A., Hames, W., Koeberl, C., Bunch, T., Storzer, D., Renne, P. R., and Witke, J. (2004). The Quaternary impact record from the Pampas, Argentina. Earth and Planetary Science Letters, Vol. 219, pp. 211–238.

The area of the Rio Cuarto field of crater-like structures on October 7, 2018 as a composite image based on TanDEM-X DEM and Copernicus Sentinel-2 (channel 2/3/4) data. The elongated structures originally identified by Schultz and Lianza (1992) are labelled A–K.

Rio Cuarto Crater Field

Argentina

latitude = 32°52‘ S / longitude = 64°11‘ W / diameter = max. 4.50 km / age = < 0.114 Myr

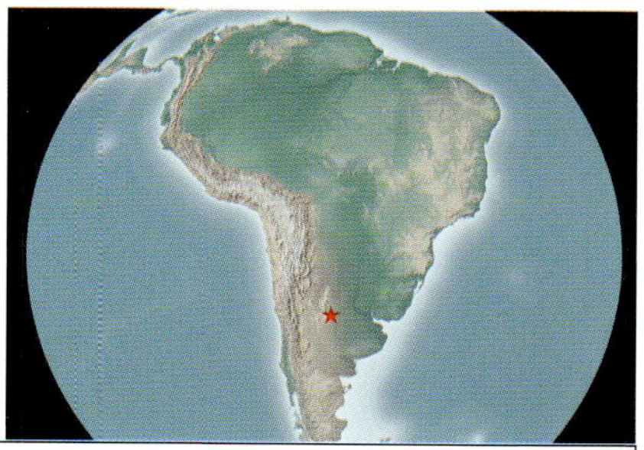

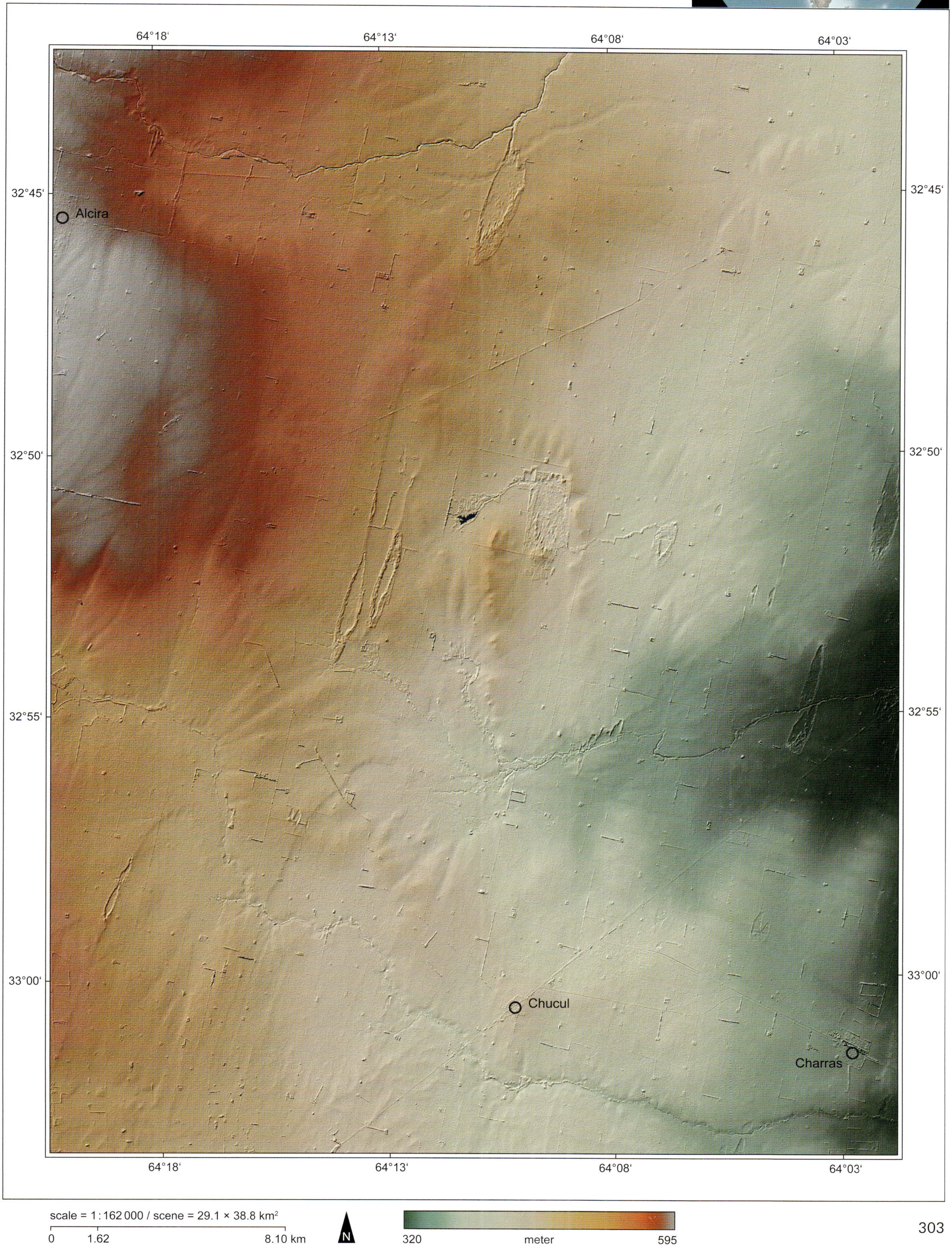

Geologic Timescale

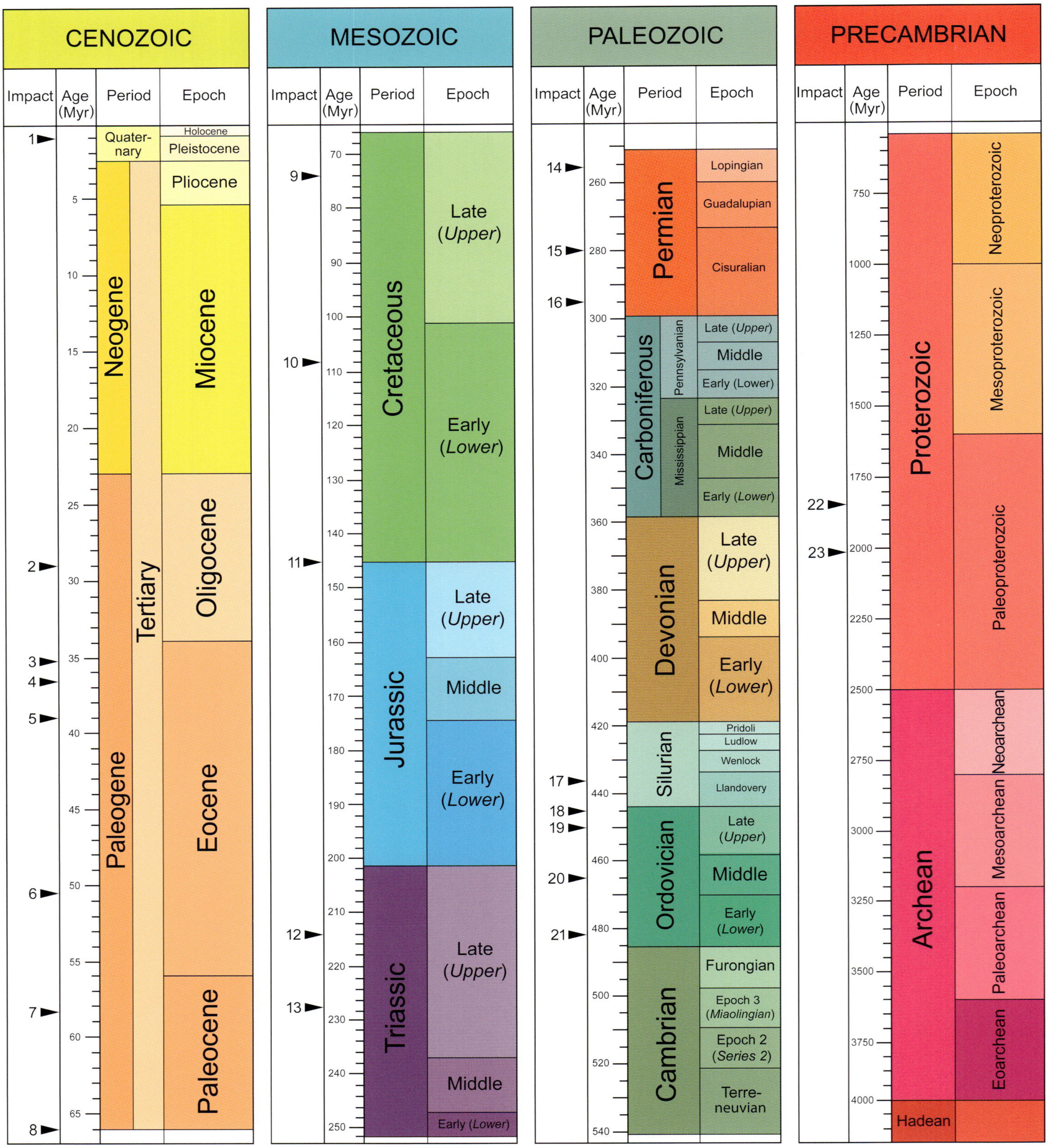

Selection of impact structures with reasonably well determined ages:

1 Bosumtwi
2 Libyan Desert Glass
3 Chesapeake Bay
4 Mistastin
5 Haughton
6 Montagnais
7 Marquez
8 Chicxulub
9 Manson
10 Steen River
11 Morokweng
12 Manicouagan
13 Saint Martin
14 Araguainha
15 Douglas Crater Field
16 Clearwater West
17 Slate Islands
18 Pilot
19 Charlevoix
20 Clearwater East
21 Carswell
22 Sudbury
23 Vredefort

Modified after (terminology different in chronostratigraphic chart given in italics):
Geologic Time Scale v. 5.0, Geological Society of America, https://doi.org/10.1130/2018.CTS005R3C. L.E.
and
International Chronostratigraphic Chart v2018/08, International Commission on Stratigraphy
End and Beginning of Holocene (0–0.012 Myr) and end of Pleistocene (0.012 Myr) not drawn to scale

Manfred Gottwald
Thomas Kenkmann
Wolf Uwe Reimold

TERRESTRIAL IMPACT STRUCTURES

The TanDEM-X Atlas

2

Asia, Australia, Europe

Manfred Gottwald
Thomas Kenkmann
Wolf Uwe Reimold

TERRESTRIAL IMPACT STRUCTURES

The TanDEM-X Atlas

2

Asia, Australia, Europe

Verlag Dr. Friedrich Pfeil · München 2020
ISBN 978-3-89937-261-8

Bibliografische Information Der Deutschen Bibliothek

Die Deutsche Bibliothek verzeichnet diese Publikation in der Deutschen Nationalbibliografie;
detaillierte bibliografische Daten sind im Internet über
http://dnb.dnb.de abrufbar.

front cover
Popigai, Wolfe Creek, Ries Crater (from left to right).
Top row: TanDEM-X topographic maps; bottom row Sentinel-2 RGB images,
fused with theTanDEM-X digital elevation model.

back cover
Top: El'gygytgyn (photo: U. Raschke);
bottom: Kara-Kul (copyright: J.-L. Dauvergne, provided by S. Bouley).

print support by UNI FREIBURG and Wolf Uwe Reimold and Manfred Gottwald

Druckvorstufe: Verlag Dr. Friedrich Pfeil, München
Druck: PBtisk a.s., Příbram I – Balonka

ISBN 978-3-89937-261-8

Printed in the European Union

Verlag Dr. Friedrich Pfeil, Wolfratshauser Straße 27, 81379 München, Germany
Tel.: +49 (0)89 5528600-0 • Fax: +49 (0)89 5528600-4 • E-Mail: info@pfeil-verlag.de
www.pfeil-verlag.de

Contents

7.4 Asia

Popigai in August 2018 – TanDEM-X/Copernicus Sentinel-2.

Asian Impact Structures – Overview

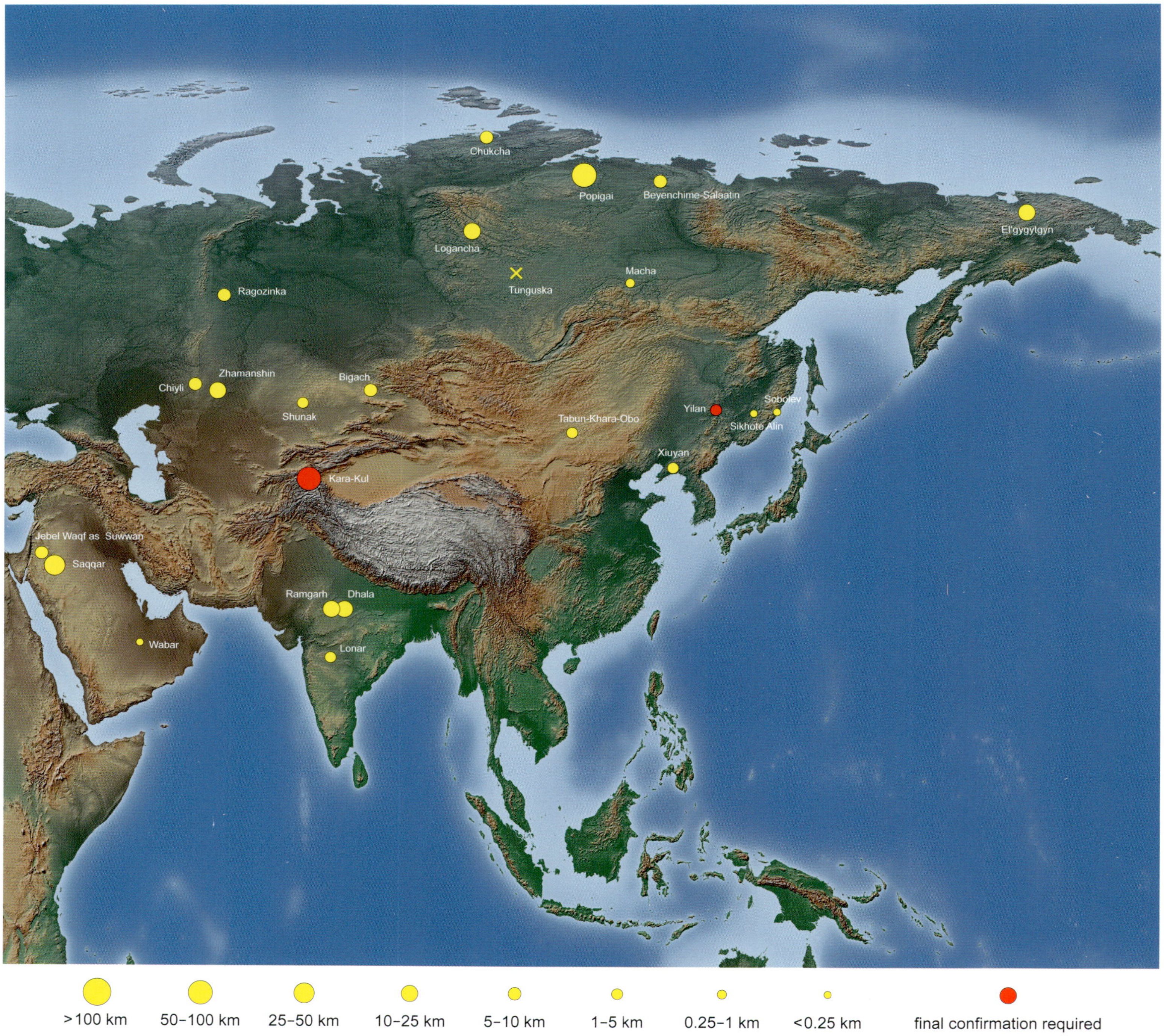

Table: Asian Impact Structures – Overview

ID	Name	Country / Region	Latitude	Longitude	Diameter (km)	Age (Myr)	Topographic Visibility
1	Beyenchime-Salaatin	Russia	71°03'29"N	121°41'23"E	8	< 66	yes
2	Bigach	Kazakhstan	48°34'33"N	82°02'11"E	8	5 ± 3	yes
3	Chiyli	Kazakhstan	49°10'32"N	57°50'01"E	5.5	46 ± 7	yes
4	Chukcha	Russia	75°42'22"N	97°50'46"E	6	< 70	yes
5	Dhala	India	25°17'55"N	78°08'33"E	12	1700–2500	yes
6	El'gygytgyn	Russia	67°29'31"N	172°03'37"E	18	3.58 ± 0.04	yes
7	Jebel Waqf as Suwwan	Jordan	31°02'54"N	36°48'24"E	6.1	< 37	yes
8	Logancha	Russia	65°29'54"N	95°57'15"E	14	23–66	yes
9	Lonar	India	19°58'36"N	76°30'32"E	1.88	0.570 ± 0.047	yes
10	Macha Crater Field	Russia	60°05'07"N	117°39'08"E	max. 0.3	0.007315 ± 0.000080	yes
11	Popigai[1]	Russia	71°38'N	111°11'E	100	35.7 ± 0.2	yes
12	Ragozinka	Russia	58°42'17"N	61°47'50"E	9	50	yes (buried)
13	Ramgarh	India	25°20'03"N	76°37'28"E	10	164.8 ± 1.3	yes
14	Saqqar[2]	Saudi Arabia	29°35'N	38°42'E	34	70–410	no (buried)
15	Shunak	Kazakhstan	47°12'35"N	72°45'41"E	2.8	< 34	yes
16	Sikhote Alin Crater Field[3]	Russia	46°09'36"N	134°39'12"E	max. 0.026	0.000072	yes
17	Sobolev[4]	Russia	46°18'22"N	137°52'24"E	0.053	0.000175 ± 0.000025	yes
18	Tabun-Khara-Obo	Mongolia	44°07'52"N	109°39'17"E	1.3	145–163	yes
19	Wabar Crater Field[5]	Saudi Arabia	21°30'12"N	50°28'20"E	max. 0.116	0.000290 ± 0.000038	yes
20	Xiujan	China	40°21'52"N	123°27'39"E	1.8	0.050 ± 0.005	yes
21	Zhamanshin	Kazakhstan	48°21'38"N	60°56'12"E	13	0.75–1.10	yes
Further confirmation required							
22	Kara-Kul	Tajikistan	39°04'N	73°26'E	52	> 5–50	yes
23	Yilan	China	46°23'03"N	129°18'40"E	1.85	> 0.01	yes
Strongest airburst in modern times							
–	Tunguska[a]	Russia	63°53'10"N	101°54'25"E	n.a.	0.000111	n.a.

a Airblast without a confirmed crater

Position references for selected impact structures:

1 Masaitis, V. L. (1999). Impact structures of northeastern Eurasia: The territories of Russia and adjacent countries. Meteoritics and Planetary Science, Vol. 34, pp. 691–711.

2 Kenkmann, T., Afifi, A. M., Stewart, S. A., Poelchau, M. H., Cook, D. J., and Neville, A. S. (2015). Saqqar: A 34 km diameter impact structure in Saudi Arabia. Meteoritics and Planetary Science, Vol. 50, pp. 1925–1940.

3 Fantucci, R., Di Martino, M., and Serra, R. (2011). Tree-ring dating of meteorite fall in Sikhote-Alin, Eastern Siberia – Russia. International Journal of Astrobiology, Vol. 11, pp. 37–42.

4 Khryanina, L. P. (1981). Sobolevskiy meteorite crater (Sikhote-Alin' Range). International Geology Review, Vol. 23, p. 1–10.

5 Gnos, E., Hofmann, B. A., Halawani, M. A., Tarabulsi, Y., Hakeem, M., Al Shanti, M., Greber, N. D., Holm, S., Alwmark, C., Greenwood, R. C., and Ramseyer, K. (2013). The Wabar impact craters, Saudi Arabia, revisited. Meteoritics and Planetary Science, Vol. 48, pp. 2000–2014.

Beyenchime-Salaatin
Russia

Thomas Kenkmann

Beyenchime-Salaatin, also known as *Beenchime-Salaaty*, is located in the northeastern part of the Siberian platform in Russia. The crater structure is morphologically well visible as an 8 km wide depression that is surrounded by a 50–70 m high rim. The structure has a somewhat polygonal outline with rather straight rim segments. Country rocks are predominantly flat-lying Cambrian carbonate rocks, with a thickness of 1000 to 1200 m. Steep faults locally dissect and displace the well-stratified target rocks. These faults are also responsible for the partly polygonal outline of the structure.

The rim is composed of uplifted and up-thrusted Cambrian carbonate rocks. Intense brittle deformation and enhanced fracturing extend, however, further outside for about 0.5–1 km. Upthrusted blocks of the lower stratigraphic units were observed in the northeastern sector of the rim, where an unusually large thrust sheet of 2–3 km size was reported. Inside the crater depression are isolated hills composed of allochthonous breccia. The lithic breccia in the inner part of the structure was interpreted as allochthonous fallback material. It is largely covered with Quaternary deposits. Breccia was also mapped along the rim and outside the crater and contains fragments of all types of the target rocks, including an admixture of Permian sandstones, which are otherwise completely eroded in the region. Some clasts contain shatter cones.

Neither geophysical nor borehole data, nor radiometric age determinations are available. The structure should – at this size – have a central uplift, but it is not in evidence. The age of formation for Beyenchime-Salaatin must be younger than Permian, based on the occurrence of these rocks in impact breccia. Due to its state of preservation, a Cenozoic age has been assumed.

A first Russian publication on the structure by M. Mikhailov dates back to 1979, with only little information provided later. Recently, Ulli Raschke, formerly of the Museum für Naturkunde Berlin, and co-workers visited Beyenchime-Salaatin and returned some samples.

Location/Access

Beyenchime-Salaatin is situated in the remote northeastern part of the Siberian tundra, about 190 km west of the river Lena and 220 km south of Olenyok Gulf of the Laptew Sea, with the prominent bulge formed by the Lena delta.

Selected References

Masaitis, V. L. (1999). Impact structures of northeastern Eurasia: The territories of Russia and adjacent countries. Meteoritics and Planetary Science, Vol. 34, pp. 691–711.

Mikhailov, M. V., Shurygin, A. G., and Khariusov, L. S. (1979). Beenchime-Salaaty meteoritic crater (in Russian). Doklady AN SSSR, Vol. 245, pp. 911–914.

The Beyenchime-Salaatin impact structure on August 26, 2018 as a composite image based on TanDEM-X DEM and Copernicus Sentinel-2 (channel 2/3/4) data.

Beyenchime-Salaatin

Russia

latitude = 71°03‘29“ N / longitude = 121°41‘23“ E / diameter = 8 km / age = < 66 Myr

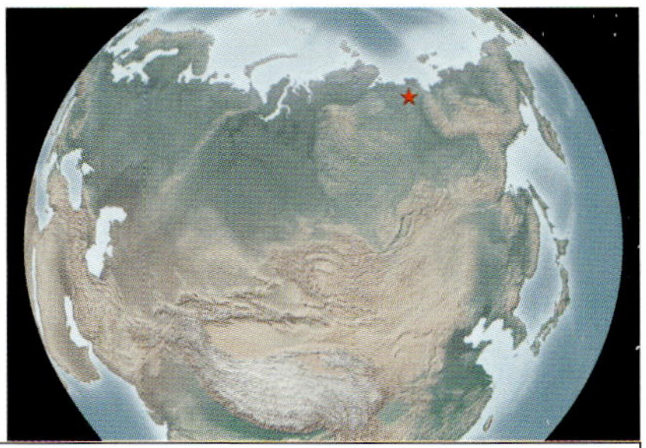

scale = 1 : 128 000 / scene = 23.1 × 30.8 km²

0 1.28 6.40 km

N

65 meter 260

Bigach
Kazakhstan

Thomas Kenkmann

Bigach is a morphologically well-expressed 8 km wide depression in an otherwise hilly region. The rim rises about 30–70 m above the flat interior that contains Quaternary sand and gravel from a southwest-northeast trending river that dissects the crater rim and crosses the structure. The southern rim area is slightly higher than the northern section.

The structure is located in a fold belt and has a complex geology. Country rocks are Devonian and Carboniferous sandstones and siltstones, as well as mafic to felsic volcanic rocks, including tuffs. Strata strike northwest and are cut by numerous faults. In valleys, the folded beds are superposed by Tertiary clays of up to 300 m thickness. The country rocks along the rim are strongly fractured and, in places, brecciated. Poorly developed shatter cones were found in this area. Blocks form terrace structures along the inner slope of the rim. Beyond the rim, within a radius of up to 9 km from the center of the structure, several outcrops of ejected breccia can be found. The ejecta contain Paleozoic megablocks up to 20 m in size that are embedded in a chaotic assemblage of finer fragments. In some places, Miocene clays underlie this breccia and provide a constraint on the age of the impact event. Secondary cratering into these clays caused by ballistic ejection of large Paleozoic blocks was demonstrated.

A gravity survey indicated a gravity low over the area of the structure. The magnetic data are compatible with the presence of a central uplift beneath Quaternary deposits in the central area. Two boreholes were drilled into the depression and reached the fractured crater floor beneath an up to 400 m thick pile of post-basin formation beds of predominantly Quaternary age, and also some breccias. Bigach is estimated to be a rather young impact structure.

A first Russian report on Bigach dates from 1986 by N. Kiselev and Y. Korotuschenko. As for so many impact structures of the former Soviet Union, literature is limited to only a few publications from the 1980s and is only available in Russian.

Location/Access

Bigach is located some 60 km west of Lake Zaysan of eastern Kazakhstan. Mountainous dry steppe and agricultural land interchange in the area. The crater is named after the town of Bigach that is situated a few kilometers north of the impact structure.

Selected References

Kiselev, N. P., and Korotuschenko, Y. G. (1986) The Bigach astrobleme, Eastern Kazkhstan (in Russian). Meteoritika, Vol. 45, pp. 119–121.

Masaitis, V. L. (1999). Impact structures of northeastern Eurasia: The territories of Russia and adjacent countries. Meteoritics and Planetary Science, Vol. 34, pp. 691–711.

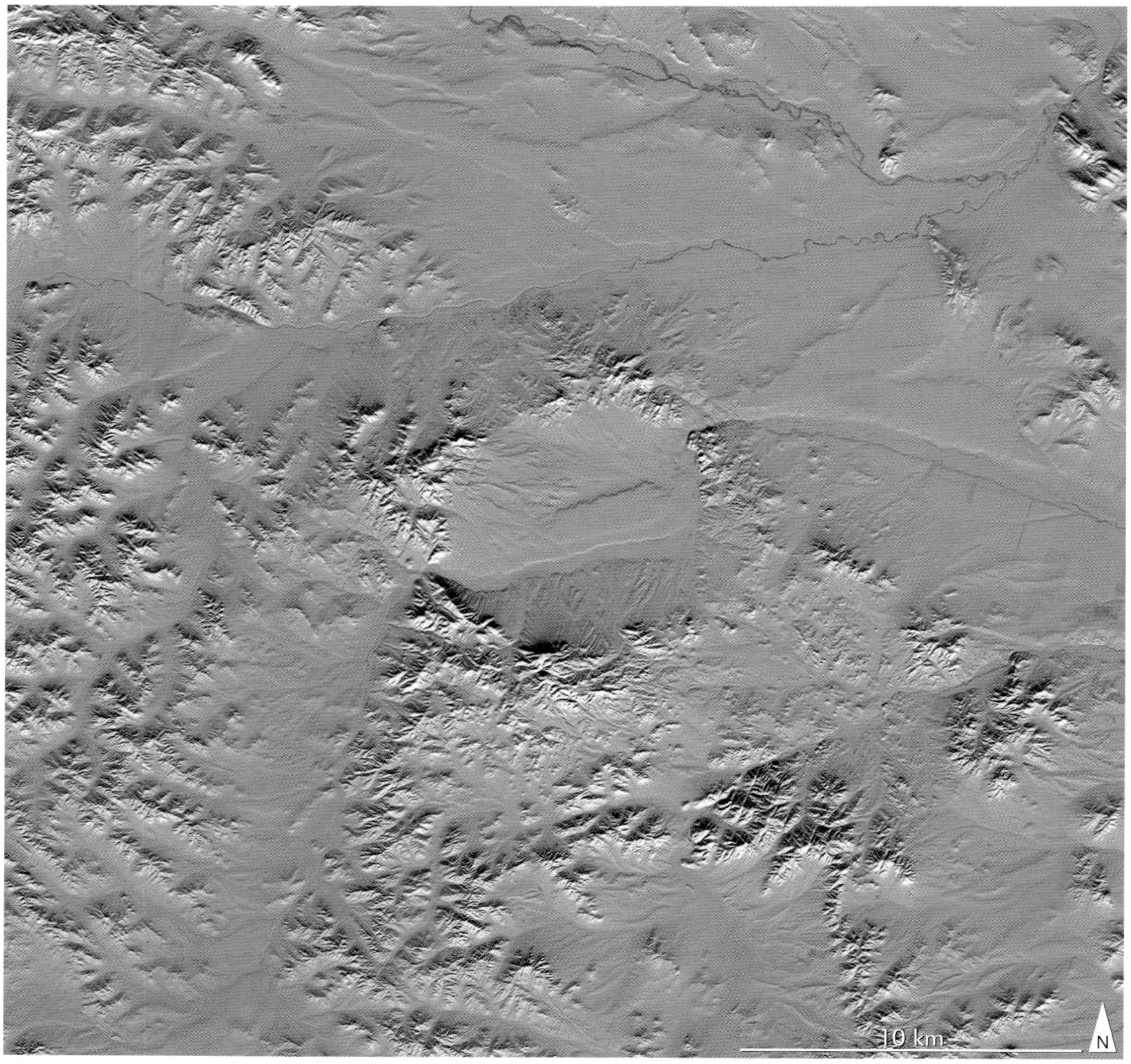

The snow-covered Bigach impact crater on January 3, 2020 as a composite image based on TanDEM-X DEM and Copernicus Sentinel-2 (channel 2/3/4) data.

Bigach

Kazakhstan

latitude = 48°34‘33“ N / longitude = 82°02‘11“ E / diameter = 8 km / age = 5 ± 3 Myr

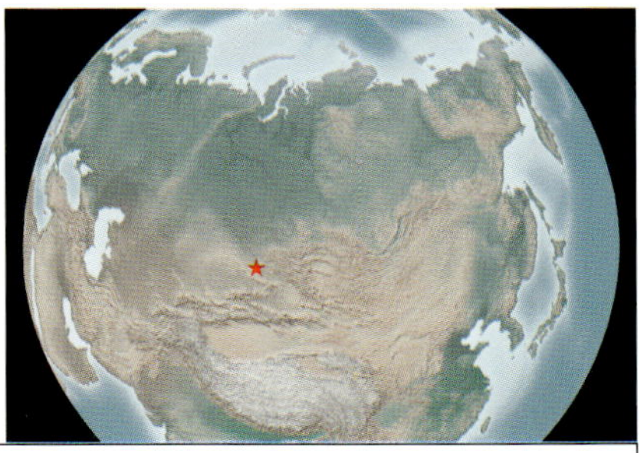

scale = 1 : 136 000 / scene = 24.5 × 32.6 km²

0 1.36 6.80 km

N

540 meter 825

Chiyli
Kazakhstan

Thomas Kenkmann

Chiyli Dome, also known as *Shiyli* or *Shili*, is an eroded, 5.5 km wide impact structure in the western part of Kazakhstan. It shows a central uplift of about 1 km diameter. Three-quarters of the crater are exposed, but the eastern sector is buried beneath Pliocene to Quaternary deposits. The structure is located about 250 km southeast of the much larger **Zhamanshin** crater.

The target was composed of Devonian and Carboniferous basement that was overlain by well bedded Cretaceous and Cenozoic cover rocks. The oldest rocks of Lower and Upper Cretaceous age are exposed in the circular central uplift. The surrounding ring syncline, the axis of which has a diameter of about 1.6 km, comprises rocks of Paleogene age. These rocks are partly brecciated and contain evidence of shock metamorphism in the form of shatter cones and planar deformation features (PDFs) in quartz. Centripetal movement during the formation of the central uplift caused the strata to form radial folds and faults, which, in turn, caused a serrated outline of the central uplift. It was suggested that the Chiyli impact event took place into a shallow, about 350 m deep, marine basin.

From the sedimentary succession of the deformed and undeformed strata of the target it has been deduced that the impact must have occurred in the Eocene, at about 46 ± 7 Myr ago.

Initially the dome was considered a salt diapir with oil and gas potential. This prompted intense geophysical exploration and drilling. In 1989 it became apparent from the detection of shock deformation features that the dome is indeed a complex impact crater.

Location/Access

The city of Embi, which has an airstrip and railway station, is located 45 km southeast of the Chiyli structure. From Embi, use Highway A 26 in northwesterly direction to Quandyaghasch. After about 45 km, the crater structure can be found about 4 km north of the highway.

Selected References

Vishnevsky, S. A. (2007). Shiyli Dome, Kazakhstan: Origin of central uplift by elastic response. Bridging the Gap II: Effect of Target Properties on the Impact Cratering Process, Saint-Hubert, Canada, LPI Contribution No. 1360, pp. 125–126.

Vishnevsky, S. A., and Korobkov, V. F. (1989). Chiyli Dome, an eroded impact structure in western Primugodzharye (in Russian). Akademiya Nauk SSSR, pp. 1–51.

The Chiyli impact structure on May 5, 2018 as a composite image based on TanDEM-X DEM and Copernicus Sentinel-2 (channel 2/3/4) data. The position of Chiyli is indicated by cross hairs.

Chiyli

Kazakhstan

latitude = 49°10'32" N / longitude = 57°50'01" E / diameter = 5.5 km / age = 46 ± 7 Myr

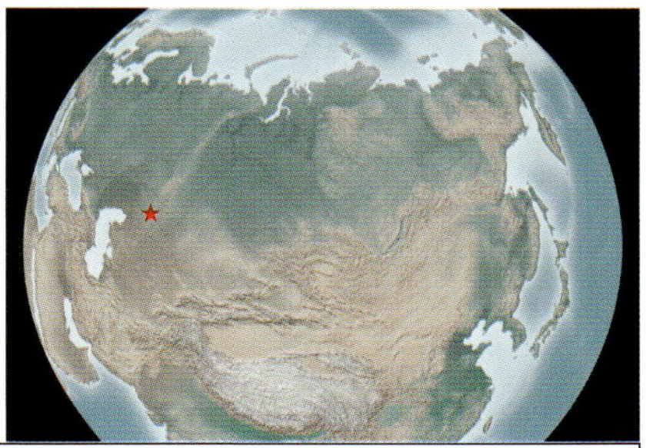

scale = 1:86 000 / scene = 15.5 × 20.7 km²

0 0.86 4.30 km

N

230 meter 345

Chukcha
Russia

Thomas Kenkmann

The Chukcha, or *Chykcha*, depression is situated in the northern tundra of Siberia. It is a roundish, 6 km wide basin, with gentle slopes to its rim, a flat interior, and a 1 km wide central uplift. Its central elevation rises about 30 m above the surrounding terrain. The annular depression, in turn, is 200–220 m deep with respect to the level of the surrounding region. The ring basin is filled with about 100 m of post-impact deposits.

Target rocks comprise silicic and carbonaceous basement rocks, into which granites and gabbros were intruded prior to the impact event. The sedimentary cover is partly preserved in the crater depression. The central uplift exposes chaotically deformed rocks, mega-breccia, and thrust slices. Quartz grains of the central uplift exhibit planar fractures (PFs) and planar deformation features (PDFs) corresponding to shock pressures of less than 10 GPa. Shatter cones have not been detected. Shock barometry indicates several hundred meters of erosion must have occurred since the time of impact. The area around the depression is devoid of any impact-induced deformation.

The age of Chukcha has been estimated based on age constraints from the basement rocks and the crater fill. The structure must be younger than 70 Myr. Studies about the Chukcha impact structure are scarce. Sergei Vishnevsky reported his findings in the early 1990s. Since then no modern descriptions have been published.

Location/Access

Chukcha is situated on the Taymyr peninsula in the northern tundra of Siberia, and is the northernmost unambiguously confirmed impact structure in the world. No infrastructure is available for accessing this structure in such a remote region.

Selected References

Vishnevsky, S. A. (1995). The Chykcha impact crater, Taymyr peninsula: Heavily eroded astrobleme of Cretaceous-Paleogene age (abstract). Meteoritics, Vol. 30, p. 591.

The Chukcha impact structure on August 25, 2018 as a composite image based on TanDEM-X DEM and Copernicus Sentinel-2 (channel 2/3/4) data.

Chukcha

Russia

latitude = 75°42‘22“ N / longitude = 97°50‘46“ E / diameter = 6 km / age = < 70 Myr

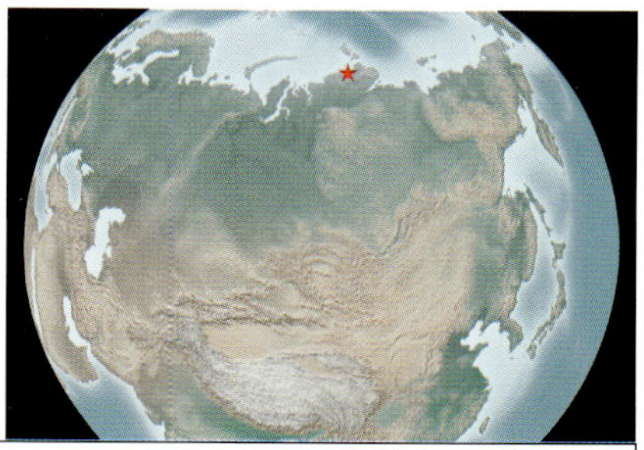

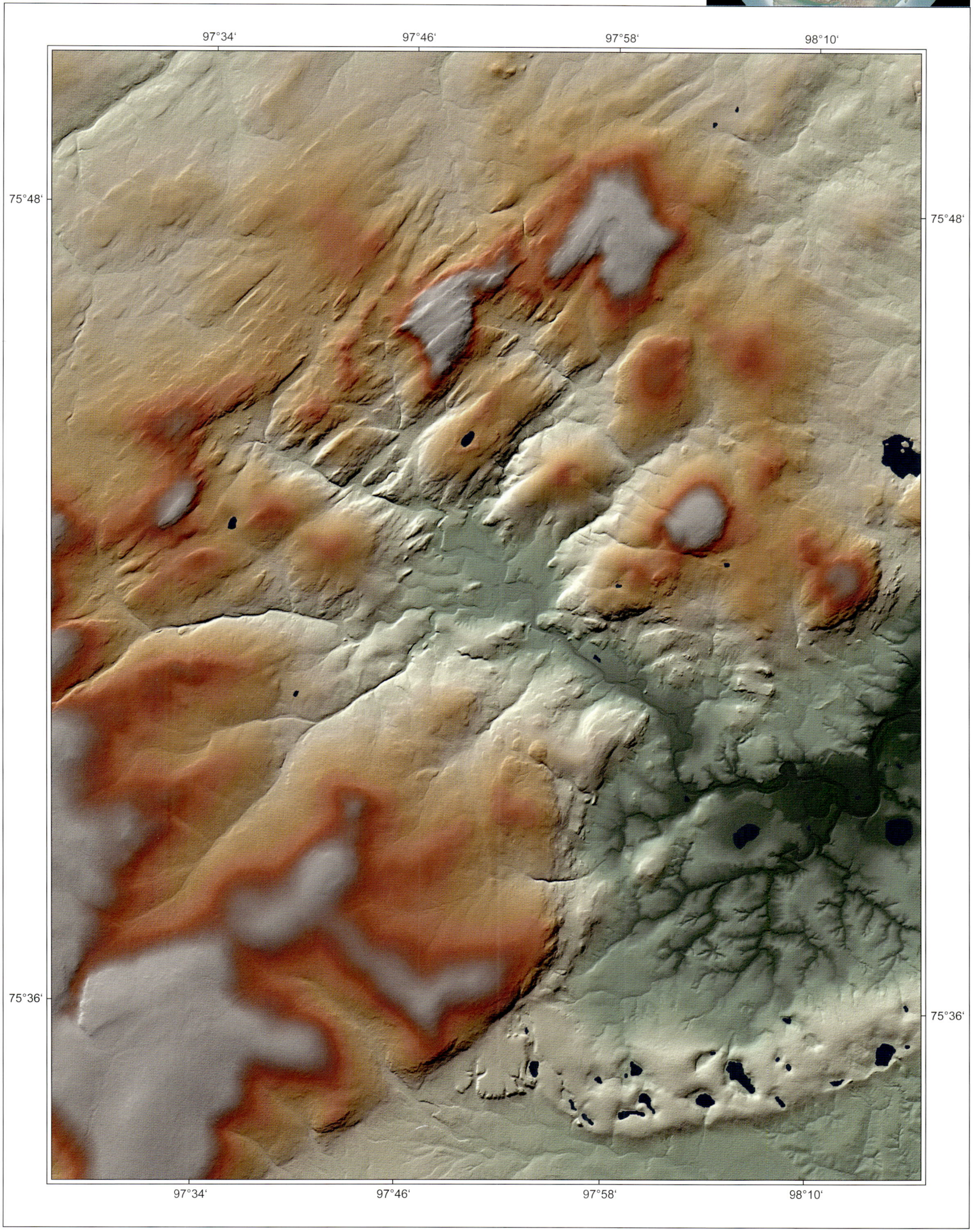

scale = 1: 132 000 / scene = 23.7 × 31.6 km²
0 1.32 6.60 km
N
35 meter 340

Dhala
India

Thomas Kenkmann

Dhala is a deeply eroded impact structure in India that is hardly visible in satellite images. A striking feature is a flat-top mesa asymmetrically disposed in the central part of the structure. This central elevated area is built up by sediments that cover the structure and, thus, are younger than the impact. The present-day apparent diameter of the structure is estimated at 12 km on the basis of the mapped distribution of impact breccias.

The structure is located on the Bundelkhand craton, which is one of five Archean shields in India. Target rocks at Dhala are mainly 2500 Myr old granitoids, with additional, very large quartz veins and mafic intrusions. The sedimentary cover comprises sandy siltstones with conglomerate lenses and an intercalated shale-sandstone unit that is exposed at the top of the central hill. These sedimentary strata belong to the 1700 Myr old Vindhyan Supergroup.

The Dhala structure comprises two subdued rings around the central elevated area, which include pre-Vindhyan metasediments and monomict granitoid breccia, respectively. There are also scattered outcrops of impact melt breccia in a zone of some 12 km^2. Many clasts within the impact melt breccia exhibit diagnostic shock metamorphic features, including multiple sets of planar deformation features (PDFs) in quartz and feldspar, and granular textured zircon, besides occurrences of coesite, and the thermal alteration features ballen quartz and checkerboard feldspar. The chemical composition of the impact melt rock is similar to that of the granitoid target rocks. Drilling revealed that the coherent impact melt sheet is up to 130 m thick. An ^{187}Os-^{188}Os ratio of 0.133 and Cr versus Ir concentration systematics indicate a probable admixture of approximately 0.3 wt% of an extraterrestrial component (chondrite or iron meteorite), left over from the projectile, in the Dhala impact melt breccia.

Attempts to date the impact with radiometric techniques have so far not been successful. Dating by the ^{40}Ar-^{39}Ar step-heating method indicated that the Dhala impact melt rock was affected by a strong thermal/hydrothermal overprint at ca. 1000 Myr and suffered further loss of radiogenic argon around 500 Myr ago. The U-Pb data yielded ages of 2550–2560 Myr that relate to the age of the granitoid basement. Thus, the Dhala impact event must have occurred between formation of the target rocks and deposition of the overlying Vindhyan Supergroup – between 1700 and 2500 Myr. The entire region was overprinted by a thermal event during Grenvillian times at about 1000 million years ago.

Initially the Dhala structure was interpreted as a "cauldron structure" of "cryptovolcanic" origin. The impact origin was established in 2005 by Jayanta Pati and co-workers.

The central elevated area (CEA, in the background) in the innermost part of the Dhala impact structure. The CEA comprises sedimentary rocks of the Dhala Formation and Vindhyan Supergroup. In the foreground monomict granitoid breccia, and in between deposits of post-impact sediments of the annular trough crop out between the CEA and the monomict breccia crop out. The hammer is 28 cm long and its head points towards the north (image courtesy: Jayanta Pati, University of Allahabad).

Location/Access

The Dhala structure is situated some 50 km west-southwest of the town of Jhansi, Madhya Pradesh State, India, about 400 km south of Delhi. Access from Delhi via Agra to Jhansi is on highways AH 1 and AH 43, and from Jhansi westwards towards the city of Shivpuri on Highway NH 25. A regional road branches from NH 25 to the center of the Dhala structure.

The central elevated area (CEA) of the Dhala structure viewed from the southeast. A large quartz vein occurs in the foreground, where the scene corresponds to a width of about 3 m. In between occur whaleback outcrops of monomict impact breccia (image courtesy: Jayanta Pati, University of Allahabad).

Part of a large outcrop of monomict breccia (length of hammer, for scale, 28 cm) comprising granitoid clasts in a clastic matrix of granitic composition. The hammer head points towards the north (image courtesy: Jayanta Pati, University of Allahabad).

Dhala

India

latitude = 25°17‘55“ N / longitude = 78°08‘33“ E / diameter = 12 km / age = 1700 - 2500 Myr

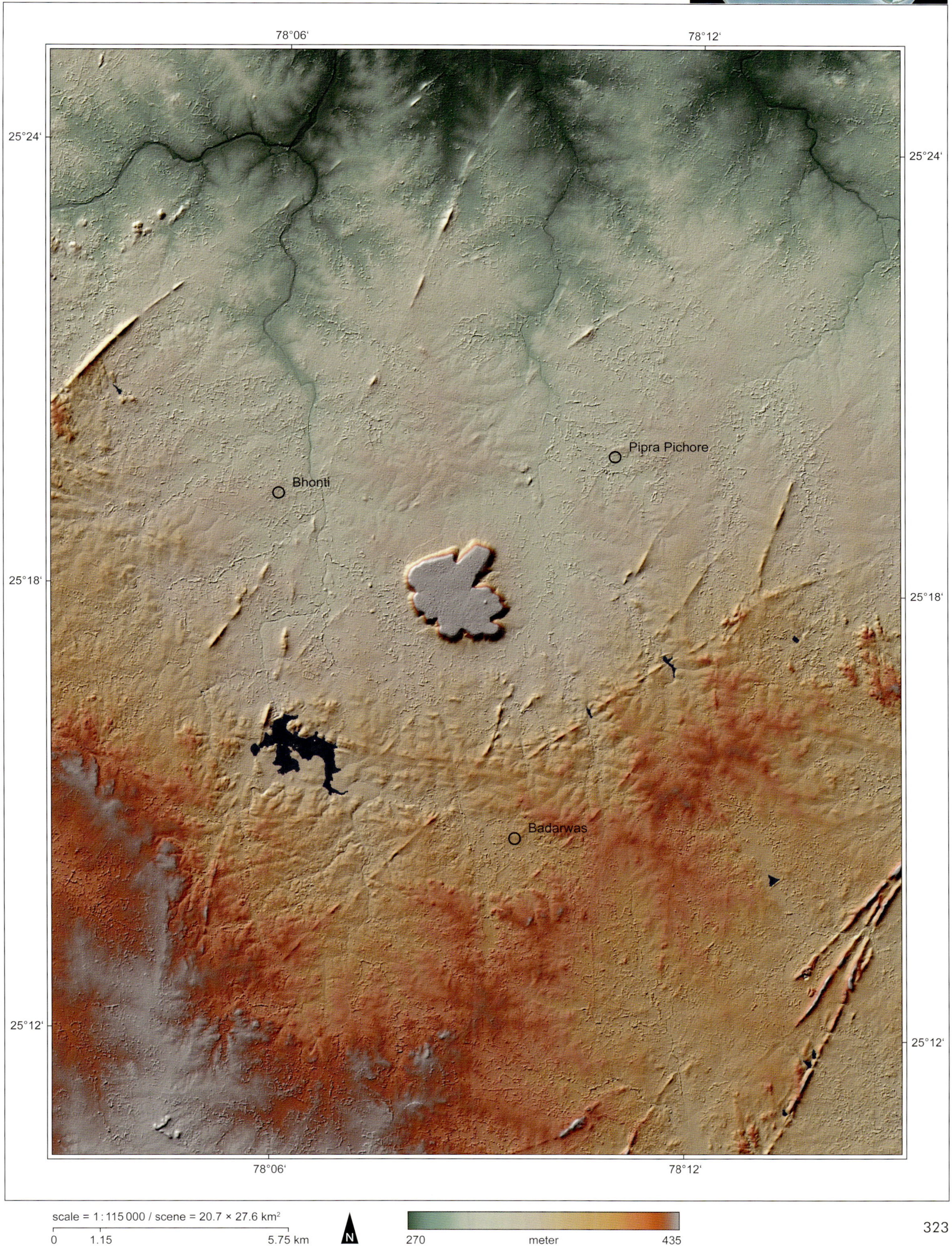

Impact melt breccia (diameter of the coin, for scale, 2.5 cm) with lithic (granitoid), mineral (feldspar and quartz), and impact melt clasts, all set in a mostly devitrified melt matrix (image courtesy: Jayanta Pati, University of Allahabad).

Dahla impact melt breccia, 13 × 7 × 5 cm in size (photo courtesy: Ralf-Thomas Schmitt, Museum für Naturkunde Berlin, sample from Jayanta Pati, University of Allahabad).

Selected References

Pati, J. K., Reimold, W. U., Koeberl, C., and Pati, P. (2008). The Dhala structure, Bundelkhand craton, Central India – Eroded remnant of a large Paleoproterozoic impact structure. Meteoritics and Planetary Science, Vol. 43, pp. 1383–1398.

Pati, J. K., Jourdan, F., Armstrong, R., Reimold, W. U., and Prakash, K. (2010). First SHRIMP U-Pb and ^{40}Ar-^{39}Ar chronological results from impact melt breccia from the Paleoproterozoic Dhala impact structure, India. Geological Society of America Special Paper, Vol. 465, pp. 571–591.

Pati, J. K., Qu, W. J., Koeberl, C., Reimold, W. U., Chakarvorty, M., and Schmitt, R. T. (2017). Geochemical evidence of an extraterrestrial component in impact melt breccia from the Paleoproterozoic Dhala impact structure, India. Meteoritics and Planetary Science, Vol. 52, pp. 722–736.

The Dhala impact structure on September 19, 2019 as a composite image based on TanDEM-X DEM and Copernicus Sentinel-2 (channel 2/3/4) data.

El'gygytgyn
Russia

Thomas Kenkmann

El'gygytgyn in Russia's most distal, Arctic, northeastern part is a well-preserved but hardly accessible complex impact structure of 18 km diameter. The crater has a morphologically visible crater rim with an average rim height of 150 m. The crater depression is to a large extent occupied by the 170 m deep and 12 km wide Lake El'gygytgyn. As the crater form has been asymmetrically filled from the west and north, the lake occupies, in particular, the southeastern part of the basin. The lake is a first-class archive for paleoclimate reconstruction for the high latitudes – for an extended timespan of several million years.

The impact structure is located in the central part of the Late Mesozoic Ochotsk-Chukotsky Volcanic Belt (OCVB). The crater was formed in mostly felsic volcanic rocks. These are mainly rhyolites, and subordinate andesites and dacites, that occur as both dense siliceous lavas and tuffs. Felsic ignimbrites occur widespread, and minor basalt as well as ultramafic components were likely part of the target geology. ^{40}Ar-^{39}Ar dating of some volcanic rocks yielded ages of 86–89 Myr for the main volcanic eruption episode of the OCVB. The surface outcrop conditions are restricted to the rim of the crater, as large parts of the crater interior are submerged and a system of lacustrine terraces exists to an elevation of up to 80 m above lake level.

El'gygytgyn impact melt breccia, 7 × 5.5 × 3 cm in size (photo courtesy: Ralf-Thomas Schmitt, Museum für Naturkunde Berlin, sample from Ulli Raschke and Patrice Zaag, formerly Museum für Naturkunde Berlin).

The crater rim is dissected by gullies and is entirely eroded by river valleys in the southeast. In contrast, the western rim is elevated up to 400 m above the lake. The rocks along the rim do not contain any evidence of shock metamorphism but show intense fracturing and, locally, brecciation. The intensity of fracturing decreases with radial distance from the crater. The rim has an asymmetric profile in cross-sectional view, with the inner wall being steeper than the outer flank. An outer concentric ring seems to be present at a radial distance of 15.5–16 km from the center of the crater and leads to deflection of tributary creeks into a more concentric direction. Future studies may test whether this outer concentric ring could be a remnant of a rampart of a fluidized ejecta blanket.

A seismic profile across the lake showed that the post-impact sediments locally reach a thickness of 400 m. The underlying rocks appear brecciated to a depth of 500–600 m. An uplift is indicated in the center of the crater structure. It develops into a ring structure of about 7–7.5 km in diameter at shallower depth. However, the topography of the ring is buried by overlying fallback breccia.

A borehole drilled by the International Continental Scientific Drilling Program (ICDP) was located slightly offset from the crater center. This core reveals a section through the crater structure. The post-impact lake deposits reach a thickness of 315 m. The uppermost impactite unit consists of a mixture of lacustrine sediments and reworked impact lithologies to a depth of 328 m. Polymict lithic impact breccia, mostly suevites, characterize the sequence to 385 m. They contain both volcanic and impact melt clasts, and shocked minerals in a fine-grained clast-dominated matrix. The distinction between original volcanic melt fragments and impact melt is difficult. Beneath this unit is a 40 m thick sequence of brecciated volcanic rocks of felsic and basaltic composition. The lowermost unit to the bottom of the core at 517 m below lake level comprises monomict-brecciated rhyo-dacitic ignimbrite.

El'gygytgyn has a negative residual Bouguer gravity anomaly of −10 mGal of 18 km diameter. A nested, 2 km wide positive anomaly of +2 mGal corresponds to the central uplift in the central part of the crater structure. Radiometric dating of the impact by the ^{40}Ar-^{39}Ar step-heating method applied to impact glasses indicated that the impact event took place at 3.58 ± 0.04 Myr ago.

El'gygytgyn was discovered in 1933 from an aircraft and was first considered a huge volcanic crater. Suggestions in the Russian literature that El'gygytgyn might be of impact origin date back to 1963 and 1970. Robert Dietz and John McHone studied Landsat imagery of the structure and concluded in 1976 that El'gygytgyn was probably the largest Quaternary impact crater preserved on Earth. An expedition to the crater, led by Eugene Gurov, confirmed the impact hypothesis for the El'gygytgyn structure in 1978, after the team had discovered shock metamorphosed rocks, impact glasses, and impact melt rock. Further investigations followed by this group. Between October 2008 and May 2009, the ICDP carried out a deep drilling campaign at Lake El'gygytgyn, with twin objectives – namely to provide a paleoclimatic record for this Arctic region and to investigate a unique impact structure formed in – mostly felsic – volcanic rocks. This highly successful and technologically challenging drilling campaign recovered, from four holes at two sites, a total of 642.3 m drill core.

Panoramic view of the El'gygytgyn impact crater, taken in late July 2011 (image courtesy: Ulli Raschke, formerly Museum für Naturkunde Berlin).

Location/Access

Lake El'gygytgyn is located one hundred kilometers north of the Arctic Circle in remote Chukotka, northeastern Siberia. Access to the structure requires thorough expedition planning. The nearest town with an airstrip is Pevek at the coast of the East Siberian Sea some 250 km further north. From there only helicopter access to the crater structure is an option.

Selected References

Dietz, R. S., and McHone, J. F. (1976). El'gygytgyn: Probably the world's largest meteorite crater. Geology, Vol. 4, pp. 391–392.

Gurov, E. P., Koeberl, C., Reimold, W. U., Brandstätter, F., and Amare, K. (2005). Shock metamorphism of siliceous volcanic rocks of the El'gygytgyn impact crater (Chukotka, Russia). In: Large Meteorite Impacts III. Kenkmann, T., Hörz, F., and Deutsch, A. (eds.), Geological Society of America Special Paper, Vol. 384, pp. 391–412.

Gurov, E. P., Koeberl, C., and Yamnichenko, A. (2007). El'gygytgyn impact crater, Russia: Structure, tectonics and morphology. Meteoritics and Planetary Science, Vol. 42, pp. 307–319.

Koeberl, C. Pittarello, L., Reimold, W. U., Raschke, U., Brigham-Grette, J., Melles, M., and Minyuk, P. (2013). El'gygytgyn impact crater, Chukotka, Arctic Russia: Impact cratering aspects of the 2009 ICDP drilling project. Meteoritics and Planetary Science, Vol. 48, pp. 1108–1129.

Layer, P. W. (2000). Argon-40/Argon-39 age of the El'gygytgyn event, Chukotka, Russia. Meteoritics and Planetary Science, Vol. 35, pp. 591–599.

Raschke, U., Schmitt, R. T., McDonald, I., Reimold, W. U., Mader, D., and Koeberl, C. (2015). Geochemical studies of impact breccias and country rocks from the El'gygytgyn impact structure, Russia. Meteoritics and Planetary Science, Vol. 50, pp. 1071–1088.

Raschke, U., Schmitt, R. T., and Reimold, W. U. (2013). Petrography and geochemistry of impactites and volcanic bedrock in the ICDP drill core D1c from Lake El'gygytgyn, NE Russia. Meteoritics and Planetary Science, Vol. 48, pp. 1251–1286.

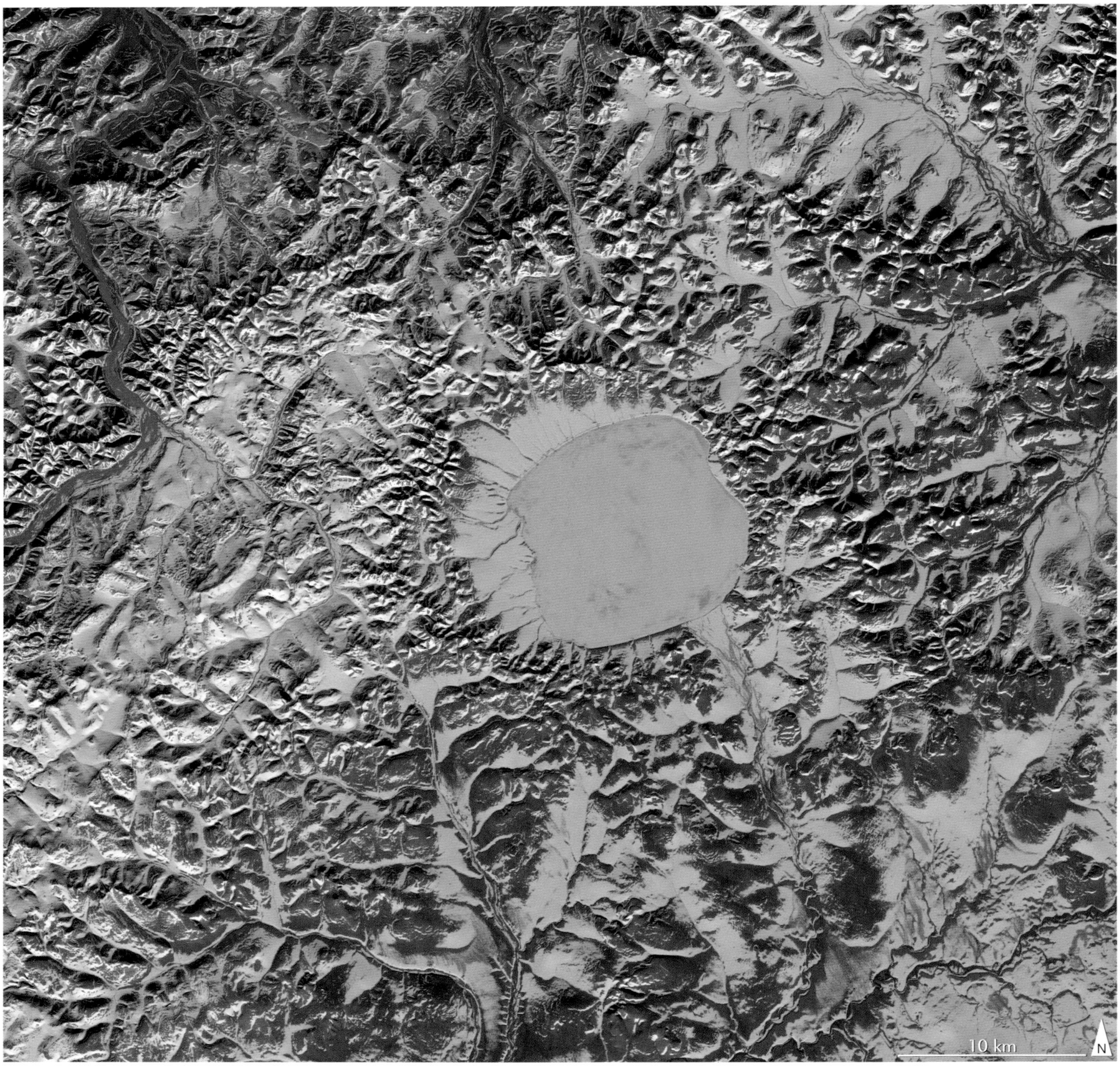

The El'gygytgyn impact crater on June 14, 2018 as a composite image based on TanDEM-X DEM and Copernicus Sentinel-2 (channel 2/3/4) data. Even at the end of spring, the scene gives a wintry impression.

El‘gygytgyn

Russia

latitude = 67°29‘31“ N / longitude = 172°03‘37“ E / diameter = 18 km / age = 3.58 ± 0.04 Myr

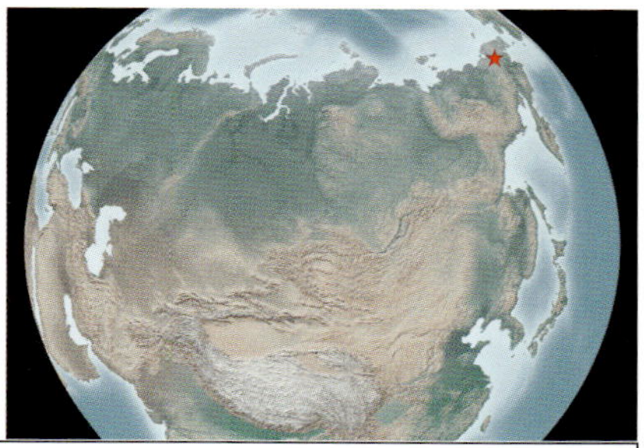

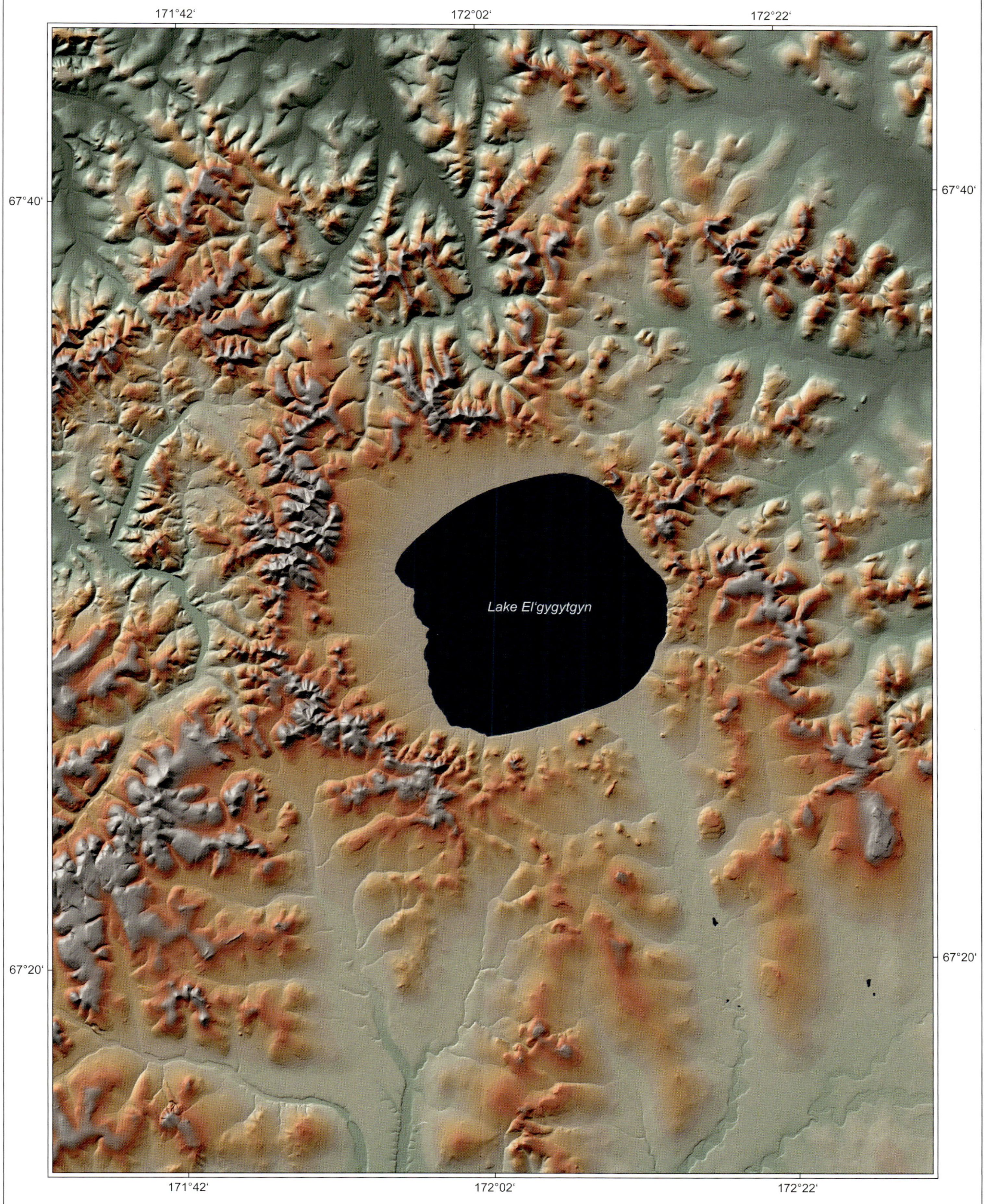

scale = 1:230 000 / scene = 41.4 × 55.2 km²

0 2.30 11.50 km

N

255 meter 935

Jebel Waqf as Suwwan
Jordan

Thomas Kenkmann

Jebel Waqf as Suwwan is a well-exposed, complex impact structure of 6.1 km diameter in the desert of eastern Jordan. It is located at the western shoulder of the Wadi Sirhan basin, where the sedimentary strata dip very gently towards the northeast. The structure is situated only a kilometer west of the escarpment of the Eastern Jordanian Chalk Plateau that here forms a promontory. Regressive erosion of the escarpment appears to have been reduced by the presence of the crater structure, which protected the escarpment from a further eastward shift.

Originally the crater diameter was about 7 km. After erosion by some 300–400 m, the ejecta and the crater fill breccia have been removed, exposing the sub-surface of the deformed crater. Today, the crater presents a superb exposure of the central uplift and the crater rim. In terms of morphology, the Jebel Waqf as Suwwan structure exhibits a ring-like elevated outer terrace area that is paved with chert and is of about 5.5 km diameter, and a 1 km wide central uplift that features a depression in its center. This central uplift is expressed by a collar of hills that rises 20–40 m above the surrounding plain.

The topographically elevated rim of Jebel Waqf as Suwwan represents the outermost terrace of the crater rim section, which slumped into the crater depression. This area is composed of white marls and massive chert beds of Eocene age, the youngest strata affected by the impact event. They also provide an upper age limit for the cratering event of about 37 Myr. The crater rim itself is located outside of the elevated area and is well exposed by a distinct circumferential fault that commonly dips towards the crater interior. The moat between central uplift and crater rim is largely covered by alluvial wadi sediments. The Cretaceous rocks of the central uplift are mainly cherts and limestones. These strata are divided into 100 m sized blocks that are internally folded and bent. The central uplift has a southwest-northeast trending symmetry axis. The southwestern sector of the central uplift is built up by normally dipping strata and radial folds with outward plunging fold axes, as opposed to the northeastern sector, where strata and fold axes are overturned. The central uplift vergence indicates a top-to-the-northeast shearing that is explained by a horizontal component of the very oblique impact trajectory. According to this, the impactor came from the southwest, defining the up-range sector of the crater. Strata in the moat appear to dip very gently within the cross-range sectors but show particularly steep dips in the down-range sector (with respect to the impact direction). The down-range (northeast) sector displays more intense deformation than the other sectors. Faults parallel to the trajectory (southwest-northeast) are dominant, especially in the up-range and down-range sectors.

A first regional geological reconnaissance of Jebel Waqf as Suwwan was conducted in the late 1960s as part of a German regional mapping project of the territory of Jordan. During this survey the unusual structural disturbance in an otherwise undeformed sedimentary environment characterized by flat-lying strata was recognized. The disturbance was interpreted as a structure of "cryptovolcanic" origin. In 2006, Elias Salameh of the University in Amman first proposed an impact origin for Jebel Waqf as Suwwan based on shatter cone findings. In 2008, besides abundant shatter cones in limestones and some rarer ones in sandstone from the central uplift, a few sets of planar deformation features (PDFs), planar fractures (PFs), and feather features (FFs) were documented by Wolf Uwe Reimold and the Jordan team. This unambiguously confirmed the impact hypothesis. Later, a German-Jordanian consortium analyzed the crater in detail. Geological mapping was conducted by Thomas Kenkmann and team, and geophysical analysis was carried out by Till Heinrichs and co-workers.

The discovery of this first impact structure in the Middle East initiated the First Arab Impact Cratering and Astrogeology Conference (AICAC) in Amman in 2009. It has, since, been followed by three further such events in Morocco and Algeria, with the purpose to address and fertilize impact cratering studies in the Arab world.

Panoramic view of the Jebel Waqf as Suwwan impact structure from the eastern crater rim, looking over the central uplift and annular moat of the structure. The bright colored rocks near the center of the structure are Turonian limestones that are surrounded by a collar of chert of Santonian age (photo: T. Kenkmann).

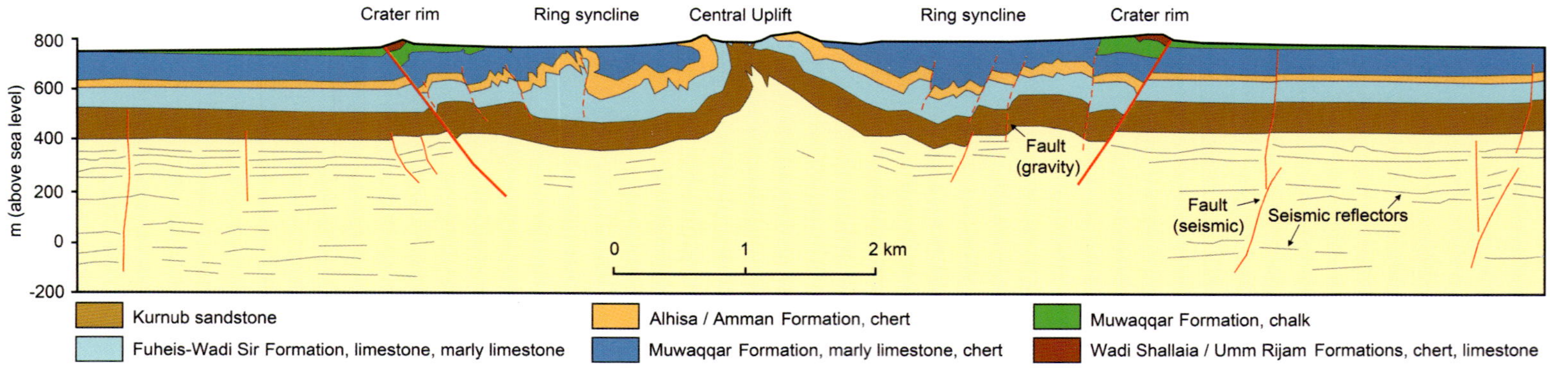

North-south cross-section through the Jebel Waqf as Suwwan impact structure. The construction relies on field and reflection seismic data (modified after Kenkmann et al., 2017).

Jebel Waqf as Suwwan

Jordan

latitude = 31°02‘54“ N / longitude = 36°48‘24“ E / diameter = 6.1 km / age = < 37 Myr

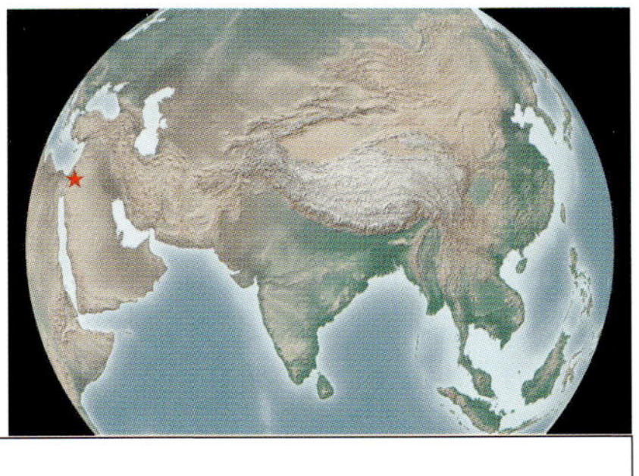

scale = 1: 145 000 / scene = 26.1 × 34.8 km²

0 1.45 7.25 km

N

650 meter 985

Exposed normal fault defining the southern rim of the Jebel Waqf as Suwwan impact structure. The rim fault is striking concentrically and is dipping towards the crater center (photo: T. Kenkmann).

Intense deformation of red claystone and yellowish marly limestone of Cenomanian age (Naur Formation) near the center of the Jebel Waqf as Suwwan structure (photo: T. Kenkmann).

Shatter cones formed in Upper Cretaceous chert of the Alhisa/Amman Formation, from the central uplift of the Waqf as Suwwan impact structure (photo: T. Kenkmann).

Location/Access

The crater is situated in the desert of eastern Jordan about 20 km west of the border to Saudi Arabia. Travel to Waqf as Suwwan requires four-wheel drive, replacement tires, and off-road experience to access the structure. Most visitors would come from Amman via Highway 40 to Azraq, and from there head southward on Highway 5 for about 70 km. Shortly after a police and service station a track branches off to the east. It continues to run southeast for about 50 km, right into the structure.

An intricately formed shatter cone in marly limestone from the Jebel Waqf as Suwwan impact structure. Coin for scale about 2 cm in diameter (photo: W. U. Reimold).

Selected References

Heinrichs, T., Salameh, E., and Khoury, H. (2014). The Waqf as Suwwan crater, Eastern Desert of Jordan: aspects of the deep structure of an oblique impact from reflection seismic and gravity data. International Journal of Earth Sciences, Vol. 103, pp. 233–252.

Kenkmann, T., Reimold, W. U., Khirfan, M., Salameh, E., Khoury, H., and Konsul, K. (2010). The complex impact crater Jebel Waqf as Suwwan in Jordan: effects of target heterogeneity and impact obliquity on central uplift formation. In: Large Meteorite Impacts and Planetary Evolution IV. Gibson, R. L., and Reimold, W. U. (eds.), Geological Society of America Special Paper, Vol. 465, pp. 471–487.

Kenkmann, T., Sturm, S., Krüger, T., Salameh, E., Al-Raggad, M., and Konsul, K. (2017). The structural inventory of a small complex impact crater: Jebel Waqf as Suwwan, Jordan. Meteoritics and Planetary Science, Vol. 52, pp. 1351–1370.

Salameh, E., Khoury, H., and Schneider, W. (2006). Jebel Waqf as Suwwan, Jordan, a possible impact crater – a first approach. Zeitschrift der deutschen Gesellschaft für Geowissenschaften, Vol. 157, pp. 319–325.

Salameh, E., Khoury, H., Reimold, W. U., and Schneider W. (2008). The first large meteorite impact structure discovered in the Middle East: Jebel Waqf as Suwwan. Meteoritics and Planetary Science. Vol. 43, pp. 1681–1690.

The Jebel Waqf as Suwwan impact structure on September 15, 2018 as a composite image based on TanDEM-X DEM and Copernicus Sentinel-2 (channel 2/3/4) data.

Logancha
Russia

Thomas Kenkmann

Logancha is an eroded, about 14 km wide impact structure in the central part of the Lower Triassic basalt plateau of eastern Siberia. The structure involves a pronounced circular depression that is 400 m lower than the surrounding ring of hills. However, the southwestern part of the rim has been eroded away, and here the circular catchment area drains towards the Vivi River in the west. On the slopes of the gently dipping depression, a number of basalt lava sheets is exposed along deeply incised creeks. Target rocks also comprise basaltic tuffs, which are intruded by dolerites.

Aerial photography has documented a number of radial and concentric faults around the depression. In the center of the basin, a group of hills within an approximately 4 km wide area consists of disturbed blocks of Upper Permian coal-bearing siltstones and claystones, and Lower Triassic slates and sandstones. The deformed beds dip steeply in various directions. Shock deformation features include shatter cones in the central area. Planar deformation features (PDFs) in quartz and isotropization of plagioclase grains (diaplectic plagioclase glass) have been described from thin sections.

According to gravity and magnetic anomaly maps, the diameter of Logancha is 14 km. However, due to the considerable incision of the river into breccias and brecciated bedrock lithologies of the crater interior, and due to slumping of the original crater rim, the topographic depression with a diameter of 20 km is wider than the original size of the impact structure. Based on geological observations, a Paleogene age has been suggested for the impact event.

Logancha was first mentioned in 1980 in a Russian book on the geology of astroblemes, with descriptions of all impact structures that were known to exist in the former Soviet Union as of 1978. Several studies on Logancha appeared shortly thereafter but modern publications are lacking.

Location/Access

Logancha is a remote place in Siberia in the oblast Krasnojarsk. The town of Tura is located about 240 km to the southeast of the structure and has an airstrip.

Selected References

Fel'dman, V. I., Mironovy, Y. U., Melikhov, V. R., Ivanov B. A., and Basilevskay. T. (1985). Astroblemes on trap – peculiarities of the inner structure, distinction from the impact structure on the other targets (in Russian). Meteoritika, Vol. 44, pp. 139–145.

Masaitis, V. L. (1999). Impact structures of northeastern Eurasia: The territories of Russia and adjacent countries. Meteoritics and Planetary Science, Vol. 34, pp. 691–711.

Masaitis, V. L., Danilin, V. N., Mashchak, M. S., Raikhlin, A. I., Selivanovskaty, A. V., and Shadenkov, E. M. (1980). The Geology of Astroblemes. Nedra Press, Leningrad, Russia. 231 pp.

The Logancha impact crater on February 21, 2018 as a composite image based on TanDEM-X DEM and Copernicus Sentinel-2 (channel 2/3/4) data.

Logancha

Russia

latitude = 65°29'54" N / longitude = 95°57'15" E / diameter = 14 km / age = 23 - 66 Myr

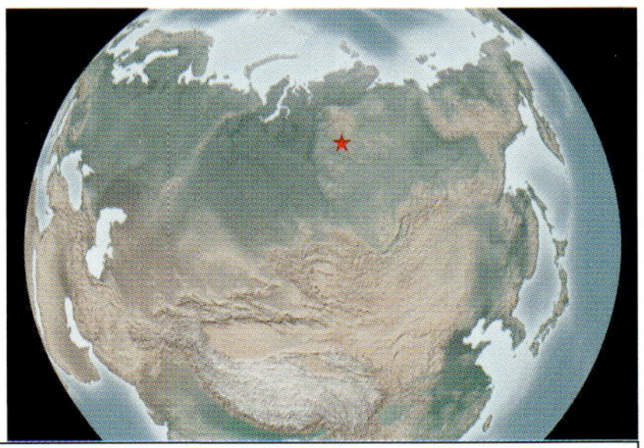

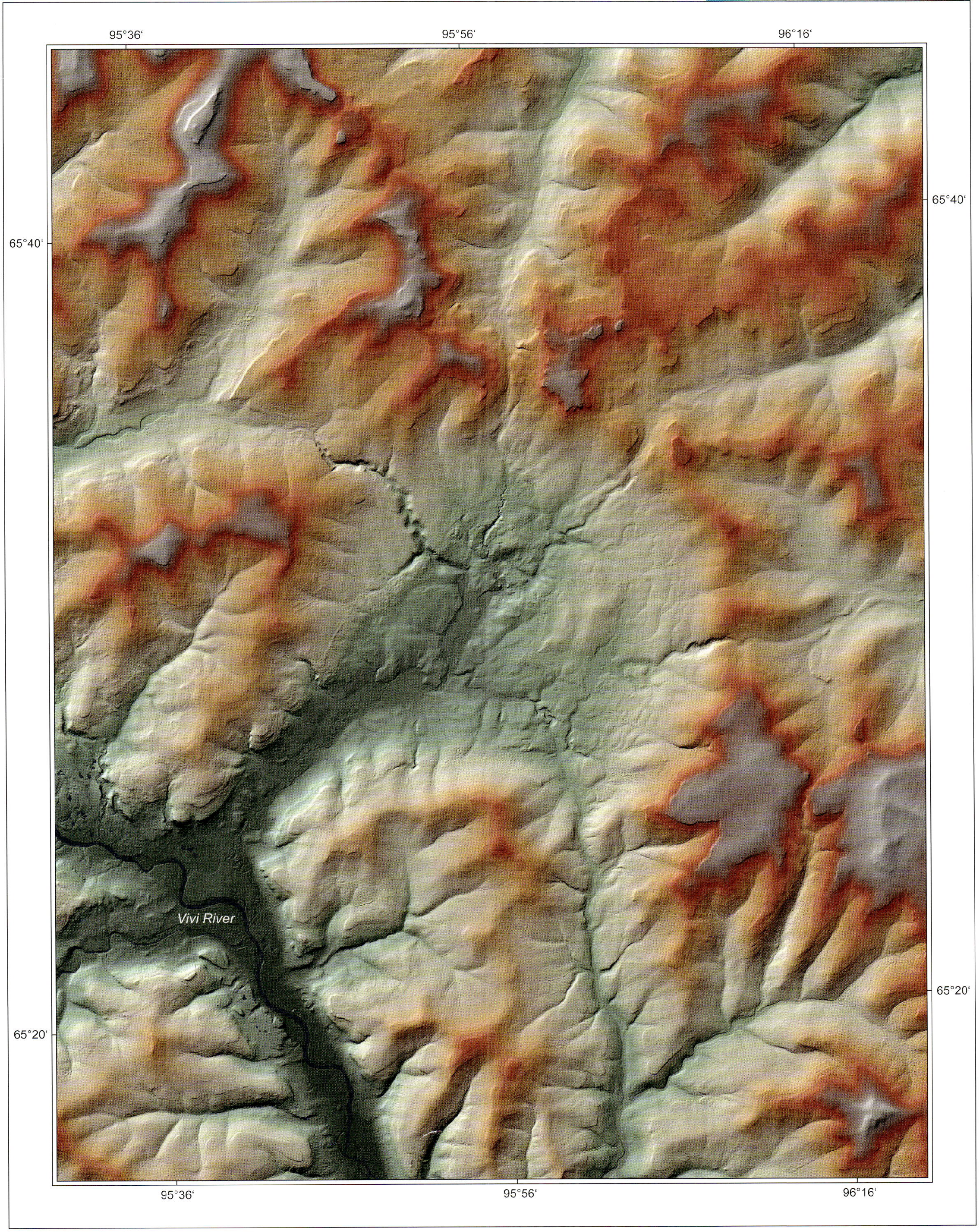

scale = 1 : 222 000 / scene = 39.9 × 53.3 km²

0 2.22 11.10 km

N

160 meter 920

Lonar
India

Thomas Kenkmann

Lonar Crater in central India is a slightly eroded, simple bowl-shaped impact structure of 1.88 km diameter. Lonar is special, as it is one of very few terrestrial impact craters formed in basalt, which makes it a sought-after planetary laboratory. This crater also has a well-preserved ejecta blanket. Overall, Lonar has received a lot of attention, as it is considered an analog structure for small craters on basaltic planetary surfaces.

The Lonar impact occurred in the region of the ca. 66 Myr old Deccan Trap volcanic province. The target rocks for this impact were exclusively basalts that form a whole series of thick flows that are superimposed onto each other. At least six such flows can be distinguished in the Lonar crater rim, each with thicknesses of 10–25 m. Each eruptive phase starts with pahoehoe lava flows and grades upward into massive, fine-grained basalt. The upper portions of a layer often show columnar jointing, vesiculation, and color change due to weathering.

The rim of the almost circular Lonar crater rises some 30 m above the surrounding plains. The crater depression is 120 m deeper than the crest. The crater interior is covered by a shallow, saline lake of less than 6 m water depth. Drilling revealed that the crater cavity is filled with more than 225 m of breccia that is overlain by 30–100 m of unconsolidated lake sediments. Along the inner wall of the crater rim, basaltic layers are dipping radially outward. Steepness of dip increases upslope. Near the rim crest, layers are overturned and form the most proximal part of the ejecta blanket. A circumferential recumbent fold is partly preserved. In addition to that, a normal fault is observed along the crater wall. Locally, the disintegration of the proximal ejecta blanket to a chaotic brecciated block field can be observed to start already at the rim crest. The emplacement of the continuous ejecta blanket onto the surrounding plain indicates a ground-hugging debris flow, as the underlying soil was incorporated into the ejecta and no stratification or size sorting have been detected. The ejecta do not show a systematic decrease in mass with radial distance from the crater but form thickened ramparts at their most distal extent. This is also known from the **Ries** and **Bosumtwi** craters, and from Martian impact craters, and suggests that ejecta fluidization processes were involved in deposition. The continuous ejecta blanket was only weakly affected by later fluvial incision and erosion. Portions of the ejecta blanket are overlain by a thin veneer of aerodynamically and rotationally sculpted glassy impact spherules, in particular at the eastern and western rim, as well as in a depression 700 m north-northeast of the crater that is known as *Little Lonar*. Little Lonar is an elliptical depression of 300 m width. The structure has a rim of more than 10 m elevation. It is currently a matter of debate whether Little Lonar is a crater formed simultaneously with the main crater, represents an unusually large secondary crater, or whether it is at all of natural origin.

The basaltic target rocks at Lonar are mainly composed of plagioclase and pyroxene. The lack of quartz in these rocks, the most important diagnostic mineral for the documentation of shock deformation, required to find shock features in plagioclase and pyroxene. Experimental studies showed that plagioclase is converted to diaplectic feldspar glass, known as maskelynite, and eventually transforms to vesiculated feldspar glass at higher shock pressures. These shock features could be documented at Lonar in the breccia lens beneath the lake and in trenches on the rim. Clasts of this breccia also show shatter cones.

Coarse ejecta on the outer slope of Lonar's northwestern crater rim (image courtesy: Amar Agarwal, University of Freiburg).

The pristine crater morphology indicates that Lonar is the result of a geologically young impact event. The most recent and probably most accurate age determination with the ^{40}Ar-^{39}Ar step-heating technique applied to four impact melt rock samples yielded an age of 0.570 ± 0.047 Myr for the impact event. Geochemical analysis of impact melt rock and spherules shows some elevated abundances of siderophile elements. The ratios of these elements indicated a contribution from a chondritic projectile to basaltic target rock.

The current remanent magnetization signature of the basaltic target rocks at Lonar consists of the original magnetization inherited during cooling of the basaltic lava 66 Myr ago and a recent overprint. Only subtle effects from the impact on the rock magnetic properties seem to exist. A gravity anomaly of +2.5 mGal coincides with Lonar crater. For a long time Lonar was considered to be of volcanic origin due to the omnipresence of basaltic rocks and the morphology of the crater.

Panoramic view of the Lonar impact crater taken from the southeastern crater rim (image courtesy: Amar Agarwal, University of Freiburg).

Lonar
India

latitude = 19°58‘36“ N / longitude = 76°30‘32“ E / diameter = 1.88 km / age = 0.570 ± 0.047 yr

scale = 1 : 83 000 / scene = 15.0 × 20.0 km²
0 0.83 4.15 km
N
485 meter 625

It was even suggested that the Deccan Trap basalts originated from this crater. Tom LaTouche modified this view and suggested in 1912 a steam explosion, because he realized that the crater landform had to be much younger than the plateau region formed by the volcanic rocks of the Deccan Traps. The hypothesis of a "cryptovolcanic" origin was supported by many geoscientists for a long time. However, in 1964 Eugene LaFond and Robert Dietz suggested that Lonar is a meteorite impact crater. Their argument was based on crater morphology, structure, and the presence of breccias. Unequivocal evidence in favor of an impact origin was provided by V. Nayak in 1972 and almost simultaneously by Kurt Frederiksson and colleagues in early 1973. The former documented glasses formed by shock, and the latter reported shatter cones and shocked feldspar. As Lonar is one of very few known terrestrial impact craters formed in basalt, it still plays a pivotal role for comparative studies of terrestrial and planetary craters.

Location/Access

Lonar Crater is located in the Buldhana District of Maharashtra State in central India. The town of Lonar is built on the eastern ejecta blanket close to the crater rim. The largest city with an airport close to Lonar is Aurangabad, about 120 km to the west. From Aurangabad the traveler drives on Highways 30, 177 and 183 to Sultanpur. The crater lies only about 10 km to the south of this town.

Selected References

Fredriksson, K., Dube, A., Milton, D., and Balasundaram, M. (1973). Lonar Lake, India: an impact crater in basalt. Science, Vol. 180, pp. 862–864.

Fudali, R., Milton, D., Fredriksson, K., and Dube, A. (1980). Morphology of Lonar crater, India: comparisons and implications. Moon and Planets, Vol. 23, pp. 493–515.

Gupta, R. D., Banerjee, A., Goderis, S., Claeys, P., Vanhaecke, F., and Chakrabarti, R. (2017). Evidence for a chondritic impactor, evaporation-condensation effects and melting of the Precambrian basement beneath the 'target' Deccan basalts at Lonar crater, India. Geochimica et Cosmochimica Acta, Vol. 215, pp. 51–75.

Jourdan, F., Moynier, F., Koeberl, C., and Eroglu, S. (2011). ^{40}Ar-^{39}Ar age of the Lonar crater and consequence for the geochronology of planetary impacts. Geology, Vol. 39, pp. 671–674.

Kieffer, S. W., Schaal, R. B., Gibbons, R., Hörz, R., Milton, D. J., and Dube, A. (1976). Shocked basalt from Lonar impact crater, India, and experimental analogues. Proceedings of the 7th Lunar Science Conference, Houston, Texas, Pergamon Press, New York, pp. 1391–1412.

LaFond, C., and Dietz, R. S. (1964). Lonar Crater, India, a meteorite crater? Meteoritics, Vol. 2, pp. 111–116.

LaTouche, T., and Christie, W. (1912). The geology of the Lonar Lake. Records of the Geological Survey of India, Vol. 41, pp. 266–275.

Maloof, A. C., Stewart, S. T., Weiss, B. P., Soule, S. A., Swanson-Hysell, N. L., Louzada, K. L., Garrick-Bethell, I., and Poussart, P. M. (2010). Geology of Lonar Crater, India. Bulletin of the Geological Society of America, Vol. 122, pp. 109–126.

Nayak, V. K. (1972). Glassy objects (impactite glasses?): A possible new evidence for meteoritic origin of the Lonar Crater, Maharashtra State, India. Earth and Planetary Science Letters, Vol. 14, pp. 1–6.

Macha Crater Field

Russia

Thomas Kenkmann

A group of five morphologically prominent, small impact craters is located in the basin of the Macha River, a tributary of the Lena River in western Yakutia. The craters have sizes of 300, 180, 90, 70, and 60 m, and maximum depths of 40, 26, 14, 12, and 19 m, respectively. The four largest craters are partly filled with water. The two largest craters overlap and form a double depression that is occupied by a lake. This crater strewn field has a total dimension of 750 × 500 m and is elongated in northwest-southeast direction. The largest craters cluster in the southeastern part of the strewn field ellipse and indicate a relatively steep impact from the northwest.

The shape of the largest (300 m) crater is irregular due to an extended embayment in its eastern part. Its rim rises about 8 m above lake level. The second largest crater is surrounded, from northwest to west, by an ejecta layer of variable thickness ranging from 1.5 to 3 m and a width ranging from 8 to 17 m. An alluvial fan modified the western part of this crater. North of it, with overlapping ejecta, lies the 70 m crater that is surrounded by a rim up to 3.5 m high and up to 17 m wide, which is covered by ejecta. This crater is filled with water. A perfectly circular 90 m crater is situated some 100 m northwest of the previous one. The inner crater wall has steep slopes but no rim around the crater has been preserved. The smallest (60 m) crater has a cone-like shape and is 19.5 m deep, is filled with a little pond, and is surrounded by a raised ejecta-covered rim up to 2 m high and 8–10 m wide.

The bolides impacted a sequence of Late Proterozoic sandstone, quartzite, and limestone that is overlain by up to 80 m thick Early Quaternary sand of eolian or fluvial-lacustrine origin. Only the two largest craters excavated the uppermost strata of the Proterozoic platform sediments below the sands, the others excavated only loose Quaternary sand.

Ejecta of the craters contain cataclased and brecciated fragments of the Proterozoic sedimentary rocks, as well as quaternary pebbles and gravels of quartzites and sandstones. Sometimes charred wood and charcoal fragments occur mixed into the sandy ejecta that cover a fossil soil horizon. Brecciated fragments reveal signs of shock metamorphism, such as narrow-spaced planar deformation features (PDFs) with an amorphous (glassy) phase. Quartz crystals show slightly lowered refractive indices. The occurrence of stishovite has also been reported. A shock pressure up to 12–15 GPa can be estimated for the shocked quartz. Impact melt rock has not been detected in the craters. A few metallic particles were found and show significant abundances of iridium and chromium, but the composition and nature of the projectile are still unknown. The age of the craters has been determined to 7315 ± 80 years by the radiocarbon method on charcoal and charred wood samples from ejecta.

The Macha craters were recognized for the first time by Eugene Gurov and Elena Gurova from aircraft in 1979. Field investigations of the craters were carried out from 1982 to 1983, and later in 1987.

Location/Access

The craters are located in the western Yakutian taiga that has very limited infrastructure. The crater field occurs about 19 km north of Nochtuisk village and the river Macha. The town of Olekminsk that has an airport lies about 170 km to the west.

Selected Reference

Gurov, E. P., and Gurova, E. P. (1998). The group of Macha craters in western Yakutia. Planetary Space Science, Vol. 46, pp. 323–328.

Macha Crater Field

Russia

latitude = 60°05'07" N / longitude = 117°39'08" E / diameter = max. 300 m / age = 7315 ± 80 yr

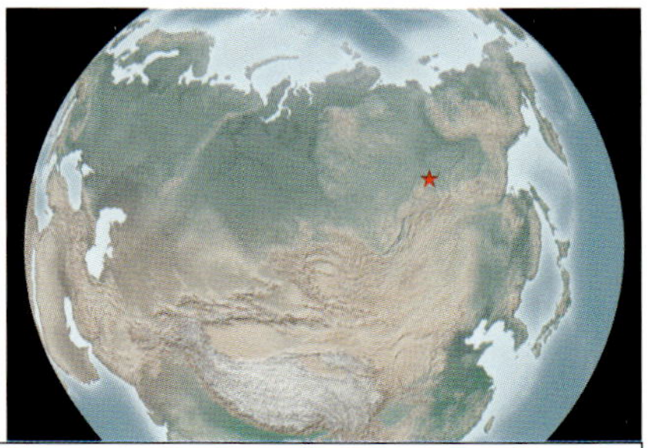

scale = 1:75 000 / scene = 13.6 × 18.1 km²

0 0.75 3.80 km

N

195 meter 430

Popigai
Russia

Thomas Kenkmann

The 100 km diameter Popigai impact structure is located on the northern Siberian plateau, some 300 km southeast of the Gulf of Khatanga and 400 km south of the Gulf of Olenyok, both of which belong to the arctic Laptew Sea. The impact structure forms a large circular depression surrounded by a plateau that rises 200 m above it. The morphology of the structure has been significantly modified by erosion and the deposition of Pliocene-Quaternary deposits since the time of impact. And yet, all different types of impact ejecta are preserved.

The Popigai projectile targeted Archean gneisses and schists of the Anabar Shield. The basement rocks are covered by a 1000 m thick sequence of platform carbonates and terrigenous rocks of Upper Proterozoic age. These are, in turn, overlain by 500 m of predominantly carbonate rocks of Cambrian age, 120–130 m of continental siliciclastic deposits with coal interbeds of Permian age, and 20–30 m of basaltic tuffs of the Lower Triassic. All these rocks are penetrated by sills and dikes of Triassic dolerite. The uppermost target involves Cretaceous deposits.

The Popigai structure has a gravity anomaly of –35 mGal, one of the largest gravity anomalies associated with a terrestrial impact structure. Superimposed on the gravity low is a concentric ring-shaped high, about 45 km in diameter, that coincides with uplifted Archean basement. Magnetic data indicate a –300 nT anomaly over the structure. These geophysical anomalies and excessive drilling have revealed that the Popigai structure represents one of the few peak-ring craters on Earth. The ring is most likely developed from the collapse of an overheightened central peak.

An inner annular depression of 20–25 km diameter is filled with impact breccia and is surrounded by a circular uplift of crystalline basement of about 45 km diameter, which represents the peak ring. This peak ring itself is covered by impact melt rock and suevite, and is surrounded by an annular trough. In a radial cross-section, this annular trough is asymmetric with the deepest part close to the peak ring and progressively shallower sections occur further outside, where the crater floor is built up by the suite of sedimentary rocks of the target, and passes over into the outer terrace zone. The annular trough is filled with 1–1.5 km thick polymict breccia. The breccia consists of variably sized fragments of diverse Archean, Proterozoic, Paleozoic and Mesozoic rocks that are cemented by a fine clastic matrix. The composition and structure of the breccia can be best examined in large cliffs on the banks of the Popigai River, upstream from the confluence with the Rassokha River.

The structural rim of the impact structure is expressed by short ridges parallel to the margin of the structure in the western and northern sectors. Deformed basement rocks are exposed in the northwestern part of the structure, whereas in the southeast, they are buried beneath 1 km of impactites. The relative depth of the surrounding depression varies from 1.2–1.5 km in the northwest sector to 1.7–2.0 km in the southeastern sector of the structure. At the edges of the crater rim, ploughed-up radial trenches up to 25 km in length and ranging from 1 to 3 or from 10 to 15 km in width are developed. The depth of these trenches reaches several hundreds of meters. In the annular trough, sedimentary rocks of the target are intensely deformed. They contain thrusts and other faults and folds. Some are completely allochthonous blocks embedded in polymict lithic impact breccia. The horizontal displacements along the thrusts with a top-outward shear sense may reach several kilometers.

Popigai impact melt rock, 9.5 × 6 × 6 cm in size (photo courtesy: Ralf-Thomas Schmitt, Museum für Naturkunde Berlin, sample obtained from the VSEGEI, St. Petersburg, Russia).

The suevite ejecta blanket can be up to 500–700 m in thickness. Suevites also form small lenses within the lithic polymict breccia, with which they often have gradual transitions. Suevites may be further subdivided based on their composition and texture, and due to the variability of the manner in which the material was transported and deposited. Both sharp meandering boundaries and gradual transitions between the varieties of suevites have been observed. Suevites consist of fragments and bombs of impact glass and clasts of rocks and minerals, all cemented by finer-grained material of the same lithologies. Although sorting of clastic debris is absent, fragments of 1–5 cm sizes are predominant. Clasts are mainly composed of shocked gneisses. In clasts of basement rocks, diaplectic minerals, fused glasses, and the high-pressure polymorphs of quartz, coesite, and stishovite have been reported, as well as diamondiferous materials after carbon. In fact, the Popigai impact structure has in the past been termed "Russia's Treasury" because of the abundance of impact-formed diamondiferous material. The diamondiferous phases, unfortunately, occur at microscopic size, so that this deposit does not represent gemstone quality material.

Massive occurrences of impact melt rock that in the past have been denoted "tagamite" in the Russian literature, are widespread, particularly on the hills

A drill core section of Popigai suevite, 12 × 5.5 cm in size (photo courtesy: Ralf-Thomas Schmitt, Museum für Naturkunde Berlin, sample obtained from the VSEGEI, St. Petersburg, Russia).

Popigai

Russia

latitude = 71°38' N / longitude = 111°11' E / diameter = 100 km / age = 35.7 ± 0.2 Myr

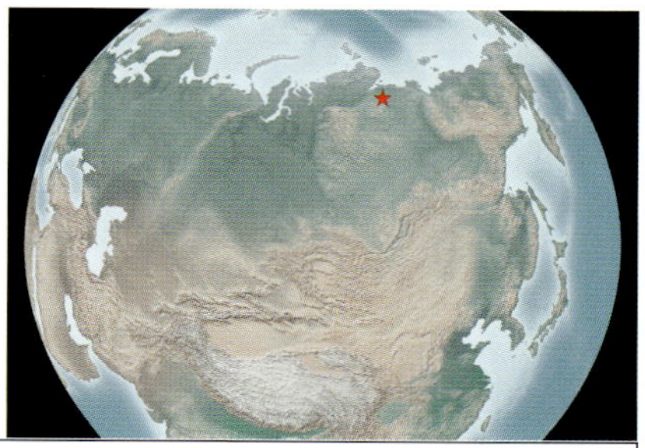

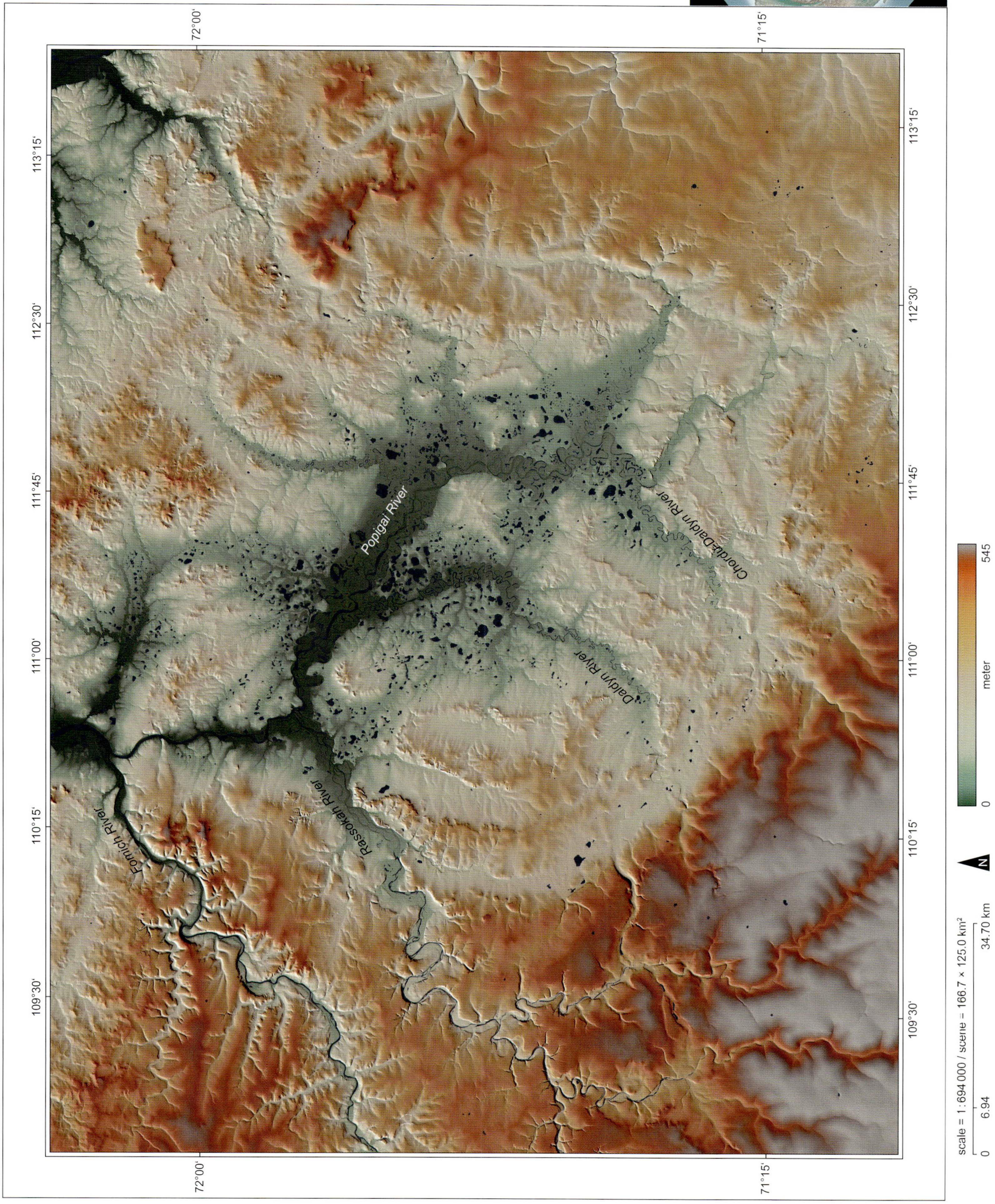

in the western part of the structure. Impact melt rocks form either sub-horizontal, sheet-like bodies close to the crater floor, or lenticular and branching bodies in the outer annular trough, where they are associated with thick sequences of suevites and mega-breccia. In the southwestern part of the annular depression, a continuous sheet of impact melt rock of about 15 km length and 150–600 m thickness was mapped by an extensive drilling program. Dikes and veins of impact melt rock are often injected into other breccia types.

It is especially interesting that the impact melt rock, as well as glass lumps within the suevite, contain impact diamonds formed after graphite from crystalline target rocks. In their overall composition, both impact melt rock and glasses from the suevite are close to the average composition of target gneisses. The total volume of impact melt preserved after erosion has been estimated at 1750 km^3, and the actual impact melt rock occurrences can account for about 950 km^3 of this. Most of the impact melt rock is situated in the northwestern, southwestern, and southeastern sectors. Allochthonous polymict breccia has a maximum thickness in the northeast part of the crater. Based on these data an oblique impact from the southwest has been proposed. Geochemical traces of the projectile were identified in the impact melt rock due to enrichment in Ni, Co, Cr and platinum group elements, compared against the composition of the target. These findings have been interpreted to suggest that the impactor was an ordinary chondrite.

Radiometric dating of impact melt rocks using the ^{40}Ar-^{39}Ar step heating technique yielded an Eocene age of 35.7 ± 0.2 Myr for the impact event. This nearly correlates with the **Chesapeake Bay** impact event. Upper Eocene marine sediments around the world contain evidence for at least two closely spaced impact layers. The tektite layer correlates with the Chesapeake Bay impact. The other layer contains clinopyroxene-bearing spherules, so-called "microkrystites". It is assumed that this globally occuring layer originated from the Popigai impact event.

Popigai's circular depression was first noticed in the late 1940s when the northern part of the Siberian platform became subject of geological studies. An impact origin was first proposed in the early 1970s by Victor Masaitis and co-workers from the Karpinsky Geological Research Institute in St. Petersburg, then Leningrad. Since then vast geological, geophysical, and drilling investigations have been carried out. More than 50 boreholes were sunk, commonly to 200–400 m depth but sometimes reaching down to 1000–1500 m.

Location/Access

The area of Popigai is situated in the Siberian taiga far from any significant infrastructure. The impact site may be reached via the airport of Norilsk city in the Krasnoyarsk district, with an onward flight to Khatanga harbour that is situated 300 km southwest of Popigai. The closest settlement possessing an airstrip is Sakylach located circa 100 km to the east of the structure. Access requires extensive logistical expedition planning and authorization by a number of agencies. A helicopter is required for visiting the structure. The most fascinating and convenient places for geological observation of the target rocks and various impact lithologies are located in the western and northern parts of the structure, where steep topography and extensive exposures exist, in particular along the Popigai, Rassokha, and Chordu-Daldyn rivers. Travelling within the crater is easiest with boats along the rivers.

Selected References

Bottomley, R. J., Grieve, R. A. F., York, D., and Masaitis, V. L. (1997). The age of the Popigai impact event and its relation to events at the Eocene/Oligocene boundary. Nature, Vol. 388, pp. 365–368.

Deutsch, A., Masaitis, V. L., Langenhorst, F., and Grieve, R. A. F. (2000). Popigai, Siberia – well preserved giant impact structure, national treasury, and world's geological heritage. Episodes, Vol. 23, pp. 3–11.

Glass, B. P. (2002). Upper Eocene impact ejecta/spherule layers in marine sediments. Chemie der Erde – Geochemistry, Vol. 62, pp. 173–196.

Koeberl, C., Masaitis, V. L., Shafranovsky, G. I., Gilmour, I., Langenhorst, F., and Schrauder, M. (1997). Diamonds from the Popigai impact structure, Russia. Geology, Vol. 25, pp. 967–970.

Masaitis, V. L. (1999). Impact structures of northeastern Eurasia: The territories of Russia and adjacent countries. Meteoritics and Planetary Science, Vol. 34, pp. 691–711.

Masaitis, V. L., Mashchak, M. S., Naumov, M. V., and Raikhlin, A. I. (1994). Large Astroblemes of Russia. VSEGEI, St. Petersburg, 34 pp.

Masaitis, V. L., Mashchak, M. S., and Naumov, M. V. (2004). Popigai Impact Crater. Guide of Geological Excursions. VSEGEI St. Petersburg, 56 pp.

Pilkington, M., Pesonen, L. J., Grieve, R. A. F., and Masaitis, V. L. (2002). Geophysics and petrophysics of the Popigai impact structure, Siberia. In: Impacts in Precambrian Shields, Plado, J. and Pesonen, L. J. (eds.), Impact Studies, Springer-Verlag, Berlin-Heidelberg, pp. 87–107.

Tagle, R., and Claeys, P. (2005). An ordinary chondrite impactor for the Popigai crater, Siberia. Geochimica et Cosmochimica Acta, Vol. 69, pp. 2877–2889.

Ragozinka

Russia

Thomas Kenkmann

Ragozinka, a 9 km diameter impact structure, is located on the eastern slope of the Ural Mountains, Russia. It is almost completely buried under Paleogene to Quaternary sediments but still makes an impression on the local topography. A circular upland, open to the north, rises 50 m above a depression that coincides with the buried crater.

The target was composed of folded Devonian and Carboniferous rocks that are partly covered by Mesozoic and Cenozoic sediments that dip gently to the northeast. These Jurassic to Paleogene siliciclastics have a total thickness of about 100–200 m. All these rock types can be found as blocks, fragments, and finely dispersed material in allochthonous breccia that fills the buried crater depression and occurs outside the crater as ejecta. The depression is surrounded by a ring of brecciated Paleozoic rocks, which is interpreted as a remnant of the crater rim. These remnants are, in places, covered by polymict breccia.

A shatter cone formed in limestone, 9 × 6.5 × 3 cm in size, from the ▷ Ragozinka impact structure (photo courtesy: Ralf-Thomas Schmitt, Museum für Naturkunde Berlin, sample obtained from the VSEGEI, St. Petersburg, Russia).

Ragozinka

Russia

latitude = 58°42‘17“ N / longitude = 61°47‘50“ E / diameter = 9 km / age = 50 Myr

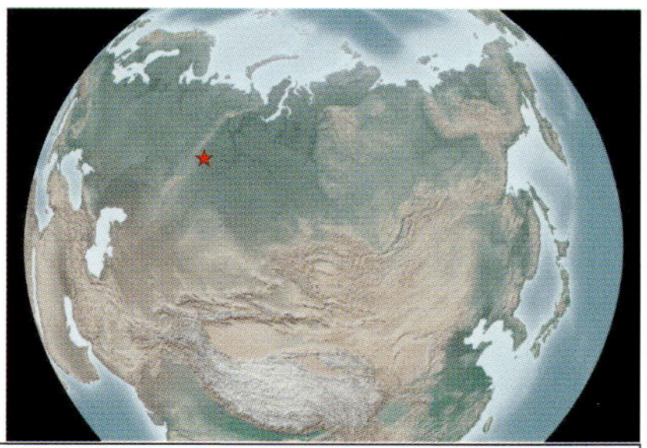

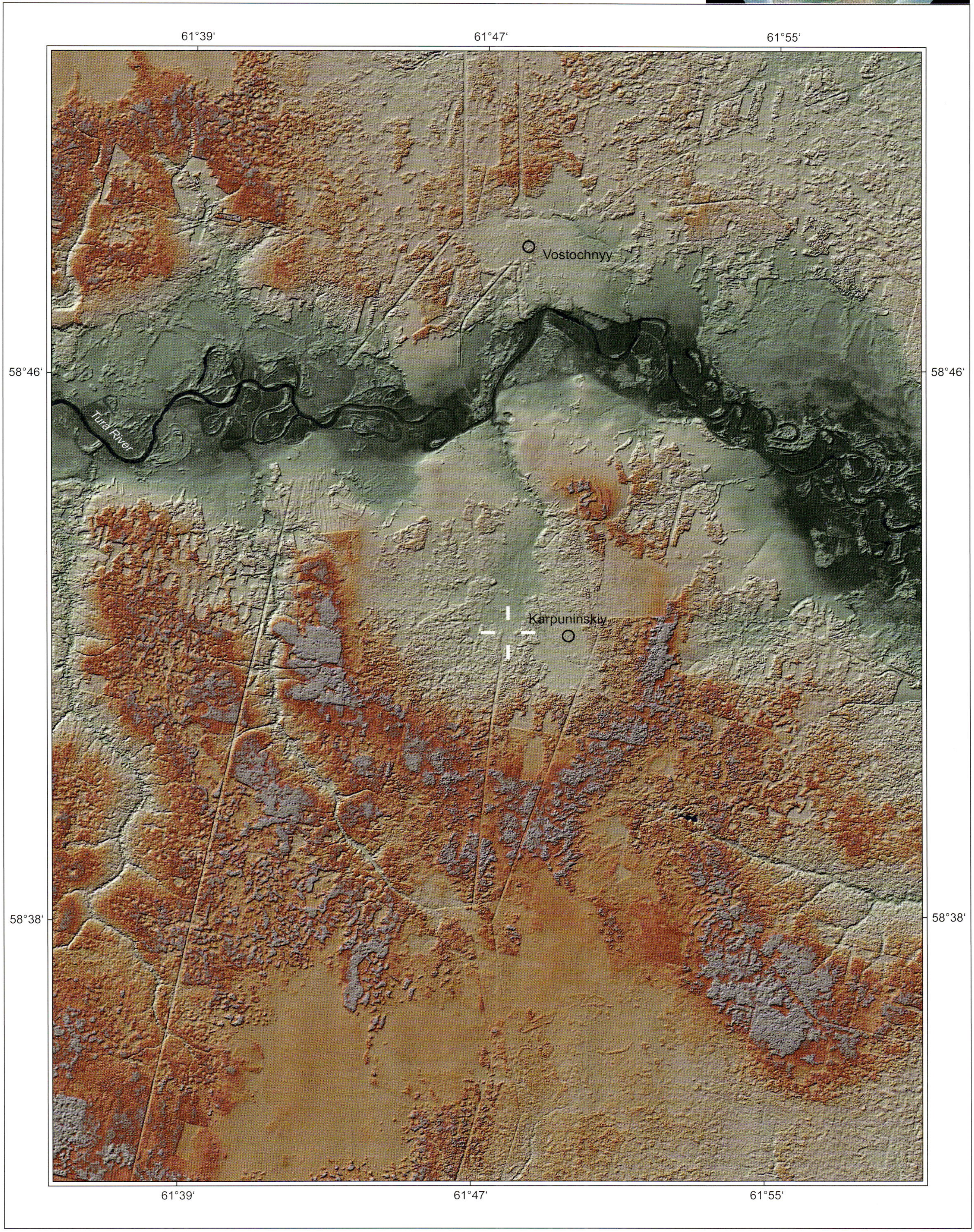

scale = 1: 128 000 / scene = 23.0 × 30.7 km²

0 1.28 6.40 km

N

65 meter 180

A few outcrops of breccia exist close to the crater and document shock metamorphism in the form of shatter cones, or planar deformation features (PDFs) in quartz parallel to various crystallographic directions. Boreholes drilled into the depression indicate that the breccia in the crater is covered by a more than 300 m thick sequence of post-impact sedimentary rocks. In the lower part of the section, numerous fragments of Paleozoic rocks are found. Marine molasse deposits from the Ural Mountains, including turbidites, were deposited inside the crater cavity. They suggest that the crater was formed in a shallow marine environment about 50 Myr ago, in the Early Eocene. The depression is characterized by a negative magnetic anomaly and a gravity low, which is surrounded by a weak gravity high. Resistivity data indicate that the crater is 550–600 m deep. A central uplift of 100 m height and 1200 m diameter was determined from geophysical surveying. The central uplift, in turn, is surrounded by an annular trough.

The structure was discovered and confirmed in 1986 by Sergei Vishnevsky and V. Lagutenko. Little work has been published since then on Ragozinka, even though it has been an object of long debate. Most of the published work was reported in Russian only.

Location/Access

Ragozinka is located 220 km north of Jekaterinburg in the oblast Swerdlowsk. The area of the structure is accessible by road and train.

Selected References

Masaitis, V. L. (1999). Impact structures of northeastern Eurasia: The territories of Russia and adjacent countries. Meteoritics and Planetary Science, Vol. 34, pp. 691–711.

Vishnevsky, S. A. (1999). Ragozinka: Eocenian marine impact crater on the middle Urals, Russia, Meteoritics and Planetary Science, Vol. 34 Supplement, pp. A117.

Vishnevsky, S. A., and Lagutenko, V. N. (1986). The Ragozinka Astrobleme: An Eocene crater in the central Urals (in Russian). Doklady Akademii Nauk SSSR, Vol. 14, 42 pp.

Ramgarh

India

Thomas Kenkmann

The Ramgarh structure is a well-known morphological landmark in the flat sedimentary terrain of south-eastern Rajasthan, India. The structure displays an annular collar, 200 m high and almost 4 km wide, which in the past has been considered the rim of a crater structure but in fact represents the remnant of the central uplift. This geomorphological phenomenon has caused a sensation among Indian geologists since the 19th century, and many hypotheses for its origin, including meteorite impact, have been proposed since then. However, unequivocal evidence was not presented until very recently.

The structure is situated in the center of the Vindhyan Basin that represents a large Proterozoic depocenter within India's Bundelkhand craton. The impact structure was formed in flat-lying sediments. The oldest strata of the central uplift, the Upper Rewa sandstone of the Proterozoic Upper Vindhyan Group, are exposed at the geographic center of the structure and show vertical dips. The next younger unit, the Ganugarh Shale Formation, occupies the central depression and comprises heavily deformed shales and sandstones, as well as limestones, that are partially dismembered into megablocks. The sandstones of the central depression show a weak shock metamorphic overprint. Planar fractures (PFs) and feather features (FFs) were recently found in quartz grains from in situ sandstones of the central part of the Ramgarh structure – thus confirming its impact origin.

The pronounced collar of the Ramgarh structure is formed by competent Lower Bhander Formation sandstone that also belongs to the Upper Vindhyan Group. The sandstones of the central uplift contain heavy mineral rich seams and are impregnated by Fe-Mn ores, which may indicate an impact-induced hydrothermal mineralization overprint.

As pointed out above, the annular morphological collar does not mark the rim of a 3.5–4 km wide impact crater, as was assumed in the previous literature. Instead, the collar represents the outer part of a central uplift of a complex impact structure of about 10 km diameter. The apparent crater rim is exposed along a local river bed at 5 km distance from the crater center as a low-angle normal fault with ramp-and-flat structure, and can be traced as lineaments in remote sensing imagery. Spectral remote sensing data, the local drainage network, and available gravity data delineate the circular structure with an even slightly larger diameter and indicate that the crater was originally somewhat larger than 10 km in diameter. With the subdued morphological appearance of the crater rim and the pronounced collar of the central uplift Ramgarh is similar to the **Gosses Bluff** impact structure in Australia.

The central uplift of Ramgarh is somewhat rectangular in shape and is dissected by numerous faults that are seemingly cogenetic with the formation of the central uplift. The central uplift has a bilateral symmetry along a southwest-northeast axis, where a large dextral strike-slip fault documents a strong horizontal shear component. This direction corresponds to the impact trajectory along a southwest-northeast line. Further structural characteristics indicative for an oblique impact from the southwest are a stack of concentrically striking thrusts in the up-range sector, a narrow-spaced radial fault swarm in the down-range region, and oblique, fault-bounded

Panoramic view from the southwestern collar, looking across the central uplift of the Ramgarh impact structure and its southwestern outlet with dammed lakes. The center of the crater is behind the far end of the lake (photo: T. Kenkmann).

Ramgarh

India

latitude = 25°20'03" N / longitude = 76°37'28" E / diameter = 10 km / age = 164.8 ± 1.3 Myr

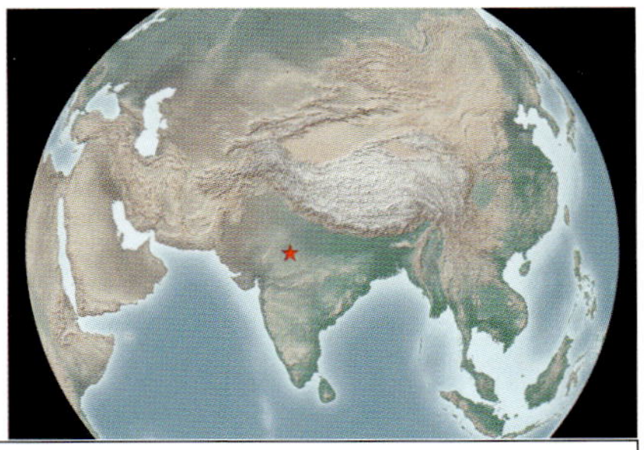

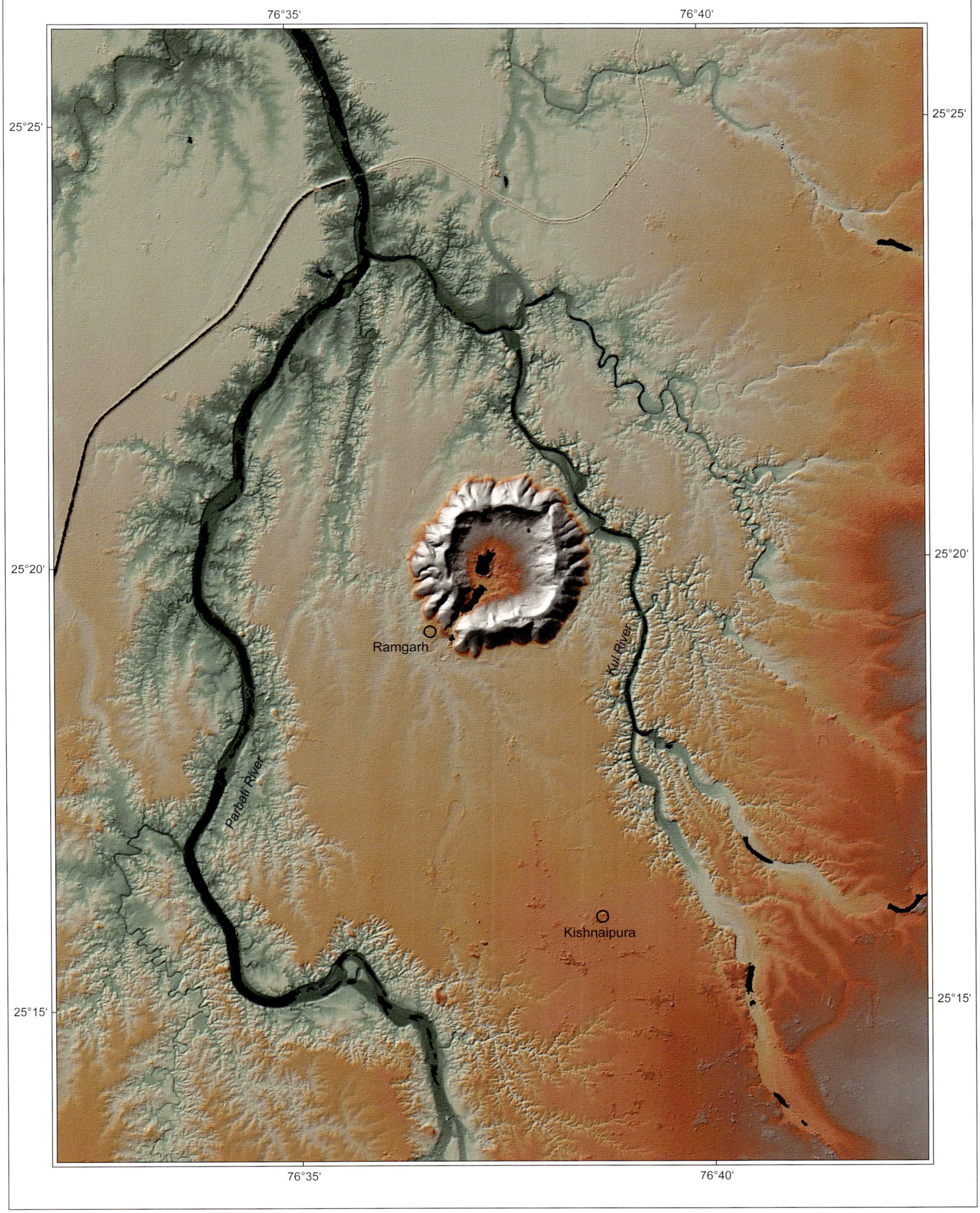

scale = 1:99000 / scene = 17.7 × 23.6 km²

0 0.99 4.95 km

N

170 meter 450

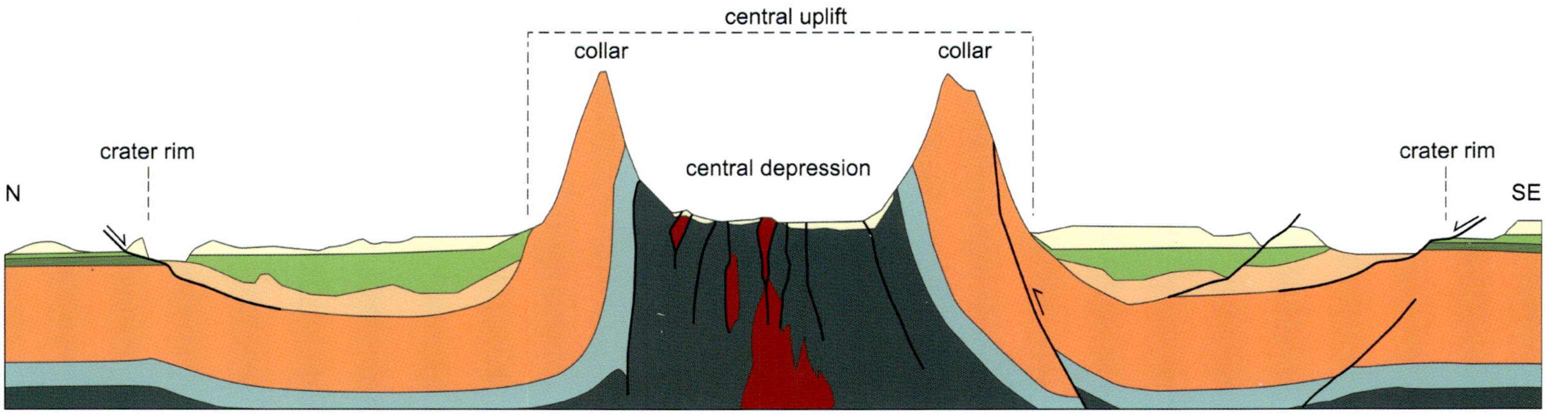

Schematic, vertically exaggerated cross-section through the Ramgarh impact structure. Ramgarh is a classical complex crater with a large central uplift that morphologically comprises a collar and a central depression. The subdued ring syncline and a suppressed crater rim are mostly covered by Quaternary deposits (modified after Kenkmann et al., 2020).

blocks in the cross-range sectors. This pattern is also comparable to that found at the Gosses Bluff crater in Australia.

Adjacent to the crater rim fault occur tilted strata of the Upper Bhander Formation, which are interpreted as slump blocks that moved into the crater depression during crater modification. The Upper Bhander sandstone that constitutes the youngest formation of the Vindhyan Supergroup has been dated at 1000–1070 Myr by the U-Pb zircon method. The Ramgarh impact event most likely occurred into a shallow water environment, as indicated by massive soft-sediment deformation and convolute bedding observed in sediments at and near the crater rim and the deposition of diamictites above the rim and within the crater moat. The lack of a visible morphological crater rim and the presence of the low-angle normal faults at the crater rim give further hints that the uppermost target was unconsolidated at the time of impact. Gastropods embedded in the diamictite suggest an Upper Middle Jurassic age of the impact.

Research on Ramgarh has been intense but the results remained ambiguous so that the structure has not been listed as a confirmed impact structure until recently. Geological interpretations of the structure included scenarios of a dome-like uplift induced by a combination of a magmatic intrusion and tectonic forcing, an origin due to explosive emplacement of a kimberlite pipe, diapirism, or an origin by an impact event. In 1972 A. Crawford was the first who favored an impact scenario, however, without being able to provide evidence. Since then an impact scenario was supported by most of the Indian researchers of this structure. Several attempts to prove the impact hypotheses have been carried out in the past decades. Although the finding of shock features including planar deformation features (PDFs), and PFs in quartz grains was repeatedly claimed, the evidence of shock metamorphism remained vague.

Parallel set of feather feature lamellae (FFs) in quartz emanating from a planar fracture that is decorated by ore mineral particles. The straight lamellae have a spacing of 4 µm. The sample is from the western part of the central depression of Ramgarh's central uplift (photo: T. Kenkmann).

Most of the previous studies interpreted the prominent 3.5 km wide morphological ring as the rim of the crater. Recently, Thomas Kenkmann, Gerwin Wulf and Amar Agarwal, however, showed that the morphological collar of Ramgarh is only the edge of the central uplift of this complex impact structure.

Location/Access

Ramgarh is situated close to the eastern border between Rajasthan and Madhya Pradesh, some 110 km northeast of the city of Kota and 350 km south-southwest of Delhi. The name-giving village of Ramgarh is located just south of the collar of the central uplift. Access to the structure is via the towns of Baran (28 km southwest of the structure center) and Mangrol (11 km west of the structure center). In Baran a visitor can find suitable accommodation.

Selected References

Crawford, A. R. (1972). Possible impact structure in India. Nature, Vol. 237, p. 96.

Kenkmann, T., Wulf, G., and Agarwal, A. (2020). Ramgarh, Rajasthan, India: a 10 km diameter complex impact crater. Meteoritics and Planetary Science, Vol. 55, pp. 936–961.

Misra, S., Srivastava, P. K., and Arif, M. (2019). Remote sensing, structural and rock magnetic analyses of the Ramgarh structure of SE Rajasthan, Central India – further clues to its impact origin and time of genesis. In: Tectonics and Structural Geology: Indian Context, Mukherjee S. (ed.), Springer Geology, pp. 327–352.

Rana, S., and Agarwal, V. (2016). Microscopic evidences for the impact origin of Ramgarh structure, Rajasthan, India. Journal of Indian Geophysical Union, Vol. 20, pp. 544–550.

Sharma, H. S. (1973). Ramgarh structure, India. Nature, Vol. 242, pp. 39–40.

Sisodia, M. S., Lashkari, G., and Bhandari, N. (2006). Impact origin of the Ramgarh structure, Rajasthan: Some new evidences. Journal of Geological Society of India, Vol. 67, pp. 423–431.

Ramgarh impact structure on December 21, 2018 as a composite image based on TanDEM-X DEM and Copernicus Sentinel-2 (channel 2/3/4) data.

Saqqar
Saudi Arabia

Thomas Kenkmann

Saqqar, meaning "Falconer" in Arabic, is a shallow-buried impact structure in the western part of the Nafud Basin of northern Saudi Arabia. The Nafud Basin is part of the Greater Arabian basin. The land surface is almost completely covered by Quaternary wadi deposits, and Tertiary and Cretaceous sediments. Deformation by the Saqqar impact affected Devonian siliciclastic strata but not the unconformably overlying Upper Cretaceous and younger beds. Hence, the age of the impact is bracketed between 70 and 410 Myr. Deformed Devonian strata can be found in the southernmost part of the structure, the only section where the crater structure is somewhat exposed at surface.

The subsurface structure has been well resolved by 2D seismic reflection profiles and six hydrocarbon exploration wells. First-order structural features are an eroded central uplift with a stratigraphic uplift of at least 2000 m with respect to the stratigraphic level outside of the structure, and a ring syncline. The crater rim is defined by circumferential normal faults and indicates an apparent crater diameter of 34 km. The form of the crater structure deviates slightly from a circle and seemingly contains some straight segments. The central uplift and ring syncline correspond to a Bouguer gravity high and an annular ring-like gravity low, respectively. The wells were drilled into the central uplift. The deepest borehole exceeded 2 km depth. Sandstone core samples from these wells show abundant evidence for shock metamorphism in the form of planar deformation features (PDFs) and planar fractures (PFs) that are locally part of feather features (FFs). In addition, some shocked feldspar grains and strongly deformed mica flakes have been found. The recorded shock pressures range from 5 to 15 GPa. The preserved level of shock and the absence of an allochthonous crater fill suggest that Saqqar was eroded by 1–2 km between the Devonian and the Cretaceous. Assuming such an erosion level and taking into account the dip of the concentric normal fault, an original rim diameter of Saqqar of approximately 36 km is derived.

The desert region around Saqqar has been explored by Saudi Aramco since the 1980s in search for hydrocarbon reservoirs. The structure was mapped as a subsurface circular feature and potential impact crater on the basis of 2D reflection seismic profiles, gravity and magnetic data. The documentation of unequivocal shock features in drill core samples in 2015 by Thomas Kenkmann and scientists from Saudi Aramco proved the origin of the Saqqar structure as the result of a hypervelocity impact event.

Location/Access

Saqqar is situated in a low-relief desert environment. The nearest village is Meegowa, 38 km to the northeast of the center of the structure. Agricultural irrigation dominates the northern part of the structure. While most parts of the impact structure are buried, exposures along its southern margin include overturned beds of Devonian age. To reach the southern rim requires 4-wheel-drive transport to cover a stretch of 55 km southwest of Meegowa and from Highway 80.

Selected References

Kenkmann, T., Afifi, A. M., Stewart, S. A., Poelchau, M. H., Cook, D. J., and Neville, A. S. (2015). Saqqar: A new 34 km diameter impact crater in Saudi Arabia. Meteoritics and Planetary Science, Vol. 50, pp. 1925–1940.

Neville, A. S., Cook, D. J., Afifi A. M., and Stewart, S. A. (2014). Five buried crater structures imaged on reflection seismic data in Saudi Arabia. GeoArabia, Vol. 19, pp. 17–44.

Stewart, S. A. (2015). Circular geological structures outcropping in the sedimentary basins of Saudi Arabia. Journal of Asian Earth Sciences, Vol. 106, pp. 95–118.

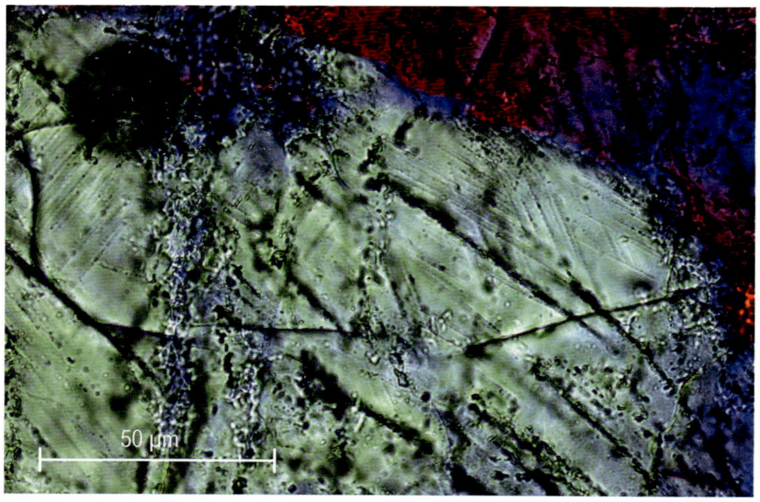

Shocked quartz grain under crossed polarizers and with a lambda plate inserted. The grain exhibits widely spaced planar fractures (PFs) and narrowly-spaced planar deformation features (PDFs). The thin section is from a sample from 457 m depth, from a well drilled by Saudi Aramco into the central uplift of the Saqqar impact structure (photo: T. Kenkmann).

Shocked quartz grain from the Saqqar impact structure with decorated PDFs in three different crystallographic orientations (photographed in cross polarized light). The sample was obtained from 194 m depth in a well drilled into the central uplift (photo: T. Kenkmann).

Saqqar
Saudi Arabia

latitude = 29°35‘ N / longitude = 38°42‘ E / diameter = 34 km / age = 70 - 410 Myr

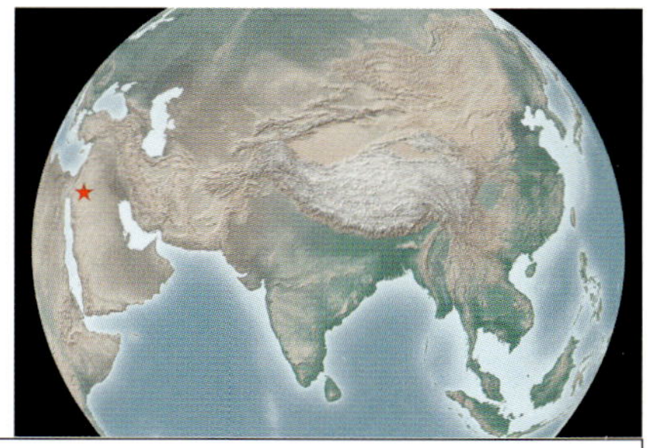

scale = 1 : 337 000 / scene = 60.6 × 80.9 km²

0 3.37 16.85 km

N

540 meter 865

Shunak
Kazakhstan

Thomas Kenkmann

Shunak is a well-preserved impact crater of 2.8 km diameter. The structure is characterized by a 2 km wide, flat, circular depression that is surrounded by an up to 400 m high rim. The radial drainage pattern of the inner slope of the rim has an outlet to the southeast where the crater rim has been eroded away. Strata along the inner slope of the western rim are overturned.

The target rocks are folded Devonian rhyolites. The crater cavity has been drilled. The lower part of the cavity comprises a more than 100 m thick sequence of blocks and fragments of volcanic rocks with small particles of melt (glass) and fractured quartz grains. This unit passes into overlying allochthonous crater fill breccia. Brecciated and cataclased rhyolites also occur on the inner slope of the rim and occasionally display shock features such as shatter cones of up to 25 cm length. These impact breccias are overlain by an up to 240 m thick sequence of post-impact crater-lake deposits. The sediments are predominantly clays, with boulders of country rock rhyolite. Quartz grains with planar deformation features (PDFs) have been found in reworked material. Fossils indicate that these deposits are of Oligocene-Miocene, and in the upper part, of Pliocene age. Thus, biostratigraphy indicates that the Shunak impact occurred < 34 Myr ago in the Oligocene or earlier.

A first report on Shunak by V. Fel'dman, published in Russian, dates back to 1976. This was supplemented by several studies over the following decade. No contemporary work exists on Shunak.

Location/Access

The structure is situated in the steppe of eastern Kazakhstan, to the northwest of Lake Balkhash. The village of Moyynty is located 45 km to the east of the structure.

Selected References

Fel'dman, V. I., Granovsky, L. B., and Lomonosov, M. V. (1978). Meteoritic crater Shunak, the central Kazakhstan, U.S.S.R. (abstract). 9th Lunar and Planetary Science Conference, Lunar and Planetary Institute, Houston, Texas, pp. 312–313.

Marakushev, A. A. (ed.) (1981) Impactites (in Russian). Moscow University Press, Moscow, Russia. 240 pp.

Masaitis, V. L. (1999). Impact structures of northeastern Eurasia: The territories of Russia and adjacent countries. Meteoritics and Planetary Science, Vol. 34, pp. 691–711.

Masaitis, V. L., Danilin, V. N., Mashchak, M. S., Raikhlin, A. I., Selivanovskaya, T. V., and Shadenkov, E. M. (1980). The Geology of Astroblemes. Nedra Press, Leningrad, Russia, 231 pp.

The Shunak impact crater on September 29, 2018 as a composite image based on TanDEM-X DEM and Copernicus Sentinel-2 (channel 2/3/4) data.

Shunak
Kazakhstan

latitude = 47°12‘35“ N / longitude = 72°45‘41“ E / diameter = 2.8 km / age = < 34 Myr

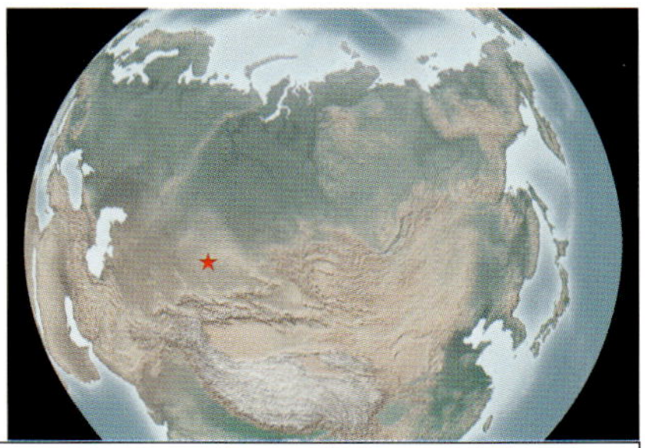

scale = 1:76 000 / scene = 13.6 × 18.1 km^2

0 0.76 3.80 km

N

690 meter 1135

Sikhote Alin Crater Field
Russia

Thomas Kenkmann

The Sikhote Alin impact is the largest meteorite fall on Earth that was demonstrably witnessed by people. The fall produced a crater strewn field, from which a large mass of the projectile could be recovered. The event happened during daytime on February 12, 1947, in the Sikhote-Alin Mountains, in the Primorje region in the Far East of the former Soviet Union.

It was calculated that the meteoroid entered the upper atmosphere at a speed of about 14 km/s, coming from the north-northeast and descending at an angle of 41°. Its luminosity constantly increased and its brightness eventually exceeded that of the sun. The bolide broke apart in the atmosphere in a violent explosion, at an altitude of 5–6 km, in an airburst. The bright flash and the extremely loud sound of the fall were observed and heard within a radius of at least 300 km around the point of impact. The wake of the bolide remained in the sky for several hours and its length was estimated at 32 km. According to modelling, the bolide fragments decelerated strongly during atmospheric passage but kept a fraction of their cosmic velocity and impacted on the ground at a velocity between 0.5 and 4 km/s.

The impacting bodies formed an elliptical crater strewn field of 0.3×0.5 km width, with 122 individual crater pits. Twenty-three craters were larger than 9 m in diameter. The largest crater has a diameter of about 26 m and is 6 m deep. This crater resulted from the more or less simultaneous impact of at least 4 or 5 large meteorites. In contrast, another crater, denoted as No. 4, measures 19.2 m in diameter and is 4.6 m deep. This crater has a distinctly conical shape and was formed by a single meteorite mass. In the surroundings of the larger craters trees were felled in a radial pattern.

The overall mass of the Sikhote-Alin meteorite has been estimated at 90 to 100 tons. In numerous expeditions about 26 tons of meteorite were collected. The meteorite strewn field is much larger than the crater strewn field and extends comparatively further in up-range direction. The meteorite is classified as an iron meteorite and belongs to the meteorite group IIAB, members of which have a coarse octahedrite structure. The composition of the meteorite is approximately 93 wt% iron, 5.9 wt% nickel, 0.42 wt% cobalt, 0.46 wt% phosphorus, and 0.28 wt% sulfur, besides trace amounts of iridium and many other elements. Minerals present include taenite, plessite, troilite, chromite, kamacite, and schreibersite.

A 165 g slab of the Sikhote Alin meteorite, on display at the University of Arizona Mineral Museum (image courtesy: The University of Arizona Mineral Museum).

The postage stamp of the painting by P. J. Medvedev, issued by the Soviet Union on November 20, 1957 to commemorate the 10th anniversary of the Sikhote-Alin meteorite shower.

Two types of meteorite pieces were recovered: The first type was formed early in the descent at cosmic velocity. These pieces show depressions on their surfaces resembling thumb prints, which are known as "regmaglypts". These fragments were aerodynamically formed by atmospheric ablation and show fusion crusts. The other type involves sharp-edged pieces called "shrapnel", and shows evidence of violent fragmentation either during the main atmospheric explosion or due to disruption upon impact.

The flight path of the Sikhote-Alin bolide was observed by many people. Among them was the artist Pjotr J. Medvedev, who witnessed the fall while painting the landscape from his house. He drew the huge wake of the bolide that formed when the meteoroid passed through the lower atmosphere. At the 10th anniversary of the Sikhote-Alin meteorite shower, the Soviet Union issued a stamp based on Medvedev's painting. The impact crater strewn field and its surroundings were initially studied by four expeditions of the Soviet Committee on Meteorites in 1947–1950. At that time about 8500 samples were collected, with their total weight exceeding 23 tons. Further expeditions were carried out later, between 1967 and 1973.

Location/Access

The bolide impacted in the Sikhote-Alin Mountains in the Primorje region of Far Eastern Russia. Lutschegorsk is the closest town that is located some 40 km northwest of the crater strewn field. This town is connected by Highway M60 to Wladiwostok 400 km further south at the coast of the Japan Sea.

Selected References

Artemieva, N. A., and Shuvalov, V. V. (2001). Motion of a fragmented meteoroid through the planetary atmosphere. Journal of Geophysical Research, Vol. 106, pp. 3297–3309.

Krinov, E. L. (1959). The main conditions of the fall of a meteoritic shower (in Russian). In: The Sikhote-Alin Iron Meteorite Shower, Fesenkov, V. G., and Krinov, E. L. (eds.), Publishing House of the USSR Academy of Sciences, Vol. 1, pp. 99–156.

Krinov, E. L. (1971). New studies of the Sikhote-Alin iron meteorite shower. Meteoritics, Vol. 6, pp. 127–138.

Krinov, E. L. (1974). Fragmentation of the Sikhote-Alin meteoritic body. Meteoritics, Vol. 9, pp. 255–262.

Sikhote Alin Crater Field

Russia

latitude = 46°09'36" N / longitude = 134°39'12" E / diameter = max. 26 m / age = 72 yr

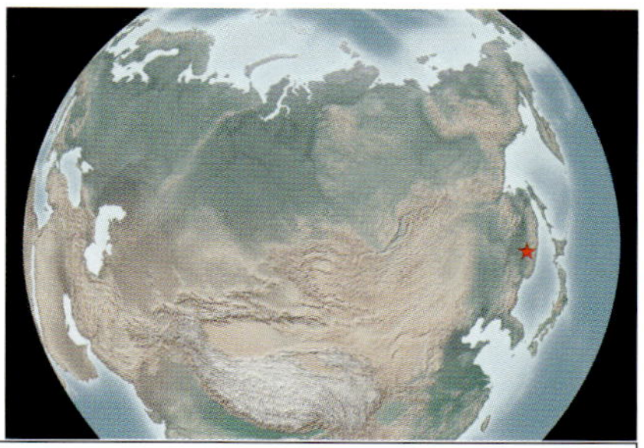

scale = 1 : 89 000 / scene = 16.0 × 21.3 km²
0 0.89 4.45 km
N
80 meter 480

Sobolev
Russia

Thomas Kenkmann

Sobolev is a small meteorite impact crater in the Sikhote-Alin Mountain Range in Russia's Far Eastern Federal District. It measures only 53 meters in diameter and has a depth of 9 m. The crater is located on the southern slope of a small valley formed by the Goreliy creek, a tributary of the Sobolevka River. The crater pit is exposed on a south-dipping slope of a 340 m high mountain.

Target rocks are Cretaceous and Tertiary quartz porphyries of subvolcanic origin partly overlain by floodplain deposits of the Goreliy creek. The crater has a wall only on its eastern side. The deepest point of the pit is offset to the east-northeast from the geometric center. In-situ brecciation of the target, as well as allochthonous breccia, have both been reported to contain shatter cones. Buried soil was excavated beneath the overturned flap at the outer slope of the raised crater rim. Allochthonous ejecta extend about 50 m up-slope of the crater rim and are locally exposed. The crater is surrounded by concentric fractures. Micro-slivers of meteoritic material and magnetic spherules with enhanced Ni-content have been reported. The age of Sobolev crater must be late Quaternary, as the crater was partly formed in floodplain deposits of the Goreliy creek. An age of 150–200 years has been estimated based on its relatively fresh appearance.

Seismic investigations have shown that the velocity of the elastic waves is decreased beneath the crater and crater rim, with respect to the environs around the structure. The area of enhanced fracturing and jointing has a diameter of more than 160 m, that is three times the crater diameter. Also the magnetic properties were changed upon impact. The remanent component of the magnetization disappears in the brecciated rocks.

The crater was discovered in 1949 by V. Yarmolyuk in the course of a geological survey. In 1974–1975 L. Khryanina and co-workers carried out a geophysical and geological survey of the structure. Their findings, reported several years later, still represent the relevant reference for this crater.

Location/Access

The crater pit is situated only 15 km west and inland from the coast of the Pacific/Japan Sea in the district of Primorje, about 250 km east of the Sikhote Alin strewn field. It is not yet proven whether Sobolev has been formed simultaneously with this strewn field. The crater can be reached best from the Terney-Samarga road. Sobolev's exact position is only poorly documented. The relevant references only provide a short description of its location including an unreferenced map.

Selected References

Khryanina, L. P. (1981). Sobolevskiy meteorite crater (Sikhote Alin' Range). International Geology Review, Vol. 23, pp. 1–10.

Khryanina, L. P. (1982). Indications of meteorite impact in Sobolevka meteorite crater. International Geology Review, Vol. 24, pp. 1019–1026.

Yarmolyuk, V. A. (1951). The Sobolevka crater (in Russian). Priroda, Vol. 6, pp. 40–42.

Sobolev

Russia

latitude = 46°18'22" N / longitude = 137°52'24" E / diameter = 53 m / age = 175 ± 25 yr

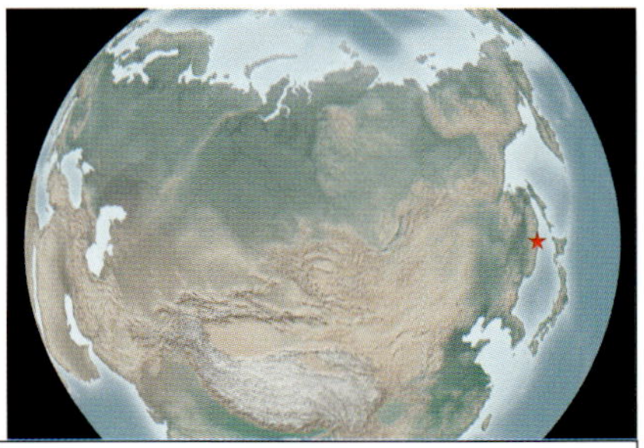

scale = 1:105 000 / scene = 18.8 × 25.1 km²

0 1.05 5.25 km

N

25 meter 925

Tabun-Khara-Obo
Mongolia

Thomas Kenkmann

Tabun-Khara-Obo is a well-exposed, 1.3 km diameter, simple bowl-shaped impact crater in the eastern Gobi desert of Mongolia. The crater rim rises some 20 to 30 meters above the crater floor. A layer of lake deposits, up to 171 m thick, covers the central depression.

The crater is situated in metamorphic rocks of Proterozoic age. This target includes schists, strongly foliated gneisses, and amphibolites. The surrounding sedimentary basins are filled with Cretaceous sediments. A large-scale, northeast-trending strike-slip shear zone with sinistral shear sense is the dominant geological feature of the region and is responsible for a penetrative northeast-trending foliation of the rocks. This foliation caused a strongly anisotropic mechanical behavior of the target rocks during the impact event and affected the crater structure and morphology. The outer wall collapsed preferentially along northeasterly trending foliation planes, resulting in a slightly oval shape of the final crater. Basement rocks with steeply dipping northeast-southwest foliation are mainly exposed in the northwestern and southeastern sectors of the crater. Overturning is not indicated along the crater rim.

Breccias cemented by carbonates occur along the northeastern-eastern crater rim. Monomict breccias form narrow lenses, whereas polymict breccias form layers outside of the eastern rim of the crater. Otherwise, no ejecta blanket has been preserved. Drilling into the center of the structure in 2008 to a depth of 206 m, with 135 m core recovery, revealed indications of shock metamorphism that had not been found at surface. From the top, the following lithologies were encountered: 3 m of eolian sand, 171 m of mud, carbonate and evaporates from an ancient lake, 11 m of polymict breccia with clasts of schist and gneiss, and 19 m of monomictly brecciated quartz-feldspar-mica schist that grades into in-situ shattered schist. Quartz grains of breccia from different depth intervals display planar deformation features (PDFs).

The age of the structure is not yet known but Early Cretaceous fossils found close to the rim in a remnant of the post-impact sediment cover suggest that the impact occurred in Late Jurassic time.

An impact origin for Tabun-Khara-Obo was first proposed by the Russian geologists O. Suetenko and L. Shkerin in the early 1970s, based on morphological and geophysical data. The final proof came in 2009 when Tsolmon Amgaa and Christian Koeberl documented shock features after drilling the structure.

Location/Access

Tabun-Khara-Obo is located in Mongolia's Dornogovi Province, about 470 km south-southeast of the capitol of Mongolia, Ulan Bator, and about 95 km south-southwest of the city of Sainshand on the Transmongolian railway. From Sainshand it is a 120 km drive on the Sainshand-Zuunbayan Road to take the visitor directly to the Tabun-Khara-Obo crater.

Selected References

Amgaa, T., and Koeberl, C. (2009a). Geology and petrography of Tabun Khara Obo crater. 72nd Annual Meeting of the Meteoritical Society, Abstract #5019.

Amgaa, T., and Koeberl, C. (2009b). Impact origin of Tabun Khara Obo crater, Mongolia, confirmed by drill core studies (abstract). Geological Society of America Annual Meeting, Portland, Oregon, Vol. 41, p. 533.

McHone, J. F., and Dietz, R. S. (1976). Tabun Khara Obo crater, Mongolia: Probably meteoritic. Meteoritics, Vol. 11, pp. 332–333.

Shkerin, L. M. (1976). Geological structure of the crater-like feature Tabun Khara Obo (south east Mongolia) (in Russian). Meteoritika, Vol. 35, pp. 97–102.

Suetenko, O. D., and Shkerin, L. M. (1970). A proposed meteorite crater in southeastern Mongolia (in Russian). Astronomicheskii Vestnik, Vol. 4, pp. 261–263.

The Tabun-Khara-Obo impact crater on July 16, 2018 as a composite image based on TanDEM-X DEM and Copernicus Sentinel-2 (channel 2/3/4) data.

Tabun-Khara-Obo
Mongolia

latitude = 44°07‘52“ N / longitude = 109°39‘17“ E / diameter = 1.3 km / age = 154 ± 9 Myr

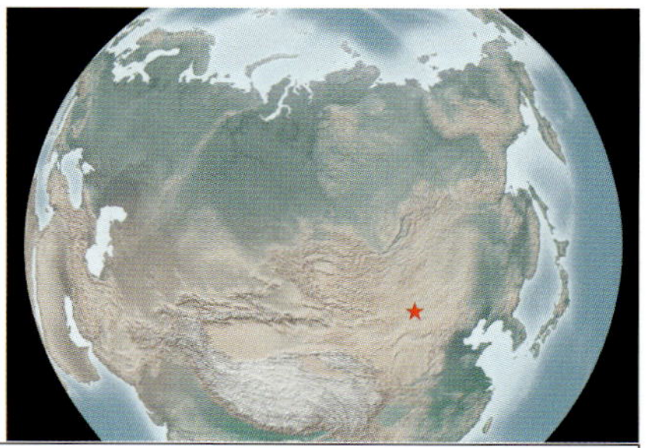

scale = 1 : 74 000 / scene = 13.4 × 17.8 km²

0 0.74 3.70 km

N

755 meter 920

Wabar Crater Field

Saudi Arabia

Thomas Kenkmann

The Wabar impactor struck the middle of the largest contiguous sand desert in the world, the Rub' al-Khalī desert of Saudi Arabia. The impact formed at least three small craters with diameters of 116, 64, and 11 m in an area of about 500 × 1000 m. Each crater forms a gentle depression that is nearly completely filled with sand and is surrounded by a raised rim. The depth of the largest crater was measured as 14.5 m. The appearance of the craters changes with time as dunes move over the crater field and temporarily cover the craters.

The porous sandy target experienced a shock metamorphic overprint upon impact that led to crushing and rock pulverization, bleaching, and welding of particles by melt formation. Also planar deformation features (PDFs) and the high-pressure minerals coesite and stishovite have been found in shocked material from Wabar. Impactite lithologies predominantly occur on the southeastern crater rims and are almost absent on the northern and western sides of the craters. This asymmetry suggests that the impact was oblique, with the incoming objects arriving from the northwest at an angle between 20° and 45° from the horizontal. Along the crater rims and beyond occur lumps up to a meter in size of white sandstone-like material and blackish, vesiculated, slag-like glasses. The lumps are cohesive because of shock lithification. Two different melt types, a black and a white melt, have been described from the impact site. They contain, on average, 4.5 and less than 1 % meteoritic component, respectively. Meteoritic iron was selectively mixed at high temperatures with high-silica melt derived from the target, resulting in high Fe/Ni ratios for the black melt and low Fe/Ni ratios for projectile droplets.

The total mass of iron meteorite fragments recovered from the Wabar area since their discovery has been estimated at about 2.55 tons. The largest fragment has a mass of 2.04 tons. The Wabar meteorite is classified as a group IIIAB iron meteorite with a Ni content of 7.45 wt%.

The age of the impact craters is not known. While in 1977 fission track dating on impact melt droplets gave an age of 6400 years, a more recent study using thermoluminescence dating of shocked quartz suggested a much younger age of 290 ± 38 years. This complies with two historic poems from the year 1704 about a bright fireball that was observed 620 km south-southeast from the Wabar site.

The Wabar craters were first reported by Harry St. John Philby in 1933 in his book "The Empty Quarter", in which he described his travels through the Rub' al-Khalī desert. He had heard from bedouins about a place where iron could be found. Upon Philby's visit to the impact site the large iron meteorite was invisible. It was, however, exposed when another expedition arrived in 1965. Subsequently, in 1966, Thomas Abercrombie mentioned three craters, found by an Aramco team at the Wabar site, in his National Geographic article about Saudi Arabia. In 1994 and 1995 three expeditions, including one of the impact pioneer Eugene Shoemaker, analyzed the crater strewn field in more detail. A recent visit in 2008 by Edwin Gnos and his team again found the Wabar structures to a large degree sand-covered.

The 11 m crater in the Wabar crater field as seen from the top of a seif dune in March 2008. Between 2009 and 2012 this crater was completely covered by the dune (image courtesy: Edwin Gnos, Natural History Museum, Geneva).

Crater Philby-B in the Wabar crater field as seen from the top of a seif dune (image courtesy: Edwin Gnos, Natural History Museum, Geneva).

Location/Access

The Wabar craters lie in the Empty Quarter of Saudi Arabia – the Rub' al-Khalī desert. To reach the craters, extensive expedition planning is required. The two largest recovered iron fragments, a 2040 kg and a 210 kg specimen, are exhibited at the Natural History Museum in Riyadh.

Wabar Crater Field

Saudi Arabia

latitude = 21°30‘12“ N / longitude = 50°28‘20“ E / diameter = max. 116 m / age = 290 ± 38 yr

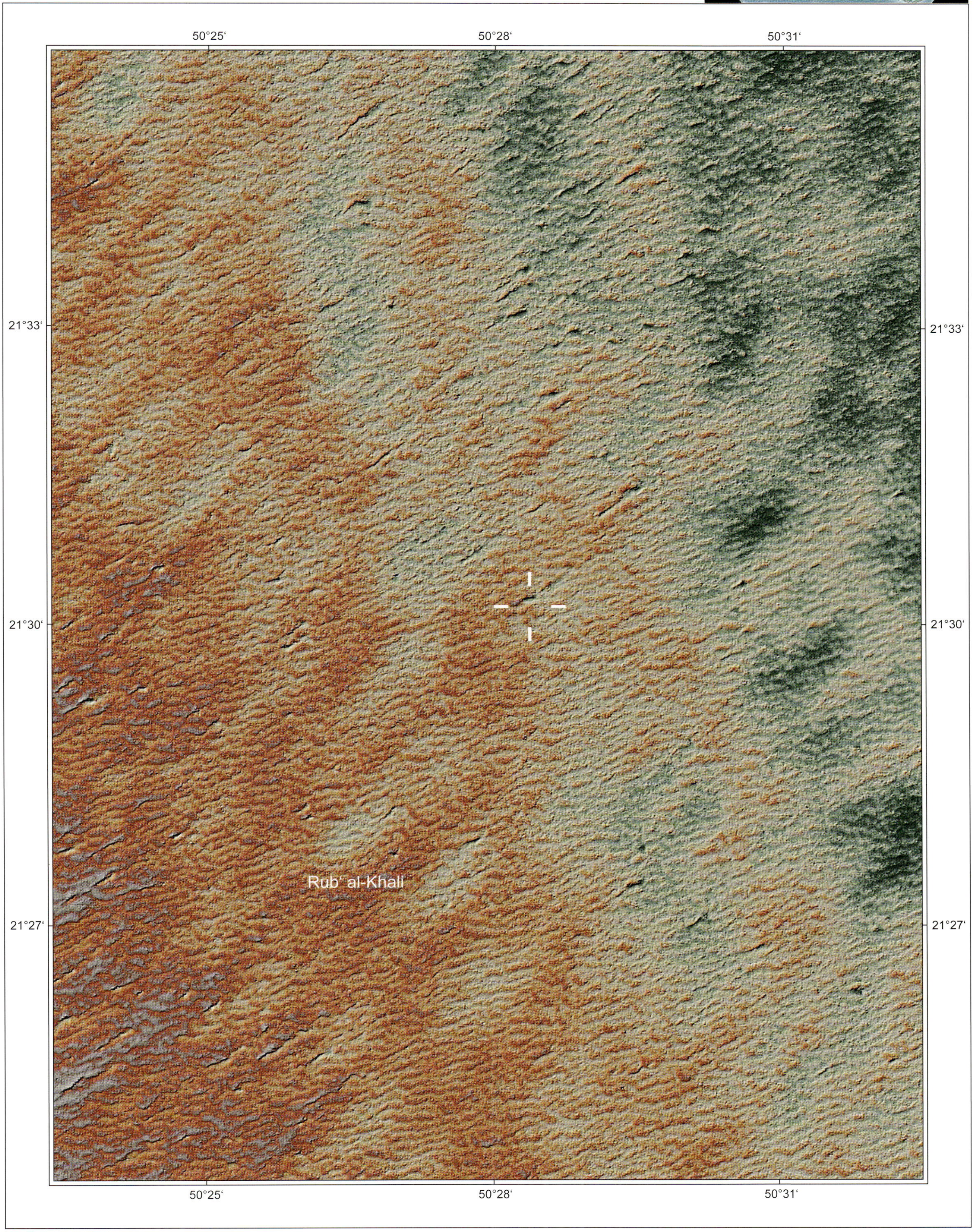

scale = 1 : 87 000 / scene = 15.6 × 20.8 km²

0 0.87 4.35 km

N

170 meter 215

Wabar: black lapilli of impact glass (image courtesy: Edwin Gnos, Natural History Museum, Geneva).

Wabar: black lapilli of impact glass (image courtesy: Edwin Gnos, Natural History Museum, Geneva).

Selected References

Abercrombie, T. J. (1966). Saudi Arabia. Beyond the sands of Mecca. National Geographic, Vol. 1291, pp. 1–53.

Chao, E. C. T., Fahey, J. J., and Littler, J. (1961). Coesite from Wabar crater, near Al Hadida, Arabia. Science, Vol. 133, pp. 882–883.

Gnos, E., Hofmann, B. A., Halawani, M. A., Tarabulsi, Y., Hakeem, M., Al Shanti, M., Greber, N. D., Holm, S., Alwmark, C., Greenwood, R. C., and Ramseyer, K. (2013). The Wabar impact craters, Saudi Arabia, revisited. Meteoritics and Planetary Science, Vol. 48, pp. 2000–2014.

Hamann, C., Hecht, L., Ebert, M., and Wirth, R. (2013). Chemical projectile-target interaction and liquid immiscibility in impact glass from the Wabar craters, Saudi Arabia. Geochimica et Cosmochimica Acta, Vol. 121, pp. 291–310.

Hörz, F., See, T. H., Murali, A. V., and Blanchard D. P. (1989). Heterogeneous dissemination of projectile materials in the impact melts from Wabar crater, Saudi Arabia. Proceedings of the 19th Lunar and Planetary Science Conference, Houston, Texas, Cambridge University Press, Cambridge, pp. 697–709.

Philby, H. S. J. (1933). The Empty Quarter. London, Constable and Company Ltd., 433 pp.

Wynn, J. C., and Shoemaker, E. M. (1998). The day the sands caught fire. Scientific American, Vol. 279, pp. 3645.

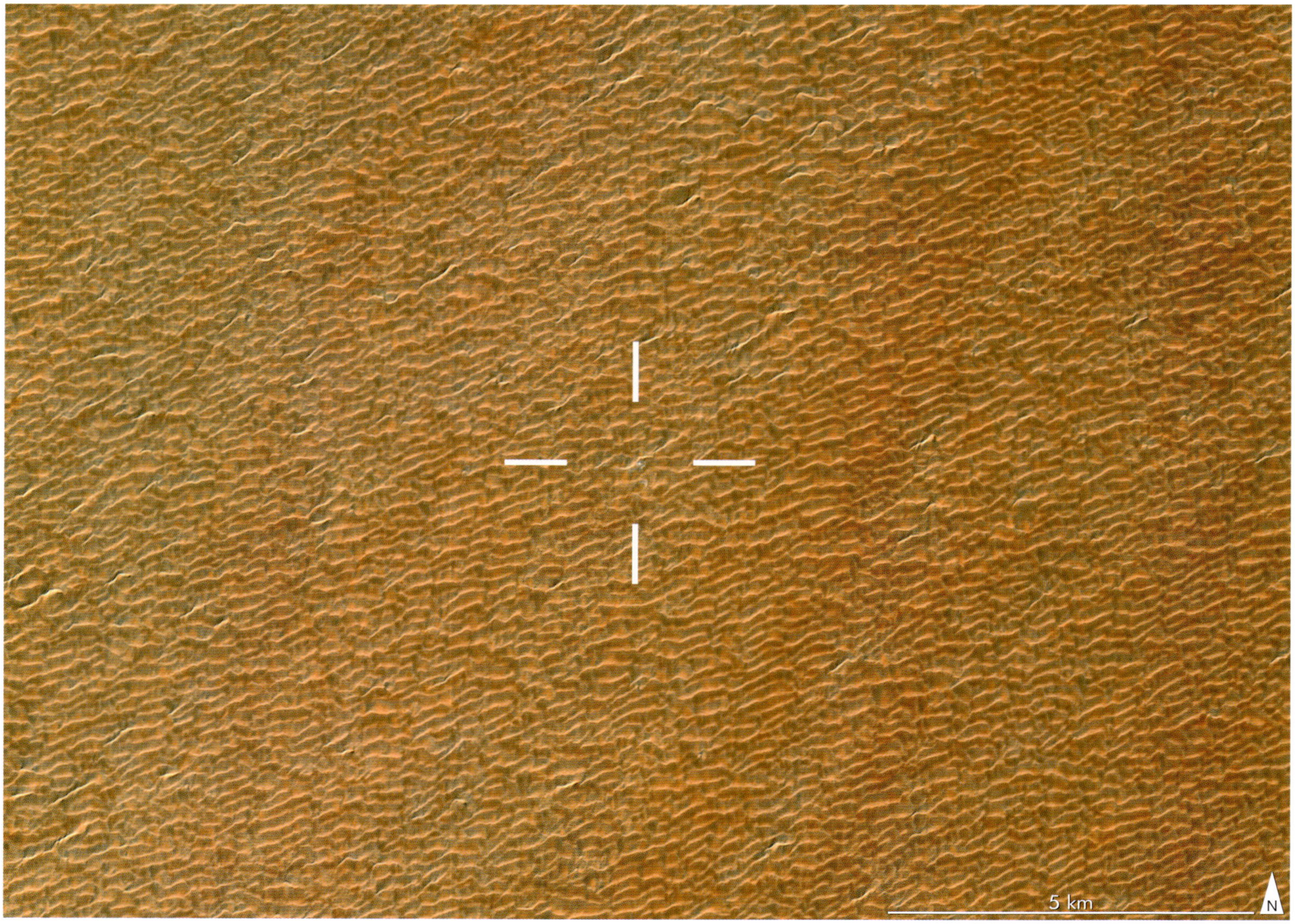

The Wabar crater field, hidden in a sea of sand, on October 2, 2018 as a composite image based on TanDEM-X DEM and Copernicus Sentinel-2 (channel 2/3/4) data.

Xiuyan
China

Thomas Kenkmann

Xiuyan is currently the only unambiguously confirmed impact crater in China. The crater is situated in northeast China in a hilly region of the central part of the Liaodong Peninsula. The crater belongs to the category of simple craters and consists of a well preserved bowl-shaped cavity with a raised rim. The rim-to-rim diameter measures about 1.8 km. The rim elevation above the crater floor amounts to 150 m, on average.

The target rocks comprise various types of Proterozoic metamorphic rocks including granulite, amphibolite, gneiss, and marble. An intrusive granite and diorite body, which was formed in the Triassic period, is located to the west of the crater. However, exposure of bedrock is very limited in and around the crater, owing to thick soil cover. Moreover, the central crater floor is covered by a thick pile of lacustrine sediments. The limited exposed bedrock along the crater rim shows heavy fracturing and in situ monomict breccia.

A central borehole revealed the structure of the crater. Beneath 107 m of lacustrine sediments, unconsolidated crystalline breccia with up to 30 cm sized fragments dominates to a depth of 260 m. Here, the breccia character changes to a polymict type down to the crater floor at ca. 150 m depth. The polymict breccia consists of a clastic matrix and clasts of granulite, gneiss, hornblendite, and glasses. Clasts reveal strong shock metamorphism. Shock-produced high-pressure polymorphs such as coesite and reidite have been found in some clasts. Quartz grains in clasts of granulite contain multiple sets of planar deformation features (PDFs) of various crystallographic orientations. They suggest a shock pressure of some 20 GPa. Fractured and monomict brecciated marble was encountered at 295 m depth in the drill core.

Radiocarbon dating of the marshy to lacustrine sediments inside the crater revealed an age of about 50000 years. The true age of the impact event may be somewhat higher though, as the maximum age limit for the radiocarbon dating method is close to this value.

Xiuyan has drawn geologists' attention since the 1990s. A decade later an impact origin was first proposed but it took another few years, until 2007, before Ming Chen and co-workers found the definite proof of shock metamorphism.

Location/Access

The crater is named after the city of Xiuyan located 18 km west-southwest of the crater. Within the agriculturally used crater depression lies the village Luoquangou Li that can be reached from the northeast following the valley along which the crater is drained. The crater rim is forested and exhibits only a few exposures.

Selected References

Chen, M., Xiao, W., and Xie, X. (2010). Coesite and quartz characteristic of crystallization from shock-produced silica melt in the Xiuyan crater. Earth and Planetary Science Letters, Vol. 297, pp. 306–314.

Chen, M., Koeberl, C., Xiao, W., Xie, X., and Tan, D. (2011). Planar deformation features in quartz from impact-produced polymict breccia of the Xiuyan Crater, China. Meteoritics and Planetary Science, Vol. 46, pp. 729–736.

Chen, M., Yin, F., Li, X., Xie, X., Xiao, W., and Tan, D. (2013). Natural occurrence of reidite in the Xiuyan crater of China. Meteoritics and Planetary Science, Vol. 48, pp. 796–805.

Impact breccia composed of fragments of gneiss, amphibolite and marble at a location at the foot of the inner crater rim (photo courtesy: Ming Chen, Guangzhou Institute of Geochemistry, Chinese Academy of Sciences).

A panoramic view of the Xiuyan impact crater in autumn (photo courtesy: Ming Chen, Guangzhou Institute of Geochemistry, Chinese Academy of Sciences).

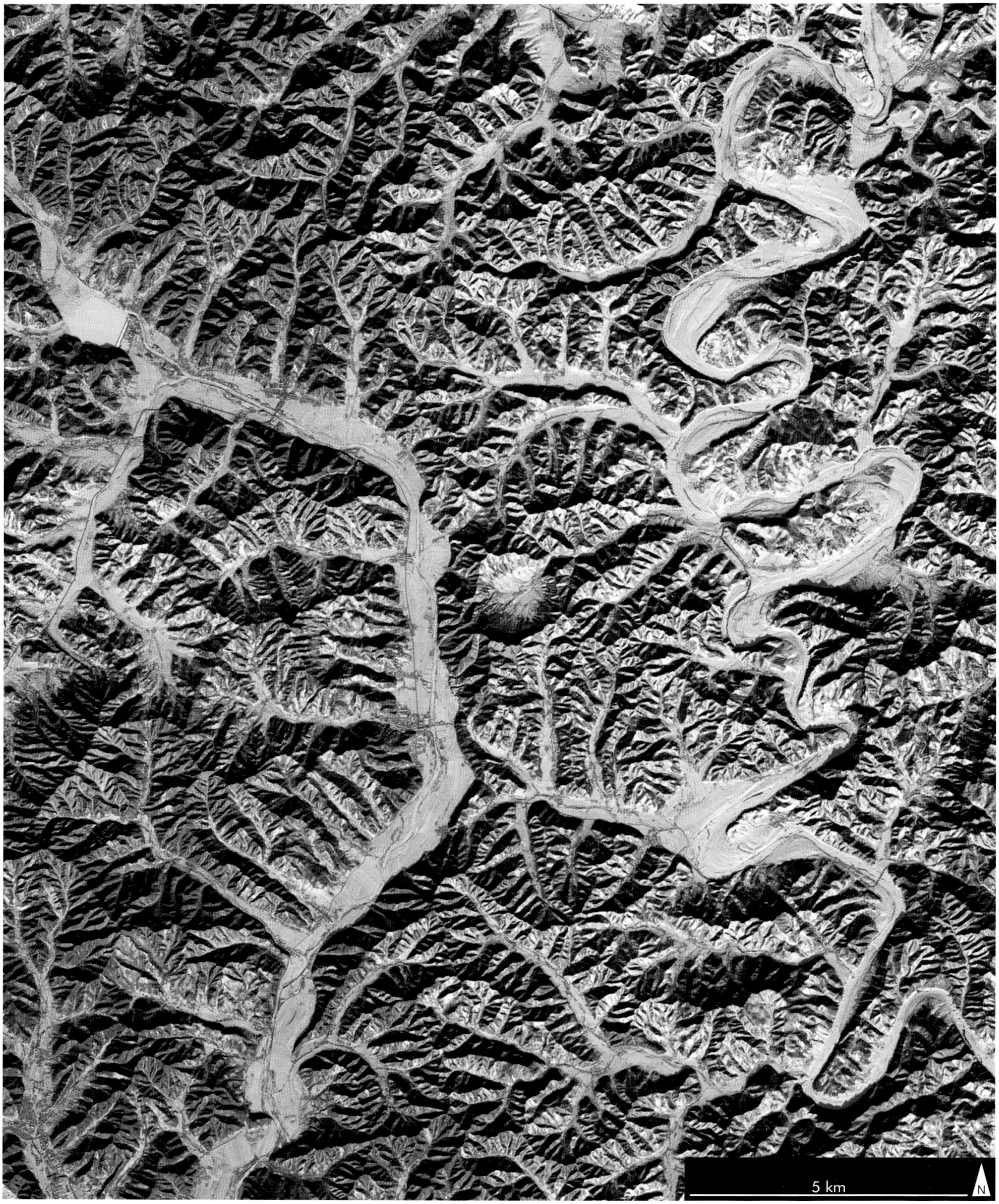

The Xijuan impact crater and environs on January 9, 2018 as a composite image based on TanDEM-X DEM and Copernicus Sentinel-2 (channel 2/3/4) data.

Xiuyan

China

latitude = 40°21'52" N / longitude = 123°27'39" E / diameter = 1.8 km / age = 50 000 ± 5000 yr

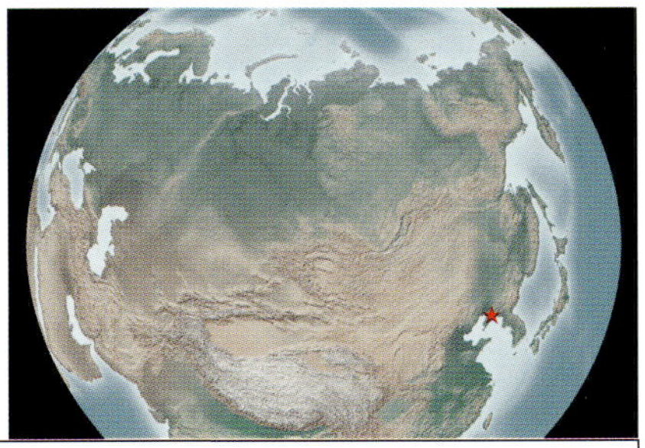

scale = 1 : 62 000 / scene = 11.2 × 15.0 km²

0 0.62 3.10 km

N

30 meter 435

Zhamanshin
Kazakhstan

Thomas Kenkmann

Zhamanshin, an impact structure with a diameter of 13 km, is situated in Kazakhstan's semiarid northern Aral Sea area. The crater structure is visible in the local topography. Zhamanshin has been famous for its impact glasses that occur dispersed to the south of the crater.

The target has a two-layer structure. The competent basement is composed of folded Paleozoic and Precambrian schists and a variety of other crystalline rocks of volcanic and sedimentary origin. These rocks are overlain by about 150–200 m of Tertiary sediments including clays, sandstone, and marls, and some Cretaceous limestones. All these strata have rather horizontal attitude. The interior of the crater structure is filled with Quaternary deposits and loess.

The crater has a nested morphology. An inner crater was excavated into basement rocks. It has a diameter of 5–6 km, a maximum depth of 600 m, and wall slopes that dip inward at 20°. This inner crater also contains a small central uplift of 500–1000 m diameter with a maximum elevation of 250 m and has an elevated rim of 5.5 km diameter, where basement lithologies are uplifted by some 50–60 m. The outer, 13 km wide crater is shallower and the depression is formed in the Cenozoic and Mesozoic sediments. This depression has more gently dipping slopes and contains outward-directed thrusts. The outer depression is defined in the northern, western and southern sectors of the structure. The floor of the inner crater depression is filled with an up to 600 m thick pile of allochthonous breccia, which – in turn – is covered by post-impact Pleistocene lacustrine sediments and loess that locally reach 100 m thickness. The allochthonous breccia mainly contains fragments and blocks of Precambrian and Paleozoic basement provenance. Autochthonous breccia is dominant in the center of the structure.

Quartz with multiple sets of planar deformation features (PDFs), as well as diaplectic glass, coesite, and fused SiO_2 (lechatelierite) occur in allochthonous breccia. Impact melt rock is present throughout the structure – mostly in the form of decimeter to meter-sized blocks. The melt shows quench textures.

To the southeast of the crater occur abundant aerodynamically shaped impact glasses dispersed over an area of 2 km^2. They are black, ropy, irregular or dumbbell shaped glassy bodies. Their appearance has been referred as "tektite-like". The most abundant group of glasses has SiO_2 contents of about 70–80 wt%, is vesiculated, and forms droplet aggregates. This group has been named "Zhamanshinite". Another group shows a layered structure, often with alternating layers of glass and partly crystallized layers, and has an average SiO_2 content of 52–57 wt%. Such glasses have been termed "Irghizites", after the nearby Irgiz river. This group is similar to a group of tektites, the so-called "Muong-Nong" type of Indochinite tektites, which belong to the huge Australasian tektite strewn field. Based on compositional and morphological similarities between Irghizites, Zhamanshinites, and Australasian tektites, it was originally considered that the Zhamanshin impact crater could be a possible source of the Australasian tektites. The idea was that the loess within the crater was the parent material of both the Zhamanshin glasses and Australasian tektites. However, it later became apparent that the glasses from Zhamanshin differ in chemical composition from Australasian tektites, and, thus, the link between the Zhamanshin impact structure and the Australasian strewn field was discarded. The wide spatial separation between the impact structure and the tektite strewn field, as well as the moderate size of the Zhamanshin structure, made a relation to the Australasian tektite strewn field even more unlikely.

A slab of Zhamanshin impact melt rock, 11.5 × 7.5 × 3.5 cm in size (photo courtesy: Ralf-Thomas Schmitt, Museum für Naturkunde Berlin, sample obtained from the VSEGEI, St. Petersburg, Russia).

Numerous age determinations of the various glasses were conducted in the 1980s and 1990s. This included application of the K-Ar and the ^{40}Ar-^{39}Ar step-heating techniques, as well as of the fission track method. The results cluster around 0.75 and 1.1 Myr and provide a rough age for the impact event.

The Zhamanshin site drew the attention of geologists already in the late 1930's because of the presence of Paleozoic rocks in a wide area of occurrence of Cenozoic rocks. About 30 years later the so-called "Zhamanshin natural landmark" was studied by L. Kiryukhin and co-workers. First ideas about an impact emerged but the structure was still considered a tectonic uplift and the glasses to be of volcanic origin. In 1975 P. Florenskiy documented a variety of shock metamorphic features and confirmed the impact origin of the Zhamanshin impact structure.

Location/Access

Zhamanshin is situated about 185 km north of the limit of the Aral Sea in a dry steppe to desert environment. Notably the extent of the Aral Sea has dramatically shrunk in recent decades. Two Kazakhstan highways, the A26 and M32, converge as close as 5 km from the structure. The nearby town of Yrgyz, located only 35 km to the northeast of the crater, has an airstrip.

Selected References

Bouşka, V., Povondra, P., Florenskiy, P. V., and Kanda, Z. (1981). Irghizites and zhamanshinites: Zhamanshin crater, USSR. Meteoritics, Vol. 16, pp. 171–184.

Florenskiy, P. V., Short, N., Winzer, S. R., and Fredriksson, K. (1977). The Zhamanshin structure: Geology and petrography (abstract). Meteoritics, Vol. 12, pp. 227–228.

Florenskiy, P. V., and Dabizha, A. I. (1980). Meteoritniy krate'r Zhamanshin (in Russian), Moscow, Nauka Press, pp. 1–127. Translation in NASA Technical Memorandum NASA TM-88531, 1987.

Glass, B. P. (1979). Zhamanshin crater, a possible source of Australasian tektites? Geology, Vol. 7, pp. 351–353.

Koeberl, C., and Fredriksson, K. (1986). Impact glasses from Zhamanshin crater (U.S.S.R.): chemical composition and discussion of origin. Earth and Planetary Science Letters, Vol. 78, pp. 80–88.

Masaitis, V. L., Boiko, Y. I., and Izokh, E. P. (1984). Zhamanshin impact crater (Western Kazakhstan): Additional geological data. 15th Lunar and Planetary Science Conference, Lunar and Planetary Institute, Houston, Texas, pp. 515–516.

Zhamanshin

Kazakhstan

latitude = 48°21'38" N / longitude = 60°56'12" E / diameter = 13 km / age = 0.75 - 1.10 Myr

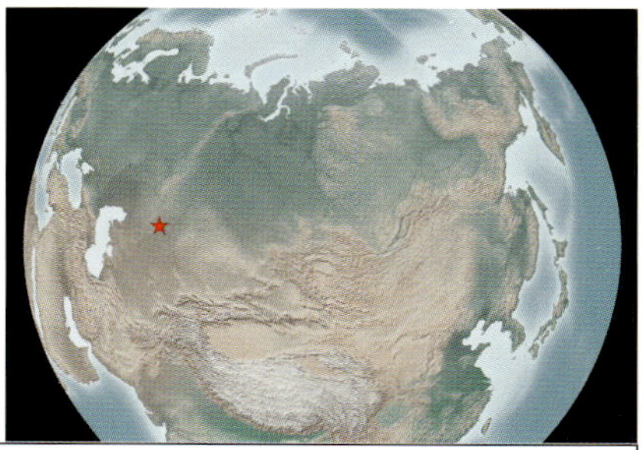

scale = 1 : 143 000 / scene = 25.7 × 34.2 km^2

0 1.43 7.15 km

N

90 meter 305

The Zhamanshin impact structure on September 29, 2018 as a composite image based on TanDEM-X DEM and Copernicus Sentinel-2 (channel 2/3/4) data.

Impact Structures – Further Confirmation Required

Kara-Kul
Tajikistan

Thomas Kenkmann

Kara-Kul is a proposed 52 km diameter, complex impact structure that forms a large depression in the 4000–7000 m high mountains of the alpine Pamir orogen. The structure is partly occupied by the 24 km wide Kara-Kul Lake. The lake is largely separated into an eastern and western part by a peninsula and islands, which have been interpreted as the morphological expression of a large central uplift that was seemingly elongated in north-south direction. The central uplift rises 250–300 m above lake level. The elevated crater rim lies about 1000 m above the floor of the depression and about 120–200 m above the surrounding terrain.

The target is composed of slightly metamorphosed and intensely folded Paleozoic sedimentary rocks, which are intruded by granites of 190–230 Myr ages. The rocks of the possible central uplift and of the inner crater rim are disturbed, and widespread breccias and fractured rocks occur. Monomict brecciation partly led to pulverization of various target lithologies, including limestones and granites. Quartz grains of brecciated granites have been reported to exhibit various degrees of shock metamorphism, in particular in the form of planar deformation features (PDFs) in different crystallographic directions. A shock pressure of 15–17 GPa was derived for these shock features. Impactites like suevite or polymict lithic breccia have not been detected so far, and a recent search by a French-Polish team for shatter cones and shock metamorphosed rocks has been unsuccessful as well.

A post-impact succession of Quaternary and Plio-Pleistocene lacustrine sediments of Lake Kara-Kul sets a minimum age constraint for the alleged impact of somewhat less than 5 Myr. An upper age limit is given by ages related to the mountain building and for granites of the target at about 50 Myr. The structure was identified by Eugene Gurov and co-workers and was investigated by their team with several expeditions from 1987 to 1991.

Further research on this apparent large impact structure is desired, and it is noted that, to date, no full scientific publication with verifiable data/observation has been published.

Location/Access

The nearest airstrip to Kara-Kul is at the town of Murgab some 110 km southeast of the structure. From there, mountain road M41 heads towards Kara-Kul Lake. Other options to reach the impact structure are via M41 from the cities of Dschalabat or Osch that are located 210 km and 170 km north of the crater, respectively.

Selected References

Gurov, E. P. (1988). The Kara-Kul Lake depression in the Pamirs – A Probable Astrobleme (abstract). 8th Soviet-American Microsymposium, Moscow, Russia, pp. 37–39.

Gurov, E. P., Gurova, H. P., Rakitskaya, R. B., and Yamnichenko, A. Y. (1993). The Karakul depression in the Pamirs – the first impact structure in central Asia (abstract). 24th Lunar and Planetary Science Conference, Lunar and Planetary Institute, Houston, Texas, pp. 591–592.

Two impressions of the proposed Kara-Kul impact structure in the Pamir mountain range. Photos taken during the August 2011 expedition to Kara-Kul Lake, 3900 m, Tajikistan, with David Baratoux, Lenka Baratoux, Sylvain Bouley, Francois Colas, Jean-Luc Dauvergne and Anna Losiak (copyright: Jean-Luc Dauvergne, supplied by Sylvain Bouley).

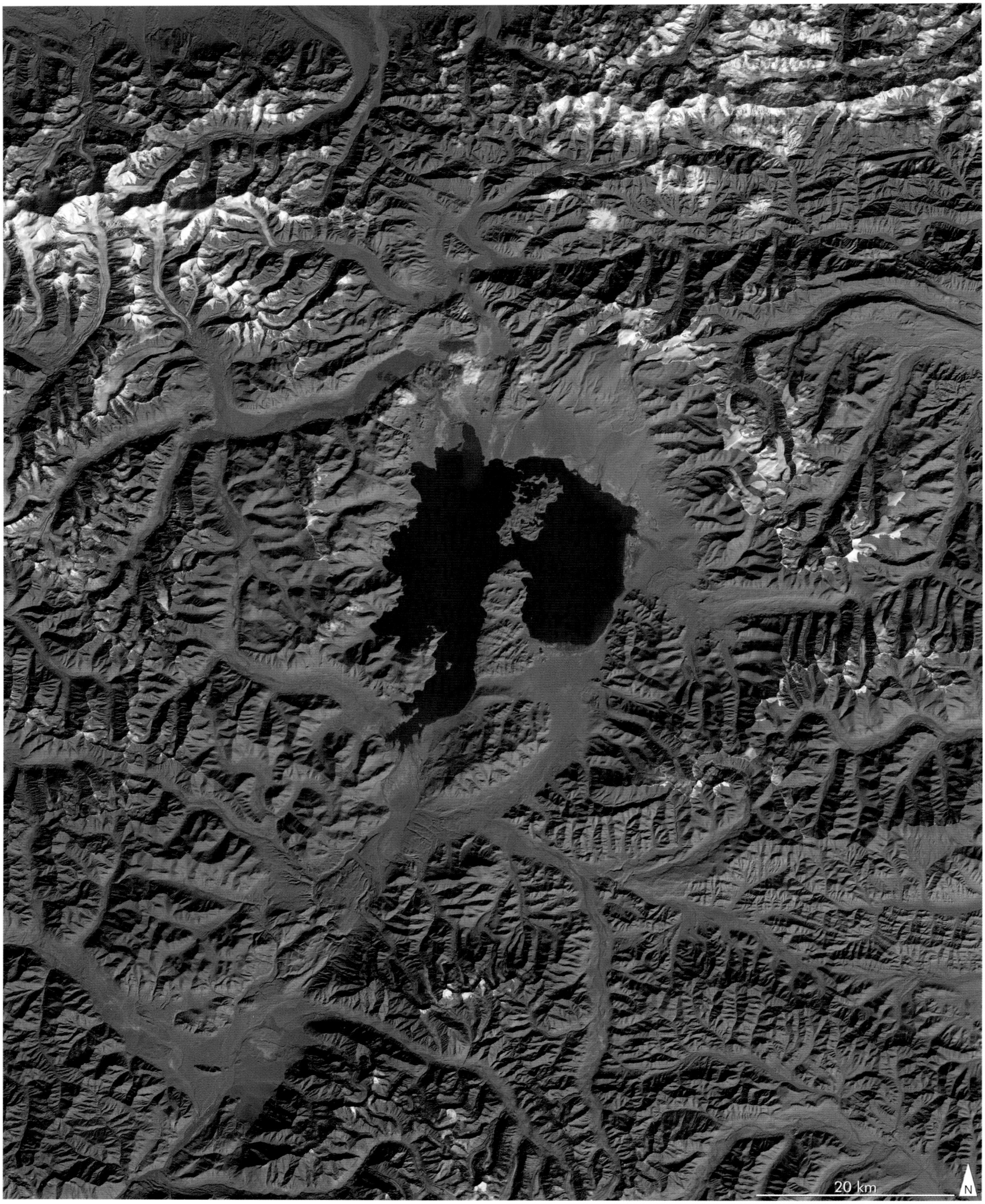

The Kara-Kul impact structure on September 3, 2018 as a composite image based on TanDEM-X DEM and Landsat-8 (channel 2/3/4) data.

Kara-Kul

Tajikistan

latitude = 39°04‘ N / longitude = 73°26‘ E / diameter = 52 km / age = > 5 - 50 Myr

scale = 1 : 474 000 / scene = 85.3 × 113.7 km²

0 4.74 23.70 km

N

3060 meter 6760

Yilan
China

Manfred Gottwald

Yilan is a circular structure in northeast China, in a hilly area of the southwestern Xiaoxinganling mountain range. The structure has a diameter of 1.85 km and appears as a simple bowl-shaped feature. The rim is largely well-preserved but has been considerably eroded in the southern part. It rises up to about 150 m above the relatively flat floor of the structure and about 70–120 m above the surrounding terrain.

The Yilan structure is formed in Cretaceous granites. Its interior is mainly covered with forest. The fill of the structure comprises lake deposits that superpose a granitic breccia. ^{14}C dating suggests that the lake vanished about 10 000 years ago, which provides a lower limit for the age of the Yilan structure. The disappearance of the lake has been attributed to the loss of the southern crater rim.

Ming Chen and co-workers claimed the detection of Planar Deformation Features (PDFs) in quartz from fine-grained samples collected from the floor of the structure, and, consequently referred to Yilan as an impact crater. However, the published material of observed microdeformation features does not provide unequivocal support for this assertion so far.

Given the very recent report about the discovery of Yilan as a possible impact crater, as well as the spurious evidence of possible shock deformation, we decided to include this structure in the section about structures, for which further confirmation is required. Particularly, unequivocal proof of shock effects needs to be accomplished.

Location/Access

Yilan crater is located in northeast China in the Heilongjiang Province and is named after the city of Yilan, located 20 km to the east-southeast at the confluence of the Shongua River and Mudanjiang River. Rural roads lead from Yilan over the Shongua River to the structure.

Selected Reference

Chen, M., Xie, X., Xiao, W., and Tan, D. (2019). Yilan crater, a newly identified impact structure in northeast China. Chinese Science Bulletin, online first, doi: 10.1360/TB-2019-0704.

The Yilan impact structure on December 14, 2019 as a composite image based on TanDEM-X DEM and Landsat-8 (channel 2/3/4) data.

Yilan
China

latitude = 46°23‘03“ N / longitude = 129°18‘40“ E / diameter = 1.85 km / age = > 10 000 yr

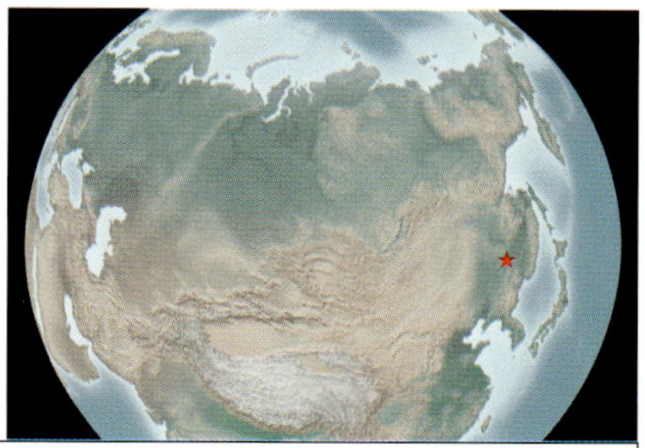

scale = 1 : 97 000 / scene = 17.5 × 23.3 km²
0 0.97 4.85 km
N
88 meter 649

Strongest Airburst in Modern Times

Tunguska

Russia

Thomas Kenkmann

The explosion of a large bolide that had entered the atmosphere over an unpopulated part of Siberia in the early morning of June 30, 1908 is widely known as the *Tunguska Event*. Although the event most likely did not form an impact crater, it is listed here, as this is an impact event of modern times, where an extraterrestrial projectile of considerable size collided with Earth and exploded in the Earth's atmosphere.

Eye witnesses and geophysical data agree that the bolide traversed the atmosphere in an oblique path over 450 km from southeast to northwest at an azimuth of 99–115°. At the final flash point at a height of 8–9 km above ground, its brightness exceeded that of the Sun. The tremendous blast wave flattened taiga forests over a 2150 km^2 area. It was seismically recorded as far to the west as Germany. The destroyed forest zone had the shape of a butterfly, symmetrical to the flight trajectory. The point where the blast wave hit the surface first was indicated by a "telegraph forest", where all trees remained standing – but with their crowns torn off. In the surroundings where the blast wave hit the Earth at an oblique angle, trees fell in a radial pattern. The effect of the blast wave was superposed by a thermal pulse that burned large forest areas. The event also caused a magnetic storm. A local disturbance of the Earth's magnetic field started with a few minutes delay after the explosion and prevailed for 4 hours. The energy of the blast was estimated at the equivalent of 3–5 megatons of TNT explosive.

The origin of the Tunguska blast has been explained by impact into the atmosphere of either a meteorite or a comet. No large fragments of meteorite were ever identified in the Tunguska blast area. Field studies carried out in 1979 and into the 1980s showed that peat samples from the Tunguska site include amalgamated polycrystalline micro-aggregates of diamond, lonsdaleite, graphite and troilite, which contain trace elements such as Ir that could be indicative for a meteoritic component. Submillimeter-sized metallic spheres extracted from soil in the Tunguska region also contain noble metals at cosmic proportions. Such meteoritic debris was also found in an ice core drilled at the South Pole and indicates that the Tunguska object indeed exploded in the atmosphere so that the debris could be distributed globally. The Tunguska Event was also associated with "white" nights over large parts of Russia and northern Europe that occurred in the direct aftermath of the Tunguska event.

The Tunguska event was modelled by Natalia Artemieva and Valery Shuvalov. Their numerical simulation from 2007 satisfactorily describes the damage caused by Tunguska and proposes the following scenario: A cosmic body of 50–80 m diameter and low strength typical for stony meteorites enters the upper atmosphere at a cosmic velocity of 20 km/s at an angle of 30–45°. At an altitude of 20 km, the impactor starts to flatten, disrupts and evaporates strongly; the vaporized impactor material completely decelerates at an altitude of 8–10 km and releases all its energy into the atmosphere. The mixture of hot air and vaporized impactor material is buoyant and accelerates back along the wake, while an atmospheric shock wave reaches the terrestrial surface. The blast wave causes the observed forest damage with its butterfly pattern and the comparatively weak damage in its center. All the extra-terrestrial impactor material is sucked backward into the wake and reaches high altitudes. Possibly the dispersal of hot condensed impactor material caused the white nights observed in Northern Eurasia shortly after the event.

In 1927 Leonid Kulik led the first expedition to the site thought to represent the center of the blast. They, as others in the following decades, did not record any signs of an impact crater formed during the event. In 2009, an Italian expedition team led by Luca Gasperini proposed that an elliptical lake of 300 m diameter, named *Lake Cheko*, resulted from the event. The location of the lake fits with the trajectory of the Tunguska projectile and the lake has a funnel-like bathymetry. Gareth Collins and a number of other impact researchers countered that the age of the lake is still unknown and no traces of meteoritic material have been found in or around Lake Cheko. Morphologically there is neither evidence for an uplifted crater rim with an overturned flap, nor for ejected material in its surroundings. An elliptical shape is also uncommon for impact craters. Moreover, aerial photographs of the lake from 1938 show large trees standing along the rim. The age of these trees ought to predate 1908. They should have been destroyed during an impact at this location. Thus, a causal relationship between Lake Cheko and the Tunguska Event is not supported at this time.

This postage stamp showing Leonid Kulik was issued by the Soviet Union in 1958 to commemorate the 50th anniversary of the Tunguska event.

Location/Access

The Tunguska Event occurred in the eastern Siberian taiga close to the Stony Tunguska River, in an area that today belongs to the Krasnojarsk region. This area is still largely as unpopulated as it was in 1908. The nearest infrastructure is found some 60 km to the south-southeast in the town of Vanawara that has an airstrip. A local operator provides tours from Krasnojarsk via Vanavara to the epicenter of the event.

Selected References

Artemieva, N. A., and Shuvalov, V. V. (2007). 3D effects of Tunguska Event on the ground and in the atmosphere. 38th Lunar and Planetary Science Conference, Lunar and Planetary Institute, Houston, Texas, Abstract #1537.

Collins, G. S., Artemieva, N. A., Wünnemann, K., Bland, P. A., Reimold, W. U., and Koeberl, C. (2008). Evidence that Lake Cheko is not an impact crater. Terra Nova, Vol. 20, pp. 165–168.

Florensky, K. P. (1963). Previous results of the Tunguska meteoritic complex expedition of 1961. Meteoritika, Vol. 23, pp. 3–29. (in Russian).

Ganapathy, R. (1983). The Tunguska explosion of 1908: discovery of meteoritic debris near the explosion site and at the South Pole. Science, Vol. 220, pp. 1158–1161.

Gasperini, L., Bonatti, E., Albertazzi, S., Forlani, L., Accorsi, C. A., Longo, G., Ravaioli, M., Alvisi, F., Polonia, A., and Sacchetti, F. (2009). Sediments from Lake Cheko (Siberia), a possible impact crater for the 1908 Tunguska event. Terra Nova, Vol. 21, pp. 489–494.

Kolesnikov, E. M., Boettger, T., and Kolesnikova, N. V. (1999). Finding of probable Tunguska cosmic body material: isotopic anomalies of carbon and hydrogen in peat. Planetary and Space Science, Vol. 47, pp. 905–916.

Vasilyev, N. V. (1998). The Tunguska meteorite problem today. Planetary and Space Science, Vol. 46, pp. 129–150.

Tunguska
Russia

latitude = 60°53'10" N / longitude = 101°54'25" E / diameter = n.a. / age = 111 yr

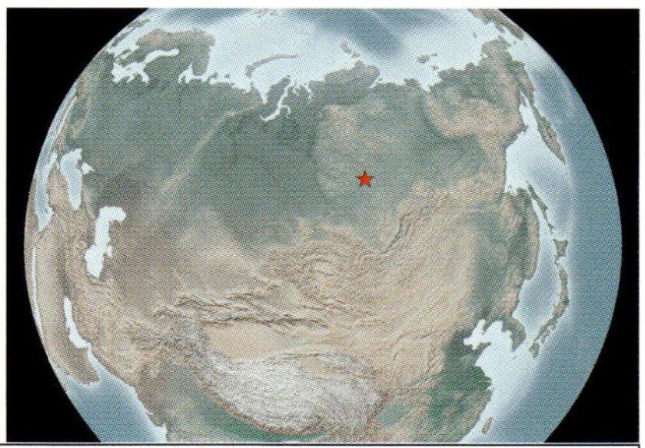

scale = 1:473 000 / scene = 85.3 × 113.7 km²
0 4.73 23.65 km
N
230 meter 600

The area that was affected by the Tunguska airblast on May 30, 2018 as a composite image based on TanDEM-X DEM and Copernicus Sentinel-2 (channel 2/3/4) data. The position of the airblast epicenter is indicated by cross hairs.

7.5 Australia

Acraman in September 2018 – TanDEM-X/Copernicus Sentinel-2

Australian Impact Structures – Overview

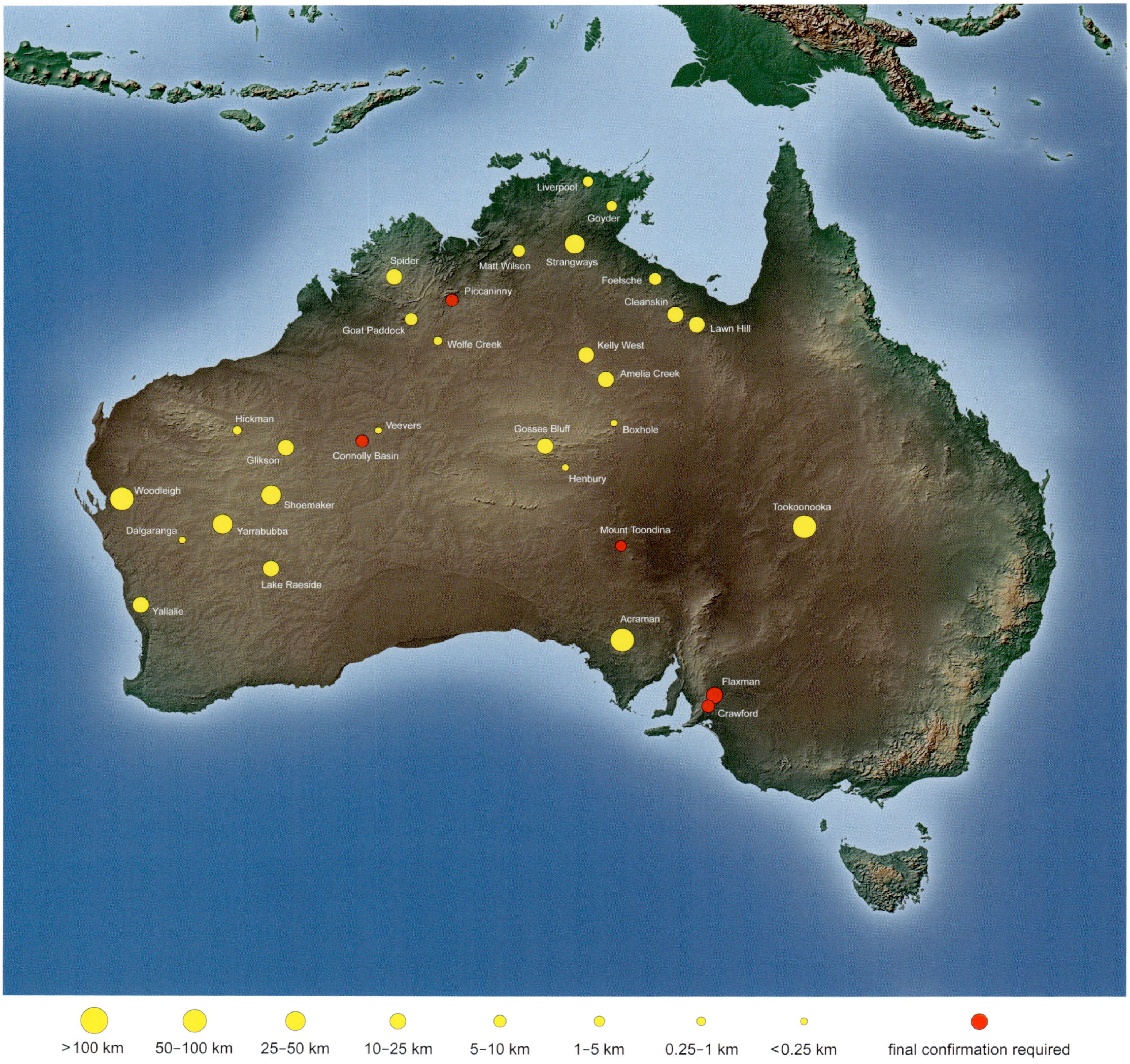

Table: Australian Impact Structures – Overview

ID	Name	Country/Region	Latitude	Longitude	Diameter (km)	Age (Myr)	Topographic Visibility
1	Acraman	South Australia	32°01'S	135°27'E	90	580	yes
2	Amelia Creek[1]	Northern Territory	20°51'S	134°53'E	20	600–1660	no
3	Boxhole	Northern Territory	22°36'45"S	135°11'43"E	0.17	0.0177 ± 0.0123	yes
4	Cleanskin	Northern Territory	18°10'00"S	137°56'30"E	15	540–1400	yes
5	Dalgaranga	Western Australia	27°38'06"S	117°17'20"E	0.024	0.27	yes
6	Foelsche	Northern Territory	16°40'15"S	136°47'02"E	6	541–981	yes
7	Glikson[2]	Western Australia	23°58'40"S	121°33'22"E	19	< 508 ± 5	no
8	Goat Paddock	Western Australia	18°20'02"S	126°40'36"E	5	56–64	yes
9	Gosses Bluff	Northern Territory	23°49'10"S	132°18'27"E	22	142.5 ± 0.8	yes
10	Goyder	Northern Territory	13°28'31"S	135°02'24"E	3	150–1400	yes
11	Henbury Crater Field	Northern Territory	24°34'18"S	133°08'52"E	max. 0.180	0.0042 ± 0.0019	yes
12	Hickman	Western Australia	23°02'13"S	119°41'00"E	0.26	0.02–0.10	yes
13	Kelly West	Northern Territory	19°55'42"S	133°57'12"E	14	> 541	yes
14	Lake Raeside[3]	Western Australia	28°44'54"S	120°56'55"E	11	< 295	no (buried – Lake Raeside)
15	Lawn Hill	Queensland	18°41'19"S	138°39'06"E	20	472 ± 8	yes
16	Liverpool	Northern Territory	12°23'45"S	134°02'50"E	2	540–1000	yes
17	Matt Wilson	Northern Territory	15°30'00"S	131°10'51"E	7.5	1400–1500	yes
18	Shoemaker	Western Australia	25°52'S	120°53'E	30	1630 ± 5	yes
19	Spider	Western Australia	16°44'27"S	126°05'21"E	13	600–900	yes
20	Strangways	Northern Territory	15°12'S	133°34'E	40	646 ± 42	yes
21	Tookoonooka[1]	Queensland	27°07'S	142°50'E	60	128 ± 5	no (buried)
22	Veevers	Western Australia	22°58'12"S	125°22'20"E	0.07	0.012 ± 0.008	yes
23	Wolfe Creek	Western Australia	19°10'18"S	127°47'43"E	0.88	0.120 ± 0.009	yes
24	Woodleigh[1]	Western Australia	26°03'S	114°40'E	60	265–318	no (buried)
25	Yallalie	Western Australia	30°26'40"S	115°47'16"E	12	83.6–89.8	yes (buried)
26	Yarrabubba[1]	Western Australia	27°11'S	118°50'E	30	2246 ± 17	no
Further confirmation required							
27	Connolly Basin	Western Australia	23°32'03"S	124°45'38"E	9	55–75	yes
28	Crawford[1]	South Australia	34°43'S	139°02'E	8.5	32–38	no
29	Flaxman[1]	South Australia	34°37'S	139°04'E	10	32–38	no
30	Mount Toondina	South Australia	27°56'41"S	135°21'31"E	4	66–144	yes
31	Piccaninny	Western Australia	17°25'31"S	128°26'18"E	7	< 360	yes

Position references for selected impact structures:

1 Haines, P. W. (2005). Impact cratering and distal ejecta: the Australian record. Australian Journal of Earth Sciences, Vol. 52, pp. 481–507.

2 MacDonald, F. A., Wingate, M. T. D., and Mitchell, K. (2005). Geology and age of the Glikson impact structure, Western Australia. Australian Journal of Earth Sciences, Vol. 52, pp. 641–651.

3 Glikson, A., Hickman, A., and Crossley, R. (2016). Evidence for a shock-metamorphic breccia within a buried impact crater, Lake Raeside, Yilgarn Craton, Western Australia. Australian Journal of Earth Sciences, Vol. 63, pp. 99–109.

Acraman
South Australia

Thomas Kenkmann

Topographically, the Acraman impact structure represents a gentle, circular, 30 km wide depression in the Gawler Ranges of the northern Eyre Peninsula, South Australia. This depression is partially filled by Lake Acraman, a somewhat polygonal and 20 km wide playa lake that has several islands. The lake is surrounded by alluvial beds. Outcrops of bedrock in the central depression are rare. Those on the islands show intensely shattered rocks with shatter cones. They also contain quartz grains with multiple sets of planar deformation features. It has been thought that the islands are remnants of the eroded central uplift of the structure.

Target rocks are exclusively crystalline and comprise flat-lying Early Mesoproterozoic dacitic lavas of the Gawler Range Volcanics of about 1590 Myr age. The structure is deeply eroded and is estimated to have undergone as much as 3–5 km of degradation. The original crater floor and rim are not preserved and, therefore, the reconstruction of the original rim diameter relies on geophysical data or the application of scaling relationships characteristic for impact structures. For Acraman the range of published diameters is particularly large and extends from 35 km, a size roughly corresponding to the morphological depression, up to 160 km, with a diameter of 90 km having been most widely accepted.

Acraman has an aeromagnetic signature in the form of a circular, 20 km wide magnetic low with a subdued magnetic relief. Dating of an altered melt-rock vein in the central uplift with K-Ar and Ar-Ar techniques provided a minimum age of 450 Myr for the impact event. However, the Acraman impact event is connected to a distal ejecta horizon occurring 240–540 km away from Acraman. This ejecta layer occurs in the 580 Myr Bunyeroo Formation of the Adelaide Fold Belt. Consequently, this latter age is currently regarded as the best available age estimate for the Acraman impact.

The impact origin for Acraman was postulated by G. Williams in 1986 based on the discovery of shatter cones, planar deformation features (PDFs) in quartz, and veins of devitrified melt in isolated, shattered dacite outcrops on islands near the center of the lake.

Location/Access

Lake Acraman is located about 80 km northeast of the town of Poochera and the Eyre Highway A1 that connects Perth with Adelaide. The only infrastructure of the area are some unpaved roads and the settlement Yardea about 40 km south of the lake. Yardea is considered to be located closest to the edge of the eroded crater structure.

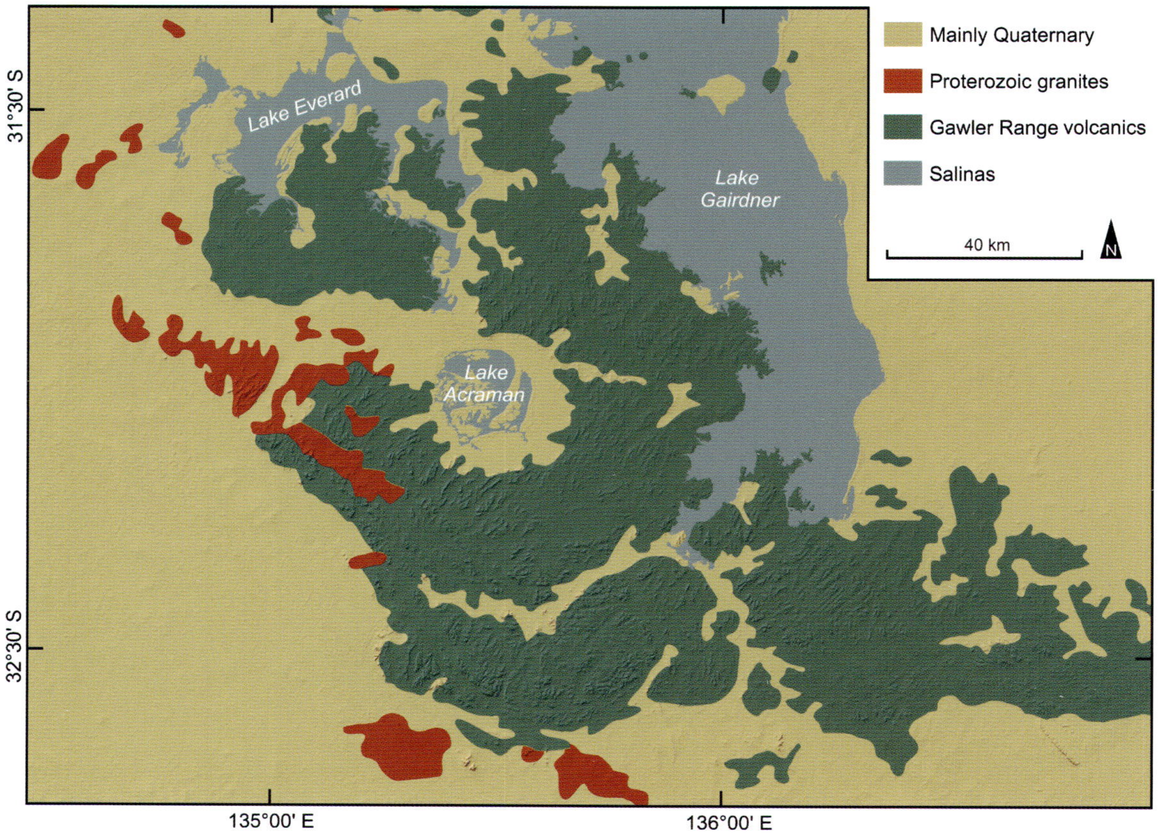

Schematic geological map of the Acraman impact structure, underlain by a TanDEM-X DEM (modified after Williams and Wallace, 2003).

Exposure of a possible clast-rich melt dike near the center of the Acraman impact structure. Hammer for scale ca. 40 cm long (photo courtesy: Peter Haines, Geological Survey of Western Australia, Perth).

Brecciated felsic volcanics are exposed near the center of the Acraman impact structure. Hammer for scale ca. 40 cm long (photo courtesy: Peter Haines, Geological Survey of Western Australia, Perth).

Acraman

South Australia

latitude = 32°01' S / longitude = 135°27' E / diameter = 90 km / age = 580 Myr

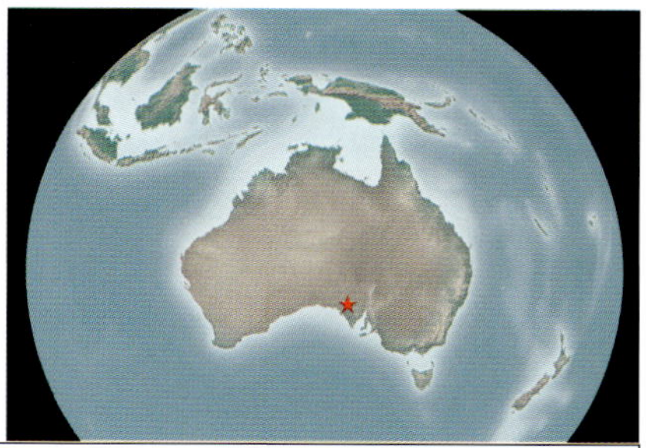

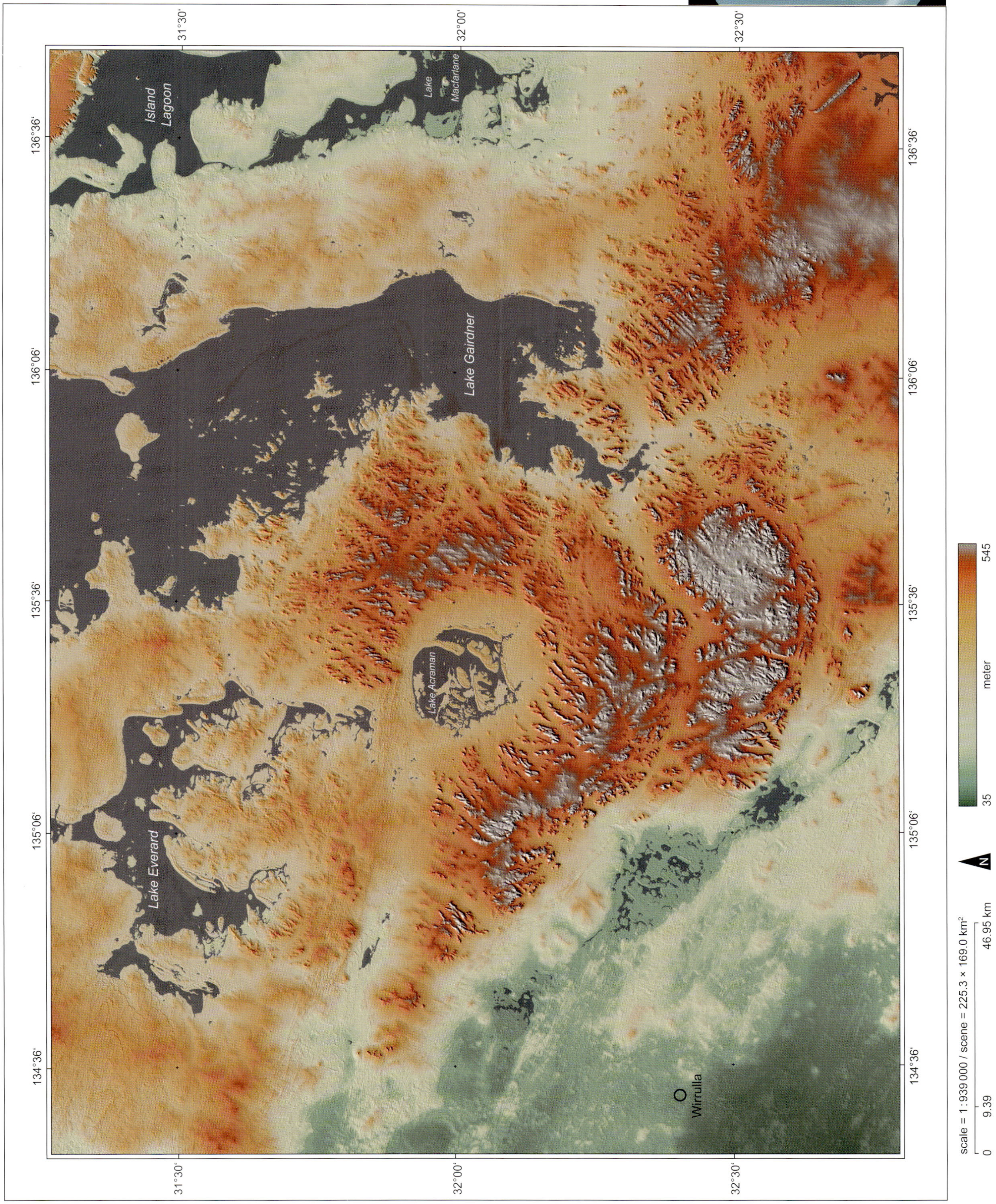

The Acraman impact structure with its playa lakes is a landmark for astronauts aboard the International Space Station. This photo was captured on October 2, 2014 (photo: NASA-ISS041-E-61822).

Selected References

Baldwin, S. L., McDougall, I., and Williams, G. E. (1991). K-Ar and ^{40}Ar-^{39}Ar analyses of melt rock from the Acraman impact structure, Gawler Ranges, South Australia. Australian Journal of Earth Sciences, Vol. 38, pp. 291–298.

Haines, P. W. (2005). Impact cratering and distal ejecta: the Australian record. Australian Journal of Earth Sciences, Vol. 52, pp. 481–507.

Shoemaker, E. M., and Shoemaker, C. S. (1996). The Proterozoic impact record of Australia. AGSO Journal of Australian Geology and Geophysics, Vol. 16, pp. 379–398.

Williams, G. E. (1986). The Acraman impact structure: Source of ejecta in Late Precambrian shales, South Australia. Science, Vol. 233, pp. 200–203.

Williams, G. E., Schmidt, P. W., and Boyd, D. M. (1996). Magnetic signature and morphology of the Acraman impact structure, South Australia. AGSO Journal of Australian Geology and Geophysics, Vol. 16, pp. 431–442.

Williams, G. E., and Wallace, M. W. (2003). The Acraman asteroid impact, South Australia: magnitude and implications for the late Vendian environment. Journal of the Geological Society, Vol. 160, pp. 545–554.

Amelia Creek

Northern Territory

Thomas Kenkmann

Amelia Creek has no obvious topographic expression, as the morphology of the region is largely governed by regional tectonics. The crater structure is located in the hilly fold belt of the Davenport Ranges, Northern Territory. This region displays complex, refolded folds with basin and dome structures.

The target rocks of the Amelia Creek structure are folded sedimentary and volcanic rocks of the Paleoproterozoic Hatches Creek Group. Deformation related to the impact is present over an elongated area of 20×12 km, with shatter cones restricted to the south-southwest portion, where their occurrences are associated with south-southwest oriented thrust sheets. Despite pervasive shatter coning, typical of the central region of complex impact structures, no structural uplift is apparent. In addition to shatter cones, evidence of low-level shock in sandstone includes sets of planar fractures (PFs) with feather features (FFs). The asymmetric distribution of shocked rocks was explained by a very oblique impact from the north-northeast. However, the structural complexity of the target area makes a proper analysis of the impact trajectory very difficult.

Amelia Creek was investigated by means of aeromagnetic, advanced space-borne thermal emission and reflection radiometer (ASTER), and X-band synthetic aperture radar (X-SAR) data. These remote sensing investigations constrained an area of anomalous deformation, in which smooth regional trends are disrupted by arcuate features to a distance of 10 km to the north and south of the occurrence of shock metamorphosed rocks. Based on geological reasons, the age of Amelia Creek was bracketed between 600 and 1660 Myr. On the one hand, a consistent shatter cone orientation indicates that impact occurred after the regional folding that took place between 1660 and 1700 Myr. And on the other hand, relics of Cambrian and Late Neoproterozoic strata overlying the impact-deformed Hatches Creek Group were seemingly not affected by the impact event.

Amelia Creek's impact origin was identified in 2003 by Francis Macdonald and Ken Mitchell, who discovered abundant shatter cones.

Amelia Creek

Northern Territory

latitude = 20°51‘ S / longitude = 134°53‘ E / diameter = 20 km / age = 600 - 1660 Myr

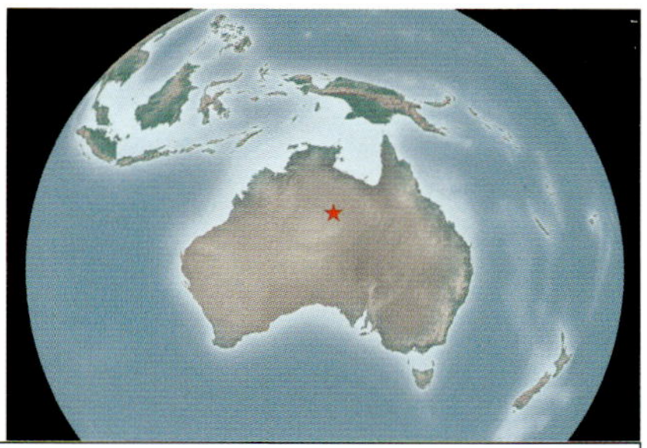

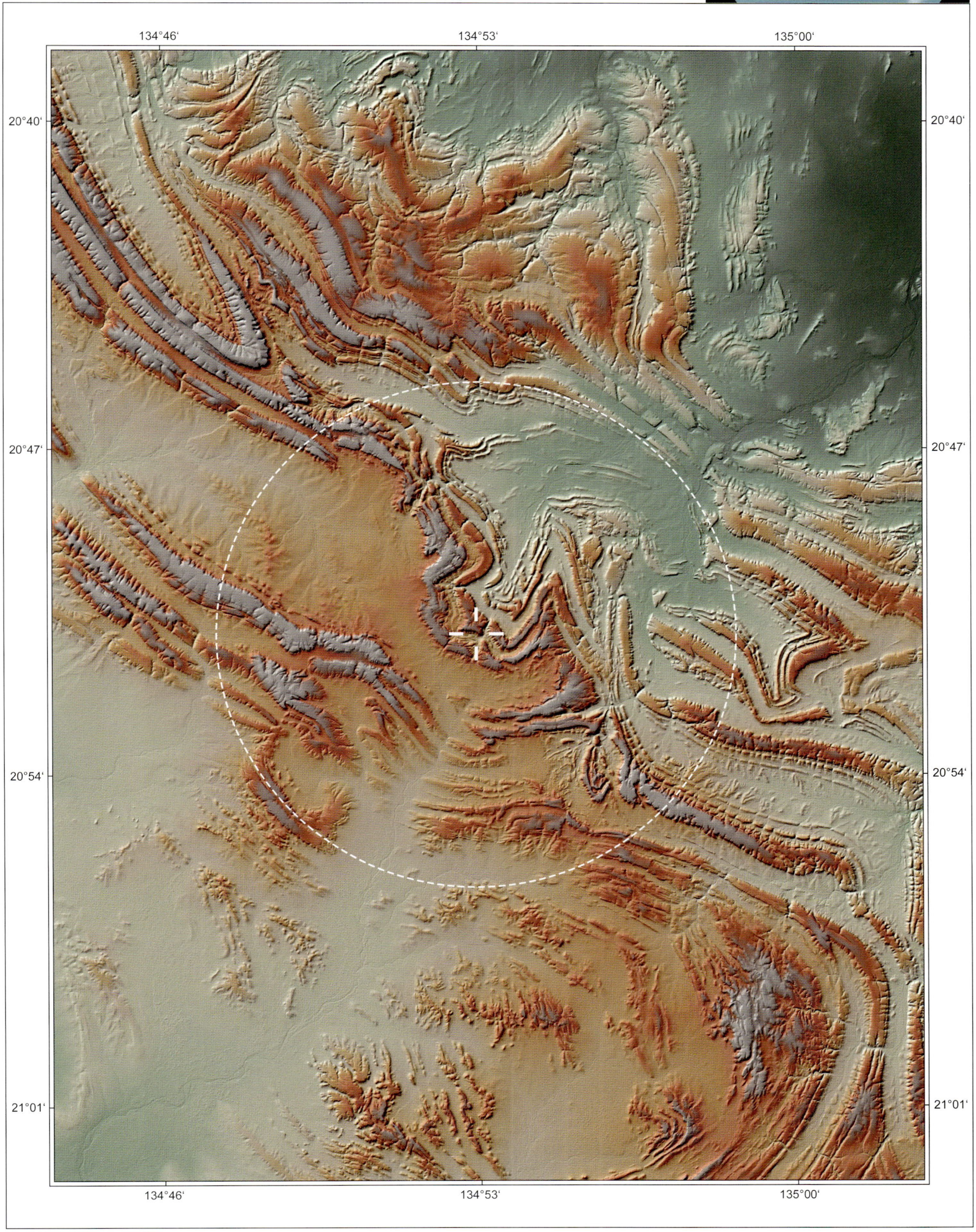

scale = 1:185 000 / scene = 33.3 × 44.4 km²

0 1.85 9.25 km

N

345 meter 615

Location/Access

The structure is located some 60 km east of Stuart Highway that connects Alice Springs with Darwin. Some 40 km south of Wycliffe Roadhouse a dirt road branches off to the east that allows to drive close to the structure. The area of Amelia Creek has no infrastructure.

Selected References

Haines, P. W. (2005). Impact cratering and distal ejecta: the Australian record. Australian Journal of Earth Sciences, Vol. 52, pp. 481–507.

Macdonald F. A., Mitchell K., and Stewart A. J. (2005). Amelia Creek: a Proterozoic impact structure in the Davenport Ranges, Northern Territory. Australian Journal of Earth Sciences, Vol. 52, pp. 631–640.

Boxhole

Northern Territory

Thomas Kenkmann

The Boxhole meteorite crater is a small and morphologically well-preserved simple crater of 170 m diameter. The raised rim is best developed on the southern side, where it displays inverted stratigraphy. The northern part of the crater is leaning against the slope of a 10 m high ridge. The ejecta blanket around the crater is partly preserved. It is thickest along the crest of the southern rim and absent on the northern rim. The target rocks are composed of Proterozoic schist and foliated gneiss of the Arunta Province that, at the time of impact, were overlain by soil. A large number of meteorites were recovered from the crater and its surroundings. The largest piece weighed 82 kg. In total, 280 kg of an iron meteorite, a medium octahedrite belonging to the chemical group IIIAB, were collected. In addition, Ni-Fe spherules of meteoritic origin were found in the soil around the crater. Meteoritic material around Boxhole crater is very much less abundant than, e.g., at the **Henbury** craters located 300 km to the southwest of Boxhole.

It was suggested that Boxhole and the Henbury craters could represent the result of a paired fall. From the asymmetry of the ejecta blanket of Boxhole – with absence of ejecta to the north – Eugene Shoemaker inferred that this side represented the up-range direction and that the projectile travelled from north to south and impacted at a relatively low angle. If this is correct, Boxhole and Henbury cannot originate from the same event, as the Henbury projectile clearly approached from the southwest. Two independent and incompatible age determinations have been reported for Boxhole: a cosmogenic ^{14}C terrestrial exposure age for the meteorite of 5400 ± 1500 years is in conflict with a ^{10}Be-^{26}Al exposure age of 30000 years for a quartz ejecta block seemingly uncovered by the impact.

Boxhole was first reported in 1937 by Cecil Madigan. The impact origin was confirmed by the discovery of iron meteorite fragments around the crater.

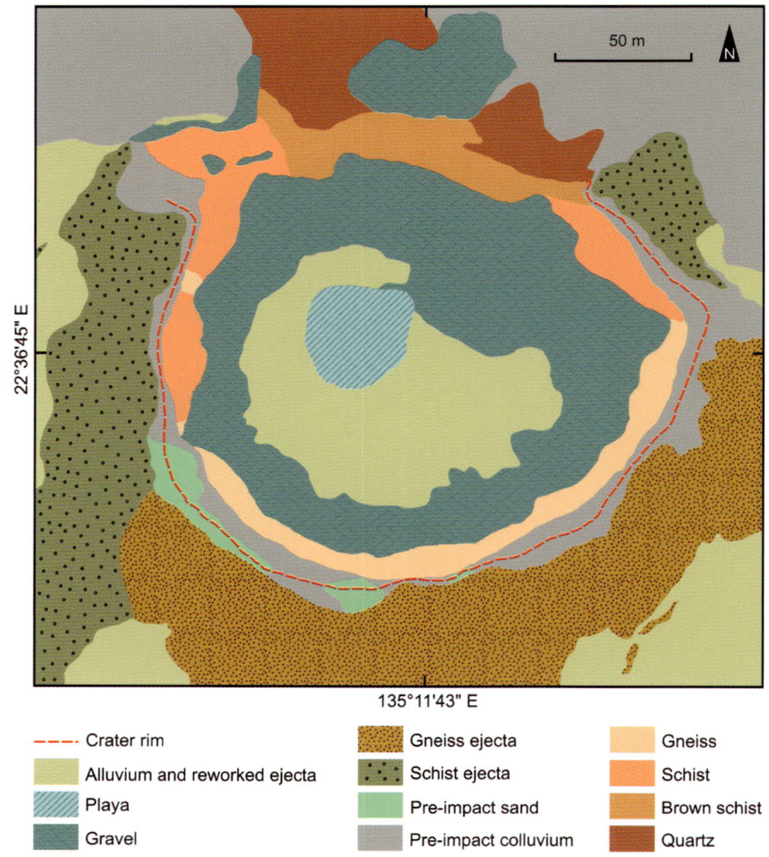

Schematic geological map of the Boxhole crater (modified after Shoemaker et al., 2005).

View onto the Boxhole meteorite impact crater. Car, at the far edge, for scale (photo courtesy: Peter Haines, Geological Survey of Western Australia, Perth).

Location/Access

The Plenty Highway (12) branches off from Stuart Highway 55 km north of Alice Springs. After 150 km on this unpaved road, another dirt road runs northward to the Boxhole meteorite crater. A cattle station lies close to the crater.

Selected References

Haines, P. W. (2005). Impact cratering and distal ejecta: the Australian record. Australian Journal of Earth Sciences, Vol. 52, pp. 481–507.

Hodge, P. W., and Wright, F. W. (1970). Meteoritic spherules in the soil surrounding terrestrial impact craters. Nature, Vol. 225, pp. 717–718.

Shoemaker, E. M., Roddy, D. J., Shoemaker, C. S., and Roddy, J. K. (1988). The Boxhole meteorite crater, Northern Territory, Australia. 19th Lunar and Planetary Science Conference, Lunar and Planetary Institute, Houston, Texas, pp. 1081–1082.

Shoemaker, E. M., Shoemaker, C. S., Nishiizumi, K., Kohl, C. P., Arnold, J. R., Klein, J., Fink, D., Middleton, R., Kubik, P. W., and Sharma, P. (1990). Ages of Australian meteorite craters – A preliminary report (abstract). Meteoritics, Vol. 25, p. 409.

Shoemaker, E. M., Macdonald, F. A., and Shoemaker, C. S. (2005). Geology of five small Australian impact craters. Australian Journal of Earth Sciences, Vol. 52, pp. 529–544.

Boxhole
Northern Territory

latitude = 22°36'45" S / longitude = 135°11'43" E / diameter = 170 m / age = 17 700 ± 12 300 yr

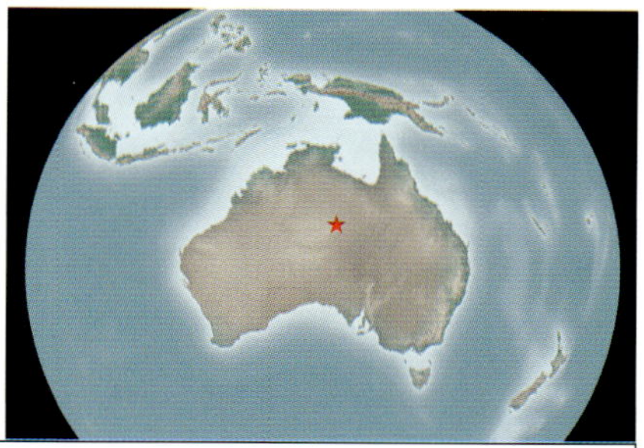

scale = 1:70 000 / scene = 12.6 × 16.8 km²
0 0.70 3.50 km
N
445 meter 520

Cleanskin

Northern Territory

Thomas Kenkmann

Cleanskin is a circular structure in northern Australia that measures approximately 15 km in diameter. It is located close to the border between the Northern Territory and Queensland. The terrain within the Cleanskin structure consists of low ridges and valleys with local relief of about 50 m. The ridges are built up by moderately-dipping sandstones and – locally – by massive lithic breccias that are strongly lithified. The western half of the structure is poorly defined because of overlying undeformed Early Cretaceous strata that cap the impact-deformed strata. Anomalous fold and fault patterns in otherwise mildly deformed rocks were first noted by Ian Sweet during a mapping campaign for the Northern Territory Geological Survey in 2001.

The inner several kilometers of the structure form a dome with at least three repetitions of the same two members of sandstone of presumed Early Mesoproterozoic age, which are separated by reverse faults. These faults form a circumferential pattern and are unrelated to the faults of the regional deformation, which display east-northeast and west-northwest orientations.

Apart from its circular nature and the presence of a central uplift, proof that Cleanskin is an impact site relies on the presence of shatter cones, extensive breccia formations, and microscopic shock deformation in quartz grains. The latter includes feather features (FFs), planar fractures (PFs), and also planar deformation features (PDFs).

Cross-cutting sets of PDFs in a quartz grain of a sandstone sample taken about 5 km southeast of the center of the Cleanskin impact structure (photo: T. Kenkmann).

The Cleanskin impact structure in August 2018 as a composite image based on TanDEM-X DEM and Copernicus Sentinel-2 (channel 2/3/4) data.

Cleanskin

Northern Territory

latitude = 18°10'00" S / longitude = 137°56'30" E / diameter = 15 km / age = 540 - 1400 Myr

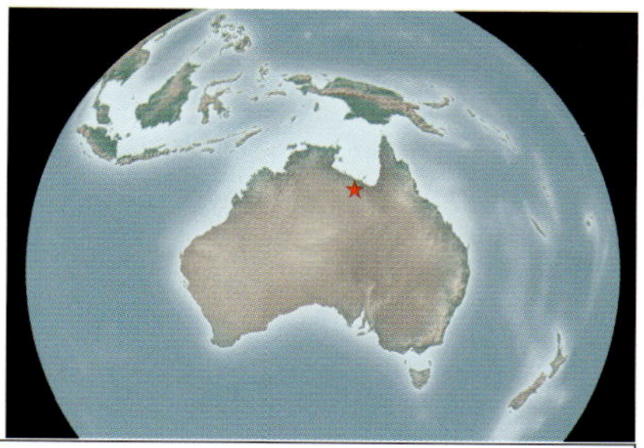

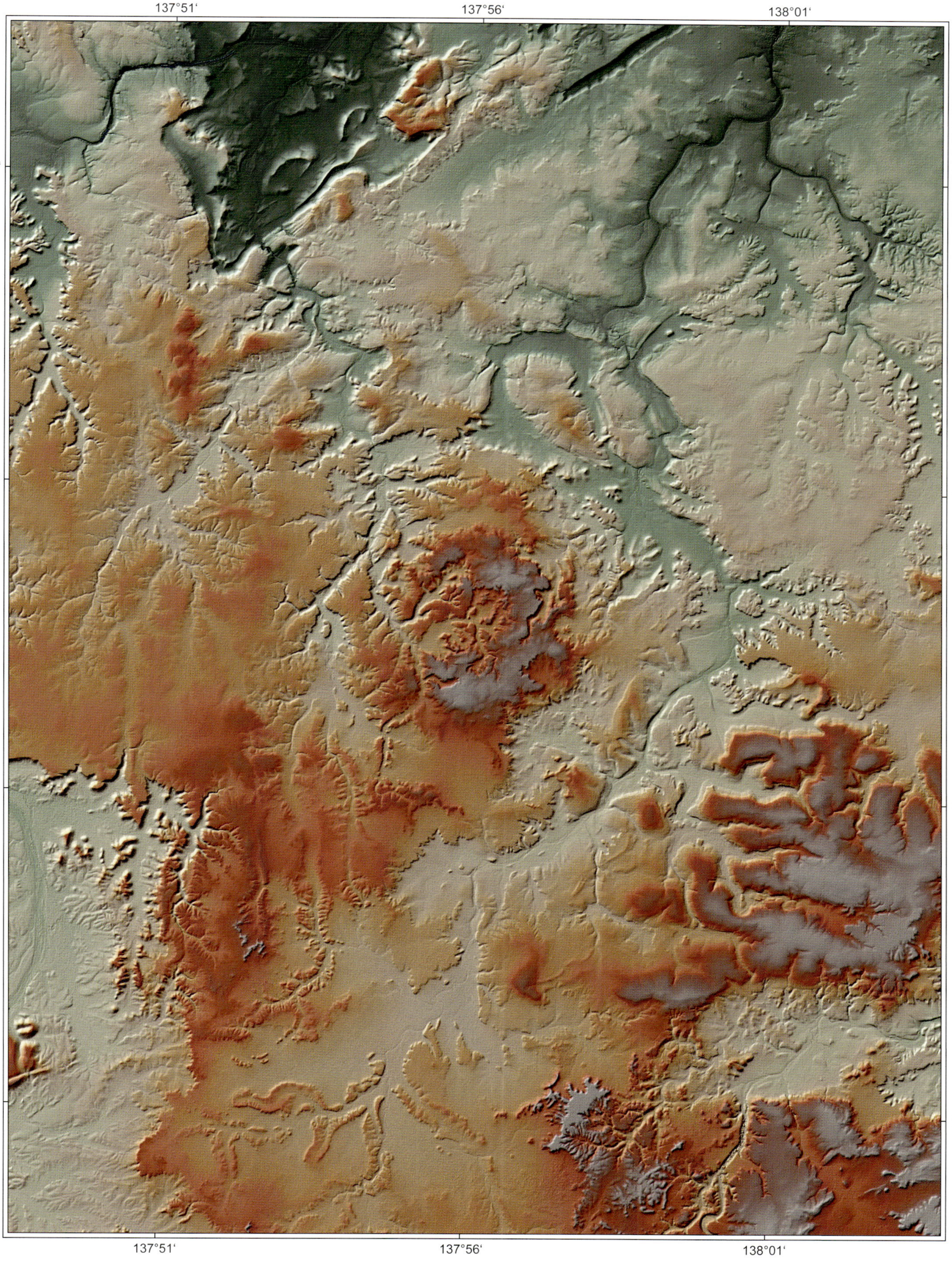

scale = 1: 150 000 / scene = 27.0 × 35.9 km²

0 1.50 7.50 km

N

90 meter 290

It has been estimated that the sedimentary rocks of the Cleanskin target, predominantly sandstones and mudstones, were deposited between 1400 and 1500 Myr ago. The impact clearly occurred prior to the Early Cretaceous, as sedimentary rocks of this age blanket the structure. The impact is inferred to have taken place also well before the Cambrian, because the Cambrian rocks lie at approximately the same erosion level outside the structure as the present day surface, indicating that the impact site must have been deeply eroded prior to Cambrian deposition. Thus, the impact is likely to have occurred between about 1400 Myr and the Early Cambrian.

In 2012 evidence for an impact origin of Cleanskin was presented by Peter Haines, Ian Sweet and Kent Mitchell in the form of unindexed shock features. Since then, the crystallographic orientations of these abundant PDFs, PFs, and FFs have been indexed by Thomas Kenkmann and Peter Haines.

Location/Access

Cleanskin lies in a remote region of the Northern Territory. Without suitable infrastructure in the region, access is severely restricted. A helicopter is required to reach the impact structure.

Selected References

Dunster, J. N., Haines, P. W., and Munson, T. J. (2014). Meteorites and impact structures in the Northern Territory. Northern Territory Geological Survey, Record 2014-007, 48 pp.

Haines, P. W., Sweet, I. P., and Mitchell, K. (2012). Cleanskin structure, Northern Territory and Queensland, Australia: evidence for an impact origin. Meteoritics and Planetary Science, Abstract. #5176.

Kenkmann, T., Haines, P. W., Sweet, I. P., and Mitchell, K. (2019). Shock deformation and faulting in the Cleanskin impact structure, Northern Territory, Australia. Large Meteorite Impacts and Planetary Evolution VI, Brasilia, Abstract #5014.

Dalgaranga

Western Australia

Thomas Kenkmann

The small Dalgaranga meteorite impact crater is situated in the Archean granite-greenstone terrain of the Yilgarn Craton. The crater is 24 m wide and approximately 3 m deep.

Target rock is a quartz-rich, medium-grained granite that is weathered to varied degree. On the southwestern side, the granite appears relatively fresh, whilst on the eastern and northern sides of the crater the granite has been lateritized. Outside the crater, the granite is overlain by weakly stratified Quaternary deposits, which are actually strongly weathered granite. A distinctive horizon, about 20–50 cm thick, rests sharply on the granite. Its contact allows defining the crater rim and hinge line.

Within the crater, a breccia unit is well exposed and consists mainly of weathered granite, but also contains fragments of the surficial unit and laterite. This breccia unit is extensively exposed around the southwest half of the crater wall. In the crater center the breccia is buried under two meters of Quaternary colluvium. The crater rim shows strata dipping locally with steep inclination. The ejecta display an inverted stratigraphy with granite ejecta overlying ejecta from surficial target material. The ejecta distribution has a pronounced bilateral symmetry and a "forbidden zone" up-range. Consequently, the projectile must have impacted from the south-southwest. Based on the preservation status of the ejecta blanket, Eugene and Carolyn Shoemaker concluded that Dalgaranga crater was formed less than 3000 years ago. More recent cosmogenic nuclide measurements based on ^{10}Be-^{26}Al isotopes, however, measured an exposure age of 0.27 Myr for granite bedrock near the crater rim.

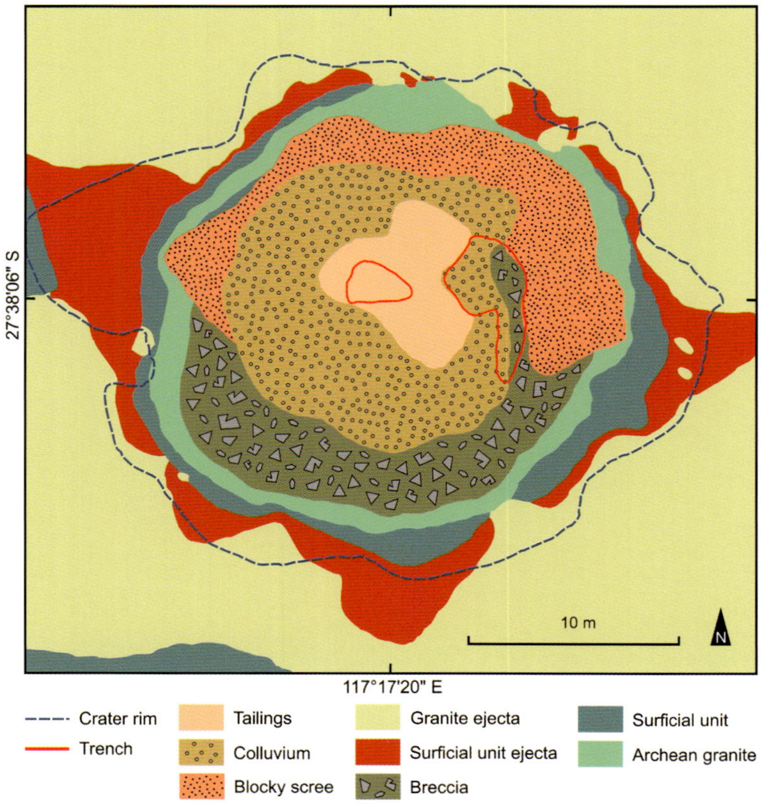

Schematic geological map of the Dalgaranga crater (modified after Shoemaker et al., 2005).

Abundant weathered, microscopic meteorite particles were found in the soil around the crater. Much of the meteoritic material is corroded due to terrestrial weathering. The impactor was a stony-iron meteorite, a so-called "mesosiderite".

Dalgaranga was the first impact crater recognized in Australia. It holds the record of being the smallest single crater on this continent. Dalgaranga was first discovered by aboriginal people in 1921 and two years later meteorite fragments of the projectile were collected inside and outside of the crater. A first report in the scientific literature did, however, not appear until 1938. Twenty years later systematic collections of meteoritic material from the crater were carried out and yielded more than 200 additional specimens. Harvey Nininger collected more than 9 kg of mesosiderite meteorites from shallow trenches in the crater. Later, Eugene and Carolyn Shoemaker mapped the crater in detail.

Location/Access

Dalgaranga is situated in the outback of Western Australia, some 50 km west of the Great Northern Highway. Unsealed mining roads branching off from Mt. Magnet lead close to Dalgaranga.

Selected References

Bevan, A. (2012a). Western Australian Impact Crater Field Excursion, 20–29 August 2012. 75th Annual Meeting of the Meteoritical Society Cairns, Queensland, Australia, 55 pp.

Hamacher, D. W., and O'Neill, C. (2013). The discovery and history of the Dalgaranga meteorite crater, Western Australia. Australian Journal of Earth Sciences, Vol. 60, pp. 637–646.

Nininger, H. H., and Huss, G. I. (1960). The unique meteorite crater at Dalgaranga, Western Australia. Mineralogical Magazine, Vol. 32, pp. 619–639.

Shoemaker, E. M., Macdonald, F. A., and Shoemaker, C. S. (2005). Geology of five small Australian impact craters. Australian Journal of Earth Sciences, Vol. 52, pp. 529–544.

Dalgaranga
Western Australia

latitude = 27°38'06" S / longitude = 117°17'20" E / diameter = 24 m / age = 0.27 Myr

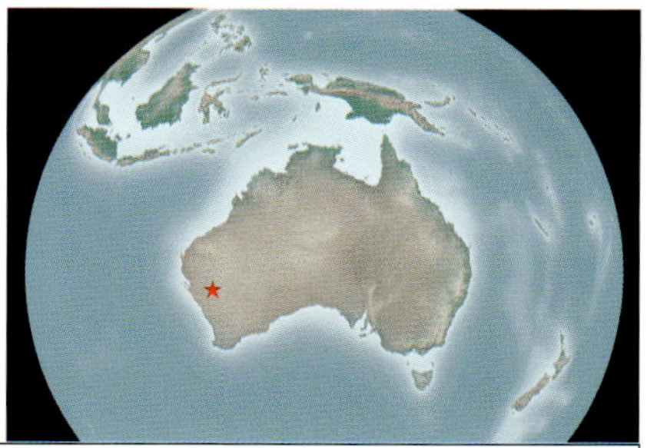

Foelsche
Northern Territory

Thomas Kenkmann

The circular Foelsche structure is easily recognizable in aerial photography by changes in vegetation and topography, and a concentric drainage system. Morphologically, Foelsche shows an annular moat with concentric drainage that isolates a central up-warp. Outside of the circular valley occurs hilly, table-mountain dominated land. The structure is interpreted as a partly buried, complex impact crater of about 6 km diameter. The central hill is formed by flat-lying Neoproterozoic Bukalara Sandstone, a deposit that filled the crater after its formation.

The target comprised flat-lying un-metamorphosed Late Paleoproterozoic to Early Mesoproterozoic sedimentary and igneous rocks of the McArthur Basin. Outside Foelsche, structural disturbance is generally restricted to some faulting. Impact lithologies crop out beneath the post-impact deposits along the annular moat. These impact breccias were mapped as conglomeratic sandstones but seem to represent redeposited fallback breccia, or material eroded from the central uplift. This unit contains abundant shocked clasts with planar deformation features (PDFs), planar fractures (PFs), and feather features (FFs) in quartz.

The presence of a central uplift of about 2 km diameter was inferred from geophysical data alone, as it appears to be completely buried by the post-impact Bukalara Sandstone. The down-faulted crater rim is partly exposed. The structure has a strong concentric aeromagnetic signature. The anomaly has a sharp outer limit at approximately 5.5 km diameter, inside of which lies a distinct annular magnetic low of 4 km width. Closer to the center there is a somewhat discontinuous annular magnetic high of roughly 2.5 km width that surrounds a central magnetic low. The aeromagnetic anomaly related to the breccia infill and central uplift is slightly smaller than the diameter inferred from the concentric rim fault.

The Foelsche impact structure was formed prior or simultaneous to the deposition of the Neoproterozoic infill of Bukalara Sandstone and must therefore be of Neoproterozoic age. The good preservation of the structure beneath the post-impact infill suggests that it was likely buried soon after impact and then only recently exhumed.

The crater was discovered by Peter Haines in 2002. The circular aeromagnetic anomaly helped to identify the structure.

Megabreccia at the base of the crater-fill deposits of the Foelsche impact structure (image courtesy: Peter Haines, Geological Survey of Western Australia, Perth).

Tilted and fractured sandstone along the northern rim of the Foelsche impact crater (image courtesy: Peter Haines, Geological Survey of Western Australia, Perth).

Bukalara Sandstone of the Foelsche impact crater. These deposits filled the crater depression after impact. The sandstone contains clasts with shocked quartz grains. Hammer for scale ca. 35 cm (image courtesy: Peter Haines, Geological Survey of Western Australia, Perth).

Shocked quartz clast with PDFs of the Foelsche crater fill sediments. Note that the quartz overgrowth is entirely unshocked indicating post-impact cementation of the rock (image courtesy: Peter Haines, Geological Survey of Western Australia, Perth).

Foelsche

Northern Territory

latitude = 16°40'15" S / longitude = 136°47'02" E / diameter = 6 km / age = 541 - 981 Myr

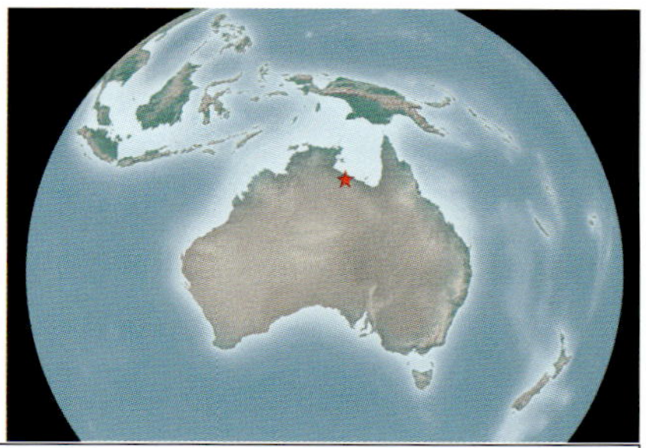

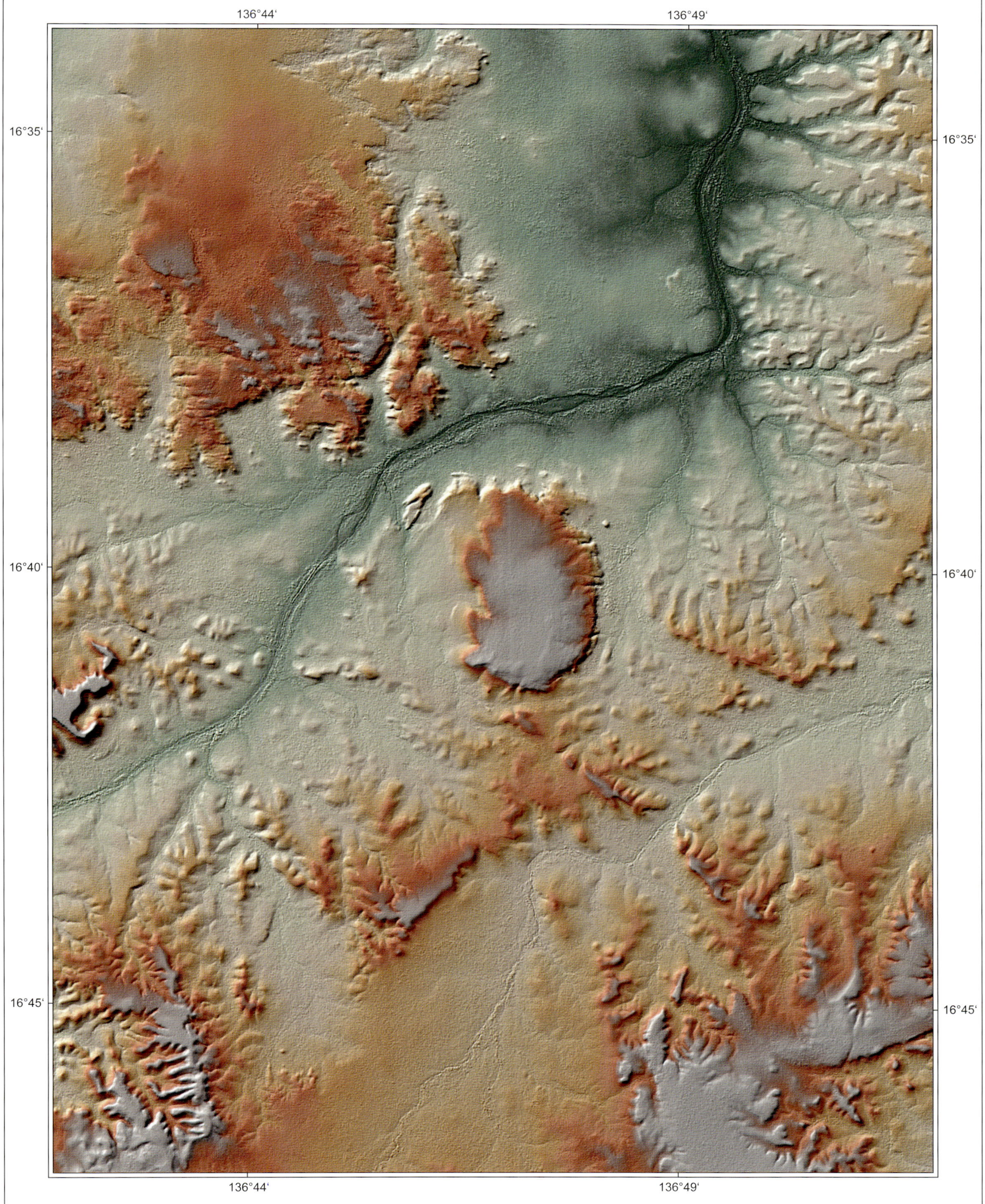

scale = 1 : 101 000 / scene = 18.2 × 24.2 km^2

0 1.01 5.05 km

N

45 meter 185

The Foelsche impact structure on October 8, 2018 as a composite image based on TanDEM-X DEM and Copernicus Sentinel-2 (channel 2/3/4) data.

Location/Access

Foelsche is situated near the Gulf of Carpentaria in northern Australia, some 20 km from Savannah Way. Large parts of the Northern Territory are Aboriginal homeland or National Park land. Thus, access to Foelsche requires special permits.

Selected References

Haines, P. W. (2005). Impact cratering and distal ejecta: the Australian record. Australian Journal of Earth Sciences, Vol. 52, pp. 481–507.
Haines, P. W., and Rawlings, D. J. (2002). The Foelsche structure, Northern Territory, Australia: an impact crater of probable Neoproterozoic age. Meteoritics and Planetary Science, Vol. 37, pp. 269–280.

Glikson

Western Australia

Thomas Kenkmann

The Glikson structure in the Little Sandy Desert of central Western Australia has a diameter of roughly 19 km, which is defined by subdued structural characteristics. The structure is morphologically not apparent. The area is flat, devoid of vegetation, and is crossed by east-west trending longitudinal dunes. In terms of geology, the Glikson structure lies in flat to gently dipping, 800 Myr old sedimentary rocks of the western Officer Basin. These deposits are mainly Neoproterozoic sandstones, siltstones and conglomerates, which are intruded by dolerite sills. Rare exposures belong to a circular central uplift, the deformation of which is consistent with an impact origin.

Shatter cones, planar fractures (PFs), and feather features (FFs) in quartz grains are present in the rocks of the highly deformed central

Glikson
Western Australia

latitude = 23°58'40" S / longitude = 121°33'22" E / diameter = 19 km / age = < 508 ± 5 Myr

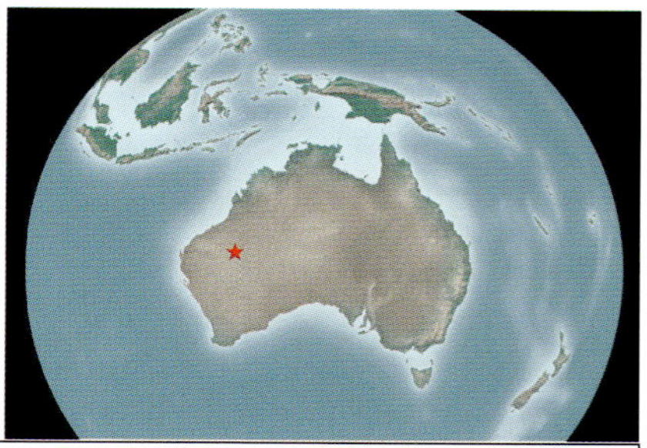

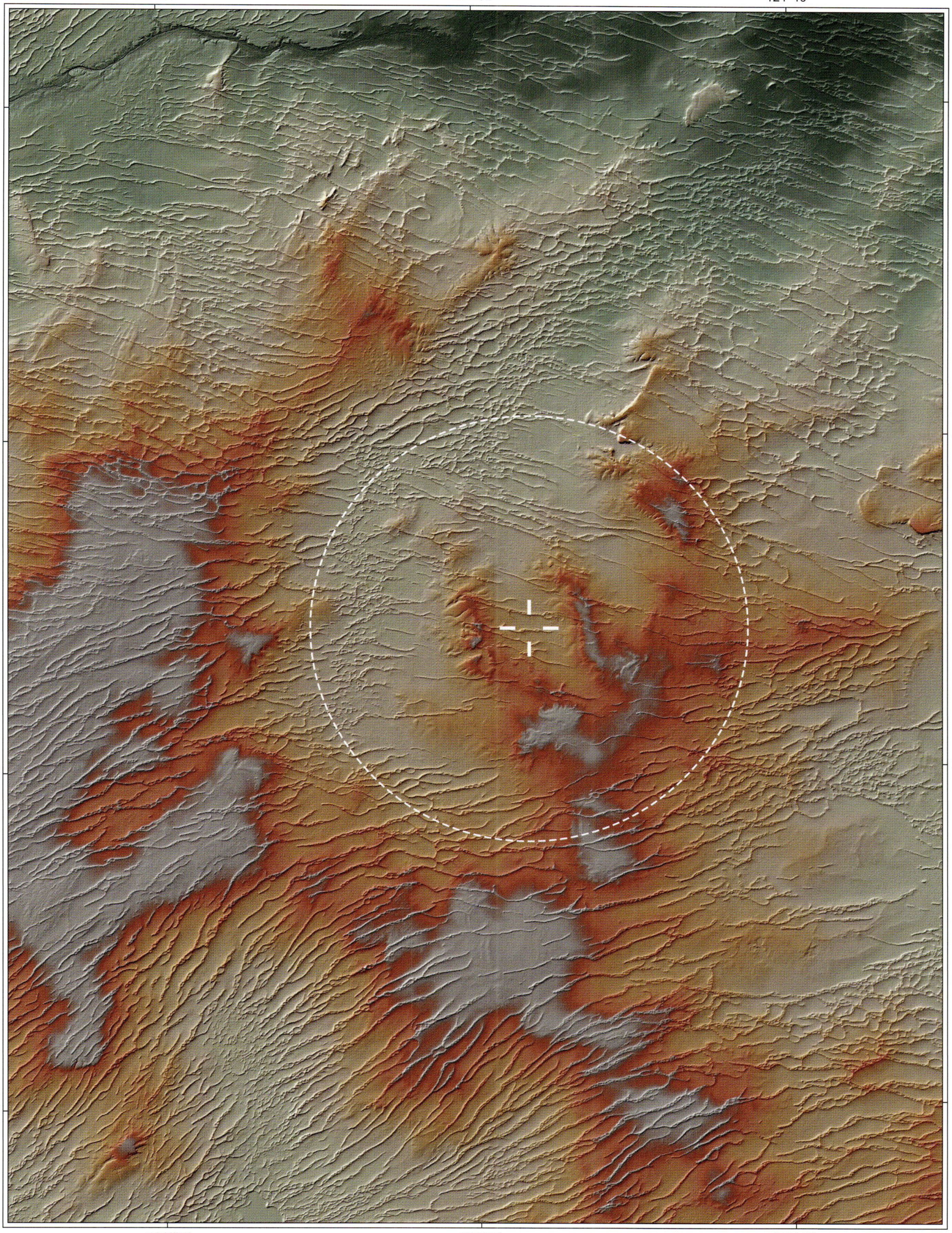

scale = 1:224 000 / scene = 40.3 × 53.7 km²
0 2.24 11.20 km
N
450 meter 605

The Glikson impact structure on June 30, 2018 as a composite image based on TanDEM-X DEM and Copernicus Sentinel-2 (channel 2/3/4) data. Cross hairs and a dashed circle indicate position and extent of the morphologically invisible crater structure.

region and indicate a low shock metamorphic overprint. Likewise, the macro-deformation inventory of the central rise is characteristic for impact craters. It includes chaotic strike of strata in the center and tight radial folding with steeply plunging fold axes in the periphery of the central uplift and further outside.

Glikson has a pronounced concentric aeromagnetic signature with a 10 km inner diameter and a 14 km outer diameter. The circular magnetic signature is most likely the result of truncation and folding of mafic sills in the ring syncline around the central peak. The sills are thought to correlate with dolerites that occur about 25 km to the north of Glikson, and for which an age of 508 ± 5 Myr was determined by U-Pb dating of zircon with the sensitive high-resolution ion microprobe (SHRIMP). This age would provide an upper limit for the impact event, if the sills within the structure are indeed a correlative of the dolerites outside the impact structure.

Detection of the Glikson impact structure was possible due to its strong ring-shaped aeromagnetic anomaly. Eugene and Carolyn Shoemaker named the structure in 1997 – in the absence of any geographic landmarks – in the honor of Andrew Glikson, an Australian geologist who investigated several impact structures.

Location/Access

Glikson is situated in the Little Sandy Desert of central Western Australia. The Canning Stock Route, some 60 km southeast of the structure, is the nearest infrastructure. The Canning Stock Route is an outback track for four-wheel drive vehicles that runs from Halls Creek in the Kimberley region of Western Australia to Wiluna in the mid-west region.

Selected References

Haines, P. W. (2005). Impact cratering and distal ejecta: the Australian record. Australian Journal of Earth Sciences, Vol. 52, pp. 481–507.

Macdonald, F. A., Wingate, M. T. D., and Mitchell, K. (2005). Geology and age of the Glikson impact structure, Western Australia. Australian Journal of Earth Sciences, Vol. 52, pp. 641–651.

Shoemaker, E. M., and Shoemaker, C. S. (1997a). Glikson, a probable impact structure, Western Australia. 28th Lunar and Planetary Science Conference, Lunar and Planetary Institute, Houston, Texas, Abstract #1669.

Goat Paddock

Western Australia

Thomas Kenkmann

Goat Paddock forms a wide, slightly oval, plain of little more than 5 km width. The structure is situated in the King Leopold Ranges of the southern Kimberley Basin. This mountainous region is in stark morphological contrast to the smooth plain of the structure. About 100–150 m high cliffs surround this plain and form the remnants of the crater rim.

Target rocks are moderately tilted strata of the Proterozoic Pentecost Sandstone. At the crater rim, however, strata have been upturned and overturned, intensely fractured, and injected by clastic dikes. Crater wall breccia occurs close to the overturned rim and along the bowl-shaped crater floor. Shock features are abundant in the impact breccia that fills the central part of the depression. This includes melt-bearing breccias, silica glass, shatter cones, and shock metamorphosed quartz grains with planar deformation features (PDFs) and planar fractures (PFs).

A mining company drilled two boreholes into the crater depression and revealed that the structure was infilled with about 200 m of lacustrine sediments. These post-impact deposits have been dated by biostratigraphy to be of Early Eocene age. They constrain the minimum age of the impact. Taking the preservation state and the moderate erosion that could be inferred from the lack of any proximal ejecta into account, the maximum age cannot be considerably older. Boreholes located in the center and midway out to the crater wall, respectively, intersected basement of shocked sandstone at essentially identical depths, which implies a flat crater floor instead of the presence of a central uplift. The absence of a central uplift in a 5 km diameter crater either suggests an unusually competent target or a very oblique impact. The slight elongation of the crater in a north-south direction supports the hypothesis of an oblique impact from either the north or the south. Owing to the flat-floored morphometry that is typical for complex craters and the lack of any central uplift, Goat Paddock appears to represent a transitional type between simple and complex craters. Recent studies of Lunar craters show that this category is more frequent than previously thought. For the Earth about 3 percent of the currently known impact structures can be regarded as transitional types.

The structure was first noticed during a reconnaissance survey of the Geological Survey of Western Australia in 1965. Exploration drilling in 1972 failed to locate an anticipated kimberlite pipe. J. Harms and co-workers identified major structural features of the impact crater, and identified shatter cones and shocked quartz grains, and proposed the formation of the structure by impact. They can therefore be regarded as the discoverers of the impact structure. Still they preferred the term "cryptoexplosion crater" instead of "impact crater", as they did not find meteoritical material.

Location/Access

The Great Northern Highway encircles the mountainous Kimberley region to the south. The distance from the closest point along this highway to Goat Paddock is about 50 km. Large parts of the Kimberley region are Aboriginal Homeland or Pastoral Lease, and access to the structure requires special permits. The best exposures of impactites are along steep-sided canyons crossing the crater rim.

Selected References

Haines, P. W. (2005). Impact cratering and distal ejecta: the Australian record. Australian Journal of Earth Sciences, Vol. 52, pp. 481–507.

Harms, J. E., Milton, D. J., Ferguson, J., Gilbert, D. J., Harris, W. K., and Goleby, B. (1980). Goat Paddock cryptoexplosion crater, Western Australia. Nature, Vol. 286, pp. 704–706.

Milton, D. J., and Macdonald, F. A. (2005). Goat Paddock, Western Australia: an impact crater near the simple-complex transition. Australian Journal of Earth Sciences, Vol. 52, pp. 691–698.

The Goat Paddock impact crater on September 30, 2018 as a composite image based on TanDEM-X DEM and Copernicus Sentinel-2 (channel 2/3/4) data.

Goat Paddock

Western Australia

latitude = 18°20'02" S / longitude = 126°40'36" E / diameter = 5 km / age = 56 - 64 Myr

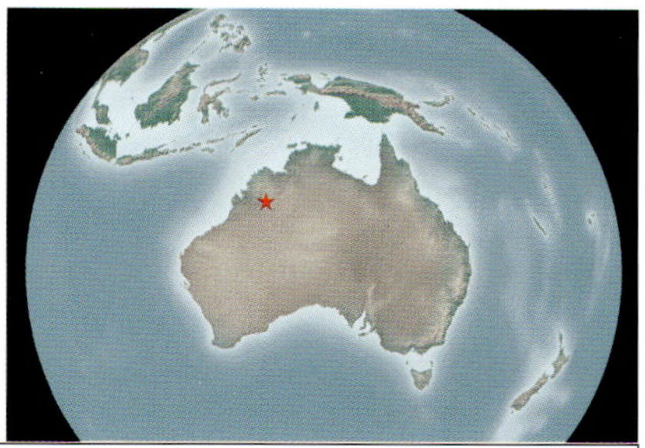

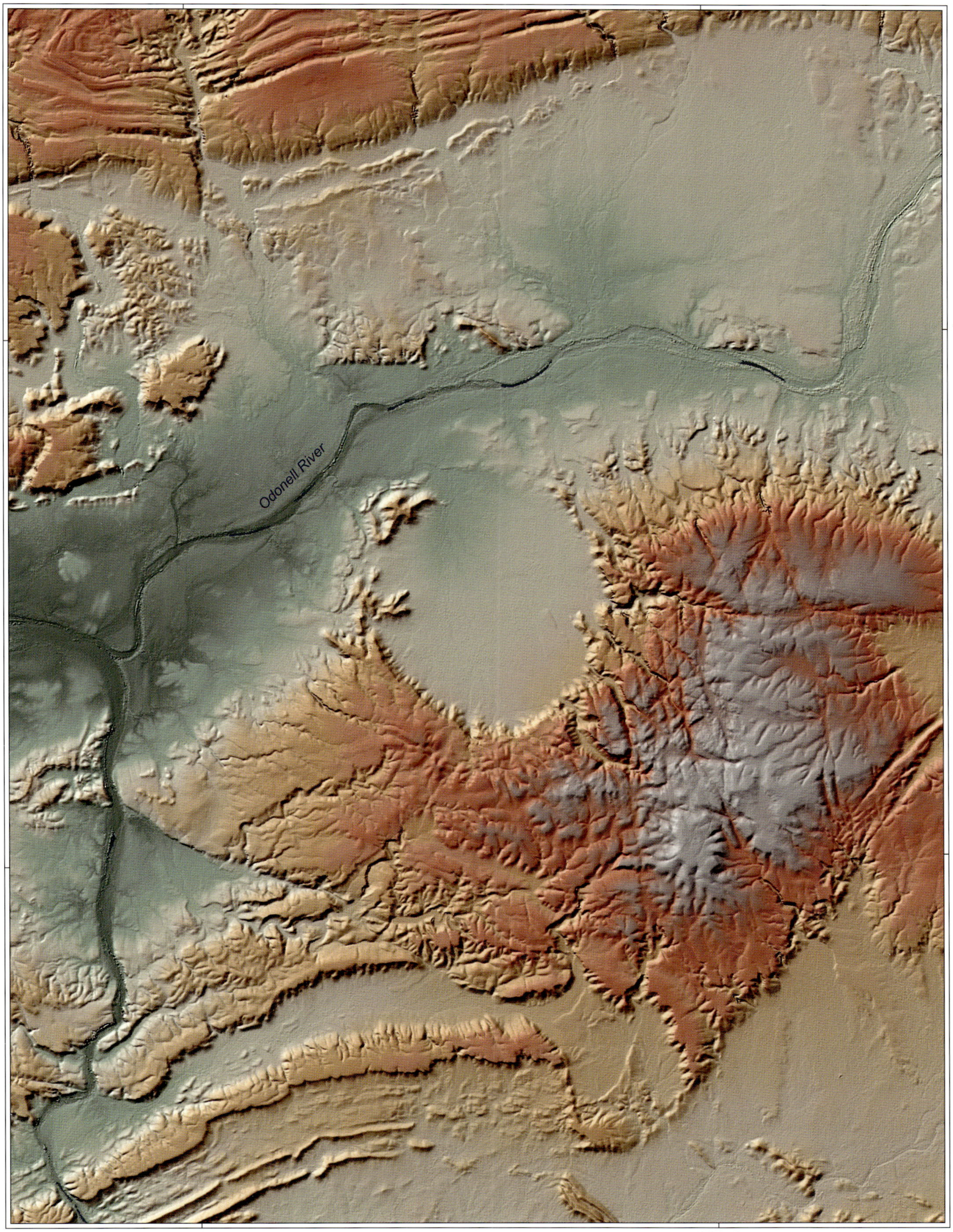

scale = 1 : 106 000 / scene = 19.1 × 25.4 km²

0 1.06 5.30 km

N

145 meter 500

Gosses Bluff
Northern Territory

Thomas Kenkmann

Gosses Bluff is located some 150 km west of Alice Springs and 50 km west of the Hermannsburg Aboriginal settlement. The structure is situated on the Missionary plain, a very broad, flat-floored, east-west trending syncline south of the West MacDonnell Ranges, in the Amadeus Basin.

The impact occurred in mainly flat-lying sedimentary strata dominated by sandstones of Ordovician to Devonian age. The crater rim has no morphological expression and was completely removed by erosion, similar to what has recently been concluded for the **Ramgarh** structure. The full extent of the impact structure with a diameter of 22 km is visible only on aerial photography or space-borne imagery. The outer limit of the structure is also marked by a circular negative gravity anomaly,

The impressive landmark of Gosses Bluff is a collar of prominent mountains, about 4.5 km wide, that is elevated above the environs by some 180 m. This collar and its inner morphological depression form the central uplift of the crater structure.

The morphological collar of Gosses Bluff, here seen from the south, represents the outer part of the central uplift (photo: T. Kenkmann).

Gosses Bluff has probably the best exposed central uplift of all impact structures on Earth. Shocked rocks can be found everywhere. Shatter cones, some of them larger than one meter, are present all over the central depression and the collar. The most impressive ones with preserved apices occur in the northern part of the collar, the up-range sector of the structure. Other shock features include planar fractures (PFs) and planar deformation features (PDFs) in quartz of sandstone and in quartz clasts of pseudotachylitic melt breccia veins. Higher shock levels are recorded south of the collar in impact melt rock and suevitic breccia, at Mount Pyroclast – a group of hills that are some 40–50 m high and are located about 4 km south of the central uplift. Diaplectic glasses as well as lechatelierite are abundant in these breccias. Mount Pyroclast is the only exposed part of the annular breccia trough.

Gosses Bluff: Panoramic view over the central depression and the crater center that is surrounded by the heavily faulted and shock-deformed collar formed by the Mereenie Sandstone Formation (photo: T. Kenkmann).

Most strata in the central uplift show very steep dips and are strongly deformed. The stratigraphic sequence is roughly intact, insofar as the oldest strata occur in the center and are surrounded by successively younger units. The oldest exposed lithologies are of Ordovician age, belong to the Upper Larapinta Group, and form the central depression. These lithologies are the so-called "Stairway Sandstone" and "Stokes Siltstone" that contain interlayered limestone. Some oolitic limestones occur. The next younger unit of Late Ordovician age is the purple-red Carmichael Sandstone that forms the lower slopes of the inner collar. The Lower and Upper Mereenie Sandstones of Late Ordovician to Early Devonian age both form the prominent collar. The Lower Mereenie Sandstone is of eolian origin and forms thick quartzitic layers that build most of the steep

Gosses Bluff

Northern Territory

latitude = 23°49'10" S / longitude = 132°18'27" E / diameter = 22 km / age = 142.5 ± 0.8 Myr

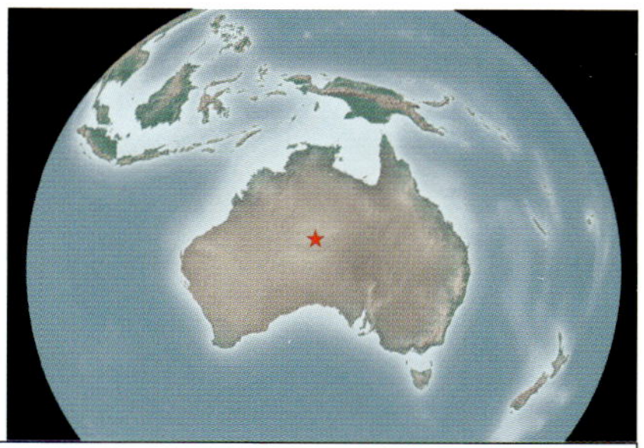

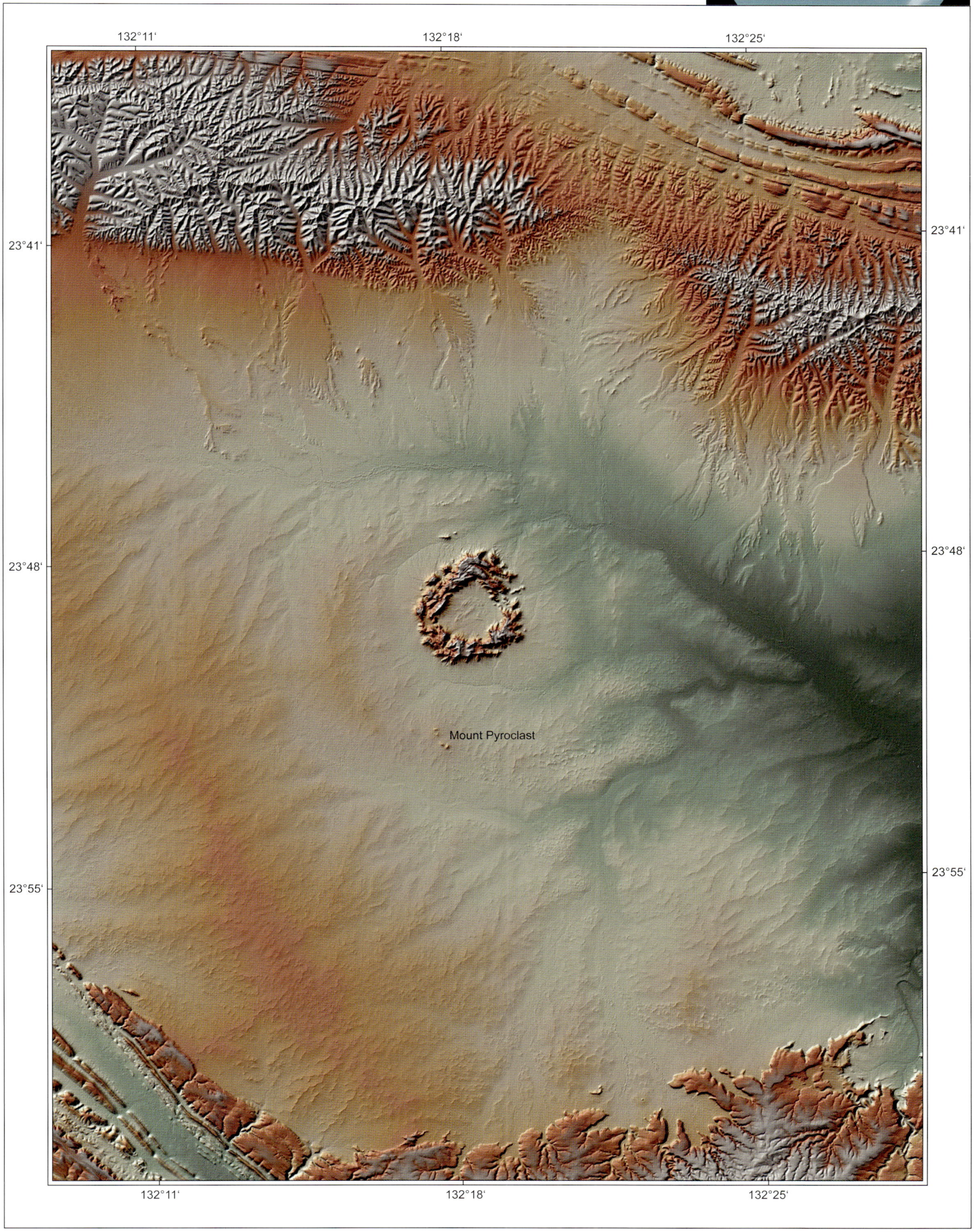

scale = 1 : 189 000 / scene = 34.0 × 45.3 km²

0 1.89 9.45 km

N

635 meter 1045

Shatter cones with mostly similar apex orientation in Lower Mereenie Sandstone. This Late Ordovician formation builds up the inner part of the collar of the central uplift of the Gosses Bluff impact structure (photo: T. Kenkmann).

cliffs of the inner collar. The Devonian Parke Siltstone forms the outer slopes of the collar.

The collar has a somewhat triangular shape with the western half resembling the eastern half. The symmetry axis trends north-northeast to south-southwest and marks the trajectory of the oblique impact. In the northern up-range sector, uptilted blocks form a stack of east-northeast to west-southwest trending slices that are thrust onto each other. They are separated by faults that mostly correspond to local valley incisions. Faults are marked by massive lithic fault breccia occurrences. In contrast, in the down-range direction the slices of uplifted Mereenie Sandstone are radially oriented to the center of the structure. In this area exist completely allochthonous patches and recumbent nappe sheets of Lower Mereenie Sandstone and megabreccia that were ejected from the uptilted slices of the down-range sector. They cover the steeply dipping to overturned beds of Upper Mereenie Sandstone and Parke Siltstone. The occurrence of these rocks suggests that the summit area of the collar is close to the ancient crater floor and that the amount of erosion of this part of the structure is small. In contrast, outside the central uplift the level of erosion is most likely more than 200 m. A moderately oblique impact from the north-northeast to the south-southwest is supported by the bilateral symmetry of the structure of the central uplift, the preferential overturning of the collar in the down-range sector, the orientation of shatter cone axes, and the preferred occurrence of melt rock at Mount Pyroclast and melt veins in the down-range sector of the crater.

The discovery of Gosses Bluff by Ernest Giles dates back to 1872. Recognized as a geological anomaly, the structure was interpreted in the 1950s as a salt diapir or as an igneous intrusion. Later, K. Crook and P. Cook assumed a "cryptoexplosion" origin, triggered either by an extraterrestrial impact or a violent endogenic event. However, in 1967 Robert (Bob) Dietz, working on shatter cones, classified Gosses Bluff as an impact crater. Considerable geological analysis was carried out subsequently by Daniel Milton and co-workers. ^{40}Ar-^{39}Ar dating of impact melt rock by Milton and his team yielded an age of 142.5 ± 0.8 Myr. The crater has also been investigated at depth by means of extensive shallow drillings and two 1383 m and 2652 m deep boreholes. They were drilled into the center of the crater structure in search for oil. Geophysical analysis included seismic, magnetic and gravity surveying. Reflection seismics indicated that the structural disturbance reaches down to 3.5 km depth. Recent structural mapping of the central uplift, under the auspices of Curtin University, Perth, was carried out by Thomas Kenkmann and collaboration partners from Curtin and Imperial College.

Location/Access

Gosses Bluff is easily accessible from the 150 km distant city of Alice Springs via Larapinta Drive and Namatjira Drive. Both roads are tarred and in good condition. A 6 km long access road into the center of the structure is unsealed and requires four-wheel drive. The crater structure is part of the Tnorala Conservation Reserve. Walking is permitted only on designated tracks. Overnight stay is forbidden. The nearest campgrounds are found at Red Bank Gorge (50 km distance) or Glen Helen (70 km) at Namatjira Drive in the West MacDonald National Park.

Selected References

Cook, P. J. (1968). The Gosses Bluff crypto-explosion structure. Journal of Geology, Vol. 76, pp. 123–139.

Crook, K. A. W., and Cook, P. J. (1966). Gosses Bluff – diapir, crypto-volcanic structure or astrobleme? Geological Society of Australia Journal, Vol. 13, pp. 495–516.

Dietz, R. S. (1967). Shatter cone orientation at Gosses Bluff astrobleme. Nature, Vol. 216, pp. 1082–1084.

Kenkmann, T., Rae, A. S. P., Cavosie, A. J., Cox, M. A., Timms, N., and Miljkovic, K. (2018). The central uplift of Gosses Bluff, Northern Territory, Australia. 81st Annual Meeting of the Meteoritical Society, Moscow, Russia, Abstract #6077.

Milton, D. J., and Sutter, J. F. (1987). Revised age for the Gosses Bluff impact structure, Northern Territory, Australia, based on ^{40}Ar-^{39}Ar dating. Meteoritics, Vol. 22, pp. 281–289.

Milton, D. J., Glikson, A. Y., and Brett, R. (1996). Gosses Bluff – a latest Jurassic impact structure, central Australia. Part 1: Geological structure, stratigraphy, and origin. AGSO Journal of Australian Geology and Geophysics, Vol. 16, pp. 453–486.

Milton, D. J., Barlow, B. C., Brown, A. R., Moss, F. J., Manwaring, E. A., Sedmik, E. C. E., Young, G. A., and Van Son, J. (1996). Gosses Bluff – a latest Jurassic impact structure, central Australia. Part 2: Seismic, magnetic, and gravity studies. AGSO Journal of Australian Geology and Geophysics, Vol. 16, pp. 487–527.

The Gosses Bluff impact structure on January 24, 2018 as a composite image based on TanDEM-X DEM and Copernicus Sentinel-2 (channel 2/3/4) data.

Goyder
Northern Territory

Thomas Kenkmann

The Goyder structure, named after the nearby Goyder River, can be identified as a 3 km wide circular feature with a central hilly region that displays some arcuate and bent, southwest-northeast trending ridges. This terrain is surrounded by a vegetated area, and an outer annulus that, in part, forms a low ridge. Outcrop in the structure is restricted to the central region and the outer annulus. These divisions relate to lithological variation and the differential resistance of the lithologies to weathering. The present morphology of the structure is not a primary feature. The ring structure and central hilly zone are interpreted to be parts of the central uplift of a deeply eroded, complex impact structure. The rest of the structure is not clearly defined.

The Goyder impact structure is situated in Mesoproterozoic sedimentary rocks of the northwestern McArthur Basin in the Northern Territory. Rocks are mainly sandstones with intercalated mudstone-siltstone layers. Outside the structure, the strata are flat-lying. Within the central uplift, beds are heavily fractured or brecciated, tend to dip at moderate angles, and are dissected into numerous blocks bounded by faults. The fault pattern resembles that of the **Spider** impact structure, but is less pronounced. The structure displays a weak bilateral symmetry about a north-northwest to south-southeast axis that might be indicative for an oblique impact from north-northwest. This is most clearly seen in the central hilly region of the structure. Shatter cones were found close to the center of the structure, a little offset from the structure's symmetry axis towards the north. Petrographic evidence of shock metamorphism includes abundant planar fractures (PFs), feather features (FFs) and planar deformation features (PDFs) in quartz grains of sandstone of the impact structure.

Pervasive shatter coning in quartzitic sandstone of the central uplift of the Goyder impact structure (image courtesy: Peter Haines, Geological Survey of Western Australia, Perth).

Intensely fractured Mesoproterozoic sandstone of the central uplift of the Goyder impact structure. Note that the fractures were later sealed with quartz (image courtesy: Peter Haines, Geological Survey of Western Australia, Perth).

The original diameter is uncertain but has been estimated at 9–12 km, based on scaling relationships for impact craters. Likewise, the age of Goyder is poorly constrained, due to the deep level of erosion and as there are no allochthonous breccia infill or impact melt rocks preserved that might be datable. The maximum possible age is given by the 1400 Myr age of the target rocks, whereas a minimum age is defined by flat-lying Cretaceous rocks exposed near Goyder that overlie the Precambrian strata and are not affected by impact deformation.

The circular structure was identified on aerial photographs by Peter Haines in the course of a geological mapping campaign in Arnhem Land, northern Australia, by the Geological Survey of the Northern Territory. Later, in 1991, he visited and sampled the structure and obtained bona fide impact evidence.

Location/Access

Goyder is located about 40 km east of the Central Arnhem Road that branches off from Stuart Highway. Special permits from the Aboriginal owners of Arnhem Land are required to visit the site.

Selected References

Haines, P. W. (1996). Goyder impact structure, Arnhem Land, Northern Territory. AGSO Journal of Australian Geology and Geophysics, Vol. 16, pp. 561–566.

Haines, P. W. (2005). Impact cratering and distal ejecta: the Australian record. Australian Journal of Earth Sciences, Vol. 52, pp. 481–507.

Goyder

Northern Territory

latitude = 13°28'31" S / longitude = 135°02'24" E / diameter = 3 km / age = 150 - 1400 Myr

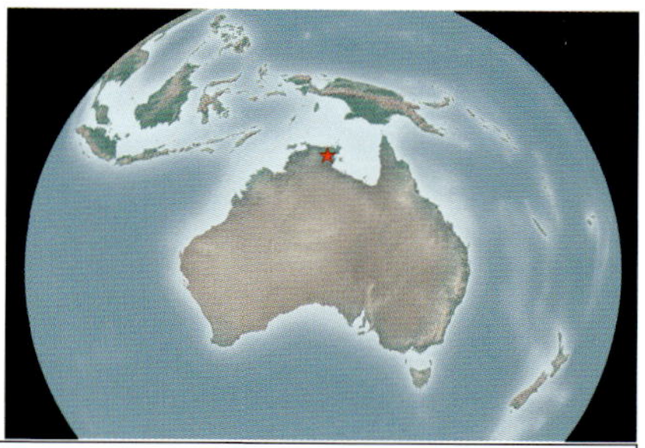

scale = 1:90 000 / scene = 16.1 × 21.5 km²

0 0.90 4.50 km

N

50 meter 145

The Goyder impact structure on September 16, 2018 as a composite image based on TanDEM-X DEM and Copernicus Sentinel-2 (channel 2/3/4) data.

Henbury Crater Field
Northern Territory

Thomas Kenkmann

The Henbury site exhibits 12 to 15 crater pits ranging from 6 to 180 m in width and up to 15 m depth. The larger craters are clustered closely together. The very small pits are hardly visible today. Since their formation about 4000–5000 years ago, the original sharp outlines of the craters have been softened by wind and rain into broad undulating mounds.

Bedrock at Henbury is composed of beds of sandstones and intercalated shales of late Precambrian age overlain by pediment gravel. The strata belong to the Pertataka Formation that strikes east-west and dips with about 35° to the south. A few craters show relics of ejecta rays; each of them is made up by fragments from a single type of bedrock. The bolide, pieces of which struck the Earth at Henbury, has been classified as a platinum group element-rich type IIIAB iron meteorite. Fragments of the meteorite consisting of nickel-iron alloys were scattered over a wide area. Several hundred kilograms of these fragments have been recovered. Magnetic traverses across 12 craters revealed no larger magnetic anomalies. Thus, no larger intact meteorite pieces are expected to be buried within the craters. The impact event was dated to have occurred 4200 ± 1900 years ago based on the cosmogenic ^{14}C terrestrial age of the meteorite.

The first scientific investigations of the site date back to 1931 and 1932, when Arthur Alderman carried out a cursory geological study. This was followed by an expedition by Robert Bedford that collected a large amount of meteorite fragments and discovered a meteorite strewn field stretching to the northeast at a distance of 1–5 km from the crater field. This was sufficient to declare the largest crater pits a Natural Reserve. In 1963 Truman Kohman and Parmatma Goel dated the impact event to prehistoric times. The Henbury Meteorite Conservation Reserve has existed since 1983. It not only includes the large pits but also the smaller craters and the meteorite strewn field to the northeast.

Location/Access

Henbury is situated some 150 km south of Alice Springs and is easily accessible, even with a standard two-wheel-drive vehicle. The turn-off from Stuart Highway, at 135 km from Alice Springs, onto the unpaved Ernest Giles Road is signposted. After 10 km on this gravel road, a 5 km long access road to the Henbury Meteorite Conservation Reserve's entrance follows. This reserve is open all year round, but roads may be closed after heavy rainfall.

Selected References

Alderman, A. R. (1931). The meteorite craters at Henbury, Central Australia, with an addendum by L. J. Spencer. Mineralogical Magazine, Vol. 23, pp. 19–32.

Hodge, P. W. (1965). The Henbury meteorite craters. Smithsonian Contributions to Astrophysics, Vol. 8, pp. 199–201.

Kohman, T. P., and Goel, P. S. (1963). Terrestrial ages of meteorites from cosmogenic ^{14}C. In: Radioactive Dating. International Atomic Energy Agency, Vienna, pp. 395–411.

Milton, D. J. (1968). Structural Geology of the Henbury Meteorite Craters, Northern Territory, Australia. Contributions to Astrogeology, Geological Survey Professional Paper 599-C, United States Government Printing Office, Washington, 17 pp.

Milton, D. J., and Michel, F. C. (1965). Structure of a ray crater at Henbury, Northern Territory, Australia. U.S. Geological Survey Professional Paper 525-C, C5–C11.

Henbury Crater Field

Northern Territory

latitude = 24°34'18" S / longitude = 133°08'52" E / diameter = max. 180 m / age = 4200 ± 1900 yr

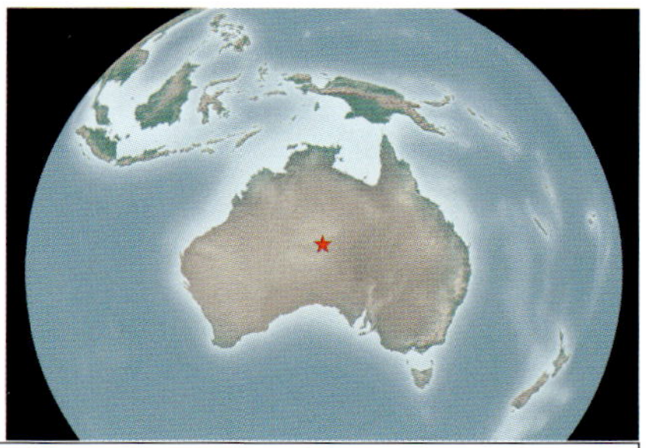

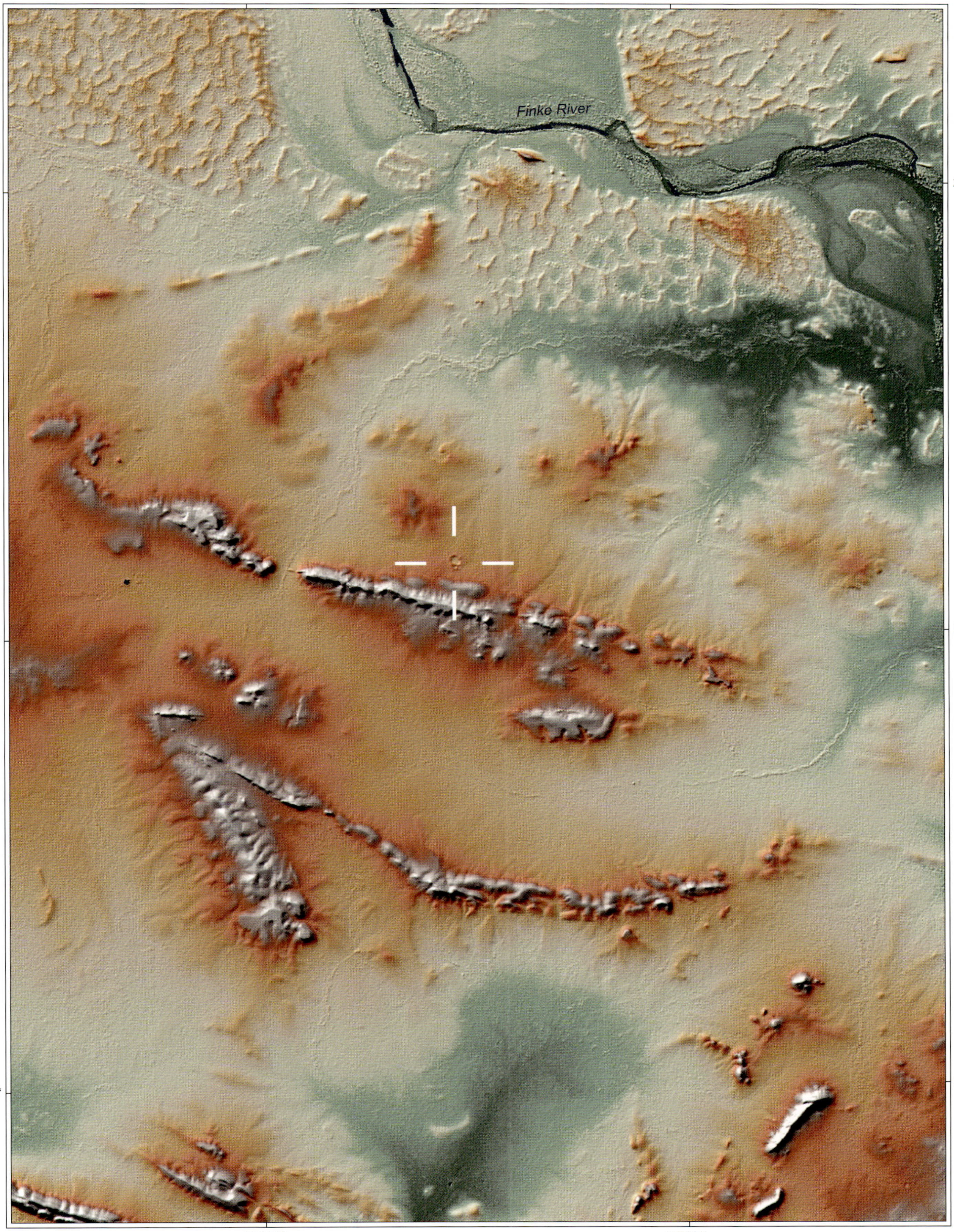

scale = 1:83 000 / scene = 14.9 × 19.9 km²

0 0.83 4.15 km

N

420 meter 530

Panoramic view of the largest of the Henbury craters. This crater is actually formed by two overlapping craters with a width, shown here, of about 150 m (photo: T. Kenkmann).

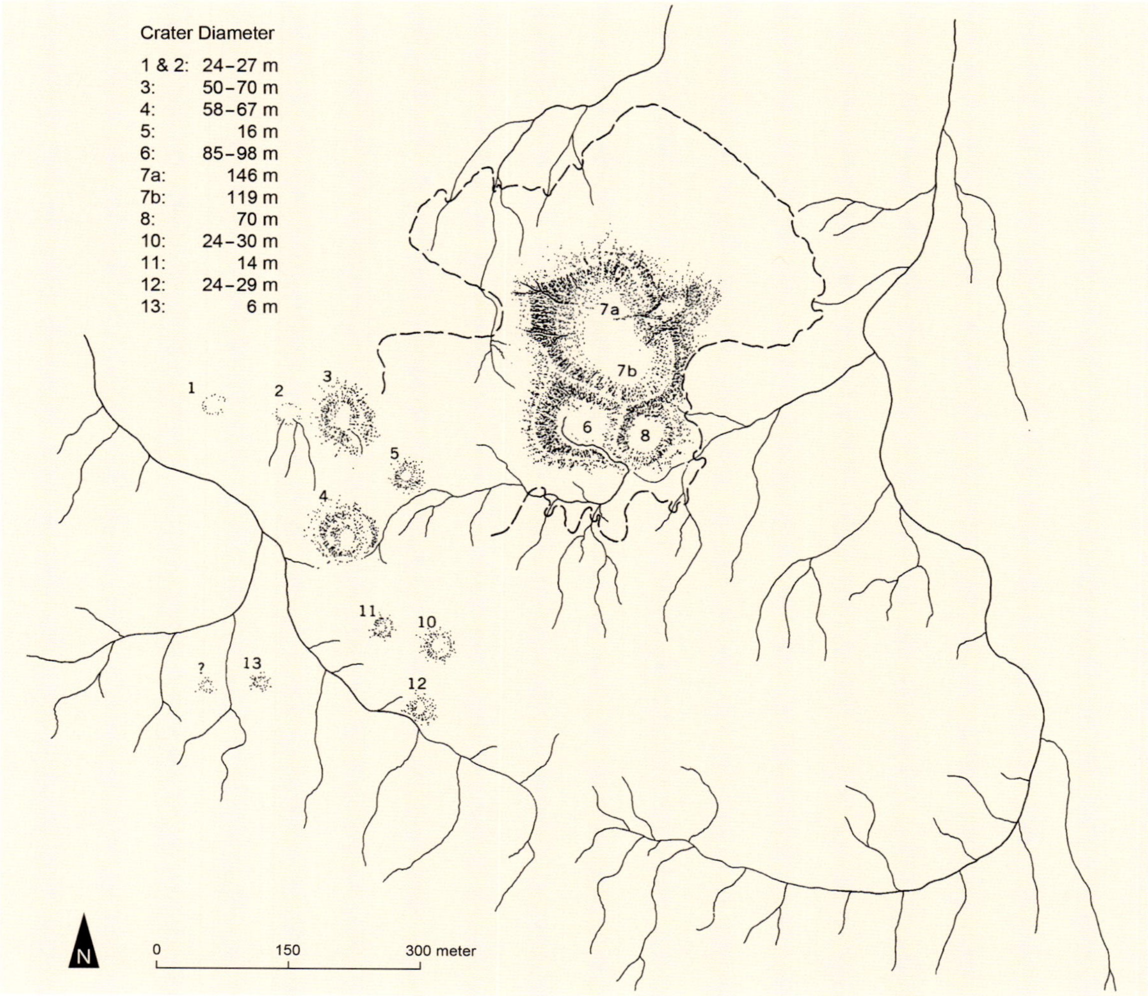

Schematic map of the outlines of the craters in the Henbury crater field. The dashed line marks the approximate outer limit of ejecta occurrence around the larger craters (modified after Milton, 1968).

The Henbury meteorite crater strewn field observed from the south (photo: T. Kenkmann).

Hickman
Western Australia

Thomas Kenkmann

The small, 260 m wide and 30 m deep Hickman crater is located on the Pilbara craton of Western Australia. The small crater is a striking, circular morphological feature. It has a flat floor that is often covered by water after heavy rainfall, and a raised rim. The southwestern rim sector has been removed by erosion. The remaining rim parts rise 15–20 m above the surrounding plateau and 20–30 m above the flat crater floor. The inner walls form steep, unstable slopes – with numerous slump blocks on them, and in the talus around the base. Target rocks at Hickman were predominantly rhyolite and banded iron formation of the early Paleoproterozoic Hamersley Basin. In the region of the crater, the dips of the strata are generally 15–25° to the south-southwest. Around the rim, bedding in the iron formation and crude flow layering in the rhyolites have been disrupted, steepened, and in places overturned. Beyond the upturned rim, angular fragments of rhyolite form an apron around the crater. Isolated fragments occur up to 300 m north of the crater. This debris has been interpreted as part of an ejecta blanket. The western section contains strongly oxidized iron fragments that are interpreted as possible remnants of iron meteorites.

Further hints for an impact origin include intense fracturing and shattering, and brecciation of the target rocks. Fractures show slickensides and radiating plumose structures. Recent drilling into the central crater revealed that the alluvium deposited on the crater floor overlies a lithic breccia. Quartz and other mineral grains in the crater basement and breccia show intense irregular and sub-planar fracturing. Unequivocal indicators for shock metamorphism though are still lacking. However, the crater floor breccia displays a strong concentration in platinum group elements (PGE) and in Nickel, which provides strong hints at the presence of contamination with extraterrestrial material. The high Ni and PGE inter-element ratios are most consistent with a chemical class IAB iron projectile.

The precise age of the impact is not known. Based on the drainage patterns and state of preservation of the rim, the impact may have happened a few 10 000 years to possibly 100 000 years ago.

The Hickman impact crater was discovered in 2007 by Arthur Hickman of the Geological Survey of Western Australia by using Google Earth. Geological analysis was carried out by Andrew Glikson and Arthur Hickman, and recently Peter Haines initiated drilling into the center of the crater.

The Hickman meteorite impact crater seen from the southern crater rim (photo courtesy: Peter Haines, Geological Survey of Western Australia, Perth).

The Hickman meteorite impact crater seen from the western crater rim (photo courtesy: Peter Haines, Geological Survey of Western Australia, Perth).

The Hickman impact crater and surroundings on October 6, 2018 as a composite image based on TanDEM-X DEM and Copernicus Sentinel-2 (channel 2/3/4) data. The position of the crater is indicated by cross hairs.

Location/Access

Hickman is situated in remote, hilly country high on the dissected plateau of the Ophthalmia Range of the Pilbara area. Access to the Hickman structure requires off-road experience and a good four-wheel drive vehicle. Visitors use the Marble Bar Road east of the town of Newman, situated on the Great Northern Highway, and then continue on a dirt road parallel to the iron ore railway line before heading into the bush for another 35 km along very rough and slow tracks.

Selected References

Bevan, A. (2012a). Western Australian Impact Craters Field Excursion 20–29 August 2012. 75th Annual Meeting of the Meteoritical Society Cairns, Queensland, Australia, 55 pp.

Glikson, A. Y., Hickman, A. H., and Vickers, J., (2008). Hickman Crater, Ophthalmia Range, Western Australia: evidence supporting a meteorite impact origin. Australian Journal of Earth Sciences, Vol. 55, pp. 1107–1117.

Haines, P. W., (2014). Collaborative scientific drilling at Hickman Crater. In GSWA 2014 extended abstracts: Promoting the prospectivity of Western Australia. Geological Survey of Western Australia, Record 2014/2, pp. 1–4.

Haines, P. W. (2017). Drilling to basement at Hickman Crater, Western Australia. In: International Workshop on Shock Metamorphism in Terrestrial and Extra-terrestrial Rocks. June, 26–29, 2017, Curtin University, Perth.

Hickman

Western Australia

latitude = 23°02'13" S / longitude = 119°41'00" E / diameter = 260 m / age = 20 000 - 100 000 yr

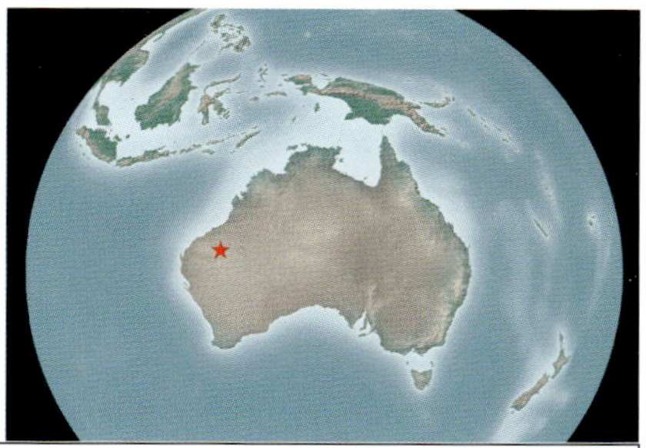

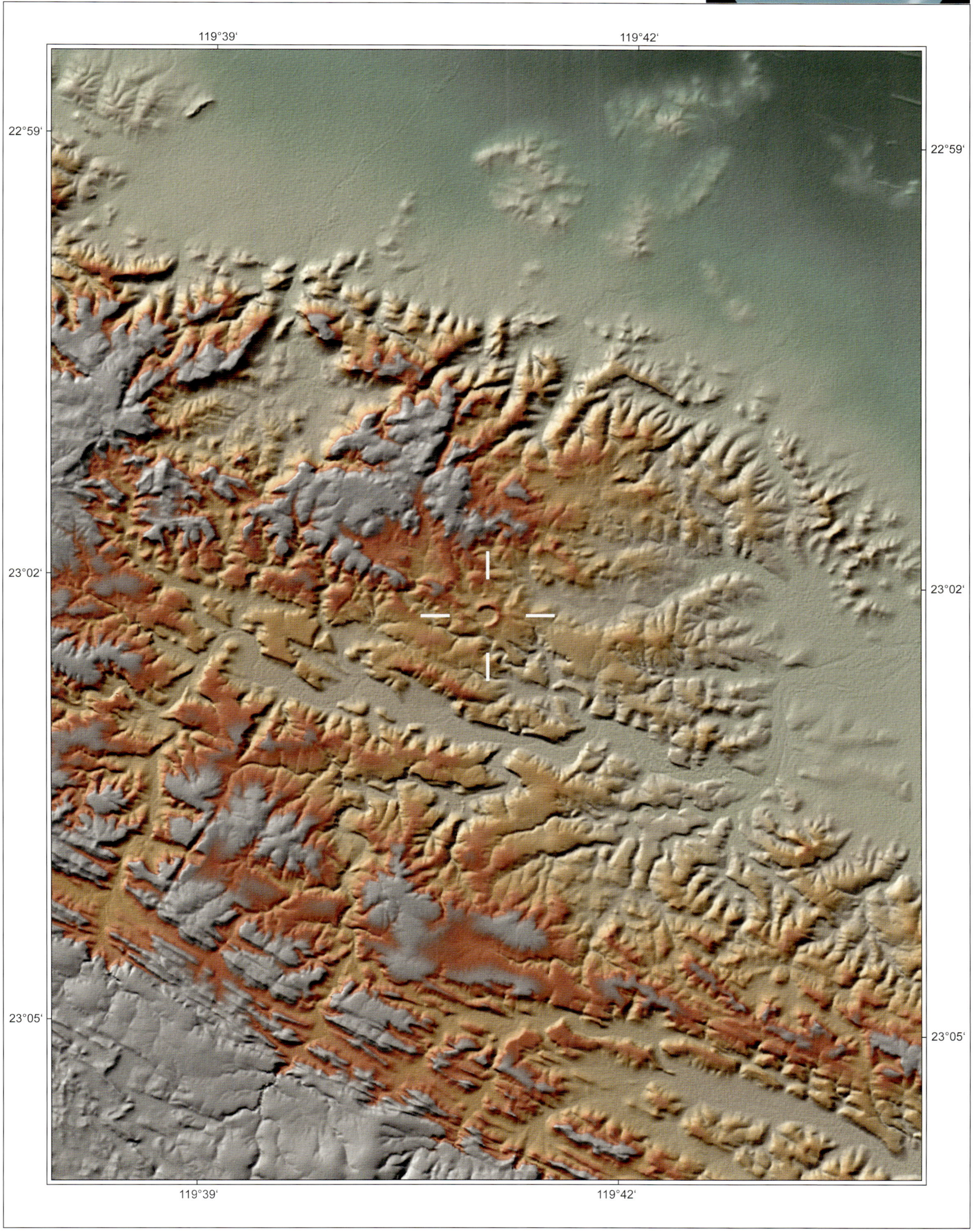

scale = 1 : 59 000 / scene = 10.6 × 14.1 km²

0 0.59 2.95 km

N

445 meter 700

Kelly West
Northern Territory

Thomas Kenkmann

Morphologically, the exposure of the Kelly West structure is limited to a 2 km diameter monadnock of deformed quartzitic sandstone standing about 10–20 m above the surrounding desert plain. Shatter cones are distributed over the entire exposure. This outcrop represents an eroded central uplift, which is elongated in northeast-southwest direction. The rest of the crater structure is not exposed. The diameter of the structure has been estimated variably at 8 to 20 km.

The target rocks belong to the Late Paleoproterozoic Hatches Creek Group, which is predominantly composed of quartzitic sandstones, besides some shale and conglomerate beds. These strata were folded at about 1660–1700 Myr ago, prior to impact. Remnants of undeformed Middle Cambrian strata drape the eroded central uplift and indicate that the impact must have occurred in the Precambrian era. The uplift exhibits tilting of the sandstones. The shatter cone axes in these rocks are pointing outward and indicate rotational movements during crater collapse. Structural characteristics of the central uplift include outward-plunging radial folds, fault bounded blocks, and duplication of strata, along with intense fracturing and brecciation. A breccia with clasts of shocked sandstone and fragments of a lithic impact breccia unconformably overlies the deformed crater rocks. Quartz grains from this breccia show planar fractures (PFs). This breccia is considered a colluvial deposit that was formed by post-impact reworking of the exposed crater.

Phil Tonkin was the first who suggested an impact origin for Kelly West in 1973 after discovering abundant shatter cones in the quartzite. Eugene and Carolyn Shoemaker mapped the site in detail in 1989.

Location/Access

Kelly West is located about 27 km west of Stuart Highway, which connects Alice Springs with Darwin. It is possible to get close to the central hill on desert dirt roads.

Selected References

Haines, P. W. (2005). Impact cratering and distal ejecta: the Australian record. Australian Journal of Earth Sciences, Vol. 52, pp. 481–507.

Shoemaker, E. M., and Shoemaker, C. S. (1996). The Proterozoic impact record of Australia. AGSO Journal of Australian Geology and Geophysics, Vol. 16, pp. 379–398.

Tonkin, P. E. (1973). Discovery of shatter cones at Kelly West near Tennant Creek, Northern Territory, Australia. Geological Society of Australia Journal, Vol. 20, pp. 99–102.

The Kelly West impact structure on July 17, 2017 as a composite image based on TanDEM-X DEM and Copernicus Sentinel-2 (channel 2/3/4) data. Cross hairs illustrate the location of Kelly West.

Kelly West

Northern Territory

latitude = 19°55'42" S / longitude = 133°57'12" E / diameter = 14 km / age = > 541 Myr

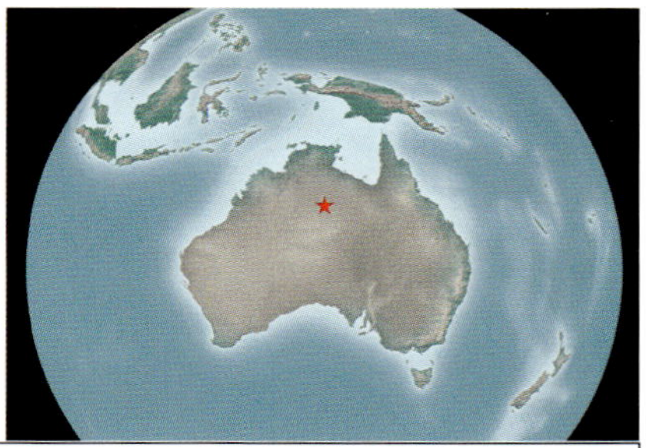

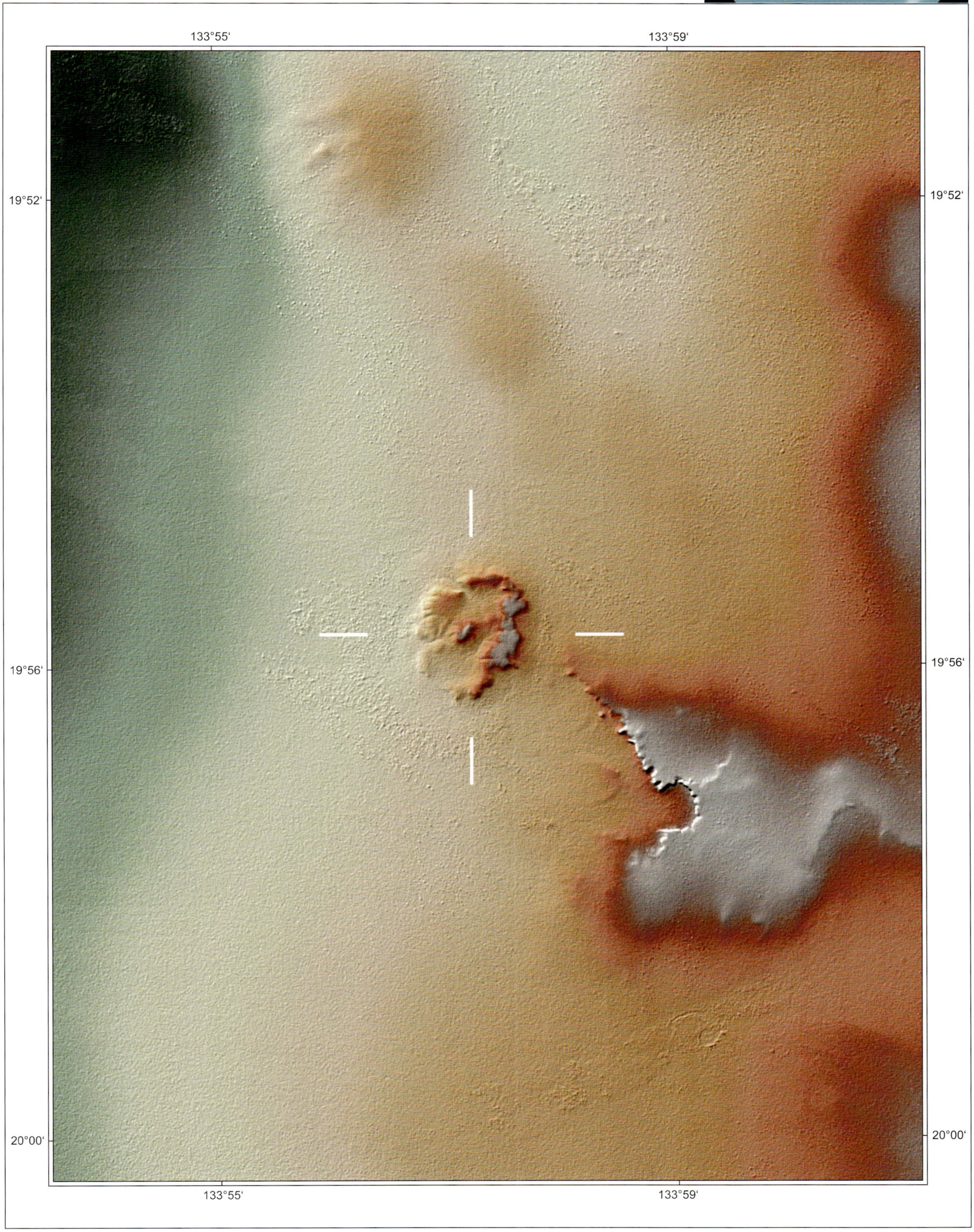

scale = 1:74 000 / scene = 13.3 × 17.7 km²

0 0.74 3.70 km

N

285 meter 345

Lake Raeside
Western Australia

Thomas Kenkmann

Lake Raeside is an impact structure on the Yilgarn craton in southwestern Australia, which lacks any morphological expression. A large part of it is buried beneath the hypersaline Lake Raeside, which is devoid of any circularity.

The structure was drilled on a well-defined magnetic anomaly. In the borehole a large breccia body was encountered at the base of a Cenozoic paleo-channel beneath Lake Raeside. Even after recovery of 95 m of impact breccias, the bottom of the breccia had still not been reached. Clearly, this indicated the presence of a large impact structure.

The breccia is composed of poorly consolidated fragments of granitoid rocks and metamorphosed mafic rock within a matrix of fine- to medium-grained brecciated material. Microscopic examination revealed pervasive grain-scale deformation with quartz grains displaying well-developed intersecting sets of planar deformation features (PDFs), diaplectic silica glass, and diaplectic plagioclase glass. The latter is partly devitrified. These glasses indicate shock pressures at least as high as 30 GPa and suggest that an allochthonous breccia body was encountered in the borehole.

Airborne total magnetic intensity data have been interpreted to suggest an outline of a probably 11 km diameter crater structure. The age of the Lake Raeside impact has been suggested to post-date the Permian glaciations of Western Australia. Thus, it must have occurred less than about 295 Myr ago.

The buried structure was discovered in the course of mineral exploration drilling in search for kimberlites. The first international report on Lake Raeside appeared in 2016 by Andrew Glikson and co-workers.

Location/Access

The buried structure is situated 60 km west of the town of Leonora. Lake Raeside covers parts of the region.

Selected Reference

Glikson, A., Hickman, A., and Crossley, R. (2016). Evidence for a shock-metamorphic breccia within a buried impact crater, Lake Raeside, Yilgarn Craton, Western Australia. Australian Journal of Earth Sciences, Vol. 63, pp. 99–109.

The area of the buried Lake Raeside impact structure on October 6, 2018 as a composite image based on TanDEM-X DEM and Copernicus Sentinel-2 (channel 2/3/4) data. Cross hairs and a dashed circle indicate position and diameter of the buried crater structure.

Lake Raeside
Western Australia

latitude = 28°44‘54“ S / longitude = 120°56‘55“ E / diameter = 11 km / age = < 295 Myr

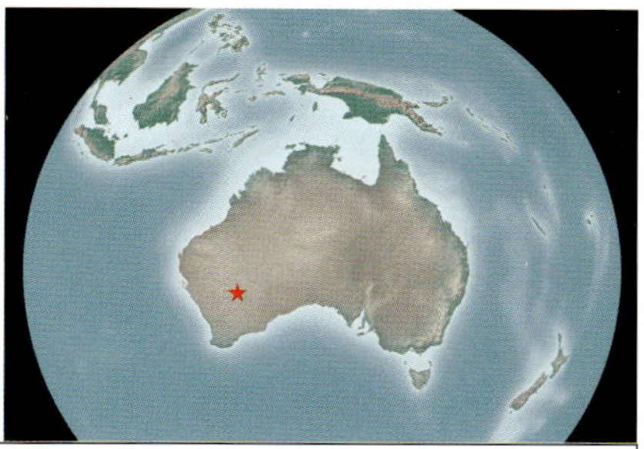

scale = 1: 190 000 / scene = 34.2 × 45.6 km²
0 1.90 9.50 km
N
360 meter 465

Lawn Hill
Queensland

Thomas Kenkmann

The Lawn Hill structure is a partly exposed impact crater with outcrops that occur predominantly in the topographically flat, but structurally uplifted central area and along the rim that partly rises 60 to 100 m above the center of the structure. The ring-like moat between rim and central uplift is covered by post-impact limestones of the mid Cambrian. From the present topographic remnants, the crater diameter was estimated at 20 km.

Target rocks from this crater structure are clastic sedimentary rocks, mainly sandstones, siltstones, and shales of the Lawn Hill Formation, with some tuff beds and older formations of the McNamara Group. These rocks were folded before impact. Beds of the central uplift have steep dips and are radially folded. Multiple repetition of strata occurs. Shatter cones occur abundantly in the rocks of the central uplift. Suevite is preserved on top of the central uplift and along the southwestern crater rim. The suevite consists of silicic impact-melt clasts embedded in a finer breccia matrix.

At the southwestern edge of the crater occurs a large zinc and lead ore body, the so-called "Century deposit", which had been actively quarried. This ore predates the impact and was truncated by the crater rim fault. Parts of the ore body were displaced and slumped towards the crater annulus along low-angle normal faults dipping towards the crater center. The ore is overlain by breccia, and Cambrian carbonates and Proterozoic shale.

The Lawn Hill impact structure with the Century Mine in the southwest on October 5, 2018 as a composite image based on TanDEM-X DEM and Copernicus Sentinel-2 (channel 2/3/4) data.

Lawn Hill was first identified as an impact structure by A. Stewart and K. Mitchell in 1987, who described shatter cones and microscopic shock features in quartz, such as multiple sets of planar deformation features (PDFs). The structure was later mapped in detail by Eugene and Carolyn Shoemaker. Current investigations focus on the world-class zinc metal deposit along the southwestern crater rim.

According to Stewart and Mitchell, and the Shoemakers, the Cambrian carbonates, although brecciated, were considered to have been deposited post-impact in the structural moat surrounding the central uplift. Brecciation of these rocks was attributed to solution collapse. Thus, these workers suggested that the impact occurred in the Late Precambrian or very Early Cambrian. In contrast, Vicki Darlington and co-workers indicated by ^{40}Ar-^{39}Ar dating of impact-related melt particles a time of impact in the Ordovician at 472 ± 8 Myr ago. At this time the area was an active depocenter. Terry Lees and his team, who restored the southwestern crater rim, suggested that the Cambrian rocks were partially consolidated prior to impact, became fluidized on impact, and then resurged into the crater from surrounding areas. The resurge process resulted in a fivefold thickening of the Cambrian carbonates in the annulus.

Location/Access

Southwest of Lawn Hill, open-pit mining activity took place until 2015 in the Century zinc, lead and silver mine. Therefore, access is via reliable infrastructure, including the Century Mine airport.

Selected References

Darlington, V., Blenkinsop, T., Dirks, P., Salisbury, J., and Tomkins, A. (2016). The Lawn Hill annulus: An Ordovician meteorite impact into water-saturated dolomite. Meteoritics and Planetary Science, Vol. 51, pp. 2416–2440.

Haines, P. W. (2005). Impact cratering and distal ejecta: the Australian record. Australian Journal of Earth Sciences, Vol. 52, pp. 481–507.

Lees, T. C., Murphy, F. C., Tomkins, A. G., and O'Donohue, D. (2019). Displacement of the Proterozoic century ore deposit at the edge of an Ordovician meteorite impact crater, Queensland. Economic Geology, Vol. 114, pp. 427–439.

Shoemaker, E. M., and Shoemaker, C. S. (1996). The Proterozoic impact record of Australia. AGSO Journal of Australian Geology and Geophysics, Vol. 16, pp. 379–398.

Stewart, A., and Mitchell, K. (1987). Shatter cones at the Lawn Hill circular structure, northwestern Queensland: presumed astrobleme. Australian Journal of Earth Sciences, Vol. 34, pp. 477–485.

Lawn Hill

Queensland

latitude = 18°41‘19“ S / longitude = 138°39‘06“ E / diameter = 20 km / age = 472 ± 8 Myr

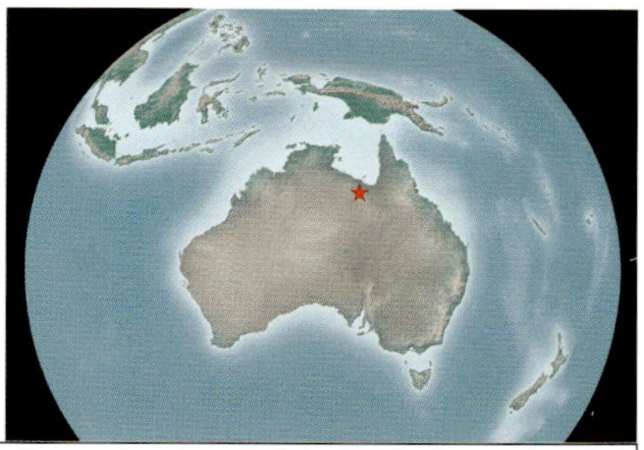

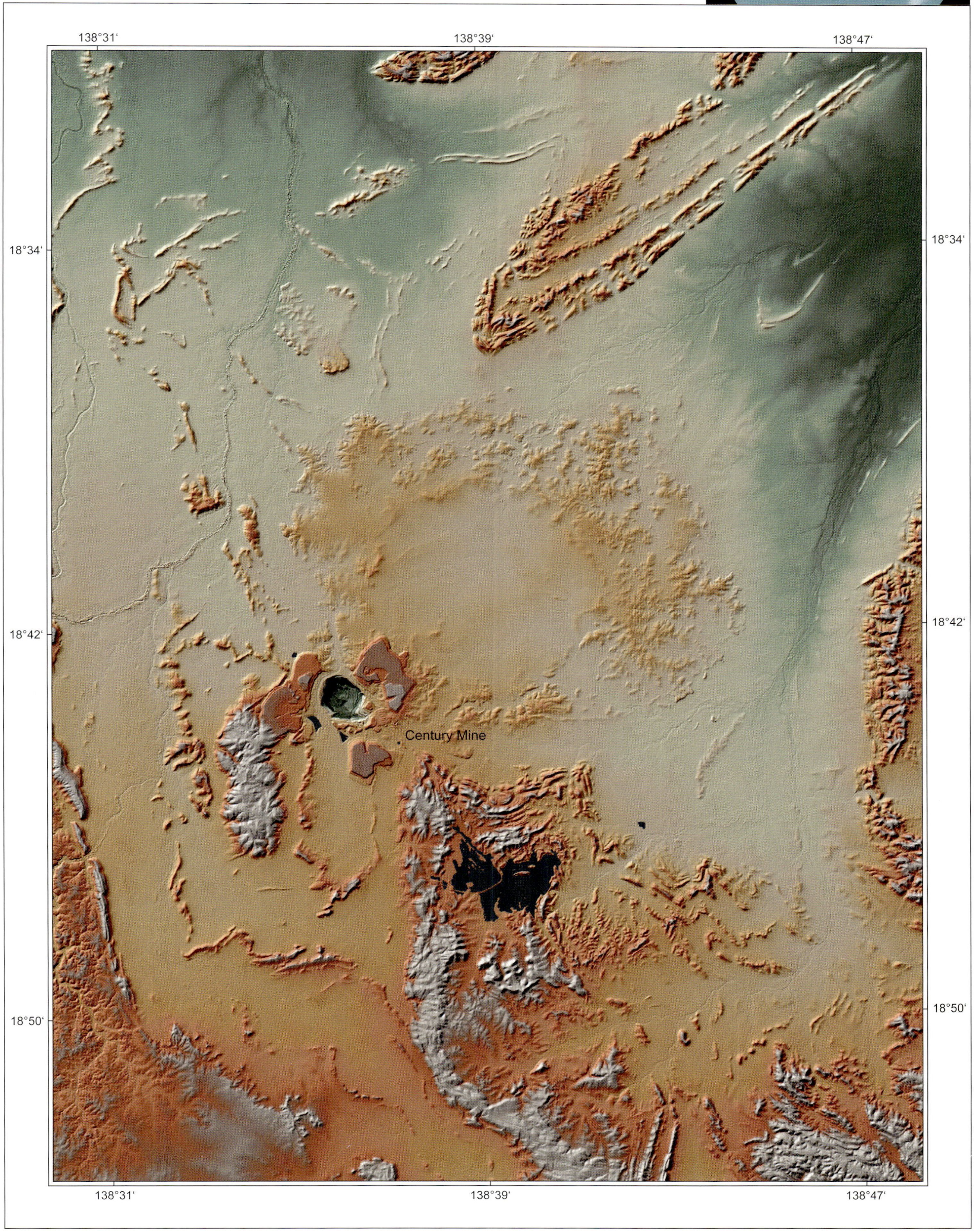

scale = 1 : 180 000 / scene = 32.5 × 43.3 km²

0 1.80 9.00 km

N

-160 meter 300

Liverpool
Northern Territory

Thomas Kenkmann

Liverpool is an only partially exhumed impact crater of about 1.6 km diameter. At a first glance, it has the classical shape of a simple crater with a circular raised rim. However, the elevated rim is not the original crater rim, but is formed by allochthonous breccia that dips consistently inward. The breccia is interpreted as a crater fill breccia. As the breccia represents a competent rock, the sealed breccia nowadays forms a circular high. The original crater diameter is estimated to have been somewhat larger at about 2 km. The Liverpool impact structure is slightly more elongated, by roughly 6 %, in the northeast-southwest direction than in the northwest-southeast direction. This difference and systematic deviations in strata tilt may reflect the direction of impact from the southwest.

The target stratigraphy involves relatively flat-lying quartzose sandstone of Early Precambrian age. Two distinct breccia units were mapped. A lower monomict breccia consists of upturned and locally overturned target rocks and meter-scale sandstone blocks in a matrix of crushed sandstone and smaller angular clasts. Quartz grains of this sandstone exhibit planar fractures (PFs) decorated with fluid inclusions. These rocks may represent moderately displaced uppermost target and the immediate crater floor breccia. A more voluminous breccia rests on the first breccia type, with a sharp and often grooved and slickensided contact that dips concentrically inward with 40–50°. Clasts in the upper breccia reach meter size. The upper breccia was likely emplaced as a sequence of discrete slumps emanating from the crater rim. Both units together have been referred as the *Liverpool Breccia*. The breccia is topped by fossil-free post-impact sediments deposited in the crater cavity.

The age of the Liverpool structure is constrained only by the age of the target rocks and the post-impact formation. As fossils are lacking in post-impact crater fill sediments, age estimates vary strongly and range from the Cretaceous to the Proterozoic.

Eugene and Carolyn Shoemaker, as well as Peter Haines, suggested a Neoproterozoic age for the post-impact sediments. If this is correct and taking into account the degree of preservation of the crater, this would imply that impact occurred close to the time of burial in the Neoproterozoic. Thus, Liverpool would perhaps be the world's oldest well-preserved simple impact crater.

The structure, known as *Yingundji* by the Aboriginal landowners, was first noticed by geologists during a 1960s geological mapping campaign. A possible origin as a meteorite impact crater was mentioned on a topographic map in 1965. The actual confirmation of the impact origin dates back to 1970. Detailed mapping was carried out by Eugene and Carolyn Shoemaker in 1996.

Location/Access

Liverpool crater is located in the Arnhem Land Aboriginal Reserve of the Northern Territory, and special permits are required for entry into this area. The estuary of the Liverpool River and the Bay of Manigrida are about 40 km northeast of the structure. Meandering streams as well as swampland form the environs of the crater on the floodplain of the Liverpool River. Access to the structure is difficult, as it is located in an area with little infrastructure, about 130 km east of Jabiru, where the Arnhem Highway meets the Kakadu Highway.

Selected References

Brett, R., Guppy, D. J., and Milton, D. J. (1970). Two circular structures of impact origin in Northern Territory, Australia (abstract). Meteoritics, Vol. 5, p. 184.

Haines, P. W. (2005) Impact cratering and distal ejecta: the Australian record. Australian Journal of Earth Sciences, Vol. 52, pp. 481–507.

Shoemaker, E. M., and Shoemaker, C. S. (1997b). Notes on the geology of Liverpool Crater, Northern Territory, Australia. 28th Lunar and Planetary Science Conference, Lunar and Planetary Institute, Houston, Texas, Abstract #1662.

Shoemaker, E. M., Macdonald, F. A., and Shoemaker, C. S. (2005). Geology of five small Australian impact craters. Australian Journal of Earth Sciences, Vol. 52, pp. 529–544.

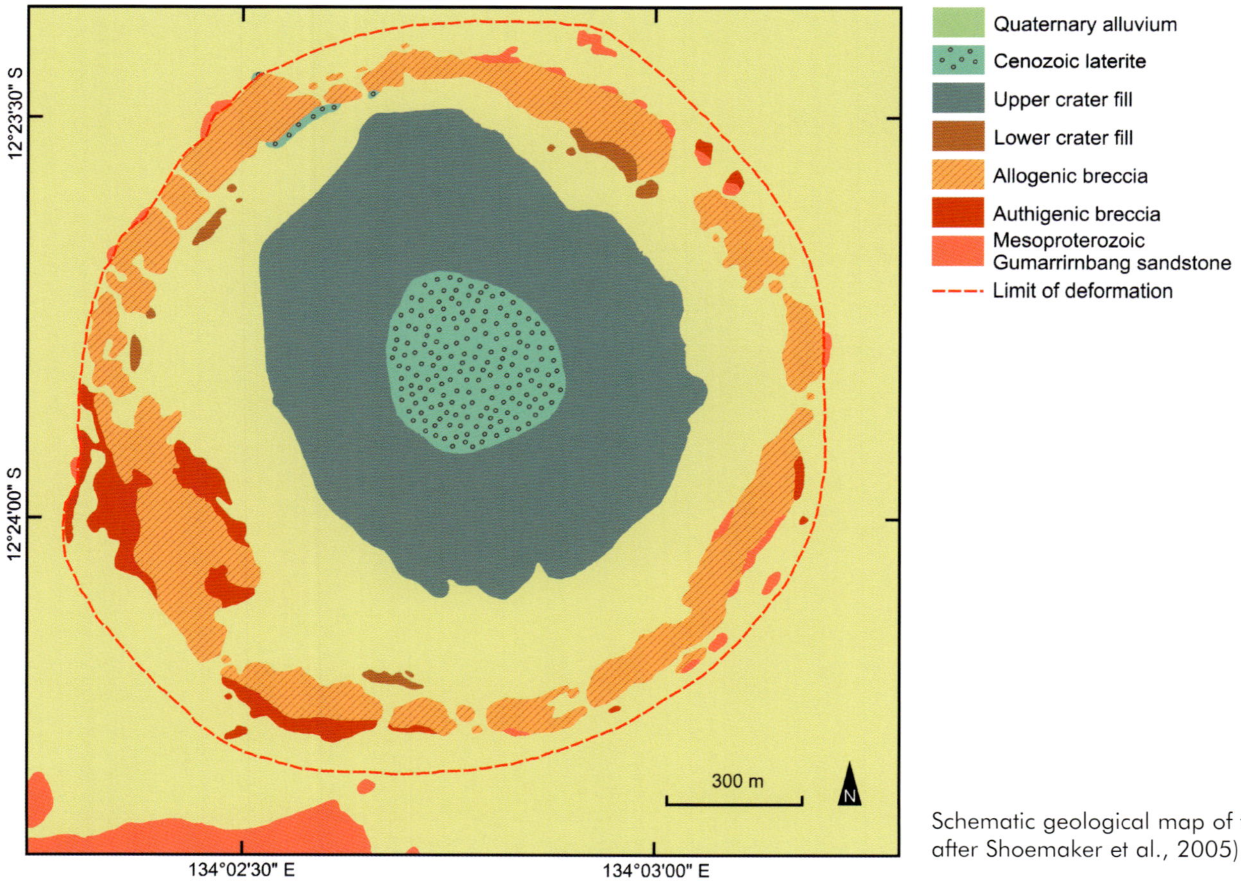

Schematic geological map of the Liverpool crater (modified after Shoemaker et al., 2005).

Liverpool

Northern Territory

latitude = 12°23‘45“ S / longitude = 134°02‘50“ E / diameter = 2 km / age = 540 - 1000 Myr

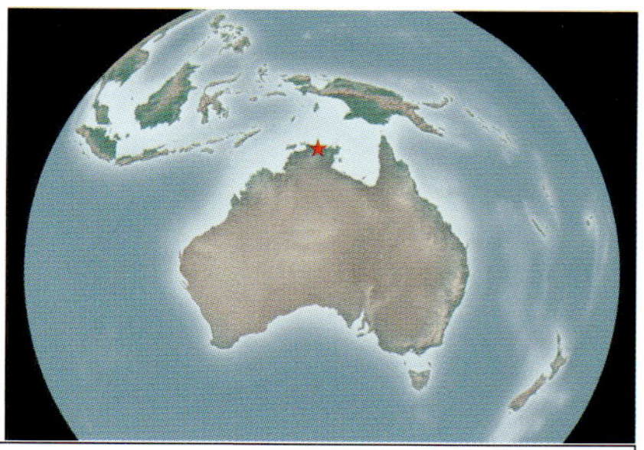

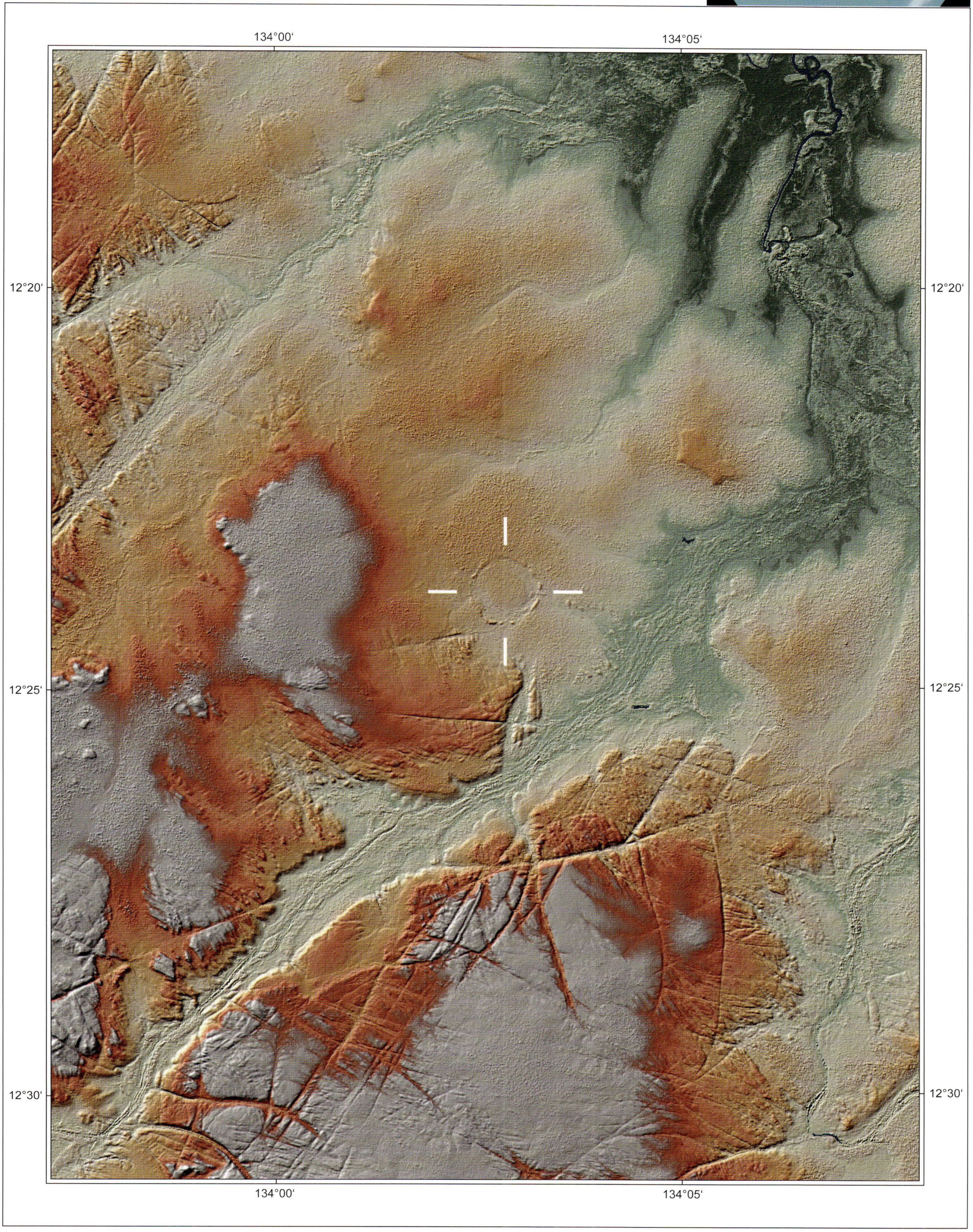

scale = 1: 108 000 / scene = 19.4 × 25.8 km²

0 1.08 5.40 km

N

0 meter 180

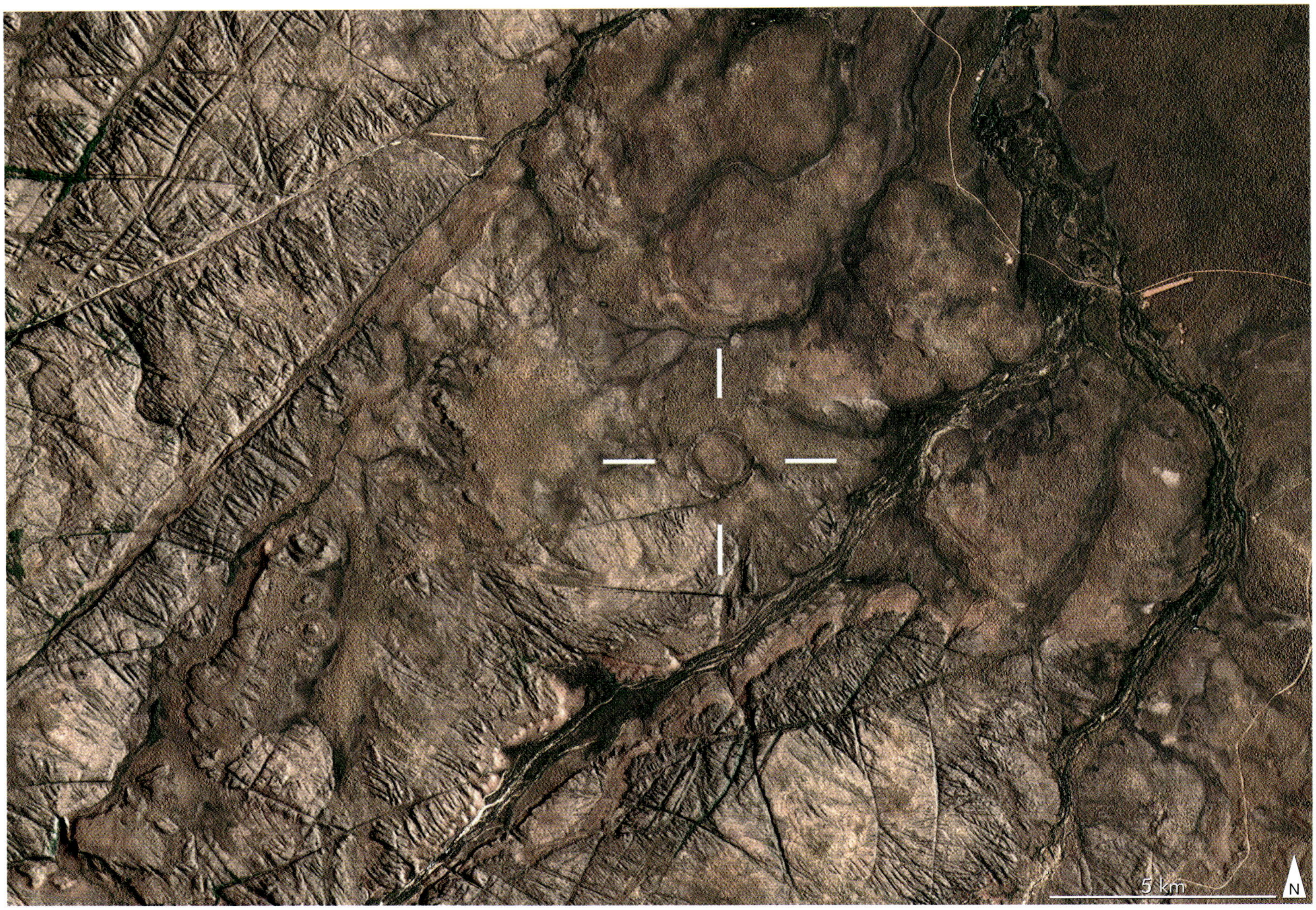

The Liverpool impact crater and surroundings on October 20, 2015 as a composite image based on TanDEM-X DEM and Copernicus Sentinel-2 (channel 2/3/4) data. The position of Liverpool is indicated by cross hairs.

Matt Wilson
Northern Territory

Thomas Kenkmann

Matt Wilson, located in Gregory National Park, is a unique complex crater. Whilst most impact structures are circular in shape, Matt Wilson displays a pronounced elliptical form of 7.5 × 6.3 km extent. Its long axis, trending northeast-southwest, corresponds to the impact direction, which is consistent with the structural observations from the structure. The projectile came from the northeast at a very oblique angle of incidence.

The impact structure is located in the Victoria Basin that is built up of Proterozoic sediments. These are mainly sandstones and siltstones that are unmetamorphosed, flat-lying, and tectonically undeformed. The oldest outcropping rocks are quartzitic sandstones of the Wondoan Hill Formation. They form prominent morphological ridges in the innermost part of the impact structure. This unit is conformably overlain by a 70 m thick sequence of mudstones and siltstones. These strata often form the lower part of escarpments and are, therefore, mostly covered by talus. The next younger stratigraphic unit, the Jasper Gorge Sandstone of probable Mesoproterozoic age, is the major lithology exposed in the crater. This formation consists of massive sandstone beds at the base, which form prominent escarpments of the National Park, with a sequence of thinly bedded siltstones and sandstones above them. A 10 m thick, white sandstone layer is a good marker bed within this sequence. The crater is deeply eroded, to slightly less than a kilometer depth, so that no crater fill breccias or ejecta can be expected anymore.

The age of Mount Wilson can only be based on stratigraphic evidence and is rather uncertain. Given the high degree of erosion, an Early Mesoproterozoic age of about 1400–1500 Myr is favored.

Matt Wilson belongs to the group of complex craters. First-order structural characteristics are: (1) The crater rim is defined by a monocline, along which flat-lying strata of Jasper Gorge Sandstone suddenly change to inward-dipping attitude. The crater rim monocline delineates the crater's elliptical shape. (2) A syncline with circumferential strike parallel to the elliptical outline is the next prominent feature of the crater, following in direction of the interior. The white sandstone layer marks the youngest strata in this syncline and traces the fold axis of this ring syncline. The inner limb of the ring syncline gradually changes into the central uplift. (3) In the central uplift the oldest exposed strata belong to the Wondoan Hill Formation, which indicates a relatively low uplift of roughly 200 m. Shape and structure of the central uplift are symmetrical to the long axis of the ellipse and elongated along the structure's symmetry axis, but doming and uplift are asymmetric. In the flanks of the central uplift radial anticlines and synclines with outward-plunging fold axes are common and can readily be seen on satellite images. Textbook exposures of radial transpression ridges and lateral thrust ramps also occur in the periphery of the central uplift. These structural features indicate convergent inward flow and increasing constriction toward the center of the crater. This type of deformation is known to occur during the gravitational collapse of a crater.

Matt Wilson

Northern Territory

latitude = 15°30'00" S / longitude = 131°10'51" E / diameter = 7.5 km / age = 1400 - 1500 Myr

scale = 1 : 101 000 / scene = 18.1 × 24.2 km²

0 1.01 5.05 km

N

30 meter 310

Mesoproterozoic Jasper Gorge Sandstone forms prominent cliffs at Matt Wilson. Here, a lateral thrust ramp caused by convergent movements during crater modification leads to a duplication of the formation (photo: T. Kenkmann).

View toward the center of the central uplift of the Matt Wilson impact structure from the southeast. The morphological ridges are formed by massive quartzitic sandstone of the Wondoan Hill Formation. Stacking of strata indicates a transport direction top to the southwest and indicates an impact from the northeast (photo: T. Kenkmann).

The core of the central uplift shows a different style of deformation. Steep parallel cliffs of quartzitic sandstone of the Wondoan Hill Formation stand in almost upright position – forming a sequence of imbricated thrusts. Each of the cliffs is bounded by faults that contain various types of lithic breccias. The preferred stacking of thrust sheets indicates an up-range to down-range transport of rock. This motion is caused by remnant momentum transferred to the target from a projectile impacting at an acute angle.

The massive quartzitic sandstones of the central uplift, as well as the lithic fault breccias between the thrust slices, show a shock metamorphic overprint, including an abundance of planar fractures (PFs) and feather features (FFs) in quartz, and rare planar deformation features (PDFs). Joints and fracture surfaces often show faint striated and undulating surfaces that may represent incipient shatter coning.

It is estimated that the projectile struck the Earth's surface at a low angle of incidence, most likely below 15°. Impacts usually occur at moderately oblique angles of around 45°, which results in the formation of circular impact structures. Only when the angle of impact becomes smaller than 15°, the shape of the crater becomes elliptical. From a statistical point of view, about 7 % of all extraterrestrial projectiles should impact at an angle below 15°. However, within the current record of confirmed impact structures, only Matt Wilson falls into the category of pronouncedly oblique impacts. Thus, elliptical craters are highly under-represented on Earth, probably because the prolonged passage through the atmosphere may filter out such impacts. Moreover, such craters are shallower and, thus, more prone to erosion than deeper, circular structures. Furthermore, a lower degree of shock metamorphism and a less pronounced radial symmetry could hamper the identification of such elliptical structures.

The Matt Wilson structure was discovered in 2005 based on aerial photography inspection and the occurrence of a faint ring-shaped negative aeromagnetic anomaly. Ian Sweet and co-workers named the structure after a nearby tributary of the Victoria River. A few years later, in 2009, the crater was mapped in detail by Thomas Kenkmann and Michael Poelchau. They discovered PDFs, and the effects of a low-angle oblique impact on crater formation could be thoroughly investigated.

Location/Access

The Matt Wilson structure is located in Gregory National Park, 15 km north of the Victoria River Roadhouse. This roadhouse is situated at the Victoria River Crossing, a key landmark on the Victoria Highway that connects Katharine to the east with Timber Creek to the west. No access roads into the structure exist, and there is no on-site infrastructure. However, the Victoria River Roadhouse has a campground, and scenic helicopter flights over the National Park can be arranged with Coolibay Air from there.

Selected References

Kenkmann, T., and Poelchau, M. H. (2009). Low-angle collision with Earth: The elliptical impact crater Matt Wilson, Northern Territory, Australia. Geology, Vol. 37, pp. 459–462.

Sweet, I. P., Haines, P. W., Mitchell, K., and Glikson, A. Y. (2005). Matt Wilson Structure: record of an impact event of possible early Mesoproterozoic age, Northern Territory. Australian Journal of Earth Sciences, Vol. 52, pp. 675–688.

The Matt Wilson impact structure on October 9, 2018 as a composite image based on TanDEM-X DEM and Copernicus Sentinel-2 (channel 2/3/4) data.

Shoemaker
Western Australia

Thomas Kenkmann

Shoemaker, also known as *Teague Ring*, is an old, deeply eroded impact structure of roughly 30 km diameter. The impact event is thought to have occurred at the turn of the Paleoproterozoic to the Mesoproterozoic – however, no satisfactory dating results on impact lithologies are available yet. The structure is well-defined in terms of topography by two concentric rings of low hills. A series of playa lakes in the center of the structure and between the inner and outer rings are responsible for the peculiar appearance of the crater in aerial views. The structure is situated at the southern edge of the Paleoproterozoic Earaheedy Basin, close to the exposed northern margin of the Archean Yilgarn Craton. It interrupts prominent lines of hills belonging to the northwest-trending Frere Range, which rise about 50–70 m above the surrounding plain.

The geological structure is characterized by a 12 km wide central uplift that consists of fractured Archean basement granitoids of syenitic composition. These granitoids are hydrothermally altered, particularly on the eastern side of the central uplift. This uplift is surrounded by a complex ring syncline of about 20–25 km width that is formed in Paleoproterozoic sedimentary rocks, among them banded iron formations. Outermost occurs the outer ring anticline. Geophysical characteristics of the structure include a negative gravity anomaly coincident with the central uplift and positive circular trends in both magnetic and gravity anomaly patterns that can be correlated with the inner ring syncline and outer ring anticline. There seems to be an east-west asymmetry in the magnetic intensity.

Several attempts to date the impact event led to conflicting results. A whole rock Rb-Sr isochron obtained from granite of the central uplift yielded 1630 ± 5 Myr. This is the most cited age of the impact. However, it remains doubtful whether this age represents a target rock age or is related to the impact event. More recent K-Ar measurements on illite and smectite revealed much younger ages of 694 ± 25 and 568 ± 20 Myr that were interpreted to date the hydrothermal alteration of the granite.

A $6.5 \times 5.5 \times 5$ cm piece of monomict impact breccia from the central basin of the Shoemaker impact structure (photo courtesy: Ralf-Thomas Schmitt, Museum für Naturkunde Berlin, sample from Dieter Stöffler).

The Shoemaker impact structure, indicated by the ring of playa lakes on July 3, 2017 as a composite image based on TanDEM-X DEM and Copernicus Sentinel-2 (channel 2/3/4) data.

Shoemaker

Western Australia

latitude = 25°52' S / longitude = 120°53' E / diameter = 30 km / age = 1630 ± 5 Myr

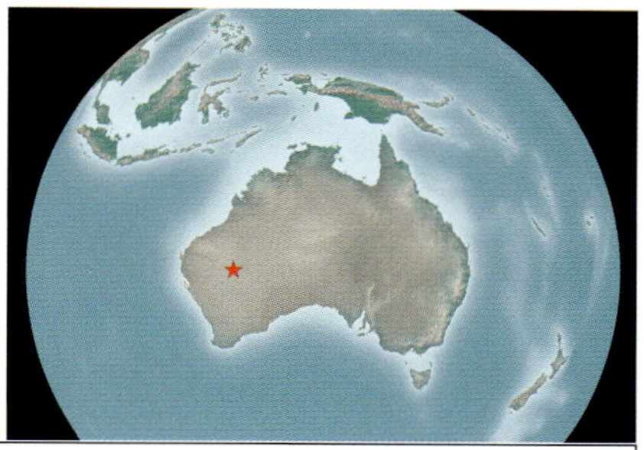

The structure was first described by H. Butler in 1974, who named it the *Lake Teague* ring structure. He suggested it might have formed as either the result of granite intrusion or an impact event. In 1980 J. Bunting and co-workers systematically mapped the area and provided evidence for shock metamorphism, including shatter cones and planar deformation features (PDFs) in quartz from the central granite. They also reported pseudotachylitic breccia. However, they concluded that the structure was caused by a "cryptoexplosive" event. This view changed when Eugene and Carolyn Shoemaker conducted fieldwork in 1986 and 1995 and confirmed the presence of shock metamorphism. This provided the definite proof that the structure had been formed by an impact event. They mapped the crater in detail. Franco Pirajno and colleagues continued Shoemaker's work with detailed investigations of the structure, which was then followed by precise 1 : 100 000 scale mapping by the Geological Survey of Western Australia. These workers suggested that 2–3 km of overlying rock may have been removed by erosion. Hence, the original crater diameter was probably larger than 30 km.

Today this impact structure is named in honor of Eugene Shoemaker. He died in a car accident on July 18, 1997 in the course of a geological field trip that included another visit to this impact structure. This commemoration honors this renowned Earth and planetary geologist for his pioneering work in impact and planetary geology.

Location/Access

From the Great Northern Highway that connects Perth with Port Hedland, the Goldfields Highway branches off at Meekatharra. This highway runs for almost 200 km to the former mining town of Wiluna. From there it is about 80 km to Millrose Homestead, and along station tracks for another 60 km northward to reach the structure. Driving to Shoemaker impact crater requires a four-wheel-drive car and outback experience.

Selected References

Bunting, J. A., de Laeter, J. R., and Libby, W. G. (1980). Evidence for the age and cryptoexplosive origin of the Teague Ring structure, Western Australia. Western Australia Geological Survey Annual Report 1979, pp. 125–129.

Butler, H. (1974). The Lake Teague ring structure, Western Australia: an astrobleme? Search, Vol. 5, pp. 536–537.

Pirajno, F., Hawke, P., Glikson, A. Y., Haines, P. W., and Uysal, T. (2003). Shoemaker impact structure, Western Australia. Australian Journal of Earth Sciences, Vol. 50, pp. 775–796.

Shoemaker, E. M., and Shoemaker, C. S. (1996). The Proterozoic impact record of Australia. AGSO Journal of Australian Geology and Geophysics, Vol. 16, pp. 379–398.

Spider

Western Australia

Thomas Kenkmann

The slightly elongated Spider structure with a size of 11 × 13 km is located on the scenic Kimberley Plateau in Western Australia. Its central uplift is marked by a striking system of imbricated, radiating thrusts. This peculiar appearance is where the name of the structure was derived from. Spider forms a conspicuous landmark on aerial images.

The target rocks are shallow marine sandstone and subordinate siltstone that were deposited in the Paleoproterozoic Kimberley Basin. The oldest stratigraphic unit exposed in this impact structure is the Warton Sandstone with an age of 1704 ± 7 Myr based on U-Pb dating of diagenetic xenotime. The Warton Sandstone is overlain by the Elgee Siltstone, a succession of siltstone and mudstone with sandstone intervals but rather poor exposure. The youngest Proterozoic unit exposed in the Spider structure is the Pentecost Sandstone. The Spider structure lies across the east-west oriented axis of the broad, deeply eroded Mount Barnett Syncline, the limbs of which dip very gently, at a few degrees, towards the north and south.

The Warton Sandstone Formation is exposed in the center of the impact structure. Deformation and the shock metamorphic overprint of this rock are intense. This sandstone shows planar deformation features (PDFs) and abundant planar fractures (PFs), many of which are part of feather features (FFs). Shock effects propagate from quartz grains into the surrounding quartz cement, suggesting that the rocks were completely lithified and of low porosity when the impact occurred. Pentecost Sandstone builds most parts of the peculiar morphological ridges. The deformation varies from "weakly deformed" to "shocked". These rocks show abundant transgranular fractures, PFs, FFs, and a few PDFs. Recently, planar microstructure and twins have also been observed in zircon and xenotime grains of the central uplift. Shatter cones are ubiquitously present in the central dome and its vicinity. Between the thrusts occur narrow zones with monomict and polymict fault breccias.

Because the outline of the crater structure is poorly defined, investigations have focused on the central uplift. It has a north-northwest trending axis of symmetry. Two sets of imbricate thrust stacks surround the core of the structure, indicating north-south shortening with a preferred material transport of top-to-the-south direction. Dips of beds within the imbricate stacks vary from 20 to 80° and do not show any systematic change from north to south. The traces of each thrust are bent and their strikes shift gradually. The change in

The characteristic morphological ridges of the Spider impact structure are built up by the Proterozoic Pentecost Sandstone. The formation is dipping into northerly direction. The multiple repetition of strata is formed by imbrication stacking (photo: T. Kenkmann).

Spider
Western Australia

latitude = 16°44'27" S / longitude = 126°05'21" E / diameter = 13 km / age = 600 - 900 Myr

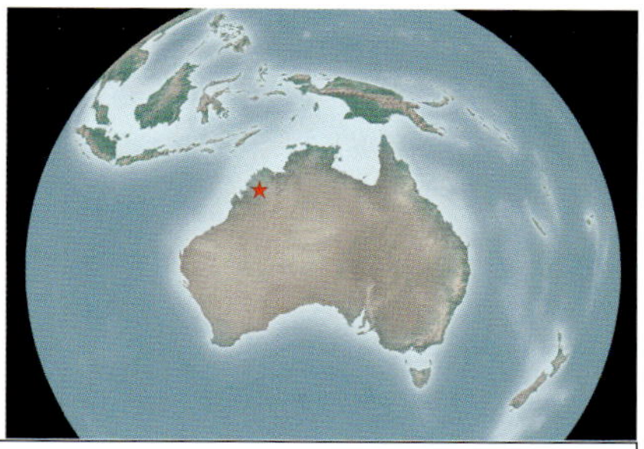

scale = 1 : 83 000 / scene = 15.0 × 20.0 km²

0 0.83 4.15 km

N

365 meter 565

Shatter cones in Warton Sandstone, the formation that builds up the center of the Spider impact structure (photo: T. Kenkmann).

orientation of the thrust slices from up-range to down-range can be modeled by assuming an oblique impact from the north-northwest, which transferred horizontal momentum to the target, while interfering with a radially symmetrical collapse flow field. The imbricate stacks initiated up-range and migrated down-range, simultaneously with the uplift of the central core.

Spider's age relies on assumptions concerning regional events. According to this, the impact occurred later than the Yampi orogeny that dates to about 900 Myr ago and earlier than regional glacial erosion at about 600 Myr. This would assign a Neoproterozoic age to Spider.

Spider was first identified on aerial photographs by John Harms during reconnaissance mapping of the Kimberley region in the late 1950s. He did not make the connection to an impact scenario. In a publication in 1980, Harms and co-workers still classified Spider as a "cryptoexplosion" structure, in accord with one of the contemporary doctrines on structures of this kind. Eugene and Carolyn Shoemaker mapped the site in 1984–1985 and published their findings shortly after. They identified Warton Sandstone piercing through the younger units in the center of the dome. They concluded that an oblique impact would be the best explanation for the structure and its imbricate thrusting, caused by a low angle projectile approaching from the north or northwest. Later on, the structure was investigated by remote sensing methods. Recently, Morgan Cox and Aaron Cavosie from Curtin University, Perth, conducted structural analysis of the central uplift together with Thomas Kenkmann.

Location/Access

The Spider impact structure is located on the Kimberley Plateau, Western Australia. The Gibb River-Wyndham Road is the main infrastructure in this area, connecting the towns of Derby to the west and Kununurra to the east. Using this dirt road requires four-wheel drive and is only recommended in the dry season from May to October, as there are some severe river crossings. Mount Barnett Roadhouse lies about 20 km from the impact structure. Access to the structure is restricted, as it is located in the Mount Barnett pastoral lease, a declared Aboriginal homeland.

The Spider impact structure and environs on September 25, 2018 as a composite image based on TanDEM-X DEM and Copernicus Sentinel-2 (channel 2/3/4) data.

Selected References

Cox, M. A., Cavosie, A. J., Miljković, K., Bland, P. A., Kenkmann, T., and Hoskins, Z. N. P. (2018). Characterization of shock deformation at the Spider impact structure, Western Australia. 81st Annual Meeting of the Meteoritical Society, Moscow, Russia, Abstract #6248.

Harms, J. E., Milton, D. J., Ferguson, J., Gilbert, D. J., Harris, W. K., and Goleby, B. (1980). Goat Paddock cryptoexplosion crater, Western Australia. Nature, Vol. 286, pp. 704–706.

Kenkmann, T., Poelchau, M. H., and Wulf, G. (2014). Structural geology of impact craters. Journal of Structural Geology, Vol. 62, pp. 156–182.

Shoemaker, E. M., and Shoemaker, C. S. (1985). Impact structures of Western Australia. Meteoritics, Vol. 20, pp. 754–756.

Shoemaker, E. M., and Shoemaker, C. S. (1996). The Proterozoic impact record of Australia. AGSO Journal of Australian Geology and Geophysics, Vol. 16, pp. 379–398.

Strangways

Northern Territory

Thomas Kenkmann

Strangways is a large complex impact structure with a diameter of probably more than 40 km. The structure is eroded but also partly buried beneath younger strata. The crater has a faint morphological expression with a ring pattern of competent sandstones forming concentric ridges that alternate with flat-floored valleys. The target consists of a granitic basement overlain by sedimentary rocks of Meso- to Neoproterozoic age of the McArthur Basin in northern Australia. Most of the central and southern parts of the structure are buried beneath Cretaceous, and possibly older, Phanerozoic beds. The northern part of the structure is partially covered by undeformed plateau volcanics of Early Cambrian age.

Strangways has a large, 8 km wide, central area, which is part of the central uplift structure. It exposes granitic basement in its inner part. Based on scaling relationships, the stratigraphic uplift and vertical movement in the central uplift have been estimated at 2.4–2.7 km. Parts of the basement in the central uplift area are covered by post-impact deposits. Morphologically the central area is encircled by a collar of competent quartzites and sandstones. This collar of upturned and overturned beds has a diameter of at least 20 km; it still belongs to the central uplift. Most of the beds are overturned and suggest that the central uplift was gravitationally unstable. The annular depression surrounding the central uplift is not well-defined and its impact-related deformation is hard to distinguish from tectonic deformation. Likewise, a crater rim cannot be precisely defined. The impact structure seems to be truncated in the east by a tectonic fault that was active after the impact.

There is a large variety of size estimates for this crater structure. Early estimates suggested a diameter of 20–25 km but remote sensing analysis favored 26 to 29 km. However, based on the 20 km size of the collapsed central uplift with overturned beds, the diameter of the structure that is clearly visible in aeromagnetic datasets and spaceborne imagery should be at least 40 km.

Near the center of the crater structure occur outcrops of suevite that rest on brecciated and strongly fragmented granite. The suevite contains shock-melted, partly melted, and unmelted clasts with planar deformation features (PDFs) in quartz, as well as diaplectic glasses embedded in a particulate matrix. Clasts are quartzite, shale, and granite and can be up to a meter in size. The granite breccia grades downward into shattered or coherent granite. Shock-melted granite occasionally also occurs. Further shock metamorphic overprint in the form of shatter cones has been recorded in quartzites and sandstones forming the concentric ridges further outward.

Geochemical analysis of the suevite indicates a contamination by the impacting asteroid. Nickel is enriched in melt rocks by factors of 10 to 60 relative to the background abundances of country rocks. Iridium is enriched generally by more than a factor of 20. Other enriched elements are Cr, Os and Pd.

The age of this impact is still controversial. The overlap of undeformed Early Cambrian plateau volcanics in the northern part of the structure indicates a Precambrian age for the Strangways impact event. Attempts to radiometrically date shock-melted granite from the crater center by the ^{40}Ar-^{39}Ar dating method led to ambiguous plateau ages of 470 Myr, which are in conflict with the age of the overlapping volcanic rocks. A more recent dating study using laser spot fusion argon dating on two impact melt samples and a highly shocked breccia clast produced a combined age of 646 ± 42 Myr, consistent with the stratigraphic superposition of the structure by Cambrian volcanic rocks.

The Strangways structure was first mapped by P. Dunn in 1963. Douglas Guppy and co-workers suggested an impact origin in 1971 and identified its prominent central uplift, shatter cones, and impact melt rocks. Later, detailed geological mapping by Eugene and Carolyn Shoemaker led to further improvement of the understanding of this impact structure.

Location/Access

As with many other crater structures in Australia, Strangways is difficult to reach and is situated on Aboriginal Land. Thus, access to the structure requires special permits. Located west of the Strangways River, the nearest infrastructure is Mataranka on Stuart Highway that links the city of Darwin at Australia's north coast with Port Augusta in the south. Seven kilometers south of Mataranka, Roper Highway branches off to the east. After about 50 km, it passes the Strangways structure at a distance of 25 km.

Selected References

Bottomley, R. J., York, D., and Grieve, R. A. F. (1990). ^{40}Ar-^{39}Ar dating of impact craters. Proceedings of the 20th Lunar and Planetary Science Conference, Houston, Texas, Cambridge University Press, pp. 421–431.

Guppy, D. J., Brett, R., and Milton, D. J. (1971). Liverpoool and Strangways craters, Northern Territory: two structures of probable impact origin. Journal of Geophysical Research, Vol. 76, pp. 5387–5393.

Haines, P. W. (2005). Impact cratering and distal ejecta: the Australian record. Australian Journal of Earth Sciences, Vol. 52, pp. 481–507.

Morgan, J. W., and Wandless, G. A. (1983). Strangways crater, Northern Territory, Australia: Siderophile element enrichment and lithophile element fractionation. Journal of Geophysical Research, Supplement, Vol. 88, pp. A819–A829.

Shoemaker, E. M., and Shoemaker, C. S. (1996). The Proterozoic impact record of Australia. AGSO Journal of Australian Geology and Geophysics, Vol. 16, pp. 379–398.

Spray, J. G., Kelley, S. P., and Dence, M. R. (1999). The Strangways impact structure, Northern Territory, Australia: geological setting and laser probe ^{40}Ar-^{39}Ar geochronology. Earth and Planetary Science Letters, Vol. 172, pp. 199–211.

Zumsprekel, H., and Bischoff, L. (2005). Remote sensing and geographic information system analyses of the Strangways impact structure, Northern Territory. Australian Journal of Earth Sciences, Vol. 52, pp. 621–630.

The Strangways impact structure in September 2018 as a composite image based on TanDEM-X DEM and Copernicus Sentinel-2 (channel 2/3/4) data.

Strangways

Northern Territory

latitude = 15°12‘ S / longitude = 133°34‘ E / diameter = 40 km / age = 646 ± 42 Myr

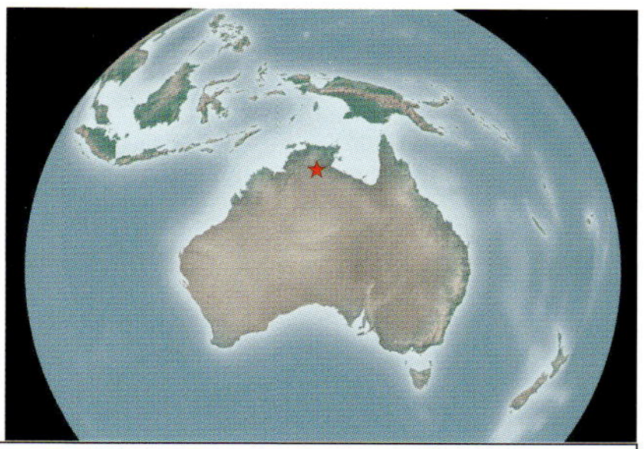

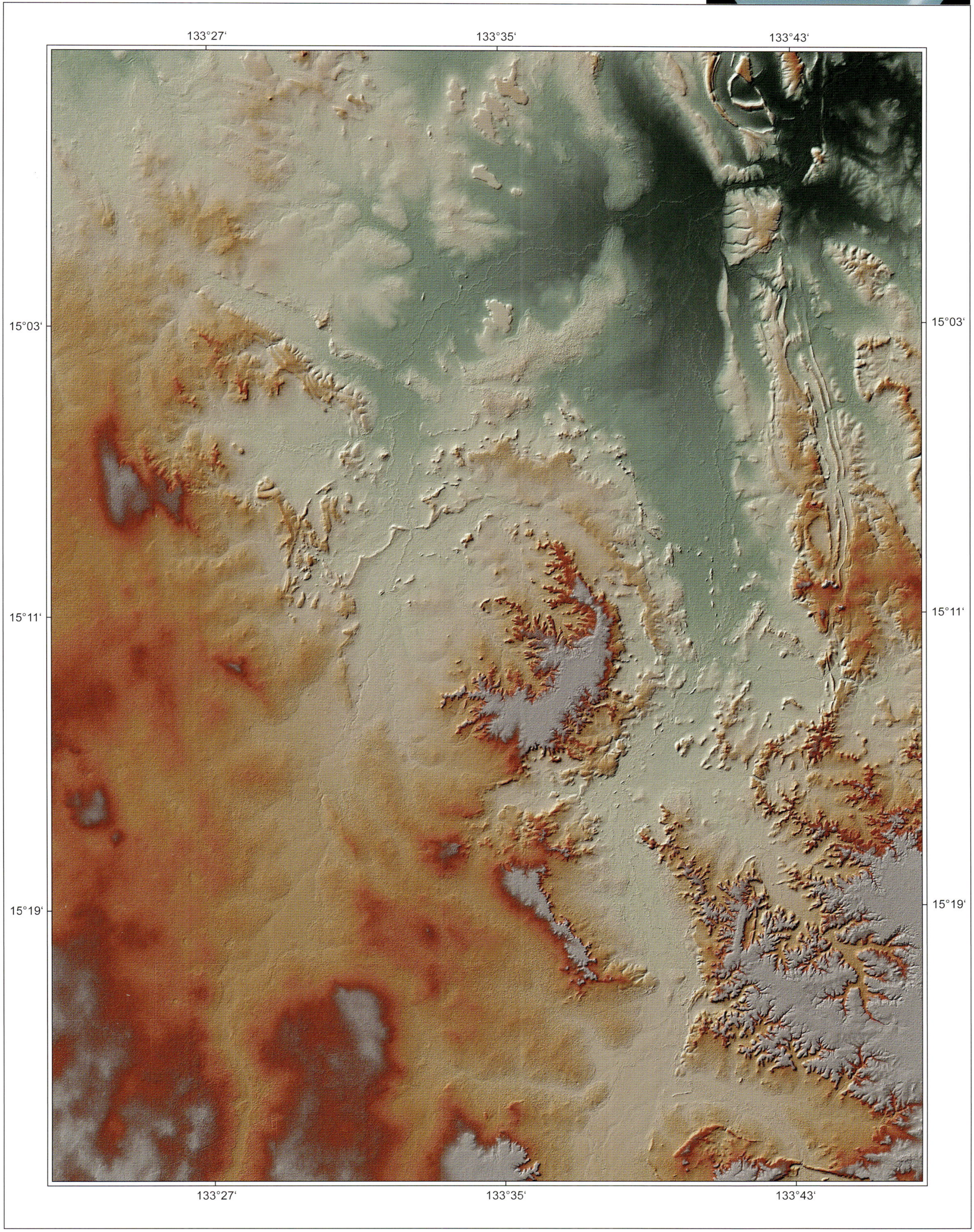

scale = 1 : 238 000 / scene = 42.8 × 57.1 km²

0 2.38 11.90 km

N

65 meter 200

Tookoonooka
Queensland

Thomas Kenkmann

Tookoonooka is a large, buried, circular impact structure in flat-lying Mesozoic strata of the Eromanga Basin, entirely without any morphological expression. The crater is buried beneath 900 m of Cretaceous and Cenozoic strata. Knowledge about the crater was derived from geophysical investigation and drilling. Seismic profiles indicate a circular disturbance consisting of a concentric arrangement of anticlines and synclines around a 22 km diameter central dome. This complex showed uplift of quartzite and phyllitic schist of probable Ordovician age by at least 450 m. The diameter of the crater structure has been reported variably between 55 and 66 km. Although the regional field was not subtracted in that study, the central uplift was shown to correlate with a strong negative Bouguer gravity anomaly of −20 mGal. Aeromagnetic data show a large positive anomaly of +100 nT that corresponds to the central uplift.

Within the area of the crater a number of exploration wells exist, and several small oilfields are situated marginal to the crater. The first borehole, Tookoonooka 1, reached the metasedimentary basement of the central uplift. Samples from this borehole showed planar deformation features (PDFs) in quartz, and intense brittle deformation. Another borehole 25 km to the southeast revealed a 115 m thick body of reworked impact melt breccia. It contains abundant devitrified melt clasts mixed with sedimentary clasts.

The impact occurred into the sedimentary Eromanga Basin. Sedimentation actually continued after the impact event. Fossil evidence from the earliest sediments deposited within the crater, directly after the event, provided biostratigraphic ages of 128 ± 5 Myr, 125 ± 1 Myr and 112–115 Myr. Because the impact occurred during sediment deposition, the ejecta blanket is preserved within the basin. The ejecta distribution extends up to 130 km from the crater structure.

Another seismic disturbance of slightly smaller size, named *Talundilly*, exists some 400 km northeast of Tookoonooka in the same basin. This structure, proposed to be possibly related to Tookoonooka, still awaits investigation.

Tookoonooka was discovered by seismic investigations during petroleum exploration. John Gorter and co-workers reported a possible impact structure in 1989, but an impact origin had already been suspected in 1982 when first regional seismic surveys had been performed. The impact confirmation was possible after the structure had been drilled and samples became available for petrographic analysis.

Location/Access

Tookoonooka lies about 400 km west of Charleville. It is accessible via the Diamantina and Cooper Developmental Roads. However, as the structure is deeply buried, no morphological expression awaits the visitors.

Selected References

Bron, K. A., and Gostin, V. (2012). The Tookoonooka marine impact horizon, Australia: Sedimentary and petrologic evidence. Meteoritics and Planetary Science, Vol. 47, pp. 296–318.

Gorter, J. D., Gostin, V. A., and Plummer, P. S. (1989). The enigmatic subsurface Tookoonooka Complex in south-west Queensland: its impact origin and implications for hydrocarbon accumulations. In: The Cooper and Eromanga Basins, Australia, O'Neil, B. J. (ed.), Petroleum Exploration Society of Australia, Adelaide, pp. 441–456.

Gostin, V. A., and Therriault, A. M. (1997). Tookoonooka, a large buried Early Cretaceous impact structure in the Eromanga Basin of southwestern Queensland, Australia. Meteoritics and Planetary Science, Vol. 32, pp. 593–599.

Haines, P. W. (2005). Impact cratering and distal ejecta: the Australian record. Australian Journal of Earth Sciences, Vol. 52, pp. 481–507.

The area of the buried Tookoonooka impact structure in July 2019 as a composite image based on TanDEM-X DEM and Copernicus Sentinel-2 (channel 2/3/4) data. Cross hairs and a dashed circle indicate position and extent of the structure.

Tookoonooka

Queensland

latitude = 27°07' S / longitude = 142°50' E / diameter = 60 km / age = 128 ± 5 Myr

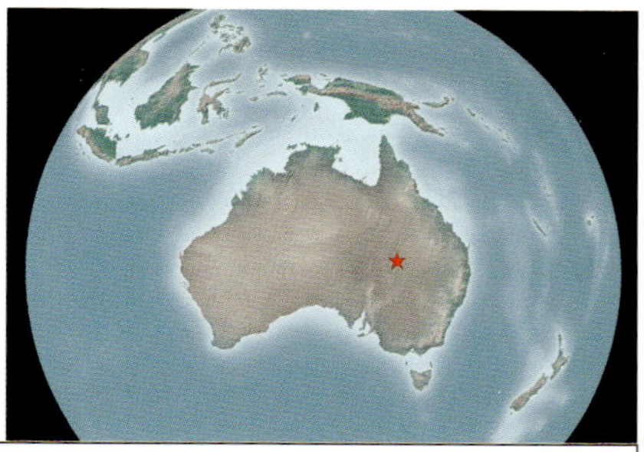

scale = 1 : 677 000 / scene = 121.9 × 162.5 km²

0 6.77 33.85 km

N

50 meter 350

Veevers
Western Australia

Thomas Kenkmann

Veevers is an exceptionally well preserved, simple bowl-shaped meteorite crater of only 70 m diameter. It is regarded as one of the least eroded terrestrial impact craters known. The crater exhibits a bowl-shaped cavity that is partly filled with eolian sand blown in from the southwest, a raised rim, and an ejecta blanket surrounding the crater. The latter has a larger extent to the northeast of the crater.

Exposed target rock is Cenozoic laterite of 10 m thickness. An upper, dark mottled zone is the marker of the original target surface. Laterite is exposed in the crater walls and dips concentrically outward, due to rim uplift during crater excavation. Dips are generally steepest low on the exposed crater walls and become shallower close to the hinge line and ejecta unit. Three distinctive outward-plunging anticlines have been recognized. The lower part of the inner slope is covered by Quaternary colluvium. The Veevers ejecta consist of blocks and nodules of laterite. The contact between the ejecta and upturned laterite is generally close to the rim crest, but lies somewhat below the crest on the western wall. Half-meter-sized blocks of laterite are exposed on the southern rim, where the ejecta are thicker and appear to be better preserved. The ejecta form ray-like lobes that commonly radiate from discrete high points on the crater rim.

A magnetic survey revealed that there is no large meteorite buried in the crater. The mass of a projectile required to form this crater was estimated at somewhere between about 100 and 1000 t, depending on the fraction of the original cosmic velocity remaining upon impact. The scarcity of meteorites, and absence of iron-rich shale balls, impactites, and melt spherules indicates that the impactor was strongly decelerated by atmospheric drag.

A cosmogenic nuclide exposure age of less than 20000 years was obtained for crater wall laterites. However, Eugene and Carolyn Shoemaker suggested a younger age of less than 4000 years on the basis of the pristine form of the ejecta blanket. The crater was discovered in 1975 by Anthony Yeates and co-workers. Later Eugene and Carolyn Shoemaker studied it in detail.

Ejecta (laterite, 5 × 5 × 3.5 cm in size) from the southeast crater rim of the Veevers impact crater (photo courtesy: Ralf-Thomas Schmitt, Museum für Naturkunde Berlin, sample from Dieter Stöffler).

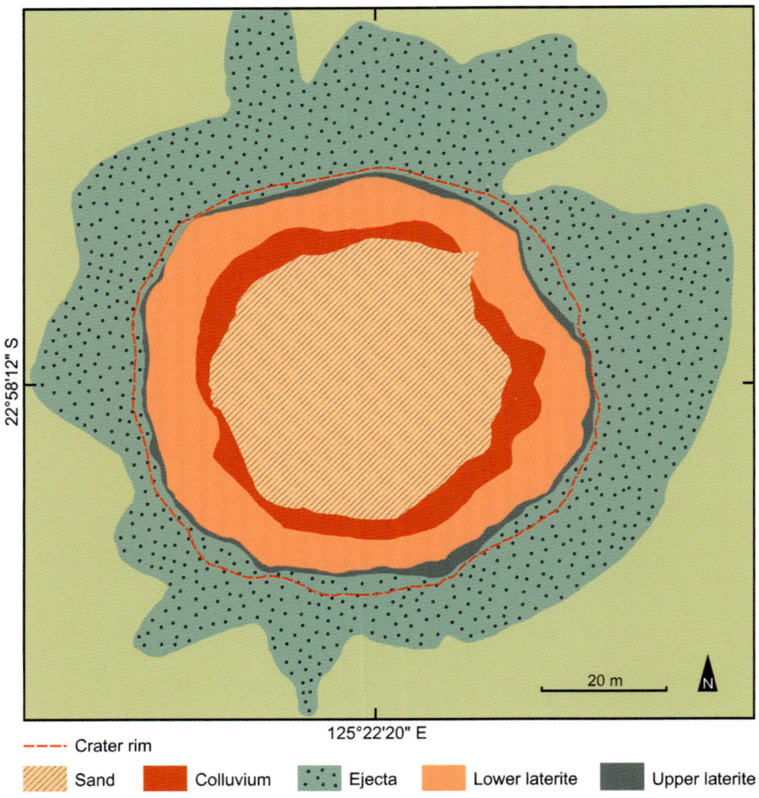

Geological map of the Veevers crater (modified after Shoemaker et al., 2005).

Location/Access

Veevers meteorite crater is located in the Great Sandy Desert of Western Australia, some 80 km northeast of the proposed **Connolly Basin** impact structure and at the same distance from the Canning Stock Route. The crater's area has native title and requires permits to visit. Access would be via desert tracks, namely the Gary Highway, which departs northward at Everard Junction from the Gunbarrel Highway. After about 80 km a 15 km long track leads eastward to the crater.

Selected References

Haines, P. W. (2005). Impact cratering and distal ejecta: the Australian record. Australian Journal of Earth Sciences, Vol. 52, pp. 481–507.

Shoemaker, E. M., Macdonald, F. A., and Shoemaker, C. S. (2005). Geology of five small Australian impact craters. Australian Journal of Earth Sciences, Vol. 52, pp. 529–544.

Yeates, A. N., Crowe, R. W. A., and Towner, R. R. (1976). The Veevers crater: a possible meteoritic feature. BMR Journal of Australian Geology and Geophysics, Vol. 1, pp. 77–78.

Veevers

Western Australia

latitude = 22°58'12" S / longitude = 125°22'20" E / diameter = 70 m / age = 12 000 ± 8000 Myr

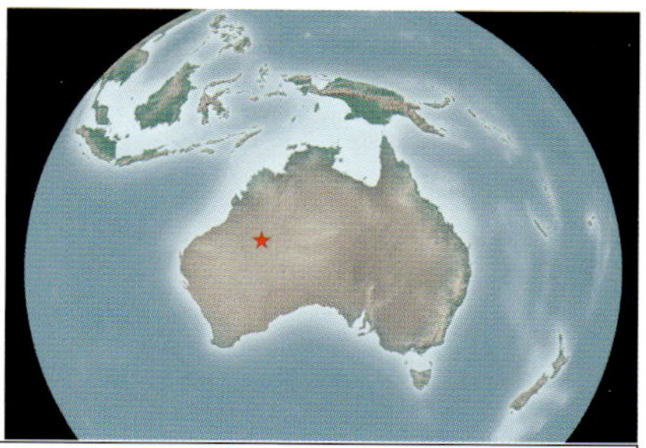

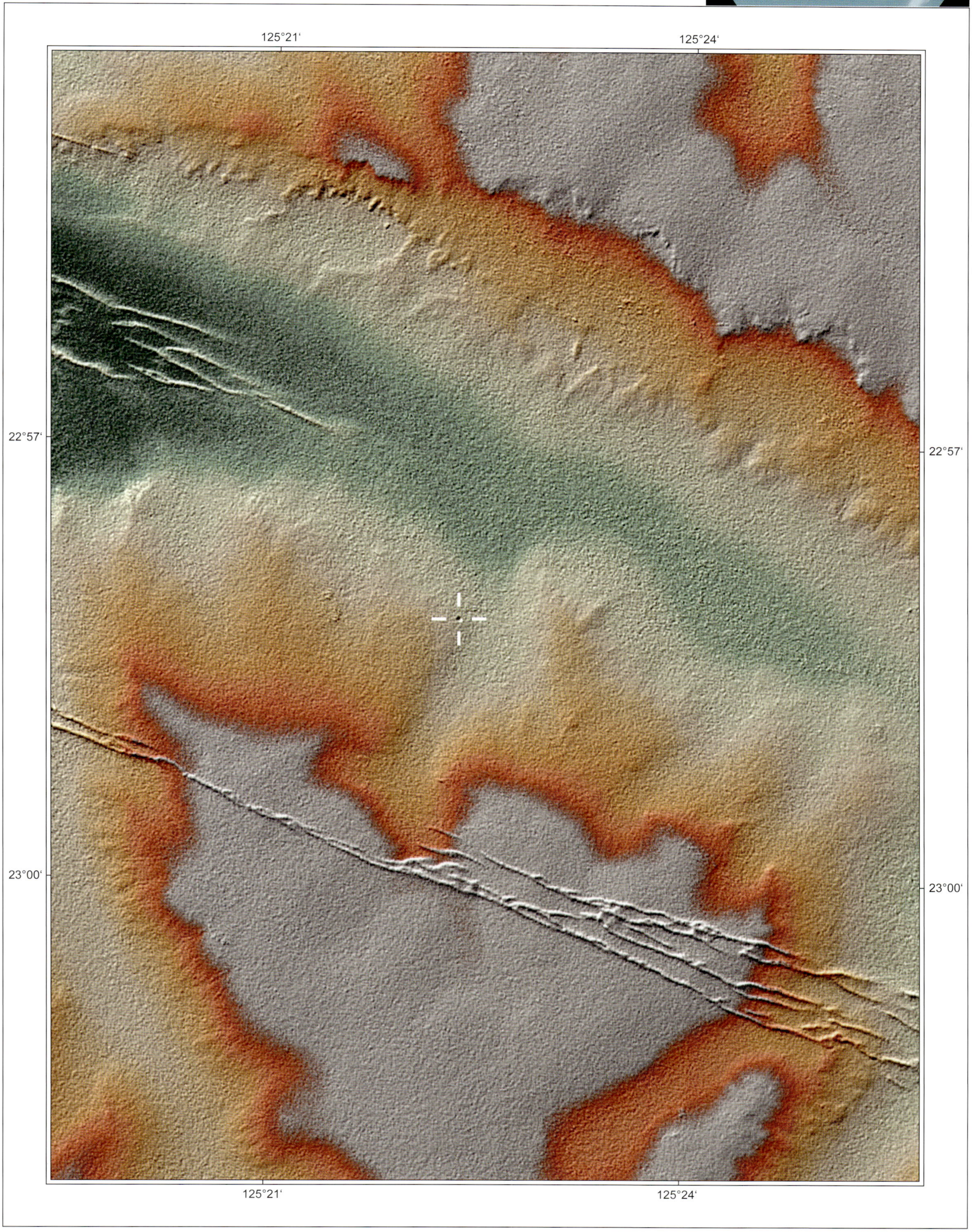

scale = 1 : 59 000 / scene = 10.7 × 14.2 km^2

0 0.59 2.95 km

N

325 meter 375

Wolfe Creek
Western Australia

Thomas Kenkmann

Wolfe Creek crater is a relatively young, small, simple bowl-shaped impact crater with an average diameter of 880 m. It was formed about 0.12 Myr ago by an iron meteorite, a medium octahedrite, chemical group IIIAB, and with a mass of more than 50000 tons. A small number of iron meteorites has been found in the vicinity of the crater, as well as larger, so-called "shale-balls" – roundish objects of iron oxide – weathering product after the original meteorite, some weighing as much as 250 kg.

The crater is well preserved and forms an obvious landmark in the plains at the edge of the Great Sandy Desert in Western Australia. It was formed in flat-lying Devonian sandstones that are overlain by a layer of Miocene laterites. Based on ^{10}Be and ^{26}Al analyses, the crater's erosion rates were calculated at 2 mm per hundred years, which yields an exposure age of some 300000 years. This would also suggest 6 m of erosion estimated for the interval since the time of impact. However, a more recent dating based on cosmogenic nuclides arrived at a considerably younger age of 120000 ± 9000 years. Erosion particularly affected the roughly 50 m high crater rim and the surrounding ejecta blanket, whilst alluvium was washed into the interior of the crater. Precipitated salt crystals cause the whitish appearance of the soil in the central depression, where a specific salt-tolerant flora developed in a circular area.

Using gravity and magnetic data the subsurface structure of the central depression could be modelled and revealed that the interpreted depth below surface to the crater floor varies from 120 m to 320 m. Wolfe Creek shows a distinct aeromagnetic anomaly with a multi-ringed pattern.

The depression is predominantly filled with fallback breccia and slumped material. Findings of shatter cones have occasionally been reported from the southwestern crater rim. Bedrocks can be studied best along the inner slope and along the crest of the rim. These rocks are tilted to various degrees in outward direction. Relatively gentle dips of less than 45° characterize the northeastern crater rim, and strongest overturning of 135° is displayed at the southwestern rim. The way the rocks have been differentially uplifted along the crater rim and their attitude that deviates from radial symmetry has allowed to unravel the direction of approach of the meteoritic projectile. Most researchers agree that the impactor came from a northeast to

Exposed beds of the overturned flap of the Wolfe Creek crater rim. Strike of bedding planes is roughly concentric with respect to the crater center (photo: T. Kenkmann).

Wolfe Creek impact crater: The flat crater depression is filled with alluvial deposits (photo: T. Kenkmann).

Wolfe Creek

Western Australia

latitude = 19°10‘18“ S / longitude = 127°47‘43“ E / diameter = 880 m / age = 0.120 ± 0.009 Myr

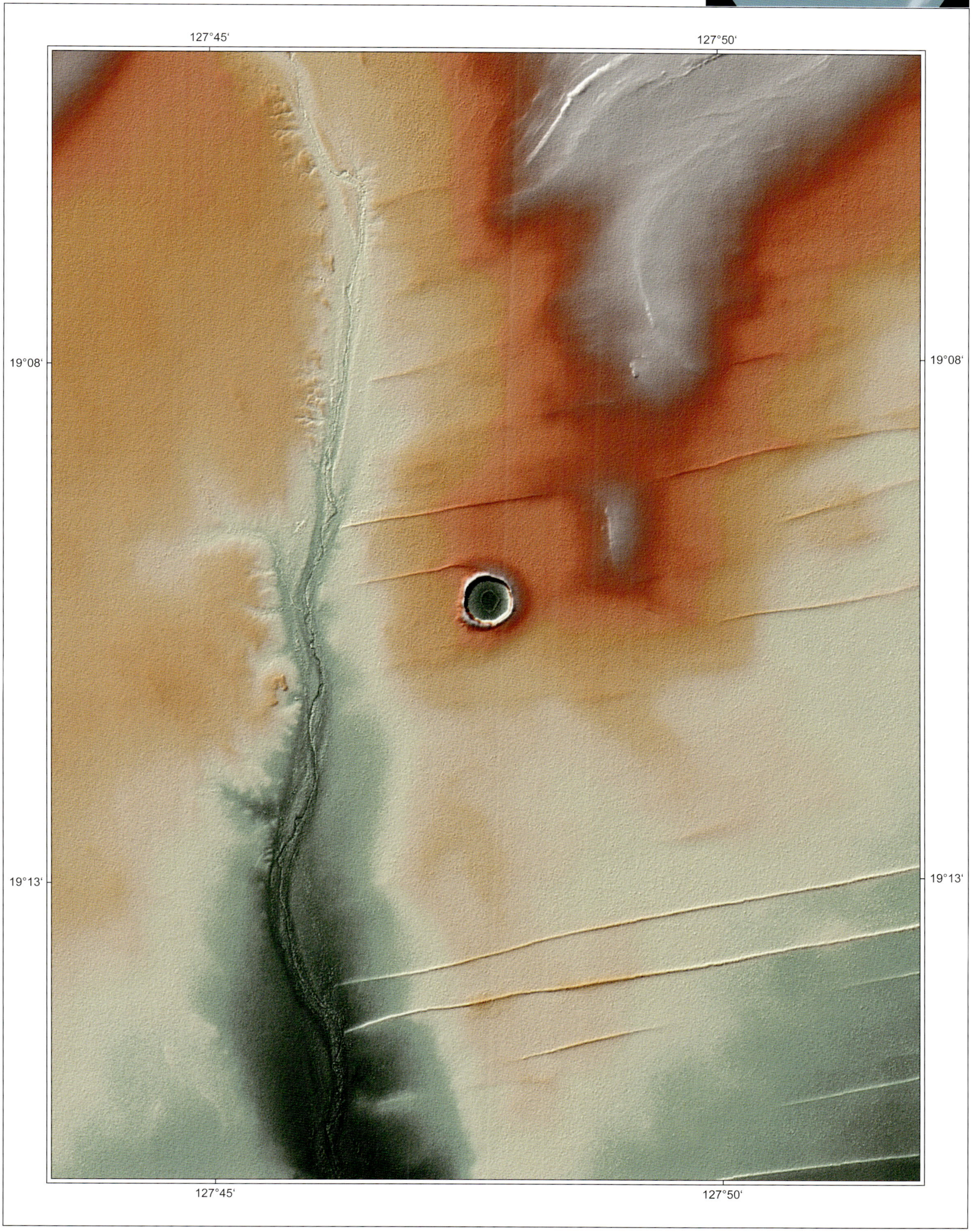

scale = 1 : 83 000 / scene = 15.0 × 20.0 km²

0 0.83 4.15 km

N

320 meter 405

easterly direction. This finding is also supported by the apparent concentration of shatter cones, the occurrence of meteoritic material, and the preferential preservation of ejecta in the southwestern sector that represents the down-range sector of the crater.

Wolfe Creek meteorite crater was noted during an aerial survey in 1947. The subsequent reconnaissance campaign resulted in a first scientific publication by F. Reeves and R. Chalmers in 1949. A nearby creek, which was named after Robert Wolfe, a prospector and storekeeper during the gold rush that established the town of Halls Creek, lent its name to the impact structure. However, for the local Darju aboriginal people the crater has been known for a long time as *Kandimalal*. The crater plays a significant role in their *Dreamtime*.

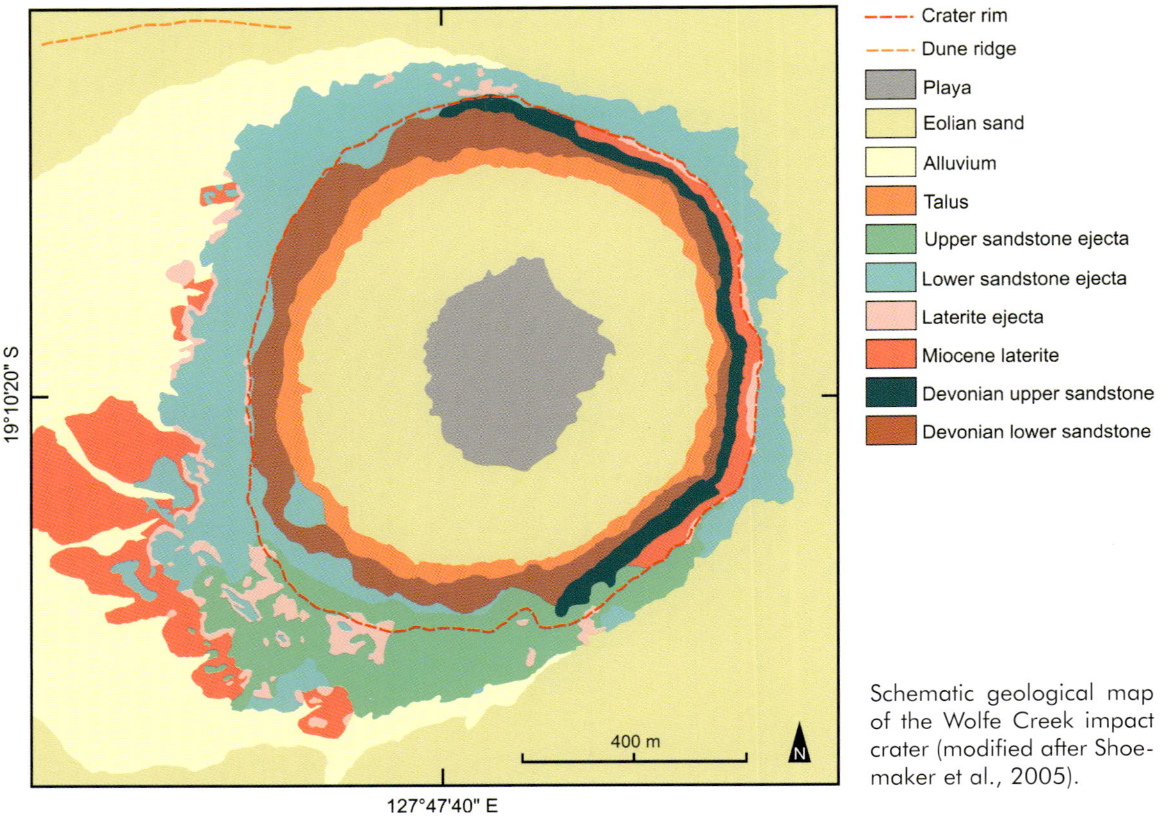

Schematic geological map of the Wolfe Creek impact crater (modified after Shoemaker et al., 2005).

Location/Access

Wolfe Creek Crater National Park lies 145 km south of the town of Halls Creek. Access is via Tanami Road that branches off the Great Northern Highway about 15 km south of Halls Creek. The gravel road is only accessible to conventional vehicles during the dry season from May to October. The trip to the crater is a two to three hour drive. All travel within the National Park is on foot. A basic camp-ground is situated close to the crater. Because the best views of Wolfe Creek are obtained from above, aerial round trips are organized from Halls Creek's regional airport.

Ejecta boulders along the crater rim. Image width in the foreground about 4 m (photo: T. Kenkmann).

Selected References

Barrows, T. T., Magee, J., Miller G., and Fifield, L. K. (2019). The age of the Wolfe Creek meteorite crater (Kandimalal), Western Australia. Meteoritics and Planetary Science, Vol. 54, pp. 2686–2697.

Hawke, P. J. (2003). Geophysical investigation of the Wolfe Creek meteorite crater. Geological Survey of Western Australia Record 2003/10, 18 pp.

O'Neill, C., Heine, C., Glikson, A. Y., and Haines, P. W. (2005). Reconstructing the Wolfe Creek meteorite impact; deep structure of the crater and effects on target rock. Australian Journal of Earth Sciences, Vol. 52, pp. 699–709.

Poelchau, M. H., and Kenkmann, T. (2008). Asymmetric signatures in simple craters as an indicator for an oblique impact direction. Meteoritics and Planetary Science, Vol. 43, pp. 2059–2072.

Reeves, F., and Chalmers, R. O. (1949). The Wolfe Creek crater. The Australian Journal of Science, Vol. 11, pp. 154–156.

Shoemaker, E. M., MacDonald, F. A., and Shoemaker, C. S. (2005). Geology of five small Australian impact craters. Australian Journal of Earth Sciences, Vol. 52, pp. 529–544.

The Wolfe Creek impact crater on September 12, 2018 as a composite image based on TanDEM-X DEM and Copernicus Sentinel-2 (channel 2/3/4) data.

Woodleigh
Western Australia

Thomas Kenkmann

The large Woodleigh impact structure is completely covered by 60–200 m of Cretaceous strata. It is located on the Gascoyne Platform of the Southern Carnarvon Basin, Western Australia. Knowledge about the crater structure is entirely based on geophysical data and two boreholes that penetrate the structure.

The size and shape of Woodleigh is only constrained by its geophysical signatures. Woodleigh has a distinctive, ringed gravity signature, with a central positive anomaly of 8 km diameter corresponding to the stratigraphic uplift expressed by a central uplift structure. This is surrounded by an annular low of 20 km diameter, which – in turn – is surrounded by a less pronounced ring-like gravity high. Data given for the crater diameter depend on definition and interpretation of geophysical data and range from 40 to 120 km, with 60 km representing the most likely diameter.

Drilling of core Woodleigh 1 was carried out in the center of the structure. It penetrated about 160 m of uplifted, shock-metamorphosed granitoids of the basement that occurs beneath 170 m of undisturbed Cretaceous sediments. Woodleigh 2A, situated 15 km west of Woodleigh 1, penetrated undisturbed Cretaceous and Jurassic sediments, before intersecting a conglomerate at a depth of 521–587 m, followed by 14 m of breccia and Silurian sediments that are now considered target rock. Abundant shock features were encountered in the uplifted Proterozoic rocks of the central uplift, including planar deformation features (PDFs) in quartz that indicated shock pressures of at least 20 GPa. In addition, diaplectic glass and partial isotropisation of quartz were documented. The lack of any impact melt-bearing breccias in the core from the central region suggests that the Woodleigh structure was exposed and eroded to some degree before subsequent sedimentation took place. The centrally drilled granitoid was affected by impact-induced hydrothermal alteration.

The age of Woodleigh is open to interpretation. Stratigraphic constraints suggest an age between Devonian and Early Jurassic and paleomagnetic evidence places the impact between 265 and 318 Myr. Absolute dating of a bona fide impact-generated phase is still pending. Around the turn of the 20th to the 21st century, Woodleigh obtained a degree of notoriety, as it was proposed as a likely, very large smoking-gun for a Late Devonian mass extinction event. The claims that the structure could be as large as 120 km in diameter and interpretations of age data for illite that would indicate a mass extinction related age were strongly rejected.

The buried crater structure was identified from geophysical data by Robert Iasky and co-workers. In 1999 the Geological Survey of Western Australia drilled two boreholes into the structure, which ultimately provided the petrographic proof of the impact.

Location/Access

Woodleigh is not exposed and has no morphological expression. Its center is located 45 km east of Shark Bay. The Northwest Coastal Highway passes close to the periphery of the structure.

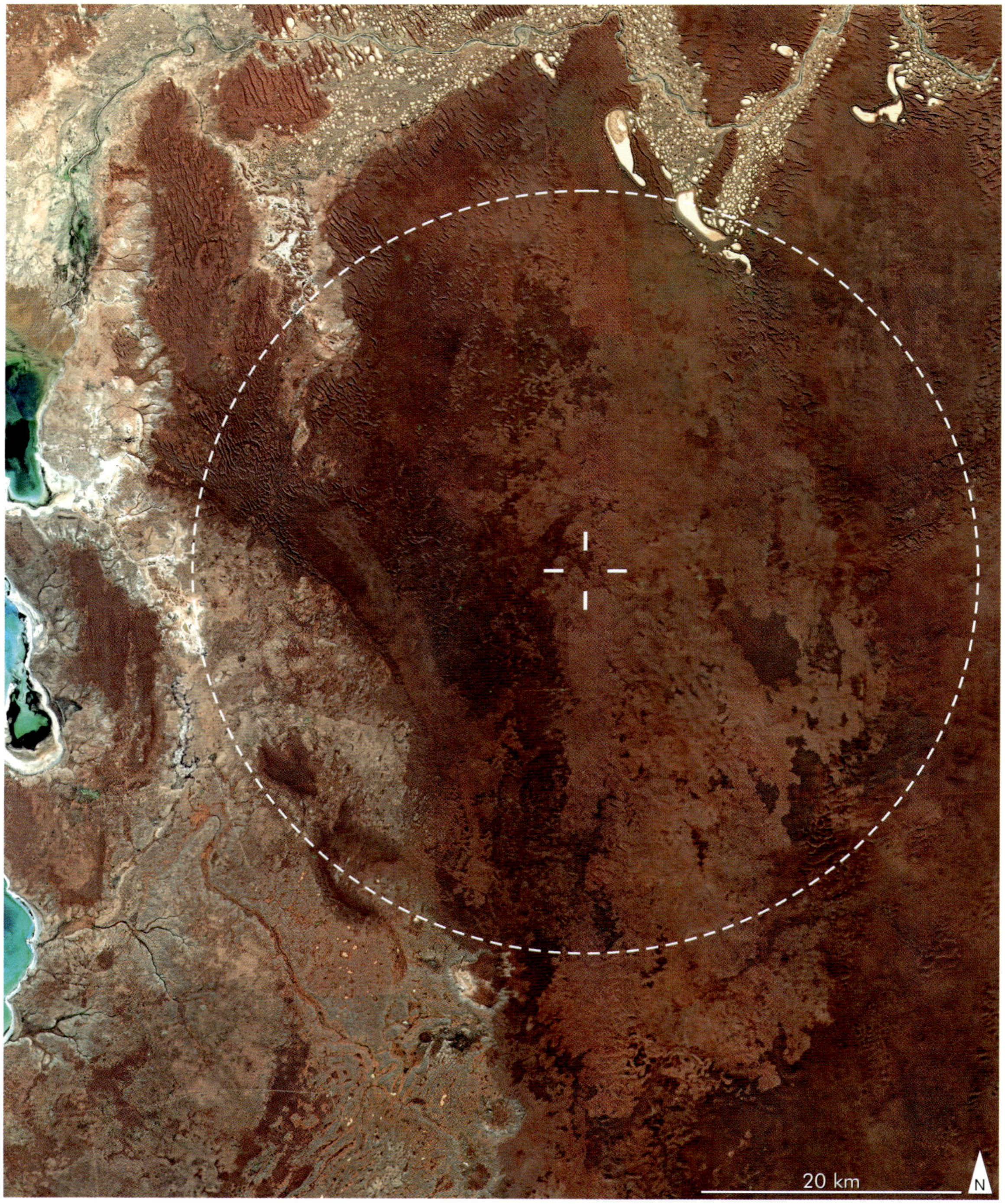

The area of the Woodleigh impact structure on June 29, 2019 as a composite image based on TanDEM-X DEM and Copernicus Sentinel-2 (channel 2/3/4) data. Cross hairs and a dashed circle indicate position and extent of the buried crater structure.

Woodleigh

Western Australia

latitude = 26°03‘ S / longitude = 114°40‘ E / diameter = 60 km / age = 265 - 381 Myr

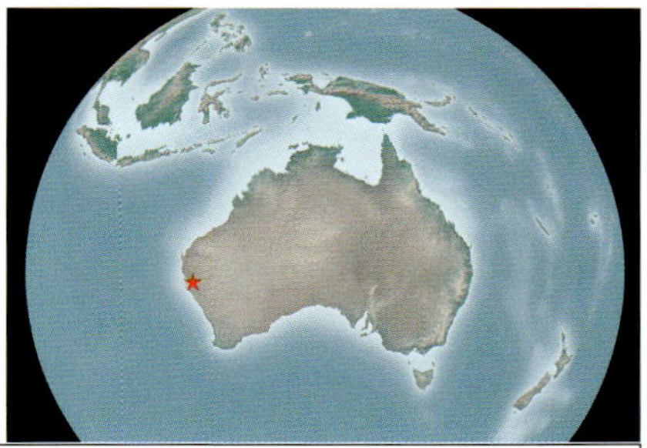

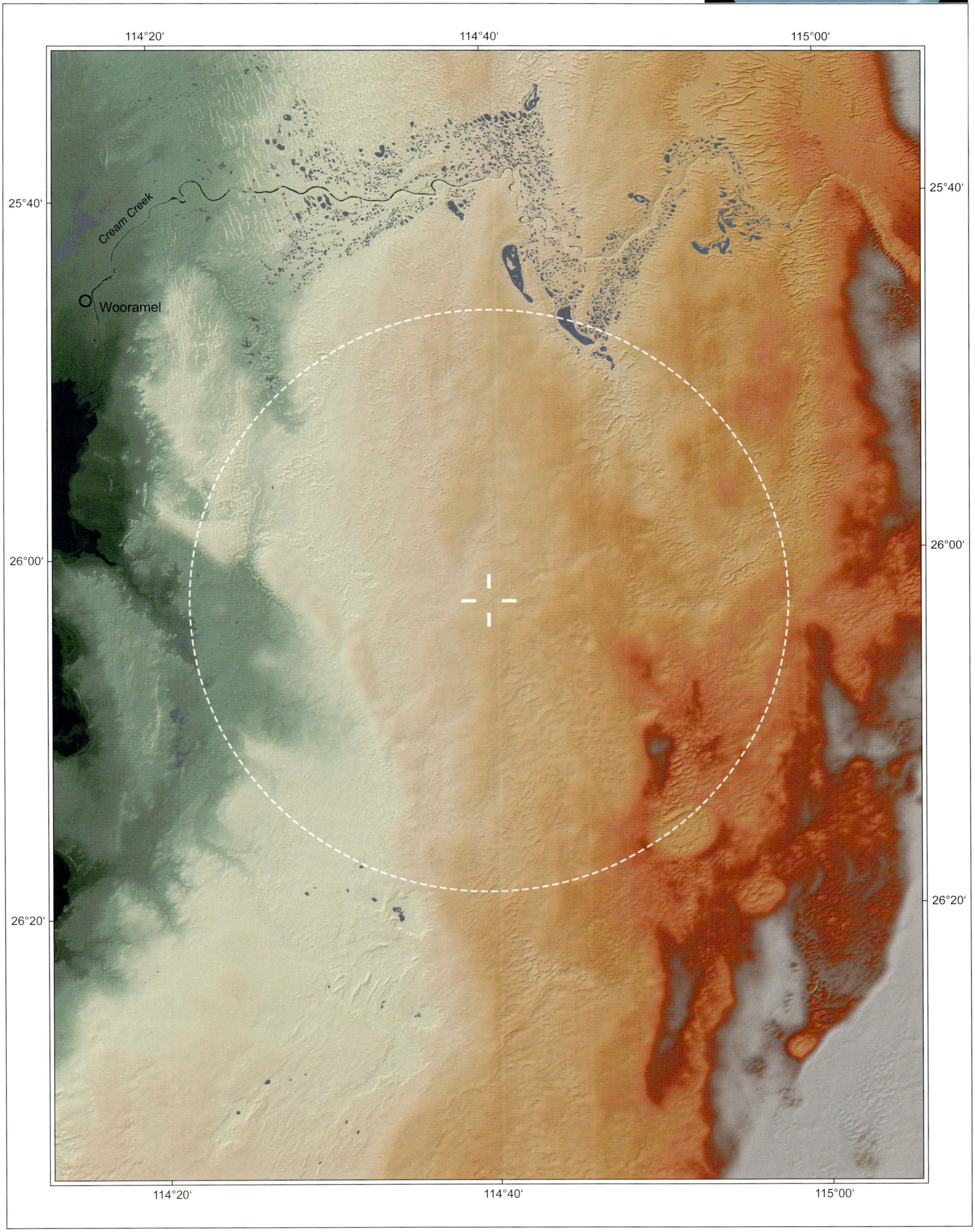

scale = 1 : 484 000 / scene = 87.2 × 116.3 km^2

0 4.84 24.20 km

N

0 meter 235

Selected References

Haines, P. W. (2005). Impact cratering and distal ejecta: the Australian record. Australian Journal of Earth Sciences, Vol. 52, pp. 481–507.

Iasky, R., Mory, A. J., and Blundell, K. A. (2001). The geophysical interpretation of the Woodleigh impact structure, Southern Carnarvon Basin, Western Australia. Western Australia Geological Survey, Report 79, 41 pp.

Mory, A. J., Iasky, R. P., Glikson, A. Y., and Pirajno, F. (2000). Woodleigh, Carnarvon Basin, Western Australia: a new 120 km diameter impact structure. Earth and Planetary Science Letters, Vol. 177, pp. 119–128.

Reimold, W. U., Koeberl, C., Hough, R. M., Mcdonald, I., Bevan, A., Amare, K., and French, B. M. (2003). Woodleigh impact structure, Australia: shock petrography and geochemical studies. Meteoritics and Planetary Science, Vol. 38, pp. 1109–1130

Uysal, I. T., Golding, S. D., Glikson, A. Y., and Mory, A. J. (2001). K-Ar evidence from illitic clays of a Late Devonian age for the 120 km diameter Woodleigh impact structure, Southern Carnarvon Basin, Western Australia. Earth and Planetary Science Letters, Vol. 192, pp. 281–289.

Yallalie

Western Australia

Thomas Kenkmann

Yallalie is a recently confirmed, not exposed impact structure of circa 12 km diameter. It is located about 200 km north of Perth in the Paleozoic-Mesozoic Perth Basin. At surface, Yallalie is expressed as a broad, circular depression that is the focus of drainage that flows centripetally into the structure. Yallalie is buried below 100–300 m of undisturbed sediment, which includes thin layers of lacustrine sediment, laterite, and recent eolian sand.

Target rocks are poorly sorted, interbedded, marine and fluvial sandstone, siltstone, shale, chalk, and glauconitic sandstone of Jurassic and Cretaceous age. The morphology of the buried impact structure can only be revealed by geophysical means. The structure is characterized by a central, 3–4 km wide, uplifted region. Reflection seismic profiles show a sharp contact between the Yallalie structure and the surrounding undisturbed sedimentary strata in the form of normal listric faults at 4.2 km radial distance and, less prominent, at 6 km, from the center. These faults likely formed during crater modification. Down to a depth of about 2 km the seismic profiles also illustrate a basin feature with chaotic reflectors. Airborne magnetic surveying displayed a conspicuous 12 km anomaly in otherwise non-magnetized surroundings. The anomaly pattern over the structure consists of several sub-circular, concentric, positive magnetic anomalies that are centered on a single magnetic high near the middle of the Yallalie structure. Yallalie has a positive gravity anomaly of +3 mGal corresponding to the width of the central uplift.

While the crater itself is buried and provides no exposures, there is an exposure of breccia 4 km to the west of the crater structure. This so-called "Mungedar breccia" represents ejecta or reworked ejecta from the crater. The breccia sits unconformably on Upper Cretaceous siliciclastic rocks. The polymict breccia contains up to meter-sized blocks of sedimentary rocks, and has a maximum thickness of about 30 m. In some of the clasts quartz and zircon with planar microdeformations were found. Quartz exhibits planar deformation features (PDFs) with different crystallographic orientations, planar fractures (PFs), and ballen texture.

The age of the impact event could be stratigraphically constrained to have occurred in the interval from 83.6 to 89.8 Myr based on the occurrence of Coniacian clasts in the breccia and undisturbed overlying Santonian to Campanian sedimentary rocks.

In 1990 Michael Dentith and colleagues first recognized Yallalie as a buried, circular structure of about 12 km diameter during oil exploration. Subsequent magnetic surveying by groups lead by Philip Hawke and Alex Bevan revealed its underground morphology. The final proof of Yallalie's impact origin came in 2017 when geoscientists from Curtin University in Perth performed microstructural investigations of the "Mungedar breccia".

Location/Access

Coming from Perth and heading north to the Yallalie impact structure takes about 200 km. The route on Brand Highway 1 runs parallel to the Darling Scarp that terminates the Perth basin to the east. At Badgingarra a right turn takes the visitor onto the Koonah road to Moora. After about 11 km, the road passes the Mungedar locality with breccia occurrence. Then further to the east appears the gentle depression of Yallalie.

Exposure of the so-called "Mungedar Breccia", about 4 km west of the Yallalie depression. This allochthonous breccia was likely deposited as ejecta or reworked ejecta from the crater (photo: T. Kenkmann).

Yallalie

Western Australia

latitude = 30°26‘40“ S / longitude = 115°47‘16“ E / diameter = 12 km / age = 83.6 - 89.8 Myr

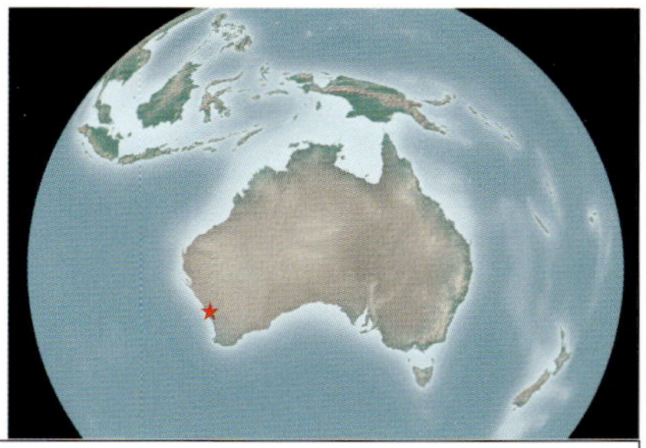

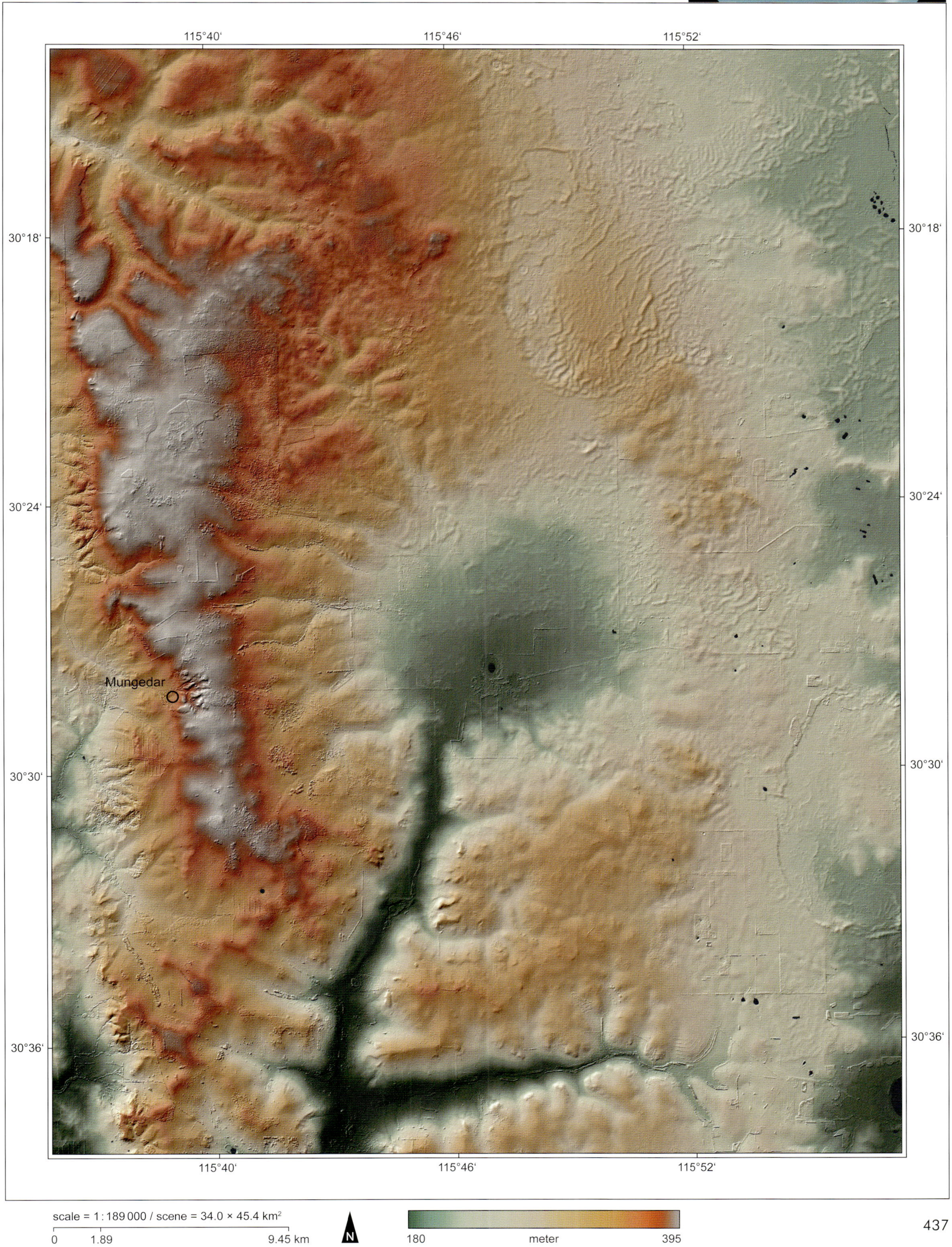

The site of the Yallalie impact structure shows a gentle circular depression, although the crater itself is not exposed (photo: T. Kenkmann).

Selected References

Bevan, A. (2012b). Yallalie impact structure. In: Western Australian Impact Craters Field Excursion, 75th Annual Meeting of the Meteoritical Society, Cairns, Queensland, Australia, pp. 12–23.

Cox, M. A., Cavosie, A. J., Ferrière, L., Timms, N. E., Bland, P. A., Miljkovic, K., Erickson, T. M., and Hess, B. (2019). Shocked quartz in polymict impact breccia from the Upper Cretaceous Yallalie impact structure in Western Australia. Meteoritics and Planetary Science, Vol. 54, pp. 621–637.

Dentith, M. C., Bevan, A. W. R., Backhouse, J., Featherstone, W. E., and Koeberl, C. (1999). Yallalie: a buried structure of possible impact origin in the Perth Basin, Western Australia. Geological Magazine, Vol. 136, pp. 619–632.

Hawke, P. J., Buckingham, A. J., and Dentith, M. C. (2006). Modelling source depth and possible origin of magnetic anomalies associated with the Yallalie impact structure, Perth Basin, Western Australia. Exploration Geophysics, Vol. 37, pp. 191–196.

Yarrabubba

Western Australia

Thomas Kenkmann

The Yarrabubba impact structure was formed in the Archean granite-greenstone terrain of the Yilgarn Craton in Western Australia. It does not have an obvious circular topographic or geological expression. Outcrops are rare and weathering of granitic rocks has produced a thick silcrete crust in this area. Erosion of the hard silcrete cap and underlying softer kaolinitinized granite led to the formation of cliffs and mesas. The original size of the structure is poorly constrained but impact-induced deformation is recognized within a minimum radius of 10 km.

Country rock in the area of the structure is primarily granite, the Yarrabubba Granite, which is cut by mafic dikes. The center of the structure at the Barlangi Rock location is marked by outcrops of a granophyre. At a distance of 15–20 km occur the Meekatharra and Barrambie greenstone belts that are composed of metamorphosed mafic, ultramafic and felsic volcanic rocks, related intrusions, and various sedimentary rocks.

Evidence for impact includes shatter cones in the granites along with planar deformation features (PDFs) in quartz grains of these rocks. The felsic granophyre, called the "Barlangi Rock", exposed in the center of the structure shows quench textures with a fine-grained matrix and elongated mineral laths. Shock features have neither been observed in the granophyre nor in xenoliths embedded in this rock. The granites are dissected by many faults with visible offsets. Fault planes contain fine-grained mylonites. Some textures along faults were interpreted as pseudotachylitic breccia. Near the contact to the granite, the granophyre is flow banded, intrudes into, and inter-fingers with, the pseudotachylitic material.

The most remarkable geophysical feature of the crater structure is provided by a 25 × 15 km zone of low aeromagnetic relief that seems to represent demagnetization of the regional granite, likely as a result of the impact. Estimates of the size of Yarrabubba vary strongly and range from 25 to 70 km.

Age determinations were not conclusive until recently. Zircon grains from the Yarrabubba Granite and the central granophyre have given Archean ages of 2600–2700 Myr and provide a maximum age, while a ^{40}Ar-^{39}Ar age of 1134 ± 26 Myr from pseudotachylitic veins in the granite can be interpreted as the lower age limit. New in situ U-Pb isotope data from zircon and monazite that recrystallized during impact-driven shock metamorphism yielded upper intercept ages of 2246 ± 17 Ma and 2229 ± 5 Ma, respectively, for the impact event. Such an age makes Yarrabubba the oldest impact structure recognized on Earth so far.

The discovery of the impact structure dates back to 1979 when Steve Williams and Will Libby from the Geological Survey of Western Australia recognized shock microdeformation in a single thin section of granite collected during a regional mapping campaign. It took, however, almost 25 years until the impact origin was documented in international literature when Francis Macdonald and co-workers published the results of their study of Yarrabubba.

Location/Access

Yarrabubba is located about 70 km southeast of Meekatharra. Access on dirt roads requires well-equipped four-wheel-drive vehicles. Outcrops are patchy, as much of the area is covered by superficial alluvium or playa lake deposits. Good granophyre outcrops exist at Barlangi. The granite occasionally contains shatter cones and veins of pseudotachylitic breccia.

Selected References

Bevan, A. (2012a). Western Australian Impact Craters Field Excursion 20–29 August 2012. 75th Annual Meeting of the Meteoritical Society Cairns, Queensland, Australia, 55 pp.

Erickson, T. M., Kirkland, C. L., Timms, N. E., Cavosie, A. J., and Davison, T. M. (2020). Precise radiometric age establishes Yarrabubba, Western Australia, as Earth's oldest recognised meteorite impact structure. Nature Communications 11, Article #300, pp. 1–8.

Macdonald, F. A., Bunting, J. A., and Cina, S. E. (2003). Yarrabubba – a large, deeply eroded impact structure in the Yilgarn Craton, Western Australia. Earth and Planetary Science Letters, Vol. 213, pp. 235–247.

Yarrabubba
Western Australia

latitude = 27°11‘ S / longitude = 118°50‘ E / diameter = 30 km / age = 2246 ± 17 Myr

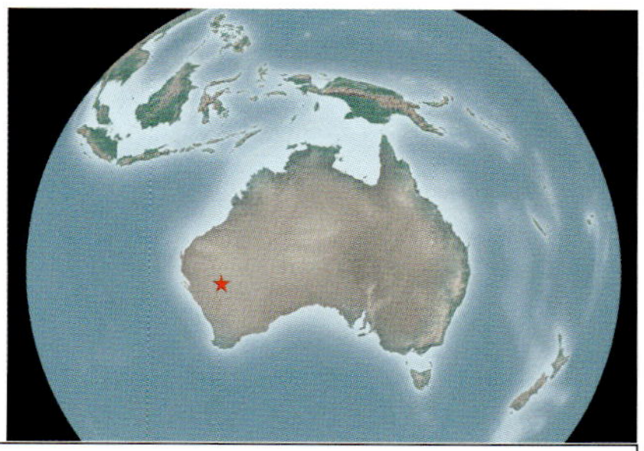

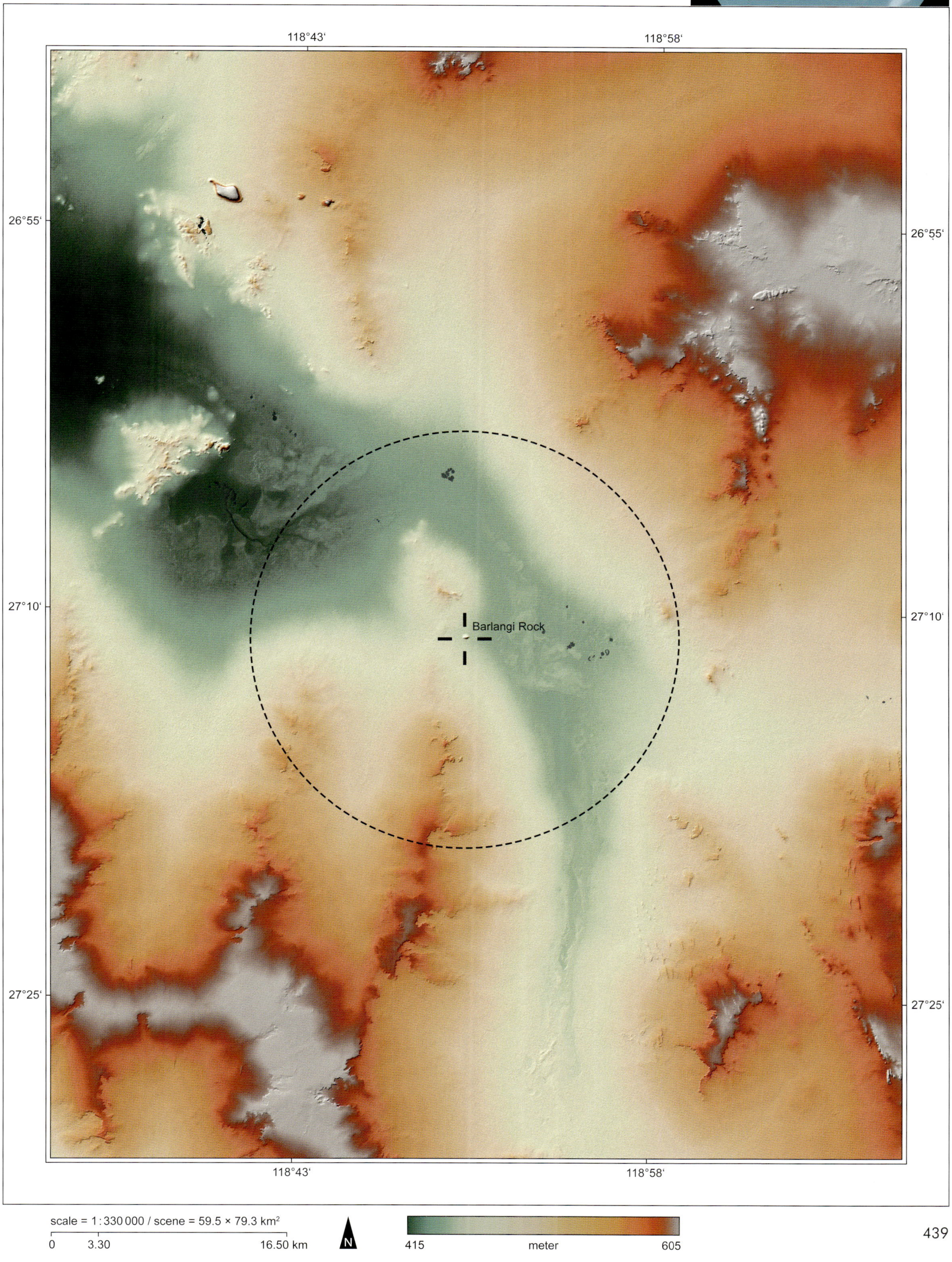

Impact Structures – Further Confirmation Required

Connolly Basin
Western Australia

Thomas Kenkmann

Connolly Basin is a gentle circular depression with a diameter of 9 km, which is situated in the Gibson Desert in central Western Australia. The structure has a topographic rim of 25–30 m height, with a radial centripetal drainage system discharging into the interior, and with an outlet to the southwest.

Within the structure the surface is largely covered with Quaternary materials, and bedrock is exposed only in scattered outcrops in a ring about 3 km across, as well as in the central uplift. The oldest exposed rocks in the structure are of Permian age and crop out in the center of the structure. They include silt- and sandstones with steep attitudes and partly chaotic strike. At about 1 km from the center, highly fractured and sheared sandstone forms an annular ring. These strata are in upright or overturned position and are thought to represent the steep sides of the central uplift. Like in other complex impact structures, the periphery of the central uplift shows tight folds, the fold axes of which plunge steeply. In contrast, the rim of the basin is formed by gently inward dipping Cretaceous sand- and siltstones. Due to erosion, no ejecta have been preserved around the structure. Breccia was locally found along the annular moat. The structure is surrounded and largely covered by Tertiary lateritic soils and Quaternary alluvium. Scattered outcrops of sediment thought to be of post-impact age in the basin are tentatively regarded to be of Upper Cretaceous or Early Tertiary age. This suggests that the impact event would have occurred around this time and that the structure is not too deeply eroded.

A seismic survey revealed disturbed reflectors in the upper kilometer below the basin structure, but it also indicated horizontal layering. Two gravity profiles across the Connolly Basin showed a weak but distinct central high with a Bouguer anomaly of +2 mGal. This anomaly is surrounded by a narrow annular gravity low, which, in turn, is surrounded by an annular gravity high of +0.3 to +0.5 mGal at a distance of about 2 km from the center. This positive anomaly dies out at 4 km distance. The gravity data were modelled as a complex impact crater with central uplift surrounded by low-density breccia and overlying higher density crater fill sandstone.

Connolly Basin was originally interpreted as the surface expression of a diapir, but the seismic profile, which indicated that strong horizontal reflectors are present at depths corresponding to 0.7 to 0.8 seconds two-way travel time beneath the margin and interior of the structure, pointed to an impact origin. Apart from the structural characteristics that perfectly match those of an eroded complex impact structure, the sandstones display intense cataclasis. They are pervasively crushed and partly pulverized to a fine-grained matrix. However, to our knowledge no impact diagnostic shock features have been presented until now. Connolly Basin was first suggested and named by Eugene and Carolyn Shoemaker in 1985.

Location/Access

Connolly Basin is located some 120 km south of the Canning Stock Road in the Australian outback. There is no infrastructure that would support access.

Selected References

Haines, P. W. (2005). Impact cratering and distal ejecta: the Australian record. Australian Journal of Earth Sciences, Vol. 52, pp. 481–507.

Shoemaker, E. M., and Shoemaker, C. S. (1986). Connolly Basin: a probable eroded impact crater in Western Australia. 17th Lunar and Planetary Science Conference, Lunar and Planetary Institute, Houston, Texas, pp. 797–798.

Shoemaker, E. M., and Shoemaker, C. S. (1989). Geology of the Connolly Basin impact structure, Western Australia. 20th Lunar and Planetary Science Conference, Lunar and Planetary Institute, Houston, Texas, pp. 1008–1009.

Shoemaker, E. M., Shoemaker, C. S., and Plescia, J. B. (1989). Gravity investigation of the Connolly Basin impact structure; Western Australia. 20th Lunar and Planetary Science Conference, Lunar and Planetary Institute, Houston, Texas, pp. 1010–1011.

The Connolly Basin structure on July 17, 2017 as a composite image based on TanDEM-X DEM and Copernicus Sentinel-2 (channel 2/3/4) data.

Connolly Basin

Western Australia

latitude = 23°32'03" S / longitude = 124°45'38" E / diameter = 9 km / age = 55 - 75 Myr

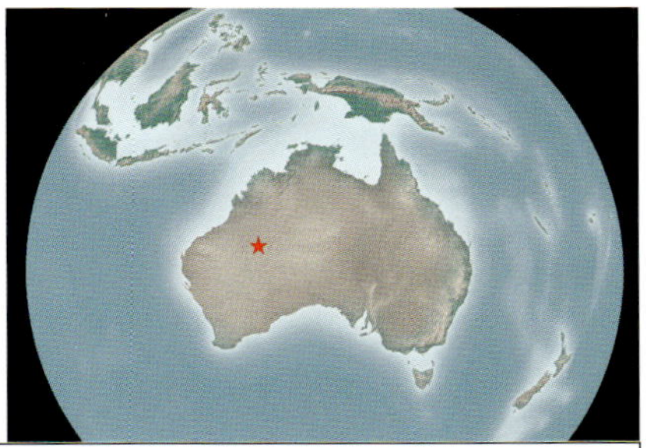

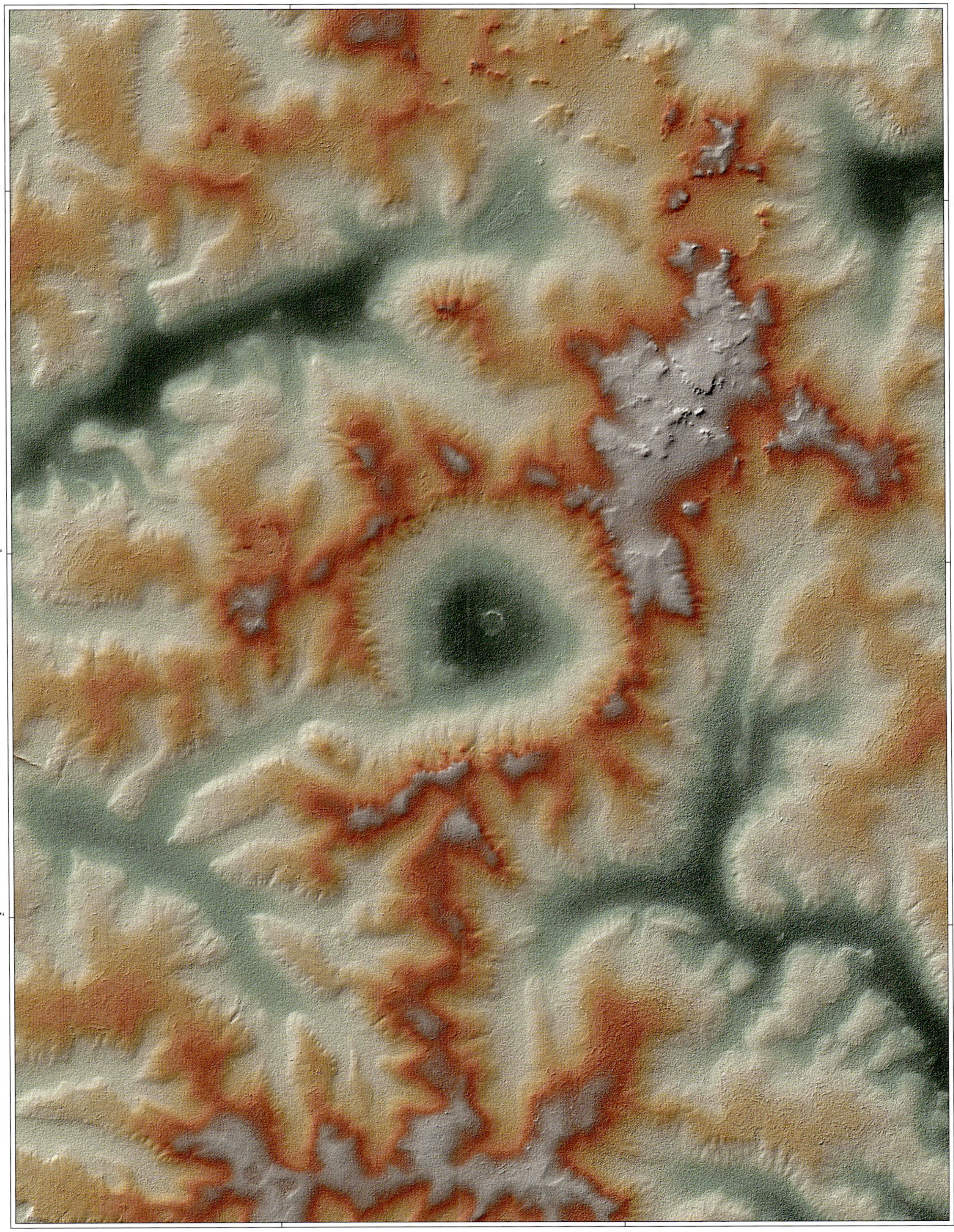

scale = 1 : 154 000 / scene = 27.7 × 36.9 km²

0 1.54 7.70 km

N

375 meter 430

Crawford and Flaxman
South Australia

Thomas Kenkmann

Crawford and Flaxman near Adelaide in South Australia are two sites where shock features were found. However, neither a circular topographic feature, nor structures suggestive of a conventional impact structure are present. Both sites are usually listed together, because it remains uncertain whether they represent separate, perhaps contemporaneous, impact craters or whether they are part of a single, much larger structure.

Flaxman consists of a north-south elongated zone of deformation that measures about 10 × 1 km in extent. Crawford, situated 12 km south of Flaxman, is shaped like a teardrop of 8.5 × 3.5 km extent. The structures were interpreted to belong to one or representing two elongated crater structures formed by highly oblique impacts from the south-southwest, with the projectiles coming in at an angle of incidence of less than 10°. However, it remains debatable whether the present morphology of the area is related to the impact event at all. At both sites outcrops are rare.

Target rocks for both craters are metamorphic and igneous rocks of the Delamerian Orogen. Evidence for shock metamorphism includes several sets of planar deformation features (PDFs) in quartz decorated with fluid inclusions, and possibly diaplectic feldspar glass (maskelynite), in shocked igneous rocks. Breccia veins with a black aphanitic matrix were tentatively interpreted as pseudotachylites, but their true nature has remained uncertain. Along with intense brittle deformation, these veins are, however, indications for stresses that were rapidly applied to these rocks. The occurrences of allochthonous breccias were interpreted in terms of being remnants of proximal ejecta. Attempts to date the breccia veins radiometrically by means of Ar-Ar dating remained unsuccessful, as the early Paleozoic metamorphic age of the target rocks was not reset.

Peter Haines, who discovered the structures in 1999, suggested a Late Eocene to Early Oligocene age for the impact event(s) based on stratigraphic relationships. However, if the crater structure(s) has/have been tectonised after impact, Flaxman and Crawford could represent tectonic slivers of the same structure, and in this case also an Ordovician or older age could be possible. So far, the evidence for impact rests solely on shock metamorphic effects. In accordance with the discoverer, we decided to list the structures in the chapter of craters that demand more investigation.

Location/Access

Both localities are situated about 50 km northeast of Adelaide in South Australia in a hummocky landscape of agricultural use but with good infrastructure. The villages of Mount Pleasant and Springton are situated in close proximity southeast and east to the structures, respectively, and can be reached via highway B10 from Adelaide.

Selected References

Haines, P. W. (2005). Impact cratering and distal ejecta: the Australian record. Australian Journal of Earth Sciences, Vol. 52, pp. 481–507.

Haines, P. W., Therriault, A. M., and Kelly S. P. (1999). Evidence for mid-Cenozoic(?) low-angle multiple impacts in South Australia. Meteoritics and Planetary Sciences, Vol. 34 (Suppl.), pp. A49–A50.

The area of the buried Crawford and Flaxman structures on August 5, 2019 as a composite image based on TanDEM-X DEM and Copernicus Sentinel-2 (channel 2/3/4) data. Cross hairs and a dashed circle indicate position and extent of each crater structure.

Crawford / Flaxman

South Australia

Crawford: latitude = 34°43‘ S / longitude = 139°02‘ E / diameter = 8.5 km / age = 32 - 38 Myr
Flaxman: latitude = 34°37‘ S / longitude = 139°04‘ E / diameter = 10 km / age = 32 - 38 Myr

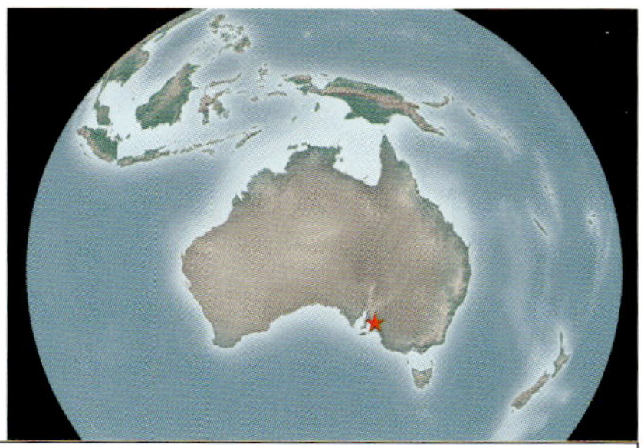

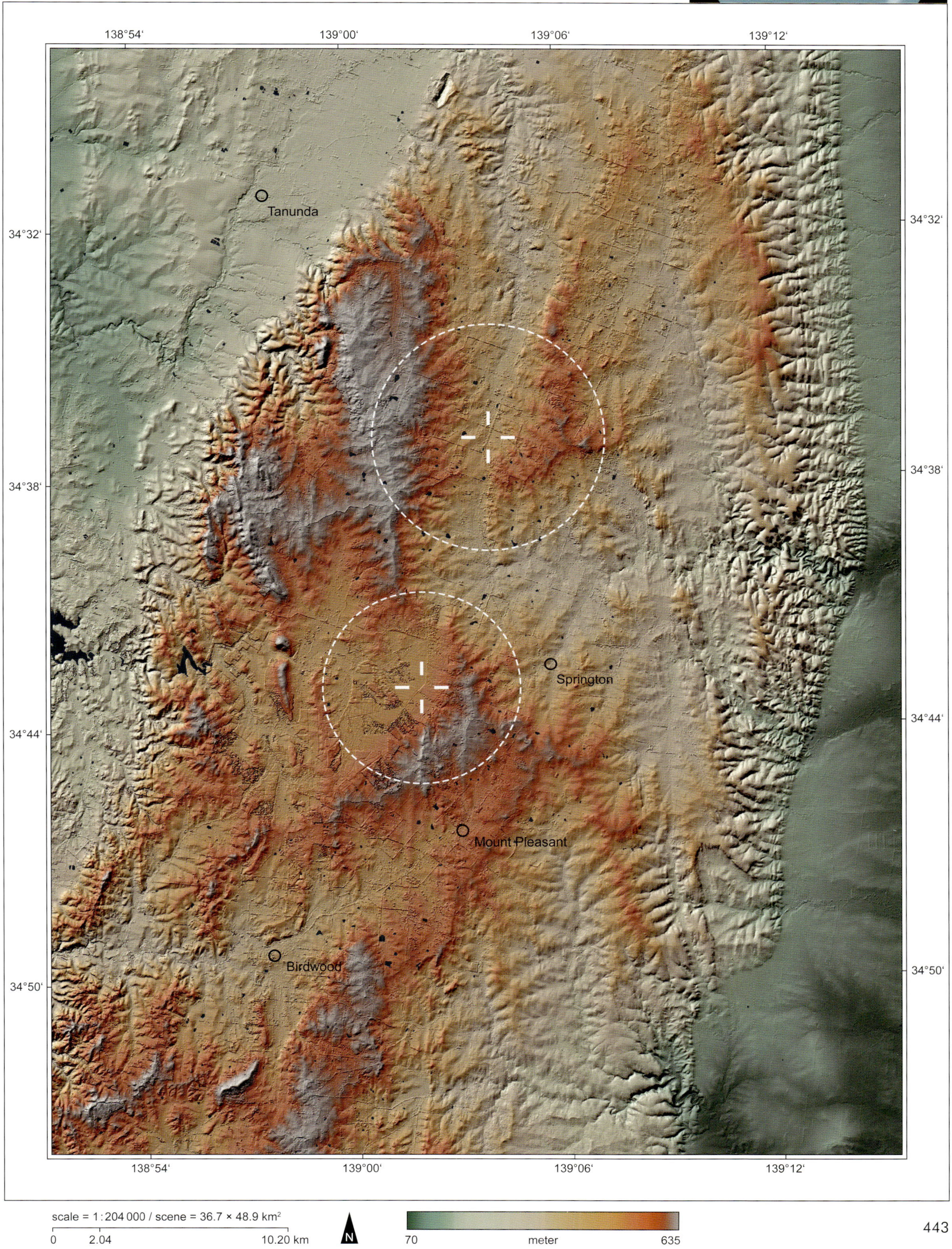

Mount Toondina
South Australia

Thomas Kenkmann

Mount Toondina is considered one of the smallest exposed, complex impact structures on earth. Its surface expression is restricted to a 1.5 km diameter circular structural uplift in an area of otherwise flat-lying Lower Cretaceous shales. Morphologically the central area forms a hilly region that surmounts the surrounding terrain by some 15 m.

The core of the uplift exposes highly deformed sedimentary rocks of Permian age that were uplifted by 400–500 m. The central uplift comprises fault bounded blocks that are internally folded; strata have steep dips. Neither the rim of the crater, nor ejecta and allochthonous impactites are preserved. This suggests that the present erosional level is below the original crater floor.

Gravity and seismic reflection data indicate that Mount Toondina is best interpreted as an eroded impact structure with a diameter of 4 km that consists of a ring-shaped structural depression surrounding the steeply rising central uplift. Seismic reflection data suggest that deformation extends to a depth of only about 800 m, which rules out that the structure represents a diapir. The structure has a gravity signature typical of complex impact structures with a positive residual anomaly of +1.0 mGal coincident with the central uplift, and a -0.5 mGal low associated with the ring depression. The age of Mount Toondina is roughly estimated to be Cretaceous, only constrained by the Early Cretaceous age of the youngest exposed possible target rocks and corresponding to the large amount of erosion.

The structural disturbance of Mount Toondina was initially interpreted as a diapir. An impact origin was first suggested by I. Youles in 1976, and supported by later field studies by Eugene Shoemaker and co-workers. Although all structural features are in favor of a small complex impact crater, the proof of the impact by documentation of shock metamorphism is to our knowledge still outstanding.

Location/Access

The Mount Toondina structure is situated 45 km south of the township of Oodnadatta in South Australia. The Stuart Highway, connecting Adelaide with Alice Springs, runs about 130 km west of the crater. Further access is provided via an old and rehabilitated seismic line. The crater area is remote and difficult to access. Careful planning and the use of four-wheel-drive vehicles are a must.

Selected References

Haines, P. W. (2005). Impact cratering and distal ejecta: the Australian record. Australian Journal of Earth Sciences, Vol. 52, pp. 481–507.

Plescia, J. B., Shoemaker, E. M., and Shoemaker, C. S. (1994). Gravity survey of the Mount Toondina impact structure, South Australia. Journal of Geophysical Research, Vol. 99E, pp. 13 167–13 179.

Shoemaker, E. M., and Shoemaker, C. S. (1988). Impact structures of Australia . 19th Lunar and Planetary Science Conference, Lunar and Planetary Institute, Houston, Texas, Abstract #1547.

Youles, I. P. (1976). Mount Toondina impact structure. Quarterly Geological Notes, The Geological Survey of South Australia, Vol. 60, pp. 10–11.

The Mount Toondina structure on October 9, 2018 as a composite image based on TanDEM-X DEM and Copernicus Sentinel-2 (channel 2/3/4) data. Cross hairs illustrate the position of Mount Toondina.

Mount Toondina

South Australia

latitude = 27°56‘41“ S / longitude = 135°21‘31“ E / diameter = 4 km / age = 66 - 144 Myr

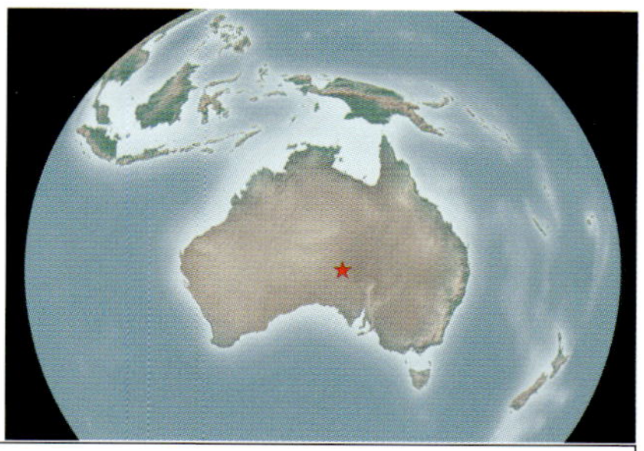

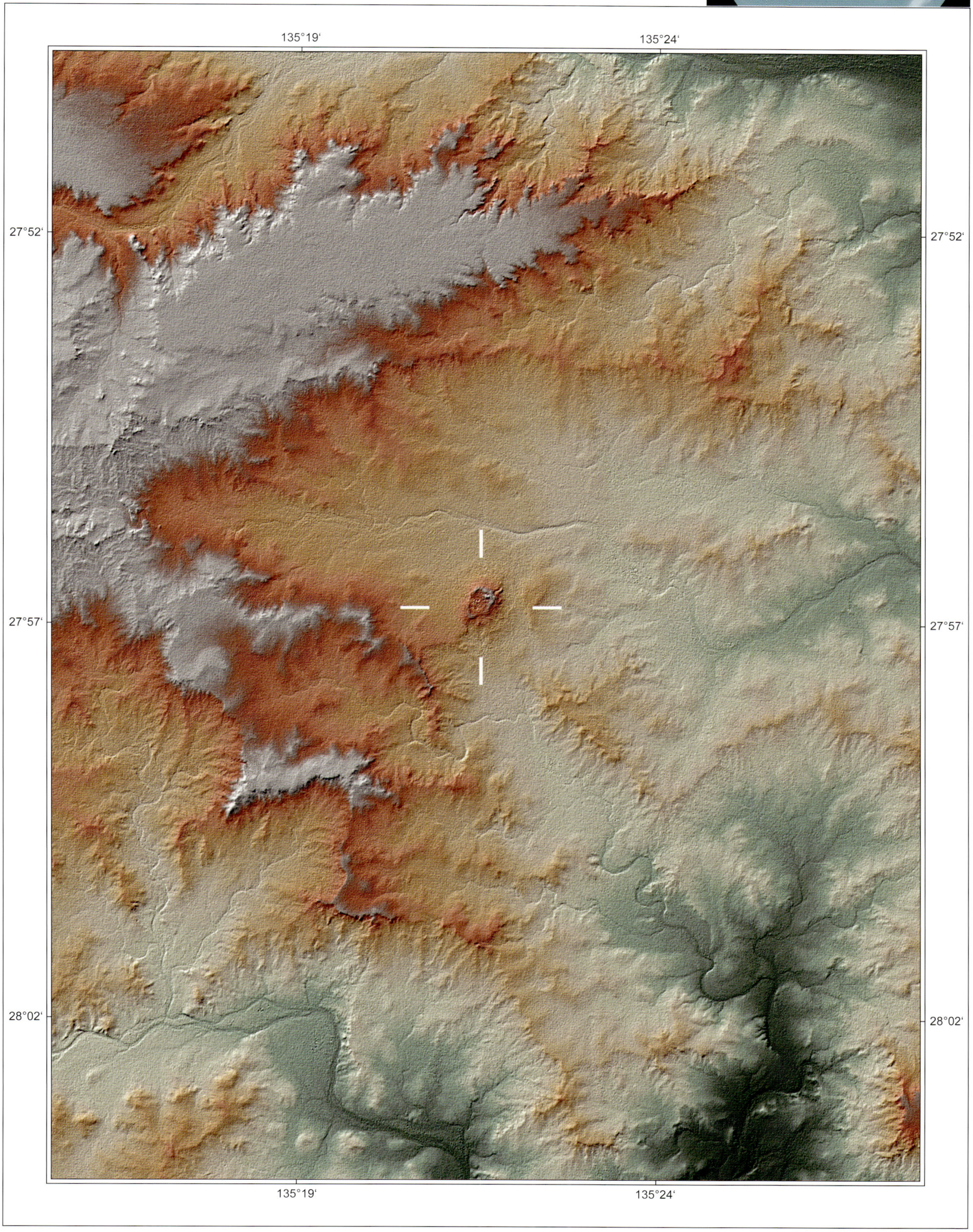

scale = 1 : 111 000 / scene = 19.9 × 26.5 km²

0 1.11 5.55 km

N

95 meter 155

Piccaninny
Western Australia

Thomas Kenkmann

The 7 km wide Piccaninny structure is located in the Purnululu National Park, in the Bungle-Bungle Range that is built up by Devonian conglomerates and sandstones in the center of the Ord Basin. Piccaninny is peculiar because it differs from the regular morphology of impact structures, where usually a depression is surrounded by a rim. At Piccaninny, most parts of the circular structure form a plateau that is framed by multicolored beehive-like domes, steep cliffs and the deeply cut Piccaninny Gorge. All these morphological landforms make Purnululu a tourist hot spot.

The structure is defined at its rim by a centrally dipping circular monocline with 5 to 15° dips that changes into a very gentle ring syncline, and a gentle central rise with a stratigraphic uplift of only 40–50 m. Most prominent is the distinct regional joint pattern. Outside Piccaninny it shows two major directions: northeast-southeast and northwest-southeast. Deeply incised gorges follow these zones of weakness. Towards the Piccaninny plateau, a systematic deflection of these joint sets towards the center of the structure can be observed, which is even obvious in space imagery.

View from the gently rising circular plateau of the Piccaninny structure towards the rim of the structure. Note the beehive-like domes that are particularly frequent near the Piccaninny gorge (photo: T. Kenkmann).

Beehive-like domes characterize the area of the Piccaninny structure. They are cut by steep faults (photo: T. Kenkmann).

On the rather flat plateau the vertical joint planes develop into faults. However, their displacements can hardly be measured and may not be large. The fault zones typically contain decimeter-wide cataclasites with powdery gouge zones and shear bands. Faults have meter-wide process zones with subsidiary fault networks. Some of the interconnected shear zones resemble breccia dikes. Along the faults minor tilting of strata can be observed. Shatter cones and other shock criteria could not be documented to date. The apparent lack of shock features either suggests a considerable amount of erosion of more than 2 km of the possible impact crater, or it would require an alternative, non-impact origin of the structure. The formation of the Piccaninny structure must post-date the age of the country rocks which are Devonian. This is the only age constraint.

Piccaninny was discovered by G. Beere in 1983. Two years later Eugene Shoemaker and his wife Carolyn first proposed an impact origin. Based on this report Piccaninny became listed as an approved impact crater in the Earth Impact Database. However, even though the gross structure indeed shows some similarities to deeply eroded complex impact structures, the origin of Piccaninny by an extraterrestrial projectile still has to be confirmed by the identification of unequivocal shock features.

Location/Access

The Piccaninny structure is located in the Purnululu National Park in Western Australia. When entering the access road to the National Park, which turns off from the Great Northern Highway close to the crossing over the Ord River between the towns of Warmun and Halls Creek, a four-wheel-drive car is prerequisite. The dirt road runs for about 50 km and

Piccaninny
Western Australia

latitude = 17°25'31" S / longitude = 128°26'18" E / diameter = 7 km / age = < 360 Myr

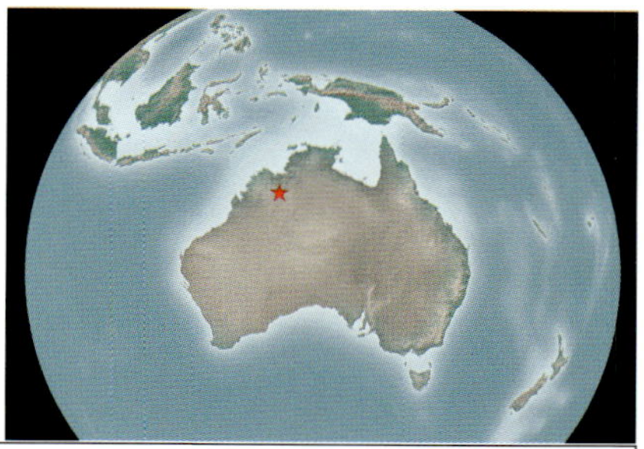

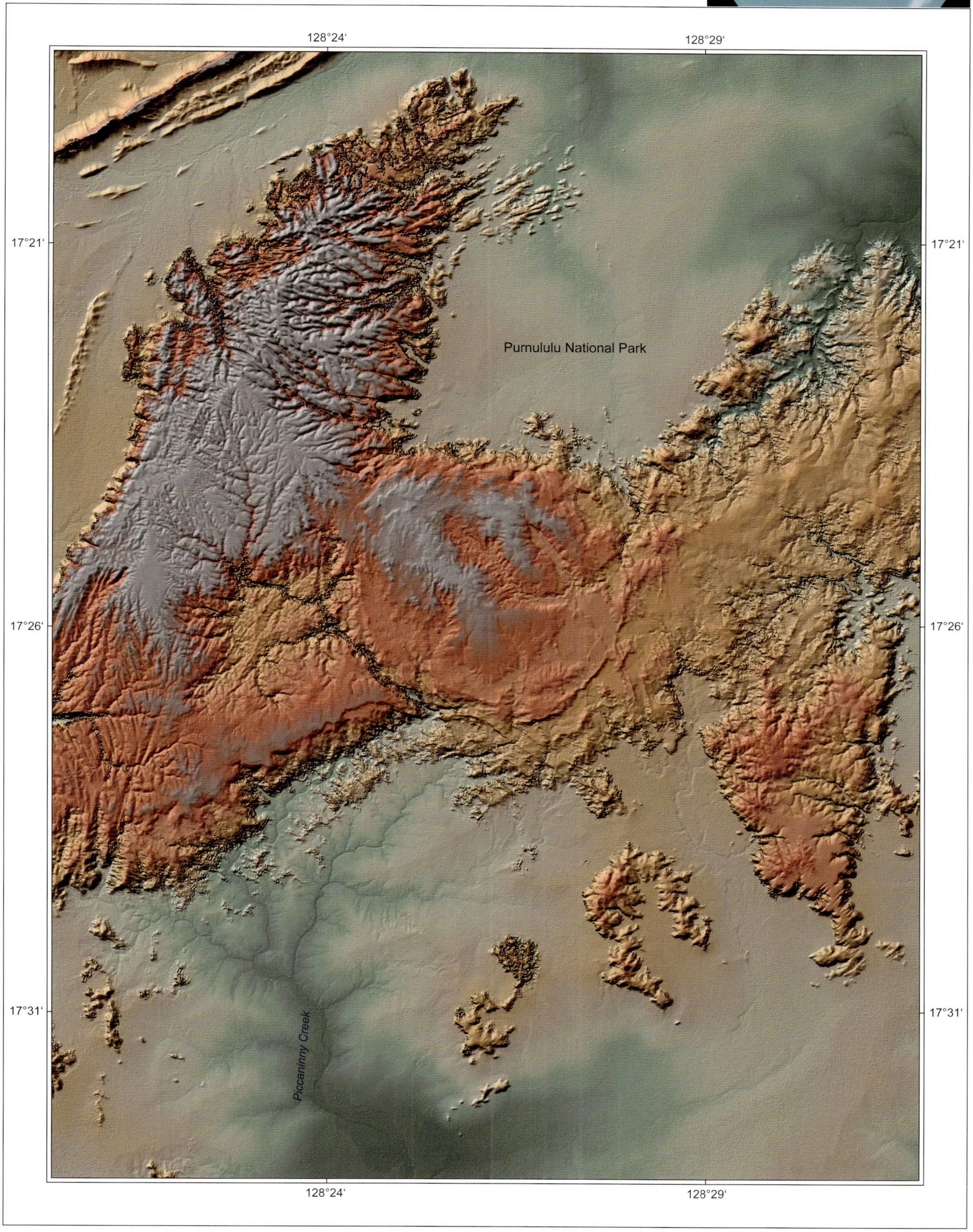

scale = 1 : 113 000 / scene = 20.3 × 27.0 km²

0 1.13 5.65 km

N

220 meter 765

Fault zone in the southern part of the Piccaninny structure, with dragging of the adjacent sandstone (photo: T. Kenkmann).

takes about 2 hours to drive, before a parking area in the southern part of the National Park is reached. From there walking along the Piccaninny creek for about 10 km takes the visitor to a site where the Piccaninny creek makes a sharp turn from a northwesterly to a southeasterly direction. The following climb onto the Piccaninny Plateau is laborious and requires a permit. The plateau itself has a very gentle topography.

Selected References

Beere, G. M. (1983). The Piccaninny structure – a cryptoexplosive feature in the Ord Basin, East Kimberley. Geological Survey of Western Australia Report 1983/6, 8 pp.

Kenkmann, T., and Poelchau, M. H. (2007). Piccaninny, WA, and Matt Wilson, NT: Two possible complex impact craters in Australia. 17th Annual Goldschmidt Conference, Cologne, Germany. Geochimica et Cosmochimica Acta, Vol. 71 Supplement, A477.

Shoemaker, E. M., and Shoemaker, C. S. (1985). Impact structures of Western Australia. Meteoritics, Vol. 20, pp. 754–756.

The Piccaninny structure on October 2, 2018 as a composite image based on TanDEM-X DEM and Copernicus Sentinel-2 (channel 2/3/4) data.

7.6 Europe

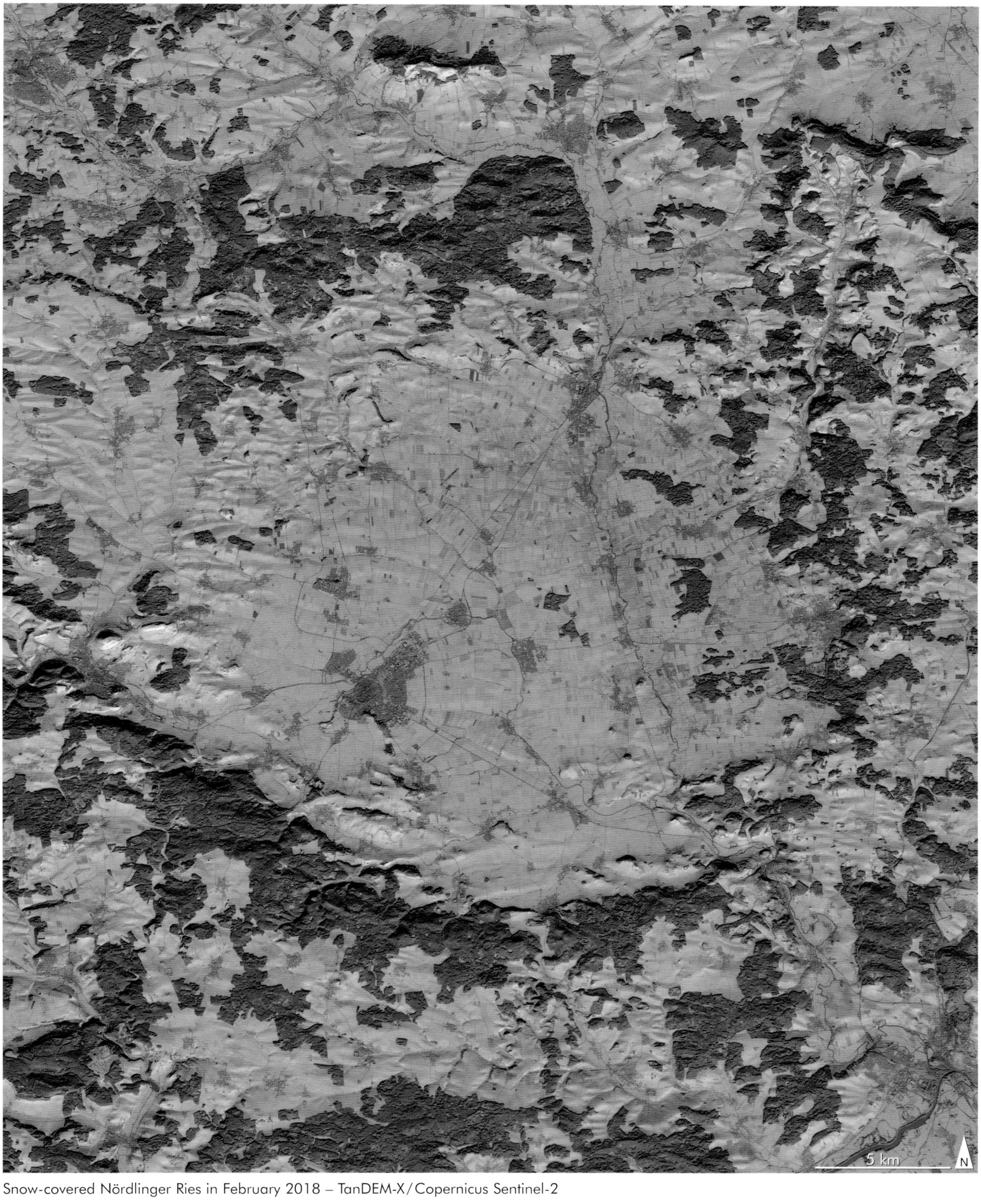

Snow-covered Nördlinger Ries in February 2018 – TanDEM-X/Copernicus Sentinel-2

European Impact Structures – Overview

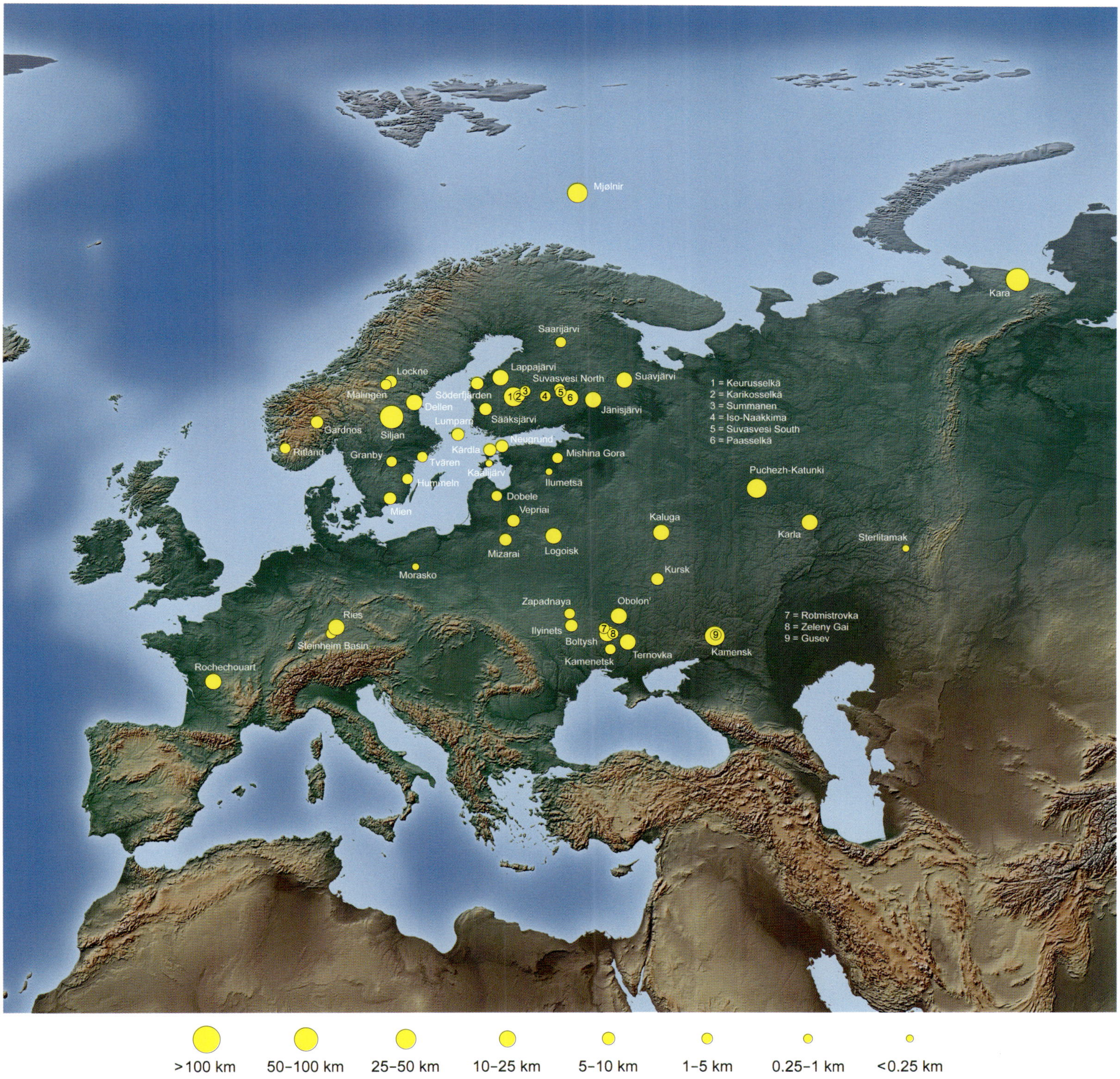

Table: European Impact Structures – Overview

ID	Name	Country/Region	Latitude	Longitude	Diameter (km)	Age (Myr)	Topographic Visibility
1	Boltysh[1]	Ukraine	48°57'30"N	32°14'23"E	24	65.17 ± 0.64	no (buried)
2	Dellen	Sweden	61°50'49"N	16°40'38"E	20	140.82 ± 0.51	yes (Lake North/South Dellen)
3	Dobele[2]	Latvia	56°34'12"N	23°17'24"E	4.5	260–340	no (buried)
4	Gardnos	Norway	60°38'29"N	09°00'39"E	5	546 ± 5	yes
5	Granby[3]	Sweden	58°25'34"N	14°55'53"E	2	470	no (buried)
6	Gusev[4,a]	Russia	48°29'N	40°32'E	3	50.36 ± 0.33	no (buried)
7	Hummeln	Sweden	57°22'05"N	16°15'10"E	1.2	463 ± 7	no (submerged – Lake Hummeln)
8	Ilumetsä Crater Field	Estonia	57°57'36"N	27°24'10"E	max. 0.08	0.0066	yes
9	Ilyinets[5]	Ukraine	49°07'N	29°06'E	8.5	445 ± 10	no (buried)
10	Iso-Naakkima[6]	Finland	62°11'46"N	27°07'48"E	3	900–1200	no (buried – Lake Iso-Naakkima)
11	Jänisjärvi	Russia	61°58'44"N	30°56'40"E	16	682 ± 4	no (submerged – Lake Janisjärvi)
12	Kaalijärv Crater Field	Estonia	58°22'22"N	22°40'10"E	max. 0.11	0.00351 ± 0.00004	yes
13	Kärdla[7]	Estonia	58°58'26"N	22°46'51"E	4	455	no (buried)
14	Kaluga[8]	Russia	54°30'N	36°12'E	15	395 ± 4	no (buried)
15	Kamenetsk[9]	Ukraine	47°46'N	32°21'E	1.1	250–540	no (buried)
16	Kamensk[4]	Russia	48°21'N	40°30'E	25	50.36 ± 0.33	no (buried)
17	Kara	Russia	69°05'N	64°20'E	65	70.3 ± 2.2	yes
18	Karikkoselkä	Finland	62°13'17"N	25°14'55"E	1.4	230–260	no (submerged – Lake Karikkoselkä)
19	Karla[10]	Russia	54°57'23"N	47°57'04"E	10	< 5	no (buried)
20	Keurusselkä[11]	Finland	62°08'N	24°37'E	30	1151 ± 10	no (Lake Keurusselkä)
21	Kursk[4]	Russia	51°42'N	36°00'E	6	250 ± 80	no (buried)
22	Lappajärvi	Finland	63°08'26"N	23°41'04"E	23	77.80 ± 0.78	no (submerged – Lake Lappajärvi)
23	Lockne	Sweden	63°00'01"N	14°49'11"E	7	461 ± 5	Yes (Lake Locknesjön)
24	Logoisk[10]	Belarus	54°15'46"N	27°47'10"E	17	29.71 ± 0.48	no (buried)
25	Lumparn	Finland	60°08'17"N	20°07'59"E	7	450–1250	no (submerged – Baltic Sea)
26	Målingen	Sweden	62°55'01"N	14°33'59"E	0.85	461 ± 5	yes (Lake Målingen)
27	Mien	Sweden	56°25'05"N	14°51'23"E	9	118.7 ± 2.3	no (submerged – Lake Mien)
28	Mishina Gora[10]	Russia	58°34'57"N	28°05'27"E	2.5	< 359	no (buried)
29	Mizarai[4]	Lithuania	54°01'N	23°54'E	5	520–560	no (buried)
30	Mjølnir[12]	Norway	73°48'N	29°40'E	40	142.0 ± 2.6	no (buried/submerged – Barents Sea)
31	Morasko Crater Field	Poland	52°29'25"N	16°53'47"E	max. 0.096	0.0057 ± 0.0007	yes
32	Neugrund[13]	Estonia	59°20'00"N	23°31'30"E	7	535	no (submerged – Baltic Sea)
33	Obolon[14]	Ukraine	49°35'48"N	32°54'33"E	18	169	no (buried)
34	Paasselkä	Finland	62°08'55"N	29°24'46"E	10	228.7 ± 3.0	no (submerged – Lake Paasselkä)
35	Puchezh-Katunki[4]	Russia	56°58'N	43°43'E	40	192–196	no (buried)
36	Ries	Germany	48°52'09"N	10°34'41"E	26	14.8	yes
37	Ritland[15]	Norway	59°14'08"N	06°25'18"E	2.7	500–540	yes
38	Rochechouart[16]	France	45°49'27"N	00°46'54"E	23	201 ± 2	no
39	Rotmistrovka[17]	Ukraine	49°08'00"N	31°44'52"E	2.7	95–145	no (buried)
40	Saarijärvi	Finland	65°17'37"N	28°23'31"E	1.5	< 520–600	no (submerged – Lake Saarijärvi)
41	Sääksjärvi	Finland	61°24'40"N	22°22'45"E	6	560 ± 12	no (submerged – Lake Sääksjärvi)
42	Siljan	Sweden	61°01'N	14°56'E	65	380.9 ± 4.6	yes
43	Söderfjärden	Finland	63°00'19"N	21°34'37"E	6.5	520–600	yes
44	Steinheim Basin	Germany	48°41'12"N	10°04'02"E	3.8	14.8	yes
45	Sterlitamak	Russia	53°36'12"N	55°35'12"E	0.0094	0.000029	no (destroyed)
46	Suavjärvi	Russia	63°07'21"N	33°22'24"E	16	2090–2700	no (Lake Suavjärvi)
47	Summanen[18]	Finland	62°39'00"N	25°22'30"E	2.6	< 1880	no (submerged – Lake Summanen)
48	Suvasvesi North	Finland	62°39'35"N	28°10'22"E	4	85	no (submerged – Lake Kuukkarinselkä)
49	Suvasvesi South	Finland	62°35'36"N	28°13'38"E	3.8	> 710	no (submerged – Lake Haapaselkä)
50	Ternovka[19]	Ukraine	48°07'48"N	33°31'12"E	11	280 ± 10	no (buried)
51	Tvären	Sweden	58°46'30"N	17°25'33"E	3.2	462 ± 5	no (submerged – Baltic Sea)
52	Vepriai[4]	Lithuania	55°05'N	24°35'E	7.5	162–174	no (buried)
53	Zapadnaya[20]	Ukraine	49°44'00"N	29°03'15"E	3.2	165 ± 5	no (buried)
54	Zeleny Gai[17]	Ukraine	48°53'17"N	32°47'55"E	3.5	80	no (buried)

Position references for selected impact structures:

a For Gusev the position in latitude is based on the claimed distance of 16 km between both impact structures in Masaitis, V. L. (1999).
1 Gurov, E. P., Kelley, S. P., Koeberl, C., and Dykan, N. I. (2006). Sediments and impact rocks filling the Boltysh impact crater. In: Biological Processes Associated with Impact Events, Cockell, C. S., Koeberl, C., and Gilmour, I. (eds.), Springer, Berlin Heidelberg, pp. 335–358.
2 Jõeleht, A., Mustasaar, M., Rooni, K., Kalvāns, A., and Popovs, K. (2015). Reflection seismic study of the Dobele impact crater, Latvia. Bridging the Gap III: Impact Cratering in Nature, Experiments and Modeling, Freiburg, Germany, Abstract #1105.
3 Henkel, H., Ekneligoda, T. C., and Aaro, S. (2010). The extent of impact induced fracturing from gravity modeling of the Granby and Tvären simple craters. Tectonophysics, Vol. 485, pp. 290–305.
4 Masaitis, V. L. (1999). Impact structures of northeastern Eurasia: The territories of Russia and adjacent countries. Meteoritics and Planetary Science, Vol. 34, pp. 691–711.
5 Gurov, E. P., Koeberl, C., and Reimold, W. U. (1998). Petrography and geochemistry of target rocks and impactites from the Ilyinets Crater, Ukraine. Meteoritics and Planetary Science, Vol. 33, pp. 1317–1333.
6 Elo, S., Kuivasaari, T., Lehtinen, M., Sarapää, O., and Uutela, A. (1993). Iso-Naakkima, a circular structure filled with neoproterozoic sediments, Pieksämäki, Southeastern Finland. Bulletin of the Geological Society of Finland, Vol. 65, pp. 3–30.
7 Kleesment, A., Konsa, M., Puura, V., Karhu, J., Preeden, U., and Kallaste, T. (2006). Impact induced and diagenetic changes in minerals in the sandy ejecta of the Kärdla crater, NW, Estonia. Proceedings of the Estonian Academy of Science, Geology, Vol. 55, pp. 189–211.
8 Masaitis, V. L. (2002). The middle Devonian Kaluga impact crater (Russia): new interpretation of marine setting. Deep-Sea Research Part II, Vol. 49, pp. 1157–1169.
9 Gurov, E., Nikolaenko, N., Helena Shevchuk, H., and Yamnichenko, A. (2017). Kamenetsk – A new impact structure in the Ukrainian Shield. Meteoritics and Planetary Science, Vol. 52, pp. 2461–2469.
10 VSEGEI. (2014). State Geological Maps 1:200,000. https://vsegei.ru/en/.
11 Pesonen, L. J., Hietala, S., Poutanen, M., Moilanen, J., Lehtinen, M., and Ruotsalainen, H. E. (2005). The Keurusselkä meteorite impact structure, central Finland: Geophysical data. 22nd Geophysics Days, Helsinki, Geophysical Society of Finland. pp. 165–169.
12 Gudlaugsson, S. T. (1993). Large impact crater in the Barents Sea. Geology, Vol. 21, pp. 291–294.
13 Suuroja, K., and Suuroja, S. (2010). The Neugrund meteorite crater on the seafloor of the Gulf of Finland, Estonia. Baltica, Vol. 23, pp. 47–58.
14 Gurov, E., Gurova, E., Chernenko, Y., Yamnichenko, A. (2009). The Obolon impact structure, Ukraine, and its ejecta deposits. Meteoritics and Planetary Science, Vol. 44, pp. 389–404.
15 Setså, R. (2011). The Ritland impact structure: Characteristics and distribution of the ejecta layer and associated Lower Paleozoic sedimentary succession. Masters Thesis (unpublished), University of Oslo, 114 pp.
16 Kelley, S. P., and Spray, J. G. (1997). A late Triassic age for the Rochechouart impact structure, France. Meteoritics and Planetary Science, Vol. 32, pp. 629–636.
17 Kelley, S., and Gurov, E. (2002). Boltysh – another end-Cretaceous impact. Meteoritics and Planetary Science, Vol. 37, pp. 1031–1043.
18 Plado, J., Hietala, S., Kreitsmann, T., Lerssi, J., Nenonen, J., and Pesonen, L. (2018). Summanen, a new meteorite impact structure in Central Finland. Meteoritics and Planetary Science, Vol. 53, pp. 2413–2426.
19 Sharpton, V. L., Krochuk, V. R., and Herrick, R. R. (2013). Characterization and morphological reconstruction of the Terny impact structure, central Ukraine. Meteoritics and Planetary Science, Vol. 48, pp. 806–818.
20 Gurov, E. P., Gurova, E. P., and Sokur, T. M. (2002). Geology and Petrography of the Zapadnaya impact crater in the Ukrainian Shield. In: Impacts in Precambrian Shields, Plado, J., and Pesonen, L. J. (eds.), Springer, Berlin Heidelberg, pp. 173–188.

Boltysh
Ukraine

Thomas Kenkmann

Boltysh is a nearly pristine complex impact crater in Ukraine, of approximately 24 km diameter. It is buried by more than 500 m of Cretaceous and younger crater-lake sediments. Thus, the crater structure does not have a distinct topographic expression. However, after subtraction of the sediments that were deposited after the impact, the impact crater forms a large circular depression in the Precambrian crystalline basement of the Ukrainian Shield, with a 6 km diameter central uplift rising some 550 m above the basement of the surrounding crater floor. The crater depression is filled with impact breccia including a massive, up to 200 m thick sheet of impact melt rock, and by a sequence of later sediments. The sedimentary infill into the crater depression has economic significance, as it contains many layers of oil shale.

Target rocks were Lower Proterozoic granites and migmatites, and minor Archean gneisses. Crystalline rocks in the central uplift are strongly shocked and show features of incipient melting and recrystallization. Quartz grains often exhibit abundant planar deformation features (PDFs), and have in places been converted to diaplectic glass, which may be recrystallized into small silica aggregates. Feldspars also display PDFs and partial isotropization. The degree of shock-induced deformation decreases in the central uplift with increasing depth.

Within the annular trough surrounding the central uplift, massive impact-generated melt rocks occupy a significant portion of the moat, whereas suevites form patches within the upper part of the impact breccia deposit and generally overlie the melt rock. The melt sheet makes up some 200 m of the allochthonous breccia fill sequence. The impact melt rocks are partially microcrystalline and partially glassy. Near the upper and lower contacts, they are vesicular. The composition of the impact melt rocks is similar to that of the granitic rocks of the target but the melt rocks show small enrichments in Ni and Cr relative to the average composition of the basement. The presence of allochthonous breccias, including suevite deposits immediately beneath the post-crater sediments, indicates minimal post-impact erosion. A polymict impact breccia also occurs outside the impact structure and has been interpreted as ejecta. It has a maximum thickness of 18 m at a distance of about 21–29 km from the crater center. This breccia is composed of fragments and blocks of granite up to 5 m in size, with shocked minerals and impact-glass particles. A suevitic breccia also covers the brecciated floor inside the crater and occurs on the outer slope of the annular trough.

Because of the economic potential of the oil shales within the post-impact sediments, detailed geophysical and drilling surveys were conducted. The crater depression is well imaged by a circular gravity low, with a weak gravity high at the center that corresponds to the central uplift. Two drillholes, located 3.5 and 2.2 km from the crater center, penetrated the melt sheet and exhumed shocked rocks. Fission-track dating of impact melt rock indicated an impact age between 96 and 105 Myr. However, comprehensive ^{40}Ar-^{39}Ar dating of a range of melt samples yielded a mean age of 65.17 ± 0.64 Myr, a value that is – within errors – identical to the age of the Cretaceous-Paleogene boundary.

A slab of impact melt rock, $8 \times 4 \times 2.5$ cm in size, from Boltysh (photo courtesy: Ralf-Thomas Schmitt, Museum für Naturkunde Berlin, sample obtained from the VSEGEI, St. Petersburg, Russia).

Boltysh was originally described as a caldera and was also interpreted as a result of tectonic movements. In 1974 Victor Masaitis proved the impact hypothesis on the basis of findings of shock metamorphic evidence. The exact position of the structure has been a subject of discussion. The most recent publication places the center of Boltysh into the city limits of Oleksandrivka in the Kirowohrad oblast (district).

Location/Access

Boltysh is situated 210 km southeast of the capital of Ukraine, Kiew, and 40 km south of the Kremenchuk Reservoir, the dammed-up part of the river Dnjepr.

Selected References

Grieve, R. A. F., Reny, G., Gurov, E. P., and Ryabenko, A. (1987). The melt rocks of the Boltysh impact crater, Ukraine, U.S.S.R. Contributions to Mineralogy and Petrology, Vol. 96, pp. 56–62.

Gurov, E. P., Kelley, S. P., Koeberl, C., and Dykan, N. I. (2006). Sediments and impact rocks filling the Boltysh impact crater. In: Biological Processes Associated with Impact Events. Cockell, C. S., Koeberl, C., and Gilmour, I. (eds.), Springer, Berlin Heidelberg, pp. 335–358.

Kelley, S. P., and Gurov, E. P. (2002). Boltysh, another end-Cretaceous impact. Meteoritics and Planetary Science, Vol. 37, pp. 1031–1043.

Masaitis, V. L. (1999). Impact structures of northeastern Eurasia: The territories of Russia and adjacent countries. Meteoritics and Planetary Science, Vol. 34, pp. 691–711.

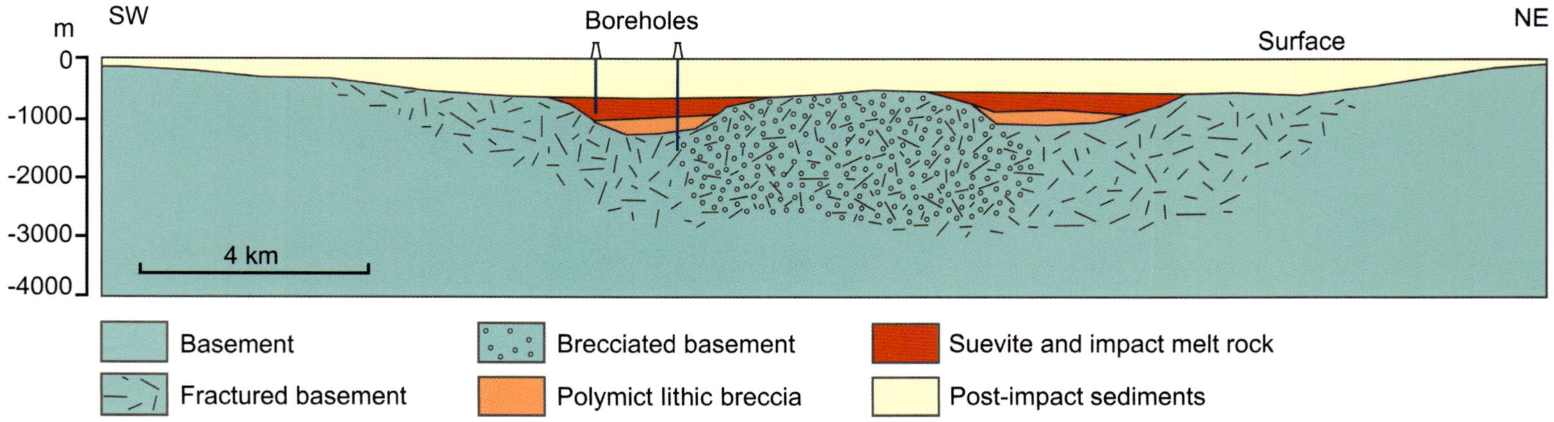

Schematic cross-section through the Boltysh impact structure (modified after Grieve et al., 1987).

Boltysh

Ukraine

latitude = 48°57‘30“ N / longitude = 32°14‘23“ E / diameter = 24 km / age = 65.17 ± 0.64 Myr

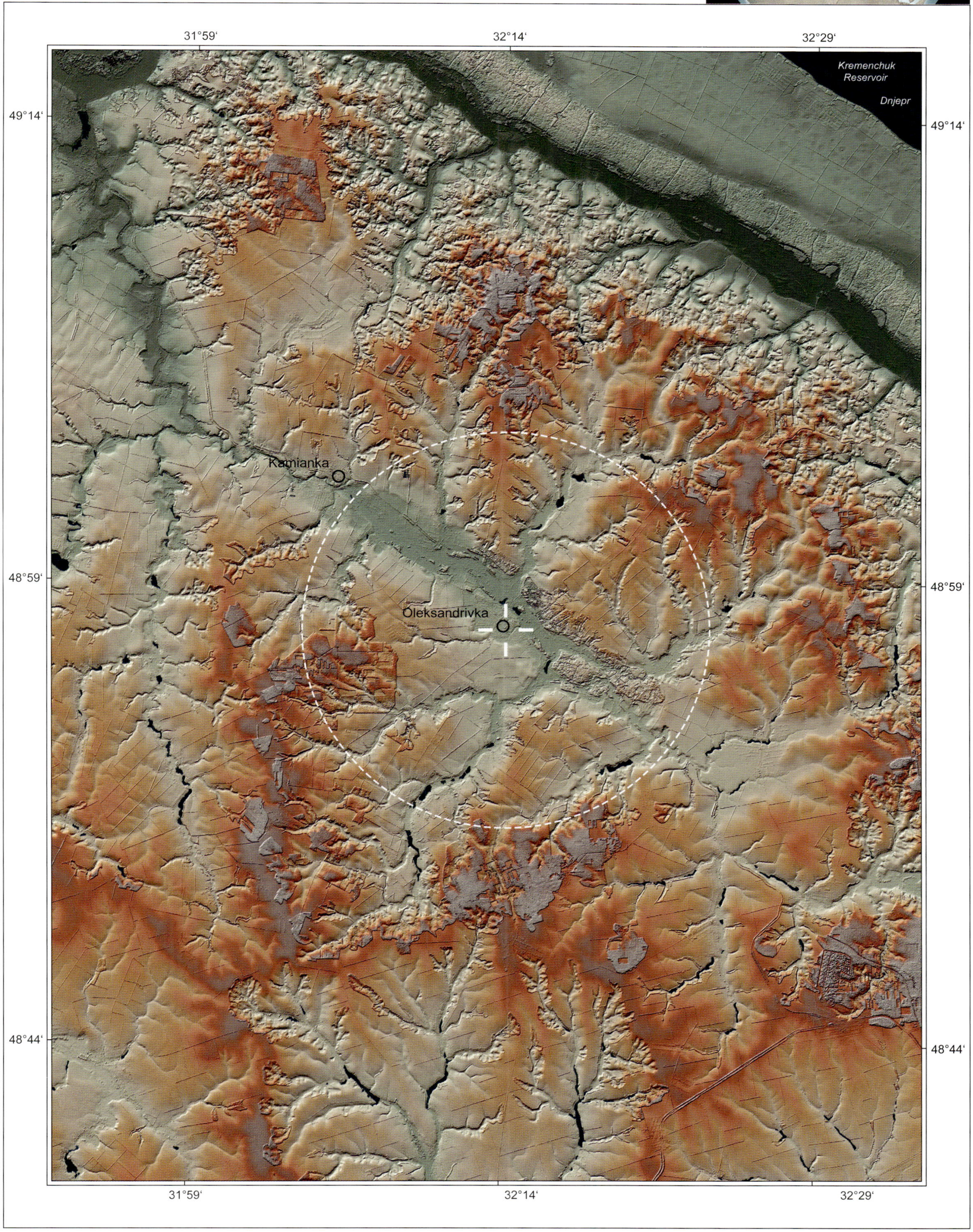

scale = 1 : 284 000 / scene = 51.2 × 68.3 km²

0 2.84 14.20 km

N

75 meter 275

Dellen
Sweden

Thomas Kenkmann

The Dellen impact structure is located in east-central Sweden. Its present diameter, based on morphology, is roughly 20 km. The roundish but somewhat irregular depression lies at a depth of about 250 m below the surrounding terrain. Two lakes, Norra and Södra Dellen, which are separated by an isthmus, occupy this depression. Dellen is a complex impact structure with a central uplift. The degree of erosion is low within the depression, so that impact melt rock as well as suevite have been preserved.

The target rocks are mainly granitoids of Paleoproterozoic age that belong to the Svecofennian phase of the Baltic Shield. A large batholith called the "Ljusdal granite", which was emplaced at 1850 Myr ago, forms the main target rock. Its composition is partly granitic, partly granodioritic. The texture of these igneous rocks was modified by regional metamorphism, prior to impact. However, a large part of the structure is buried under Quaternary moraines. Only a few in-situ outcrops exist. Fortunately, glacial erosion has exhumed crater lithologies and incorporated them into moraines as glacial boulders.

A prominent impact lithology at Dellen is glassy impact melt rock that occurs as glacial boulder trails and has been found in situ in a few places. Geophysical data indicate that the melt layer is approximately 9 km wide and 200–500 m thick. Suevite also occurs as fill of the crater depression and as allochthonous material. The melt-bearing impactites superpose lithic impact breccias and the strongly fractured and shocked crystalline basement rocks.

Over the years the glassy impact melt rocks at Dellen have been dated by different methods and with partly conflicting results. Paleomagnetic data suggested an Early Tertiary age, while measurements of whole-rock melt rock samples using the ^{40}Ar-^{39}Ar and Rb-Sr radiometric dating methods yielded Lower and Upper Cretaceous ages, respectively. The application of state-of-the-art laser ^{40}A-^{39}Ar stepwise-heating to single grains of alteration-free, homogeneous impact glass has supported a lowermost Cretaceous age of 140.82 ± 0.51 Myr for the Dellen impact.

A specimen of suevite from the Dellen impact structure, $10.5 \times 7.5 \times 6$ cm in size. Note the dark, fresh-looking schlieren of impact melt (photo courtesy: Ralf-Thomas Schmitt, Museum für Naturkunde Berlin, sample from T. Kenkmann).

Gravity measurements at the Dellen impact crater have shown a distinct negative anomaly of about −7 mGal superimposed onto the regional gravity gradient. Magnetic measurements trace outline and interior pattern of the melt sheet due to a preserved strong remanent magnetization. The resistivity of the rocks to electrical currents is also quite characteristic at Dellen. The brecciated Ljusdal granite forming the crater floor reduces the rock resistivity because conductive fluids can circulate through its fractures. The lowest resistivities were measured where the basement is overlain by highly porous suevite and other breccias. The presence of the continuous melt sheet in the inner part of the crater causes a relatively high electrical resistivity but with magnitudes that are still below those measured outside the crater.

Until the 1960s and early 1970s the Dellen structure was thought to be of volcanic origin. The peculiar glassy impact melt rock outcropping at the shores of the Dellen lakes was called "Dellenite". It was interpreted variably as tuff, tuff breccia, and devitrified and glassy lava. However, already in 1910, at a meeting of the Geological Society of Stockholm, Arvid Högbom tentatively suggested that the Dellen Lakes could have an impact origin. Nils Svensson provided the proof for the impact hypothesis in 1968, when he documented planar deformation features (PDFs) in quartz clasts in "Dellenite" impact melt rock.

Location/Access

The Dellen Lakes are situated about 300 km north-northwest of Stockholm. A fast connection via motorway E 4 and national road 84 leads to Delsbo at the southwestern margin of the impact structure. At Näsviken the visitor can leave National Road 84 and travel on Noorbovägen towards Bjuråker. This road crosses the isthmus and the crater center between the North and South Dellen lakes, where one finds boulders of impact lithologies along the shorelines of the lakes.

Selected References

Deutsch, A., Buhl, D., and Langenhorst, F. (1992). On the significance of crater ages: new ages for Dellen (Sweden) and Araguainha (Brazil). Tectonophysics, Vol. 216, pp. 205–218.

Henkel, H. (1992). Geophysical aspects of meteorite impact craters in eroded shield environment, with special emphasis on electric resistivity. Tectonophysics, Vol. 216, pp. 63–89.

Högbom, A. (1910). Geologiska Föreningen i Stockholm Förhandlingar, Vol. 32, pp. 482.

Mark, D. F., Lindgren, P., and Fallick, A. E. (2014). A high-precision ^{40}Ar-^{39}Ar age for hydrated impact glass from the Dellen impact, Sweden. In: Advances in ^{40}Ar-^{39}Ar Dating: from Archaeology to Planetary Sciences, Jourdan, F., Mark, D. F., and Verati, C. (eds.), Geological Society of London, Special Publications, Vol. 378, pp. 349–366.

Müller, N., Hartung, J. B., Jessberger, E. K., and Reimold, W. U. (1990). ^{40}Ar-^{39}Ar ages of Dellen, Janisjärvi, and Sääksjärvi impact craters. Meteoritics, Vol. 25, pp. 1–10.

Svensson, N. B. (1968b). The Dellen lakes: a probable meteorite impact in central Sweden. Geologiska Föreningen i Stockholm Förhandlingar, Vol. 90, pp. 14–316.

Dellen

Sweden

latitude = 61°50‘49“ N / longitude = 16°40‘38“ E / diameter = 20 km / age = 140.82 ± 0.51 Myr

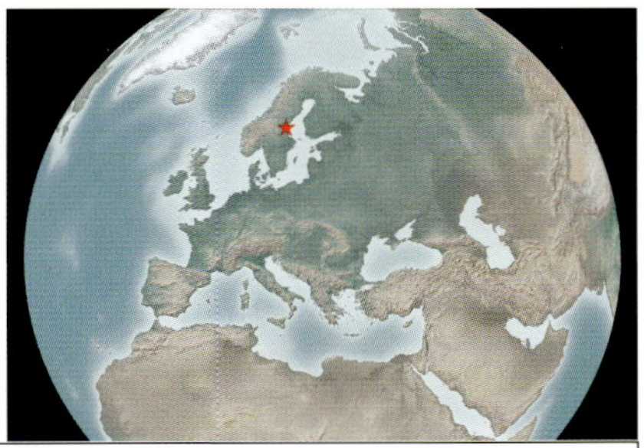

scale = 1 : 305 000 / scene = 54.9 × 73.3 km²
0 3.05 15.25 km

N

0 meter 240

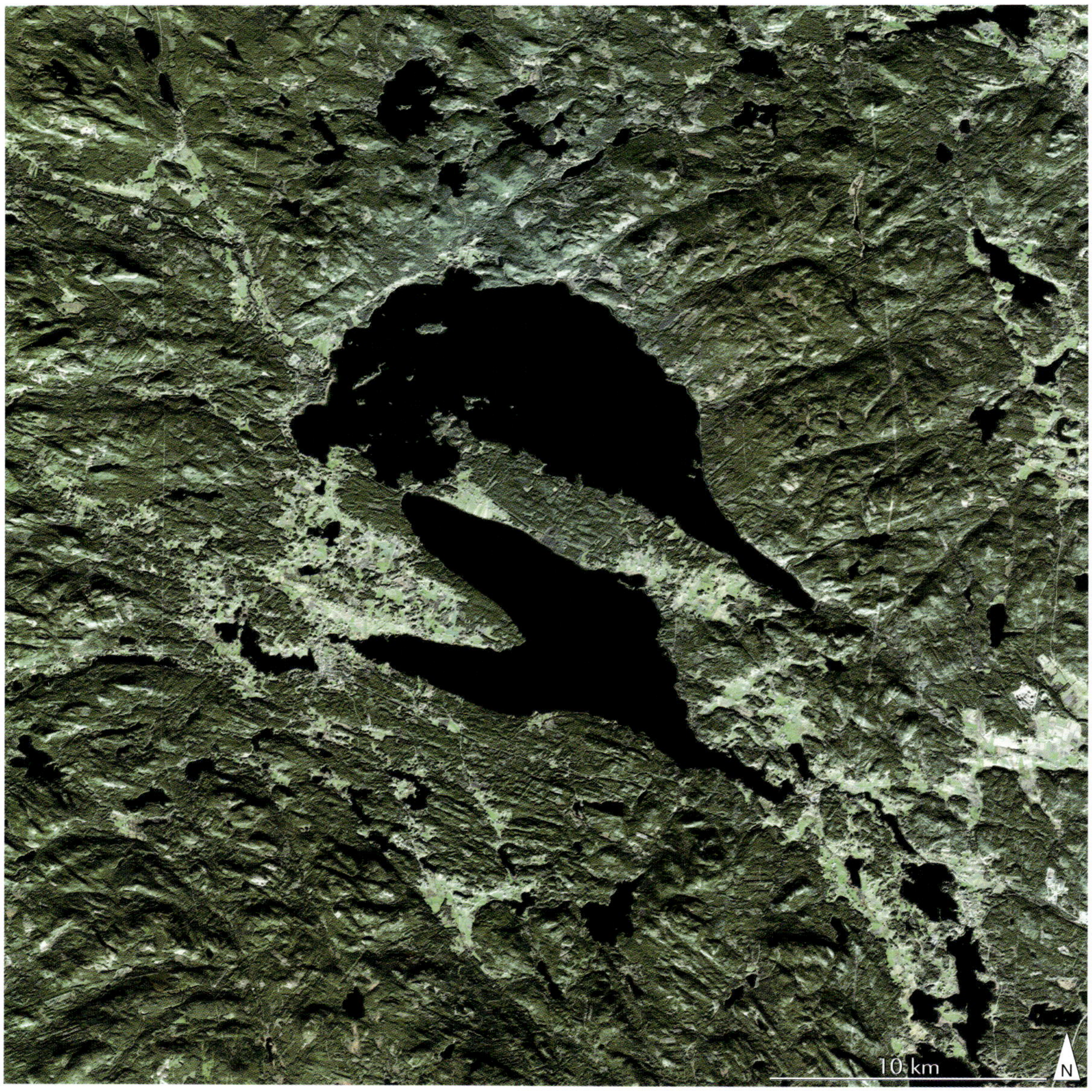

The Dellen impact structure with the lakes Norra and Södra Dellen on October 22, 2018 as a composite image based on TanDEM-X DEM and Copernicus Sentinel-2 (channel 2/3/4) data.

Dobele
Latvia

Thomas Kenkmann

Dobele is a buried impact structure with a diameter of 4.5 km without a morphological expression. The structure consists of a central uplift surrounded by an annular trough that is filled with 300 m of impact breccia.

The target rocks are Paleozoic strata, consisting of Silurian carbonates overlain by various Devonian sediments, including sandstones, sulfate, carbonate rocks, and a thin veneer of Carboniferous limestone. The thickness of the Paleozoic target rocks amounts to 750–800 m. The oldest strata, brecciated and displaced, occur in the central uplift, where the rocks show a stratigraphic uplift of 300 m. Occasionally, shatter cones have been observed in drill core samples, as well as quartz grains with planar deformation features (PDFs). Strata show multiple repetition in drill core from the central uplift. According to drilling and seismic data, the allochthonous breccias in the annular moat that surrounds the central uplift occur at about 300–350 m depth and contain large blocks of target rock. The impact crater is buried under fifty meters of variegated carbonate rocks of upper Permian and Lower Triassic age. These strata provide a minimum age for this impact event of about 260 Myr, whereas the upper age limit of 340 Myr is defined by lower Carboniferous strata of the target succession. Quarternary beds of 25 m thickness cover the entire impact structure.

Initially, the Dobele structure was suspected to be a kimberlite pipe, however, gravimetric and magnetic studies could not support this. Not much can be said about how Dobele became recognized as an impact structure. The corresponding entry in Victor Masaitis' overview paper about the impact structures in Russia and adjacent countries is still the most comprehensive report describing Dobele.

Location/Access

Dobele is located in rural land about 50 km south of the Gulf of Riga, and 70 km southwest of Riga, the capital of Latvia. The area of the Dobele municipality is reached from Riga on Highway A9.

Selected References

Abels, A., Plado, J., Pesonen, L. J., and Lehtinen, M. (2002). The impact cratering record of Fennoscandia – A close look at the database. In: Impacts in Precambrian Shields, Plado, J., and Pesonen, L. J. (eds.), Springer, Berlin Heidelberg, 339 pp.

Jõeleht, A., Mustasaar, M., Rooni, K., Kalvāns, A., and Popovs, K. (2015). Reflection seismic study of the Dobele impact crater, Latvia. In: Bridging the Gap III: Impact Cratering in Nature, Experiments and Modeling, Freiburg, Germany, Abstract #1105.

Masaitis, V. L. (1999). Impact structures of northeastern Eurasia: The territories of Russia and adjacent countries. Meteoritics and Planetary Science, Vol. 34, pp. 691–711.

Dobele

Latvia

latitude = 56°34‘12“ N / longitude = 23°17‘24“ E / diameter = 4.5 km / age = 260 - 340 Myr

scale = 1 : 120 000 / scene = 21.7 × 28.9 km²

0 1.20 6.00 km

N

5 meter 145

Gardnos
Norway

Thomas Kenkmann

The Gardnos impact structure in south-central Norway measures roughly 5 km in diameter. It has a topographic expression that is difficult to discern, as the degree of glacial erosion varies strongly over the area of the structure. In the east, the structure is truncated by a large glacial valley, which has exposed the crater's sub-surface. In the center of the structure the fractured crystalline bedrock is uplifted by roughly 400 m and forms a small topographic high of 0.5 km width. This high has been interpreted as a central uplift. A relatively flat-floored annular moat surrounds the central uplift. It is about 1 km wide and filled with allochthonous breccia and post-impact sediments.

The target rocks at Gardnos are of Precambrian age. In the area of the structure one finds mostly granitic gneisses, along with quartzites, banded gneisses, and amphibolites. For these rocks ages from 1150 to 1500 Myr have been determined. The rocks experienced strong deformation and metamorphic overprint during the Sveconorwegian orogeny around 960 to 1140 Myr ago. This orogeny was also responsible for the formation of a steep north-south trending foliation. Relatively undeformed pegmatite dikes intruded the basement sequence at about 900 Myr ago.

The Gardnos area was repeatedly glaciated and deeply eroded during the Quaternary. At that time the crater reappeared at the surface from underneath Caledonian thrusts that eventually had been eroded away. The present glacial relief is high, but is by no means related to the crater morphology. Low altitude areas in the main valley Hallingdal expose deep seated rocks of the crater's subsurface, whereas the area further up-slope exposes the crater floor and the crater fill. Glacial till covers much of the area, but local river beds have provided excellent sections through different levels of the crater.

The most remarkable impact lithology at Gardnos is the aptly termed "Gardnos Breccia," a lithic breccia consisting of light-colored, intensely shattered granitic gneiss clasts in a black matrix. The black color is due to strong grain comminution and the presence of carbon within most of the micro-fractures. The presence of carbon suggests that a carbon-rich shale may have overlain the metamorphic basement at the time of impact. In the shattered crater sub-surface the gneiss fragments form jigsaw patterns, but the continuation of the foliation documents almost no rotational movements of the clasts with respect to each other. Here the rocks were "shaken but not stirred". Closer to the crater floor clasts are more disintegrated and the matrix content increases. A series of suevitic breccias overlies the Gardnos Breccia. Petrographic inspection of these breccias showed distinctive shock metamorphic effects, mainly planar deformation features (PDFs) in quartz. Patches of impact melt breccia also occur. A minor extraterrestrial component of about 0.15 % was detected in the melt-bearing breccias, as evidenced by significantly elevated iridium and osmium contents.

A 400 m long scientific borehole was drilled 1100 m northwest of the crater's center in the outer part of the annular moat around what is believed to represent the central uplift of the crater structure. From bottom to top, about 200 m of Gardnos Breccia, 50 m suevite, and 150 m of post-impact sedimentary formations were recovered. The sedimentary rocks comprise conglomerates, sandstones, siltstones, and mudstones, and form a thick fining-upward cycle. The post-impact sedimentary succession indicates a retarded water resurge, possibly from a breach of the crater rim. This may indicate that the crater was formed in shallow water.

The impact age has to be older than that of undeformed pegmatite dikes of the crater basement. Overthrusting of the crater during the Caledonian orogeny 425 Myr ago provides a lower age limit. Attempts to date the impact event using the U-Pb radiometric method applied to zircon grains were hindered by the difficulty that the Caledonian orogeny caused a thermal pulse that resulted in partly reset ages. However, the data from a few zircon grains, including those from a grain with granular shock deformation texture, plot along a discordia line with an upper intercept age of 546 ± 5 Myr. This value was suggested to represent Gardnos' age but confirmation that this age relates to the impact event is still outstanding.

The river bed crossing the Gradnos impact structure provides good exposure of the Gardnos Breccia, lithologies of the crater floor, and suevitic breccia (photo: T. Kenkmann).

Brecciated rocks of the Gardnos area were initially regarded the result of explosive volcanism. An impact origin of the structure was first proposed in the early 1990s by J. Dons and J. Naterstad. They also confirmed the impact origin by documentation of PDFs in quartz. Further detailed studies by Bevan French and co-workers in the mid-1990s focused on the petrography and geochemistry of impactites. The analysis of the post-impact infill sequence was carried out by Erin Kalleson.

Location/Access

The Gardnos impact structure is located between the villages of Nesbyen and Gol about 150 km northwest of the city of Oslo. The structure's close proximity to Highway R7 makes access easy. The crater is situated on the western shoulder of the Hallingdal valley. A local road winds up for 3 km into the structure, where a small visitor center informs the public

Gardnos
Norway

latitude = 60°38‘29“ N / longitude = 09°00‘39“ E / diameter = 5 km / age = 546 ± 5 Myr

scale = 1 : 104 000 / scene = 18.7 × 25.0 km²
0 1.04 5.20 km
N
105 meter 1235

Typical Gardnos Breccia, a monomict lithic breccia with a dark carbon-rich matrix (photo: T. Kenkmann).

Suevitic breccia from Gardnos, with clasts of impact melt and of different crystalline target lithologies (photo: T. Kenkmann).

about the crater. The interested traveler can visit some important outcrops along the nearby river bed and along a dedicated trail.

Selected References

Dons, J. A., and Naterstad, J. (1992). The Gardnos impact structure, Norway (Abstract). Meteoritics, Vol. 27, p. 215.

French, B. M., Koeberl, C., Gilmour, I., Shirley, S. B., Dons J. A., and Naterstad, J. (1997). The Gardnos impact structure, Norway: Petrology and geochemistry of target rocks and impactites. Geochimica et Cosmochimica Acta, Vol. 61, pp. 873–904.

Gilmour, I., French, B. M., Franchi, I. A., Abbott, J. I., Hough, R. M., Newton, J., and Koeberl, C. (2003). Geochemistry of carbonaceous impactites from the Gardnos impact structure, Norway. Geochimica et Cosmochimica Acta, Vol. 67, pp. 3889–3903.

Kalleson, E., Dypvik, D., and Naterstad, J. (2007). Post-impact sediments in the Gardnos impact structure, Norway. In: The Sedimentary Record of Meteorite Impacts, Evans, K., Horton Jr., J. W., King Jr. D. T., and Morrow, J. R. (eds.), Geological Society of America Special Paper, Vol. 437, pp. 19–41.

Kalleson, E., Jahren, T., and Dypvik, D. (2008). Guide to Gardnos Meteorite Crater. Department of Geoscience, University of Oslo, 24 pp.

Granby

Sweden

Thomas Kenkmann

Granby is a small impact structure in southern Sweden, buried under Quaternary glacial till. The Granby structure is a slightly asymmetrical, bowl-shaped cavity of approximately 2 km diameter. It is over 377 m deep and has a raised rim of 70 m. Target rocks are 1650–1850 Myr old Precambrian crystalline rocks of the Trans-Scandinavian Igneous Belt that is part of the Baltic Shield.

The crater depression is surrounded by a rim of brecciated crystalline basement rocks that have been uplifted in the crater rim by up to 70 m. The depression is filled with various sedimentary rocks of Paleozoic age. A borehole drilled into the cavity revealed that the lowermost 70 m of the infill consist of a polymict breccia with angular fragments of Cambrian and Ordovician sandstone, limestone, and shale, as well as pieces of the underlying crystalline bedrock. The heterogeneous distribution of fragment sizes is most likely the result of slumping processes, so that the breccia ought to be classified as a sedimentary breccia. The breccia is overlain by dark bituminous shale, followed by limestone and mudstone.

Multiple sets of planar deformation features (PDFs) decorated with fluid inclusions were discovered in quartz grains of clasts of the breccia infill. The decorated nature of the PDFs implies a water-bearing target. The absence of melt particles gives an estimated maximum shock pressure of less than 15–20 GPa. The presence of abundant fluid inclusions on PDFs, along with the multitude of slump blocks and resurge deposits in the cavity, and the regional geological framework suggest that the Granby impact occurred into a shallow marine environment with a possible water depth ranging from 25 to 75 m.

Based on lithostratigraphy, the structure was formed in the Middle Ordovician some 470 Myr ago. Granby was long considered to be of impact origin. Geological surveying was conducted in 1982. However, only in 2009 Carl Alwmark reported the finding of shock deformation features providing the unequivocal evidence for origin by impact.

Location/Access

Granby is located in Östergötland province of southern Sweden close to the eastern shore of Lake Vättern and about 3 km from the town of Vadstena. Vadstena is located 250 km southwest of Stockholm, from where it can be reached via Highway E4.

Selected References

Alwmark, C. (2009). Shocked quartz grains in the polymict breccia of the Granby structure, Sweden – Verification of an impact. Meteoritics and Planetary Science, Vol. 44, pp. 1107–1113.

Bruun, Å., and Dahlman, B. (1982). Den palaeozoiska berggrunden (in Swedish). In: Beskrivning till berggrundskartan Hjo NO, Wikman, H., Bruun, Å., Dahlman, B., and Vidal, G. (eds.), Sveriges Geologiska Undersökning, Vol. Af120, pp. 76–109.

Granby
Sweden

latitude = 58°25'34" N / longitude = 14°55'53" E / diameter = 2 km / age = 470 Myr

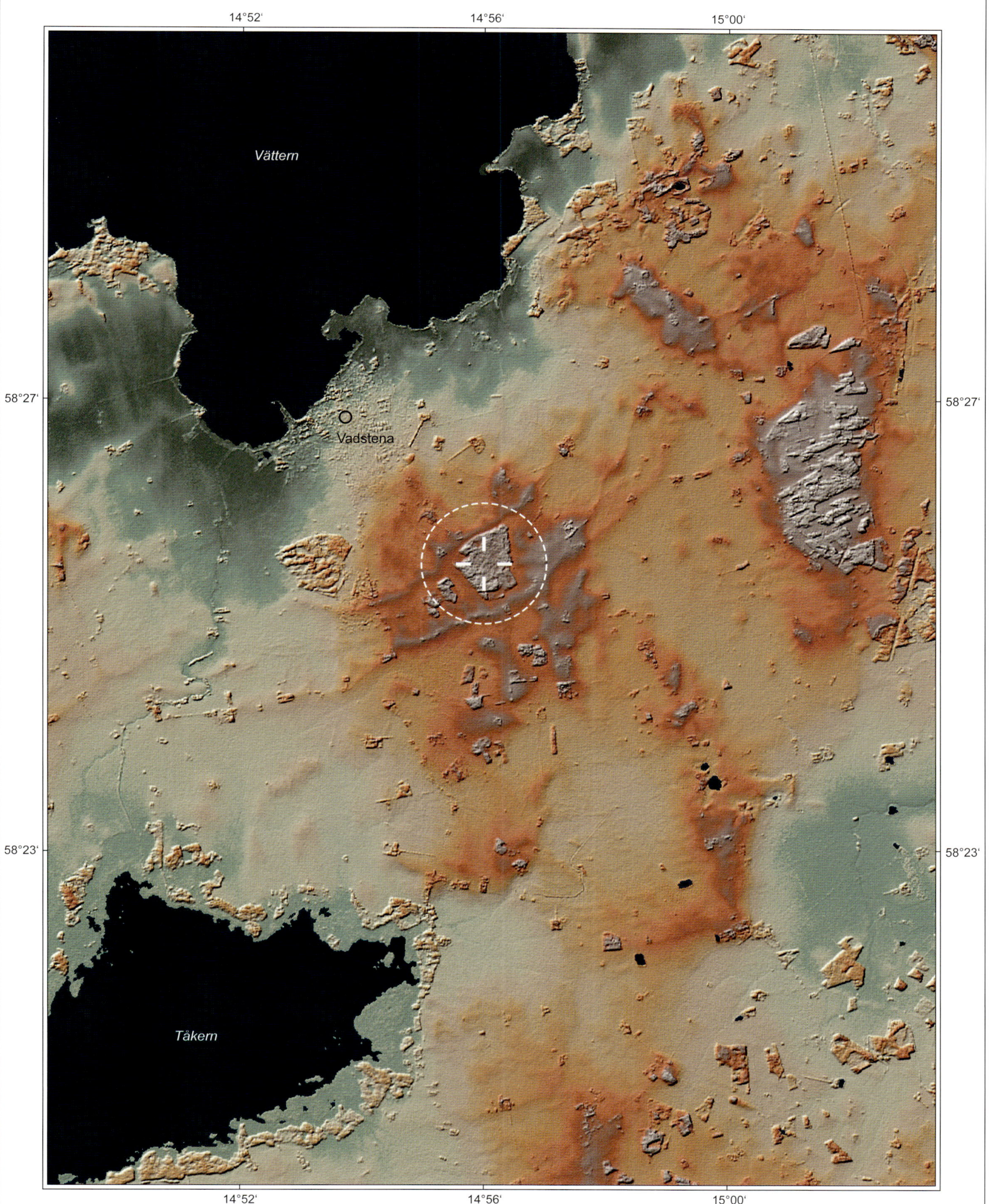

scale = 1 : 79 000 / scene = 14.2 × 18.9 km²
0 0.79 3.95 km
N
80 meter 160

Gusev and Kamensk
Russia

Thomas Kenkmann

The Gusev and Kamensk impact craters are considered to be the result of a double impact. Gusev with a diameter of 3 km is located adjacent to the northern rim of Kamensk, which has a diameter of 25 km. The respective crater centers are separated by 16 km. Both structures have been formed in the northeastern part of the Donetsk basin, are buried, and have no surface expression.

Target rocks for both crater structures are folded Carboniferous and Lower Permian rocks that are discordantly overlain by Lower Triassic and Upper Cretaceous rocks. The coal-bearing Carboniferous rocks are 3–4 km thick and comprise sandstones, siltstones, shales, and limestones. Permian strata reach 600 m in thickness and show a similar lithological variation. Triassic beds are primarily siltstone and claystone of 150 m thickness. Carboniferous, Permian and Triassic rocks are covered by Upper Cretaceous strata, with a total thickness of 300 m.

Gusev has a simple bowl shape with an extent of 4×2.5 km. The depression is elongated in east-west direction, and has a depth of 600 m. A borehole into the crater center encountered 360 m of allochthonous breccia, composed of fragments and blocks of the various sedimentary target rocks. The crater floor is covered with a 200 m thick pile of brecciated Carboniferous rocks that thins out radially from the center.

Kamensk has a central uplift with a diameter of about 5–6.5 km and a height exceeding 350 m. The stratigraphic uplift of strata above their original position, however, is much larger and has been estimated at 2–4 km. This central uplift is surrounded by a ring syncline that may be as deep as 700–800 m. The crater floor is made up of a breccia composed of displaced and crushed blocks of Carboniferous lithologies. The breccia is 150–200 m thick near the outer slope of the crater depression but in the vicinity of the central uplift the crushed and brecciated target rocks are 2 km thick. The lithic breccia of the crater floor gradually turns into polymict impact breccia. The thickness of this breccia reaches 700 m in the annular trough and still attains 100–200 m above the central uplift. Shatter cones are found in some places in the polymict breccia. Thin irregular lenses of suevite are found within the breccia layer. They contain clasts of diaplectic quartz glass and coesite-bearing quartz.

A thick layer of reworked breccia covers the rocks adjacent to both the Gusev and Kamensk impact structures. Therefore, both craters are considered contemporaneous. The resurge deposit is called the "Glubokinskaya Formation" and occupies an area of $30 \text{ km} \times 60$ km, elongated in an east-west direction. Drilling has revealed numerous radial gullies in the target rocks that are filled by what could be resurge deposits. The maximum thickness of this resurge formation in the annular trough is 200–300 m.

The fossil content of these strata suggested that the impact occurred within a shallow sea around 62–68 Myr before present. However, radiometric age dating of impact glass from a borehole into the Kamensk structure yielded a younger age for this double impact event, namely 50.36 ± 0.33 Myr.

The impact origin for both Gusev and Kamensk was postulated in the mid-1970s. Since then, the structures have been studied by 2D reflection seismics, gravity, geoelectric and magnetic surveys, as well as multiple boreholes. Most of the published literature appeared in the two decades following the discovery of these impact structures.

Location/Access

Kamensk, named after the city of Kamensk-Shakhtinsky located southwest of the crater, is situated close to the border of Russia and Ukraine in the oblast Rostow. The Seversky Donets river, a tributary of the Don river, meanders along the southern rim of the buried impact structure. River incision provides outcrops of post-impact strata. Gusev crater is also named after a nearby town and is situated 25 km northeast of the city of Kamensk-Shakhtinsky, which lies about 150 km north of the Sea of Azov. Earlier publications had placed both impact strcutures about 20 km further to the west.

Selected References

Izett, G. A., Masaitis, V. L., Shoemaker, E. M., Dalrymple, G. B., and Steiner, M. B. (1994). Eocene age of the Kamensk buried crater of Russia. In: New Developments Regarding the KT Event and Other Catastrophes in Earth History, Lunar and Planetary Institute Contribution No. 825, pp. 55–56.

Masaitis, V. L. (1976). Astroblemes in the USSR. International Geology Review, Vol. 18, pp. 1249–1258.

Masaitis, V. L. (1999). Impact structures of northeastern Eurasia: The territories of Russia and adjacent countries. Meteoritics and Planetary Science, Vol. 34, pp. 691–711.

Schmieder, M., Trieloff, M., Schwarz, W. H., Buchner, E., and Jourdan, F. (2014). Supportive comment on: "Morphology and population of binary asteroid impact craters", by K. Miljkovic, G. S. Collins, S. Mannick and P. A. Bland [Earth Planet. Sci. Lett. 363(2013) 121–132]. An updated assessment. Earth and Planetary Science Letters, Vol. 405, pp. 281–284.

Gusev / Kamensk

Russia

Gusev: latitude = 48°29‘ N / longitude = 40°32‘ E / diameter = 3 km / age = 50.36 ± 0.33 Myr
Kamensk: latitude = 48°21‘ N / longitude = 40°30‘ E / diameter = 25 km / age = 50.36 ± 0.33 Myr

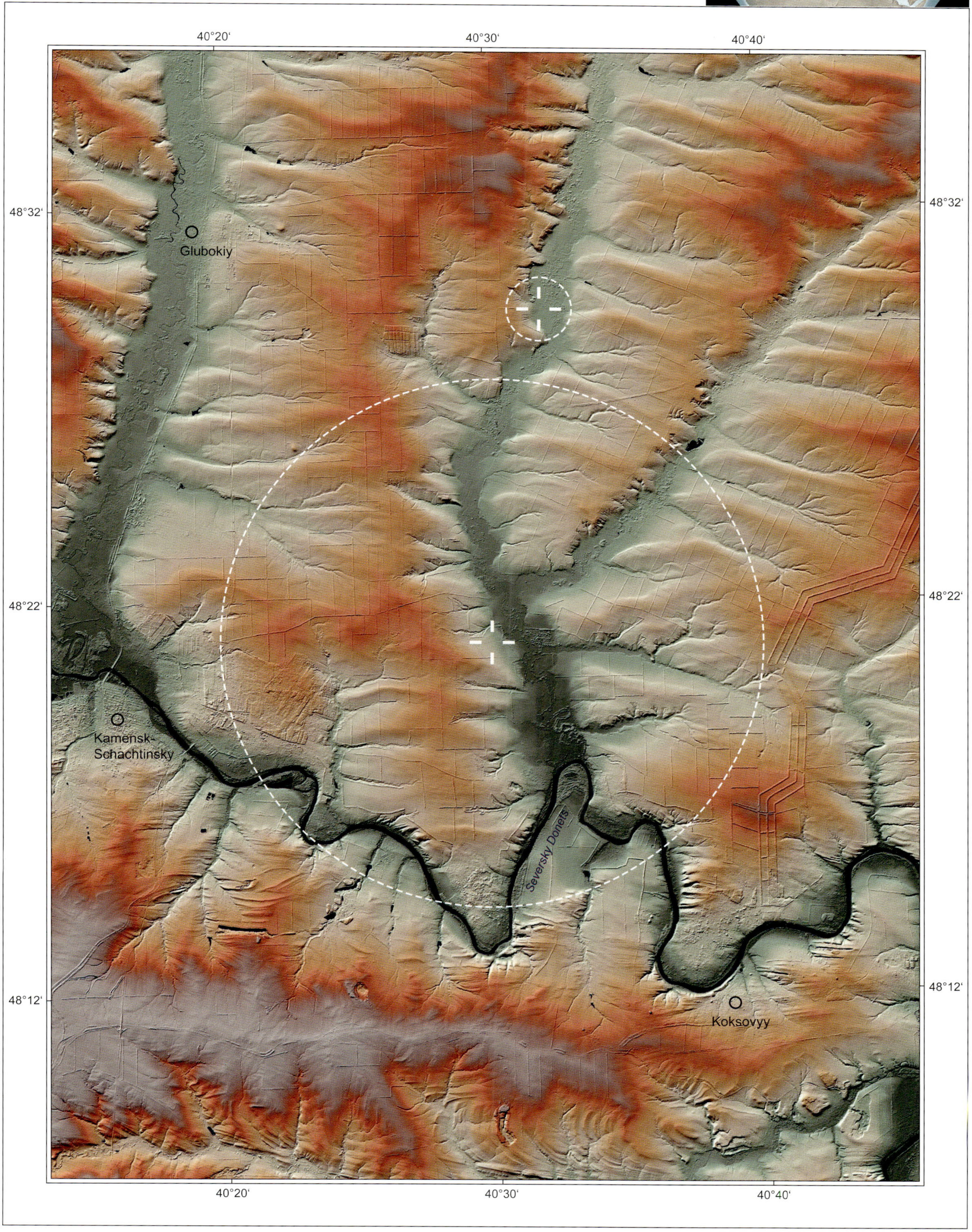

scale = 1:221 000 / scene = 39.7 × 52.9 km²

0 2.21 11.05 km

N

5 meter 225

Hummeln
Sweden

Thomas Kenkmann

The 1.2 km diameter Hummeln impact structure almost completely lies within Lake Hummeln, which is slightly larger than the structure itself. Only the southeastern shore partly exposes the rim of the crater structure. The shape of Lake Hummeln is also a result of glacial erosion. Except for the elevated crater rim that has been glacially levelled, the Hummeln impact structure is remarkably well preserved.

Target rocks are Svecofennian granites and diorites of the Baltic Shield. Within Lake Hummeln the crater structure forms a more than 160 m deep and 1.2 km wide depression in Precambrian crystalline basement. A 160 m deep drill core, recovered from the eastern part of the structure, revealed that the depression is partially filled with slumped Paleozoic sedimentary rocks, mainly Cambrian muddy sandstone and Ordovician limestone that originally covered the entire Baltic Shield. Brecciated target rocks from the crater rim are found in situ along the southern shore of the lake and as erratic boulders along the shore and in glacial deposits. Crystalline clasts show a shock metamorphic overprint in the form of decorated planar deformation features (PDFs). The impact crater has not been radiometrically dated, but the Hummeln structure is believed to have formed during the Middle Ordovician, subsequent to the deposition of the Ordovician limestones. Apparently, the impact happened coeval with the formation of a cluster of impact structures in this period, which has been considered a consequence of the break-up of the L chondrite parent body in the asteroid belt at ca. 470 Myr ago.

The peculiar circular depression of Hummeln, with preserved Paleozoic sediments, has been known for 200 years. In the 1930s and 1940s the depression was explained by either explosive volcanism or tectonics. An impact origin for the structure was proposed by Kurt Fredriksson and Frans Wickman in the 1960s. However, only in 2015 was the ultimate proof of impact presented, when Carl Alwmark and co-workers found unambiguous shock deformation features in quartz grains, which proved the hypervelocity impact origin of the structure.

Location/Access

The Hummeln structure is located in Småland province of southern Sweden, about 25 km west of the shoreline of the Baltic Sea. The largest city nearby is Oskarshamn at the Baltic Sea. From there it is a 20 km drive to Lake Hummeln, first westward on Highway 47, then for a short distance northward on Highway 711, which passes Hummeln on the western lake shore.

Selected References

Alwmark, C., Ferrière, L., Holm-Alwmark, S., Ormö, J., Leroux, H., and Sturkell, E. (2015). Impact origin for the Hummeln structure (Sweden) and its link to the Ordovician disruption of the L chondrite parent body. Geology, Vol. 43, pp. 279–282.

Lindström, M., Flodén, T., Grahn, Y., Hagenfeldt, S., Ormö, J., Sturkell, E. F. F., and Törnberg, R. (1999). The lower Palaeozoic of the probable impact crater of Hummeln, Sweden. GFF, Vol. 121, pp. 243–252.

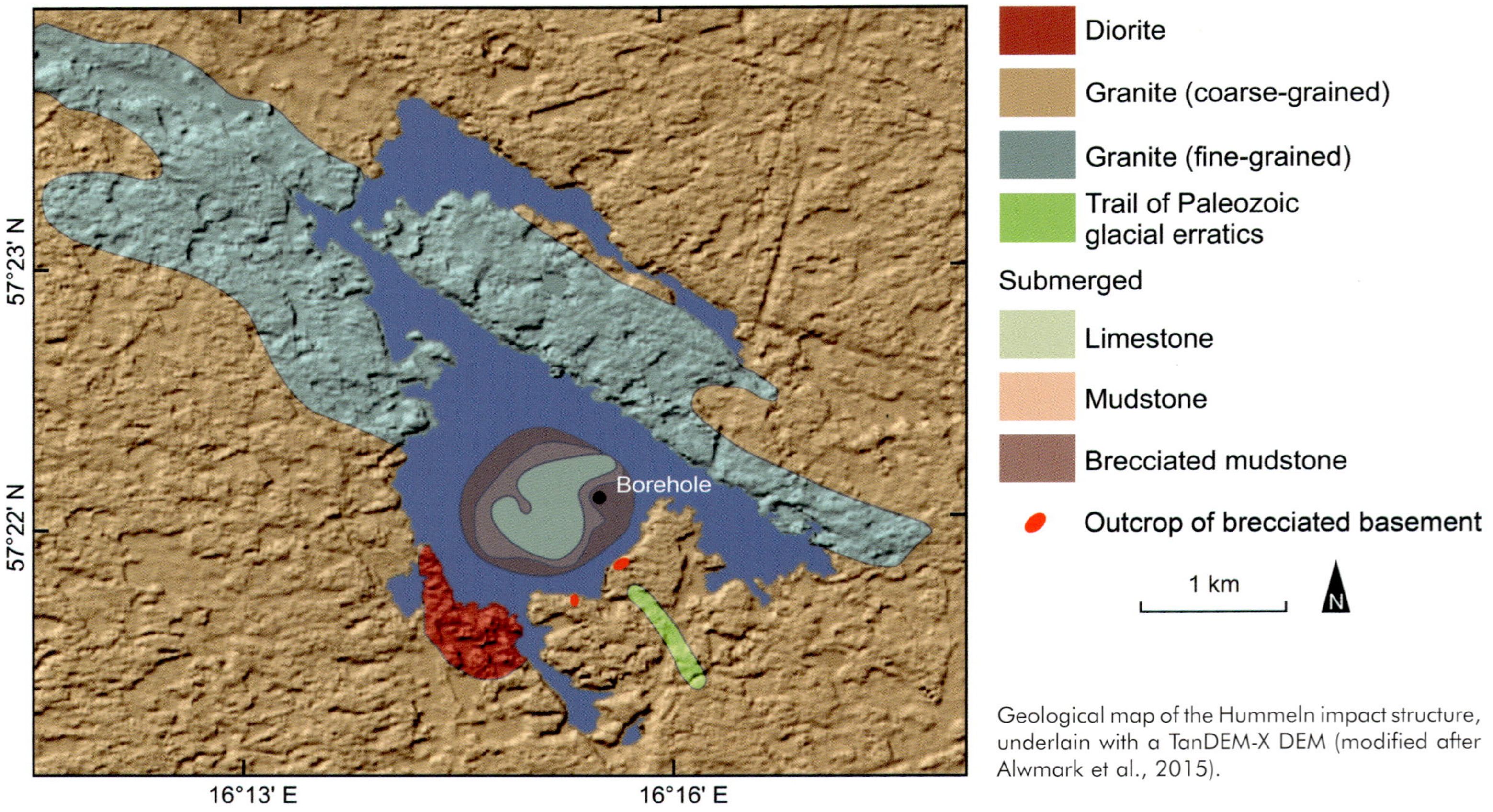

Geological map of the Hummeln impact structure, underlain with a TanDEM-X DEM (modified after Alwmark et al., 2015).

Hummeln

Sweden

latitude = 57°22'05" N / longitude = 16°15'10" E / diameter = 1.2 km / age = 463 ± 7 Myr

scale = 1 : 95 000 / scene = 17.1 × 22.7 km²

0 0.95 4.75 km

N

15 meter 195

Ilumetsä Crater Field

Estonia

Thomas Kenkmann

The Ilumetsä crater field in southeastern Estonia consists of two simple structures with rim-to rim diameters of 75–80 m and about 50 m, respectively, and with respective depths of about 12.5 and 4.5 m. Both craters have rims raised by up to 4.5 and 1.5 m, respectively, above the crater interior. Three additional smaller pits were reported by local people.

The target for this impact comprised Middle Devonian siltstones and sandstones that were overlain by a 1–2 m thick layer of Quaternary till. The craters are filled with allochthonous breccia. A small lens of gyttja and peat, at maximum some 3 m thick, occurs in the center of the largest crater. The organic-rich deposits of the crater pits have been dated with the ^{14}C method as well as palynology, which resulted in a minimum age of 6030 ± 100 years before present (BP), as a close approximation of the impact age. Glassy spherules, discovered in a 5.7 m deep peat layer approximately 6 km west-southwest of the Ilumetsä structures yielded a somewhat older ^{14}C age of around 6600 yrs BP that is considered as the age for the Ilumetsä impact. Magnetic profiling across the two Ilumetsä craters showed local positive and negative anomalies up to 20 nT, corresponding to the rim and inner parts of the structures, respectively.

The Ilumetsä impact crater pits were discovered in 1938 in the course of a geological mapping campaign. However, the results by A. Luha and co-workers of the Estonian Geological Committee, who already at that time suggested a meteorite impact origin, remained unpublished. A. Aaloe conducted further studies in the early 1960s, which included a few excavations. Later Anto Raukas studied Ilumetsä in detail and derived an updated age.

Location/Access

The Ilumetsä crater field is located about 1 km southwest of Ilumetsä station on the Tartu-Petseri railway. The crater field occurs in a forested area and can be reached on forest roads. Since 1964, two craters known as Pdrguhaud (Hell's Grave) and Sugavhaud (Deep Grave) have been placed under state protection.

Selected References

Aaloe, A. O. (1963). New data on the structure of Ilumetsa craters. Eesti NSV Teaduste Akadeemia Geoloogia Instituudi uurimused (in Russian). Vol. 11, pp. 35–43.

Plado, J. (2012). Meteorite impact craters and possibly impact-related structures in Estonia. Meteoritics and Planetary Science, Vol. 47, pp. 1590–1605.

Raukas, A., Tiirmaa, R., Kaup, E., and Kimmel, K. (2001). The age of the Ilumetsä me-eorite craters in southeast Estonia. Meteoritics and Planetary Science, Vol. 36, pp. 1507–1514.

Ilumetsä Crater Field

Estonia

latitude = 57°57'36" N / longitude = 27°24'10" E / diameter = max. 80 m / age = 6600 yr

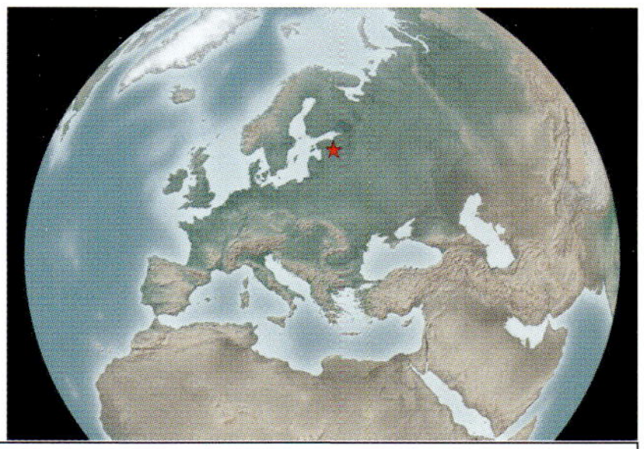

Ilyinets
Ukraine

Thomas Kenkmann

Ilyinets is one of the many impact structures found on the Ukrainian Shield. A significant part of the crater is buried under Neogene and Quaternary deposits. Impact lithologies crop out in an oval area of 3 × 4 km size. However, the crater is larger than this. Its diameter of 8–9 km is best constrained by the gravity low associated with it. The best exposure of impact lithologies occurs in the southern crater sector.

Target rocks are Archean and Early Proterozoic granites, gneisses, and amphibolites of the western part of the Ukrainian Shield. They are thinly covered by Paleozoic sediments. The structure has a central uplift of about 300 m elevation and 750 m diameter, which is surrounded by an annular trough. The central uplift is composed of brecciated and shocked granite. Shock features in these rocks indicate that they were subjected to shock pressures below 35 GPa. Shatter cones, as well as injections of melt rock, are observed. The thickness of the crater floor breccia underneath the allochthonous infill reaches 600–800 m close to the southern rim. The annular trough is filled with a variety of impactites, including lithic impact breccia of crystalline rock, suevite, and, to a lesser degree, impact melt rock. The latter forms a sheet-like body of 25–30 m thickness in the southern sector of the annular trough. Maximum thickness of the crystalline breccia is 200 m. At its base, this breccia grades into the crater floor breccias and shattered basement rocks. The upper part of the breccia contains a few glass fragments in addition to crushed crystalline rock, and thus, per definition, becomes a suevitic breccia. Diaplectic glasses and coesite have been noted in these rocks. The transition to the overlying suevite is gradational. The thickness of the suevite reaches 300 m. Two varieties of suevite with different composition and different amounts of glass have been distinguished. Both varieties are strongly affected by hydrothermal alteration.

The age of the impact event was approximated with the K-Ar method on whole-rock samples of impact melt rock, which yielded ages between 385 and 410 Myr. This age range is in agreement with the previously determined palynological age for the crater-fill sediments. More recent combined paleo-magnetic and ^{40}Ar-^{39}Ar dating by Lauri Pesonen and colleagues suggested an age of ca. 445 ± 10 Myr for the impact. Petrographic observations showed that some secondary hydrothermal alteration affected the rocks after the impact, which may have been responsible for some of the slightly younger K-Ar ages.

The crater is well explored at depth by various boreholes drilled since its first description by Victor Masaitis in 1973.

Location/Access

Ilyinets is located in central Ukraine, some 200 km south of the capital city of Kiev and 45 km southeast of the town of Vinnitsa. Access from Kiev requires driving about 200 km southward to the town of Uman. There the traveler has to turn westward on road M 12 towards Vinnitsa. After 110 km, local roads lead to the village of Illintsi in close vicinity of the Ilyinets structure.

A small piece of Ilyinets suevite, 6.5 × 5.5 cm in size (photo courtesy: Ralf-Thomas Schmitt, Museum für Naturkunde Berlin, sample obtained from the VSEGEI, St. Petersburg, Russia).

Selected References

Gurov, E. P., Koeberl, C., and Reimold, W. U. (1998). Petrography and geochemistry of target rocks and impactites from the Ilyinets Crater, Ukraine. Meteoritics and Planetary Science, Vol. 33, pp. 1317–1333.

Masaitis, V. L. (1976). Astroblemes in the USSR. International Geology Review, Vol. 18, pp. 1249–1258.

Masaitis, V. L. (1999). Impact structures of northeastern Eurasia: The territories of Russia and adjacent countries. Meteoritics and Planetary Science, Vol. 34, pp. 691–711.

Pesonen, L. J., Mader, D., Gurov, E. P., Koeberl, C., Kinnunen, K. A., Donadini, F., and Handler, R. (2004). Paleomagnetism and ^{40}Ar-^{39}Ar age determinations of impactites from the Ilyinets structure, Ukraine. In: Cratering in Marine Environments and on Ice, Dypvik, H., Burchell, M. J., and Claeys, P. H. (eds.), Springer, Berlin, Heidelberg, pp. 251–280.

Ilyinets
Ukraine

latitude = 49°07‘ N / longitude = 29°06‘ E / diameter = 8.5 km / age = 445 ± 10 Myr

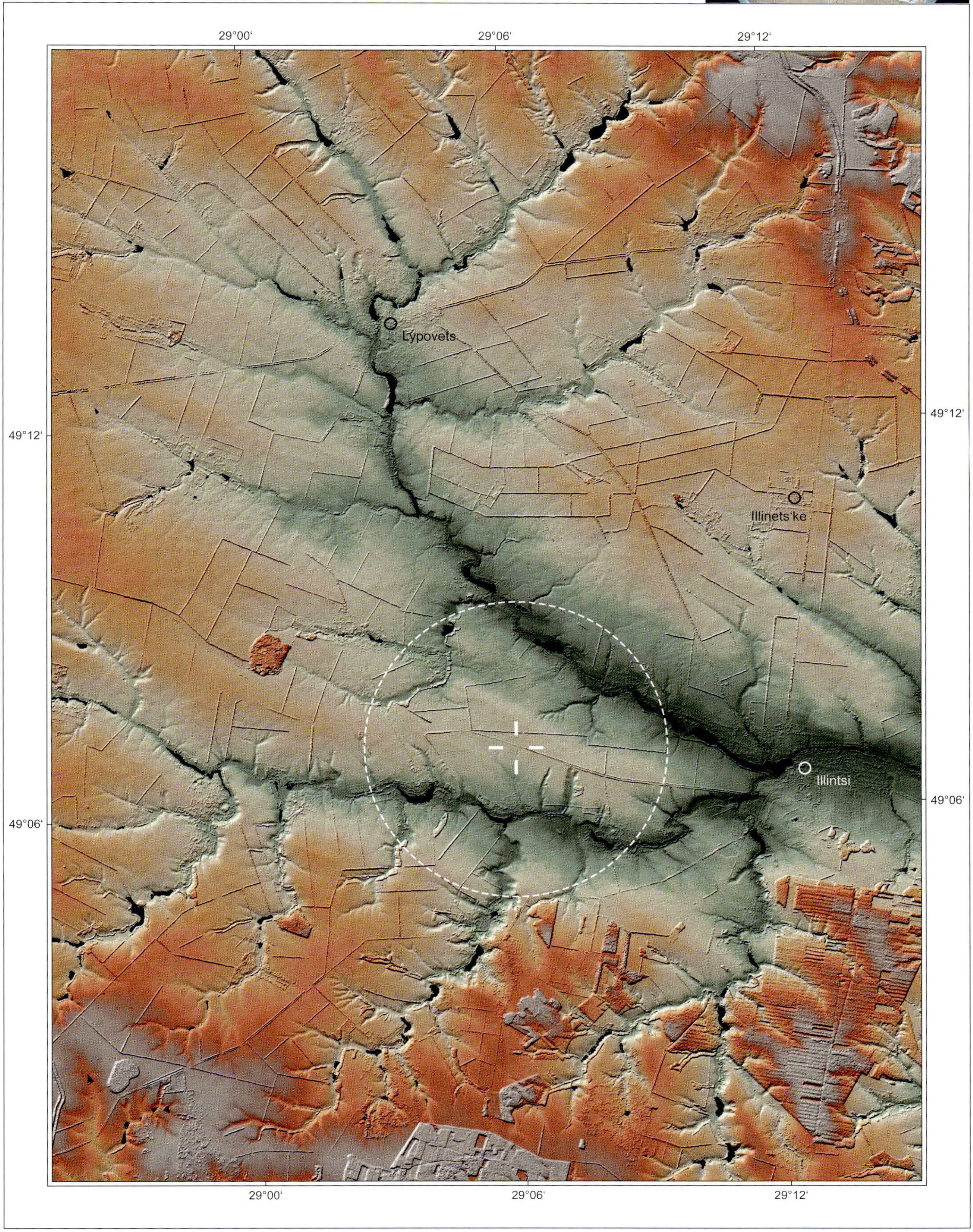

scale = 1:135 000 / scene = 24.3 × 32.4 km²

0 1.35 6.75 km

N

195 meter 340

Iso-Naakkima
Finland

Thomas Kenkmann

The Iso-Naakkima impact crater is a buried structure of 3 km diameter lacking any morphological expression. The crater depression was filled with sediments of Meso- to Neoproterozoic age and Quaternary deposits. The topography around the structure is rather flat, with undulating long and narrow moraine ridges and a chain of lakes that is tens of kilometers long and includes Lake Iso-Naakkima. The area is entirely covered by glacial till. The primary target rock is mica gneiss with occasional amphibolite intercalations and granitic veins. These rocks belong to the Svecokarelian Belt of the Baltic Shield. The rocks have magmatic and metamorphic ages of 1850 to 1930 Myr.

Iso-Naakkima was investigated by means of geophysical surveying and drilling. The crater has a pronounced circular gravity anomaly of −4 mGal. Likewise, the magnetic and electromagnetic signatures show circular patterns. Drilling into the center of the geophysical anomaly revealed, beneath 25–40 m of Quaternary glacial deposits, a 100 m thick sequence of non-metamorphosed shale, siltstone, quartz sandstone, kaolinitic clay, and conglomeratic sandstone, which rests on a basement of weathered mica gneiss. Planar deformation features (PDFs) and feather features (FFs) in quartz are widespread in samples from the basal conglomerate and in dike lithologies that crosscut the mica gneiss of Paleoproterozoioc age. The depth of the depression amounts to 160 m. A puzzling aspect of the Iso-Naakkima crater structure is that a thick breccia lens is missing in the crater depression. The lack of impact breccia may indicate a rather deep level of erosion.

A tentative paleomagnetic age for the impact event was determined on shocked samples from the central borehole at 900–1200 Myr.

The crater was discovered in 1989 during the regional gravity mapping program of the Geological Survey of Finland. The discovery was based on the striking circular negative Bouguer anomaly of the structure. Proof of impact origin became possible when the central borehole was drilled and shock metamorphic features were discovered.

Location/Access

The structure is located at the northwestern end of Lake Iso-Naakkima in southeastern Finland, between the towns of Pieksämäki and Mikkeli. National road 72, which links both towns, passes close to the eastern shoreline of Lake Iso-Naakima, only 10 km south of Pieksämäki.

Selected References

Elo, S., Kuivasaari, T., Lehtinen, M., Sarapää, O., and Uutela, A. (1993). Iso-Naakkima, a circular structure filled with Neoproterozoic sediments, Pieksämäki, Southeastern Finland. Bulletin of the Geological Society of Finland, Vol. 65, pp. 3–30.

Pesonen, L. J., Järvelä, J., Sarapää, O., and Pietarinen, H. (1996). The Iso-Naakkima meteorite impact structure: Physical properties and paleomagnetism of a drill core (abstract). Meteoritics and Planetary Science, Vol. 31, pp. A105–A106.

The area of the Iso-Naakkima impact structure with the homonymous lake and environs on August 11, 2018 as a composite image based on TanDEM-X DEM and Copernicus Sentinel-2 (channel 2/3/4) data.

Iso-Naakkima

Finland

latitude = 62°11'46" N / longitude = 27°07'48" E / diameter = 3 km / age = 900 - 1200 Myr

scale = 1 : 104 000 / scene = 18.7 × 24.9 km²

0 1.04 5.20 km

N

100 meter 205

Jänisjärvi
Russia

Thomas Kenkmann

The Jänisjärvi impact structure is located in the southeastern part of the Baltic Shield. It is mostly submerged by Lake Jänisjärvi, a somewhat irregular and elongated, 13 × 17 km in size, and up to 50 m deep lake. Its shape is different from that of other lakes in this region that are mostly elongated and formed by glacial overprint. The best estimate for the diameter of Jänisjärvi, based on the diameter of the −7 mGal gravity anomaly over the structure, is about 16 km. Whether the structure has a central uplift has remained unclear to date.

The structure is located in a region that is part of the Proterozoic Svecofennian realm of the Baltic Shield. It is located close to the border with an Archean cratonic province. Target rocks are metasedimentary rocks – crystalline schists and gneiss composed of plagioclase, biotite, quartz, muscovite, garnet, staurolite, and cordierite. The age for this metamorphism is 1885 ± 30 Myr. These metamorphic strata may have been overlain by Late Precambrian sedimentary strata at the time of impact.

A variety of impact lithologies occurs at Jänisjärvi. They include impact melt rock forming a 15–20 m thick coherent sheet that overlies suevites and allochthonous polymict breccia. The only exposures of impact melt rock occur on the three little islands in the northwest-central part, and on one promontory (peninsula) at the southwestern shore of the lake. Shock metamorphic features include shatter cones, planar deformation features (PDFs) in quartz, diaplectic quartz and feldspar glasses, fused minerals, and the SiO_2 high-pressure mineral coesite. Analysis of the magnetic properties of the impact melt rocks indicates that two essentially different types of melt exist. One type was likely formed by extremely rapid cooling of a superhot melt with initial temperature well above 2000 °C. The other type of melt was relatively cooler and is also enriched in iron.

The aeromagnetic anomaly map over the crater shows a rather subdued circular magnetic relief that corresponds to the location and size of the crater. This is in contrast to the strong relief of the surrounding Proterozoic rocks. Such a pattern is characteristic for many impact craters formed in crystalline targets and is consistent with relatively weak magnetization of the impactites.

Several attempts to date the impact melt rocks from Jänisjärvi have been made with different techniques over the years. Initial K-Ar age estimates gave an age of 718 ± 5 Myr. A subsequent study using the ^{40}Ar-^{39}Ar step-heating method yielded complex data that suggested an age of 698 ± 22 Myr. The most recent ^{40}Ar-^{39}Ar investigation provided an isochron age of 682 ± 4 Myr, with a much reduced error. Paleomagnetic data, however, are in conflict with the isotopic data and suggest ages for the rock magnetization either during the Late Sveconorwegian period (850–900 Myr) or during the Late Cambrian about 500 Myr ago.

Based on findings of melt rocks, a volcanic origin of Jänisjärvi was initially proposed in the early 20th century. An impact origin was first suggested in 1971 by Michael Dence. Findings of diaplectic mineral glasses and shatter cones by Victor Masaitis and co-workers a few years later confirmed this hypothesis.

Jänisjärvi impact melt breccia, 8.5 × 8 × 2.5 cm in size (photo courtesy: Ralf-Thomas Schmitt, Museum für Naturkunde Berlin, sample obtained from the VSEGEI, St. Petersburg, Russia).

Location/Access

Jänisjärvi is located 40 km north of the large Lake Ladoga, only 25 km from the border between Finland and the Russian Republic of Karelia. Lake Jänisjärvi can be reached by car via Highway A 129 from St. Petersburg, which is situated about 250 km to the south.

Selected References

Jourdan, F., Renne, P. R., and Reimold, W. U. (2008). High-precision ^{40}Ar-^{39}Ar age of the Jänisjärvi impact structure (Russia). Earth and Planetary Science Letters, Vol. 265, pp. 438–449.

Masaitis, V. L., Sindeev, A. S., and Staritsky, Y. G. (1976). The impactites of the Jänisjärvi astrobleme (in Russian). Meteoritika, Vol. 35, pp. 103–110.

Müller, N., Hartung, J. B., Jessberger, E. L., and Reimold, W. U. (1990). ^{40}Ar-^{39}Ar ages of Dellen, Jänisjärvi and Sääksjärvi impact craters. Meteoritics, Vol. 25, pp. 1–10.

Salminen, J. M., Donadini, F., Pesonen, L. J., Masaitis, V. L., and Naumov, M. V. (2006). Paleomagnetism and petrophysics of the Jänisjärvi impact structure, Russian Karelia. Meteoritics and Planetary Science, Vol. 41, pp. 1853–1870.

Sergienko, E. S., Kosterov, A., and Kharitonskii, P. V. (2017). Two types of impact melts with contrasting magnetic mineralogy from Jänisjärvi impact structure, Russian Karelia. Geophysical Journal International, Vol. 209, pp. 1080–1094.

Jänisjärvi

Russia

latitude = 61°58‘44“ N / longitude = 30°56‘40“ E / diameter = 16 km / age = 682 ± 4 Myr

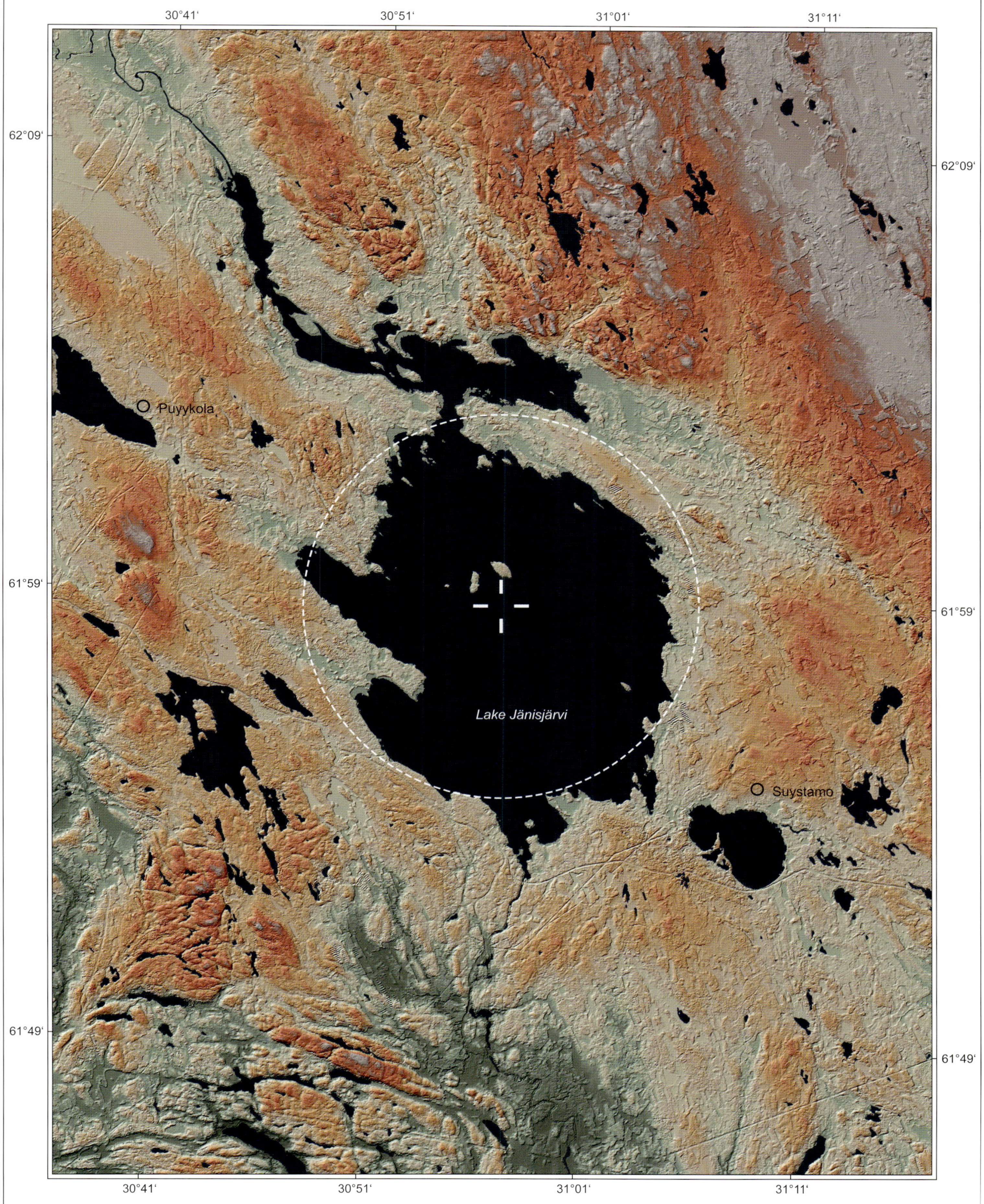

scale = 1 : 197 000 / scene = 35.5 × 47.3 km²

0 1.97 9.85 km

N

0 meter 250

Kaalijärv Crater Field
Estonia

Thomas Kenkmann

The Kaalijärv crater strewn field consists of nine craters located on Saaremaa Island in Estonia. The largest crater is 110 m wide and has a rim that rises about 6–7 m above the surrounding terrain. The crater depression is occupied with a deep pond of 9–10 m depth below the surface outside the crater. The smallest craters are 13 and 15 m wide, respectively, and the others range in size from 25 to 40 meters. The slightly elliptical strewn field extends over ca. 1 km in northwest-southeast direction. The long axis of the elliptical strewn field defines the trajectory of the impacting bolide and indicates an impact from the southeast.

Target rocks are Silurian dolomites that at the time were covered by a several meter thick glacial till. The bowl-shaped craters are partly forested and filled with ponds. Outcrops are rare. Recently, a trench was excavated through the ejecta blanket of the largest crater that shows an overturned stratigraphy characteristic of the proximal ejecta of impact craters. The rocks of the ejecta flap are crushed. Pulverized dolomite was described to occur along the inner slope of the largest crater. According to drilling, the dolomite in the largest crater was compressed to a depth of 4–5 meters.

The crater strewn field was formed by the impact of an IAB group iron meteoroid. Meteorite fragments have been found within the craters and in their surroundings. The age of the impact is still a matter of debate. Measurements by different techniques and of different materials resulted in ages between about 6400 years Before Current Epoch (BCE) and about 400 BCE. Recently, ^{14}C dating was carried out on charcoal fragments embedded in the proximal ejecta. The charcoal is believed to result from roasting of Picea branches that adhered to hot clasts within the ejecta blanket. This dating revealed an age of 3237 ± 10 years before present (BP). Saaremaa was already inhabited when the bolide hit the Earth, but there is no evidence that this event caused any changes in material culture or human habitation.

The craters aroused the interest of scientists as far back as the first quarter of the 19th century. The first systematic investigation was carried out by the German mining engineer Ivan Reinwald, who also found the first meteorite fragments and, thus, in 1928 proved the impact origin of the craters.

Location/Access

The crater strewn field on Saaremaa Island is located 2 km north of main Highway 10 that connects mainland Estonia with the town of Arensburg.

A 9.5 × 5 × 4.5 cm sample of Kaali dolomite (photo courtesy: Ralf-Thomas Schmitt, Museum für Naturkunde Berlin, sample obtained from Ivan Reinwald in 1938).

Selected References

Krinov, E. L. (1961). The Kaalijarv meteorite craters on Saarema Island, Estonian SSR. American Journal of Science, Vol. 259, pp. 430–440.

Losiak, A., Wild, E. M., Geppert, W. D., Huber, M. S., Jõeleht, A., Kriiska, A., Kulkov, A., Paavel, K., Pirkovic, I., Plado, J., Steier, P., Välja, R., Wilk, J., Wisniowski, T., and Zanetti, M. (2016). Dating a small impact crater: An age of Kaali crater (Estonia) based on charcoal emplaced within proximal ejecta. Meteoritics and Planetary Science, Vol. 51, pp. 681–695.

Reinwald, I. (1928). Bericht über geologische Untersuchungen am Kaalijärv (Krater von Sall) auf Ösel. Loodusuurijate Seltsi Aruanded, Vol. 35, pp. 30–70.

Veski, S., Heinsalu, A., Lang, V., Kestlane, Ü., and Possnert, G. (2004). The age of the Kaali meteorite craters and the effect of the impact on the environment and man: evidence from inside the Kaali craters, island of Saaremaa, Estonia. Vegetation History and Archaeobotany, Vol. 13, pp. 197–206.

Panoramic view of the largest, 110 m wide crater of the Kaali impact crater strewn field (image courtesy: Jakob Wilk, University of Freiburg).

Kaalijärv Crater Field

Estonia

latitude = 58°22'22" N / longitude = 22°40'10" E / diameter = max. 110 m / age = 3509 ± 40 yr

scale = 1:79 000 / scene = 14.2 × 18.9 km²

0 0.79 3.95 km

N

5 meter 55

Kärdla
Estonia

Thomas Kenkmann

Kärdla is a 4 km diameter crater structure in Estonia, situated close to the northeastern coast of the island of Hiiumaa. The crater does not have a prominent morphological expression.

The target represented an epi-continental shelf sequence composed of a 140 m thick Ordovician and Cambrian sedimentary succession of limestones, sandstones, siltstones, and clays overlying the Precambrian crystalline basement. The crater depression was immediately filled with allochthonous fragmental breccias, that, in turn, were covered by resurge conglomerates, turbidites, and sands eroded from the uplifted crater walls prior to burial of the structure. Except for the crater rim, the inner structure and the ejecta blanket are perfectly preserved. Drilling proved the existence of a small central uplift of approximately 100 m height and a diameter of roughly 300 m.

The ejecta blanket occurs stratigraphically in a succession of Upper Ordovician carbonate rocks as a 0.01–9.6 m thick bed of fragmented and disintegrated Cambrian and Lower Ordovician sand, silt, and clay, with minor admixture of fragments from the other lithological units of the target.

Shock features including planar deformation features (PDFs) in quartz were recorded in different settings of the crater, e.g., in the distal ejecta blanket. Remote sensing data and marine seismic reflection profiles show that Kärdla is surrounded by an elliptical, southwest-northeast elongated ring fault. The 2 to 8 km wide zone between the rim and the ring feature is built up by fractured and folded blocks that presumably belong to the ejecta blanket. Mineral parageneses of the lithic breccia and shattered crater floor material indicate hydrothermal activity that was active until 9000–10 000 years after the impact.

A 7 × 5 cm sample from core K1 that was drilled into the central depression of the Kärdla impact structure. The sample is derived from 542 m depth. The allochthonous clast-rich breccia shows injected matrix material (image courtesy: Jüri Plado, University of Tartu, Estonia).

The Kärdla structure is characterized by a negative gravity anomaly of −3 mGal that is surrounded by a positive anomaly. The negative anomaly is due to low-density post-impact sediments infilling the structure, and the low density of impact breccias, whereas the gravity high reflects the uplifted and comparatively denser crystalline rim. The local magnetic anomaly displays virtually the same pattern as the gravity anomaly. Reflection seismic and ground-penetrating radar investigations showed that the rim of Kärdla does not have a uniform height, but displays various degrees of denudation by the resurging sea. The variations in rim height may be attributed either to an oblique impact from the northeast or local differences in the physical properties of the crystalline target. The impact took place in a shallow marine sea of likely 20 to 100 m depth, at approximately 455 Myr ago. Kärdla formed during the Middle to Late Ordovician during a period, for which an enhanced flux of L chondrite asteroids has been suggested due to a major collision event in the asteroid belt at 470 ± 6 Myr.

Kärdla attracted the attention of geologists already in the mid 19th century, when Karl Eduard von Eichwald and Alexander Gustav von Schrenk in 1843 and 1854, respectively, described abnormally tilted sedimentary rocks in a quarry at Palükula/Hiiumaa. More than a century later, in 1967, drilling for groundwater at Palükula revealed crystalline rocks at an unusually shallow depth. An extensive drilling program was initiated in the 1970s, as well as gravity and ground magnetic mapping, which finally revealed the crater structure. As a result of this, the impact origin was proven by the discovery of quartz with PDFs in allochthonous breccia by Victor Masaitis.

Location/Access

Kärdla is named after the town of Kärdla at the northern shore of Hiiumaa Island in western Estonia. The structure is located onshore within the limits of Kärdla town. Hiiumaa is linked with Estonia's mainland by ferry between the villages of Rohükula and Helterma. From Helterma regional road 80 leads to Kärdla. Alternative access is provided by air between Tallin and Kärdla.

Selected References

Plado, J. (2012). Meteorite impact craters and possibly impact-related structures in Estonia. Meteoritics and Planetary Science, Vol. 47, pp. 1590–1605.

Plado, J., Pesonen, L. J., Elo, S., Purra, V., and Suuroja, K. (1996). Geophysical research on the Kärdla impact structure, Hiiumaa Island, Estonia. Meteoritics and Planetary Science, Vol. 31, pp. 289–298.

Puura, V. A., and Suuroja, K. (1992). Ordovician impact crater at Kärdla, Hiiumaa Island, Estonia. Tectonophysics, Vol. 216, pp. 143–156.

Suuroja, S., Suuroja, K., and Flodén, T. (2013). A comparative analysis of two Early Palaeozoic marine impact structures in Estonia, Baltic Sea: Neugrund and Kärdla. Bulletin of the Geological Society of Finland, Vol. 85, pp. 83–101.

Kärdla

Estonia

latitude = 58°58‘26“ N / longitude = 22°46‘51“ E / diameter = 4 km / age = 455 Myr

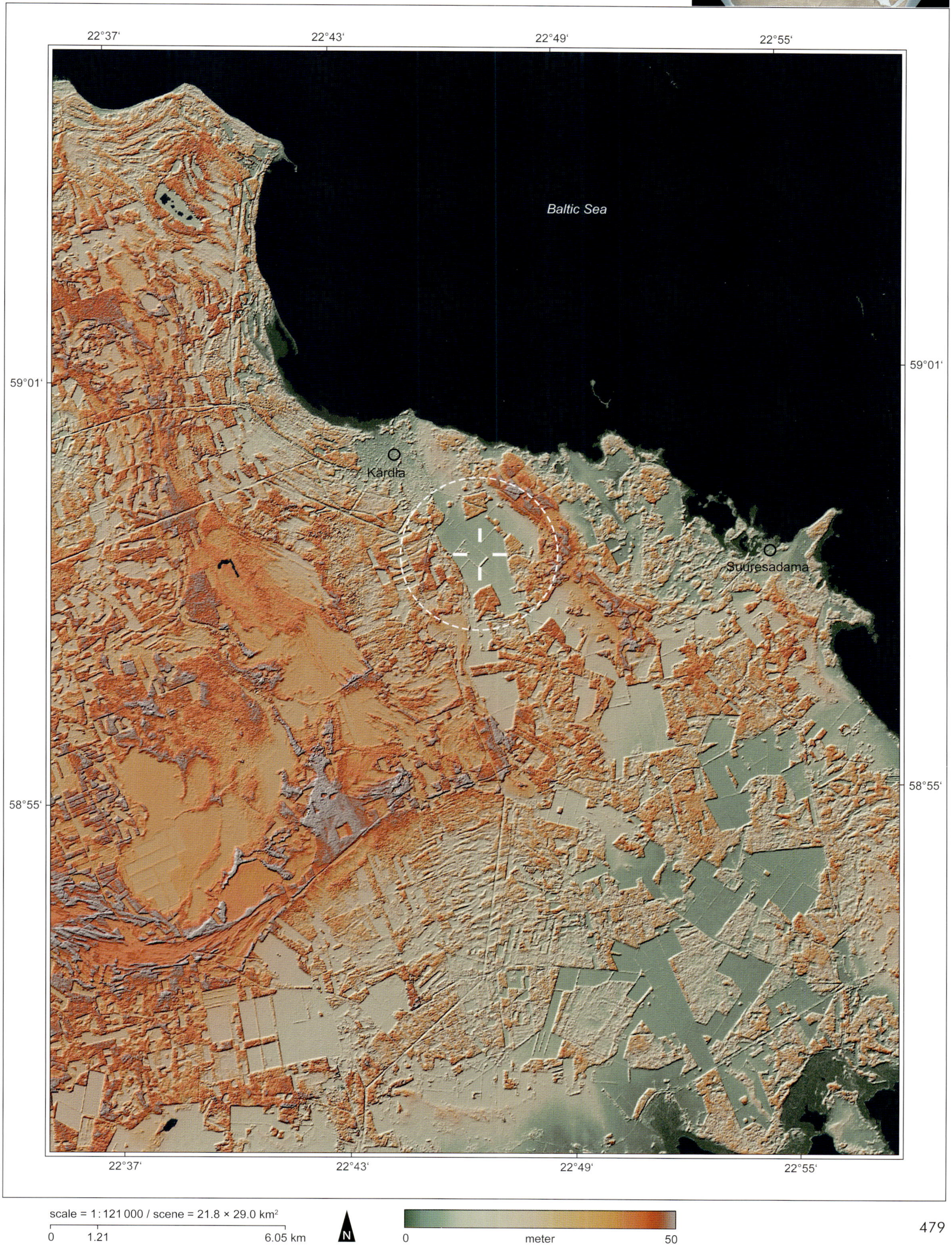

Kaluga
Russia

Thomas Kenkmann

The 15 km diameter Kaluga impact structure is situated in the central part of the Eastern European plain. The structure is buried under Middle to Upper Devonian and Carboniferous beds, with a total thickness of 800 m. The impact happened into a marine environment with a water depth of at least 300 m.

The target rocks are Archean and Lower Proterozoic biotite gneisses, granites, and schists, which were covered at the time when the impact occurred by a 125 m thick sedimentary sequence of Precambrian and Middle Devonian siliciclastic rocks. There is a distinct horizon of reworked breccia within the Middle Devonian rocks that coincides with the impact. These deposits are interpreted as an impact tsunami layer. Strata below are deformed by the impact, strata above are not affected. By removing the sediments resting on top of the crater, the structure appears as a rounded depression with a true depth of the annular moat of 450–500 m. It is filled with polymict breccias, suevite, and – to a minor degree – with irregular and up to 20 m thick bodies of impact melt rock. The polymict breccia consists of fragments and small, up to 2–4 m sized blocks of migmatitic biotite gneiss, granite, and minor schist. The clasts display abundant shock features including planar deformation features (PDFs) and planar fractures (PFs) in quartz. The presence of a central uplift is likely; it has, however, not been drilled yet. The crater is a prominent example of an impact into a shallow sea.

Seismic surveying and a dense network of boreholes in the periphery of the crater structure revealed a prominent and well-preserved crater rim with a height of 250–300 m and a width of 6 km. The lower portion of the rim is made up of imbricated, outward thrust, and inward slumped blocks of the target. At the top occur 100 m of polymict breccias, which, in turn, are covered by up to 150 m of impactoclastic breccias that were reworked by water resurging into the crater. Intense outward and inward slumping of blocks is associated with the formation of the initially outward-migrating, high-amplitude rim, collapse waves of the water column, and the subsequent resurge waves. Resurging sedimentary breccia and suevite filled the crater depression. They have a thickness of about 250 m or more. This sedimentary breccia is widely distributed in northwest Russia, Belarus, Lithuania, Latvia, and Estonia.

The age of the impact event is defined by the reworked impact-tsunami breccia bed of Middle Devonian age. Based on the bio-stratigraphic ages of over- and underlying strata the impact age is constrained to 395 ± 4 Myr.

The Kaluga structure was discovered by geophysical observations and drilling. Its recognition as an impact structure was achieved in 1969 by V. Petrov, and followed by a detailed analytical study by Victor Masaitis in 1973 and following years.

Location/Access

The buried Kaluga impact structure lies in an area with well developed infrastructure in the oblast Kaluga. The town of Kaluga 160 km southwest of Moscow is located in the northeastern part of the impact structure. Highway M3 brings the visitor from Moscow close to the buried impact structure.

Selected References

Masaitis, V. L. (1976). Astroblemes in the USSR. International Geology Review, Vol. 18, pp. 1249–1258.

Masaitis, V. L. (1999). Impact structures of northeastern Eurasia: The territories of Russia and adjacent countries. Meteoritics and Planetary Science, Vol. 34, pp. 691–711.

Masaitis, V. L. (2002). The middle Devonian Kaluga impact crater (Russia): new interpretation of marine setting. Deep-Sea Research Part II, Vol. 49, pp. 1157–1169.

Petrov, V. G. (1974). Peculiarities of the Interior of the Kaluga Circular Structure and its Location in the Moscow Syneclise (in Russian). Unpublished Ph.D. Thesis. Moscow State University, Moscow, Russia.

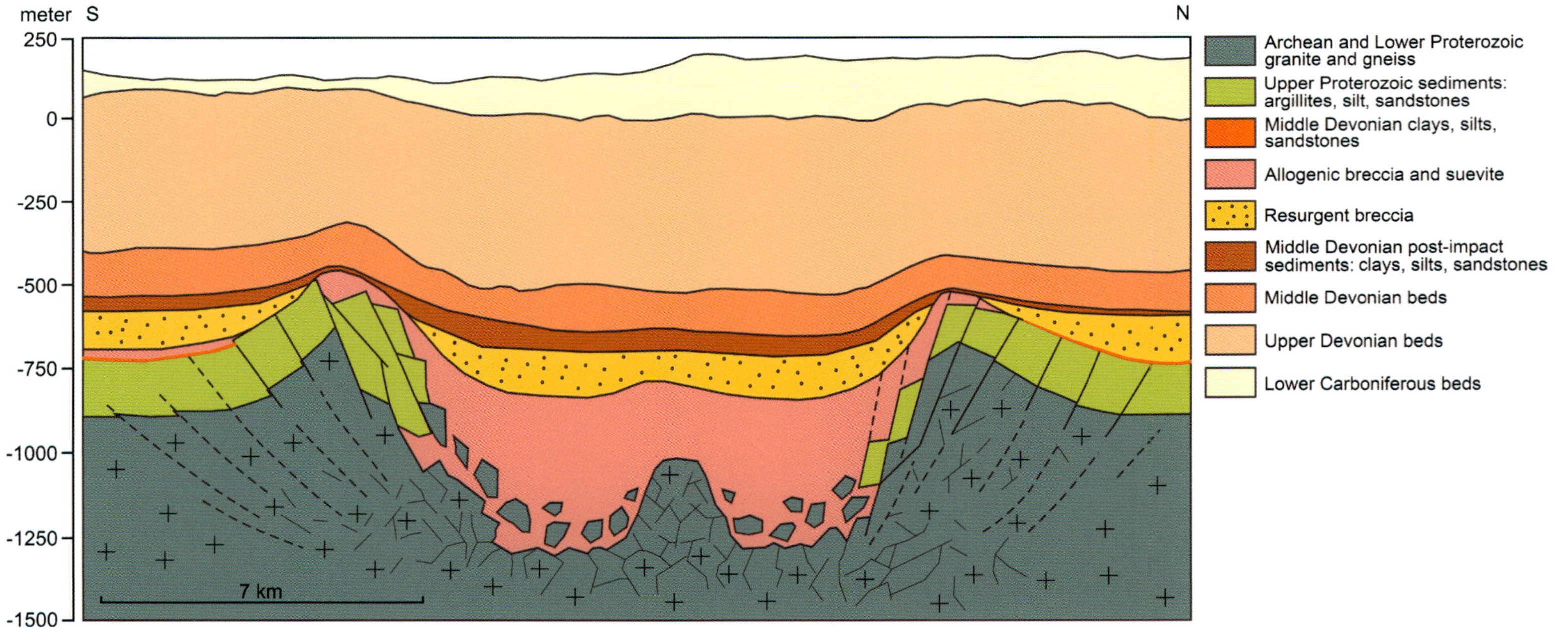

Schematic cross-section of the Kaluga impact structure (modified after Masaitis, 2002).

Kaluga

Russia

latitude = 54°30' N / longitude = 36°12' E / diameter = 15 km / age = 395 ± 4 Myr

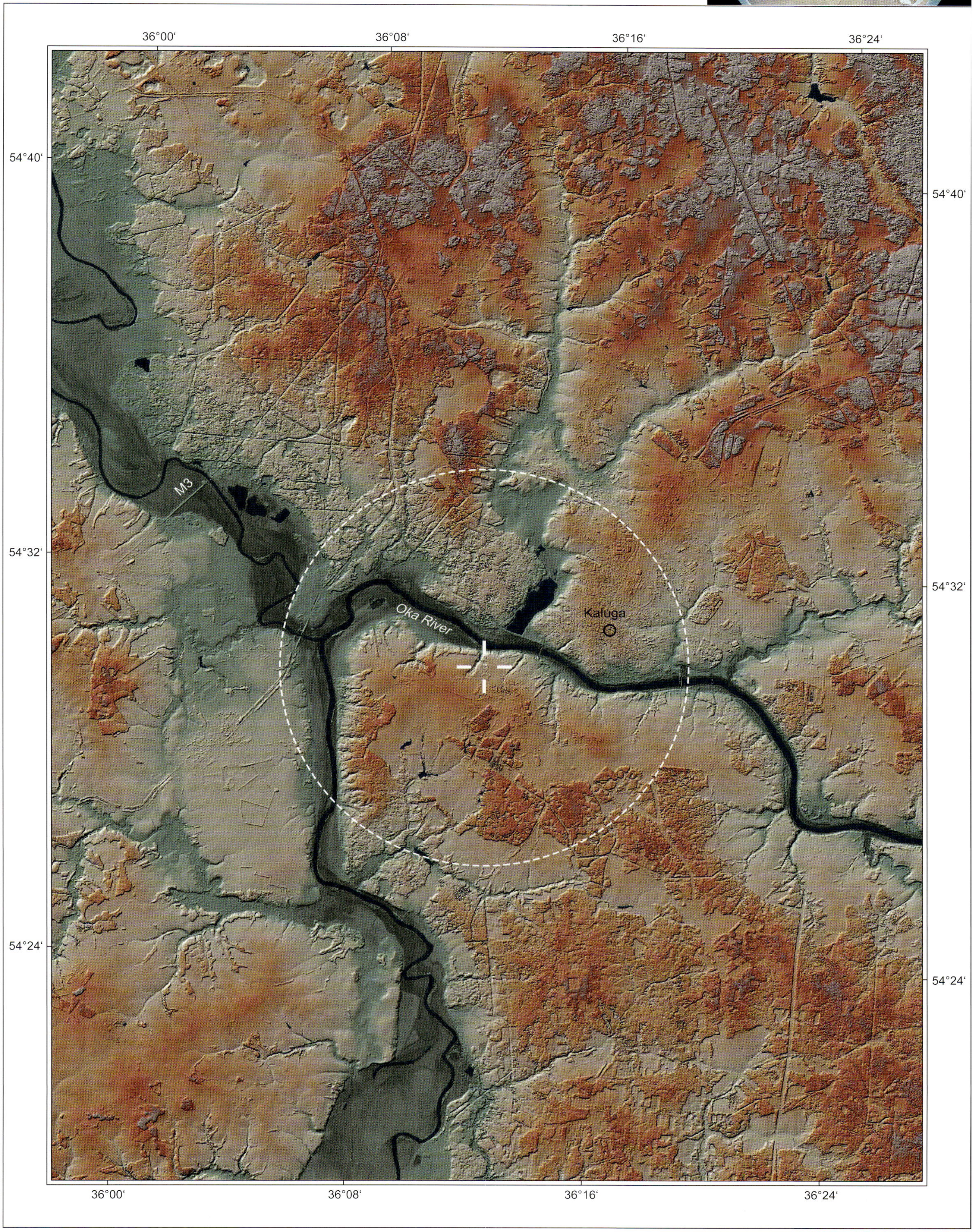

scale = 1 : 177 000 / scene = 31.9 × 42.6 km²

0 1.77 8.85 km

N

100 meter 275

Kamenetsk

Ukraine

Thomas Kenkmann

The recently discovered simple bowl-shaped impact crater Kamenetsk is a buried structure with an apparent diameter of 1.0–1.2 km and a depth of 130 m. Its original diameter was, however, larger because the structure has been affected by erosion.

The crater was formed in a crystalline target of the Kirovograd Complex, which is part of the Ukrainian Shield. The crystalline basement is composed of 2300–2400 Myr old biotite-rich gneisses. A gradual transition from brecciated crystalline rocks to overlying lithic breccias has been revealed. The lithic breccias fill the depression by up to 60 m. A series of sedimentary breccias unconformably overlies the eroded surface of these lithic breccias, and, in turn, are overlain in the central part of the Kamenetsk structure by an 18–25 m thick, lenticular body of lignite. The upper part of the post-impact sedimentary sequence is built by a series of shallow marine Miocene sediments with a thickness of 11–19 m, and 8–15 m thick Quaternary loess.

Distinctive signs of shock metamorphism were discovered in the lithic breccias. Also, weak shock metamorphism occurs occasionally in minerals of the crystalline basement. Quartz and alkali feldspar from lithic breccias contain planar deformation features (PDFs) and planar fractures (PFs) at various crystallographic orientations.

The absence of impact glasses and impact melt rocks has prevented precise age determination. A maximum age for the impact is constrained by the youngest rocks of the crystalline target – Proterozoic granites of 2000–2100 Myr age. The sedimentary infill does not contain datable rock, except for the Upper Neogene lignite sequence. Based on mean erosion rates of the Ukrainian Shield, a Paleozoic age has been estimated for the impact event.

The Kamenetsk gravity anomaly of −2 mGal was discovered in 2004 during gravity prospecting of the Kirovograd Block of the Ukrainian Shield. This prompted between 2005 and 2009 a drilling program of 15 boreholes within the extent of the gravity anomaly. Because of the wide distribution of rocks resembling tuff, the structure was initially regarded as an endogenic explosive structure of volcanic origin. Evgeny Gurov and co-workers eventually discovered shock effects in drill core samples from the structure and confirmed the impact origin.

Location/Access

The Kamenetsk impact structure is located in the basin of the Ingul River, 95 km south of the town of Kirovograd and 80 km southeast of Krywoj Rog. Highway H 14 from Kirovograd passes 15 km to the west, and local road T 1602 from Krywoj Rog about 8 km south of the crater.

Selected References

Gurov, E., Nikolaenko, N., Shevchuk, H., and Yamnichenko, A. (2017). Kamenetsk – A new impact structure in the Ukrainian Shield. Meteoritics and Planetary Science, Vol. 52, pp. 2461–2469.

Kamenetsk
Ukraine

latitude = 47°46‘ N / longitude = 32°21‘ E / diameter = 1.1 km / age = 250 - 540 Myr

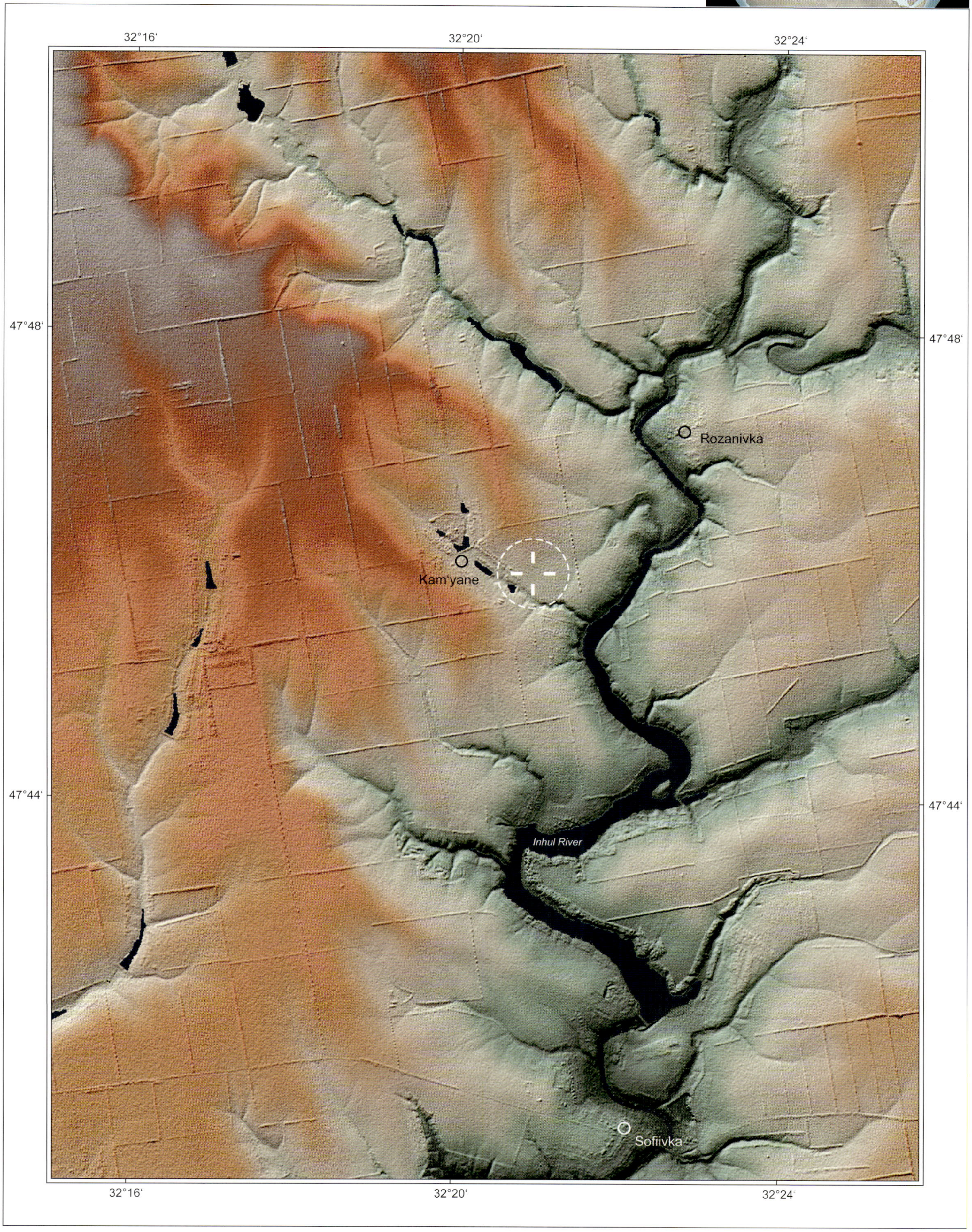

scale = 1:74 000 / scene = 13.3 × 17.7 km²
0 0.74 3.70 km
N
10 meter 125

Kara
Russia

Thomas Kenkmann

The 65 km diameter Kara impact structure is located in the northernmost part of the polar Urals, entirely on land in a tundra environment. The northeastern part of the low-relief Kara structure opens to the Kara Sea and is otherwise surrounded by a hilly terrain in the southwestern sector, which corresponds to the crater rim zone. Major outcrops occur along the river valleys, which cut into the plain close to the rim. The periphery of the crater has a very complicated structure and is best preserved in the southwestern crater sector, where overthrusts, faults, displaced blocks, and allochthonous megablocks of different target rock types are exposed.

Kara was formed in gently folded Paleozoic strata of Ordovician to Lower Permian age that are intruded by dolerite dikes. The total thickness of these sedimentary sequences is 5.5 km. The impact structure is overlain by massive Quaternary deposits. A thin lens of Pliocene sediments below the Quaternary strata is the remnant of a crater lake that formed in the Late Tertiary.

Allochthonous polymict lithic breccia and suevites overlying the disturbed Paleozoic rocks are mainly exposed in river valleys, along the periphery of the structure. These impactites reach a thickness of 2 km within the annular moat. They are composed of shocked, melted, and brecciated Lower Paleozoic rocks and dolerites. Planar deformation features (PDFs) record shock pressures of 12–15 GPa. Numerous, well-preserved shatter cones occur on clasts in different breccia types. Impactites of the Kara crater also contain impact diamonds formed due to transformation of coal under shock pressures of about 26–30 GPa and high post-shock temperature. Coesite, a high-pressure polymorph of quartz, was also found in the shocked rocks and suggests the local attainment of even higher shock pressure. Recently glass veinlets formed under ultra-high pressures and temperatures have been reported from these rocks.

The allochthonous breccias can be subdivided into two major units: A lower unit is composed of megabreccias, embedded into a block agglomerate and suevites. The vesicular glass fragments and bombs of the suevites are usually altered and transformed into aggregates of secondary clay minerals, minor calcite, zeolites, and pyrite, amongst other minerals. The lower unit also includes irregular bodies of impact melt rock reaching thicknesses of 15–20 m. The bulk composition of impact melt rock corresponds to the composition of Permian sandstones. The lenses of impact melt rock are commonly full of clasts of shocked and partly fused sandstone and siltstone. Geochemical investigations indicate that the impact melt rocks are slightly enriched in siderophile elements, possibly caused by contamination with projectile material. The total thickness of the lower unit is 0.7–0.9 km. The upper unit is composed of finer-grained lapilli stone-like suevites and fine-grained polymict lithic breccias. A gradational increase in fine-grained material upwards suggests the deposition of air-fall material. The total thickness of this unit is 0.8–1.2 km.

Drill core section of Kara suevite, 12.5 × 7.5 cm in size (photo courtesy: Ralf-Thomas Schmitt, Museum für Naturkunde Berlin, sample obtained from the VSEGEI, St. Petersburg, Russia).

A 10 km wide central uplift in the middle of the breccias has a topographic expression of roughly 2 km width. However, the stratigraphic uplift is even higher and was estimated to be 5 km. The annular moat is characterized by a negative gravity anomaly, whereas the central uplift is correlated to a positive gravity excursion. According to geophysical data, the brecciation reaches to a depth of 3.5 km. Further knowledge of the crater came from an extensive drilling program into the impactites of the annular moat and the rim of the central uplift.

Early attempts to date the impact event using the K-Ar dating method gave ages that were associated with the Cretaceous-Paleogene boundary at 66 Myr. This association could be ruled out later by Christian Koeberl and co-workers, when a precise age of 70.3 ± 2.2 Myr was determined based on the weighted mean of ^{40}Ar-^{39}Ar plateau ages for glassy impact melt rocks from Kara that had been cleaned of mineral and rock clasts.

The impact origin of Kara was considered for the first time by P. S. Voronov in 1970. Sergej Vishnevsky was the first to document

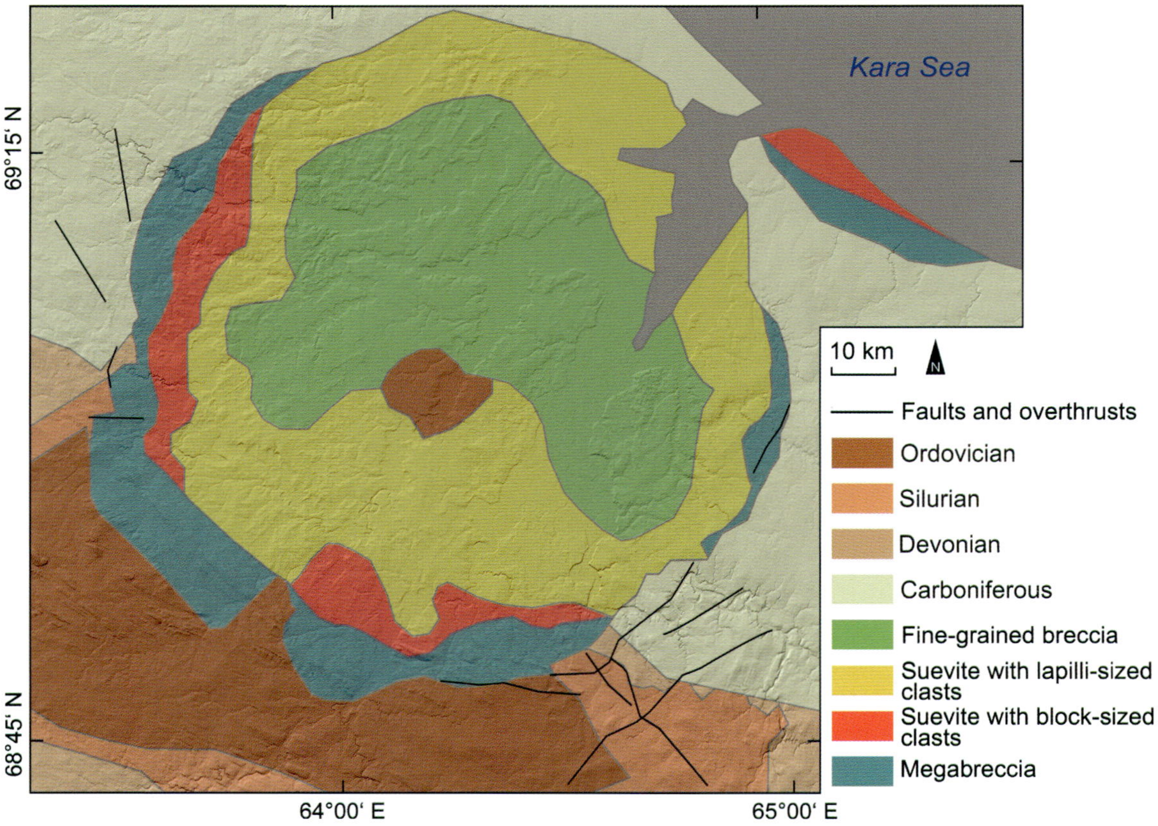

Schematic geological map of the Kara impact structure, underlain with a TanDEM-X DEM (modified after Masaitis, 1999).

Kara

Russia

latitude = 69°05‘ N / longitude = 64°20‘ E / diameter = 65 km / age = 70.3 ± 2.2 Myr

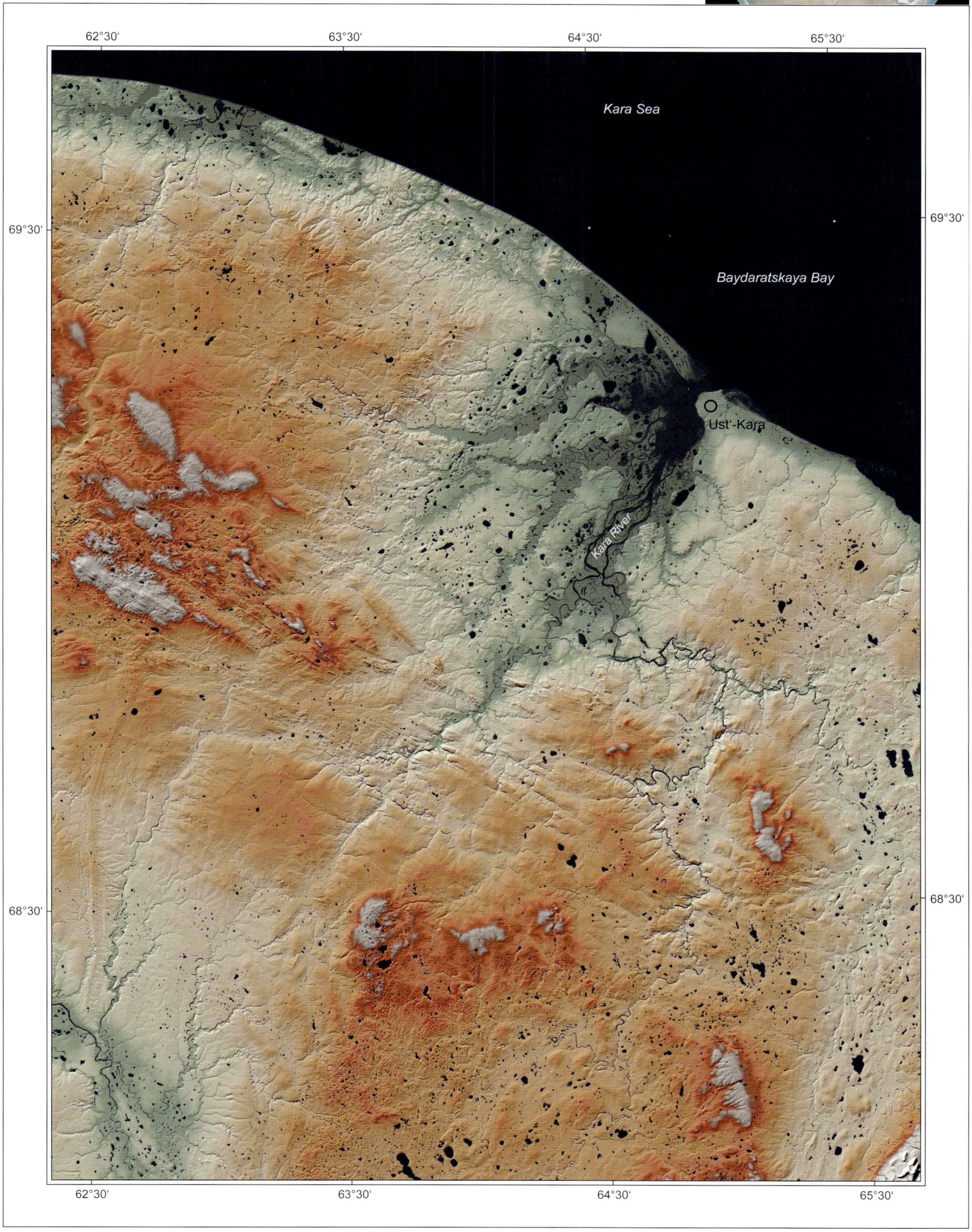

scale = 1 : 761 000 / scene = 137.0 × 182.6 km²

0 7.61 38.05 km

N

0 meter 1370

Kara impact structure: The Kara river valley with diamond-bearing suevites on both banks of the river. Both photos were taken with a drone in the framework of the Russian Foundation for Basic Research Project #17-05-00516 (photo courtesy: Alexander Zubov and Tatyana Shumilova, both Institute of Geology of Komi Scientific Center, Syktyvkar, Russia).

the high-pressure mineral coesite from this impact structure in 1977. Later, in 1980, Kara was presented in a Russian textbook on the geology of astroblemes published by Victor Masaitis and co-workers. Various publications exist on Kara, though most of them were published in Russian. A recent expedition to Kara led by Tatyana Shumilova of the Komi Scientific Center, Syktyvkar, Russia, revived research on Kara.

Close to Kara, onshore exposures can be found near the Kara River estuary. They have often been attributed to a smaller twin impact structure **Ust-Kara**, hidden mostly underwater. Its presence may be inferred from outcrops of suevites, shocked country rocks, impact melt rock and impact glasses that are distributed in a 16 km wide band southeast of Cape Polkovnik, just northeast of Kara crater. The diameter of the alleged Ust-Kara structure is poorly constrained and has been alternatively reported at 25 to 70 km. Another explanation for these reported exposures does, however, not require a second impact but would be consistent with a larger diameter for the Kara impact structure of about 100 km.

Location/Access

Kara is situated in the polar tundra without any infrastructure. It is only accessible through the Russian authorities with extensive logistical support, such as helicopter transportation.

Selected References

Koeberl, C., Sharpton, V. L., Murali, A. V., and Burke, K. (1990). Kara and Ust-Kara impact structures (USSR) and their relevance to the K/T boundary event. Geology, Vol. 18, pp. 50–53.

Masaitis, V. L. (1999). Impact structures of northeastern Eurasia: The territories of Russia and adjacent countries. Meteoritics and Planetary Science, Vol. 34, pp. 691–711.

Masaitis, V. L., Danilin, V. N., Mashchak, M. S., Raikhlin, A. I., Selivanovskaty, A. V., and Shadenkov, E. M. (1980). The Geology of Astroblemes. Nedra Press, Leningrad, Russia. 231 pp.

Shumilova, T. G., Lutoev, V. P., Isaenko, S. I., Kovalchuk, N. S., Makeev, B. A., Lysiuk, A. Yu., Zubov, A. A., and Ernstson, K. (2018). Spectroscopic features of ultrahigh-pressure impact glasses of the Kara astrobleme. Scientific Reports, Vol. 8, No. 6923.

The Kara impact structure on May 21, 2018 as a composite image based on TanDEM-X DEM and Copernicus Sentinel-2 (channel 2/3/4) data.

Trieloff, M., Deutsch, A., and Jessberger, E. (1998). The age of the Kara impact structure, Russia. Meteoritics and Planetary Science, Vol. 33, pp. 361–372.

Vishnevsky, S. A., Maslov, M. A., Pal'chik, N. A., and Ponomarev, G. Y. (1977). Coesite in rocks of the Kara structure. Akademiya Nauk SSSR Doklady, Vol. 232, pp. 446–448.

Kara impact structure: Massif of the clast-free impact melt rocks with columnar jointing. The photo was taken w th a drone in the framework of the Russian Foundation for Basic Research Project #17-05-00516 (photo courtesy: Alexander Zubov and Tatyana Shumilova, both Institute of Geology of Komi Scientific Center, Syktyvkar, Russia).

Karikkoselkä

Finland

Thomas Kenkmann

Karikkoselkä is an almost perfectly circular lake of about 1.4 km diameter in Central Finland. The lake coincides with a simple, bowl-shaped impact structure. With a depth of 26 m, this Quaternary lake is anomalously deep. The present size of the crater is 1.5 km but the original crater, now somewhat eroded, must have been larger.

Target rocks are porphyritic granites that belong to the Central Finland Granitoid Complex (CFGC) of the Svecofennian orogeny with an age of about 1880 Myr. The granites are exposed along the shore of the lake and exhibit shatter cones, in particular along the southeastern shoreline. The shatter cone hosting granites are in situ, and shatter cones range in size from a few centimeters to almost two meters. The apices of the shatter cones point toward the lake. Additional shatter cones were found in the glacial till. Moreover, the granites show intense fracturing. Quartz grains from granite display planar deformation features (PDFs) and feather features (FFs).

An airborne electromagnetic survey indicated a strong, circular anomaly over the structure that was explained by the presence of a conductive sedimentary layer below the lake bottom, and the presence of lithic impact breccias containing saline water. Based on geophysical modeling it was estimated that the bottom of the crater lies 120 m below lake level. The size of the crater is constrained by a negative gravity anomaly of −3.8 mGal. As no impact-generated melt rock has been found so far, radiometric age determination has remained impossible. Paleomagnetic data suggest either an age between 230 and 260 Myr or between 530 and 560 Myr. A study of microfossils, however, excludes the Cambrian to Late Neoproterozoic period and leaves the Early Triassic to Late Permian as the most likely time for the impact event. Karikkoselkä was proposed as an impact structure in 1996 by Martti Lehtinen and co-workers based on the evidence listed above.

Location/Access

Karikkoselkä lies about 230 km north of the capital of Finland, the city of Helsinki, in a well developed area. Highway 4 links Helsinki with the city of Jvväskylä. From there subsidiary road 6016 brings the visitor close to the lake.

Selected References

Lehtinen, M., Pesonen, L. J., Puranen, R., and Deutsch, A. (1996). Karikkoselkä – a new impact structure in Finland (abstract). 27th Lunar and Planetary Science Conference, Lunar and Planetary Institute, Houston, Texas, pp. 739–740.

Pesonen, L. J., Elo, S., Lehtinen, M., Jokinen, T., Puranen, R., and Kivekäs, L. (1999). Lake Karikkoselkä impact structure, central Finland: New geophysical and petrographic results. In: Large Meteorite Impacts and Planetary Evolution II. Dressler, B. O., and Sharpton, V. L. (eds.), Geological Society of America, Special Paper, Vol. 339, pp. 131–147.

Uutela, A. (2001). Proterozoic and early Palaeozoic microfossils in the Karikkoselkä impact crater, Central Finland. Bulletin of the Geological Society of Finland, Vol. 73, Parts 1–2, pp. 75–85.

Karikkoselkä
Finland

latitude = 62°13'17" N / longitude = 25°14'55" E / diameter = 1.4 km / age = 230 - 260 Myr

scale = 1:119 000 / scene = 21.5 × 28.6 km²

0 1.19 5.95 km

N

90 meter 260

Karla
Russia

Thomas Kenkmann

Karla, a 10 km wide impact structure in oblast Tatarstan, Russia, is located in the middle of the East European plain. The area is predominantly agricultural land, meadow and forest, and the structure is inconspicious in relief. The geologically reconstructed depth of the annular trough amounts to about 500 m. A central uplift is present and consists of brecciated carbonate rocks over an area of 600 × 800 m, which are cataclased and cut by veins of loose, fine-grained breccia.

The sequence of target rocks affected by the impact consisted of more than 390 m of horizontally stratified Middle and Upper Carboniferous limestones and dolomites, followed by 320 m of Upper Permian dolomites, limestones, and terrigenous rocks, 100 m of Middle and Upper Jurassic clays and sandstones, and finally 100 m of Cretaceous clays. These lithologies also provide the constituents of a breccia that fills the annular trough of the crater to a thickness of up to 300 m, and which overlies the crater environs as a thin veneer. The ejecta blanket is particularly well preserved to the northeast of the crater and can be traced 15 km outside the crater. Fragment sizes of this breccia deposit range from 5 to 10 cm, and the clasts are cemented with comminuted material of the same composition. However, there are also some megablocks of up to 1 km size of Upper Permian rocks preserved. They occur close to the rim of the crater and are embedded in polymict lithic breccia.

Within the area of the crater depression, the impact lithologies are buried under post-impact, Pliocene lake deposits of up to 100 m thickness. Outside the crater the impact lithologies are partly exposed and include rock fragments that contain shatter cones. The presence of Miocene age rock fragments constrains the age of the impact to less than 5 Myr.

First reports on Karla that were written in Russian were presented by Victor Masaitis in 1976 and 1980. He also presented a short review in 1999.

Shatter cones in dolomite from the Karla impact crater. The rock is part of the exposed ejecta blanket. Image width is about 20 cm (image courtesy: Michael Poelchau, University of Freiburg).

Location/Access

Karla is situated in east-central Russia some 120 km southwest of the city of Kasan. The stucture can be reached via roads A 151 and P 241.

Selected References

Masaitis, V. L. (1999). Impact structures of northeastern Eurasia: The territories of Russia and adjacent countries. Meteoritics and Planetary Science, Vol. 34, pp. 691–711.

Masaitis, V. L., Danilin, V. N., Karpov, G. M., and Raikhlin, A. I. (1976). Karla, Obolon', Rotmistrovka astroblemes in European part of the USSR (in Russian). Doklady ANSSSR, Vol. 230, pp. 174–177.

Masaitis, V. L., Danilin, V. N., Mashchak, M. S., Raikhlin, A. I., Selivanovskaya, A. V., and Shadenkov, E. M. (1980). The Geology of Astroblemes. Nedra Press, Leningrad, Russia, 231 pp.

Karla
Russia

latitude = 54°57‘23“ N / longitude = 47°57‘04“ E / diameter = 10 km / age = < 5 Myr

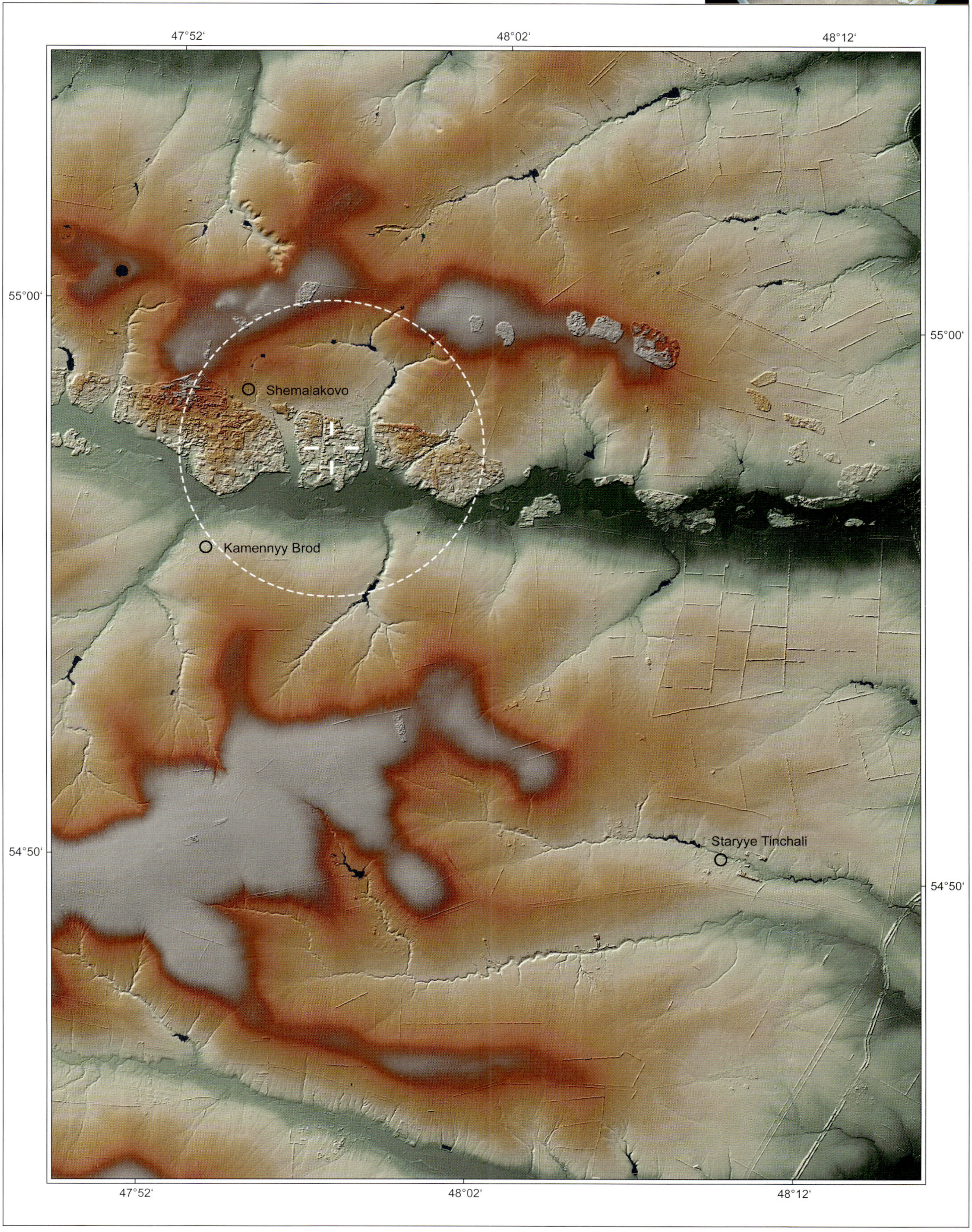

scale = 1 : 158 000 / scene = 28.5 × 38.0 km²

0 1.58 7.90 km

N

65 meter 245

Keurusselkä
Finland

Thomas Kenkmann

The Keurusselkä structure is a deeply eroded impact crater named after the large but shallow Lake Keurusselkä in Central Finland. The lake is devoid of any circularity, as its outline is determined by the topography of glacial moraines. Thus, the diameter of the structure is difficult to determine precisely. The impact structure has a faint topographic expression with some concentric ridges occurring between 10 and 30 km from the assumed center. A value of 30 km for the structure's diameter has been widely accepted.

The impact structure was formed in the Central Finland Granitoid Complex (CFGC) of the Finnish Svecofennian. The CFGC has a Precambrian age of 1870 to 1900 Myr. Exposed rocks include metavolcanic rocks, granites, granodiorites, and mica schists. Exposure of these rocks in the impact area is mainly along the shoreline of Lake Keurusselkä.

Shatter cones can be found in a 14 km wide circular area and are abundant in the central part of the structure where they have been mapped at many outcrops. They occur in various lithologies including granites, granodiorites, gneisses, and metavolcanic rocks. The best developed ones occur in the fine-grained, dense metavolcanic rocks. The extent of the circular area may represent the eroded remnants of the central uplift. Locally, pseudotachylitic breccia veinlets have been reported to occur. Shock metamorphic features include planar

A view from Jylhänniemi peninsula across Lake Keurusselkä (photo: W. U. Reimold).

Shatter cone-bearing fractures in granitoid from the shoreline at the Jylhänniemi peninsula in the central part of the Keurusselkä impact structure. Length of the pen, for scale, about 13 cm (photo: W. U. Reimold).

Keurusselkä shatter cone in gneiss; a 28 × 23 × 15 cm sized specimen (photo courtesy: Ralf-Thomas Schmitt, Museum für Naturkunde Berlin, sample from Max Hasch).

Keurusselkä

Finland

latitude = 62°08' N / longitude = 24°37' E / diameter = 30 km / age = 1151 ± 10 Myr

scale = 1 : 280 000 / scene = 50.3 × 67.1 km²

0 2.80 14.00 km

N

65 meter 250

fractures (PFs) and decorated planar deformation features (PDFs) of multiple orientations in quartz and feldspar grains of shatter coned rock. Due to the deep level of erosion, impact melt rocks or suevites are lacking in this impact structure.

A pseudotachylitic breccia veinlet, possibly formed by frictional melting and crosscutting a shatter cone surface, was dated with the stepwise heating ^{40}Ar-^{39}Ar method and yielded an age of 1151 ± 10 Myr. However, whether this age represents the impact event or some tectonic event remains to be shown.

Rock magnetic data measured in the area imply an increased magnetization and susceptibility within the area of shatter cones. The presence of magnetite and pyrrhotite was observed as carriers for the remanent magnetization.

The crater was discovered and proven by Satu Hietala and Jarmo Moilanen in 2004. They documented shatter cones in both in-situ outcrops and in glacial boulders to the south of the lake. In addition, observations of shock microstructures in quartz from glacial boulders were presented. These findings were substantiated by Ludovic Ferriere and co-workers in 2010. Martin Schmieder and his team provided the age determination, and, most recently, further work on the shatter cones was reported by Max Hasch and colleagues in 2016.

Location/Access

Keurusselkä is situated in central Finland and can be easily reached from the city of Helsinki via Lahti on Highways 4 and 24. The distance from Helsinki to the lake is about 250 km. A good location for studying shatter cones is the Jylhäänniemi peninsula at Lake Keurusselkä.

Selected References

Ferrière, L., Raiskila, S., Osinski, G. R., Pesonen, L. J., and Lehtinen, M. (2010). The Keurusselkä impact structure, Finland – Impact origin confirmed by characterization of planar deformation features in quartz grains. Meteoritics and Planetary Science, Vol. 45, pp. 434–446.

Hasch, M. Reimold, W. U., Raschke, U., and Zaag, P. T. (2016). Shatter cones at the Keurusselkä impact structure and their relation to local jointing. Meteoritics and Planetary Science, Vol. 51, pp. 1534–1552.

Hietala, S., and Moilanen, J. (2004). Keurusselkä – A new impact structure in central Finland. 35th Lunar and Planetary Science Conference, Lunar and Planetary Institute, Houston, Texas, Abstract #1619.

Raiskila, S., Salminen, J., Elbra, T., and Pesonen, L. J. (2011). Rock magnetic and palaeomagnetic study of the Keurusselkä impact structure, central Finland. Meteoritics and Planetary Science, Vol. 46, pp. 1670–1687.

Schmieder, M., Jourdan, F., Moilanen, J., Buchner, E., and Öhman, T. (2016). A late Mesoproterozoic $^{40}Ar/^{39}A$ age for a melt breccia from the Keurusselkä impact structure, Finland. Meteoritics and Planetary Science, Vol. 51, pp. 303–322.

Kursk
Russia

Thomas Kenkmann

Kursk, a 6 km wide impact structure, is located in the East European plain. It is buried under a sequence of Middle Jurassic, Cretaceous, and Quaternary deposits, with a total thickness of about 100–150 m. Thus, the impact structure has no morphological expression at the surface.

Target rocks are Archean and Proterozoic granites, gneisses, and basic igneous rocks, which were overlain by 100 meters of Middle Devonian sediments. The crater was geophysically explored and drilled at several sites. The structure is outlined by an annular gravity low, and a weak gravity high in its center, which corresponds to a central uplift. In the central region, brecciated basement was encountered at a depth of 200 m. In contrast, within the surrounding annular trough these rocks occur at a depth of 310–410 m. Here, they have a maximum thickness of 270 m. The breccias contain shocked clasts with abundant shock features in the clasts of crystalline rocks, including diaplectic mineral and fused glasses. The largest blocks in the breccias measure some tens of meters in diameter. Fragments of Upper Devonian and Lower Carboniferous strata are embedded in the lithic impact breccias. Hence, the impact age must be younger than Lower Carboniferous and older than Middle Jurassic – and has consequently been estimated at 250 ± 80 Myr. First reports about the Kursk impact date back to the 1970s. Since then, several more have been published, most of them in Russian.

Location/Access

Access to the buried Kursk impact structure, which is located only 10 km west of the city of Kursk and some 450 km south of Moscow, is easy by using the local infrastructure such as the M2 motorway.

Selected References

Masaitis, V. L. (1976). Astroblemes in the USSR. International Geology Review, Vol. 18, pp. 1249–1258.

Masaitis, V. L. (1999). Impact structures of northeastern Eurasia: The territories of Russia and adjacent countries. Meteoritics and Planetary Science, Vol. 34, pp. 691–711.

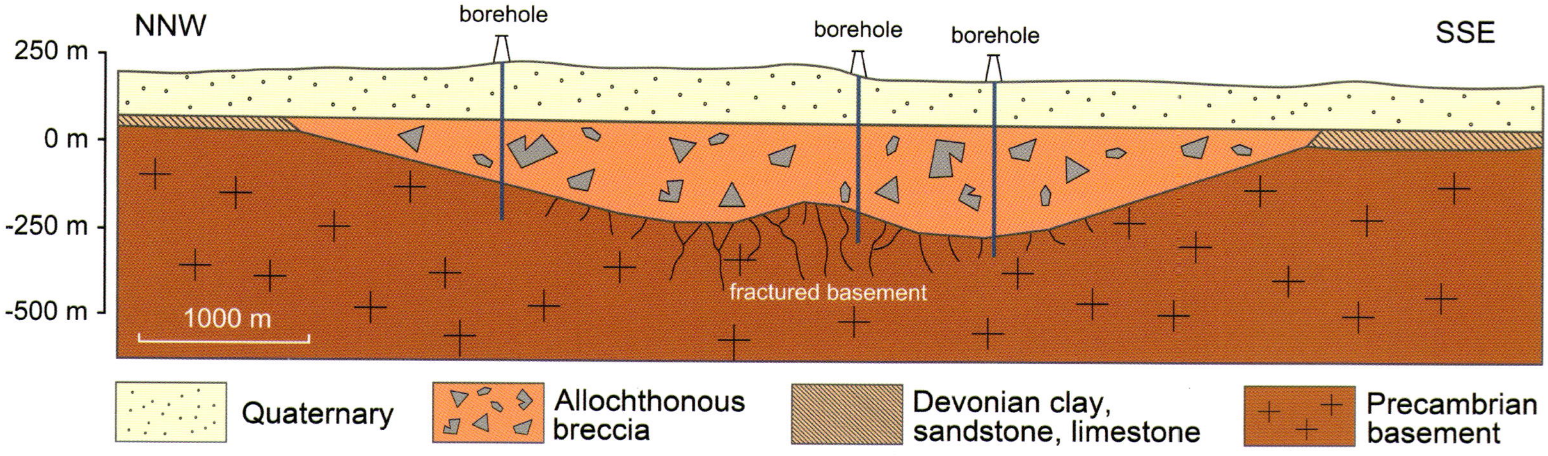

Schematic cross-section through the buried Kursk impact structure (modified after Masaitis, 1999).

Kursk
Russia

latitude = 51°42' N / longitude = 36°00' E / diameter = 6 km / age = 250 ± 80 Myr

scale = 1: 151 000 / scene = 27.1 × 36.1 km²

0 1.51 7.55 km

N

145 meter 285

Lappajärvi
Finland

Thomas Kenkmann

The 23 km diameter Lappajärvi impact structure is to a large extent located beneath Lake Lappajärvi, whose north-south elongated shape is the result of glacial overprint. The lake bed was excavated due to the low resistance of crater fill breccias and shattered bedrock to glacial erosion. The center of the geophysically mapped impact crater is close to the eastern shore of the north-central island Kärnänsaari. The gravity low associated with the crater has a diameter of 17 km. The bathymetry of the lake indicates two major troughs: one is close to the eastern shore and the other close to the western shore of the lake. These troughs represent fractures of tectonic origin that have still been active since the impact, as verified by repeated land levelling and recent earthquakes. The lake is surrounded by a rim of forested hills that is considered a relic of the original crater rim.

Target rocks are 1750–1900 Myr old Svecofennian gneisses and schists, with subordinate granite and granite pegmatite, granodiorite, and amphibolite, which are mostly covered by Quaternary till. The main impact-generated rock type occurring on the central island is a fine-grained melt rock of dacitic composition that is known as *kärnäite* – named after the large island Kärnänsaari in the northern part of the lake. The melt rock represents the remnant of an impact melt rock complex that is made up of both clast-rich and clast-poor varieties. Other impact lithologies include suevite and lithic impact breccia. Boulders of these various impactites are also found in glacial till in eskers to the south and southeast of the lake.

Geochemical analyses of the impact melt rock showed that traces of the impactor were admixed into the impact melt. Trace element analysis favored a projectile of H-chondritic composition. ^{40}Ar-^{39}Ar dating of impact melt rock and recrystallized K-feldspar melt particles of the Lappajärvi impact structure provided a weighted mean age for the impact of 76.2 ± 0.29 Myr, whereas recent U-Pb dating on shock metamorphosed zircon supports an age of 77.85 ± 0.78 Myr.

The impact melt rock was originally thought to be of volcanic origin. Kurt Fredriksson and Frans E. Wickman proposed an impact origin for this structure already in 1963. Extensive evidence for an impact origin of the Lappajärvi structure was presented by Nils Bertil Svensson in 1968 and Martti Lehtinen in 1976, who identified shock-metamorphosed rocks.

An about 25 cm wide slab of suevite from the Lappajärvi impact structure. The grey, variably bent or fluidal-structured bodies in this light-brownish altered, polymict impact breccia are all small, plastically deformed bodies of impact melt. In contrast, the matrix is composed of both small impact melt particles and clastic mineral and lithic particles (Photo: W. U. Reimold).

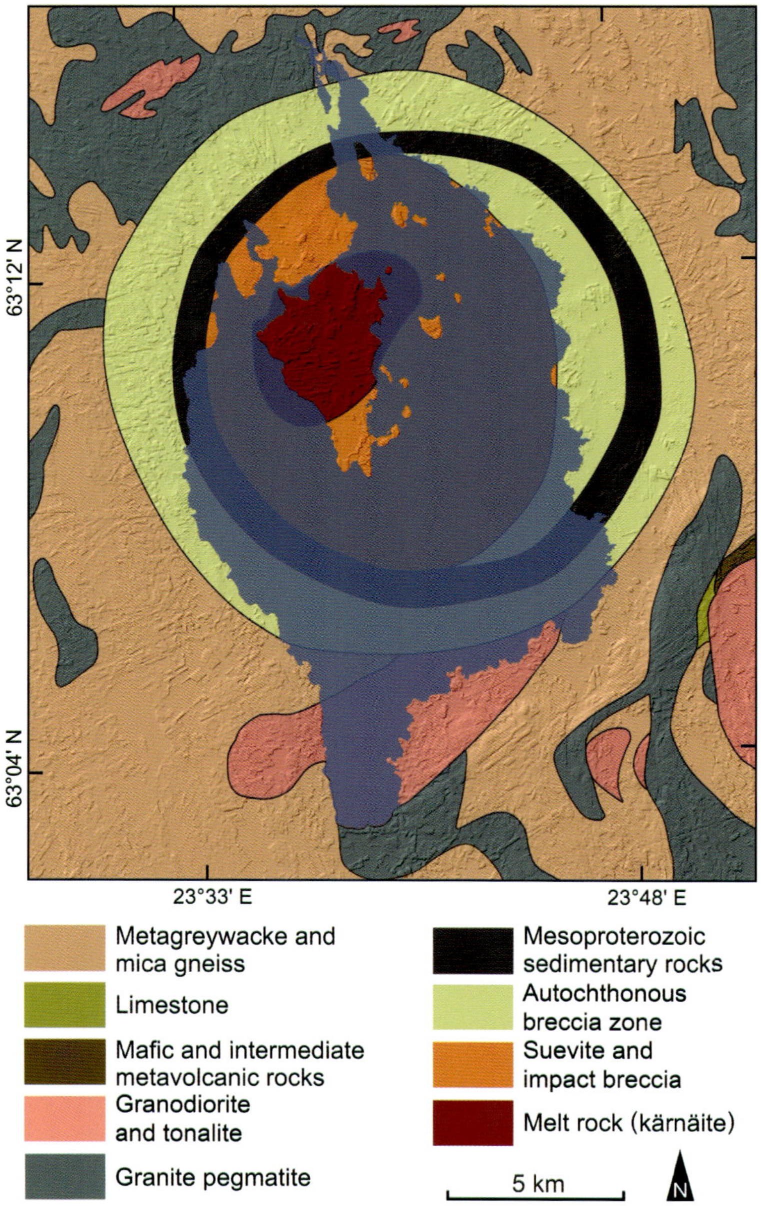

Schematic geological map of the Lappajärvi impact structure, underlain by a TanDEM-X DEM (modified after Kenny et al., 2019).

Location/Access

The Lappajärvi structure is located in western Finland, roughly 350 km north of Helsinki and 100 km east of Vaasa at Finland's west coast. Vaasa has an airport. Highways 18 and 16 lead directly from Vaasa to the southern tip of Lake Lappajärvi. Outcrops of impact lithologies can be found along the shore of Lake Lappajärvi, in particular on the island of Kärnänsaari. At the Hotelli Kivitippu, a small museum provides further information about the Lappajärvi impact and the impact cratering process in general.

Lappajärvi

Finland

latitude = 63°08'26" N / longitude = 23°41'04" E / diameter = 23 km / age = 77.80 ± 0.78 Myr

scale = 1:262 000 / scene = 47.2 × 63.0 km²
0 2.62 13.10 km
N
40 meter 225

Selected References

Lehtinen, M. (1976). Lake Lappajarvi, a meteorite impact site in western Finland. Geological Survey of Finland Bulletin, Vol. 282, 92 pp.

Kenny, G. G., Schmieder, M., Whitehouse, M. J., Nemchin, A. A., Morales, L. F. G., Buchner, E., Bellucci, J. J., and Snape, J. F. (2019). A new U-Pb age for shock-recrystallised zircon from the Lappajärvi impact crater, Finland, and implications for the accurate dating of impact events. Geochimica et Cosmochimica Acta, Vol. 245, pp. 479–494.

Reimold, W. U. (1982). The Lappajärvi meteorite crater, Finland: petrography, Rb-Sr, major and trace element geochemistry of the impact melt and basement rocks. Geochimica et Cosmochimica Acta, Vol. 46, pp. 1203–1225.

Schmieder, M., and Jourdan, F. (2013). The Lappajärvi impact structure (Finland): Age, duration of crater cooling, and implications for early life. Geochimica et Cosmochimica Acta, Vol. 112, pp. 321–339.

Svensson, N. B. (1968a). Lake Lappajärvi, Central Finland: a possible meteorite impact crater. Nature, Vol. 217, p. 438.

The area of the Lappajärvi impact structure on May 9, 2018 as a composite image based on TanDEM-X DEM and Copernicus Sentinel-2 (channel 2/3/4) data.

Lockne
Sweden

Thomas Kenkmann

The sub-circular Lockne structure in Central Sweden has a subdued morphological expression due to the preferred erosion of sedimentary strata in its center and the presence of an up to 70 m high crystalline rim. The eastern half of the crater structure is covered by Lake Locknesjön and tills related to Quaternary glaciation. No impact-induced circularity is expressed by the lake's shoreline. The central crater depression is 7 km wide. Here, crystalline basement was excavated. The crater is surrounded by an elevated crater rim with brecciated and fractured crystalline rock that has been interpreted as an overturned ejecta flap.

The Lockne crater was formed in Mid-Ordovician time, about 461 ± 5 Myr ago, in a marine environment. Dating is based on biostratigraphy using graptolites and chitinizoans of the Dalby Limestone that was deposited right after the impact. Target rocks at Lockne are crystalline rocks (Revsund granite, dolerites) of the Baltic Shield. On top of the crystalline basement, separated by a flat peneplain, rest Cambrian Alum Shale, a black mudstone enriched in organic matter, and a sequence of thickly-bedded Ordovician orthoceratite limestone that was deposited in a shelf sea with water depths of probably five hundred meters. The impact happened during the deposition of the Middle Ordovician Dalby Limestone at the western continental shelf of the Baltic Shield.

Marine-target impact craters differ from craters formed on land. The investigation of the Lockne structure has considerably contributed to our present knowledge of impact into wet environments, and this structure is regarded as a reference crater for marine impacts, in which the projectile diameter is of a similar size as the water depth. Excavated rocks are displaced by tsunami waves, and the forceful resurge of water back into the crater washes debris back into the crater cavity. Boreholes and exposures at Lockne show that the elevated crystalline rim of the crater is rootless and is underlain by Cambrian Alum Shale. The sediments surrounding the inner crater, such as the Ordovician orthoceratite limestone, were locally stripped away from the sub-Cambrian peneplain by the expanding water crater. Inside and outside of the Lockne crater resurge deposits are preserved and are crowded together in the crater cavity. The present relief of the crystalline basement displays several topographic lows along the crystalline rim, in which resurge deposits are preserved. These topographic lows were interpreted as resurge gullies that were formed and widened by eroding resurging seawater.

Lockne crater is, however, not pristine. In the western part of the crater pre-impact, impact, and post-impact rocks are incorporated into Caledonian thrust slices and were subject to folding and faulting. A nappe outlier in the central crater depression is a relic of the Caledonian nappe cover that reached a thickness of more than

The overturned ejecta flap of the Lockne crater, consisting of crystalline rock, overlies the Cambrian Alum shale, Skanska quarry. Width of photo is 12 m (photo: T. Kenkmann).

Folding of Dalby limestone occurred post-impact during the Caledonian overthrusting of the Lockne crater. Width of this image is about 3 m (photo: T. Kenkmann).

Tandsbyn Breccia of the Lockne impact structure. This exposure was photographed at the southern crater rim. Image width in the foreground is 2 m (photo: T. Kenkmann).

Monomict crystalline breccia of the Lockne impact stucture – the so-called "Tandsbyn Breccia" (photo: T. Kenkmann).

5 km. The overthrust crater itself is gently deformed. Strike of strata and trend of fold axes deviate from the standard northeast-southwest Caledonian directions. Radially oriented crater depressions that were originally understood as primary marine resurge gullies were interpreted by Thomas Kenkmann to be open tectonic folds, i. e., synclines in which resurge deposits had been preferentially preserved. The presence of the impact structure influenced the propagation of thrusts during orogenesis in a threefold way: A raised crater rim acted as an obstacle during nappe propagation, the occurrence of a central crater depression caused downward sagging of nappes, and the lack of an appropriate glide plane (Alum Shale) within the crater led to enhanced mechanical coupling and internal deformation of the nappe and the overthrusted foreland.

The structure was discovered in 1988 by Frans Wickman, who interpreted the origin of unusual breccias of the Lockne region as the products of an impact cratering event into a shallow sea. Maurits Lindstrom and co-workers refined the crater location and discovered planar deformation features (PDFs) in quartz grains from resurge deposits. Detailed geological investigation was also contributed by Jens Ormö and Erik Sturkell. Resurge deposits are enriched in chromium spinel. Chemical analysis indicated that the impactor was an L-chondrite. Further insight about the impact event came from numerical modeling by V. Shuvalov, who proposed an oblique impact from an easterly direction. Recently, it became apparent that Lockne and the small **Målingen** crater located 16 km west of Lockne are doublet craters.

Location/Access

Lockne crater is located in Jämtland, Sweden, some 15 km south of the city of Östersund. The structure is easily accessible via Highways E45 and E14. The crater is difficult to see in the landscape. However, some touristic infrastructure including a small geological museum exists, which helps the lay public to better understand the marine impact event. Relevant impact lithologies can be studied, e. g., at *Kraterstigen* near the village of Tandsbyn, where a crater trail was setup to observe resurge deposits. Another important outcrop is the Skanska quarry, which is still in operation. It exposes the crater rim topped by proximal ejecta.

Selected References

Kenkmann, T., Kiebach, F., Rosenau, M., Raschke, U., Pigowske, A., Mittelhaus, K., and Eue, D. (2007). Coupled effects of impact and orogeny: Is the marine Lockne crater, Sweden, pristine? Meteoritics and Planetary Science, Vol. 42, pp. 1995–2012.

Lindström, M., Ekvall, J., Hagenfeldt, S. E., Säwe, B., and Sturkell, E. F. F. (1991). A well-preserved Cambrian impact exposed in Central Sweden. Geologische Rundschau, Vol. 80, pp. 201–204.

Lindström, M., Shuvalov, V. V., and Ivanov, B. A. (2005). Lockne crater as a result of a marine-target oblique impact. Planetary and Space Science, Vol. 53, pp. 803–815.

Lindström, M., Ormö, J., Sturkell, E. F. F., and von Dalwigk, I. (2005). The Lockne crater: revision and reassessment of structure and impact stratigraphy. In: Impact Tectonics, Koeberl, C., and Henkel, H. (eds.), Springer, Berlin Heidelberg, pp. 357–388.

Ormö, J., and Lindström, M. (2000). When a cosmic impact strikes the sea bed. Geological Magazine, Vol. 137, pp. 67–80.

Sturkell, E. F. F. (1998). The marine Lockne impact structure, Jämtland, Sweden: a review. Geologische Rundschau, Vol. 87, pp. 253–267.

Wickman, F. E. (1988). Possible impact structures in Sweden. In: Deep Drilling in Crystalline Bedrock. Volume 1, Bodén, A., and Eriksson, K. G. (eds.), Springer, Berlin Heidelberg, pp. 298–327.

Scene of the Målingen and Lockne impact structures on July 2, 2018 as a composite image based on TanDEM-X DEM and Copernicus Sentinel-2 (channel 2/3/4) data. Cross hairs on the right side indicate Lockne, whereas those on the left side refer to Målingen.

Lockne

Sweden

latitude = 63°00'01" N / longitude = 14°49'11" E / diameter = 7 km / age = 461 ± 5 Myr

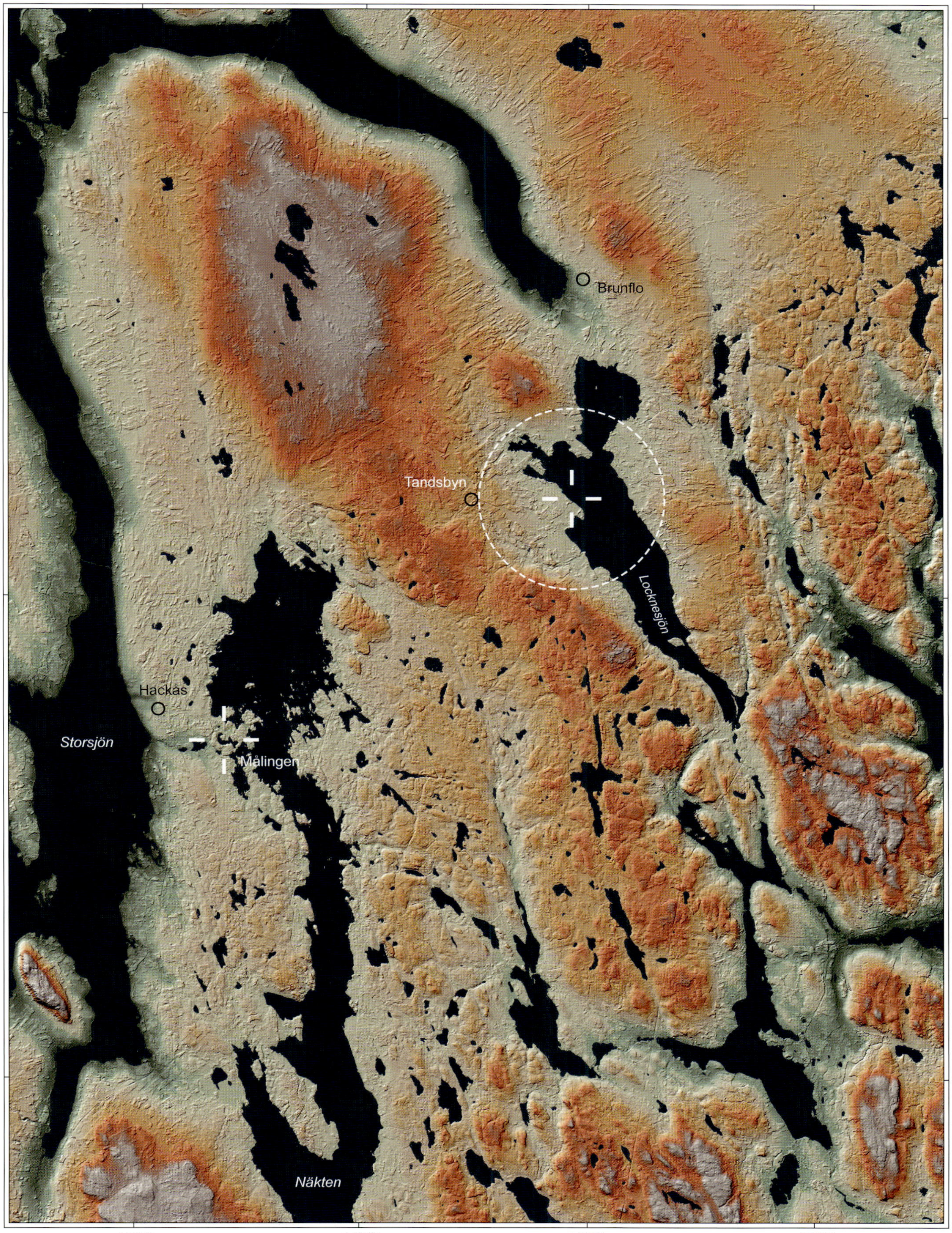

scale = 1 : 195 000 / scene = 35.0 × 46.7 km²

0 1.95 9.75 km

N

215 meter 550

Logoisk
Belarus

Thomas Kenkmann

Logoisk is a buried complex impact structure in Belarus, Eastern Europe. It has a diameter of 17 km. A central uplift of about 1.5 km width and 200 m stratigraphic uplift occurs in the central part of Logoisk.

The crater structure was formed in a target composed of crystalline basement involving granites and gneisses that were overlain by uppermost Precambrian and Devonian deposits. A 270–290 m thick sequence of Oligocene-Miocene siltstones, sands, and marls is found only in the central part of the crater structure and represents crater lake deposits. All rocks are buried under Quaternary deposits that reach a thickness of 190 m.

The rim rises 200–220 m above the surrounding terrain. The inner crater cavity is filled with allochthonous polymict breccias and suevite. The maximum thickness of the crater fill amounts to 290 m. Large megablocks of Cretaceous provenance are embedded in the breccias in the central part of the structure and also occur in peripheral areas. Impact rocks include impact melt rock, suevites, allochthonous polymict lithic, and massive, monomict lithic breccias. The impact melt rock is found in the central part of the structure covering the central uplift. It is composed of a glassy matrix and contains clasts of diaplectic mineral glass, shocked target rock, and minerals derived thereof. Coesite and stishovite occurrence has been described from clasts. The structure is asymmetric in cross-section, with the true center displaced by 1–1.5 km to the northeast of the geometric center of the structure. The age of the impact has been dated quite precisely at 29.71 ± 0.48 Myr by the ^{40}Ar-^{39}Ar method on impact melt rock.

In 1973 a negative geophysical anomaly was discovered at Logoisk. Subsequent drilling revealed rocks that were interpreted by N. Veretennikov as the result of an impact event. Meanwhile, a total of 29 boreholes has been drilled, which has provided a detailed picture of the suite of impact-related lithologies and impact-diagnostic shock effects.

Location/Access

Logoisk is situated about 40 km north of Minsk, the capital of Belarus. The structure is located close to Highway M3 connecting Minsk with the city of Lahojsk (Logoisk).

A small sample of Logoisk impact melt rock, drill core diameter = 5.5 cm (photo courtesy: Ralf-Thomas Schmitt, Museum für Naturkunde Berlin, sample obtained from the VSEGEI, St. Petersburg, Russia).

Selected References

Masaitis, V. L. (1999). Impact structures of northeastern Eurasia: The territories of Russia and adjacent countries. Meteoritics and Planetary Science, Vol. 34, pp. 691–711.

Sherlock, S. C., Kelley, S. P., Glazovskaya, L., and Ukstins Peate, I. (2009). The significance of the contemporaneous Logoisk impact structure (Belarus) and Afro-Arabian flood volcanism. Journal of the Geological Society, Vol. 166, pp. 5–8.

Veretennikov, N. V. (2010). The Logoisk impact crater. In: The Sun, the Stars, The Universe and General Relativity: International Conference in Honor of Ya. B. Zeldovich's 95th Anniversary, Minsk, Belarus, Ruffini, R., and Vereshchagin, G. (eds.), AIP Conference Proceedings, Vol. 1205, pp. 185–190.

Veretennikov, N. V., Il'kevich, G., and Makhnacha, S. (1979). Logoisk buried depression – an ancient meteorite crater (in Russian). Doklady Acad. Nauk BSSR, Vol. 23, pp. 156–160.

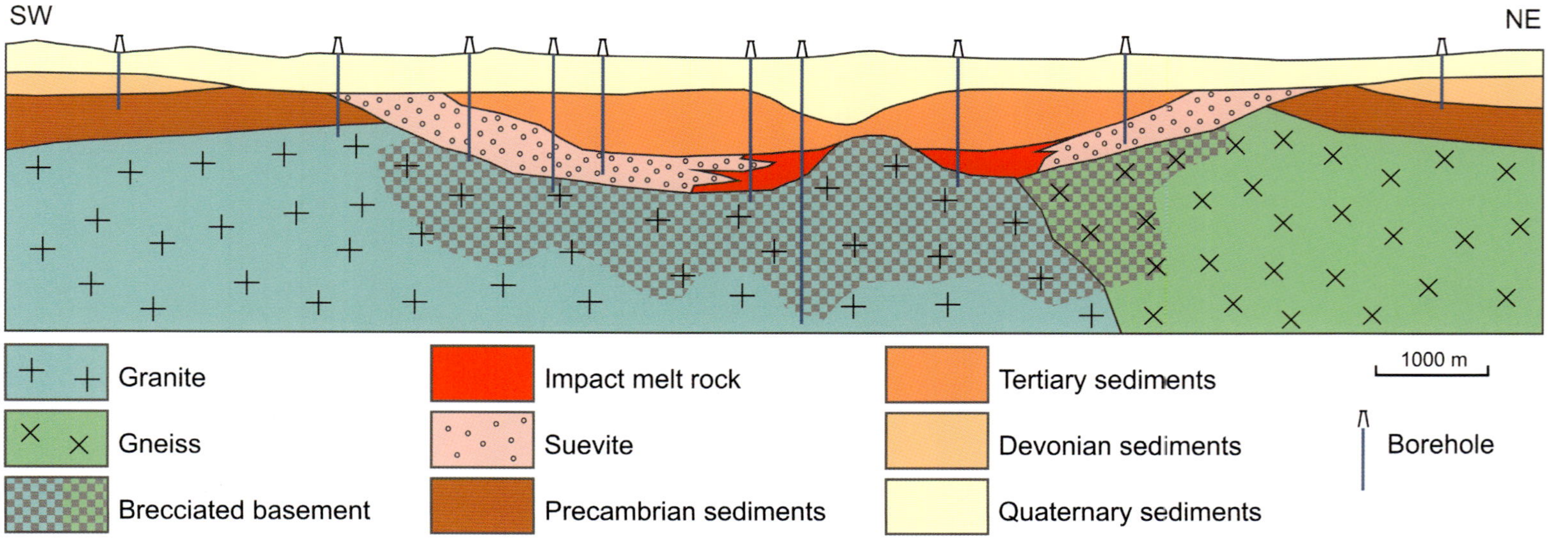

Schematic cross-section of the Logoisk impact structure (modified after Masaitis, 1999).

Logoisk

Belarus

latitude = 54°15'46" N / longitude = 27°47'10" E / diameter = 17 km / age = 29.71 ± 0.48 Myr

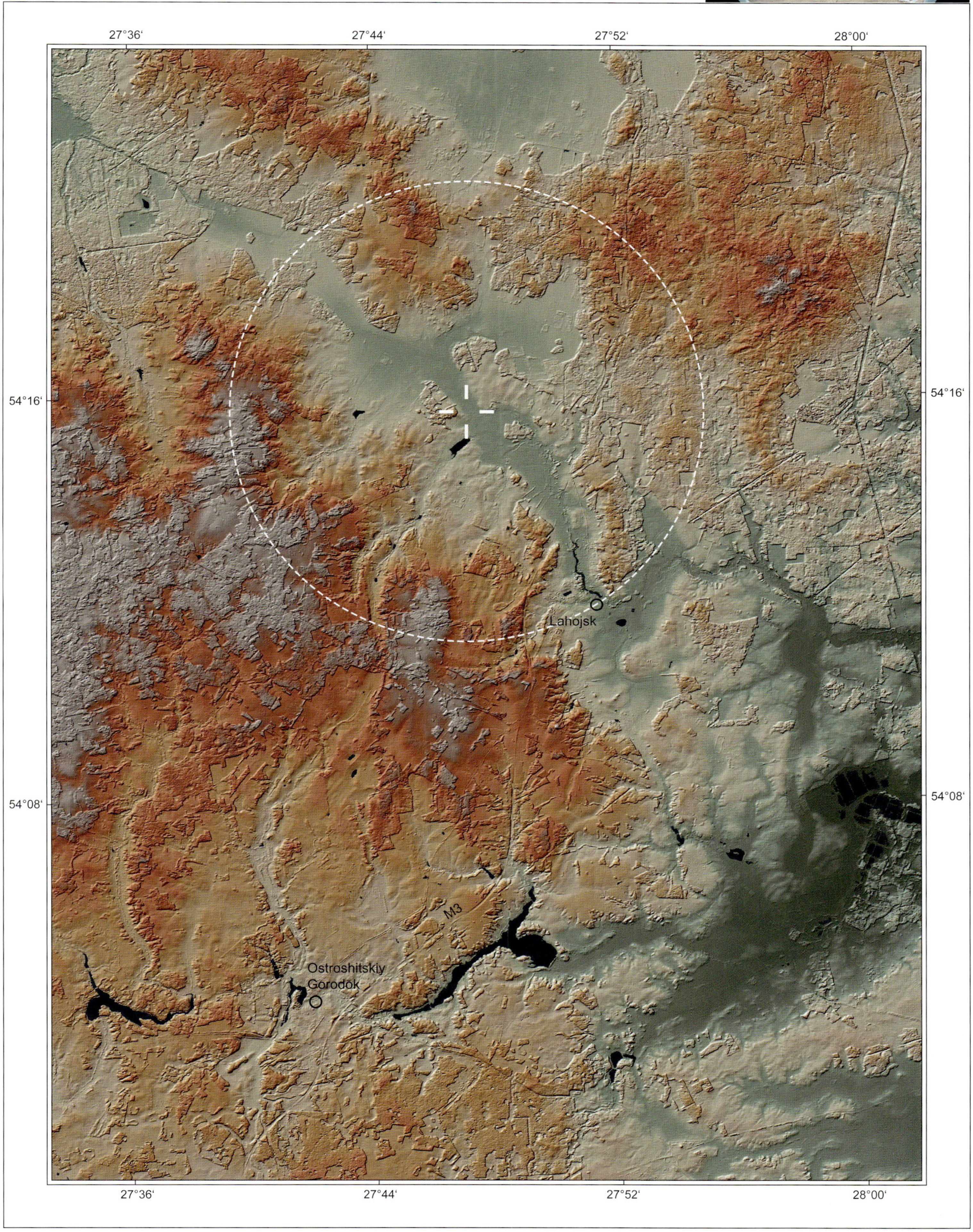

scale = 1:172 000 / scene = 31.0 × 41.4 km²

0 1.72 8.60 km

N

160 meter 355

Lumparn
Finland

Thomas Kenkmann

The Lumparn crater structure coincides with a bay in the strongly serrated outline of the main island Fasta Åland of the Åland archipelago in the Baltic Sea. The crater diameter is not well defined. Seismic and bathymetric data indicate a 7 km wide structure, whereas other reports have suggested a 9 km diameter.

Bedrock is a porphyritic to coarse-grained Rapakivi granite. A borehole drilled into the center of the bay revealed uppermost Pleistocene sediments, above Ordovician limestone of at least 40 m thickness, and crushed and brecciated Rapakivi granite underneath. At the interface between the limestone and the granite occurs a particularly fine-grained layer that contains glassy particles. Shock features were detected in the crushed granites of the drill core and along the southwestern shore of the bay. Here, in situ shatter cones were also found, together with patches of breccia and breccia dikes. The brecciated granite shows decorated planar deformation features (PDFs), planar fractures (PFs), and mosaicism in quartz and feldspar. Zircon appears granularized. Glassy fragments exist in this breccia as well, with fluidal textures. Consequently, this breccia should be regarded as a suevite.

Seismic data could not substantiate the presence of a central uplift. Post-impact tectonic movements affected the region, including the limestone deposited on top of the shattered granites. The minimum impact age of 450 Myr is constrained by the biostratigraphic age for the limestone that was not affected by the impact. The fine-grained layer just below the limestone contains redeposited acritarchs of Lower Cambrian age. An uppermost Precambrian age therefore seems likely for the impact event.

An impact origin for Lumparn was first proposed by Glen Merrill, who studied Ordovician conodonts. He concluded that the preservation of the Paleozoic section in Lumparn may be best explained by the assumption that the Lumparn basin was a meteorite impact structure. Later, Nils Svensson confirmed the impact hypothesis by documentation of shatter cones.

Location/Access

The island of Fasta Åland can be reached by ferry from either Helsinki or Turku in Finland, as well as from Stockholm in Sweden. The main town on the island, Marihamn, is located only 5 km southwest of Lumparn Bay. Related outcrops along the shore exist north of the hamlet of Hägnars.

Selected References

Abels, A., Mannola, P., Lehtinen, M., Bergman, L., and Pesonen, L. J. (1998). New observations of the properties of the Lumparn impact structure, Åland Islands, Southwestern Finland. Meteoritics and Planetary Science, Vol. 33, pp. A7–A8.

Abels, A., Bergman, L., Lehtinen, M., and Pesonen, L. J. (2000). Structural constraints and interpretations on the formation of the Soderfjärden and Lumparn impact structures, Finland. 4th ESF Workshop: Response of the Earth System to Impact Processes (IMPACT), Helsinki, p. 26.

Merrill, G. K. (1980). Ordovician conodonts from the Åland Islands, Finland. Geologiska Föreningen i Stockholm Förhandlingar, Vol. 101, pp. 329–341.

Svensson, N. B. (1993). Lumparn Bay: A meteorite impact crater in the Åland archipelago, southwest Finland. Meteoritics. Vol. 28, pp. 245.

Part of the Åland archipelago with the Lumparn bay on July 25, 2018 as a composite image based on TanDEM-X DEM and Copernicus Sentinel-2 (channel 2/3/4) data.

Lumparn
Finland

latitude = 60°08‘17“ N / longitude = 20°07‘59“ E / diameter = 7 km / age = 450 - 1250 Myr

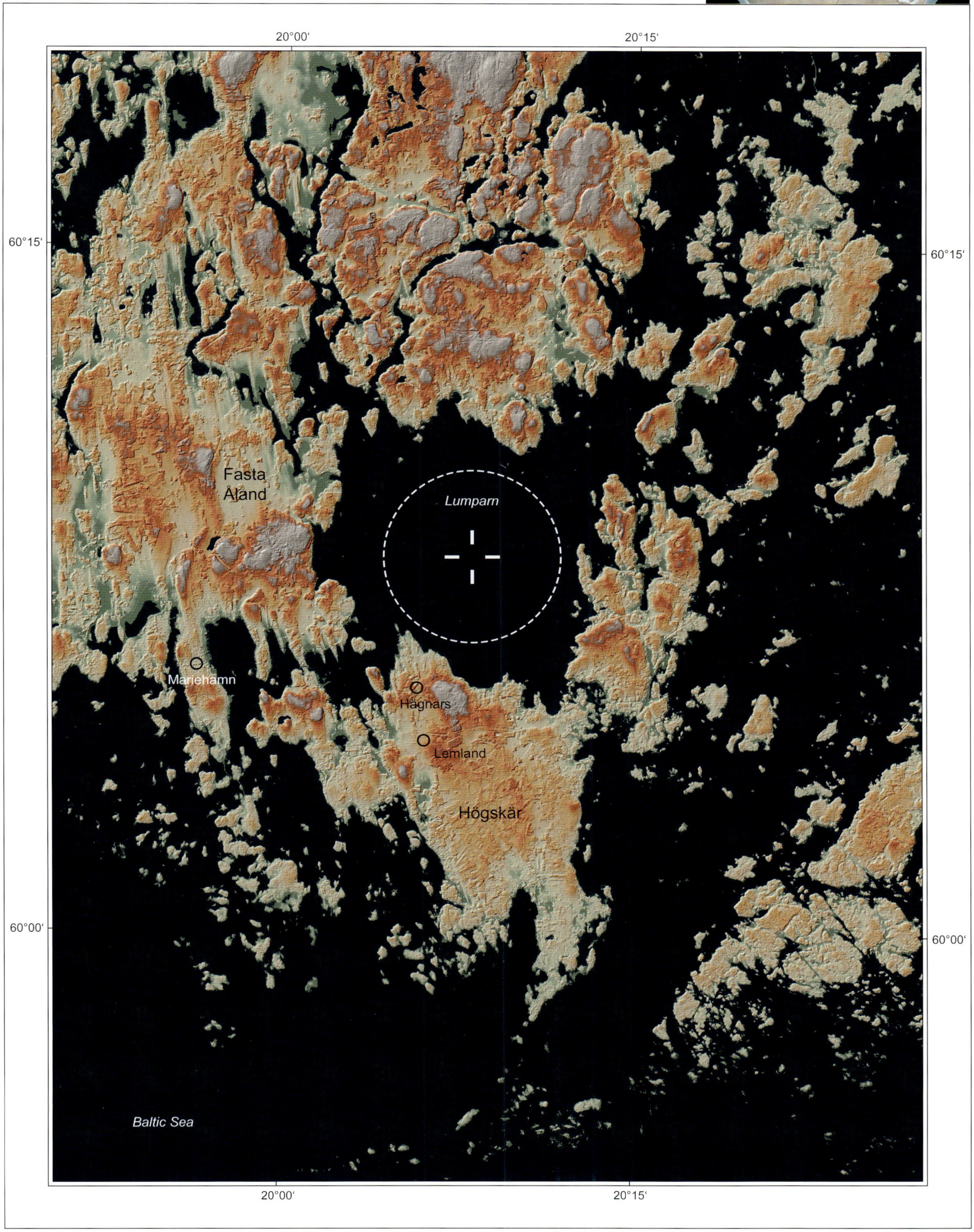

scale = 1 : 191 000 / scene = 34.5 × 45.9 km²

0 1.91 9.55 km

N

0 meter 120

Målingen
Sweden

Thomas Kenkmann

The Målingen crater structure is an approximately 0.7–1 km wide, sediment-filled, circular depression in Precambrian crystalline basement. It is located about 16 km to the southwest of the **Lockne** crater structure. Its central region is partly submerged by Lake Målingen. The impact excavated into the Precambrian crystalline basement and formed a prominent rim of fractured and brecciated crystalline rocks that is elevated above the surrounding peneplain. The sedimentary breccias within the depression have been interpreted as resurge deposits.

The Målingen structure is located on today's erosional front of the Caledonian overthrusts. The Precambrian basement rocks consist predominantly of Revsund granite. At the time of the Cambrian transgression, these basement rocks had been eroded down to a nearly horizontal peneplain that was then covered with a thin transgression conglomerate, followed by 30 m of Cambrian Alum Shale Formation and some 50 m of thickly-bedded Ordovician orthoceratite limestone. The impact occurred contemporaneous with the deposition of the Middle Ordovician Dalby Limestone in a marine environment. Tsunami waves generated during the impact and the forceful resurge led to the deposition of the sedimentary infill of the crater cavity.

In 2010 a 149 m deep borehole was drilled into the center of the structure and revealed, from bottom to top, approximately 33 m of shattered basement rocks with increasing degree of fracturing in upward direction, approximately 10 m of polymict crystalline breccia with shocked quartz grains, 97 m of slumped Cambrian mudstone, and 5 m of resurge deposits. Evidence for shock metamorphism in quartz comprises planar deformation features (PDFs).

Based on biostratigraphic dating of the Lockne crater, the Målingen structure also formed at about 461 ± 5 Myr ago in conjunction with the Lockne crater in the same marine setting. The stratigraphic and geographic relationship with Lockne suggests that the Lockne and Målingen craters are the first recognized doublet impact structure – formed by a binary asteroid – in a marine-target setting.

The anomalous geology of the Målingen structure was recognized already in the 1940s. The numerous studies on the nearby Lockne crater in the 1980s and 1990s by Maurits Lindström, Erik Sturkell, Jens Ormö, and co-workers also briefly mentioned Målingen – but it was only recently that Jens Ormö, Erik Sturkell, Carl Alwmark, and their co-workers obtained the results of the drilling campaign, which confirmed that Målingen is indeed a separate impact crater.

Location/Access

Målingen, situated between the large Storsjön and Näkten lakes of Jämtland, is easily accessible from, and directly connected to, the nearby Lockne crater via the E45 from Östersund. From Håkas turn off onto a paved road to Målingen. Outcrops of crystalline breccias and resurge deposits exist north and west of Lake Målingen.

Tilted and disintegrated limestone and claystone beds at the Målingen impact crater (photo: T. Kenkmann).

Selected References

Alwmark, C., Alwmark-Holm, S., Ormö, J., and Sturkell, E. (2014). Shocked quartz grains from the Malingen structure, Sweden – Evidence for a twin crater of the Lockne impact structure. Meteoritics and Planetary Science, Vol. 49, pp. 1076–1082.

Ormö, J., Sturkell, E., Olvak, J. N., Melero-Asensio, I., Frisk, A., and Wikström, T. (2014). The geology of the Målingen structure: A probable doublet to the Lockne marine-target impact crater, central Sweden. Meteoritics and Planetary Science, Vol. 49, pp. 313–327.

Målingen

Sweden

latitude = 62°55‘01“ N / longitude = 14°33‘59“ E / diameter = 0.85 km / age = 461 ± 5 Myr

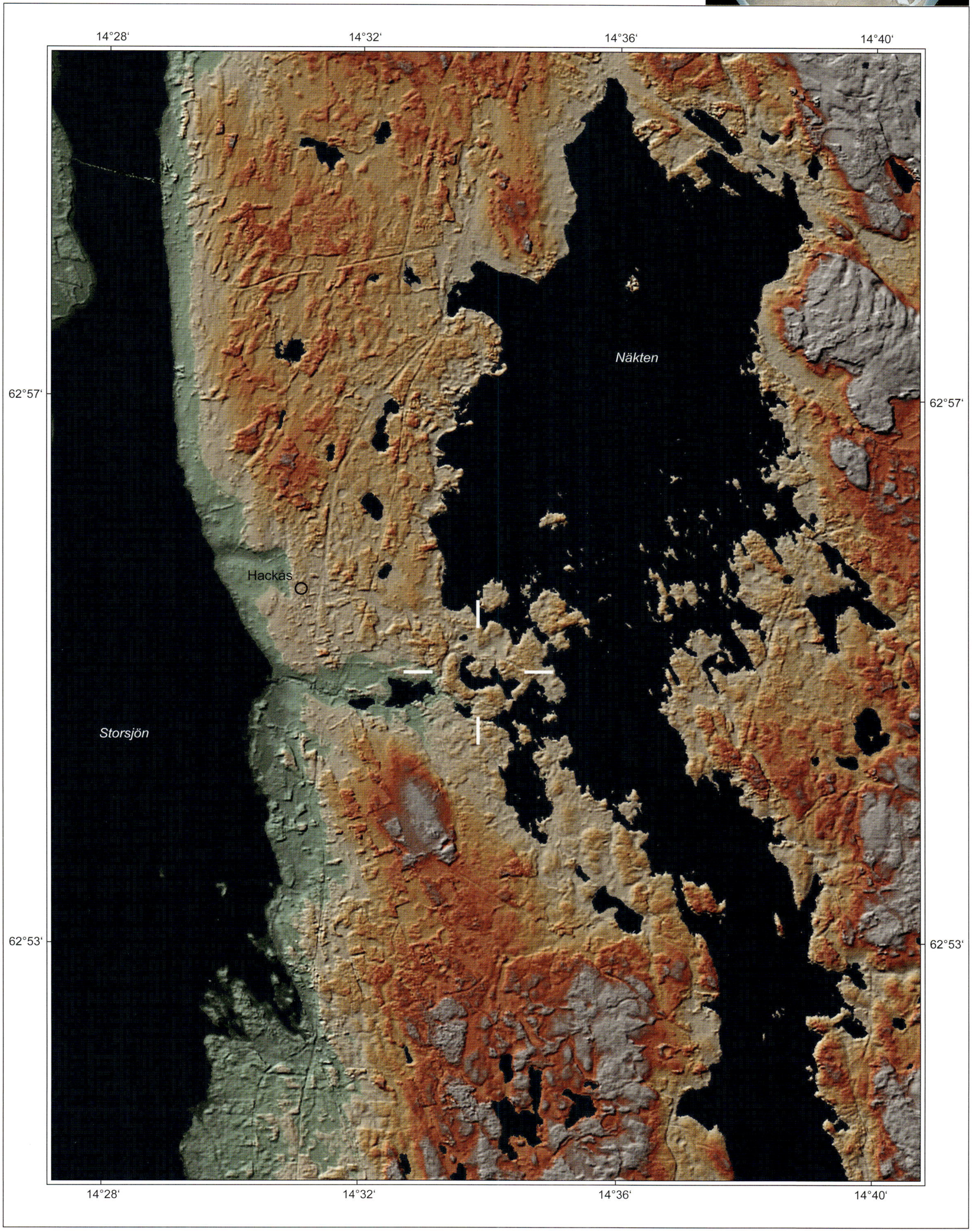

scale = 1 : 64 000 / scene = 11.5 × 15.3 km²

0 0.64 3.20 km

N

250 meter 430

Mien
Sweden

Thomas Kenkmann

The Mien impact crater in southern Sweden is largely occupied by the almost circular Lake Mien, which is roughly 40 m deep. The current topographic rim has a diameter of 6–7 km but it is estimated that the original crater had a diameter of about 9 km prior to extensive erosion. Only remnants of the crater fill and the crater floor are preserved.

Lake Mien is situated within the Trans-Scandinavian Igneous Belt of the Baltic Shield. Target rocks are Precambrian granites, gneisses and amphibolite of 1700–1800 Myr age. There is no sedimentary cover over the crystalline basement, but glacial till and glacio-fluvial deposits can be up to a few meters thick and limit the number of outcrops around the lake. Various breccias and impact melt rock occur as boulders and pebbles in glacial deposit. On the island of Ramsö in the western part of the lake, impact melt rock boulders have been encountered beneath moraine material. A shallow borehole was drilled here and penetrated a 10 m thick layer of dense impact melt rock, underlain by suevitic breccias. The bulk geochemical composition of the melt rock is granitic. It has been suggested that the crater contains a total of about 0.5 km^3 of impact melt rock. The melt rock shows a marginal enrichment (over country rock abundances) in the element iridium that might be indicative of projectile contamination, but a projectile type could not be identified from this.

The impact melt rock was initially considered of volcanic origin until coesite was found in it. Coesite is a high-pressure polymorph of quartz and documents in impact geology that the rock was affected by a strong shock wave. Impact melt rock and suevite show all types of shock metamorphism, including planar deformation features (PDFs) in quartz and feldspar, diaplectic glasses, and ballen textured silica. Based on ^{40}Ar-^{39}Ar dating of melt rock samples, the age of the impact could be constrained to a Lower Cretaceous age of 118.7 ± 2.3 Myr. Mien has a characteristic gravity anomaly with a distinct residual gravity low of −5 mGal. This anomaly is focused on the center of the impact structure. The width of the gravity low measures about 2 km. It is thought that this anomaly is due to strong fragmentation of the target rocks. The measurement of shear wave velocity in the upper part of the crust over the Mien impact structure shows a reduction in velocity over a 17 km wide zone. This indicates that target rocks are disrupted even beyond the limit of the crater.

For a long time, Lake Mien was considered of volcanic origin, although Arvid Högbom – already in 1910 – proposed a meteorite impact formation. The impact hypothesis was proven in 1965 by the documentation of coesite by Nils-Bertil Svensson and Frans Wickman, and of PDFs in quartz by Wolf von Engelhardt and Dieter Stöffler.

Along the shore of Lake Mien occur loose blocks of impactites (photo: T. Kenkmann).

A slab of Mien suevite, 12 × 10 × 2.5 cm in size (photo courtesy: Ralf-Thomas Schmitt, Museum für Naturkunde Berlin, sample from Dieter Stöffler).

Location/Access

Lake Mien is located close to the border between the provinces of Blekinge and Småland in southern Sweden. It is easily accessible, e.g., from the town of Karlshamn that is situated at the Baltic Sea about 30 km south of Lake Mien. National road 29 brings the visitor to the eastern shore of the lake.

Selected References

Åström, K. (1998). Seismic signature of the Lake Mien impact structure, southern Sweden. Geophysical Journal International, Vol. 135, pp. 215–231.

Bottomley, R. J., York, D., and Grieve, R. A. F. (1978). ^{40}Ar-^{39}Ar ages of Scandinavian impact structures: I. Mien and Siljan. Contributions to Mineralogy and Petrology, Vol. 68, pp. 79–84.

Henkel, H., and Pesonen, L. J. (1992). Impact craters and craterform structures in Fennoscandia. Tectonophysics, Vol. 216, pp. 31–40.

Schmidt, G., Palme, H., and Kratz, K. L. (1997). Highly siderophile elements (Re, Os, Ir, Ru, Rh, Pd, Au) in impact melts from three European impact craters (Sääksjärvi, Mien, and Dellen): Clues to the nature of the impacting bodies. Geochimica et Cosmochimica Acta, Vol. 61, pp. 2977–2987.

Svensson, N. B. (1969). Lake Mien, southern Sweden – a possible astrobleme. Geologiska Föreningen i Stockholm Förhandlingar, Vol. 91, pp. 101–110.

Svensson, N. B., and Wickman, F. E. (1965). Coesite from Lake Mien, southern Sweden. Nature, Vol. 205, pp. 1202–1203.

Von Engelhardt, W., and Stöffler, D. (1965). Spaltflächen im Quarz als Anzeichen für Einschläge großer Meteoriten. Naturwissenschaften, Vol. 52, pp 489–490.

Mien
Sweden

latitude = 56°25'05" N / longitude = 14°51'23" E / diameter = 9 km / age = 118.7 ± 2.3 Myr

scale = 1 : 135 000 / scene = 24.4 × 32.5 km²

0 1.35 6.75 km

N

40 meter 255

Mishina Gora
Russia

Thomas Kenkmann

Mishina Gora is a deeply eroded simple crater with an estimated diameter of 2.5 km in the northwestern part of the East European plain in Russia.

The region has a crystalline basement of Archean gneisses and granites that are overlain by a flat-lying sedimentary cover of Proterozoic and Paleozoic sediments. In the area of the crater the sedimentary sequence is 460–530 m thick and is composed of uppermost 90 m of Proterozoic sandstones and siltstones, followed by 100 m of Cambrian clays and sandstones, 150 m of Ordovician sandstones, dolomites, and limestones, and 200 m of Devonian dolomites, sandstones, and clays.

This sedimentary succession is strongly disturbed over an area of about 2–3 km in extent. In an outer zone intense faulting is noted that led to various block rotations and tilting of strata. A large Ordovician block tilted by 70° is exposed in an abandoned quarry. The bowl-shaped crater depression is filled with a lithic polymict breccia, composed of fragments and blocks up to tens of meters in size. A deep borehole into the crater center revealed a succession of 200 m of sedimentary rocks above 600 m of polymict breccia composed of sedimentary and crystalline fragments. Shatter cones, shocked quartz grains with planar deformation features (PDFs) and diaplectic glasses, and patches of altered impact glass were observed. The age of the impact event has been tentatively estimated at Late Paleozoic.

A. Shmayenok was the first to investigate the structure in 1974 but regarded it as a "cryptovolcanic" explosion pipe. Later, Victor Masaitis and co-workers identified the shock metamorphic evidence that confirmed the impact origin of Mishina Gora.

Location/Access

Mishina Gora is located close to the border between Estonia and Russia, some 15 km east of Lake Peipus (ozero Chudskoye). Gdov is the largest town nearby, about 20 km to the northwest of the structure.

Selected References

Masaitis, V. L. (1999). Impact structures of northeastern Eurasia: The territories of Russia and adjacent countries. Meteoritics and Planetary Science, Vol. 34, pp. 691–711.

Shmayenok, A. I., and Malakhovskiy, F. B. (1974). Explosion pipe near the southeast shore of Lake Chud' (in Russian). Vestnik Leningrad University, Vol. 24, pp. 97–107.

Mishina Gora monomict impact breccia, 7 × 6 × 4 cm in size (photo courtesy: Ralf-Thomas Schmitt, Museum für Naturkunde Berlin, sample obtained from the VSEGEI, St. Petersburg, Russia).

Mishina Gora

Russia

latitude = 58°34'57" N / longitude = 28°05'27" E / diameter = 2.5 km / age = < 359 Myr

Mizarai
Lithuania

Thomas Kenkmann

Mizarai with a diameter of 5 km is a buried, complex impact structure with a central uplift. It is located in the southern part of the Baltic region. No morphological expression exists.

Target rocks are Archean and Early Proterozoic amphibolites, gneisses, metagabbro, and granites. The depth of the crater is at least 230 m. The crater depression is filled by variegated sandy to clayey deposits some 150 m in thickness. At the base of this sequence occurs a coarse-grained bed containing particles of impact glass. These deposits have probably been derived from reworking of the ejecta blanket. They include fragments of primary sediments with microfossils that indicate an uppermost Precambrian to Early Cambrian age for the impact event. This sequence is covered by 70 m of Lower Triassic sediments that pinch out outside the crater. Younger post-impact deposits of 200 m thickness include Cretaceous sands and chalk, and Quaternary deposits that also cover the surrounding area and indicate that the crater morphology was peneplained at that time.

A borehole drilled into the structure's center revealed that crystalline target rocks were affected by cataclasis and brecciation to a depth of 525–600 m. Shatter cones and various shock features in minerals, as well as injections of thin veins of altered impact generated glass, were encountered in rocks of the central uplift. Small bodies of impact melt rock were found in cores close to the outer slope of the annular trough that surrounds the central uplift.

Gediminas Motuza and Rimantas Gailius noted the structure in 1978 in the course of a geophysical survey. Subsequent drilling occurred in the 1980s and provided the shock metamorphic evidence for impact. A great deal of the reference literature was published in Russian or Lithuanian only.

Location/Access

Mizarai is situated in a forested area about 6 km north of the border between Lithuania and Belarus, and about 25 km east of the border between Lithuania and Poland. The town of Druskininkai lies 5 km to the east, on the opposite side of the Neman River. This town can be reached via Highway A4 or regional road 180.

Selected References

Masaitis, V. L. (1999). Impact structures of northeastern Eurasia: The territories of Russia and adjacent countries. Meteoritics and Planetary Science, Vol. 34, pp. 691–711.

Motuza, G. B., and Gailius, P. (1978). On the supposed astroblemes in Lithuania (in Russian). In: Local Structures in Belarus and the Baltic Region, Suveizdis, P. I. (ed.), Geological Research Institute, Vilnius, Lithuania, pp. 91–94.

Mizarai

Lithuania

latitude = 54°01' N / longitude = 23°54' E / diameter = 5 km / age = 520 - 560 Myr

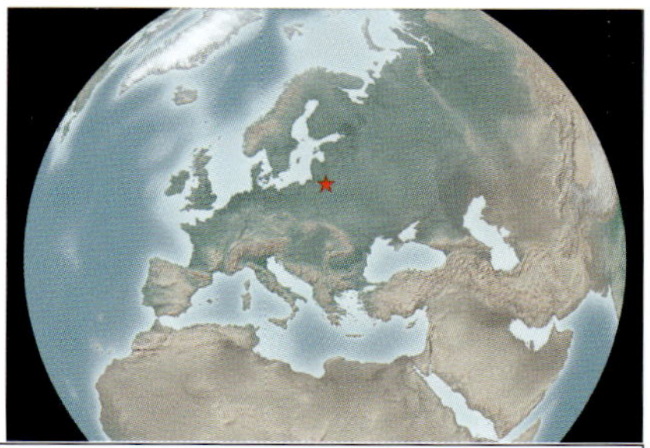

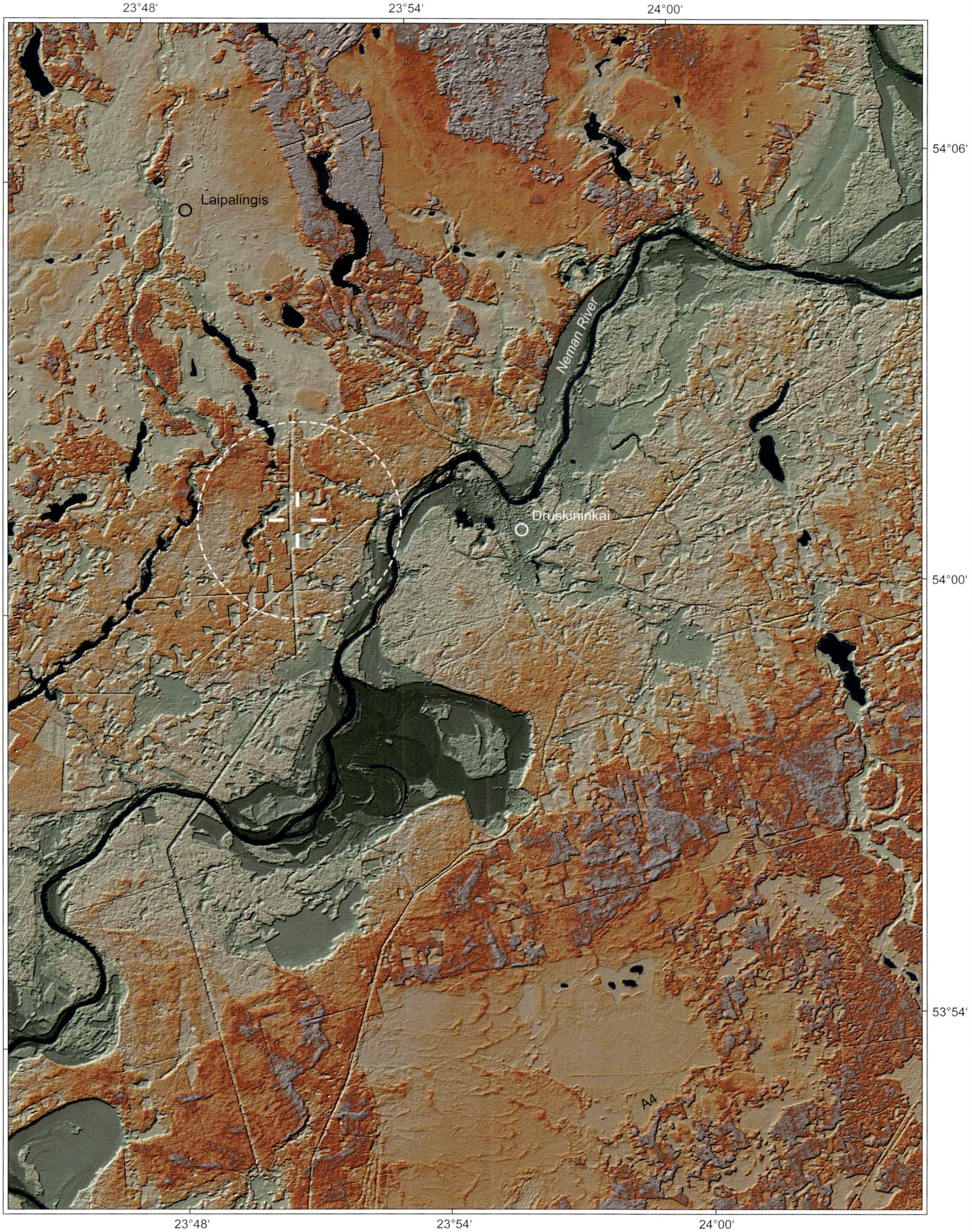

scale = 1 : 127 000 / scene = 22.8 × 30.5 km²

0 1.27 6.35 km

N

15 meter 230

Mjølnir
Norway

Thomas Kenkmann

The Mjølnir crater, 40 km in diameter, is located in the central Barents Sea. It lies buried under about 50–150 m of post-Jurassic sediments and is submerged in 350 m deep water. Key features are a prominent central high of 8 km diameter, a 4 km wide annular basin, and a 12-km-wide outer zone which, in turn, comprises an inner, low relief ring, and a marginal fault zone. Seismic imaging and interpretation have shown that the Mjølnir structure has an exceptionally shallow apparent crater depth and exhibits extensive post-impact deformation. A prominent feature of Mjølnir is the impact breccia forming the so-called "Ragnarok Formation". It is composed of a mixture of re-deposited excavated material, fall back ejecta, back wash and structurally uplifted and slumped target material with partly inverted stratigraphy. This impact breccia and ejecta deposition formed during and immediately after the Mjølnir impact. On seismic reflection profiles the Ragnarok Formation can reach thicknesses of 1.3 km.

Shock features were first found in drill cores obtained 30 km outside of the crater, where the breccia thins out to meters and less. The ejecta bearing layer consists mainly of rock fragments from the target area and is enriched in iridium, and contains Ni-rich spinel, spherules, and soot. This layer also contains shocked quartz grains and is associated with strata deformation. Ejecta have also been found in cores from elsewhere in the Barents Sea and in outcrops on Svalbard (Spitzbergen) archipelago 600 km from the impact site.

Mjølnir exhibits a distinct crater relief. This is attributed to the post-impact overburden, which is estimated to have reached a maximum of 2–2.5 km, in combination with variations in the physical properties triggered by the impact at a shallow water depth of 300–500 m in the paleo-Barents Sea. At Mjølnir, the central high developed as a prominent feature during post-impact burial, the height of the peak ring was enhanced, and the cumulative throw on the rim faults was increased.

Mjølnir has been studied intensely by means of geophysical techniques, in particular by Filipos Tsikalas, Henning Dypvik, and colleagues. The impact-induced seismic disturbance at Mjølnir has a parabolic bowl-shape at the center of the structure and turns into a shallow "broad-brim" towards the periphery. Seismic investigations also showed that the impact structure is asymmetric. The lateral extent of the shallow broad brim part of the impact-induced seismic disturbance is elongated in northward direction, while the central uplift is off-set from the geometrical center of the structure. Features like the elongated crater diameter, the crater radius/seismic-disturbance asymmetry, the offsets in brecciation and structural uplift, and finally the preferred distribution of ejecta north of the crater point to a direction of impact for Mjølnir from the south-southwest.

The biostratigraphic age of 142 ± 2.6 Myr for the Mjølnir impact is based on detailed paleontological analyses and places the impact event close to the Jurassic-Cretaceous boundary.

Extensive geological and geophysical investigations have been carried out in the Barents Sea in the past 40 years or so. This activity, including seismic analyses, gravimetric measurements, magnetic surveys, and drilling of exploration wells has been carried out by the hydrocarbon industry. In the course of this exploration activity, the large-scale seismic structure of Mjølnir was first mapped by Steinar Gudlaugsson in 1993. A systematic search for impact indicators in a core of Late Jurassic/Early Cretaceous sedimentary strata from the vicinity of the proposed Mjølnir impact structure by Henning Dypvik and co-workers led to the discovery of shock features and ultimately the confirmation of the marine impact event in 1996. Since then, this impact site has been studied in much detail by Dypvik and co-workers, as well as by Filippos Tsikalas, based on an extensive amount of geophysical information, including 2100 km of seismic reflection profiles. The best summary of the current state of knowledge provides the book *The Mjølnir impact event and its consequences. Geology and Geophysics of a Late Jurassic / Early Cretaceous marine impact event.*, published in 2010.

Location/Access

The Mjølnir Structure lies on the Bjarmeland Platform, is submerged in the Barents Sea, and is completely buried.

Selected References

Dypvik, H., Gudlaugsson, S. T., Tsikalas, F., Attrep, M. Jr., Ferrell, R. E. Jr., Krinsley, D. H., Mork, A., Faleide, J. I., and Nagy, J. (1996). Mjølnir structure: An impact crater in the Barents Sea. Geology, Vol. 24, pp. 779–782.

Dypvik, H., Tsikalas, F., and Smelror, M. (eds.) (2010). The Mjølnir Impact Event and its Consequences. Geology and Geophysics of a Late Jurassic/Early Cretaceous Marine Impact Event, Springer, Berlin Heidelberg, 318 pp.

Gudlaugsson, S. T. (1993). Large impact crater in the Barents Sea. Geology, Vol. 21, pp. 291–294.

Sandbakken, P. T., Langenhorst, F., and Dypvik, H., (2005). Shock metamorphism of quartz at the submarine Mjølnir impact crater, Barents Sea. Meteoritics and Planetary Science, Vol. 40, pp. 1363–1375.

Shuvalov, V. V., and Dypvik, H., (2004). Ejecta formation and crater development of the Mjølnir impact. Meteoritics and Planetary Science, Vol. 39, pp. 467–479.

Shuvalov, V. V., Dypvik, H., and Tsikalas, F. (2002). Numerical simulations of the Mjølnir marine impact crater. Journal of Geophysical Research, Vol. 107, E7.

Tsikalas, F., Gudlaugsson, S. T., and Faleide, J. I. (1998). The anatomy of a buried complex impact structure: the Mjølnir structure, Barents Sea. Journal of Geophysical Research, Vol. 103, pp. 30469–30483.

Tsikalas, F., and Faleide, J. I. (2007). Post-impact structural crater modification due to sediment loading: An overlooked process. Meteoritics and Planetary Science, Vol. 42, pp. 2013–2029.

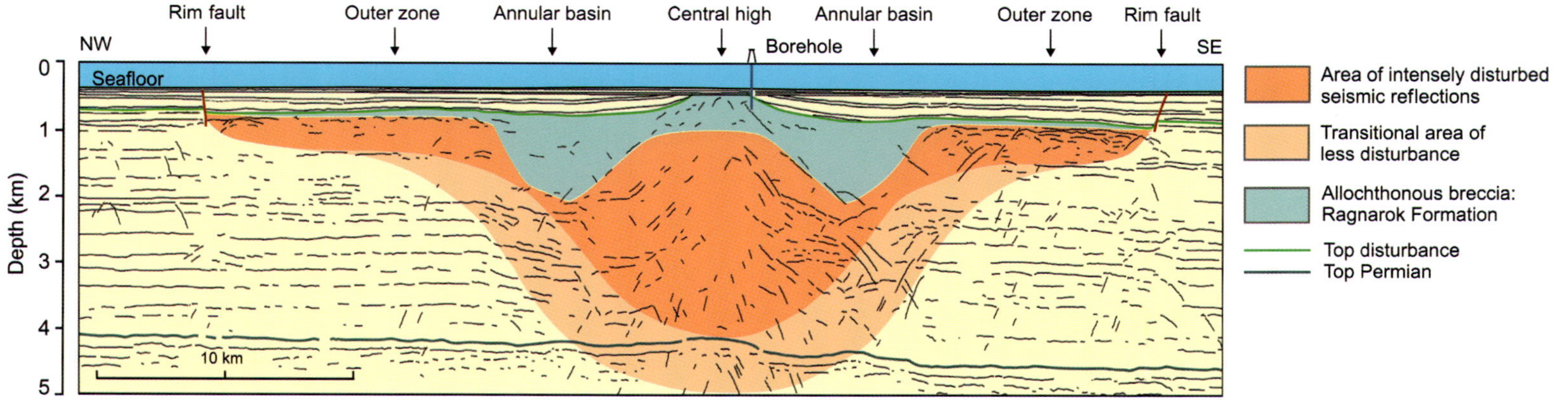

The modeled subsurface structure of the Mjølnir impact structure, based on seismic reflection data (modified after Dypvik et al., 2010).

Mjølnir

Norway

latitude = 73°48' N / longitude = 29°40' E / diameter = 40 km / age = 142.0 ± 2.6 Myr

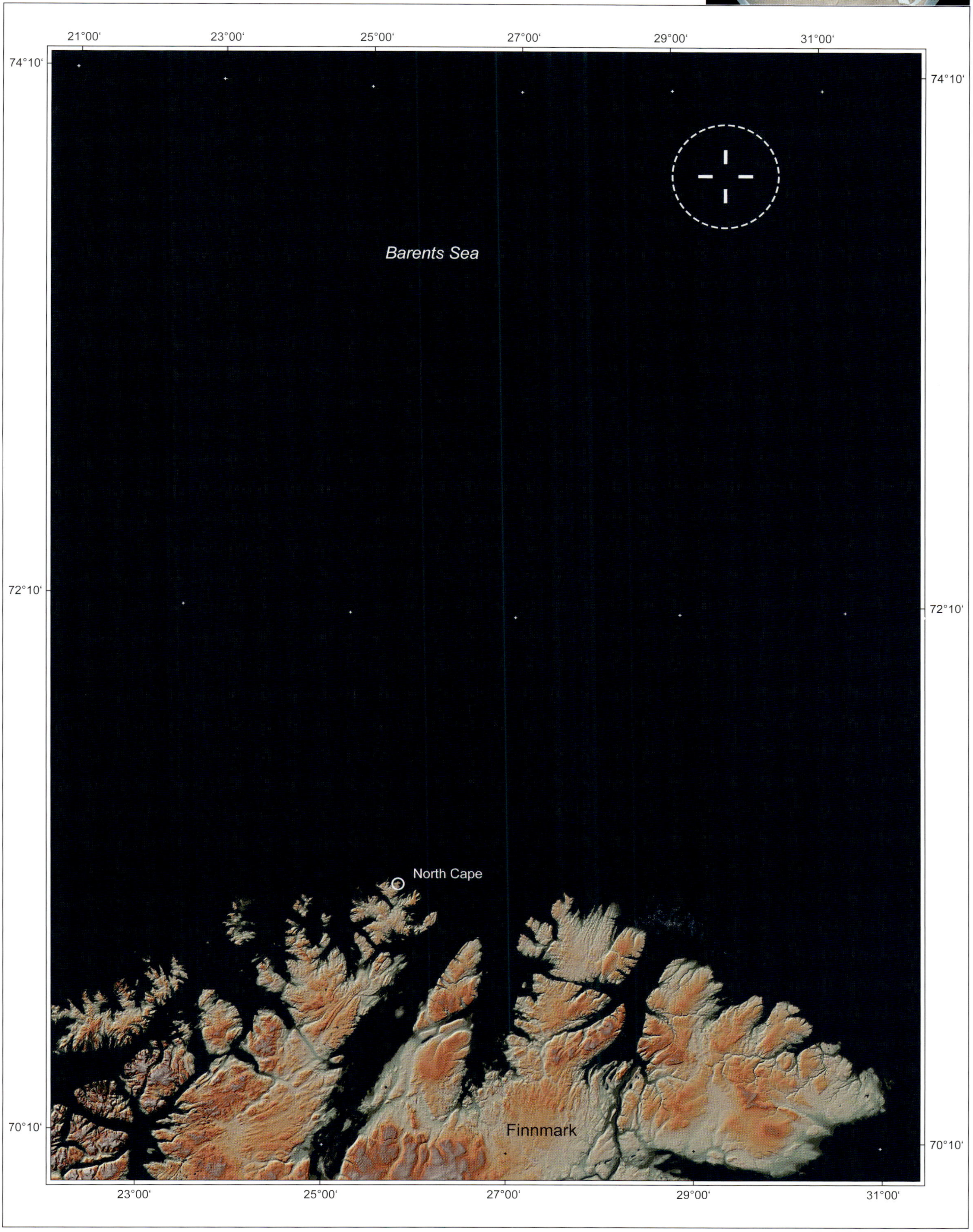

scale = 1 : 1 916 666 / scene = 345.0 × 458.0 km²

0 19.16 95.80 km

N

0 meter 1160

Morasko Crater Field
Poland

Thomas Kenkmann

The Morasko crater field is located near Poznan, Poland, in the Morasko Natural Reserve. The strewn field comprises eight craters with diameters ranging from 20 to 96 m, all of which were formed in soft glacial tills and sands. Abundant meteorites have been found in the area. Their composition and distribution suggest that the strewn field was formed by an iron projectile with a trajectory from northeast to southwest. As the Morasko impact craters are formed in glacial till, the event must be younger than the last glaciation in Poland. Today the total mass of pieces of iron that have been recovered amounts to 400 kg. Many studies have been focused on the characterization of meteorite properties.

The crater pits are almost circular and bowl-shaped. High-resolution topographic modeling of the crater field obtained from airborne LiDAR data show that the studied craters display different cross-sectional asymmetries that would have resulted from the interplay between the oblique trajectories of the incoming projectiles and the topography of the primary pre-impact glacial and post-glacial landforms of terminal moraine ridges. The oblique impacts likely influenced also the asymmetric distribution of ejecta during the excavation of the craters. Due to erosion, the crater rims have largely been degraded, and the craters have depth/diameter ratios lower than typical for simple impact craters. The largest crater has a depth of 12.1 m and a crater volume of about 39 000 m^3. Numerical modeling of the pre-atmospheric parameters of the Morasko iron meteoroid indicates that the entry mass was between 600 and 1100 tons. The meteoroid entered the upper atmosphere at an angle of 30–40° and at a velocity between 16 and 18 km/s. The largest of the Morasko craters was formed by a projectile about 1.5 m in diameter impacting at about 10 km/s.

A maximum age of the impact of 6400 years before present (BP) is provided by the age of paleosols buried underneath the impact ejecta layer. This is in agreement with the results of ^{14}C radiometric dating of the oldest organic material in the fill of the craters, which suggests that the impact had occurred earlier than 5000 years BP.

For a long time, the morphological depressions in the forest of Morasko had been considered small lakes from the glacial era. In 1914, during World War I, digging of trenches uncovered a block of meteoritic iron at a depth of five meters. Further meteorite discoveries followed, with the largest find having a mass of 78 kg. The first comprehensive study of the Morasko strewn field was conducted in the 1950s by Jerzy Pokrzywnicki, who linked the recovered meteorites with the circular depressions near Morasko and suggested their impact origin. Interdisciplinary studies in the 1970s addressed many aspects of the Morasko meteorite fall, including theoretical calculations of the trajectory, the composition and structure of meteorites, and studies of the morphology and morphometry of the craters. In the 1990s, radiometric dating methods were used to determine the age of the event. Recently, state of the-art geoscience techniques were applied to investigate the craters and to model the impact event.

Location/Access

The village of Morasko is located on the outskirts of the city of Poznań, Poland, nine kilometers from the city. One can find the crater pits in the village's beech-oak forest. Some of them are filled with water. The craters are protected and part of the Morasko Natural Reserve.

Selected References

Bronikowska, M., Artemieva, N. A., and Wünnemann, K. (2017). Reconstruction of the Morasko meteoroid impact – Insight from numerical modeling. Meteoritics and Planetary Science, Vol. 52, pp. 1704–1721.

Karaszewski, W. (1974). Geological studies of "meteorite" craters in Nördlinger Ries (West Germany) and at Morasko (Poland) (in Polish). Przeglad Geologiczny, Vol. 22, pp. 626–627.

Korpikiewicz, H. (1978). Meteoritic shower Morasko. Meteoritics, Vol. 13, pp. 311–326.

Pilski, A. S., Wasson, J. T., Muszynski, A., Kryza, R., Karwowski, Ł., and Nowak, M. (2013). Low-Ir IAB irons from Morasko and other locations in central Europe: One fall, possibly distinct from IAB-MG. Meteoritics and Planetary Science, Vol. 48, pp. 2531–2541.

Włodarski, W., Papis, J., and Szczuciński, W. (2017). Morphology of the Morasko crater field (western Poland): Influences of pre-impact topography, meteoroid impact processes, and post-impact alterations. Geomorphology, Vol. 295, pp. 586–597.

The 96 m wide main crater of the Morasko crater field (photo: Kai Wünnemann, Museum für Naturkunde, Berlin).

Morasko Crater Field

Poland

latitude = 52°29'25" N / longitude = 16°53'47" E / diameter = max. 96 m / age = 5700 ± 700 yr

scale = 1:90 000 / scene = 16.2 × 21.6 km²

0 0.90 4.50 km

N

15 meter 170

Neugrund
Estonia

Thomas Kenkmann

The 7 km diameter Neugrund impact crater is situated in the southwestern part of the Gulf of Finland; it is completely submerged. The structure involves an unusually shallow submarine plateau with a circular outline of 4.5 km diameter. The plateau is built up by competent post-impact sediments of the fill of the submarine crater structure. On the nearby island Osmussaar, up to tens of meter-sized brecciated boulders represent pieces of the crystalline rim of the Neugrund structure.

The depth of the 7 km Neugrund crater structure has not yet been assessed, as the depression is filled by post-impact sediments and allochthonous breccias. The topmost layers of the post-impact infill are Ordovician limestones and dolostones that form the central plateau over the basin. The plateau is surrounded by a 200–500 m wide and 20–70 m deep erosional canyon. Outside of the canyon, crystalline rocks form a 2.5–3 km wide annular ridge that extends 60–120 m above the target surface of crystalline bedrock. This ridge is considered the crater rim. Other striking features beyond this unusual crater structure are a ring fault with a diameter of 20 km and a 3–5 km wide circular depression between the rim wall and the fault that contains distal breccia. Shock metamorphism, in particular planar deformation features (PDFs) in quartz, was detected in rocks from the rim and the distal breccias.

The impact occurred into shallow marine sea. The underlying epi-continental shelf sediments were composed of 100 m thick Lower Cambrian clay-, silt-, and sandstones that were underlain by crystalline basement of the Baltic Shield, which had experienced a regional metamorphism and deformation in the Svecofennian orogeny at 1750–1900 Myr ago. Neugrund has a stratigraphic age of 535 Myr – based on the age of the youngest sedimentary strata involved in the impact event. The structure is associated with an event horizon in the form of a distal breccia layer with ejecta and tsunami deposits of Early Cambrian age, which has been recognized far beyond the crater. Aeromagnetic data over the Neugrund structure show a local negative anomaly, but this anomaly is superposed by regional magnetic features. An arc of positive anomalies that spatially corresponds to the rim wall surrounds the central minimum. A central uplift could not be identified by aeromagnetic surveying.

The erratic, brecciated boulders on the Island of Osmussaar were recognized already in the 1920s. Later, a 19 m thick, strongly deformed layer of breccia was drilled on Osmussaar Island. Neugrund became the subject of detailed studies in the 1980s in the course of regional geological-geophysical mapping of the sea around the former Soviet Union – however, without leading to the identification of the structure as an impact crater. Between 1996 and 2006 the father-and-son team of Kalle and Sten Suuroja undertook ten marine expeditions that included seismic reflection as well as side-scan sonar profiling, observing submarine outcrops by a video robot, diving, and sampling. They finally solved the puzzle of the origin of the submarine crater.

Location/Access

Neugrund is located 10 km offshore from the coast of mainland Estonia in the Baltic Sea. The site is located approximately 60 km southwest of the capital city of Tallinn. The erratic boulders exposed on the island of Osmussaar occur about 10 km southwest of the submerged Neugrund crater.

Selected References

Plado, J. (2012). Meteorite impact craters and possibly impact-related structures in Estonia. Meteoritics and Planetary Science, Vol. 47, pp. 1590–1605.

Suuroja, K., and Suuroja, S. (2000). Neugrund structure – the newly discovered submarine Early Cambrian impact crater. In: Impacts and the Early Earth, Gilmour I., and Koeberl C. (eds.), Springer, Berlin Heidelberg, Vol. 91, pp. 389–416.

Suuroja, S., Suuroja, K., and Flodén, T. (2013). A comparative analysis of two Early Palaeozoic marine impact structures in Estonia, Baltic Sea: Neugrund and Kärdla. Bulletin of the Geological Society of Finland, Vol. 85, pp. 83–101.

Neugrund

Estonia

latitude = 59°20‘00“ N / longitude = 23°31‘30“ E / diameter = 7 km / age = 535 Myr

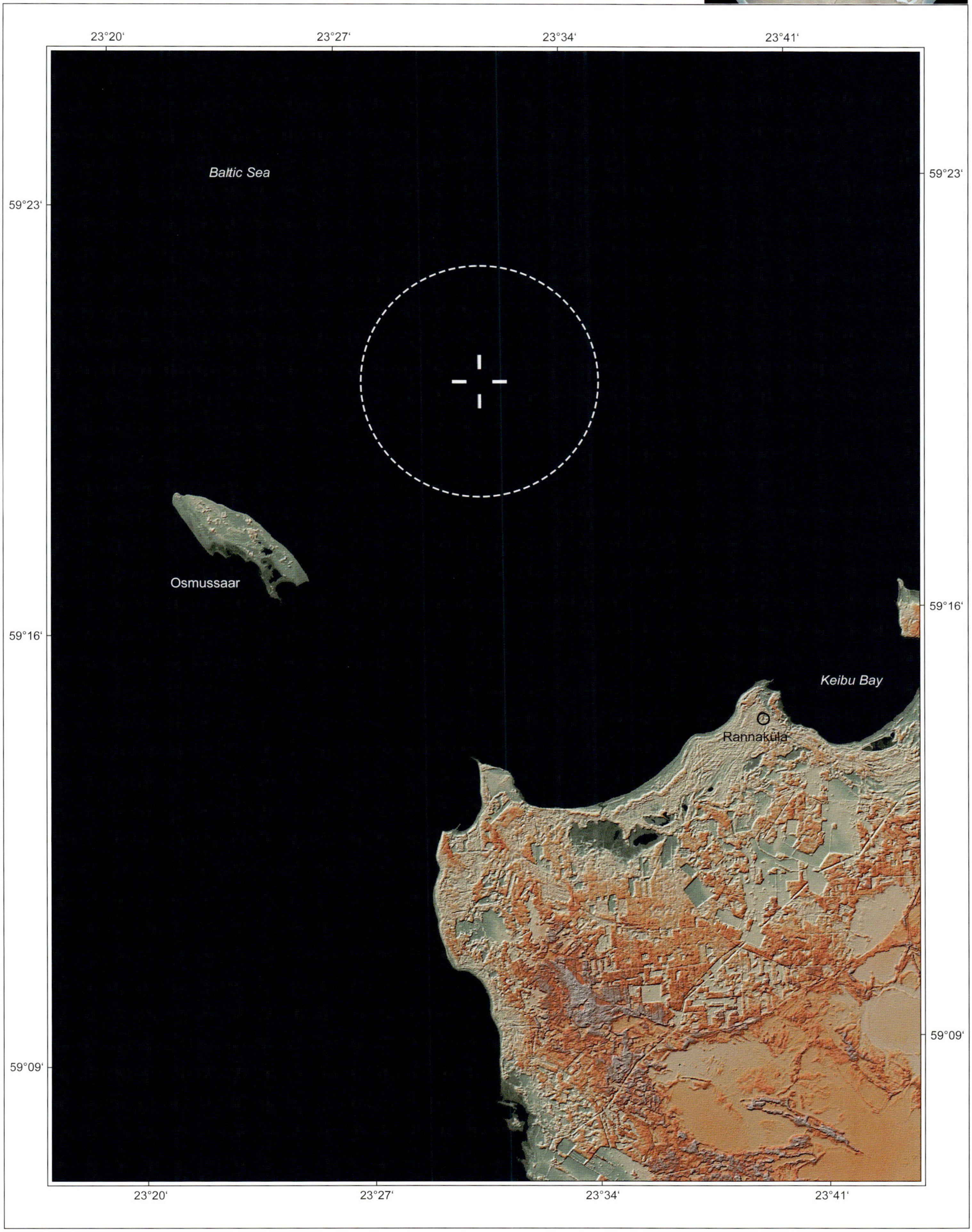

scale = 1:142 000 / scene = 25.6 × 34.1 km²

0 1.42 7.10 km

N

0 meter 35

Obolon
Ukraine

Thomas Kenkmann

Obolon, also spelled *Obolon'*, is an 18 km diameter impact structure in Ukraine. It is buried under a 300 m thick sequence of Mesozoic-Cenozoic sediments. The structure has no geomorphological expression and does not leave a trace on spaceborne and aerial imagery.

The target is composed of crystalline basement rocks and a sedimentary cover. In this region, the crystalline basement of the Ukrainian Shield is formed by Precambrian gneisses and granites. For a long time, this part of the Ukrainian Shield was land and subject to erosion rather than sedimentation. However, in the Paleozoic – from the Middle Devonian to the Carboniferous – a significant part of the shield's terrane underwent transgression. Further sedimentation occurred in the Lower Triassic.

Two boreholes drilled in the western part of the structure and one borehole at the southeastern rim have provided insight into the crater structure. The deepest part of the structure is formed in crystalline basement, whereas its peripheral zone involves platform deposits. The crater fill at Obolon consists of a sequence of breccias, including polymict lithic breccias, suevites, and clast-rich impact melt rocks of total drilled thickness of more than 70 m. Shock metamorphic effects observed in impactite samples are planar deformation features (PDFs) in quartz, diaplectic as well as fused glasses of feldspar minerals, coesite, and impact diamonds. Rare shatter cones have been described.

These impact breccias are overlain by up to 175 m of coarse-grained sandstones and sedimentary rock breccia, which, in turn, are covered by crystalline rock breccias of up to 41 m thickness. This sedimentary succession of the crater fill indicates a forceful tsunami resurge and shows that the impact occurred under shallow marine conditions. The post-impact deposits of the Obolon structure are Middle Jurassic marine deposits, mainly clays and sandstones. These strata were then buried under Upper Jurassic, Cretaceous, and Cenozoic sediments. The maximum thickness of post-impact crater infill reaches 750 m in the crater. Layers of coarse-grained ejecta and tsunami deposits from the impact event were also encountered at the boundary between the Lower Triassic and Middle Jurassic, at distances of up to 50 km from the crater.

There is a strong residual negative Bouguer gravity anomaly over the Obolon structure of −10 mGal. This negative anomaly decreases to −4 mGal in the center due to uplift of relatively denser rocks from the crystalline basement. A ring of positive gravity excursions surrounds the structure's negative anomaly and indicates structural uplift of dense crystalline rocks along the rim.

The age of the Obolon impact was determined by the K-Ar method applied to impact melt glass to 169 Myr; thus, the impact occurred in the Jurassic.

In 1947 geophysical analysis of the region by M. Chirvinskaya revealed the existence of a morphological depression of about 17 km diameter in the crystalline basement rocks. In the course of prospecting for oil shales in the mid-1960s, two deep boreholes were drilled into this depression. They encountered breccias that were initially interpreted as products of volcanic and tectonic activity. In 1976, the impact origin of the Obolon depression was confirmed by Victor Masaitis and Anton Val'ter, who inspected core samples for shock effects.

Location/Access

The Obolon impact structure is situated in Ukraine, at the left bank of the Dniepr River, where it has been dammed to form the Kremenchuk Reservoir. The buried crater is located some 65 km southwest of the city of Mirgorod.

Selected References

Gurov, E. P., and Gurova, E. P. (1995). Impact-melt composition of the Obolon crater; chlorine as a possible indicator of the submarine crater formation. Meteoritics and Planetary Science, Vol. 30, p. 515.

Gurov, E., Gurova, E., Chernenko, Y., and Yamnichenko, A. (2009). The Obolon impact structure, Ukraine, and its ejecta deposits. Meteoritics and Planetary Science, Vol. 44, pp. 389–404.

Masaitis, V. L. (1999). Impact structures of northeastern Eurasia: The territories of Russia and adjacent countries. Meteoritics and Planetary Science, Vol. 34, pp. 691–711.

Masaitis, V. L., Danilin, A. N., Karpov, G. M., and Raikhlin, A. I. (1976). Karla, Obolon' and Rotmistrovka astroblemes in the European part of the USSR (in Russian). Doklady Akademii Nauk SSSR, Vol. 230, pp. 174–177.

Val'ter, A. A., Gurov, E. P., and Ryabenko, V. A. (1976). The Obolon' fossil meteorite crater astrobleme on the northeast flank of the Ukrainian Shield (in Russian). Doklady Akademii Nauk SSSR, Vol. 232, pp. 170–173.

Obolon

Ukraine

latitude = 49°35‘48“ N / longitude = 32°54‘33“ E / diameter = 18 km / age = 169 Myr

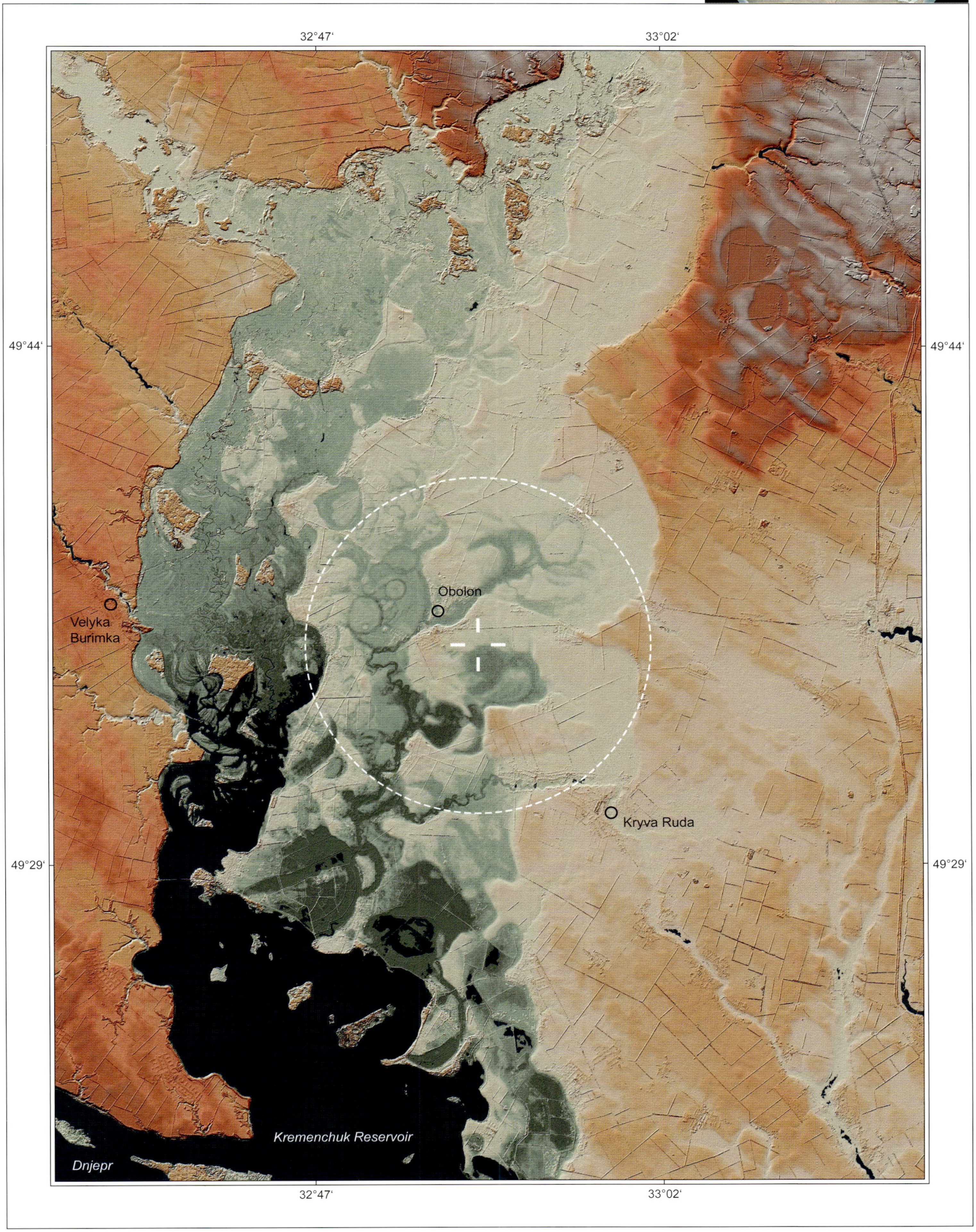

scale = 1 : 251 000 / scene = 45.2 × 60.3 km²

0 2.51 12.55 km

N

80 meter 155

Paasselkä
Finland

Thomas Kenkmann

The 10 km diameter Paasselkä impact structure is completely submerged in the 75 m deep Lake Paasselkä. The structure appears to be strongly eroded, in particular by Pleistocene glaciation. The southern and eastern shoreline of Lake Paasselkä is circular but the remainder is not reminiscent of the morphology of an impact structure.

Target rocks comprise ca. 1900 Myr old mica schist with mafic intrusions. A drilling program, carried out by the Geological Survey of Finland in 1999, recovered a core from the center of the structure, of about 200 m disintegrated target rock material that was overlain by some Quaternary deposits. Drilling neither reached an impact breccia layer, nor the crater floor. A central uplift feature was also not indicated. The target is devoid of pre-impact and post-impact sediments, although some sandstone clasts occur in the breccias.

Glacial till collected from a pit south of the crater contains scoured blocks of impact lithologies that are thought to be derived from this impact structure. Among them is fresh, clast-rich impact melt rock that contains intensely shocked (planar deformation features – PDFs) and toasted quartz grains, recrystallized silica melt, and diaplectic silica glass grains, and recrystallized feldspar glass. ^{40}Ar-^{39}Ar dating of recrystallized feldspar glass particles separated from such rock yielded a Middle to Late Triassic age of 228.7 ± 3.0 Myr for the impact. Airborne and ground magnetic data reveal a low magnetic relief over the lake, with a negative anomaly of −300 nT near the center. A negative gravimetric anomaly of −8 mGal is associated with the structure.

Paasselkä was first noted by Seppo Elo in 1979. A possible impact origin of the structure was proposed by Herbert Henkel and Lauri J. Pesonen in 1992. The impact origin was finally confirmed by Pesonen and co-workers in 1999, when shock deformation in brecciated bedrock was discovered.

Suevite, 7 × 4.4 × 3.5 cm in size, from the Paasselkä impact structure (photo courtesy: Ralf-Thomas Schmitt, Museum für Naturkunde Berlin, sample from Roald Tagle).

The area of the Paasselkä impact structure with the homonymous lake on May 21, 2018 as a composite image based on TanDEM-X DEM and Copernicus Sentinel-2 (channel 2/3/4) data.

Location/Access

The impact structure is located about 7 km east-southeast of the village of Savonranta in southeastern Finland. Fresh impact lithologies were sampled from the open Sikosärkät till pit about 700 m from the southeastern shoreline of Lake Paasselkä.

Selected References

Kuivasaari, T., Pesonen, L. J., Elo, S., Plado, J., and Lehtinen, M. (2000). Paasselkä – the ninth meteorite impact structure in Finland. 4th ESF Workshop: Response of the Earth System to Impact Processes (IMPACT), Helsinki, p. 79.

Pesonen, L. J., Kuivasaari, T., Lehtinen, M., and Elo, S. (1999). Paasselkä: A new meteorite impact structure in Eastern Finland. Meteoritics and Planetary Science, Vol. 34 Supplement, p. A90-1.

Schmieder, M., Moilanen, J., and Buchner, E. (2008). Impact melt rocks from the Paasselkä impact structure (SE Finland): Petrography and geochemistry. Meteoritics and Planetary Science, Vol. 43, pp. 1189–1200.

Schmieder, M., Schwarz, W. H., Buchner, E., Trieloff, M., Moilanen, J., and Ohman, T. A. (2010). A Middle-Late Triassic ^{40}Ar-^{39}Ar age for the Paasselkä impact structure (SE Finland). Meteoritics and Planetary Science, Vol. 45, pp. 572–582.

Paasselkä

Finland

latitude = 62°08‘55“ N / longitude = 29°24‘46“ E / diameter = 10 km / age = 228.7 ± 3.0 Myr

scale = 1: 182 000 / scene = 32.8 × 43.7 km²

0 1.82 9.10 km

N

75 meter 205

Puchezh-Katunki

Russia

Thomas Kenkmann

The large Puchezh-Katunki structure is located in the center of the Russian Plain. The structure has no topographic expression and is almost completely buried beneath Mesozoic-Cenozoic deposits. However, along the bank of the dammed Volga River that runs through the western part of the structure polymict breccia of the impact ejecta blanket is exposed. The crater is famous for the Vorotilov borehole, a 5374 m deep research drilling into the structure's center.

The undisturbed target outside of the structure is composed of Archean gneissic basement rocks at a depth of 1.5–2 km and deeper. These rocks are unconformably overlain by sediments of uppermost Precambrian (500–800 m), Devonian (650 m), Carboniferous (400 m), Permian (360 m), and Lower Triassic (110 m) ages. Limestones, dolostone and mudstones dominate the Paleozoic succession, and siliciclastic strata are abundant in the Precambrian and Mesozoic succession. The crater is covered by lake deposits that reach a thickness of up to 450 m.

Main structural features of the Puchezh-Katunki impact structure are an 8–10 km wide central uplift, surrounded by an annular depression. The axis of this annular basin, where it attains its greatest depth, has a radius of 10 km, and the depression has a radial extent of 20 km, measured from the center. Often, the corresponding diameter is quoted as the rim-to-rim diameter of 40 km, but it excludes a peripheral terrace zone that is, on average, about 20 km wide. Consequently, a diameter of 80 km has also been published for Puchezh-Katunki. The terrace zone varies in width and dips gently inward. Terraces are covered by an up to 200 m thick blanket of variegated polymict breccia that has been interpreted as deposits of the continuous ejecta blanket. This allochthonous breccia is composed of sedimentary rocks of primarily Upper Permian and Lower Triassic age that build the upper portion of the impact target. Megablocks up to 1–2 km in diameter, containing abundant faults, are embedded in the breccia and gradually pass over to the terrace zone beneath. According to reflection seismic studies, the annular depression is up to 1.8 km deep. The moat is filled with polymict breccia and megablocks that are derived from uppermost Precambrian, Devonian, Carboniferous, and, to a lesser extent, Permian strata.

A drill core section (9 cm diameter) of a polymict impact breccia vein from the Puchezh-Katunki impact structure (photo courtesy: Ralf-Thomas Schmitt, Museum für Naturkunde Berlin, sample obtained from the VSEGEI, St. Petersburg, Russia).

The central uplift was examined by the Vorotilov drilling and further boreholes that penetrated through shocked basement rocks and impact-derived deposits. The uplift is composed of blocks of Archean gneisses and amphibolites enveloped by disturbed uppermost Precambrian and Devonian strata. A central pit up to 525 m deep and 3–5 km in diameter is formed at the top of the central uplift and is filled with various types of polymict impact breccia, and finally overlain by suevite. The elevation of the central uplift over the annular depression is 1.5–1.9 km. Suevite forms an 80 km^2 wide, discontinuous, patchy blanket of, on average, 30 m thickness, with local thicknesses of as much as 116 m. Impact melt rock occurs mainly at the top of the central uplift as thin veins and is rarely up to 2–3 m thick. On the southwestern slope of the uplift, however, it may reach up to 10 m thickness. The overall composition of impact melt rock corresponds to that of the crystalline target composed of gneisses and amphibolites.

The deep Vorotilov core hole, drilled in the central pit, penetrated successively through post-impact deposits, suevites and polymict breccia, and reached crystalline impact breccia at a depth of 546 m. Crystalline rocks of the central uplift intersected by impact melt rock dikes display abundant shock features including shatter cones, planar deformation features (PDFs) in quartz and feldspar, diaplectic glasses, fused glasses (impact glass), coesite, and impact diamonds. Gradients in the degree of shock metamorphism, brecciation, post-impact annealing, and hydrothermal alteration were established in this deep section down to the depth of 5374 m. The values of shock pressure vary gradually from more than 45 GPa near the present surface of the basement to 15 GPa at −5 km; the gradient of attenuation decreases from 9 to 4 GPa per km. Shock-induced heating may locally have reached 950–1100 °C and led to complete recrystallization of rocks. Subsequent hydrothermal alteration was restricted to the central uplift and caused secondary mineralization including minerals such as smectites, chlorites, zeolites, calcite, and anhydrite. New ^{40}Ar-^{39}Ar data from step-heating analysis of five impact melt rock samples indicate an age for the formation of the Puchezh-Katunki impact structure of 192–196 Myr.

Puchezh-Katunki had puzzled geologists for more than a century, until in the early 1970s L. Firsov and Susan Kieffer suggested an impact origin. Since then, this structure, although not exposed at the surface, was intensively investigated and can today be regarded as one of the best examined impact structures on Earth. Nearly 170 boreholes penetrated through impact-derived rocks within the area of the structure. The Vorotilov deep research drilling project, conducted from 1988 to 1992, led to a series of important conclusions about rock transformations during large hypervelocity impacts and the attenuation of shock waves with depth.

Location/Access

The city of Nizhny Novgorod about 80 km south of Puchezh-Katunki is connected to Moscow via train, highways, and airport. The approximate geographic center is close to the village of Vorotilovo where the research borehole was drilled. The deep research drilling is stored in the Russian drill core facility in the city of Yaroslavl. Allochthonous polymict impact breccia is exposed in cliffs on the eastern bank of the dammed river Volga.

Puchezh-Katunki

Russia

latitude = 56°58' N / longitude = 43°43' E / diameter = 40 km / age = 192 - 196 Myr

scale = 1 : 360 000 / scene = 64.9 × 86.5 km²

0 3.60 18.00 km

45 meter 240

Selected References

Firsov, L., and Kieffer, S. W. (1973). Concerning the meteoritic origin of the Puchezh-Katunki Crater. Meteoritics, Vol. 8, pp. 223–244.

Henkel, H. (2018). Provisional translation of: Deep Drilling in the Puchezh-Katunki Impact Structure (in Russian), Masaitis, V. L., and Pevzner, L. A. (eds.), 1999, VSEGEI Press, St. Petersburg, 185 pp. Published by Henkel, H., ISBN: 978-91-519-0289-0.

Holm-Alwmark, S., Alwmark, C., Ferrière, L., Lindström, S., Meier, M. M. M., Scherstén, A., Herrmann, M., Masaitis, V. L., Mashchak, M. S., and Naumov, M. V., and Jourdan, F. (2019). An Early Jurassic age for the Puchezh-Katunki impact structure (Russia) based on ^{40}Ar-^{39}Ar data and palynology. Meteoritics and Planetary Science, Vol. 54, pp. 1764–1780.

Ivanov, B. A., Kocharyan, G. G., Kostuchenko, V. N., Kirjakov, A. F., and Pevzner, L. A. (1996). Puchezh-Katunki impact crater: preliminary data on recovered core block structure (abstract). 27th Lunar and Planetary Science Conference, Lunar and Planetary Institute, Houston, Texas, pp. 589–590.

Masaitis, V. L. (1999). Impact structures of northeastern Eurasia: The territories of Russia and adjacent countries. Meteoritics and Planetary Science, Vol. 34, pp. 691–711.

Masaitis, V. L., Mashchak, M. S., Naumov, M. V., and Raikhlin, A. I. (1994). Large Astroblemes of Russia. VSEGEI St. Petersburg, 34 pp.

Ries
Germany

Wolf Uwe Reimold

The Ries crater in Southern Germany, also known as the *Nördlinger Ries*, is one of the best preserved, complex impact craters on Earth. On surface, the structure with a diameter of 26 km represents an almost circular depression with an inner ring accentuated by some hills and a prominent crater rim that is well exhibited, with the exception of the northern/northwestern sector where it is substantially eroded. The crater structure cuts into the Swabian-Frankonian Alb, a west-southwest–east-northeast trending major escarpment in Southern Germany that was already present at the time of impact. Consequently, the southern crater rim is about 150 m higher than the northern rim.

The Ries impact occurred in 650–750 m thick sediments of Triassic to Neogene age that are underlain by crystalline basement rocks. The crystalline basement consists predominantly of pre-Variscan metamorphic rocks (ortho- and paragneisses, metabasites), with Variscan intrusive magmatites, especially granites, that belong to the Moldanubian zone of the Variscan orogeny. The overlying Triassic sediments are made up of about 50 m of Mid-Triassic Muschelkalk and 200 m thick sandstones and red mudstones of the Upper Triassic Keuper sequence. They are, in turn, overlain by Jurassic sediments, including 30 m of Liassic interlayered sandstone and mud- and claystone deposits, followed by 140 m of distinctly brownish, iron-rich sandstones of Dogger age, and a 200 m thick Malmian limestone sequence. During the Cretaceous, the Jurassic limestones underwent deep erosion and karstification. During the Upper Miocene, the southern part of the Ries became a distal part of the fore-deep of the northward propagating Alpine orogeny, the so-called "Molasse basin". The deposits of the Upper Marine Molasse transgressed into the outer parts of the Ries area, and during the Middle Miocene, the Upper Freshwater Molasse reached into the future crater region, depositing mica-rich sands and shales. The landscape before the impact was dominated by the Jurassic west-southwest–east-northeast trending escarpment of Malmian limestone that dissected the future target area.

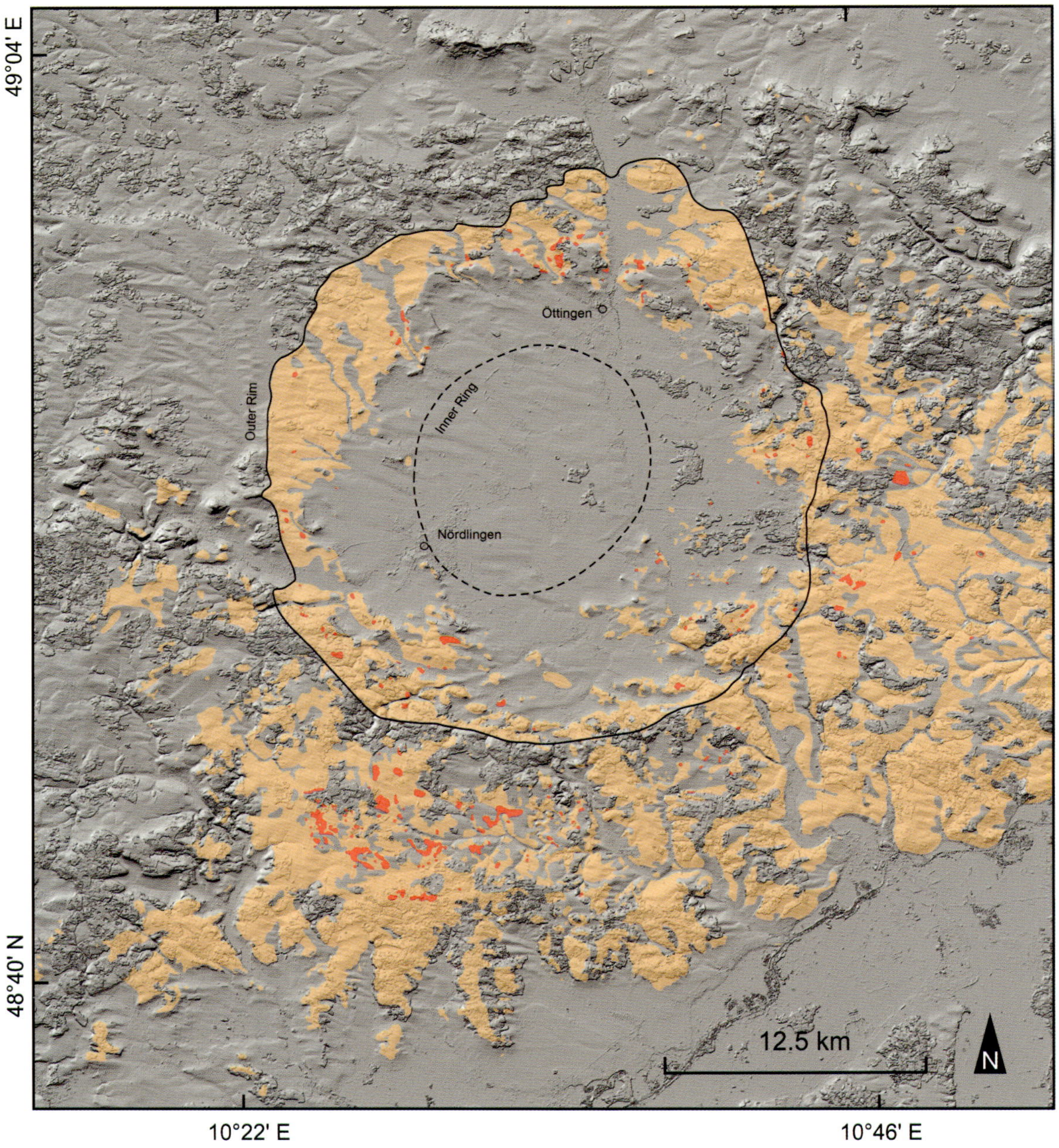

Distribution of deposits of Bunte Breccia (red) and suevite (amber) around the Ries impact crater, superimposed onto the hillshaded TanDEM-X DEM (modified after Geological Map, Bayerisches Geologisches Landesamt, 2004).

Ries

Germany

latitude = 48°52'09" N / longitude = 10°34'41" E / diameter = 26 km / age = 14.8 Myr

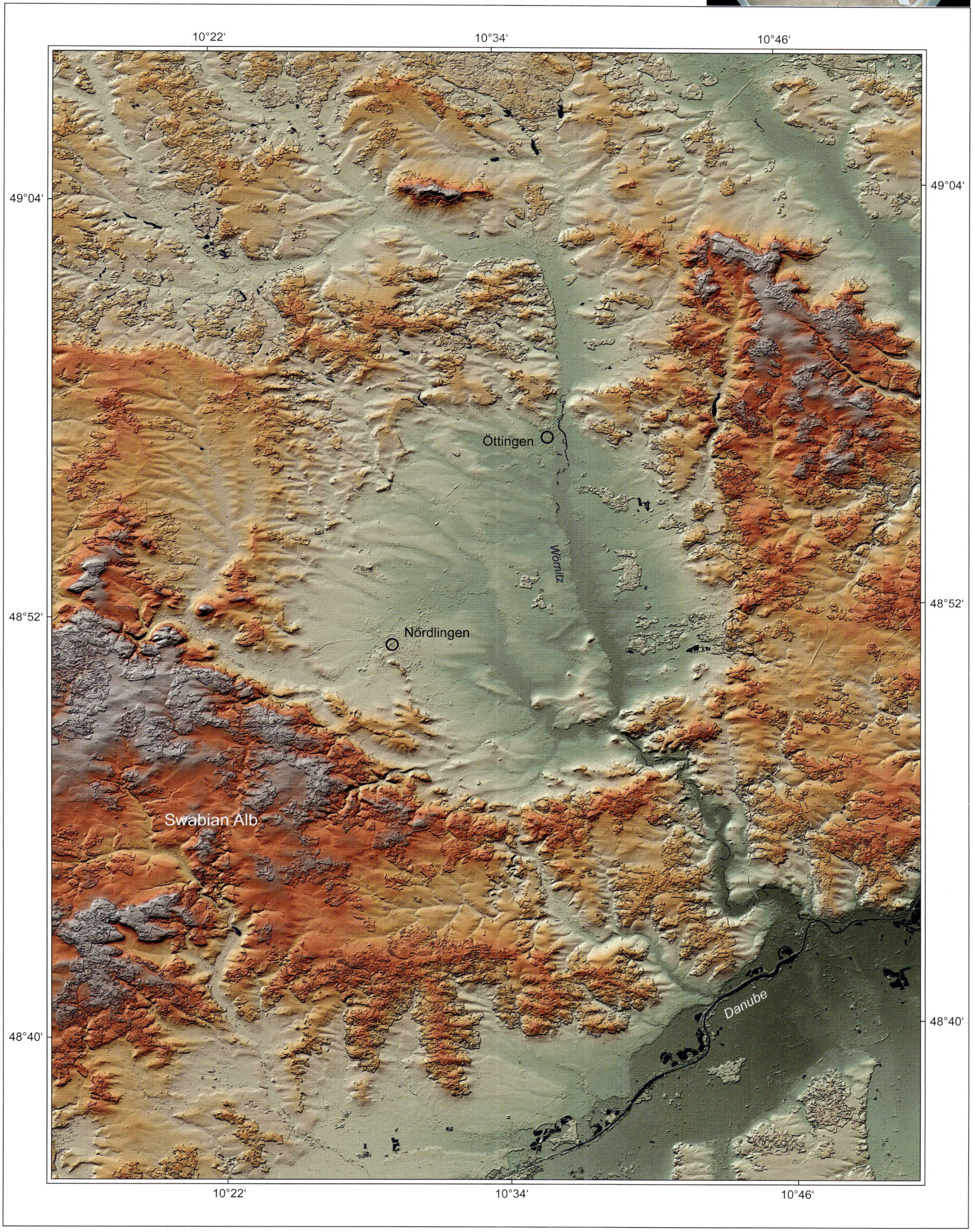

scale = 1 : 250 000 / scene = 45.0 × 60.0 km²

0 2.50 12.50 km

N

350 meter 695

Aumühle quarry in the northeastern part of the Ries impact crater. Greyish suevite overlies variegated, here predominantly reddish Bunte Breccia (photo: T. Kenkmann).

Inclined contact between the Bunte Breccia and the overlying suevite at Aumühle quarry, northeastern Ries impact crater (photo: T. Kenkmann).

The Ries has a flattish central basin with a diameter of approximately 12–13 km. This basin has a bowl shape, whose true crater floor is located in uplifted crystalline basement about 600 m below the current surface level. The basin fill comprises about 300 m of melt fragment-bearing impact breccia, the so-called "suevite" (after the Latin word "Suevia" for the region "Schwaben"), and post-impact lake deposits of 200–400 m thickness. The research borehole Nördlingen 1973 provides a profile through the shattered crater floor, suevite, and lake deposits. This basin is surrounded by the inner ring that consists of uplifted crystalline basement rocks, which are partly covered by sediments of the post-impact Ries lake. The inner ring itself is surrounded by a 7–8 km wide hummocky annulus called the megablock zone, in which large slump blocks are covered by ejecta deposits. The ejecta comprise a polymict lithic breccia called the "Bunte Breccia" and allochthonous megablocks of both crystalline and sedimentary origin. This outer rim is defined by steep concentric faults that separate the surrounding autochthonous units from the parautochthonous megablocks of the crater.

The Ries crater is the only mid-sized complex crater on Earth where actually all types of impact formations can be studied, both the proximal deposits (the crater fill and ejecta of the immediate surroundings of the crater), and the distal deposits (the ejecta in the wider environs), and even tektites several hundred kilometers away. Most notable is suevite, a polymict breccia containing impact-generated melt fragments. Suevite occurs as crater fill, the so-called "crater suevite", and in the environs of the crater, the so-called "outer suevite", and can be detected up to a distance of 30 km from the structure. Bunte Breccia, a multi-colored (variegated) breccia consisting of a mixture of all rock types of the target stratigraphy but mainly blocks and fragments of the sedimentary target rocks, occurs up to 45 km from the center and is considered as the deposit from the impact ejecta curtain. The Ries ejecta blanket contains a massive and continuous rampart structure at 1.45–2.12 crater radii from the crater center, similar to Martian impact craters. Ejecta distribution and thickness, as well as the ejecta fabric, indicate the presence of fluids during the emplacement process. In the eastern and southern parts of the crater structure isolated patches of impact melt rock within suevite have been found. Megablocks, large blocks ejected during crater formation, populate an area extending up to 20 km from the crater center. Well outside the crater, ejecta comprise the Reuter blocks of limestone that occur up to a distance of 180 km and tektites, known as moldavites.

The Ries crater is located only 45 km northeast of the **Steinheim Basin** crater, a joint genesis with which is generally accepted. Of the two scenarios invoked – a 1.5 km asteroid that formed the Ries accompanied by a smaller satellite of 150 m size, or a single bolide composed of possibly two different rock types that separated upon entry into the atmosphere – the former is by far favored and compatible

A large quarry at the eastern crater rim of the Ries impact crater, just south of the town of Wemding, exposes crater rim faults and tilted Malmian strata (photo: T. Kenkmann).

Ziswingen quarry in the southeastern part of the Ries impact crater, located near the crater rim, exposes stratified limestones and marly limestones of the Upper Jurassic, which are truncated by a large normal fault. Cliff height is about 35 m (photo: T. Kenkmann).

with numerical simulation of the impact scenario. In fact, not much is known about the composition(s) of the projectile(s) that formed these two impact structures at all. Taking into account all available age data, a best value for the timing of the Ries impact event of 14.8 Myr is obtained.

The impact changed the region surrounding the crater dramatically by displacing large amounts of material, which then blocked the pre-impact stream beds. As a consequence, a large lake, the Rezat-Altmühl Lake, comparable in size to the present-day Lake Constance, developed. During the Miocene, the Rezat-Altmühl Lake was filled with sediments deposited by the ancient Main River, covering parts of the proximal ejecta in the northeast with sand deposits. Within the crater, the Ries Lake formed and existed for approximately 2 Myr, depositing an about 300 m thick pile of

The Unterwilflingen quarry in the western megablock zone of the Ries impact crater shows the complex interplay between a crystalline megablock (at right), altered suevite containing a large block, and, on the left, variegated Bunte Breccia. Width of the quarry wall is about 30 m (photo: T. Kenkmann).

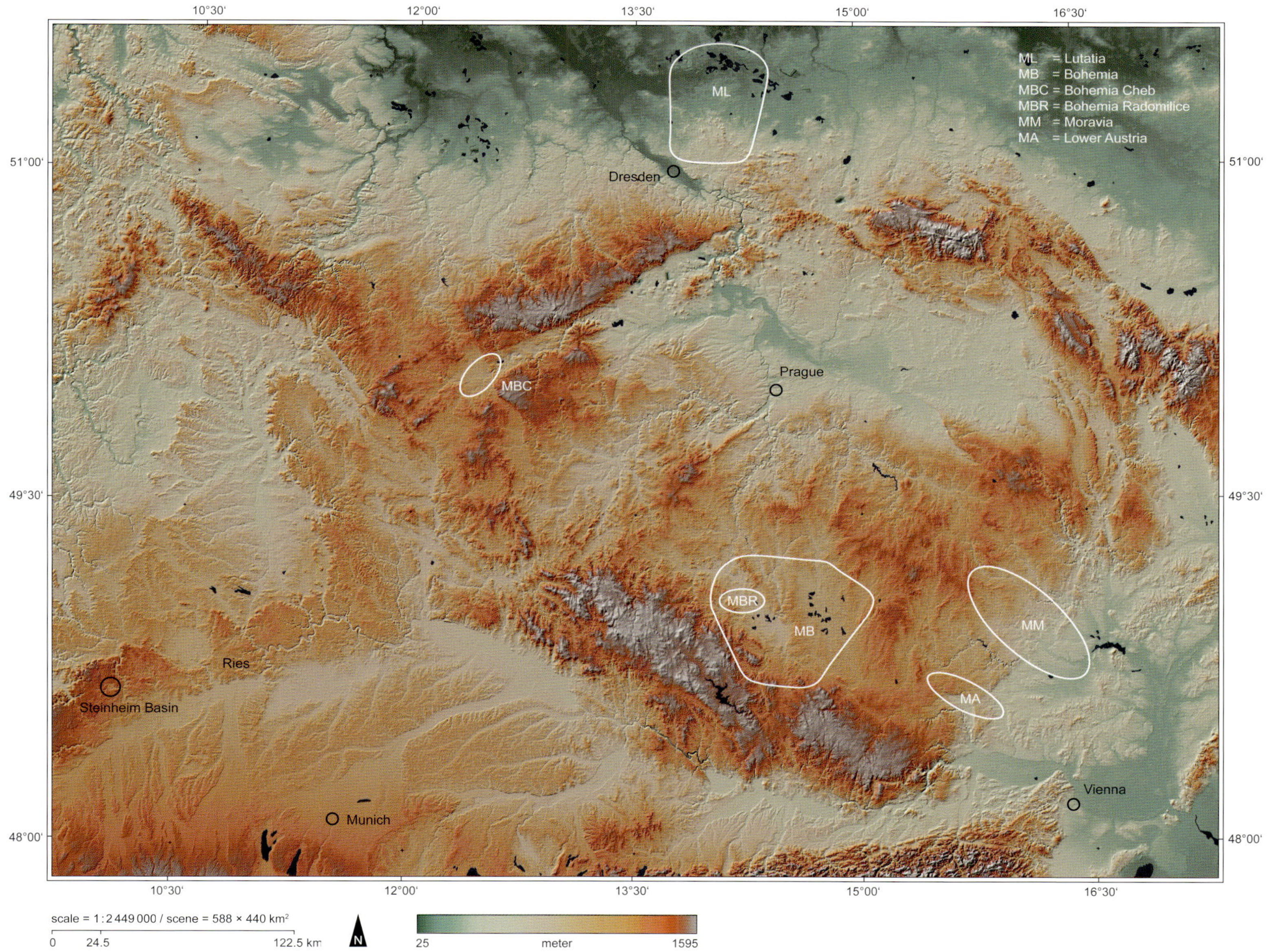

The location of the Moldavite tektite strewn field and the Ries, the source impact crater, in a physical map generated from the TanDEM-X 90m DEM.

sediment. The Ries crater and parts of its ejecta blanket were covered with terrestrial sediments, until the area was uplifted and exhumed much later during the Pleistocene.

This extended time of burial led to the good preservation of the Ries crater and its impact formations. The relief observed today was formed during the Pleistocene ice ages and is the result of erosional processes, valley formation, and karstification. The Wörnitz and Eger rivers – discharging today through the Ries depression towards the southeast – partially removed the Ries Lake clays and mudstones from the central basin, leaving behind the more resistant Ries Lake limestones. In the western and northern part of the crater, the morphologic crater rim has been degraded by post-impact erosion, which led to the current topography of the crater.

The Ries region has been geologically investigated ever since the late 18th century. Impact cratering and its universal importance did not feature on the agenda of geologists at that time, so that it was logical to draw on known volcanic processes, and later also tectonic ideas, to interpret the noted stratigraphic peculiarities and strange rock formations of the Ries. After all, the structure does lie at the northeastern continuation of the Swabian Alb range with its prominent volcanic activity. And those who favored a tectonic origin for the Ries drew on its position just north of the Alpine orogen.

In the initial absence of well-developed shatter cones that might have favored an impact origin to some already in the late 1950s, it was only the discovery of shock metamorphism, which confirmed an impact origin for the Ries crater. Shock metamorphism at Ries included – first – coesite and stishovite, the high-pressure modifications of the mineral quartz, which were discovered by Ed Chao and Eugene Shoemaker. This was soon followed by recognition of planar deformation features (PDFs) in quartz and feldspar, diaplectic quartz and feldspar glasses, and fused quartz glass, known as lechatelierite, in the later 1960s and 1970s by Wolf von Engelhardt and Dieter Stöffler, and their co-workers. In later years, shock deformation and transformation in other minerals were observed, such as shock effects in zircon, as well as impact-generated diamond, and other high-pressure phases. Incidentally, in 2016 a first report of nice shatter cones from a quarry in the western Ries was published.

In 1990, the well-appointed Ries Crater Museum was opened in Nördlingen, the main city within the boundaries of the Ries. It focuses on the Ries crater, the impact processes and evidence for them in the regional rock record; and on a wider scale, the impact history of our planet, the types of extraterrestrial projectiles that formed impact structures on Earth, the importance of impact in the context of the solar system, and for the evolution of life on Earth. Annually the museum attracts some 25000 visitors with even more paying a visit to the Ries crater because the area of the structure was declared a National Geopark in 2005. The Ries Geopark is part of the German network of geoparks dedicated to highlight and showcase landscapes of special geological significance, with the aim of educating the public and preserving natural heritage. Currently efforts are being made to upgrade the Ries Geopark to European Geopark status and eventually to World Natural Heritage. The ZERIN (Zentrum für Rieskrater und Impaktforschung Nördlingen) located close to the Ries Crater Museum is a center for impact research, with facilities for researchers and students, who have come to study the impact evidence and environmental consequences of this impact catastrophe.

The Ries Crater is the source crater for the **Moldavites**. These are centimeter-sized impact glass bodies that have been collected as far as 400 km east of the crater where they form the Central European tektite strewn field with occurrences in Bohemia, Moravia, Lower Austria, and Lutatia (Lausitz) in eastern Germany. Moldavites have been known since the late 18th century. In 1900 the Austrian geologist Eduard Suess published a monograph about Moldavites and related glasses (*Die Herkunft der Moldavite und verwandter Gläser*) where he applied the term *tektites* to this kind of glass. Moldavites are usually of greenish color and transparent with a surface marked by elliptical or circular pits and striae. Of the four large terrestrial tektite strewn fields only the Australasian field has provided more tektites than the Central European one. The Ries Crater has been identified as the source of the Moldavites because of their similar ages. In addition, the chemical composition of near-surface sediments of the Upper Freshwater Molasse supports the hypothesis that this is the parent material for the tektites.

Location/Access

The Ries is easily accessible. It is located nearly central to a triangle formed by the large cities of Stuttgart, Nürnberg, and Munich. Main access is via Freeway (Autobahn) 8, exit Augsburg-West; National Highway (Bundesstraße) 2, Donauwörth; National Highway 25, Nördlingen, or Freeway 7 exit Aalen/Westhausen; National Highway 29, Nördlingen, or Freeway 6, exit Schwabach-West and then National Highway 466, Nördlingen. The city of Nördlingen is also served by regional railway connections.

Selected References

Arp, G., Reimer, A., Simon, K., Sturm, S., Wilk, J., Kruppa, C., Hecht, L., Hansen, B. T., Pohl, J., Reimold, W. U., Kenkmann, T., and Jung, D. (2019). The Erbisberg drilling 2011: Implications for the structure and postimpact evolution of the inner ring of the Ries impact crater. Meteoritics and Planetary Science, Vol. 54, pp. 2448–2842.

Artemieva, N. A., Wünnemann, K., Krien, F., Reimold, W. U., and Stöffler, D. (2013). Ries crater and suevite revisited – Observations and modeling, Part II: Modeling. Meteoritics and Planetary Science, Vol. 48, pp. 590–627.

Hörz, F., Ostertag, R., and Rainey, D. A. (1983). Bunte Breccia of the Ries: continuous deposits of large impact craters. Review of Geophysics and Space Physics, Vol. 21, pp. 1667–1725.

Kenkmann, T., and Ivanov, B. A. (2006). Target delamination by spallation and ejecta dragging: An example from the Ries crater's periphery. Earth and Planetary Science Letters, Vol. 252, pp. 15–29.

Nördlingen Ries Crater Museum: www.riescrater-museum.de

Osinski, G. R., Grieve, R. A. F., and Spray, J. G. (2004). The nature of the groundmass of surficial suevite from the Ries impact structure, Germany, and constraints on its origin. Meteoritics and Planetary Science, Vol. 39, pp. 1655–1683.

Pösges, G., and Schieber, M. (2015). The Ries Crater Museum Nördlingen – Museums Guide. Academy Bulletin Nr. 253, Bavarian Academy for Teacher Training, Dillingen, Pfeil-Verlag, München, 128 pp.

Pohl, J., Stöffler, D., Gall, H., and Ernstson, K. (1977). The Ries impact crater. In: Impact and Explosion Cratering, Roddy, D. J., Pepin, R. O., and Merrill, R. B. (eds.), Pergamon Press, New York, pp. 343–404.

Shoemaker, E. M., and Chao, E. C. T. (1961). New evidence for the impact origin of the Ries basin, Bavaria, Germany. Journal of Geophysical Research, Vol. 66, pp. 3371–3378.

Stöffler, D. (1977). Research drilling Nördlingen 1973: polymict breccias, crater basement, and cratering model of the Ries impact structure. Geologica Bavarica, Vol. 75, pp. 443–458.

Stöffler, D., Artemieva, N. A., and Pierazzo, E. (2002). Modeling the Ries-Steinheim impact event and the formation of the moldavite strewn field. Meteoritics and Planetary Science, Vol. 37, pp. 1893–1907.

Stöffler, D., Reimold, W. U., Jacob, J., Hansen, B. K., Summerson, I. A. T., Artemieva, N. A., and Wünnemann, K. (2013). Ries crater and suevite revisited – Observations and modeling, Part I: Observations. Meteoritics and Planetary Science, Vol. 48, pp. 515–589.

Sturm, S., Wulf, G., Jung, D., and Kenkmann, T. (2013). The Ries impact, a double-layer rampart crater on Earth. Geology, Vol. 41, pp. 531–534.

Wünnemann, K., Morgan, J. V., and Jödicke, H. (2005). Is Ries crater typical for its size? An analysis based upon old and new geophysical data and numerical modeling. Geological Society of America Special Paper, Vol. 384, pp. 67–83.

The Nördlinger Ries on July 7, 2017 as a composite image based on TanDEM-X DEM and Copernicus Sentinel-2 (channel 2/3/4) data.

Ritland
Norway

Thomas Kenkmann

The Ritland impact structure is hardly recognizable from its topography alone. It forms a circular depression, about 2.7 km wide and 350 m deep, in the Sub-Cambrian peneplain of southwestern Norway. This flattish peneplain formed during a long period of exposure and erosion of the Proterozoic Baltic Shield. In the Paleozoic era the Baltic Shield was transgressed by the sea and the region became submerged. The first deposits laid down under these shallow marine conditions were the Cambrian Alum shales.

The impact occurred in Early Cambrian times when the shallow sea was already established. The crater cavity was partly filled with debris washed back into the cavity. The fill comprises Cambrian shales and Ordovician sandstone. The Caledonian orogeny reached the area of Ritland in Silurian time about 425 Myr ago. Then, the crater was overthrust by a pile of nappes that filled the still existing crater cavity. Quaternary uplift, erosion, and finally glaciation led to the current state of exposure of the structure.

At Ritland a continuous section through the crater from the shattered crater subsurface and crater floor, over the various crater-filling impact lithologies, to the post-impact sedimentary infill is preserved. The good exposures of infilling sediments have allowed a detailed reconstruction of the original crater morphology and the evolution of the post-impact infill.

The Precambrian basement rocks within the crater are intensely brecciated and fractured gneisses, and the impact-related deformation continues throughout a 100–200 m wide zone outside the morphological crater wall, as it appears around the eroded structure. The fracture patterns in the basement seem chaotic, with no obvious preferred orientation. In the lowermost, central part of the crater and along a cliff face in the crater valley, a layer of melt-bearing rocks overlying lithic basement breccia crops out. The base of the melt-bearing unit marks the transition from the underlying lithic basement breccia and shows flow structures. Quartz clasts with planar deformation features (PDFs) occur frequently within the melt-bearing unit and provided the confirmation of the impact origin of the structure. The upper part of this unit is characterized by more particulate matrix with patchy melt matrix domains, commonly observed as deformed melt slivers intermingled with small lithic clasts. The melt-bearing impactites grade into clastic deposits, representing slump and avalanche deposits formed during the modification stage of cratering. Numerical modeling of the impact event strongly supports a water depth of up to 100 m at the time of impact.

When the Caledonian nappes moved over the crater from the northwest, during the Silurian, the sediments were only weakly compacted and remained relatively incompetent. Consequently, the crater fill deposits were pushed down or squeezed aside by the nappes. The southeastern crater wall represented an immense obstacle for further nappe propagation towards the southeast, which is why the nappes became stacked into a duplex structure. When the cavity was completely filled with the nappe pile, an out-of-sequence thrust formed on top, which leveled the crater depression. The crater wall also forced the nappes to deflect their general propagation direction towards the southeastern crater sector.

An impact origin of the Ritland structure was first proposed by Fridtjof Riis in 2001 based on geological mapping of the circular structure and the deformed basement. Impact rocks with melt and shock-metamorphosed quartz grains were systematically documented by the group around Elin Kalleson and Henning Dypvyk in 2011.

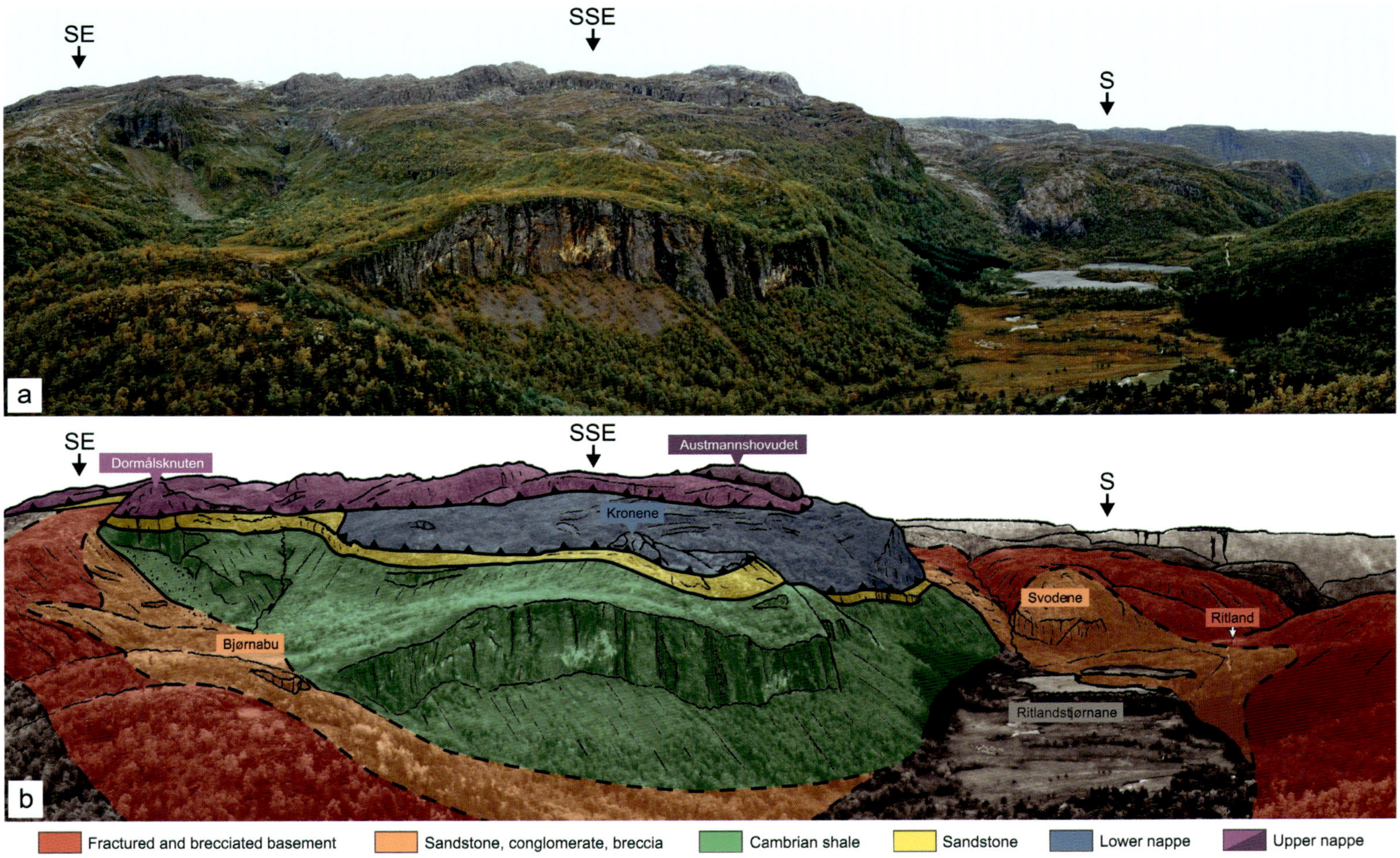

Panoramic view, about 1.5 km wide, of the Ritland impact structure (a) and geological interpretation (b) after Löffler (2013). The crater depression was partly filled with shale and sandstone, and subsequently overthrust during the Caledonian orogeny.

Ritland

Norway

latitude = 59°14'20" N / longitude = 06°25'18" E / diameter = 2.7 km / age = 500 - 540 Myr

scale = 1 : 88 000 / scene = 15.8 × 21.1 km²
0 0.88 4.40 km
N
0 meter 965

Monomict crystalline breccia of the Ritland impact structure illustrates the strong mechanical disintegration of the crater floor (photo: T. Kenkmann).

The Cambrian shale and sandstone filling of the Ritland impact structure was strongly tectonized when the region was overthrust by Caledonian nappes. The cliff face shown is about 100 m high (photo: T. Kenkmann).

Location/Access

Ritland is situated in a mountainous, formerly glaciated area called Rogaland, in the municipality of Hjelmeland, southwestern Norway. A number of scenic fjords occur in this landscape. The western rim of Ritland can be reached on a subordinate, winding road from Hjelmeland. This town is connected to the city of Stavanger via Highway E13 and ferries crossing the Stavangersund.

Selected References

Kalleson, E., Dypvik, H., Riis, F., and Nilsen, O. (2013). The melt-bearing impactites of the Ritland structure, Norway – Implications for melt formation in small impact craters. Meteoritics and Planetary Science, Vol. 48, pp. 1678–1701.

Löffler, J. (2013): The Ritland Impact Structure, southwestern Norway: post-impact Deformation and the Effect of the Crater on Caledonian Thrust Tectonics. Master Thesis (unpublished), University of Freiburg, 127 pp.

Riis, F., Kalleson, E., Dypvik, H., Krøgly, S. O., and Nilsen, O. (2011). The Ritland impact structure, southwestern Norway. Meteoritics and Planetary Science, Vol. 46, pp. 748–761.

Setså, R. (2011). The Ritland Impact Structure: Characteristics and Distribution of the Ejecta Layer and associated Lower Paleozoic sedimentary Succession. Masters Thesis (unpublished), University of Oslo, 114 pp.

Shuvalov, V. V., Dypvik H., Kalleson, E., Setsa, R., and Riis, F. (2012). Modeling the 2.7 km in diameter, shallow marine Ritland impact structure. Earth Moon Planets, Vol. 108, pp. 175–188.

The area of the Ritland impact structure on September 18, 2018 in an image based on Copernicus Sentinel-2 (channel 2/3/4) data.

Rochechouart

France

Thomas Kenkmann

The Rochechouart impact structure is one of only few confirmed impact structures in Central Europe. The structure is located at the northwestern edge of the Massif Central, France, about 40 km west of the city of Limoges. The original crater morphology has been erased by erosion. Therefore, no crater rim or remnants of a central peak are discernible. However, erosion did not cut too deep into the crater, as a variety of impact lithologies is still preserved and exposed at surface. This includes fallback breccia forming the upper part of the crater fill. From the distribution of impact breccias and from a comprehensive shock zoning study, a crater size of approximately 20–23 km diameter was derived. This is in rough agreement with geophysical data that include a 26–28 km wide, circular, negative gravity anomaly.

Target rocks include mica-rich paragneisses, leptynites, granites and granodiorites. North-south trending dikes and porphyritic leucogranites intruded into this sequence before the rocks were affected by the impact. The age of this rock suite is predominantly Upper Paleozoic and these lithologies are related to the Variscan orogeny. Satellite imaging shows various linear fracture systems, but no major differences in the pattern of joints inside and outside the impact structure. The Champagnac quarry in the northeastern part of the crater exposed a large low-angle normal fault in the basement rocks that is lubricated with fine-grained fault gouge and dark veins of a melt breccia (pseudotachylite) related to frictional melting. This quarry has been active for decades and geological evidence is changing though. Such friction-related, gravity-induced sliding occurred in acoustically fluidized rocks with low cohesion during crater modification and led to the collapse of the crater cavity. The final morphology of Rochechouart must have been relatively flat and can be reconstructed following the contact of the crater fill breccias with the crater basement rocks.

Dating attempts of the Rochechouart impact yielded ages of 186 ± 8 Myr using the whole rock Rb-Sr technique on melt rocks, and more recently 214 ± 8 Myr using the ^{40}Ar-^{39}Ar laser spot fusion dating technique on pseudotachylite. The most recent dating yielded an age of 201 ± 2 Myr. This analysis is based on ^{40}Ar-^{39}Ar dating of reset potassium feldspar separated from an impact-metamorphosed gneiss. Recently, a drilling campaign was carried out at various sites within the crater structure by the CIRIR (Center for International Research and Restitution on Impacts and on Rochechouart), a provincial/municipal organization for promoting and valorizing the scientific heritage of the Rochechouart impact initiated by Philippe Lambert. In total, 515 m of core could be recovered and expose a great variety of impact lithologies including impact melt breccia, melt-rich and melt-poor suevite, polymict lithic breccias, monomict breccia, and cataclastically deformed gneisses, as well as a variety of clastic, impact-formed dikes, and pseudotachylitic breccia. Hydrothermal sealing of fractures, e.g., by carbonates and fluorite, is abundant at Rochechouart and is particularly strong in the outer parts of the structure.

An impact origin of the Rochechouart structure was first postulated by François Kraut in 1969 and later confirmed by Lambert. They identified remnants of a thin melt rock layer and suevite breccia. Impact indicators such as shatter cones, planar deformation features (PDFs) in quartz, and isotropic (diaplectic) plagioclase proved the impact hypothesis. Today, Rouchechouart can be regarded as a well studied impact structure.

A local landmark is the castle of Rochechouart that towers on cliffs of impact breccias and, itself, has been constructed from this lithology. Another prominent building constructed from suevite is the St. George church at the historic market-place in Rochechouart. Close by is also a small but well-appointed museum about the Rochechouart impact structure and impact cratering, in general.

The castle in the small town of Rochechouart was built on a hill comprising suevitic impact breccia, and itself was originally constructed to a large extent from suevite blocks (photo: T. Kenkmann).

Rochechouart

France

latitude = 45°49'27" N / longitude = 00°46'54" E / diameter = 23 km / age = 201 ± 2 Myr

scale = 1 : 258 000 / scene = 46.4 × 61.9 km²

0 2.58 12.90 km

N

75 meter 565

The Champagnac quarry in the northeastern part of the Rochechouart impact structure exposes low-angle normal faults that displace the crystalline bedrock (photo: T. Kenkmann).

Location/Access

The crater structure is situated in a hilly landscape of west-central France some 40 km west of the city of Limoges, from where it can be reached on N141. The town of Rochechouart belongs to the Haute-Vienne Department of the region Nouvelle-Aquitaine and is the regional cultural center within the area of the crater. The center of the crater is some 4 km west of the town of Rochechouart. Other important outcrops are close to the villages of Montoume, Chassenon, and Champagnac.

Selected References

Bischoff, L., and Oskierski, W. (1987). Fractures, pseudotachylite veins and breccia dykes in the crater floor of the Rochechouart impact structure, SW-France, as indicators of crater forming processes. In: Research in Terrestrial Impact Structures, Pohl, J. (ed.), Friedrich Vieweg publishers, Braunschweig/Wiesbaden, pp. 5–29.

Kelley, S. P., and Spray, J. G. (1997). A late Triassic age for the Rochechouart impact structure, France. Meteoritics and Planetary Science, Vol. 32, pp. 629–636.

Kenkmann, T., Ivanov, B. A., and Stöffler, D. (2000). Identification of ancient impact structures: low-angle faults and related geological features of crater basements. Lecture Notes in Earth Sciences, Vol. 91, pp. 271–309.

Kraut F. (1969). Über ein neues Impaktitvorkommen im Gebiet von Rochechouart-Chassenon (Department Haute-Vienne und Chartres), Frankreich. Geologica Bavarica, Vol. 61, pp. 428–460.

Lambert, P. (1977a). The Rochechouart Crater: Shock zoning study. Earth and Planetary Science Letters, Vol. 35, pp. 258–268.

Lambert, P. (1977b). Rochechouart impact crater: Statistical geochemical investigations and meteoritic contamination. In: Impact and Explosion Cratering, Roddy, D. J., Pepin, R. O., and Merrill, R. B. (eds.), Pergamon Press, New York, pp. 449–460.

Reimold, W. U., Oskierski, W., and Huth, J. (1987). The pseudotachylite from Champagnac in the Rochechouart meteorite crater, France. Journal of Geophysical Research, Vol. 92, pp. E737–E748.

Schmieder, M., Buchner, E., Schwarz, W., Trieloff, M., and Lambert, P. (2010). A Rhaetian ^{40}Ar-^{39}Ar age for the Rochechouart impact structure (France) and implications for the latest Triassic sedimentary record. Meteoritics and Planetary Science, Vol. 45, pp. 1225–1242.

Rotmistrovka
Ukraine

Thomas Kenkmann

The Rotmistrovka structure with a diameter of 2.7 km is a simple bowl-shaped crater located on the Ukrainian Shield. It is buried beneath Cenozoic and Upper Cretaceous sediments of roughly 120 m thickness. The crater rim was eroded prior to post-impact sedimentation. Today, no morphological expression of the crater exists.

Target rocks are Proterozoic granites that become increasingly fractured and brecciated close to the crater floor. Pseudotachylitic breccia veins have been described from this shattered unit, as well as plenty of shock features including planar deformation features (PDFs) in quartz grains and small patches of mostly altered impact glass. Allochthonous breccia superposes the disintegrated granite and fills the crater depression by approximately one third. The amount of melted material increases upward within this suevite layer. An up to 200 m thick sequence of continental and marine deposits, interpreted as reworked ejecta, was washed into the depression and covers the suevites. These wash-back deposits are composed of Lower Cretaceous shales, oil shales, limestones, and sandstones. Like the rest of the crater, the crater infill is overlain by Cenomanian-Turonian sediments with discontinuous lenses of wash-back breccia. Based on the age of the overburden sediments, the age of the impact has been estimated roughly between 95 and 145 Myr. The structure was investigated by Anton Val'ter and V. Ryabenko in 1977. Several boreholes have been drilled into the structure since then.

Location/Access

Rotmistrovka, located close to the village of the same name, lies about 40 km northwest of the **Boltysh** structure and about 80 km northwest of **Zeleny Gai**, two other buried impact structures on the Ukrainian Shield. As for Boltysh and Zeleny Gai, contradicting locations can be found for Rotmistrovka in the literature.

Selected References

Masaitis, V. L. (1999). Impact structures of northeastern Eurasia: The territories of Russia and adjacent countries. Meteoritics and Planetary Science, Vol. 34, pp. 691–711.

Val'ter, A. A., and Ryabenko, V. A. (1977). Explosion Craters of the Ukrainian Shield (in Russian). Naukova Dumka Press, Kiev, Ukraine. 154 pp.

Rotmistrovka

Ukraine

latitude = 49°08'00" N / longitude = 31°44'52" E / diameter = 2.7 km / age = 95 - 145 Myr

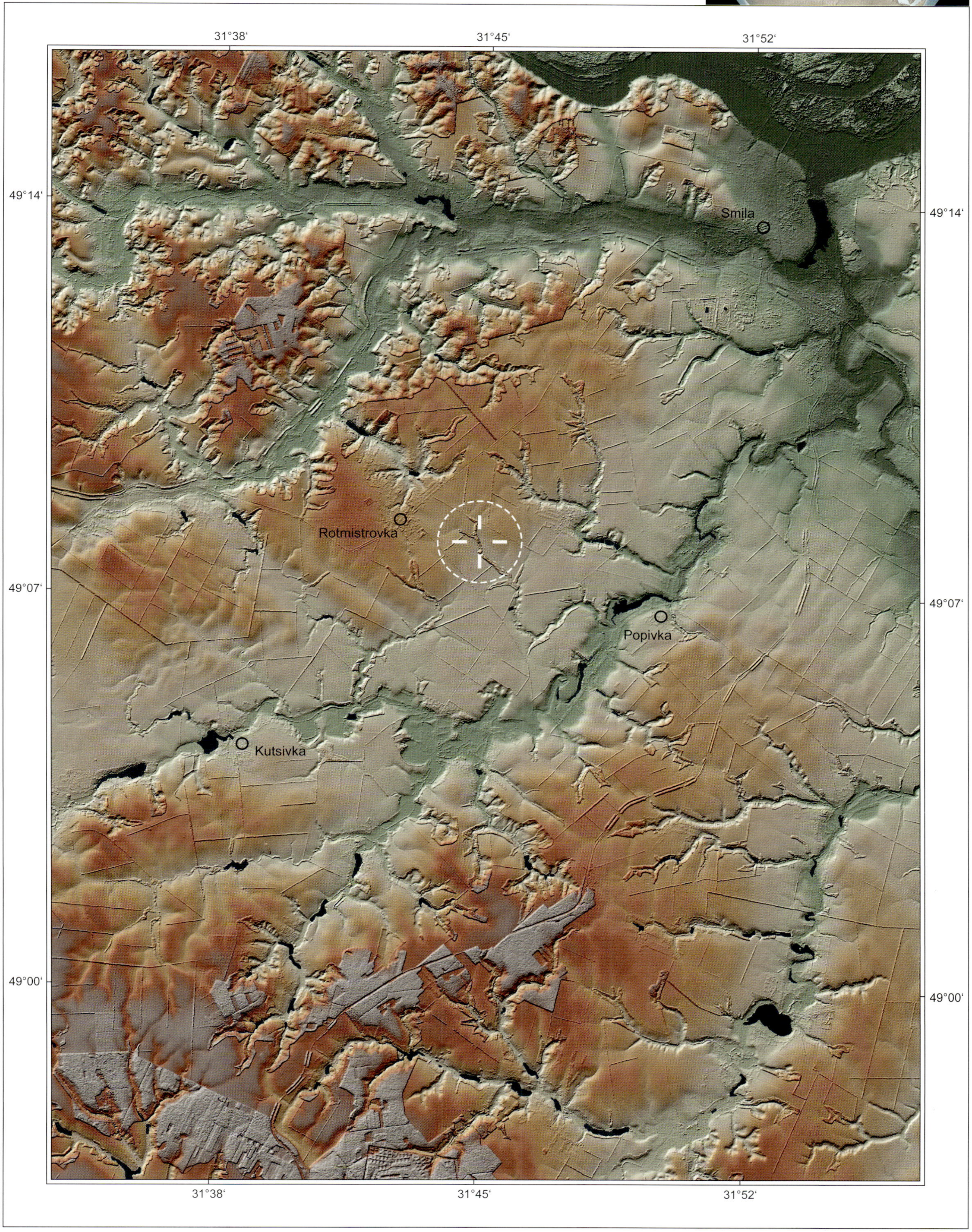

scale = 1: 156 000 / scene = 28.0 × 37.3 km²

0 1.56 7.80 km

N

40 meter 255

Saarijärvi
Finland

Thomas Kenkmann

The Saarijärvi impact structure occupies the area of the present Lake Saarijärvi but extends slightly to the north into a topographically low area. Saarijärvi is interpreted as a deeply eroded structure, whose original diameter is not well constrained but should have been larger than the present apparent diameter of 1.5 km.

The main target rock is a foliated tonalitic gneiss of Archean age. These basement rocks are cut by several generations of diabase dikes. The basement is overlain by an up to 156 m thick unit of sand- and claystones. A drill core obtained in the central part of the structure penetrated the sedimentary succession and reached into fractured basement gneiss. A thin breccia layer was found at the top of fractured basement just below the sediments. Two breccia outcrops were also discovered on the island in the central part of the lake. The uplift of these rocks is most likely unrelated to the impact and the result of later tectonic displacements.

Petrographic analysis of a breccia sample from the drill core from 156 m depth showed planar microdeformations in quartz along various crystallographic planes that were interpreted as planar deformation features (PDFs). Revisiting the published photograph, an interpretation as planar fractures (PFs) with oblique feather-like features (FFs) seems more plausible, as the spacing between the planar features is relatively large. Irrespective of this, this observation indicates that the rock was shocked. Breccia exposed on the central island also displays shock effects.

Saarijärvi is associated with an electromagnetic anomaly of polygonal outline and about 2 km width. Aeromagnetic data show a weak signature over the lake, and some of the regional linear aeromagnetic anomalies thought to be caused by mafic dikes appear to be truncated by the structure. The gravity data reveal a negative Bouguer anomaly of about −1.5 mGal over the structure.

The age of this impact is still poorly constrained. The Saarijärvi impact structure is filled with Ediacaran and Early Cambrian sedimentary rocks that are interpreted to be of pre-impact origin, suggesting an impact age younger than 520–600 Myr.

Saarijärvi was proposed as an impact structure in 1997, when shock effects in quartz were reported from a fragile granite breccia sample from the end of the drill core where the 156 m thick sedimentary rock sequence of the crater fill grades into the brecciated crater floor.

Location/Access

The Saarijärvi impact structure is located in Taivalkoski, northern Finland, and can be reached from the city of Oulu, 140 km to the southwest of the structure, via Highway 20 and regional road 800.

Selected References

Öhman, T. (2007). The origin and tectonic modification of the Saarijärvi impact structure, northern Finland. Bridging the Gap II: Effect of Target Properties on the Impact Cratering Process, Saint-Hubert, Canada, LPI Contribution No. 1360, pp. 85–86.

Öhman, T., Pesonen, L. J., Elo, S., Uutela, A., Tuiska, P., and Raitala, J. (2003). The origin and evolution of the Saarijärvi impact structure. Meteoritics and Planetary Science, Vol. 38 Supplement, p. A52.

Pesonen, L. J. (1998). The Lake Saarijarvi – A new meteorite impact structure in northern Finland. 29th Lunar and Planetary Science Conference, Lunar and Planetary Institute, Houston, Texas, Abstract #1262.

Saarijärvi

Finland

latitude = 65°17'37" N / longitude = 28°23'31" E / diameter = 1.5 km / age = < 520 - 600 Myr

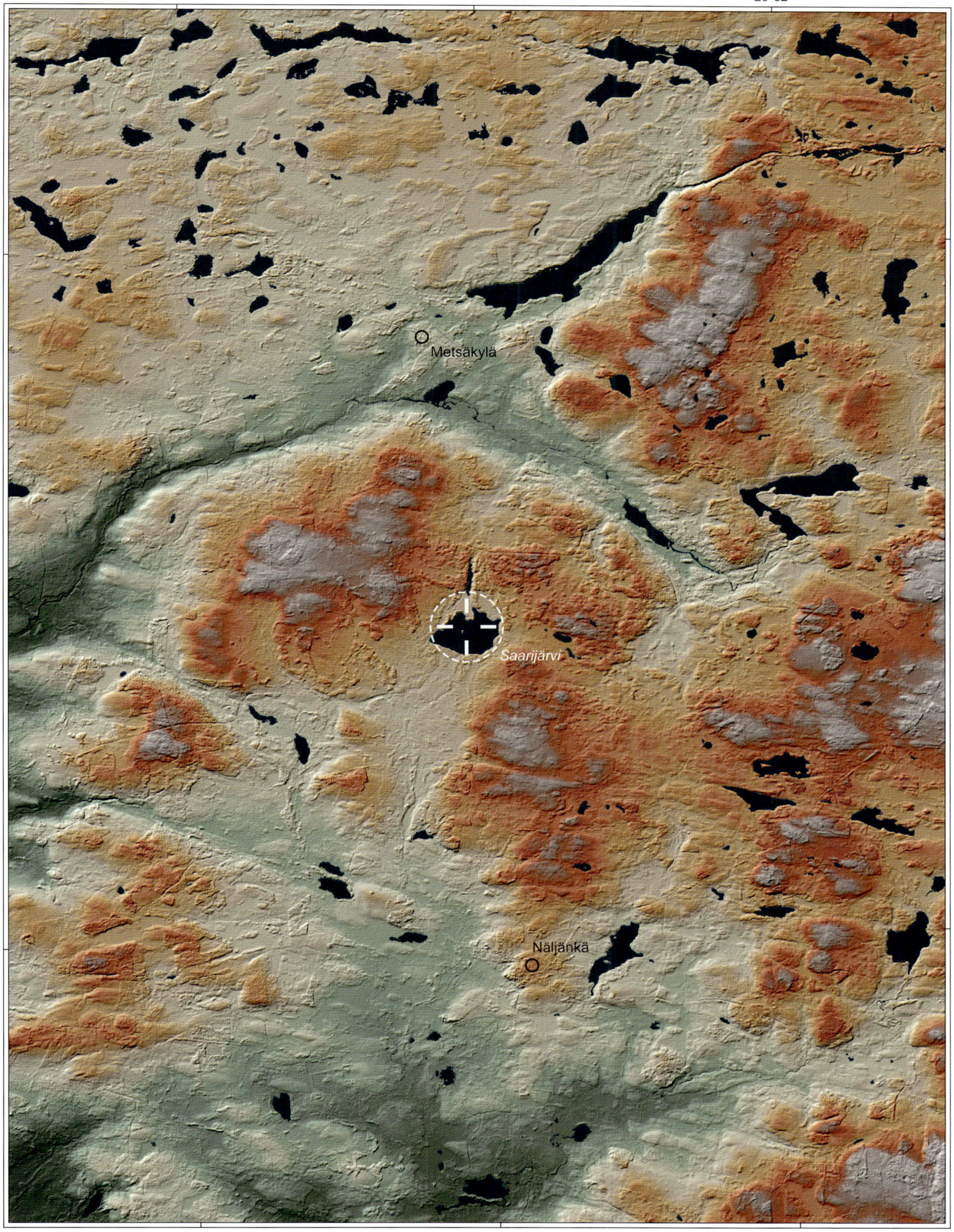

scale = 1:108 000 / scene = 19.5 × 26.0 km²

0 1.08 5.40 km

N

160 meter 315

Sääksjärvi
Finland

Thomas Kenkmann

Sääksjärvi is a deeply eroded impact structure of roughly 6 km diameter in southwestern Finland. The crater is submerged in Lake Sääksjärvi, whose northern shoreline is conspicuously semicircular. Sääksjärvi is located on Scandinavia's Baltic Shield within the Svecofennian realm. The bedrocks are mainly intermediate plutonic rocks, such as granodiorite, quartz-diorite, and tonalite, as well as gneisses, migmatites, and mica schists. The Paleoproterozoic bedrock was peneplained and is now overlain by till formed during and after the various glaciation stages.

Impactite lithologies at surface occur exclusively as glacial boulders. The boulder-rich shorelines on the eastern side of Lake Sääksjärvi exhibit various types of impact breccias. Mostly, impactites are, however, found as pebbles and trains of boulders in gravel pits some 20 km southeast of Lake Sääksjärvi. There are particularly occurrences of vesicle-rich impact melt rock, whose vesicles may be filled with agate. The direction of glacier movement measured from glacial striations indicates that Lake Sääksjärvi is the likely source of these rocks.

Drilling into the crater revealed the presence of suevite and other breccias. Of the two boreholes one was offset westward from the center of the lake and penetrated into the western margin of the crater through 105 m of brecciated mica gneiss, with zones of impact melt breccia, each up to several meters thick. The second core was drilled into the crater center and revealed abundant impact melt breccia and suevite with plenty of clasts with shock features, such as multiple sets of planar deformation features (PDFs) in quartz, isotropization of feldspar, and kinked biotite.

The impact melt rocks found as glacial boulders are enriched in platinum group elements, Ni, and Cr, compared to the regional lithologies that represent the likely target rocks for this impact. Such an enrichment is generally interpreted to result from contamination with a projectile component. Different types of projectiles have been suggested for this impact structure by different teams: a stony-iron, a magmatic iron of type IIA or IIIAB, and a non-magmatic iron meteorite of either type IA or IIIC.

Sääksjärvi displays a residual circular Bouguer anomaly of −6.5 mGal with a diameter of 5 km. There is a small positive residual anomaly within the larger gravity low, which suggests the presence of a central uplift. Aeromagnetic surveying showed that the crater is a circular region of low magnetic relief, in stark contrast to the strong regional magnetic signature outside the crater structure.

^{40}Ar-^{39}Ar age determinations on three whole-rock melt rock samples yielded an age of 560 ± 12 Myr from an isochron calculated for the combined data from these three experiments. U/Pb ages obtained from zircon grains of an impact melt rock sample revealed a slightly older age of 602 ± 17 Myr.

Sääksjärvi was identified as an ancient impact crater in 1969 by Heikki Papunen through the investigation of erratic boulders near Lake Sääksjärvi and the documentation of shock deformation features in quartz and planar microstructures in plagioclase. Later, analysis of impact melt rock samples allowed to date the impact event and to identify the trace of the meteoritic projectile.

A slice of Sääksjärvi impact melt rock, $11.5 \times 8 \times 5$ cm in size (photo courtesy: Ralf-Thomas Schmitt, Museum für Naturkunde Berlin, sample from Dieter Stöffler).

Location/Access

Lake Sääksjärvi is situated in southwestern Finland about 35 km east of the town of Pori. From Pori it is a short drive on Highway 11 to get close to the northern shore of the lake. Impactites can be found as glacial boulders along the lakeshore. Most of the investigated samples are from the location Karhiniemi in the northern part of the commune of Huittinen, some 20 km southeast of Lake Sääksjärvi, where they were obtained in a small glacial esker.

Selected References

Elo, S., Kivekas, L., Kujala, H., Lahti, S. I., and Pihlaja, P. (1992). Recent studies of the Lake Sääksjärvi meteorite impact crater, southwestern Finland (extended abstract). Tectonophysics, Vol. 216, pp. 163–167.

Müller, N., Hartung, J. B., Jessberger, E. K., and Reimold, W. U. (1990). ^{40}Ar-^{39}Ar ages of Dellen, Janisjärvi, and Sääksjärvi impact craters. Meteoritics, Vol. 25, pp. 1–10.

Papunen, H. (1969). Possible impact metamorphic textures in the erratics of the Lake Saaksjärvi area in southwestern Finland. Bulletin of the Geological Society of Finland, Vol. 41, pp. 151–155.

Schmidt, G., Palme, H., and Kratz, K.-L. (1997). Highly siderophile elements (Re, Os, Ir, Ru, Rh, Pd, Au) in impact melts from three European impact craters (Sääksjärvi, Mien, and Dellen): Clues to the nature of the impacting bodies. Geochimica et Cosmochimica Acta, Vol. 61, pp. 2977–2987.

Tagle, R., Schmitt, R. T., and Erzinger, J. (2009). Identification of the projectile component in the impact structures Rochechouart, France and Sääksjarvi, Finland; implications for the impactor population for the Earth. Geochimica et Cosmochimica Acta, Vol. 73, pp. 4891–4906.

Sääksjärvi

Finland

latitude = 61°24‘40“ N / longitude = 22°22‘45“ E / diameter = 6 km / age = 560 ± 12 Myr

scale = 1 : 180 000 / scene = 32.4 × 43.2 km²

0 1.80 9.00 km

N

5 meter 155

Siljan
Sweden

Thomas Kenkmann

The Siljan impact structure is the largest known impact structure in Western Europe. However, its size is still a matter of debate, with estimates ranging from 52 to 75 km. The structure is considerably eroded, in the order of 2 km. Its circular outline is still visible in topography, and the annular moat is covered by a number of lakes, the largest of which is Lake Siljan.

The target for this impact structure comprises crystalline basement rocks overlain by Paleozoic sediments of predominantly Ordovician, Silurian, and Devonian age. The latter are not preserved in the crater but can be assumed to have been present at the time of impact from oil maturity in the Paleozoic sedimentary strata that indicates a target sediment thickness of more than 2 kilometers. The crystalline basement is part of the Trans-Scandinavian Igneous Belt and is made up of the Dala Granite suite, which, in turn, comprises the Siljan granite and the Järna granite of 1436–1702 Myr age. Siljan granite is dominant in the outer part of the central uplift and is relatively more felsic than the Järna granite that dominates in the central and southeastern parts. The granites are cut by dikes and sills of dolerite and porphyritic intermediate to basic rocks. Seismic data indicated that 5–60 m thick dolerite sills could occur throughout the granitic basement. Dolerites from surface outcrops have been dated to 789–1098 Myr. In the western part of the crater the granites are unconformably overlain by Proterozoic metavolcanic and metasedimentary rocks. The Ordovician and Silurian strata are preserved along the annular moat of the crater and are deformed to different degrees. They are mostly disintegrated into fault-bounded blocks that range in size from tens of meters to kilometers. The dominant lithology of the Paleozoic strata is epicontinental limestone.

Tilted and disintegrated beds of the Late Ordovician Boda Formation of the eastern ring syncline of the Siljan impact structure (Ostbjörka quarry). The deformation occurred when the strata were downfaulted during crater modification. Length of the quarry wall measures some 60 m (photo: T. Kenkmann).

The central part of the Siljan structure comprises a hummocky, about 30 km wide complex of crystalline basement rocks. This complex is considered the remnant of a central uplift or perhaps a peak ring structure. The ring basin around this central complex contains the Ordovician and Silurian sedimentary strata. These strata represent downfaulted rocks and material slumped off the inner crater wall. Siljan is the type locality for so-called "radial transpression ridges" that can be observed in the periphery of a central uplift as basement highs that radially subdivide the ring syncline. This structural feature of complex impact craters results from the convergent inward movements of slumped blocks and rock units during the gravity-induced crater collapse. In such transpression ridges material is uplifted to accommodate the converging mass flow.

Strongly disintegrated blocks of the Late Ordovician Boda limestone at the Osmundsberget quarry, in the eastern ring syncline of the Siljan impact structure. Field of view is about 20 m wide (photo: T. Kenkmann).

As the crater has been eroded by about two kilometers, allochthonous crater fill breccia or ejecta do not occur and cannot be expected. The shocked area of the crater has a width of about 30 km and roughly corresponds to the size of the broad central complex. Shatter cones occur in this zone. Quartz grains of Siljan and Järna granite exhibit planar deformation features (PDFs) and planar fractures (PFs), and record a maximum shock pressure of up to 20 GPa in the centralmost part of the structure. Based on the

Siljan

Sweden

latitude = 61°01' N / longitude = 14°56' E / diameter = 65 km / age = 380.9 ± 4.6 Myr

scale = 1:528 000 / scene = 95.1 × 126.7 km²

0 5.28 26.40 km

N

40 meter 785

Downfaulted Silurian Orsa Sandstone and Late Ordovician shales of the eastern ring syncline at the Styggforsen geosite are in upright position, close to the contact with the Siljan granite (photo: T. Kenkmann).

distribution pattern of shocked quartz at Siljan, the original transient cavity was estimated at approximately 32–38 km diameter. Locally, impact breccia dikes including impact melt rock and pseudotachylitic breccia are exposed in the central uplift and were also encountered in some drill cores. Using these melt rock occurrences, the age of the impact structure was recently constrained to 380.9 ± 4.6 Myr. Whether the impact could be causally connected to the Late Devonian mass extinction event is a matter of continuing research.

A number of drill cores were extracted from the Siljan structure in a large-scale drilling project of the late 1980s, which was aimed at investigating possible resources of mantle-derived methane below and around the impact structure. This drilling included the 6337 m long Gravberg-1 core obtained outside of the northeastern central uplift, besides nine much shorter cores. One 602 m deep core was extracted just northwest of the center of the impact structure.

The circular outline in the Siljan area was noticed already in the 19th century. The impact hypothesis was first proposed by Frans Wick-

A pseudotachylitic breccia vein carrying a seemingly exotic clast (not similar to the host rock material), cutting through Järna granite. The small drill core sample of 6 cm diameter is from the 600 m deep borehole Hättberg that was drilled into the center of the Siljan structure (photo: T. Kenkmann).

man in 1963 and then backed up with findings of shatter cones and shock-deformed quartz by Nils-Bertil Svensson in 1971 and 1973. The Geological Survey of Sweden provided a detailed geological map of the Siljan impact structure.

Location/Access

The Siljan ring structure is located in Dalarna in south-central Sweden in a well developed area. Several picturesque lakes, among them the lakes Siljan, Orsasjön, Skattungen, and Oresjön, mark the annular ring of the structure. A number of small towns like Mora, Orsa, and Rättvik are situated at their shorelines. Siljan can be reached via train or via car from Stockholm or Oslo, both being about 250 km away.

Selected References

Bodén, A., and Eriksson, K. G. (eds.) (1988). Deep Drilling in Crystalline Bedrock. Volume 1: The Deep Gas Drilling in the Siljan Impact Structure, Sweden and Astroblemes. Springer, Berlin Heidelberg, 364 pp.

Holm, S., Alwmark, C., Alvarez, W., and Schmitz, B. (2011). Shock barometry of the Siljan impact structure, Sweden. Meteoritics and Planetary Science, Vol. 46, pp. 1888–1909.

Jourdan, F., and Reimold, W. U. (2012). Age of the Siljan impact structure. 75th Annual Meeting of the Meteoritical Society, Cairns, Australia, Abstract #5093.

Kenkmann, T., and von Dalwigk, I. (2000). Radial transpression ridges: A new structural feature of complex impact craters. Meteoritics and Planetary Science, Vol. 35, pp. 1189–1201.

Reimold, W. U., Fischer, L., Müller, J., Kenkmann, T., Schmitt, R. T., Altenberger, U., Kowitz, A. (2015). Impact generated pseudotachylitic breccia in drill core BH-5 Hättberg, Siljan impact structure, Sweden. GFF 137, Vol. 2, pp. 141–162.

Reimold, W. U., Kelley, S. P., Sherlock, S. C., Henkel, H., and Koeberl, C. (2005). Laser argon dating of melt breccias from the Siljan impact structure, Sweden: Implications for a possible relationship to Late Devonian extinction events. Meteoritics and Planetary Science, Vol. 40, pp. 591–607.

Svensson, N. B. (1971). Probable meteorite impact crater in central Sweden. Nature, Vol. 229, pp. 90–92.

Wickman, F. E. (1980). The impact center of the Siljan structure, central Sweden. Geologiska Föreningen i Stockholm Förhandlingar, Vol. 102, pp. 105–109.

Panoramic view from the southwestern crater rim at Gesundaberget, across Lake Siljan and the island Sollerön towards the center of the Siljan impact structure located beyond Sollerön. (photo: T. Kenkmann).

The Siljan impact structure on May 30, 2018 as a composite image based on TanDEM-X DEM and Copernicus Sentinel-2 (channel 2/3/4) data.

Söderfjärden

Finland

Thomas Kenkmann

Söderfjärden is a 6.5 km diameter impact structure on Finland's west coast. It forms a circular plain with a diameter of 5 to 5.5 km. The plain is elevated only a few meters above the sea level. It is surrounded by a discontinuous ring of hills with a maximum elevation of 55 m above the plain. This ring is 20 to 30 m higher than the surrounding landscape.

Target rocks are Paleoproterozoic porphyritic granitoids, which represent the unexposed crater floor and the partly exposed rim. The crater fill consists of Cambrian sediments that are overlain by Quaternary material. The Cambrian clastic sediments within the crater are thought to have been deposited in a very shallow marine environment, although sedimentary features of resurge deposits such as a basal breccia unit and graded bedding are lacking at Söderfjärden. It is assumed that the infill was not a violent process because the crater cavity was dammed by the elevated crater rim. The inner crater became only gradually submerged while slow erosion of the rim took place. The rock found in the central uplift, 40 m below the glacial overburden, consists of layers varying from slightly mechanically altered bedrock to unidentifiable, finely ground breccia that appears to be locally melt-bearing. Core from boreholes drilled into the central uplift and into the surrounding moat displayed shocked quartz grains containing planar deformation features (PDFs).

Söderfjärden shows a distinct negative gravity anomaly of −6 mGal corresponding to the extent of the circular depression. The central uplift is marked by a positive excursion within the overall negative gravity signal. Electromagnetic data display a distinct negative anomaly, which is approximately congruent with relatively subdued magnetic rock properties. Both anomalies have a somewhat polygonal outline caused by impact-induced straight faults terminating the sedimentary infill of the crater. A sharp magnetic anomaly in the center of the structure is possibly caused by impact melt rock or suevite associated with the central uplift. Seismic measurements revealed a maximum thickness of 350 m for the seismically slower cover rocks encircling the central uplift.

The age of Söderfjärden is only poorly constrained. Current lithostratigraphic estimates range from 520 to 600 Myr, based on the age of the sedimentary infill of the crater cavity, and assuming an impact into a shallow marine environment. However, if the impact occurred instead onshore, the age could be considerably higher and perhaps placing the Söderfjärden impact into Proterozoic times. Attempts to date impact breccia radiometrically using whole rock K-Ar yielded an age of 1200–1300 Myr – but this may well be a mixed age.

The gravity anomaly at Söderfjärden was recognized already in 1978, along with its peculiar almost circular shape, by Lennart Laurén and his team. However, its origin was still a matter of debate at that time. The impact hypothesis was substantiated by Jyrki Lehtovaara after the structure had been drilled and rocks became available for microscopic inspection for shock deformation.

Location/Access

Söderfjärden is located on the west coast of Finland. Its interior is in agricultural use, whereas the elevated crater rim is covered by forest. From the city of Vaasa, about 10 km to the north of the structure, Highway 3 leads to the village of Vikby. From there, regional road 673 reaches Söderfjärden after about 5 km.

Selected References

Abels, A., Pesonen, L. J., Deutsch, A., Bischoff, L., and Lehtinen, M. (1998). The Söderfjärden impact structure, W. Finland: Multiple data integration with implications for morphometry, cratering process, and post-impact sedimentary history. 24th Lunar and Planetary Science Conference, Lunar and Planetary Institute, Houston, Texas, Abstract #1264.

Laurén, L., Lehtovaara, J., Bostrom, R., and Tynni, R. (1978). On the geology and the Cambrian sediments of the circular depression at Söderfjärden, western Finland. Geological Survey of Finland Bulletin, Vol. 297, pp. 1–81.

Lehtovaara, J. J. (1985). ^{40}K-^{40}Ar dating of the Söderfjärden crater, Vaasa, western Finland. Geologiska Föreningen i Stockholm Förhandlingar, Vol. 107, pp. 1–6.

Lehtovaara, J. J. (1992). Söderfjärden: a Cambrian impact crater in western Finland. Tectonophysics, Vol. 216, pp. 157–161.

The Söderfjärden impact structure (top left) on October 28, 2018 as a composite image based on TanDEM-X DEM and Copernicus Sentinel-2 (channel 2/3/4) data.

Söderfjärden

Finland

latitude = 63°00'19" N / longitude = 21°34'37" E / diameter = 6.5 km / age = 520 - 600 Myr

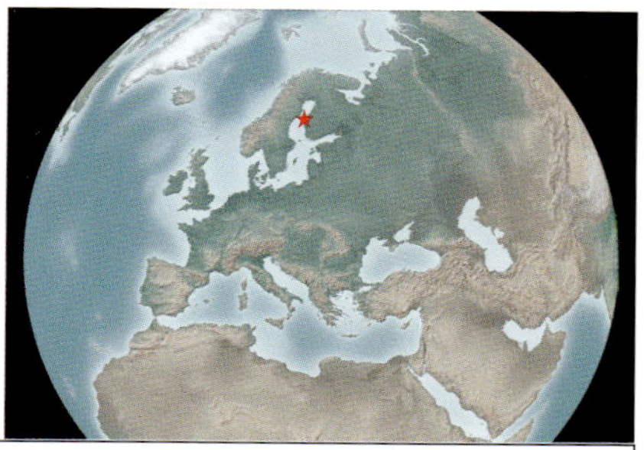

scale = 1:97 000 / scene = 17.4 × 23.2 km²

0 0.97 4.85 km

N

0 meter 75

Steinheim Basin
Germany

Thomas Kenkmann

The Steinheim Basin is a 3.8 km diameter, complex impact crater located on the northeastern Swabian Alb range of Southern Germany. The small city of Steinheim am Albuch in the Federal State of Baden-Württemberg is situated within the crater structure. The Steinheim Basin crater structure has a rather well preserved morphology. It represents one of the smallest complex impact craters on Earth. The crater has an approximately 1 km wide central uplift that rises some 50 m above the surrounding annular moat. Due to erosion the crater rim is not elevated above the surrounding terrain. Ejecta are lacking outside the crater as well. Particularly the southern rim area has been removed by the formation of the Stubental, an east-west trending valley that was formed during the glacial period of the Pleistocene. The crater, however, is still visible, as the basin floor forms a distinct depression in the area of the annular moat around the central uplift.

The Steinheim Basin structure was most likely formed simultaneously with the much larger **Ries Crater** of 26 km diameter by an oblique, east-northeast directed impact of a double asteroid about 14.8 Myr ago. The larger asteroid, approximately 1.5 km in size, formed the Ries, and the smaller satellite of about 150 m size the Steinheim Basin. The distance between both impact structures is 45 km. An alternative scenario has been suggested whereby a single bolide was split into two pieces upon entry into the Earth's atmosphere. However, it could be shown that this scenario is incompatible with the relatively large distance between the craters.

The target rocks of the Steinheim Basin impact crater consist of a layered sequence of sedimentary rocks that are gently inclined towards the south. They comprise – from top to bottom – approximately 380 m of Upper Jurassic limestone, 210 m of Mid-Jurassic shale and sandstone, 90 m of Lower Jurassic shale, 250 m of Upper Triassic shale and sandstone, and 250 m of Lower Triassic and Permian sandstone. The top of the crystalline basement lies at a depth of about 1200 m. Observations from drilling indicate that the crystalline basement was not displaced by the impact. But all other strata were uplifted and strongly deformed into large blocks that were displaced with respect to each other. The stratigraphic uplift amounts to about 500 m so that Mid-Jurassic strata crop out in the center of the uplift. The crater floor is overlain by a layer of rocks that were first fractured during the impact, then lifted as ejecta, and that subsequently fell back into the crater cavity. Some of these ejected boulders contain shatter cones. In fact, the Steinheim structure is well-known for the aggregates of intricate shatter cones formed in limestone. Fallback breccia of the annular moat reaches thicknesses of 20–70 m. Steinheim is associated with a negative Bouguer gravity anomaly of −2 mGal.

Like many other impact structures, Steinheim Basin was first regarded as the product of magmatic processes; more so, as the structure is situated on the Swabian Alb range with its prominent volcanoes. An early proposal suggested that there was a laccolith underneath Steinheim,

Upper Jurassic limestone beds show tight folding and overturning near the rim of the crater floor. Burgstall hill, southern Steinheim Basin impact structure. Width of the photo is about 4 m (photo: T. Kenkmann).

Panoramic view of the Steinheim Basin impact structure. View from east to west (left to right), with the Klosterberg, the central uplift structure, in the middle (photo: M. Gottwald).

Steinheim Basin
Germany

latitude = 48°41'12" N / longitude = 10°04'02" E / diameter = 3.8 km / age = 14.8 Myr

scale = 1:73 000 / scene = 13.1 × 17.5 km²

0 0.73 3.53 km

N

465 meter 780

Monomict breccia of upper Jurassic limestone at the Burgstall hill at the southern rim of the Steinheim Basin impact structure (photo: T. Kenkmann).

The Steinheim Basin is not only famous as a well-preserved complex impact crater. It also became a key locality for biologists in 1867 owing to the fossil-rich, 50 m thick lake sediments that form the uppermost strata in the crater. These sediments were deposited in a fresh-water lake that filled the crater cavity shortly after the impact. In 1867, only a few years after the publication of Darwin's *Origin of Species*, Franz Hilgendorf published a phylogenetic tree of Miocene planorbid snails from the Steinheim Basin crater-lake. This tree provided fossil evidence in particular for transitional forms between taxa and elucidated that one species may branch into two new species. His tree of planorbid snails is seen today as the first phylogenetic tree of fossils and one of the first fossil lines of evidence supporting Darwin's descendant theory. However, this finding was widely ignored by Darwin. The lake deposits are an eldorado for paleontologists, who have recovered an extremely rich and diverse fauna and flora that can also be admired in a small but well-appointed museum in the village of Sontheim close to the southern crater rim. Exhibits provide ample information about the Steinheim impact, the impact process in general, and the post-impact lake stage and associated fauna and flora.

an onion-shape magmatic intrusion into the shallow crust, that did not reach the surface but lifted up the rocks above the intrusion, as one would observe in the central uplift. Another proposal regarded Steinheim as a volcano with a resurgent dome in the center. But as volcanic rocks are completely absent at Steinheim, the structure was classified as a so-called "cryptovolcanic" feature. In 1912 Walter Kranz developed a blast theory to explain the crater morphology and the strong shattering of the rocks in the Ries and Steinheim structures. Although this hypothesis comes closer to the actual impact scenario, it was still assumed that the explosion was driven by an endogenic source, namely large quantities of steam that formed when groundwater reacted with ascending magma.

After having visited Meteor Crater in Arizona the German geologist Otto Stutzer presumed in 1936 that the Steinheim Basin structure might be an impact crater. This hypothesis became a well-founded theory when Robert Dietz, one of the pioneers of modern plate tectonics and impact crater research alike, published his work on shatter cones in 1959. He found these conical fracture surfaces with divergent striations not only at Steinheim but in several other structures of the United States, labelled "cryptovolcanic". He concluded that a meteorite impact was responsible for their formation. Shatter cones are now recognized as unequivocal evidence for impact. Further proof for shock metamorphic overprint, including planar deformation features (PDFs) in the mineral quartz, and partially melted sandstones along with Fe-Ni-Co rich breccias, was obtained when a number of drill cores were extracted from the central "Klosterberg" hill. The name Klosterberg testifies to the settlement of a monastery founded as early as 1190 AD right on top of the central uplift.

Location/Access

The Steinheim Basin is easily accessible. It is located 7 km west of the city of Heidenheim an der Brenz in Southern Germany (state of Baden-Württemberg). Access is via Freeway (Autobahn) 7, exit Heidenheim; or via National Highway (Bundesstraße) 466 to Steinheim am Albuch.

Selected References

Buchner, E., and Schmieder, M. (2010). Steinheim suevite – A first report of melt-bearing impactites from the Steinheim Basin (SW Germany). Meteoritics and Planetary Science Vol. 45, pp. 1093–1107.

Dietz, R. S. (1959). Shatter cones in cryptoexplosion structures (Meteorite impact?). Journal of Geology, Vol. 67, No. 5, pp. 496–505.

Heizmann, E. P. J., und Reiff, W. (2002). Der Steinheimer Meteorkrater (in German). Pfeil-Verlag, München, 160 pp.

Reiff, W. (1977). The Steinheim Basin – An impact structure. In: Impact and Explosion Cratering, Roddy, D. J., Pepin, R. O., and Merrill, R. (eds.), New York, Pergamon Press, pp. 309–320.

Reiff, W. (1979): Guide-book to the Steinheim Basin Impact Crater. Geologisches Landesamt Baden-Württemberg, Stuttgart, 32 pp.

Sontheim Meteorkrater-Museum: https://www.steinheim-am-albuch.de/tourismus-a-freizeit/meteorkrater-museum

The Steinheim Basin (lower left, indicated by cross hairs) and the Ries (right) impact structures on April 7, 2018 as a composite image based on TanDEM-X DEM and Copernicus Sentinel-2 (channel 2/3/4) data.

Sterlitamak

Russia

Manfred Gottwald

The Sterlitamak impact is one of the few meteorite falls with crater formation, for which eyewitness reports exist. The event occurred shortly before midnight on May 17, 1990 close to the city of Sterlitamak at the southern tip of the Ural mountains. It represents the smallest known single impact cratering event.

The impacting body produced a crater with a diameter of 9.4 m and a depth of about 4.5–5 m. The crater cavity consisted of a steep funnel 3 m deep that terminated in a conical funnel. A rim was present, which was characterized to have a width of 60–70 cm, and radial ejecta extended out to more than 10 m in ray-like fashion. The ejecta rays were shorter but thicker on the southern side than north and east of the crater. Rim and ejecta were composed of brown loam.

Eyewitness accounts describing a fireball with a brightness of about −5 mag allowed to shed light on the flight path of the meteoroid and its velocity. The projectile went approximately from south to north, the incidence angle was estimated at 45°, and the shine of the fireball was visible until impact. This constrained the velocity at entry to 11.2–17.8 km/s. Modeling the impact by taking an iron projectile with and without mass-loss into account yielded approximate estimates for the total mass of the meteoroid of more than 1500 kg, a size of about 0.8–0.9 m with slab geometry and a velocity at impact of 2–3 km/s. Initial excavation of the crater delivered two meteorite fragments of 6.5 kg and 3 kg from a depth of 8 m. They were classified as an iron meteorite belonging to the meteorite group IIIAB with an octahedrite structure. One year after the impact, the largest fragment weighing 315 kg with dimensions of 50 × 48 × 28 cm was found at a depth of about 12 m. In total, the recovered mass amounts to 325 kg. Most of the meteorite mass remained in the crater.

The original small impact crater has not been preserved. Already 6 days after the impact, when a scientific team arrived at the site, the rim had been disturbed by visitors and part of the crater cavity filled with material from the rim. Excavation work with heavy machinery destroyed the Sterlitamak crater completely and modified the topography in its immediate environs. Today a pond with a diameter of about 45 m can be found where once the projectile had hit the ground.

Location/Access

The bolide impacted about 20 km west of the city of Sterlitamak in the distant European part of Russia in the Republic of Bashkortostan. From there the impact site can be reached on rural roads. Ufa, 120 km north of Sterlitamak, has an international airport and is connected via Highway P314. In Ufa the largest fragment of the Sterlitamak meteorite is on display in the Museum of Archeology and Ethnography.

Sterlitamak is another case with poorly documented location. The exact position has not been provided in the literature. It could only be inferred from reports of eyewitnesses and visitors to the impact site. The pond that lies where the original crater once had been can be found on farmland 2 km west of the village of Severnaya.

Selected References

Ivanov, B. A., and Petaev, M. I. (1992). Mass and impact velocity of the meteorite formed the Sterlitamak crater in 1990 (abstract). 23rd Lunar and Planetary Science Conference, Lunar and Planetary Institute, Houston, Texas, pp. 573–574.

Petaev, M. I., Kisarev, Y. L., Mustafin, S. A., Shakurov, R. K., Pavlov, A. V., and Ivanov, B. A. (1991). Meteorite Sterlitamak – A new crater forming fall (abstract). 22nd Lunar and Planetary Science Conference, Lunar and Planetary Institute, Houston, Texas, pp. 1059–1060.

The 6.5 kg fragment of the Sterlitamak meteorite that fell on late May 17, 1990 in Russia. The cube for scale has a 1 cm side length (image courtesy: Laurence Garvie, Arizona State University, Center for Meteorite Studies).

Sterlitamak

Russia

latitude = 53°36'12" N / longitude = 55°35'12" E / diameter = 9.4 m / age = 29 yr

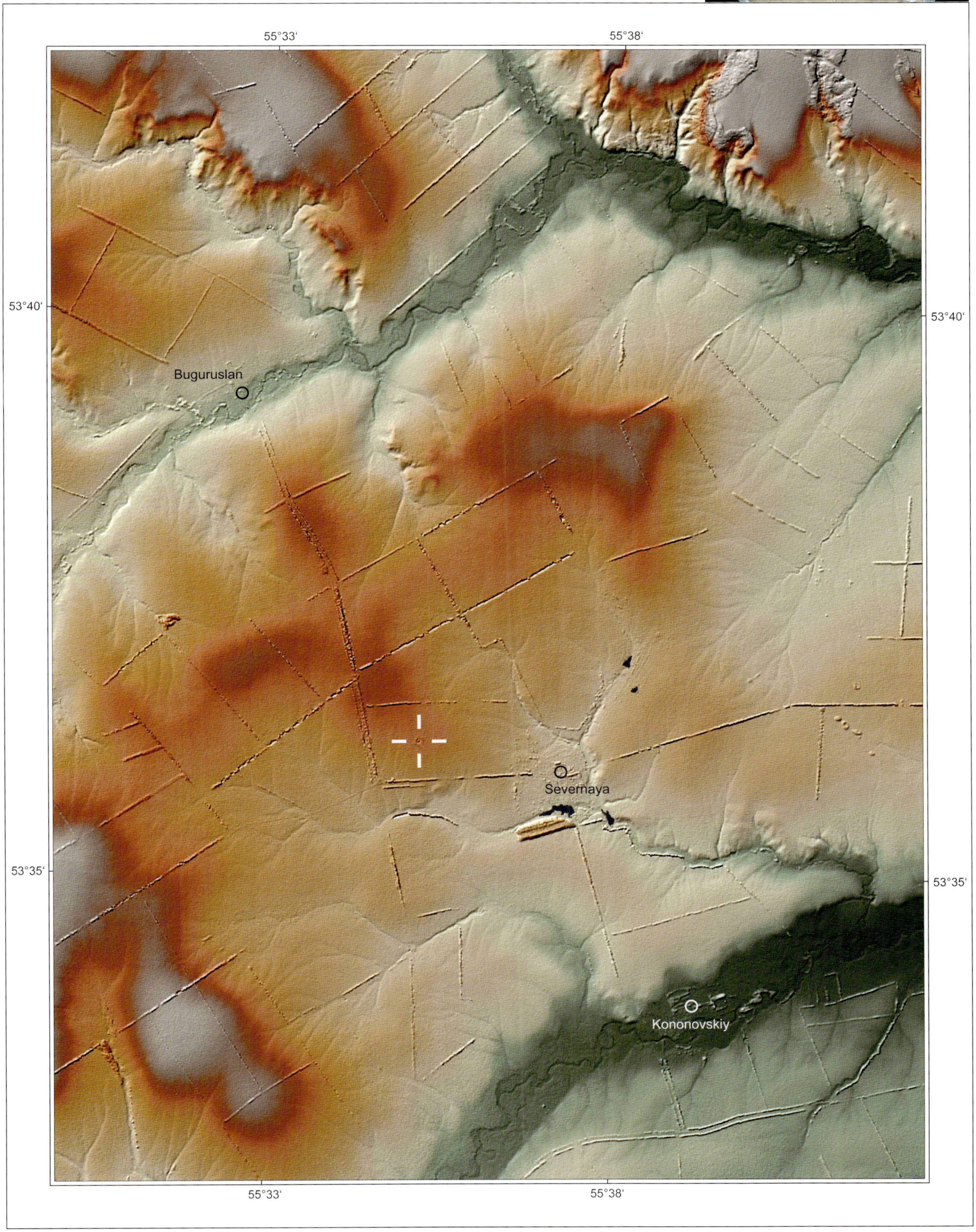

scale = 1:77 000 / scene = 13.9 × 18.5 km²

0 0.77 3.85 km

N

150 meter 295

Suavjärvi
Russia

Thomas Kenkmann

The Suavjärvi structure is a circular feature of about 16 km diameter in a hilly and swampy area of northwestern Russia close to the Finnish border. At the center of this area lies the small Lake Suavjärvi. Arcuate lakes and narrow ridges in the north and west are possible faint morphological imprints of the crater structure.

Suavjärvi is situated within the Karelian craton. The basement of the craton is composed of Late Archean granite-greenstone belt rocks comprising granitoid-gneiss complexes and supracrustal rocks ranging in age from 2620 to 3140 Myr. These units are overlain by a metamorphosed volcano-sedimentary succession dated at 2100–2300 Myr. Most of the area of Suavjärvi is composed of various granitoids and granitic gneiss of Archean and Early Proterozoic age but Quaternary deposits up to 10–15 m thick cover large parts of the area.

The structure contains highly disturbed and partly ore impregnated mega-breccias with a patchy distribution. There are three areas where monomict and polymict breccias are exposed, namely in the northeast and the southwest along the shore of Lake Aeningalampi, and in close proximity to the latter but closer to the center of the area. The polymict breccias consist of blocks ranging from a few meters to more than hundred meters in size. The rock types of these blocks are various types of gneisses, and, to a lesser degree, trondhjemite, greenschist, and quartzites.

Some quartz grains in granitoid blocks from the polymict breccia, in particular of the northeastern breccia occurrence, display sets of planar fractures (PFs) and planar deformation features (PDFs). In plagioclase, the bending of twin planes, the disappearance of twin lamellae, and PDFs were observed. Kink bands are common in biotite. The northeastern breccia is impregnated with various sulfide ore minerals including pyrite, pyrrhotite, chalcopyrite, pentlandite, galena, bornite, and sphalerite; also traces of silver have been noted.

Suavjärvi's circular appearance is visible in the geophysical signature. The residual gravity anomaly ranges from −4 to −10 mGal. This negative anomaly extends southward from the structure. The airborne magnetic pattern shows a circular area of low magnetic relief that interrupts the linear anomalies of the regional pattern.

The age of this structure is not well constrained. No direct dating of impact lithologies has been possible. Only stratigraphic considerations have been referred. Thus, an upper age limit of about 2700 Myr has been quoted, based on age determinations on granitoids that occur as fragments within the mega-breccias. A lower limit of 2150 ± 60 Myr has been given in the literature, whereby the cited error is of unclear significance. It is based on unreferenced dating of dikes cutting through rocks affected by the impact and Pb-Pb isotopic analysis of carbonates from the overlying metamorphosed volcano-sedimentary formation that is unaffected by the impact. It appears that this impact structure is indeed very old. However, whether Suavjärvi could represent a structure older than **Vredefort** and the even older **Yarrabubba** structure, which until now have been considered the oldest confirmed terrestrial impact structures, remains to be shown convincingly.

Sulfide mineralization of the area led to early geological exploration in the 1930s. A circular feature at Suavjärvi had been noted on the Cosmogeological Map of the eastern part of the Baltic Shield in 1981. In 1983–1984, this circular feature was studied by geologists of the Karpinsky Geological Research Institute in St. Petersburg. The work by Mikhail Mashchak and J. Orlova resulted in the detection of shock metamorphic features in 1986. These findings were eventually published in full form in 2012.

Location/Access

Suavjärvi is part of the Republic Karelia in northwestern Russia, some 60 km west-northwest of the town of Medvejegorsk. The structure can be reached on a road connecting Medvejegorsk with the village Padany, which is located at Lake Segozero to the northeast of the structure.

Selected References

Mashchak, M. S., and Orlova, J. V. (1986). Shock deformations in lower Proterozoic breccias of the Lake Suav'yarvi region (central Karelia) (in Russian). Meteoritika, Vol. 45, pp. 137–141.

Mashchak, M. S., and Naumov, M. V. (1996). The Suavjärvi structure: An early Proterozoic impact site on the Fennoscandian shield (abstract). 27th Lunar and Planetary Science Conference, Lunar and Planetary Institute, Houston, Texas, pp. 825–826.

Mashchak, M. S., and Naumov, M. V. (2012). The Suavjärvi impact structure, NW Russia. Meteoritics and Planetary Science, Vol. 47, pp. 1644–1658.

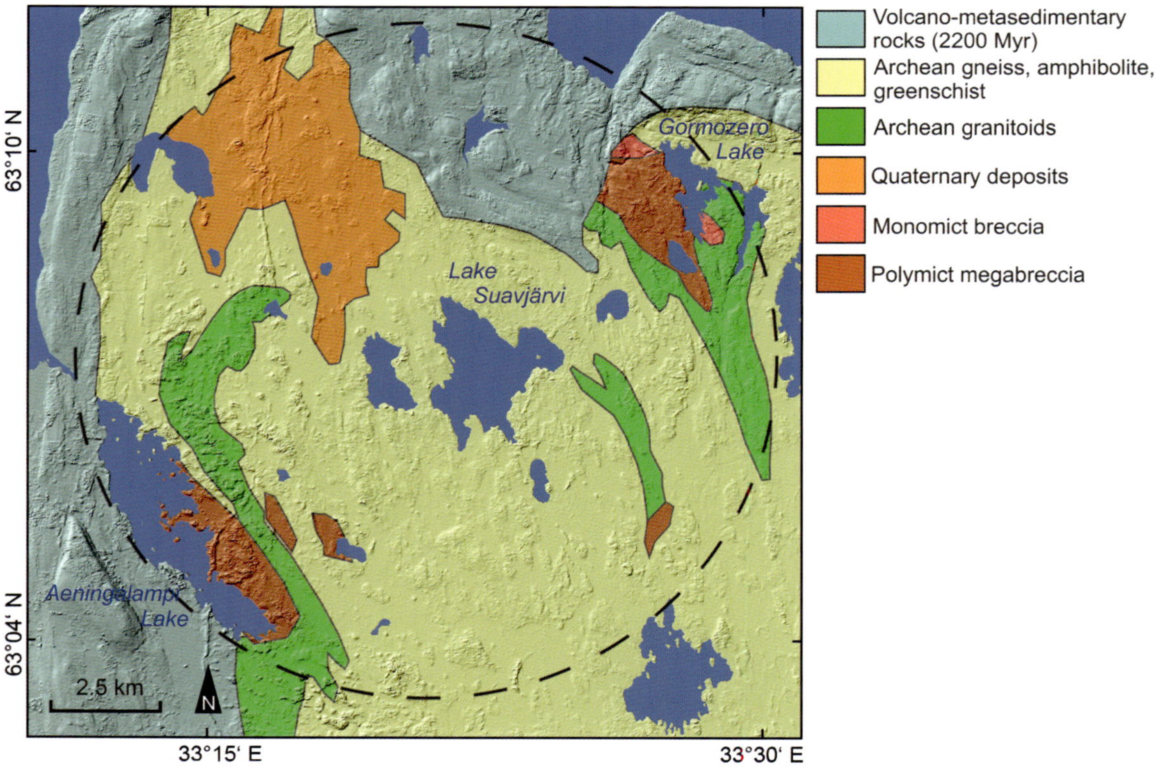

Geological map of the Suavjärvi impact structure, underlain with a TanDEM-X DEM. The dashed circle outlines the apparent circular feature in space-borne imaging (modified after Mashchak and Naumov, 2012).

Suavjärvi

Russia

latitude = 63°07‘21“ N / longitude = 33°22‘24“ E / diameter = 16 km / age = 2090 - 2700 Myr

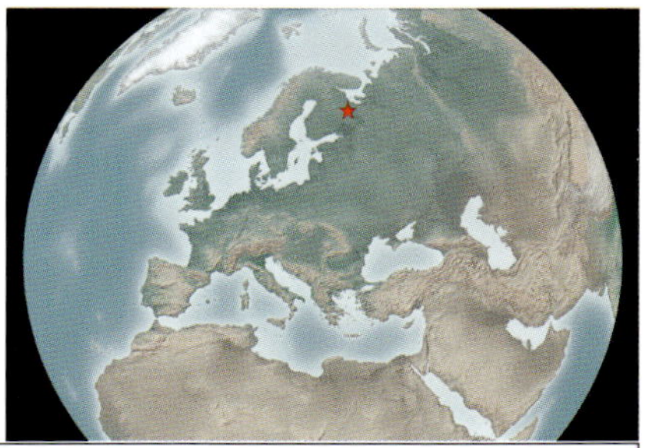

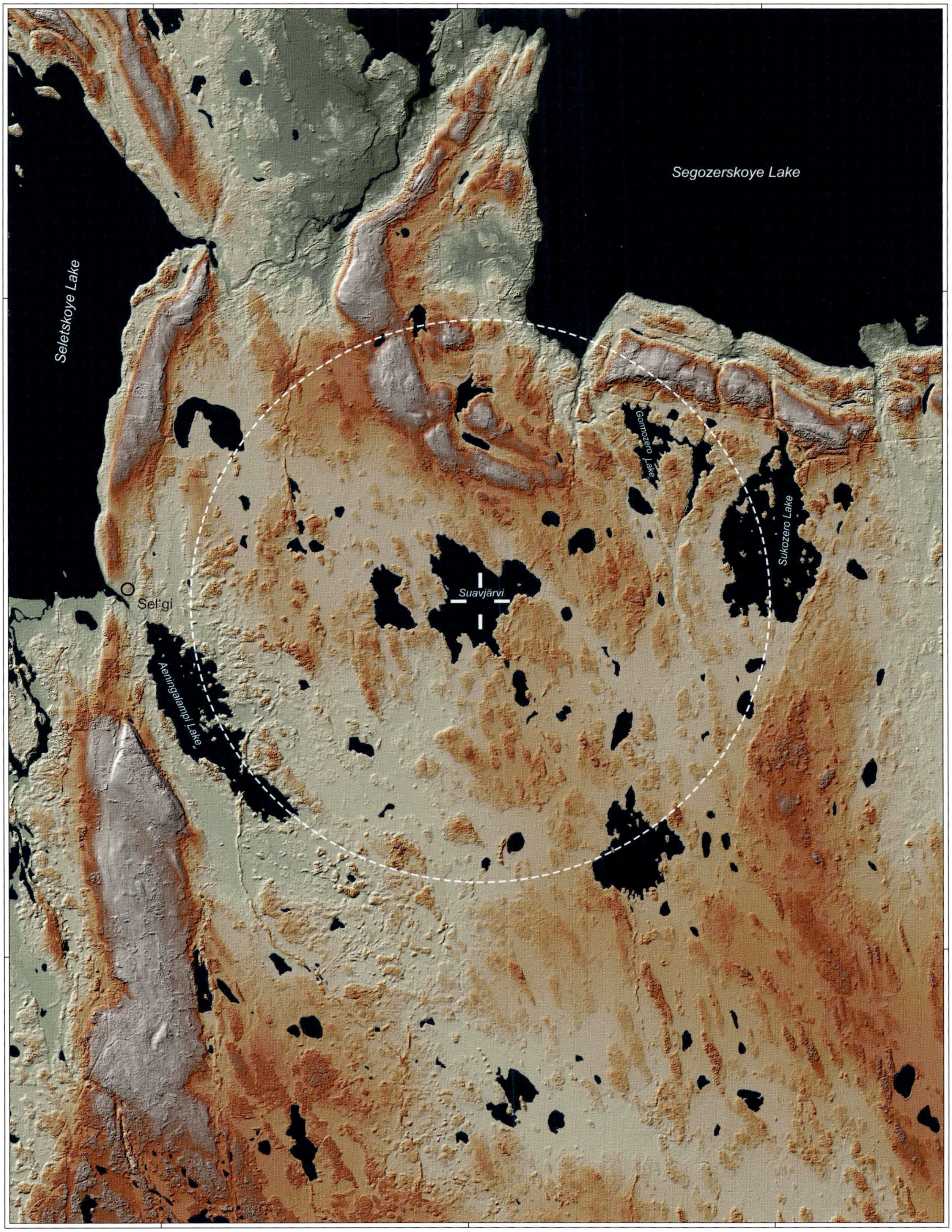

scale = 1 : 143 000 / scene = 25.7 × 34.2 km²

0 1.43 7.15 km

N

115 meter 285

Summanen
Finland

Thomas Kenkmann

Summanen is a recently discovered impact structure of 2.6 km diameter in central Finland. It is submerged below Lake Summanen, which lacks a circular shape due to glacial scouring. Compared to other lakes, Lake Summanen is relatively deep. Its maximum depth of 41 m corresponds to the position of the crater. The structure was identified by recognition of a circular aero-electromagnetic anomaly.

Target rocks are Paleoproterozoic granitoids of the Baltic Shield. In the Summanen area, the plutonic basement displays a weak magmatic foliation and has a porphyritic texture. Zircon U-Pb dating of biotite-hornblende granite gave an age of 1880 Myr. Quaternary deposits in the Lake Summanen area consist of sandy basal till, silt, clay, gravel, sand, and peat.

Shock features were mostly found in glacial boulders on the shore of the island Lamposaari, situated east-southeast of the center of the structure, and on the mainland to the southeast of the lake. This direction corresponds to the glacial drift. This includes shatter cones formed in granite, and monzodiorite and aplitic granite also contain quartz with multiple sets of planar deformation features (PDFs), planar fractures (PFs), and feather features (FFs).

The central part of Lake Summanen is characterized by a circular magnetic anomaly of low relief. Such a pattern is typical for many impact structures in Fennoscandia, and may indicate the lack of highly magnetic impact melt rock or suevite. Alternatively, such a pattern may be due to shock-induced demagnetization, post-impact oxidation of ferromagnetic minerals, low magnetization of post-impact sediment infill, or a random orientation of magnetization vectors.

Summanen's age is presently unknown. An upper age limit is given by the approximately 1880 Myr old target rocks. Considering the limited life-time of a relatively small, simple bowl-shaped crater, a Late Paleozoic to Cenozoic age might be reasonable.

Aerogeophysical studies were carried out in 1995 by the Geological Survey of Finland and led to the discovery of the circular structure in the early 2000s. The impact could recently be confirmed by Jüri Plado and co-workers, who found shatter cones and other shock deformation features.

Location/Access

Lake Summanen is located 9 km southeast of the nearest city, Saarijärvi, about 300 km north of Helsinki. From Helsinki, Highway 4 runs 250 km northward to Jyväskylä. Thereafter, Highway 4 turns into Highway 13, which leads after another 50 km directly to Lake Summanen and the town of Saarijärvi.

Selected References

Plado, J., Hietala, S., Kreitsmann, T., Lerssi, J., Nenonen, J., and Pesonen, L. (2018). Summanen, a new meteorite impact structure in Central Finland. Meteoritics and Planetary Science, Vol. 53, pp. 2413–2426.

Monomict breccia from a less than 1 cubic meter sized boulder at Lake Summanen. The sample shown measures 10 × 5 cm in size (photo: Satu Hietala, Geological Survey of Finland).

Summanen

Finland

latitude = 62°39'00" N / longitude = 25°22'30" E / diameter = 2.6 km / age = < 1880 Myr

scale = 1 : 146 000 / scene = 26.3 × 35.1 km²

0 1.46 7.30 km

N

70 meter 280

Suvasvesi North and Suvasvesi South
Finland

Thomas Kenkmann

Suvasvesi North and Suvasvesi South are two craters of similar size in close vicinity to each other. Recent dating showed, however, that they were not formed as a doublet. Both structures are submerged in Lake Suvasvesi that consists of two circular parts: Kuukkarinselkä in the north and Haapaselkä in the south. The bathymetry of the northern part shows a 90 m deep basin that coincides with the Suvasvesi North structure with a diameter of about 3.5–4 km. With a diameter of 3.8 km, Suvasvesi South, submerged in Haapaselkä, is of similar size.

Suvasvesi North is straddling the border between the 2700 Myr old Archean and the 1880 Myr old Svecofennian terranes of the Baltic Shield. The suture between these terranes runs through the structure with a northwest-southeast direction. It also has some effect on the bathymetry and the outline of the lake. The target rocks at Suvasvesi North are granites, migmatites and schists. A borehole drilled into the center of Suvasvesi North penetrated through 80 m of impactites, including allochthonous lithic breccia, suevite, and impact melt rock. Impact lithologies are only known from the drill core and have not been found along the shore. Impact melt rock contains clasts of various degrees of shock metamorphism – quartz with crosscutting sets of planar deformation features (PDFs) and planar fractures (PF) in zircon, as well as the ballen quartz thermal alteration texture in quartz.

The residual gravity anomaly of the structure after subtraction of the regional trend shows a negative anomaly of −2.8 mGal. Suvasvesi North was originally found during a regional aeromagnetic mapping campaign that indicated a circular area of low magnetic relief with a diameter of about 3.5 km. A later survey with higher spatial resolution revealed a central negative anomaly caused by reversely magnetized impact breccia and suggested the presence of a simple bowl-shaped crater structure.

Suvasvesi South is located in 1880 Myr old Svecofennian rocks. Little is known about the composition of the target here, as the structure has not been drilled. Nearest outcrops exist on several small islands east of Suvasvesi South. They contain shatter cones of up to 50 cm length in medium-grained mica gneiss and gneissose granodiorite. The rocks exposed there are heavily fractured with subvertical north-northwest to south-southeast and east-northeast to west-southwest trending fracture cleavage. A tectonic origin of those fractures is likely, as a major tectonic suture occurs 10 km to the north.

The Suvasvesi South crater was confirmed by glacially deposited impactite boulders that were found at the shoreline and in a gravel pit further south. The boulders of shatter cones and impact melt rock cannot originate from the Suvasvesi North structure, because material transported from it by ice drift would not have passed into this southern area. Microscopic studies of impact melt rock samples and shatter cone specimens revealed the presence of decorated planar deformation features (PDFs) in quartz grains and diaplectic feldspar glass, so-called "maskelynite".

The airborne magnetic anomaly of Suvasvesi South has essentially no relief, similar to that of the northern structure, but it is devoid of a central anomaly. This is an indicator that the southern structure probably does not have a continuous layer of impact melt rock. As for Suvasvesi North, the southern counterpart was initially identified based on the circular magnetic anomaly contrasting with the magnetic pattern of the region. In 2002 impactites were found at the southern shore of Lake Haapaselkä and in a gravel pit 5 km south of the structure.

The Suvasvesi structures are spaced at about 7 km from center to center. Somewhat surprisingly, ^{40}Ar-^{39}Ar dating of impact melt rock chips recovered from the Suvasvesi North drill core yielded a Cretaceous age of about 85 Myr, whereas a Proterozoic ^{40}Ar-^{39}Ar minimum age of 710 Myr was obtained for a melt rock sample from Suvasvesi South. Hence, these two structures can not represent a crater doublet.

Location/Access

Lake Suvasvesi lies in a well developed area in central Finland. The Suvasvesi North impact structure is situated beneath the lake's northern part, whereas Suvasvesi South is submerged in its southern part. The nearest town is Kuopio some 35 km northwest of the crater structures. The distance of the lake to Helsinki in the south is about 350 km. Highway 4 brings the visitor after about 130 km from Helsinki to Lusi, where Highway 5 branches off, which leads after another 220 km to Kuopio.

Selected References

Donadini, F., Plado, J., Werner, S. C., Salminen, J., Pesonen, L. J., and Lehtinen, M. (2006). New evidence for impact from the Suvasvesi South structure, central east Finland. In: Biological Processes Associated with Impact Events, Cockell C., Koeberl C., and Gilmour I. (eds.), Springer, Berlin Heidelberg, pp. 287–307.

Lehtinen, M., Pesonen, L. J., Stehlik, H., and Kuulusa. M. (2002). The Suvasvesi South structure, Central Finland: New evidence for impact. 33rd Lunar and Planetary Science Conference, Lunar and Planetary Institute, Houston, Texas, Abstract #1188.

Öhman, T., Badjukov, D., Raitlala, .., Petrova, T., and Stehlik, H. (2003). Impactites of the Paaselka and Suvasvesi South craters, Finland. 34th Lunar and Planetary Science Conference, Lunar and Planetary Institute, Houston, Texas, Abstract #1571.

Pesonen, L. J., Lehtinen, M., Deutsch, A., Elo, S., and Lukkarinen, H. (1996). New geophysical and petrographic results of the Suvasvesi N impact structure, Finland (abstract). 27th Lunar and Planetary Science Conference, Lunar and Planetary Institute, Houston, Texas, pp. 1021–1022.

Schmieder, M., Schwarz, W. H., Trieloff, M., Buchner, E., Hopp, J., Tohver, E., Pesonen, L. J., Lehtinen, M., Moilanen, J., Werner, S. C., and Öhman, T. (2016). The two Suvasvesi impact structures, Finland: Argon isotopic evidence for a "false" impact crater doublet. Meteoritics and Planetary Science, Vol. 51, pp. 966–980.

Werner, S. C., Plado, J., Pesonen, L. J., Janle, P., and Elo, S. (2002). Potential fields and subsurface models of Suvasvesi North impact structure, Finland. Physics and Chemistry of the Earth, Vol. 27, pp. 1237–1245.

Suvasvesi North / South

Finland

North: latitude = 62°39‘35“ N / longitude = 28°10‘22“ E / diameter = 4 km / age = 85 Myr
South: latitude = 62°35‘36“ N / longitude = 28°13‘38“ E / diameter = 3.8 km / age = > 710 Myr

scale = 1 : 136 000 / scene = 24.4 × 32.5 km²
0 1.36 6.80 km
N
85 meter 210

Ternovka
Ukraine

Thomas Kenkmann

The 11 km diameter Ternovka structure, also known as *Terny*, is a deeply eroded complex impact crater located in the southern part of the Ukrainian Shield. Its age has been estimated at 280 ± 10 Myr. The central uplift of the structure is 2.5 km across. The structure is the site of extensive iron ore mining.

Target rocks are folded and ore-bearing Proterozoic schists and quartzites, as well as amphibolites, syenites and granitoids that have partially been subjected to alkaline metasomatism. Ore formation is considered to have occurred prior to impact, but the impact event has brought ore close to surface within the central uplift strata – thereby facilitating ore extraction and the development of a mining district. The exposed crater subsurface is composed of large, radially displaced blocks. Brecciation attenuates with depth and its lateral disappearance constrains the size of the crater structure. Numerous irregular sub-vertical lenses of vesicular impact melt rock, up to 10 m thick, are found in the central uplift. Most of these pinch out downward. Two types of impact melt rock with different composition – obviously derived from melting of different target rocks – have been distinguished. Allochthonous crystalline breccia occurs in the annular trough around the central uplift and grades upward into suevite. Breccias contain diaplectic glasses and high-pressure polymorphs such as coesite, stishovite, and diamond. Fragments often display shatter cones.

A first publication on the Ternovka structure by V. Tikhonov and co-workers stems from 1968. Other studies have appeared in the following years, published mainly in Russian. More recently, a detailed investigation was published by Virgil Sharpton and co-workers.

Location/Access

Ternovka is situated 25 km northwest of Krywoj Rog in the Ternivsky mining district. Two very large iron ore producing quarries allow to observe different types of brecciated and displaced rocks and impactites. However, access requires special permits.

Selected References

Masaitis, V. L. (1999). Impact structures of northeastern Eurasia: The territories of Russia and adjacent countries. Meteoritics and Planetary Science, Vol. 34, pp. 691–711.

Masaitis, V. L., Mashchak, M. S., and Sokolova, I.Y. (1980). High-pressure silica phases in the Ternovka astrobleme (in Russian). Doklady Akademii Nauk SSSR, Vol. 255, pp. 709–713.

Sharpton, V. L., Krochuk, V. R., and Herrick, R. R. (2013). Characterization and morphological reconstruction of the Terny impact structure, central Ukraine. Meteoritics and Planetary Science, Vol. 48, pp. 806–818.

Suevite from the Ternovka impact structure, 8 × 6.5 × 4.5 cm in size. Note the darker greenish altered melt clasts with fluidal texture (photo courtesy: Ralf-Thomas Schmitt, Museum für Naturkunde Berlin, sample obtained from the VSEGEI, St. Petersburg, Russia).

Ternovka

Ukraine

latitude = 48°07'48" N / longitude = 33°31'12" E / diameter = 11 km / age = 280 ± 10 Myr

scale = 1:112 000 / scene = 20.2 × 26.9 km²

0 1.12 5.60 km

N

-305 meter 230

Tvären
Sweden

Thomas Kenkmann

Tvären impact crater is completely submerged in an unusually deep, circular, about 4 km wide bay of the Baltic Sea at the southeastern coast of Sweden. The outline of the bay is circular and has only a narrow connection to the Baltic Sea. Tvären crater formed under shallow marine conditions.

The impact occurred into Svecofennian gneisses of granitic to tonalitic composition and of Proterozoic age. These rocks are exposed on land in the surrounding area. At the time of impact, however, these basement rocks were covered by Cambrian and Ordovician sediments. Boulders of such Paleozoic rocks occur nowadays along the shoreline of Tvären Bay. Quartz grains in crystalline rock boulders reveal shock features. Two boreholes were drilled into the structure and investigation of the cores shed light on the origin of the crater. The stratigraphic sequence of one of the cores comprises post-impact Quaternary sediments of the crater fill, and Paleozoic sediments, mainly sandstone, shale and limestones. Slump breccia of Paleozoic limestone mixed with crystalline rock clasts indicates that the impact occurred under shallow marine conditions. This unit was deposited when a forceful resurge of the sediment-laden water column washed over the elevated crater rim. The crater floor is covered with a basal crystalline rock breccia, which grades below into brecciated basement.

Tvären is characterized by a magnetic signature that disrupts regional anomaly trends. Based on the size of the geophysical anomaly, the original crater is estimated to have had a flat bowl shape with a diameter of about 3.2 km, and a maximum depth of 800 m. There are no geophysical indications for the presence of a melt rock layer. The structure is probably eroded by 300 m. Based on the fossil content of graptolites and chitinizoans, the age of the impact is constrained to 462 ± 5 Myr. Thus, it has been suggested that Tvären belongs to the group of Late Ordovician impact craters that resulted from a major collision event in the asteroid belt at 470 Myr.

Kurt Fredriksson and Frans E. Wickman were the first to suggest, in 1963, that the unusually deep Tvären Bay could be an impact crater. Seismic investigations in 1986 by Tom Flodén and co-workers supported this hypothesis, and studies of the two drill cores, in which shock features were found by Maurits Lindström, provided the final proof.

Location/Access

The Tvären structure is located at the southeast coast of Sweden, within an almost completely closed bay occurring 23 km east of Nyköping and 75 km south of Stockholm. From Stockholm Highway E4 brings the visitor southward to Vagnhärad, where county road 219 leads close to Tvären. From Nyköping a county road provides direct access to the crater area.

Selected References

Flodén, T., Tunander, P., and Wickman, F. E. (1986). The Tvären Bay structure, an astrobleme in southeastern Sweden. Geologiska Föreningen i Stockholm Förhandlingar, Vol. 108, pp. 225–234.

Frisk, A. M., and Ormö, J. (2007). Facies distribution of post-impact sediments in the Ordovician Lockne and Tvären impact craters; indications for unique impact-generated environments. Meteoritics and Planetary Science, Vol. 42, pp. 1971–1984.

Henkel, H., Ekneligoda, T. C., and Aaro, S. (2010). The extent of impact induced fracturing from gravity modeling of the Granby and Tvären simple craters. Tectonophysics, Vol. 485, pp. 290–305.

Lindström, M., Flodén, T., Grahn, Y., and Kathol, B. (1994). Post-impact deposits in Tvären, a marine Middle Ordovician crater south of Stockholm, Sweden. Geological Magazine, Vol. 131, pp. 91–103.

Ormö, J., and Blomqvist, G. (1996). Magnetic modelling as a tool in the evaluation of impact structures, with special reference to the Tvären Bay impact crater, SE Sweden. Tectonophysics, Vol. 262, pp. 291–300.

The area of the Tvären impact structure on April 24, 2019 as a composite image based on TanDEM-X DEM and Copernicus Sentinel-2 (channel 2/3/4) data.

Tvären

Sweden

latitude = 58°46'30" N / longitude = 17°25'33" E / diameter = 3.2 km / age = 462 Myr

scale = 1:91 000 / scene = 16.3 × 21.7 km²

0 0.91 4.55 km

N

0 meter 130

Vepriai
Lithuania

Thomas Kenkmann

The Vepriai impact structure on the Baltic Shield has a diameter of 7.5 km and is buried beneath 70–100 m of Quaternary sediments. There is no morphological expression for this structure. Target rocks in this area of the Baltic Shield comprise horizontally layered sediments of predominantly Paleozoic age. The sedimentary sequence above the crystalline basement starts with a thin veneer of uppermost Precambrian sandstones, followed by 100 m of Cambrian siliciclastics, 300 m of Ordovician and Silurian carbonate rocks and clays, and a 125–275 m thick sequence of Devonian sandstones, siltstones, clays, and carbonate rocks. These latter rocks most certainly formed the target surface at the time of impact. Drillholes outside the impact structure penetrated the crystalline basement at the depth of 700 m.

The cratering event disturbed and excavated the target rocks down to the Cambrian beds. The depth of the crater depression is about 525 m. A central uplift is suspected but has not been confirmed. The southeastern slope of the depression is steeper than the northwestern side, and it is structured by slump terraces. Towards the edge of the crater depression target rocks that were uplifted by 20–50 m can be found. An allochthonous sedimentary breccia, more than 400 m thick, fills the depression and is composed of fragments and blocks of different stratigraphic units. Quartz fragments are partly shocked and contain planar deformation features (PDFs). Small shatter cones and impact glass shards were observed in the breccia. Relics of ejecta exist on undisturbed Devonian beds along the southeastern rim. The polymict breccia of the crater cavity is covered by crater lake deposits, namely Upper Jurassic black clays, reaching a maximum thickness of 162 m in the center of the depression. The crater has been extensively drilled. Geological data suggest that the impact occurred in Middle Jurassic time. Vepriai was described for the first time by Gediminas Motuza and R. Gailius in 1978. Only a few more studies of this impact structure have been published since then.

Location/Access

Vepriai is located in the Baltic region some 65 km northwest of the capital of Lithuania, Vilnius, in agricultural land and 60 km northeast of Kaunas. From Kaunas it is a 50 km drive on Highway A6 to the exit at Pageliziai, from where a regional road leads to the village of Vepriai. The buried impact structure is located on the southern side of the Šventoji river without a morphological expression on surface.

Selected References

Masaitis, V. L. (1999). Impact structures of northeastern Eurasia: The territories of Russia and adjacent countries. Meteoritics and Planetary Science, Vol. 34, pp. 691–711.

Masaitis, V. L., Danilin, V. N., Mashchak, M. S., Raikhlin, A. I., Selivanovskaty. A. V., and Shadenkov, E. M. (1980). The Geology of Astroblemes. Nedra Press, Leningrad, Russia. 231 pp.

Motuza, G. B., and Gailius, R. P. (1978). On the supposed astroblemes in Lithuania (in Russian). In: Local Structures in Belarus and the Baltic Region, Suveizdis, P. I. (ed.), Geological Research Institute, Vilnius, Lithuania, pp. 91–94.

Vepriai

Lithuania

latitude = 55°05‘ N / longitude = 24°35‘ E / diameter = 7.5 km / age = 162 - 174 Myr

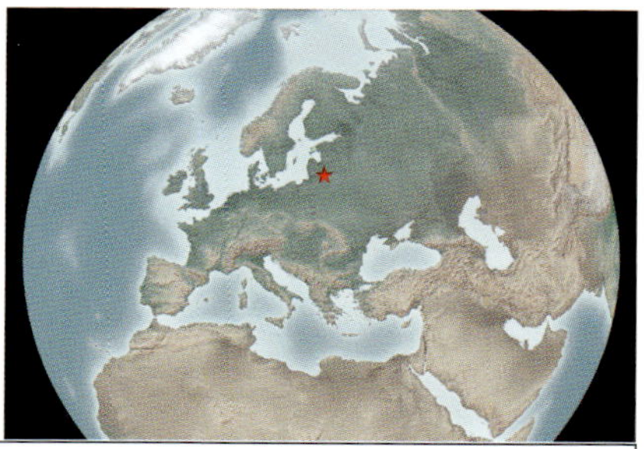

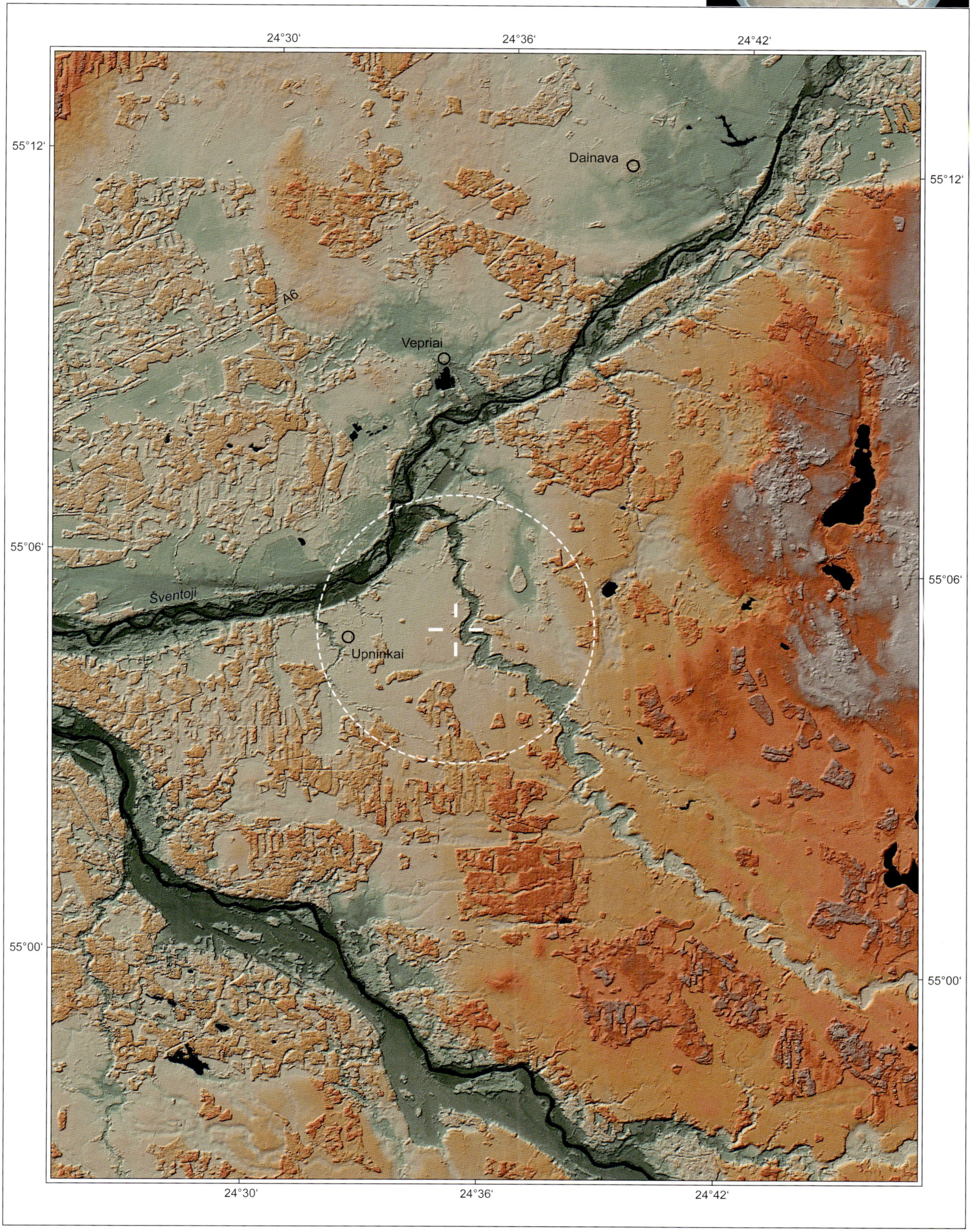

scale = 1:130 000 / scene = 23.5 × 31.3 km²

0 1.30 6.50 km

N

35 meter 185

Zapadnaja
Ukraine

Thomas Kenkmann

Zapadnaya is an elongated, 3.2 × 2.3 km sized, complex impact structure located on the Ukrainian Shield. It is buried under 40 m of loose siliciclastic sediments of Neogene and Quaternary age. An uplifted horst-like block of cataclased basement rocks separates the crater into two unequal depressions and complicates its structure. The crater forms a subdued morphological depression with a southeast trending drainage system.

Target rocks are Early Proterozoic granites, minor migmatites, and gneisses of the Ukrainian Shield. The crater has a central uplift that is surrounded by a 220 m deep ring syncline, which is filled with shocked crystalline breccia. The crater floor itself consists of fractured and brecciated granites. Shock features in these rocks, such as rare planar deformation features (PDFs) in quartz and feldspar, indicate shock pressures of about 5–10 GPa. Dikes of impact melt rock up to 10 m thick were injected into rocks of the central uplift. Allochthonous crater fill breccia covers the central uplift by some 60 m but reaches thicknesses of 120–130 m in the surrounding annular trough. Here, basal polymict crystalline breccia changes into suevites with melt particles, of 80 m thickness. Suevite is absent above the central uplift and pinches out towards the edge of the depression. Suevite and impact melt rock contain small impact diamonds produced by the transformation of graphite flakes. The Zapadnaja impact structure is defined by an elongated gravity low of −4.2 mGal over an area of 4.7 × 3.9 km. K-Ar dating of clast-free impact-melt rock provided a Middle Jurassic age for the impact of 165 ± 6 Myr.

The impact origin of the structure was first recognized in 1975 by Anton Val'ter and Eugene Gurov, who studied core samples from a circular negative gravity anomaly – at that time of unknown origin. Their discovery of impact diamonds lead to extensive drilling. Since then, only a few papers in the Ukrainian and Russian literature have been published on this structure.

Location/Access

Zapadnaja is located 130 km southwest of Kiev in agricultural land. As the structure is buried, there is not much to see on surface. As with many of the buried impact structures on the Ukrainian Shield, the published locations of Zapadnaya are not fully consistent. They differ by 2–3 crater radii.

Selected References

Gurov, E. P., Gurova E. P., and Sokur, T. M. (1999) Impact diamonds in the rocks of the Zapadnaya astrobleme (in Russian). Mineralni resursi Ukraini, Vol. 3, pp. 30–32.

Gurov, E. P., Gurova, E. P., and Sokur, T. M. (2002). Geology and petrography of the Zapadnaya impact crater in the Ukrainian Shield. In: Impacts in Precambrian Shields, Plado, J., and Pesonen, L. J. (eds.), Springer, Berlin Heidelberg, pp. 173–188.

Masaitis, V. L. (1999). Impact structures of northeastern Eurasia: The territories of Russia and adjacent countries. Meteoritics and Planetary Science Vol. 34, pp. 691–711.

Sokur, T. M. (1999) Shock metamorphism of quartz in rocks from the Zapadnaya impact structure (in Russian). Geologicheskii Zhurnal Vol. 1, pp. 87–93.

Zapadnaya

Ukraine

latitude = 49°44'00" N / longitude = 29°03'15" E / diameter = 3.2 km / age = 165 ± 5 Myr

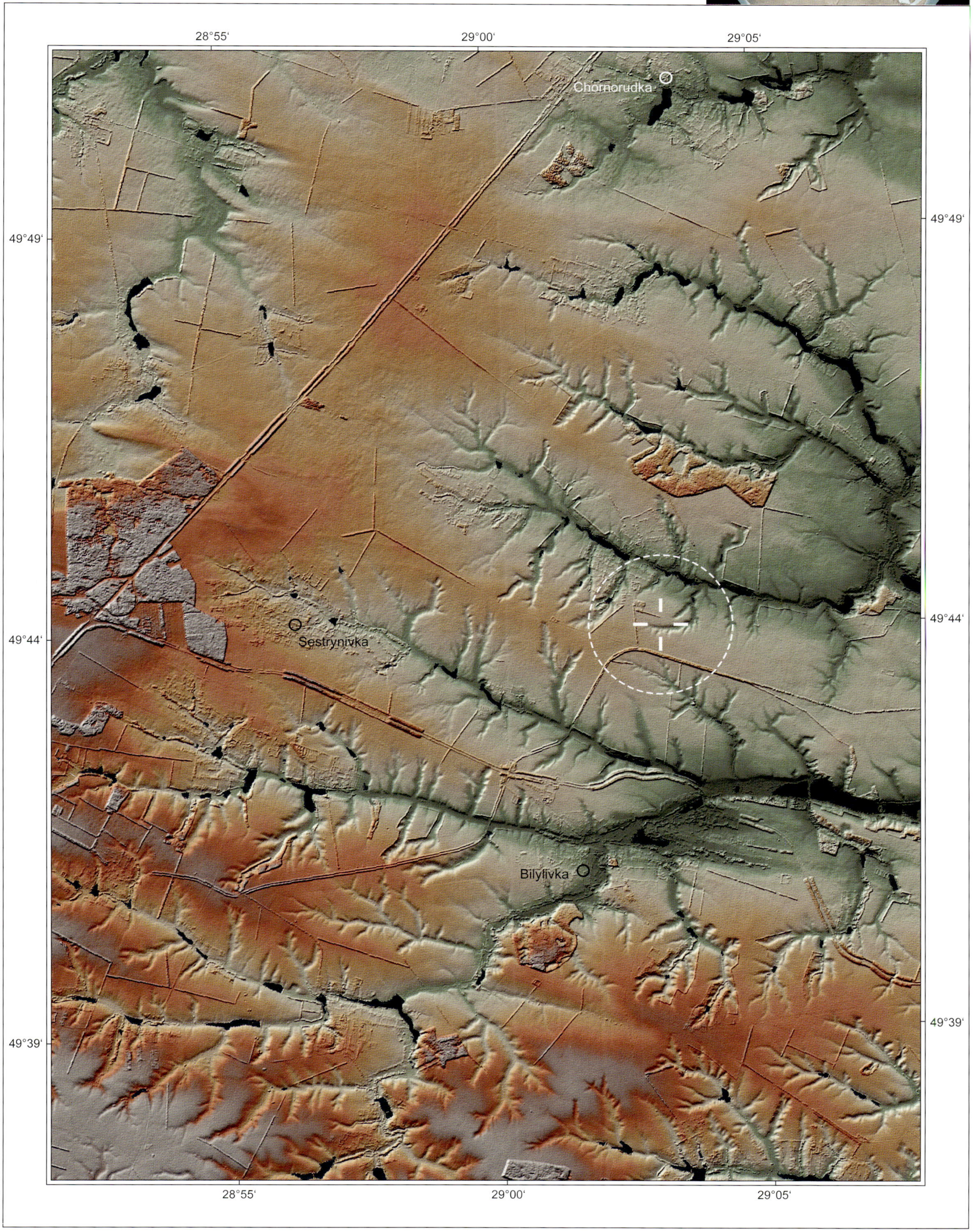

scale = 1:108 000 / scene = 19.5 × 26.6 km²

0 1.08 5.40 km

N

210 meter 325

Zeleny Gai
Ukraine

Thomas Kenkmann

The roughly 3.5 km wide Zeleny Gai impact structure on the Ukrainian Shield is buried under 50 m of Quarternary deposits and, thus, is without obvious surface expression. The outer limit of the structure is poorly defined. Target rocks were Archean granites, migmatites, and gneisses of the Ukrainian Shield. The crater diameter suggests that the crater size is close to the transition between a simple and complex crater morphology. There are indeed indications for the presence of a gentle central uplift.

A 380 m deep borehole sunk near the center of the structure reached fractured granite with shatter cones and planar deformation features (PDFs) in quartz. This slightly uplifted crater floor is overlain by a 120 m thick unit of polymict impact breccias that is composed of angular fragments and blocks of crystalline rocks, with abundant shocked mineral clasts and shards of vesicular impact glasses. After the formation of the impact structure in the Upper Cretaceous, some 80 Myr ago, a crater lake existed for a while and led to the deposition of 115 m of lacustrine sediments of Paleogene age. The lake deposits contain boulders and gravel lenses of redeposited, shocked crystalline rocks, probably mainly representing debris flows off the crater wall and possibly from the central uplift. Quaternary sediments represent the topmost unit and buried the entire structure.

A first report on the structure by Anton Val'ter and co-workers dates back to 1976. The structure was conspicuous by the presence of a circular gravity low. The analysis of core samples from the 380 m deep borehole documented shock effects and proved the impact theory.

Location/Access

Contrasting locations have also been provided for Zeleny Gai. Maps showing its position in relation to the Boltysh and Rotmistrovka impact structures place Zeleny Gai about 40 km east of Boltysh and about 20 km south of the Kremenchuk water reservoir. The town of Olexandrija lies about 30 km to the southeast of the structure. As the crater is buried, there is not much to see on surface.

Selected References

Masaitis, V. L. (1999). Impact structures of northeastern Eurasia: The territories of Russia and adjacent countries. Meteoritics and Planetary Science, Vol. 34, pp. 691–711.

Val'ter, A. A., Bryansky, V. P., Ryabenko, V. A., and Lazarenko, E. (1976). On the explosion (meteoritic) origin of Zeleny Gai structure on the Ukrainian Shield (in Russian). Doklady ANSSSR, Vol. 229, pp. 160–162.

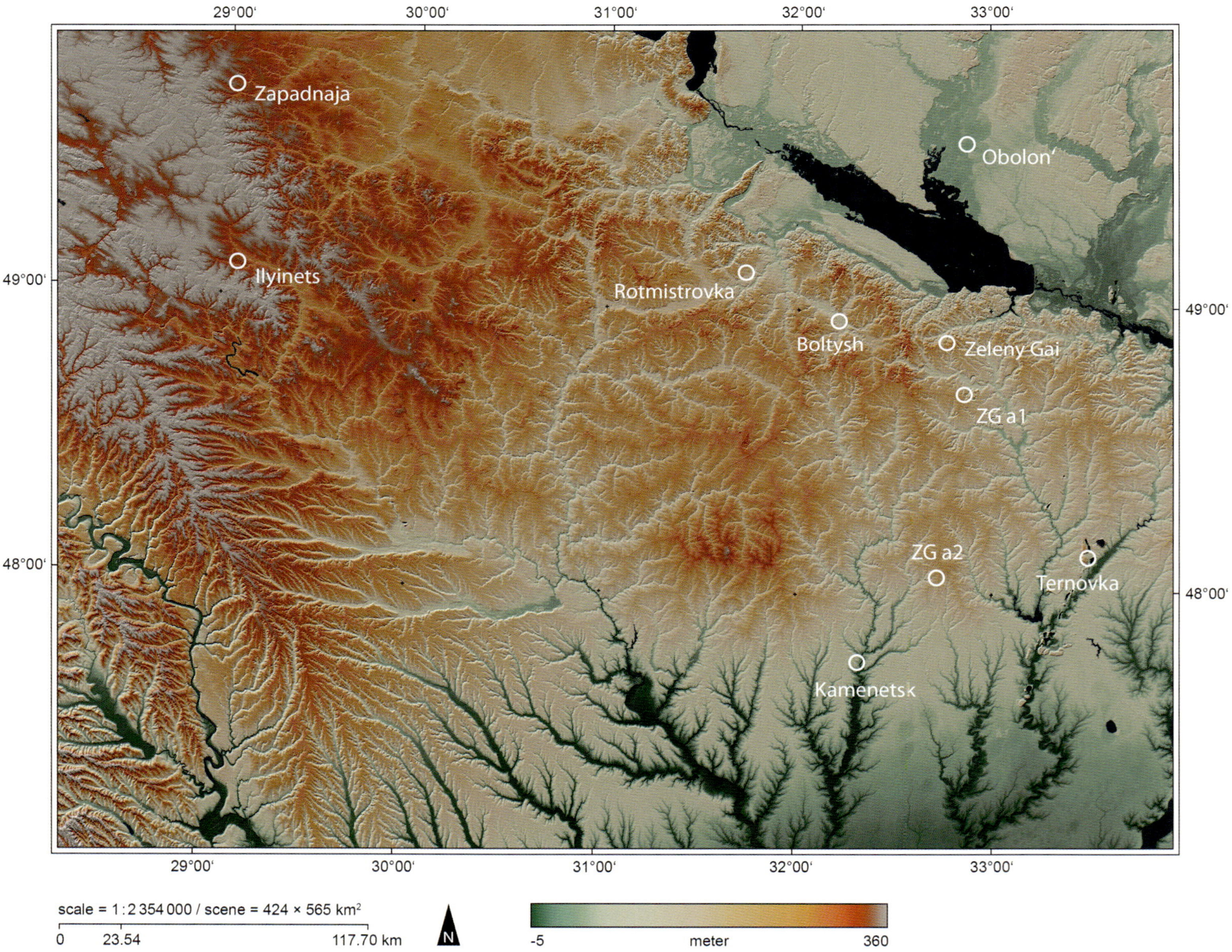

The buried impact structures on the Ukrainian Shield. For Zeleny Gai two alternative positions (ZG a1 and ZG a2) are quoted in the literature, and both are marked.

Zeleny Gai

Ukraine

latitude = 48°53'17" N / longitude = 32°47'55" E / diameter = 3.5 km / age = 80 Myr

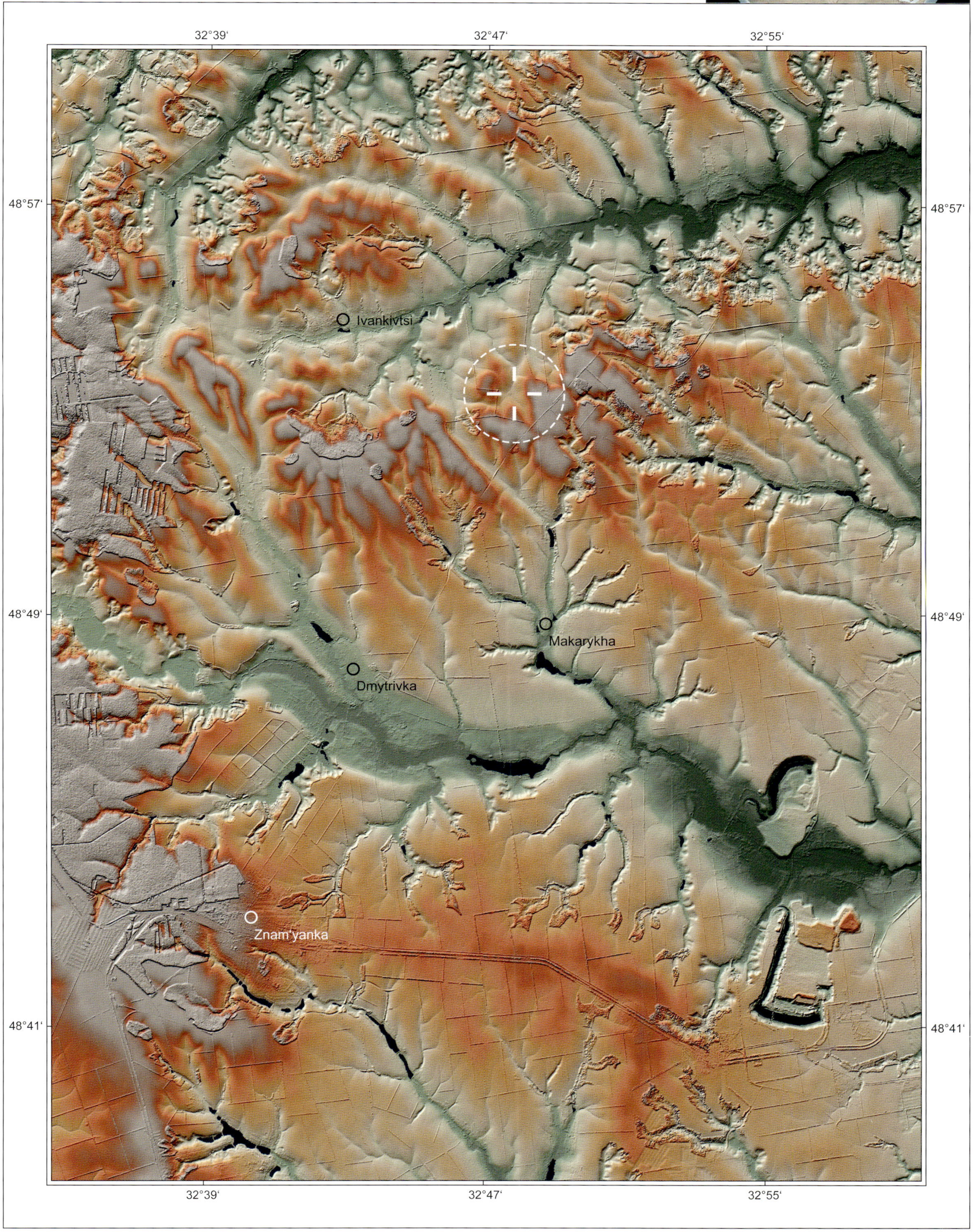

scale = 1:170 000 / scene = 30.5 × 40.7 km²

0 1.70 8.50 km

N

65 meter 265

Geologic Timescale

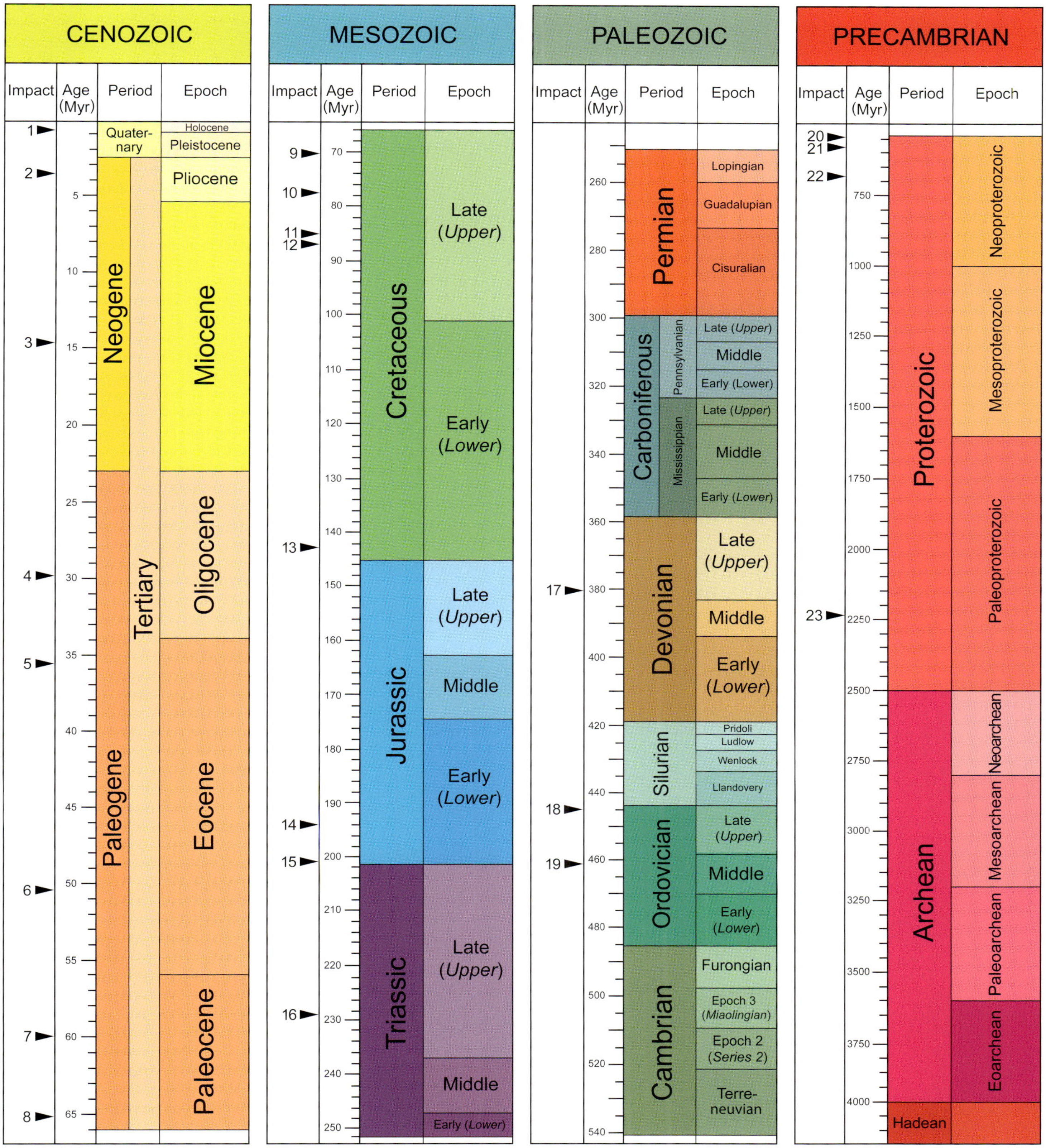

Selection of impact structures with reasonably well determined ages:

1 Zhamanshin
2 El'gygytgyn
3 Ries/Steinheim Basin
4 Logoisk
5 Popigai
6 Gusev/Kamensk
7 Goat Paddock
8 Boltysh
9 Kara
10 Lappajärvi
11 Suvasvesi North
12 Yallalie
13 Gosses Bluff
14 Puchezh-Katunki
15 Rochechouart
16 Paasselkä
17 Siljan
18 Ilyinets
19 Lockne/Målingen
20 Gardnos
21 Acraman
22 Jänisjärvi
23 Yarrabubba

Modified after (terminology different in chronostratigraphic chart given in italics):
Geologic Time Scale v. 5.0, Geological Society of America, https://doi.org/10.1130/2018.CTS005R3C. L.E.
and
International Chronostratigraphic Chart v2018/08, International Commission on Stratigraphy
End and Beginning of Holocene (0–0.012 Myr) and end of Pleistocene (0.012 Myr) not drawn to scale

Selected References

Aaloe, A. O. (1963). New data on the structure of Ilumetsa craters. Eesti NSV Teaduste Akadeemia Geoloogia Instituudi uurimused (in Russian). Vol. 11, pp. 35–43.

Abels, A., Mannola, P., Lehtinen, M., Bergman, L., and Pesonen, L. J. (1998). New observations of the properties of the Lumparn impact structure, Åland Islands, Southwestern Finland. Meteoritics and Planetary Science, Vol. 33, pp. A7–A8.

Abels, A., Pesonen, L. J., Deutsch, A., Bischoff, L., and Lehtinen, M. (1998). The Söderfjärden impact structure, W. Finland: Multiple data integration with implications for morphometry, cratering process, and post-impact sedimentary history. 24th Lunar and Planetary Science Conference, Lunar and Planetary Institute, Houston, Texas, Abstract #1264.

Abels, A., Bergman, L., Lehtinen, M., and Pesonen, L. J. (2000). Structural constraints and interpretations on the formation of the Soderfjärden and Lumparn impact structures, Finland. 4th ESF Workshop: Response of the Earth System to Impact Processes (IMPACT), Helsinki, p. 26.

Abels, A., Plado, J., Pesonen, L. J., and Lehtinen, M. (2002). The impact cratering record of Fennoscandia – A close look at the database. In: Impacts in Precambrian Shields, Plado, J., and Pesonen, L. J. (eds.), Springer, Berlin Heidelberg, 339 pp.

Abercrombie, T. J. (1966). Saudi Arabia. Beyond the sands of Mecca. National Geographic, Vol. 1,291, pp. 1–53.

Abrantes Jr., F. R. (2016). O Permo-Triássico da Bacia do Parnaíba, norte do Brasil: implicações paleoambientais, paleoclimáticas e paleogeográficas para o Pangea Ocidental. Ph.D. thesis, Universidade Federal do Pará, 130 pp.

Alderman, A. R. (1931). The meteorite craters at Henbury, Central Australia, with an addendum by L. J. Spencer. Mineralogical Magazine, Vol. 23, pp. 19–32.

Allix, A. (1951). Note et correspondence à propos des cratères météoriques. Revue de Géographie de Lyon, Vol. 26, pp. 357–359.

Alvarez, L. W., Alvarez, W., Asaro, F., and Michel, H. V. (1980). Extraterrestrial cause for the Cretaceous-Tertiary Extinction. Science, Vol. 208, pp. 1095–1108.

Alwmark, C. (2009). Shocked quartz grains in the polymict breccia of the Granby structure, Sweden – Verification of an impact. Meteoritics and Planetary Science, Vol. 44, pp. 1107–1113.

Alwmark, C., Alwmark-Holm, S., Ormö, J., and Sturkell, E. (2014). Shocked quartz grains from the Målingen structure, Sweden – Evidence for a twin crater of the Lockne impact structure. Meteoritics and Planetary Science, Vol. 49, pp. 1076–1082.

Alwmark, C., Ferrière, L., Holm-Alwmark, S., Ormö, J., Leroux, H., and Sturkell, E. (2015). Impact origin for the Hummeln structure (Sweden) and its link to the Ordovician disruption of the L chondrite parent body. Geology, Vol. 43, pp. 279–282.

Alwmark, C., Bleeker, W., LeCheminant, A., Page, L., and Scherste'n, A. (2017). An early Ordovician ^{40}Ar-^{39}Ar age for the ~50 km Carswell impact structure, Canada. Geological Society of America Bulletin, Vol. 129, pp. 1442–1449.

Amgaa, T., and Koeberl, C. (2009a). Geology and petrography of Tabun Khara Obo crater. 72nd Annual Meeting of the Meteoritical Society, Abstract #5019.

Amgaa, T., and Koeberl, C. (2009b). Impact origin of Tabun Khara Obo crater, Mongolia, confirmed by drill core studies (abstract). Geological Society of America Annual Meeting, Portland, Oregon, Vol. 41, p. 533.

Anderson, R. R., and Hartung, R. B. (1992). The Manson impact structure; Its contribution to impact materials at the Cretaceous/Tertiary Boundary. Proceedings of the 22nd Lunar and Planetary Science Conference, Houston, Texas, pp. 101–110.

Anderson, R. R., Witzke, B. J., and Roddy, D. J. (1996). The drilling of the 1991–1992 Geological Survey Bureau and U.S. Geological Survey Manson impact structure research cores. Geological Society of America Special Paper, Vol. 302, pp. 45–88.

Andreoli, M. A. G., Ashwal, L. D., Hart, R. J., Smith, C. B., Webb, S. J., Tredoux, M., Gabrielli, F., Cox, R. M., and Hambleton-Jones, B. B. (1995). The impact origin of the Morokweng Ring Structure, Southern Kalahari, South Africa (abstract). Centennial Geocongress of the Geological Society of South Africa, Johannesburg, South Africa, pp. 541–554.

Arp, G., Reimer, A., Simon, K., Sturm, S., Wilk, J., Kruppa, C., Hecht, L., Hansen, B. T., Pohl, J., Reimold, W. U., Kenkmann, T., and Jung, D. (2019). The Erbisberg drilling 2011: Implications for the structure and postimpact evolution of the inner ring of the Ries impact crater. Meteoritics and Planetary Science, Vol. 54, pp. 2448–2842.

Artemieva, N. A., and Shuvalov, V. V. (2001). Motion of a fragmented meteoroid through the planetary atmosphere. Journal of Geophysical Research, Vol. 106, pp. 3297–3309.

Artemieva, N. A., and Shuvalov, V. V. (2007). 3D effects of Tunguska Event on the ground and in the atmosphere. 38th Lunar and Planetary Science Conference, Lunar and Planetary Institute, Houston, Texas, Abstract #1537.

Artemieva, N. A., Wünnemann, K., Krien, F., Reimold, W. U., and Stöffler, D. (2013). Ries crater and suevite revisited – Observations and modeling, Part II: Modeling. Meteoritics and Planetary Science, Vol. 48, pp. 590–627.

Åström, K. (1998). Seismic signature of the Lake Mien impact structure, southern Sweden. Geophysical Journal International, Vol. 135, pp. 215–231.

Attar, A. (1981). La structure de Tin Bider (plateau du Tinhert, Algérie). Un impact météoritique. Unpublished report, Sonatrach, 15 pp.

Avice, G., Meier, M. M. M., Marty, B., Wieler, R., Kramers, J. D., Langenhorst, F., Cartigny, P., Maden, C., Zimmermann, L., and Andreoli, M. A. G. (2015). A comprehensive study of noble gases and nitrogen in "Hypatia", a diamond-rich pebble from SW Egypt. Earth and Planetary Science Letters, Vol. 432, pp. 243–253.

Baldwin, S. L., McDougall, I., and Williams, G. E. (1991). K-Ar and ^{40}Ar-^{39}Ar analyses of melt rock from the Acraman impact structure, Gawler Ranges, South Australia. Australian Journal of Earth Sciences, Vol. 38, pp. 291–298.

Banet, A. C., and Fenton, J. P. G. (2008). An examination of the Simpson core test wells suggests an age for the Avak impact feature near Barrow, Alaska. In: The Sedimentary Record of Meteorite Impacts, Evans, K., Horton, Jr., J.W., King, D. T., and Morrow, J. R. (eds.), Geological Society of America Special Paper, Vol. 437, pp. 139–145.

Baratoux, D., Niang, C. A. B., Reimold, W. U., Salah, M. S., Jessell, M. W., Boamah, D., Faye, G., Bouley, S., and Vanderheaghe, O. (2019). Bosumtwi impact structure, Ghana: Evidence for fluidized emplacement of the ejecta. Meteoritics and Planetary Science, Vol. 54, pp. 2541–2556.

Barbour, Jr. E., and Corrêa, W. A. G. (1981). Geologia da Estrutura de Vargeão, SC. São Paulo, Consórcio CESP/IPT, 33 pp. (unpublished report, Paulipetro).

Barringer, B. (1967). Historical notes on the Odessa meteorite crater. Meteoritics, Vol. 3, pp. 161–168.

Barringer, D. M. (1910). Meteor Crater in northern central Arizona. Paper presented at the National Academy of Sciences, Princeton University, 16 November 1909, 24 pp. (plus 18 plates and 3 maps).

Barringer, D. M. Jr. (1928). A new meteor crater. Proceedings of the Academy of Natural Sciences of Philadelphia, Vol. 80, pp. 307–311.

Barrows, T. T., Magee, J., Miller G., Fifield, L. K. (2019). The age of the Wolfe Creek meteorite crater (Kandimalal), Western Australia. Meteoritics and Planetary Science, Vol. 54, pp. 2686–2697.

Bayerisches Geologisches Landesamt. (2004). Geologische Karte 1 : 100 000 des Impaktraters Nördlinger Ries mit Kurzerläuterungen auf der Rückseite in deutscher und englischer Sprache, 1 map.

Beals, C. S. (1960). A probable meteorite crater of Precambrian age at Holleford, Ontario. Publication of the Dominion Observatory Ottawa, Vol. 24, pp. 117–142.

Beals, C. S., Ferguson, G. M., and Landau, A. (1956). A search for analogies between lunar and terrestrial topography on photographs of the Canadian Shield. Journal of the Royal Astronomical Society of Canada, Vol. 50, pp. 203–250.

Beals, C. S., Innes, M. J. S., and Rottenberg, J. A. (1960). The search for fossil meteorite craters. Ottawa Dominion Observatory Contributions, Vol. 4, pp. 1–31.

Beals, C. S., Dence, M. R., and Cohen, A. J. (1967). Evidence for the impact origin of Lac Couture. Publications of the Dominion Observatory Ottawa, Vol. 31, pp. 409–426.

Beatty, J. K., Collins Petersen, C., Chaikin, A. (1999). The New Solar System, Sky Publishing Corporation & Cambridge University Press.

Becq-Giraudon, J. F., Rouzeau, O., Goachet, E., and Solages, S. (1992). Impact hyperveloce d'une météorite géante a l'origine de la dépression circulaire d'Aorounga au Tchad (Afrique). Comptes Rendus de l'Académie des Sciences, Paris, Vol. 315, pp. 83–88.

Beere, G. M. (1983). The Piccaninny structure – a cryptoexplosive feature in the Ord Basin, East Kimberley. Geological Survey of Western Australia Report 1983/6, 8 pp.

Belyanin, G. A., Kramers, J. D., Andreoli, M. A. G., Greco, F., Gucsik, A., Makhubela, T. V., Przybylowicz, W. J., and Wiedenbeck, M. (2018). Petrography of the carbonaceous, diamond-bearing stone "Hypatia" from southwest Egypt: A contribution to the debate on its origin. Geochimica et Cosmochimica Acta, Vol. 223, pp. 462–492.

Beran, A., and Koeberl, C. (1997). Water in tektites and impact glasses by FTIR spectrometry. Meteoritics and Planetary Science, Vol. 32, pp. 211–216.

Bevan, A. (2012a). Western Australian Impact Crater Field Excursion, 20–29 August 2012. 75th Annual Meeting of the Meteoritical Society Cairns, Queensland, Australia, 55 pp.

Bevan, A. (2012b). Yallalie impact structure. In: Western Australian Impact Craters Field Excursion, 75th Annual Meeting of the Meteoritical Society, Cairns, Queensland, Australia, pp. 12–23.

Bibbins, A. B. (1926). A small meteor crater in Texas. Engineering and Mining Journal Press, Vol. 121, p. 932.

Biren, M. B., and Spray, J. G. (2011). Shock veins in the central uplift of the Manicouagan impact structure: Context and genesis. Earth and Planetary Science Letters, Vol. 303, pp. 310–322.

Biren, M. B., van Soest, M., Wartho, J.-A., and Spray, J. G. (2014). Dating the cooling of exhumed central uplifts of impact structures by the (U-Th)/He method: A case study at Manicouagan. Chemical Geology, Vol. 377, pp. 56–71.

Bischoff, L., and Oskierski, W. (1987). Fractures, pseudotachylite veins and breccia dykes in the crater floor of the Rochechouart impact structure, SW-France, as indicators of crater forming processes. In: Research in Terrestrial Impact Structures, Pohl, J. (ed.), Friedrich Vieweg publishers, Braunschweig/Wiesbaden, pp. 5–29.

Bischoff, L., and Oskierski, W. (1988). The surface structure of the Haughton impact crater, Devon Island, Canada. Meteoritics, Vol. 23, pp. 209–220.

Bjornerud, M. G. (1998). Superimposed deformation in seconds: Breccias from the impact structure at Kentland, Indiana (USA). Tectonophysics, Vol. 290, pp. 259–269.

Black, D. F. B. (1964). Cryptoexplosive structure near Versailles, Kentucky. U.S. Geological Survey Professional Paper, Vol. 501-B, pp. B9–12.

Bland, P. A., de Souza Filho, C. R., Jull, A. J. T., Kelley, S. P., Hough, R. M., Artemieva, N. A., Pierazzo, E., Coniglio, J., Pinotti, L., Evers, V., and Kearsley, A. T. (2002). A possible tektite strewn field in the Argentinian pampa. Science, Vol. 296, pp. 1109–1111.

Bodén, A., and Eriksson, K. G. (eds.) (1988). Deep Drilling in Crystalline Bedrock. Volume 1: The Deep Gas Drilling in the Siljan Impact Structure, Sweden and astroblemes. Springer, Berlin Heidelberg, 364 pp.

Boon, J. D., and Albritton, C. C. (1936). Meteorite craters and their possible relationship to "cryptovolcanic structures". Field and Laboratory, Vol. 5, pp. 7–9.

Bootsman, C. S., Reimold, W. U., and Brandt, D. (1999). Evolution of the Molopo drainage and its possible disruption by the Morokweng impact event at the Jurassic-Cretaceous boundary. Journal of African Earth Sciences, Vol. 29, pp. 669–678.

Borovicka, J., and Spurný, P. (2008). The Carancas meteorite impact – Encounter with a monolithic meteoroid. Astronomy and Astrophysics, Vol. 485, pp. L1–L4.

Bottomley, R. J., and York, D. (1988). Age measurement of the submarine Montagnais impact crater. Geophysical Research Letters, Vol. 15, pp. 1409–1412.

Bottomley R. J., York, D., and Grieve, R. A. F. (1978). ^{40}Ar-^{39}Ar ages of Scandinavian impact structures: I. Mien and Siljan. Contributions to Mineralogy and Petrology, Vol. 68, pp. 79–84.

Bottomley, R. J., Grieve, R. A. F., York, D., and Masaitis, V. L. (1997). The age of the Popigai impact event and its relation to events at the Eocene/Oligocene boundary. Nature, Vol. 388, pp. 365–368.

Bottomley, R. J., York, D., and Grieve, R. A. F. (1990). ^{40}Argon-^{39}Argon dating of impact craters. Proceedings of the 20th Lunar and Planetary Science Conference, Houston, Texas, Cambridge University Press, pp. 421–431.

Bouşka, V., Povondra, P., Florenskiy, P. V., and Kanda, Z. (1981). Irghizites and zhamanshinites: Zhamanshin crater, USSR. Meteoritics, Vol. 16, pp. 171–184.

Brandt, D., and Reimold, W. U. (1995a). The geology of the Pretoria Saltpan impact structure and the surrounding area. South African Journal of Geology, Vol. 98, pp. 287–303.

Brandt, D., and Reimold, W U. (1995b). The petrology of the volcanic rocks occurring in the Pretoria Saltpan impact crater and surrounding area. South African Journal of Geology, Vol. 98, pp. 304–318.

Brandt, D., Reimold, W. U., Franzen, A. J., Koeberl, C., and Wendorff, L. (1998). The geophysical signature of the Roter Kamm impact crater, Namibia. Meteoritics and Planetary Science, Vol. 33, pp. 447–454.

Brandt, D., Holmes, H., Reimold, W. U., Paya, B. K., Koeberl, C., and Hancox, P. J. (2002). Kgagodi Basin: the first impact structure recognized in Botswana. Meteoritics and Planetary Sciences, Vol. 37, pp. 1765–1779.

Brenan, R. L., Peterson, B. L., and Smith, H. J. (1975). The origin of Red Wing Creek structure: McKenzie county, North Dakota. Wyoming Geological Association Earth Science Bulletin, Vol. 8, 41 pp.

Brett, R., Guppy, D. J., and Milton, D. J. (1970). Two circular structures of impact origin in Northern Territory, Australia (abstract). Meteoritics, Vol. 5, p. 184.

Bron, K. A., and Gostin, V. (2012). The Tookoonooka marine impact horizon, Australia: Sedimentary and petrologic evidence. Meteoritics and Planetary Science, Vol. 47, pp. 296–318.

Bronikowska, M., Artemieva, N. A., and Wünnemann, K. (2017). Reconstruction of the Morasko meteoroid impact – Insight from numerical modeling. Meteoritics and Planetary Science, Vol. 52, pp. 1704–1721.

Brown, P., ReVelle, D. O., Silber, E. A., Edwards, W. N., Arrowsmith, S., Jackson, L. E., Tancredi, G., and Eaton, D. (2008). Analysis of a crater-forming meteorite impact in Peru. Journal of Geophysical Research, Vol. 113, 13 pp.

Bruun, Å., and Dahlman, B. (1982). Den palaeozoiska berggrunden (in Swedish). In: Beskrivning till berggrundskartan Hjo NO, Wikman, H., Bruun, Å., Dahlman, B., and Vidal, G. (eds.), Sveriges Geologiska Undersökning, Vol. Af120, pp. 76–109.

Buchanan, P. C., Koeberl, C., and Reid, A. M. (1998). Impact into unconsolidated, water-rich sediments at the Marquez Dome, Texas. Meteoritics and Planetary Science, Vol. 33, pp. 1053–1064.

Bucher, W. H. (1933). Cryptovolcanic structures in the United States. 16th International Geological Congress, London. Report of the United States, Vol. 2, pp. 1055–1084.

Bucher, W. H. (1936). Cryptoexplosion structures caused from without or from within the Earth? American Journa of Science, Vol. 261, pp. 597–649.

Buchner, E., and Kenkmann, T. (2008). Upheaval Dome, Utah: Impact origin confirmed. Geology, Vol. 36, pp. 227–230.

Buchner, E., and Schmieder, M. (2010). Steinheim suevite – A first report of melt-bearing impactites from the Steinheim Basin (SW Germany). Meteoritics and Planetary Science Vol. 45, pp. 1093–1107.

Buchwald V. F. (1975). Handbook of Iron Meteorites: Their History, Distribution, Composition, and Structure. University of California Press, Berkeley, California, 1426 pp.

Bunch, T. E., and Cassidy, W. A. (1972). Petrographic and electron microprobe study of the Monturaqui impactite. Contributions to Mineralogy and Petrology, Vol. 36, pp. 95–112.

Bunting, J. A., de Laeter, J. R., and Libby, W. G. (1980). Evidence for the age and cryptoexplosive origin of the Teague Ring structure, Western Australia. Western Australia Geological Survey Annual Report 1979, pp. 125–129.

Buschbach, T. C., and Ryan, R. (1963). Ordovician explosion structure at Glasford, Illinois. Bulletin of the American Association of Petroleum Geologists, Vol. 47, pp. 2015–2022.

Butler, H. (1974). The Lake Teague ring structure, Western Australia: an astrobleme? Search, Vol. 5, pp. 536–537.

Campbell-Smith, W., and Hey, M. H. (1952). The silica glass from the crater of Aouelloul (Adrar, western Sahara). Bulletin de l'Institut fondamental d'Afrique noire, Vol. 14, pp. 762–776.

Carlton, R. W., Koeberl, C., Baranoski, M. T., and Schumacher, G. A. (1998). Discovery of microscopic evidence for shock metamorphism at the Serpent Mound structure, south-central Ohio: confirmation of an origin by impact. Earth and Planetary Science Letters, Vol. 162, pp. 177–185.

Carpenter, B. N., and Carlson, R. (1992). The Ames impact crater. Oklahoma Geological Survey, Vol. 52, pp. 208–223.

Carrigy, M. A., and Short, M. N. (1968). Evidence of shock metamorphism in rocks from the Steen River structure, Alberta. In: Shock Metamorphism of Natural Materials, French, B. M., and Short, N. M. (eds.), Mono Book Corp., Baltimore, pp. 367–378.

Cassidy, W. A., and Renard, M. L. (1996). Discovering research value in the Campo del Cielo, Argentina, meteorite craters. Meteoritics and Planetary Science, Vol. 46, pp. 935–949.

Catchings, R. D., Powars, D. S., Gohn, G. S., Horton, J. W., Goldman, M. R., and Hole, J. A. (2008). Anatomy of the Chesapeake Bay impact structure revealed by seismic imaging, Delmarva Peninsula, Virginia, USA. Journal of Geophysical Research, Vol. 113, 23 pp.

Caty, J. L., Chown, E. H., and Roy, D. W. (1976). A new astrobleme: Île Rouleau structure, Lake Mistassini, Quebec. Canadian Journal of Earth Sciences, Vol. 13, pp. 824–831.

Cavosie, A. J., Erickson, T. M., and Timms, N. E. (2015). Nanoscale records of ancient shock deformation: Reidite ($ZrSiO_4$) in sandstone at the Ordovician Rock Elm impact crater. Geology, Vol. 43, pp. 315–318.

Cavosie, A. J., Montalvo, P. E., Timms, N. E., and Reddy, S. M. (2016). Nanoscale deformation twinning in xenotime, a new shocked mineral, from the Santa Fe impact structure (New Mexico, USA). Geology, Vol. 44, pp. 803–806.

Chabou, M. C. (2019). Meteorite impact structures in the Arab World: An overview. In: The Geology of the Arab World – An Overview, Bendaoud, A., Hamimi, Z., Hamoudi, M., Djemai, S., Zoheir, B. (eds.), Springer Geology, pp. 455–506.

Chao, E. C. T., Fahey, J. J., and Littler, J. (1961). Coesite from Wabar crater, near Al Hadida, Arabia. Science, Vol. 133, pp. 882–883.

Chen, M., Xiao, W., and Xie, X. (2010). Coesite and quartz characteristic of crystallization from shock-produced silica melt in the Xiuyan crater. Earth and Planetary Science Letters, Vol. 297, pp. 306–314.

Chen, M., Koeberl, C., Xiao, W., Xie, X., and Tan, D. (2011). Planar deformation features in quartz from impact-produced polymict breccia of the Xiuyan Crater, China. Meteoritics and Planetary Science, Vol. 46, pp. 729–736.

Chen, M., Yin, F., Li, X., Xie, X., Xiao, W., and Tan, D. (2013). Natural occurrence of reidite in the Xiuyan crater of China. Meteoritics and Planetary Science, Vol. 48, pp. 796–805.

Chen, M., Xie, X., and Xiao, W., Tan, D. (2019). Yilan crater, a newly identified impact structure in northeast China (in Chinese with English summary). Chinese Science Bulletin, online first, doi: /10.1360/TB-2019-0704.

Chennaoui-Aoudjehane, H., El Kerni, H., Reimold, W. U., Baratoux, D., Koeberl, C., Bouley, S., and Aoudjehane, M. (2016). The Agoudal (High Atlas Mountains, Morocco) shatter cone conundrum: A recent meteorite fall onto the remnant of an impact site. Meteoritics and Planetary Science, Vol. 51, pp. 1497–1518.

Choe, B.-H., Tornabene, L. L., Osinski, G. R., and Newman, J. D. (2019). Remote predictive mapping of the Tunnunik impact structure in the Canadian Arctic using multispectral and polarimetric SAR data fusion. Canadian Journal of Remote Sensing, Vol. 44, https://doi.org/10.1080/07038992.2018.1544846

Cione, A. L., Tonni, E. P., San Cristóbal, J., Hernández, P. J., Benítez, A., Bordignon, F., and Perí, J. A. (2002). Putative meteoritic craters in Río Cuarto (Central Argentina) interpreted as eolian structures. Earth Moon Planets, Vol. 91, pp. 9–24.

Claeys, P., Chan, J. Ch.-W., and Dujardin, R. (2008). The Luizi structure: remote sensing study of a possible impact crater in Central Africa. 39th Lunar and Planetary Science Conference, Lunar and Planetary Institute, Houston, Texas, Abstract #1720.

Clayton, P. A., and Spencer, L. J. (1934). Silica Glass from the Libyan Desert. Lecture presented on the 9 November 1933 at the Annual Meeting of the Mineralogical Society. Mineralogical Magazine, Vol. 23, pp. 501–508.

Clement, J. H., and Mayhew, T. E. (1979). Newporte discovery opens new pay. Oil and Gas Journal, Vol. 77, pp. 165–172.

Cohen, A. J., Bunch, T. E., and Reid, A. M. (1961). Coesite discoveries establish cryptovolcanics as fossil meteorite craters. Science, Vol. 134, pp. 1624–1625.

Collins, F. R., and Robinson, F. M. (1967). Exploration of Naval Petroleum Reserve No. 4 and adjacent areas, northern Alaska, 1944–53 – Part 5, Subsurface geology and engineering data. U.S. Geological Survey Open File Report, Geological Survey Professional Paper 305-K, pp. 140–145.

Collins, G. S., and Wünnemann, K. (2005). How big was the Chesapeake Bay impact? Insights from numerical modeling. Geology, Vol. 33, pp. 925–928.

Collins, G. S., Artemieva, N. A., Wünnemann, K., Bland, P. A., Reimold, W. U., and Koeberl, C. (2008). Evidence that Lake Cheko is not an impact crater. Terra Nova, Vol. 20, pp. 165–168.

Cook, P. J. (1968). The Gosses Bluff crypto-explosion structure. Journal of Geology, Vol. 76, pp. 123–139.

Cordua, W. S. (1985). The Rock Elm structure, Pierce County, Wisconsin, a possible cryptoexplosion structure. Geology, Vol. 13, pp. 372–374.

Corner, B., Reimold, W. U., Brandt, D., and Koeberl, C. (1997). Morokweng impact structure, Northwest Province, South Africa: geophysical imaging and shock petrographic studies. Earth and Planetary Science Letters, Vol. 146, pp. 351–364.

Cox, M. A., Cavosie, A. J., Ferrière, L., Timms, N. E., Bland, P. A., Miljkovic, K., Erickson, T. M., and Hess, B. (2019). Shocked quartz in polymict impact breccia from the Upper Cretaceous Yallalie impact structure in Western Australia. Meteoritics and Planetary Science, Vol. 54, pp. 621–637.

Cox, M. A., Cavosie, A. J., Miljković, K., Bland, P. A., Kenkmann, T., and Hoskins, Z. N. P. (2018). Characterization of shock deformation at the Spider impact structure, Western Australia. 81st Annual Meeting of the Meteoritical Society, Moscow, Russia, Abstract #6248.

Crawford, A. R. (1972). Possible impact structure in India. Nature, Vol. 237, p. 96.

Crook, K. A. W., and Cook, P. J. (1966). Gosses Bluff – diapir, crypto-volcanic structure or astrobleme? Geological Society of Australia Journal, Vol. 13, pp. 495–516.

Crósta, A. P. (1987). Impact Structures in Brazil. In: Research in Terrestrial Impact Structures, Pohl, J. (ed.), Springer Vieweg, Wiesbaden, pp. 30–38.

Crósta, A. P., Gaspar, J. C., and Candia, M. A. F. (1981). Feições de metamorfismo de impacto no Domo de Araguainha. Revista Brasileira de Geociências, Vol. 11, pp. 139–146.

Crósta, A. P., Kazzuo-Vieira, C., and Schrank, A. (2004). Vista Alegre: a newly discovered impact crater in Southern Brazil. Meteoritics and Planetary Science, Vol. 39 (Supplement): A-28.

Crósta, A. P., Koeberl, C., Furuie, R. A., and Kazzuo-Vieira, C. (2010a). Vista Alegre, southern Brazil: a new impact structure in the Paraná flood basalts. Meteoritics and Planetary Science, Vol. 45, pp. 181–194.

Crósta, A. P., Lourenço, F. S., and Priebe, G. H. (2010b). Cerro Jarau, Rio Grande do Sul: a possible new impact structure in Southern Brazil. In: Large Meteorite Impacts and Planetary Evolution IV, Gibson, R. L., and Reimold, W. U. (eds.), Geological Society of America Special Paper, Vol. 465, pp. 173–190.

Crósta, A. P., Jourdan, F., and Koeberl, C. (2012a). ^{40}Ar-^{39}Ar dating of the Vista Alegre crater, Brazil. 34th International Geological Congress, Brisbane, Australia, 1 p.

Crósta, A. P., Kazzuo-Vieira, C., Pitarello, L., Koeberl, C., and Kenkmann, T. (2012b). Geology and impact features of Vargeão Dome, southern Brazil. Meteoritics and Planetary Science, Vol. 47, pp. 51–71.

Crósta, A. P., Reimold, W. U., and Vasconcelos, M. A. R. (2019). Cerro Do Jarau and São Miguel do Tapuio: Two newly confirmed large impact structures in Brazil. 50th Lunar and Planetary Science Conference, Lunar and Planetary Institute, Houston, Texas, Abstract #3042.

Crósta, A. P., Reimold, W. U., Vasconcelos, M. A. R., Hauser, N., Oliveira, G. J. G., Mazivieiro M. V., and Góes, A. M. (2019a). Impact cratering: The South American record – Part 1. Geochemistry – Chemie der Erde, Vol. 79, pp. 1–61.

Crósta, A. P., Reimold, W. U., Vasconcelos, M. A. R., Hauser, N., Oliveira, G. J. G., Maziviero, M. V., and Góes, A. M. (2019b). Impact cratering: The South American record – Part 2. Geochemistry – Chemie der Erde, Vol. 79, pp. 191–220.

Currie, K. L. (1965) The geology of the New Quebec crater. Canadian Journal of Earth Sciences, Vol. 2, pp. 141–160.

Currie, K. L. (1969). Geological notes on the Carswell circular structure, Saskatchewan. Canadian Geological Survey of Canada Paper 67–32, 60 pp.

Currie, K. L. (1970). New Canadian cryptoexplosion crater at Lake St Martin, Manitoba. Nature, Vol. 226, pp. 839–841.

Currie, K. L. (1971). Geology of the resurgent cryptoexplosion crater at Mistastin Lake, Labrador. Geological Survey of Canada, Bulletin 207, 62 pp.

Currie, K. L., and Dence, M. R. (1963). Rock deformation in the rim of the New Quebec crater, Canada. Nature, Vol. 198, p. 153.

D'Orazio, M., Folco, L., Zeoli, A., and Cordier, C. (2011). Gebel Kamil: the iron meteorite that formed the Kamil Crater (Egypt). Meteoritics and Planetary Science, Vol. 46, pp. 1179–1196.

Darlington, V., Blenkinsop, T., Dirks, P., Salisbury, J., and Tomkins, A. (2016). The Lawn Hill annulus: An Ordovician meteorite impact into water-saturated dolomite. Meteoritics and Planetary Science, Vol. 51, pp. 2416–2440.

De Mille, G. (1960). Elbow structure of south-central Saskatchewan. Journal of the Alberta Society of Petroleum Geologists, Vol. 8, pp. 154–162.

Dence, M. R. (1968). Shock zoning at Canadian craters: Petrography and structural observations. In: Shock Metamorphism of Natural Materials, French, B. M., and Short, N. M. (eds.), Mono Book Corp., Baltimore, pp. 169–184.

Dence, M. R., and Popelar, J. (1972). Evidence for an impact origin for Lake Wanapitei, Ontario, In: New Developments in Sudbury Geology, Guy-Bray, J. V. (ed.), Geological Association of Canada Special Paper, No. 10, 117 pp.

Dence, M. R., Innes, M. J. S., and Robertson, P. B. (1968). Recent geological and geophysical studies of Canadian craters. In: Shock Metamorphism of Natural Materials, French, B. M., and Short, N. M. (eds.), Mono Book Corp., Baltimore, pp. 339–362.

Dence, M. R., Grieve, R. A. F., and Robertson, P. B. (1977). Terrestrial impact structures: Principal characteristics and energy considerations. In: Impact and Explosion Cratering, Roddy, D. J., Pepin, R. O., and Merrill, R. B. (eds.), Pergamon Press, New York, pp. 247–275.

Dentith, M. C., Bevan, A. W. R., Backhouse, J., Featherstone, W. E., and Koeberl, C. (1999). Yallalie: a buried structure of possible impact origin in the Perth Basin, Western Australia. Geological Magazine, Vol. 136, pp. 619–632.

Deptuck, M. A., and Campbell, D. C. (2012). Widespread erosion and mass failure from the ~51 Ma Montagnais marine bolide impact off southwestern Nova Scotia, Canada. Canadian Journal of Earth Sciences, Vol. 49, pp. 1576–1594.

Deutsch, A., and Koeberl, C. (2006). Establishing the link between the Chesapeake Bay impact structure and the North American tektite strewn field: The Sr-Nd isotopic evidence. Meteoritics and Planetary Science, Vol. 41, pp. 689–703.

Deutsch, A., Buhl, D., and Langenhorst, F. (1992). On the significance of crater ages: new ages for Dellen (Sweden) and Araguainha (Brazil). Tectonophysics, Vol. 216, pp. 205–218.

Deutsch, A., Masaitis, V. L., Langenhorst, F., and Grieve, R. A. F. (2000). Popigai, Siberia – well preserved giant impact structure, national treasury, and world's geological heritage. Episodes, Vol. 23, pp. 3–11.

Dewing, K., Pratt, B. R., Hadlari, T., Brent, T., Bédard, J., and Rainbird, R. H. (2013). Newly identified "Tunnunik" impact structure, Prince Albert Peninsula, northwestern Victoria Island, Arctic Canada. Meteoritics and Planetary Science, Vol. 48, pp. 211–223.

Dietz, R. (1960). Meteorite impact suggested by shatter cone in rock. Science, Vol. 131, pp. 1781–1784.

Dietz, R. S. (1947). Meteorite impact suggested by the orientation of shatter cones at the Kentland, Indiana, disturbance. Science, Vol. 105, pp. 42–43.

Dietz, R. S. (1959). Shatter cones in cryptoexplosion structures (Meteorite impact?). Journal of Geology, Vol. 67, No. 5, pp. 496–505.

Dietz, R. S. (1966). Shatter cones at the Middlesboro structure, Kentucky. Meteoritics, Vol. 3, pp. 27–29.

Dietz, R. S. (1967). Shatter cone orientation at Gosses Bluff astrobleme. Nature, Vol. 216, pp. 1082–1084.

Dietz, R. S., and Butler, L. W. (1964). Shatter-cone orientation at Sudbury, Canada. Nature, Vol. 204, pp. 280–281.

Dietz, R. S., and French, B. M. (1973). Two probable astroblemes in Brazil. Nature, Vol. 244, pp. 561–562.

Dietz, R. S., and McHone, J. F. (1976). El'gygytgyn: Probably the world's largest meteorite crater. Geology, Vol. 4, pp. 391–392.

Dietz, R. S., and Lambert, P. (1980). Shock metamorphism at Crooked Creek cryptoexplosion structure, MO (abstract). Meteoritics, Vol. 15, pp. 281–282.

Dietz, R. S., and McHone, J. F. (1991). Astroblemes recently confirmed with shatter cones. Meteoritics, Vol. 26, pp. 332.

Donadini, F., Plado, J., Werner, S. C., Salminen, J., Pesonen, L. J., and Lehtinen, M. (2006). New evidence for impact from the Suvasvesi South structure, central east Finland. In: Biological Processes Associated with Impact Events, Cockell, C., Koeberl, C., and Gilmour, I. (eds.), Springer, Berlin, Heidelberg, pp. 287–307.

Donofrio, R. R. (1997). Survey of hydrocarbon-producing impact structures in North America: Exploration results to date and potential for discovery in Precambrian basement rock. In: Ames Structure in Northwest Oklahoma and Similar Features. Origin and Petroleum Production, Johnson, K. S., and Campbell, J. A. (eds.), Oklahoma Geological Survey, Norman, Oklahoma, Circular 100, pp. 17–29.

Dons, J. A., and Naterstad, J. (1992). The Gardnos impact structure, Norway (Abstract). Meteoritics, Vol. 27, p. 215.

Dressler, B. (1990). Shock metamorphic features and their zoning and orientation in the Precambrian rocks of the Manicouagan Structure, Quebec. Tectonophysics, Vol. 171, pp. 229–245.

Dressler, B. O. (1982). Geology of the Wanapitei Lake area, District of Sudbury. Ontario Geological Survey Report, Vol. 213, 131 pp.

Dressler, B. O. (1984). The effects of the Sudbury event and the intrusion of the Sudbury Igneous Complex on the footwall rocks of the Sudbury structure. In: The Geology and Ore Deposits of the Sudbury Structure, Pye, E. G., Naldrett, A. J., and Giblin, P. E. (eds.), Ontario Geological Survey, Special Volume 1, pp. 97–136.

Dressler, B. O., Crabtree, D., and Schuraytz, B. C. (1997). Incipient melt formation and devitrification at the Wanapitei impact structure, Ontario, Canada. Meteoritics and Planetary Science, Vol. 32, pp. 249–258.

Dressler, B. O., Sharpton, V. L., and Schuraytz, B. C. (1998). Shock metamorphism and shock barometry at a complex impact structure: Slate Islands, Canada. Contributions to Mineralogy and Petrology, Vol. 130, pp. 275–287.

Dumont, P. (1990). Plaidoyer pour une revalorisation des photographies aériennes. Bulletin de la Société belge de Géologie, Vol. 99, pp. 57–65.

Dunster, J. N., Haines, P. W., and Munson, T. J. (2014). Meteorites and impact structures in the Northern Territory. Northern Territory Geological Survey, Record 2014-007, 48 pp.

Dypvik, H., Gudlaugsson, S. T., Tsikalas, F., Attrep, M. Jr., Ferrell, R. E. Jr., Krinsley, D. H., Mork, A., Faleide, J. I., and Nagy, J. (1996). Mjølnir structure: An impact crater in the Barents Sea. Geology, Vol. 24, pp. 779–782.

Dypvik, H., Tsikalas, F., and Smelror, M. (eds.) (2010). The Mjølnir Impact Event and its Consequences. Geology and Geophysics of a Late Jurassic/Early Cretaceous Marine Impact event. Springer Berlin Heidelberg, 318 pp.

Earth Impact Database: http://www.passc.net/EarthImpactDatabase/New website_05-2018/Index.html

Eggleton, R. E., and Shoemaker, E. M. (1961). Breccia at Sierra Madera, Texas. U.S. Geological Survey Professional Paper, Vol. 424-D, pp. D151–D153.

Eitel, M., Gilder, S. A., Spray, J. G., Thompson, L. M., and Pohl, J. (2016). A paleomagnetic study of the Manicouagan impact structure: Implications for crater formation and geodynamo effects. Journal of Geophysical Research – Solid Earth, Vol. 121, pp. 436–454.

Ekern, G. L., and Thwaites, F. (1930). The Glover Bluff structure, a disturbed area in the Paleozoic of Wisconsin. Wisconsin Academy of Sciences, Transactions, Vol. 25, pp. 89–97.

El Kerni, H., Chennaoui-Aoudjehane, H., Baratoux, D., Aoudjehane, M., Charrière, A. Ibouh, H., Rochette, P., Quesnel, Y., Uehara, M., Kenkmann, T., Wulf, G., Poelchau, M. H., Nguyen, V.B., Aboulahris, M., Makhoukhi, S., Aumaître, G., Bourlès, D., and Keddadouche, K. (2019) Geological and geophysical studies of the Agoudal impact structure (Central High Atlas, Morocco): New evidence for crater size and age. Meteoritics and Planetary Science, Vol. 54, pp. 2483–2509.

Elo, S., Kivekas, L., Kujala, H., Lahti, S. I., and Pihlaja, P. (1992). Recent studies of the Lake Saaksjärvi meteorite impact crater, southwestern Finland (extended abstract). Tectonophysics, Vol. 216, pp. 163–167.

Elo, S., Kuivasaari, T., Lehtinen, M., Sarapää, O., and Uutela, A. (1993). Iso-Naakkima, a circular structure filled with neoproterozoic sediments, Pieksämäki, Southeastern Finland. Bulletin of the Geological Society of Finland, Vol. 65, pp. 3–30.

Emrich, G. H., and Bergstrom, R. E. (1962). Des Plaines disturbance, Northeastern Illinois. Geological Society of America Bulletin, Vol. 73, pp. 959–968.

Englund, K. J., and Roen, J. B. (1963). Origin of the Middlesboro Basin, Kentucky. U.S. Geological Survey Professional Paper, Vol. 450-E, pp. E20–E22.

Erickson, T. M., Kirkland, C. L., Timms, N. E., Cavosie, A. J., and Davison, T. M. (2020). Precise radiometric age establishes Yarrabubba, Western Australia, as Earth's oldest recognised meteorite impact structure. Nature Communications 11, Article #300, pp. 1–8.

Ernst, B., and Paulson, T. (2011). New constraints on the timing of deformation at the Glover Bluff structure near Coloma, WI, from a calcite twinning analysis (abstract). Geological Society of America Annual Meeting, Minneapolis, Minnesota, Abstracts with Programs, Vol. 43, p. 188.

Evenick, J. C., Lee, P., and Deane, B. (2004). Flynn Creek impact structure: New insights from breccias, melt features, shatter cones, and remote sensing. 35th Lunar and Planetary Science Conference, League City, Texas, Abstract #1131.

Fabre, J., Kazi-Tani, N., and Megartsi, M. (1970). Le rond de l'Ouarkziz (Sahara nordoccidental), un astroblème. Comptes Rendus de l'Académie des Sciences, Paris, Vol. D270, pp. 1212–1215.

Fackelman, S. P., McElvain, T. H., Morrow, J. R., and Koeberl, C. (2007). Shatter cone exposures indicate a new bolide impact structure near Santa Fe, New Mexico. 38th Lunar and Planetary Science Conference, Lunar and Planetary Institute, Houston, Texas, Abstract #1207.

Fackelman, S. P., Morrow, J. R., Koeberl, C., and McElvain, T. H. (2008). Shatter cone and microscopic shock-alteration evidence for a post-Paleoproterozoic terrestrial impact structure near Santa Fe, New Mexico, USA. Earth and Planetary Science Letters, Vol. 270, pp. 290–299.

Fazio, A., Folco, L., D'Orazio, M., Frezzotti, M. L., and Cordier, C. (2014). Shock metamorphism and impact melting in small impact craters on Earth: Evidence from Kamil Crater, Egypt. Meteoritics and Planetary Science, Vol. 49, pp. 2175–2200.

Fel'dman, V. I., Granovsky, L. B., and Lomonosov, M. V. (1978). Meteoritic crater Shunak, the central Kazakhstan, U.S.S.R. (abstract). 9th Lunar and Planetary Science Conference, Lunar and Planetary Institute, Houston, Texas, pp. 312–313.

Fel'dman, V. I., Mironovy, Y. U., Melikhov, V. R., Ivanov, B. A., and Basilevskay, T. (1985). Astroblemes on trap – peculiarities of the inner structure, distinction from the impact structure on the other targets (in Russian). Meteoritika, Vol. 44, pp. 139–145.

Ferrière, L., Raiskila, S., Osinski, G. R., Pesonen, L. J., and Lehtinen, M. (2010). The Keurusselkä impact structure, Finland – Impact origin confirmed by characterization of planar deformation features in quartz grains. Meteoritics and Planetary Science, Vol. 45, pp. 434–446.

Ferrière, L., Lubala, F. R. T., Osinski, G. R., and Kaseti, P. K. (2011). The newly confirmed Luizi impact structure, Democratic Republic of Congo – insights into central uplift formation and post-impact erosion. Geology, Vol. 39, pp. 851–854.

Firsov, L., and Kieffer, S. W. (1973). Concerning the meteoritic origin of the Puchezh-Katunki Crater. Meteoritics, Vol. 8, pp. 223–244.

Fiske, P. S., Hargraves, R. B., Onstott, T. C., Koeberl, C., and Hougen, S. B. (1994). Pseudotachylites of the Beaverhead impact structure: Geochemical, geochronological, petrographic, and field investigations. Geological Society of America Special Paper, Vol. 293, pp. 163–176.

Flodén, T., Tunander, P., and Wickman, F. E. (1986). The Tvären Bay structure, an astrobleme in southeastern Sweden. Geologiska Föreningen i Stockholm Förhandlingar, Vol. 108, pp. 225–234.

Floran, R. J., Grieve, R. A. F., Dence, M. R., Phinney, W. C., Warner, J. L., Blanchard, D. P., and Simonds, C. H. (1978). Manicouagan impact melt, Quebec. I. Stratigraphy, petrology, and chemistry. Journal of Geophysical Research, Vol. 83, pp. 2737–2759.

Florenskiy, P. V., and Dabizha, A. I. (1980). Meteoritniy krate'r Zhamanshin (in Russian), Moscow, Nauka Press, pp. 1–127. Translation in NASA Technical Memorandum NASA TM-88531, 1987.

Florenskiy, P. V., Short, N., Winzer, S. R., and Fredriksson, K. (1977). The Zhamanshin structure: Geology and petrography (abstract). Meteoritics, Vol. 12, pp. 227–228.

Florensky, K. P. (1963). Previous results of the Tunguska meteoritic complex expedition of 1961. Meteoritika, Vol. 23, pp. 3–29. (in Russian).

Folco, L., Di Martino, M., El Barkooky, A., D'Orazio, M., Lethy, A., Urbini, S., Nicolosi, I., Hafez, M., Cordier, C., van Ginneken, M., Zeoli, A., Radwan, A.M., El Khrepy, S., El Gabry, M., Gomaa, M., Barakat, A. A., Serra, R., and El Sharkawi, M. (2010). The Kamil Crater in Egypt. Science, Vol. 329, p. 804.

Folco, L., Di Martino, M., El Barkooky, A., D'Orazio, M., Lethy, A., Urbini, S., Nicolosi, I., Hafez, M., Cordier, C., van Ginneken, M., Zeoli, A., Radwan, A.M., El Khrepy, S., El Gabry, M., Gomaa, M., Barakat, A. A., Serra, R., and El Sharkawi, M. (2011). Kamil Crater (Egypt): ground truth for small scale meteorite impact on Earth. Geology, Vol. 39, pp. 179–182.

Folco, L., Mugnaioli, E., Gemelli, M., Masotta, M., and Campanale, F. (2018). Direct quartz-coesite transformation in shocked porous sandstone from Kamil Crater (Egypt). Geology, Vol. 46, pp. 739–742.

Folco, L., Reimold, W. U., and El Barkooky, A. (2019). Impact craters and meteorites: the Egyptian record. In: The Geology of Egypt, Hamimi, Z., El Barkooky, A., Martínez-Frías, J., Fritz, H., and Abdelrahman, Y. M. (eds.), Springer Publishers, Regional Geology Reviews, pp. 415–444.

Ford, J. R. H., Orchiston, W., and Clendening, R. (2012). The Wells Creek meteorite impact site and changing views on impact cratering. Journal of Astronomical History and Heritage, Vol. 15, pp. 159–178.
Forsman, N. F., Gerlach, T. R., and Anderson, N. L. (1996). Impact origin of the Newporte structure, Williston Basin, North Dakota. American Association of Petroleum Geologists Bulletin, Vol. 80, pp. 721–730.
Fredriksson, K., Dube, A., Milton, D., and Balasundaram, M. (1973). Lonar Lake, India: an impact crater in basalt. Science, Vol. 180, pp. 862–864.
French, B. M. (1998). Traces of Catastrophe: A Handbook of Shock-Metamorphic Effects in Terrestrial Meteorite Impact Structures. LPI Contribution No. 954, Lunar and Planetary Institute, Houston, 120 pp. Available at http://www.lpi.usra.edu/publications/books/CB-954/CB-954.pdf
French, B. M., Hartung, J. B., Short, N. M., and Dietz, R. S. (1970). Tenoumer crater, Mauritania: Age and petrologic evidence for origin by meteorite impact. Journal of Geophysical Research, Vol. 75, pp. 4396–4406.
French, B. M., Underwood, J R., and Fisk, E. P. (1974). Shock metamorphic features in two meteorite impact structures, south-eastern Libya. Geological Society of America Bulletin, Vol. 85, pp. 1425–1428.
French, B. M., Koeberl, C., Gilmour, I., Shirley, S. B., Dons J. A., and Naterstad, J. (1997). The Gardnos impact structure, Norway: Petrology and geochemistry of target rocks and impactites. Geochimica et Cosmochimica Acta, Vol. 61, pp. 873–904.
French, B. M., Cordua, W. S., and Plescia, J. B. (2004). The Rock Elm meteorite impact structure, Wisconsin; geology and shock-metamorphic effects in quartz. Geological Society of America Bulletin, Vol. 116, pp. 200–218.
French, B. M., McKay, R. M., Liu, H. P., Briggs, D. E. G., and Witzke, B. J. (2018). The Decorah structure, northeastern Iowa: Geology and evidence for formation by meteorite impact. Geological Society of America Bulletin, Vol. 130, pp. 2062–2086.
Frisk, A. M., and Ormö, J. (2007). Facies distribution of post-impact sediments in the Ordovician Lockne and Tvären impact craters; indications for unique impact-generated environments. Meteoritics and Planetary Science, Vol. 42, pp. 1971–1984.
Fudali, R., Milton, D., Fredriksson, K., and Dube, A. (1980). Morphology of Lonar crater, India: comparisons and implications. Moon and Planets, Vol. 23, pp. 493–515.
Fudali, R. F., and Cassidy, W. A. (1972). Gravity reconnaissance at three Mauretanian craters of explosive origin. Meteoritics, Vol. 7, pp. 51–70.
Ganapathy, R. (1983). The Tunguska explosion of 1908: discovery of meteoritic debris near the explosion site and at the South Pole. Science, Vol. 220, pp. 1158–1161.
Gasperini, L., Bonatti, E., Albertazzi, S., Forlani, L., Accorsi, C. A., Longo, G., Ravaioli, M., Alvisi, F., Polonia, A., and Sacchetti, F. (2009). Sediments from Lake Cheko (Siberia), a possible impact crater for the 1908 Tunguska event. Terra Nova, Vol. 21, pp. 489–494.
Genest, S., Robert, F., and Duhamel, I. (2010). The Carswell impact event, Saskatchewan, Canada: Evidence for a pre-Athabasca multiring basin? In: Large Meteorite Impacts and Planetary Evolution IV, Gibson, R. L., and Reimold, W. U. (eds.), Geological Society of America Special Paper, Vol. 465, pp. 543–570.
Gent, M. R., Kreis, L. K., and Gendzwill, D. (1992). The Maple Creek structure, southwestern Saskatchewan. Summary of Investigations 1992, Saskatchewan Geological Survey, Rep. 92-4, pp. 204–208.
Giacomini, B. B., Leite, E. P., and Crósta, A. P. (2017). 3D gravimetric investigation of the Cerro do Jarau structure, Rio Grande do Sul, Brazil. Meteoritics and Planetary Science, Vol. 52, pp. 565–583.
Gibson, J. W. Jr., and Sharpton, V. L. (1989). Marquez Dome: A newly recognized complex impact structure in east Texas (abstract). AGU Spring Meeting, Baltimore, Maryland, EOS, Transactions of the American Geophysical Union, Vol. 70, p. 383
Gibson, R. L., and Reimold, W. U. (2005). Shock pressure distribution in the Vredefort impact structure, South Africa. In: Large Meteorite Impacts III, Kenkmann, T., Hörz, F., and Deutsch, A. (eds.), Geological Society of America, Special Paper, Vol. 384, pp. 329–349.
Gibson, R. L., and Reimold, W. U. (2008). The Geology of the Vredefort impact structure. Memoir 97, Council for Geoscience, Pretoria, 181 pp.
Gibson, R. L., Reimold, W. U., Baegi, M., Crosta, A. P., Shbeli, E., Eshwedi, A., and Plescia, J. (2011). Remote sensing analysis and field mapping of the Oasis structure, Libya (abstract). 2nd Arab Impact Cratering and Astrogeology Conference (AICAC II), Casablanca, Morocco, pp. 47–48.
Gilmour, I., French, B. M., Franchi, I. A., Abbott, J. I., Hough, R. M., Newton, J., and Koeberl, C. (2003). Geochemistry of carbonaceous impactites from the Gardnos impact structure, Norway. Geochimica et Cosmochimica Acta, Vol. 67, pp. 3889–3903.
Glass, B. P. (1979). Zhamanshin crater, a possible source of Australasian tektites? Geology, Vol. 7, pp. 351–353.
Glass, B. P. (2002). Upper Eocene impact ejecta/spherule layers in marine sediments. Chemie der Erde – Geochemistry, Vol. 62, pp. 173–196.
Glikson, A. Y., Hickman, A. H., and Vickers, J., (2008). Hickman Crater, Ophthalmia Range, Western Australia: evidence supporting a meteorite impact origin. Australian Journal of Earth Sciences, Vol. 55, pp. 1107–1117.
Glikson, A., Hickman, A., and Crossley, R. (2016). Evidence for a shock-metamorphic breccia within a buried impact crater, Lake Raeside, Yilgarn Craton, Western Australia. Australian Journal of Earth Sciences, Vol. 63, pp. 99–109.
Gnos, E., Hofmann, B. A., Halawani, M. A., Tarabulsi, Y., Hakeem, M., Al Shanti, M., Greber, N. D., Holm, S., Alwmark, C., Greenwood, R. C., and Ramseyer, K. (2013). The Wabar impact craters, Saudi Arabia, revisited. Meteoritics and Planetary Science, Vol. 48, pp. 2000–2014.
Gohn, G. S., Koeberl, C., Miller, K. G., Reimold, W. U., Browning, J. V., Cockell, C. S., Horton, J. W., Kenkmann, T., Kulpecz, A. A., Powars, D. S., Sanford, W. E., and Voytek, M. A. (2008). Deep drilling into the Chesapeake Bay impact structure. Science, Vol. 320, pp. 1740–1745.
Gold, D. P., Tanner, J. G., and Halliday, D. W. (1978). The Lac La Moinerie crater: A probable impact site in New Quebec. Geological Society of America, Vol. 10, pp. 44.
Goldin, T. J., Wünnemann, K., Melosh, H. J., and Collins, G. S. (2006). Hydrocode modeling of the Sierra Madera impact structure. Meteoritics and Planetary Science, Vol. 41, pp. 1947–1958
Gorter, J. D., Gostin, V. A., and Plummer, P. S. (1989). The enigmatic sub-surface Tookoonooka Complex in south-west Queensland: its impact origin and implications for hydrocarbon accumulations. In: The Cooper and Eromanga Basins, Australia, O'Neil, B. J. (ed.), Petroleum Exploration Society of Australia, Adelaide, pp. 441–456.
Gostin, V. A., and Therriault, A. M. (1997). Tookoonooka, a large buried Early Cretaceous impact structure in the Eromanga Basin of southwestern Queensland, Australia. Meteoritics and Planetary Science, Vol. 32, pp. 593–599.
Gottwald, M., Fritz, T. Helko, B., Schättler, B., and Harris, A. (2017). Remote Sensing of terrestrial impact craters: The TanDEM-X digital elevation model. Meteoritics and Planetary Science, Vol. 52, pp. 1412–1427.
Grant, J. A., and Schultz, P. H. (1993). Erosion of ejecta at Meteor Crater, Arizona. Journal of Geophysical Research, Vol. 98, pp. 15033–15047.
Grieve, R. A. F. (1978). The melt rocks at Brent crater, Ontario, Canada. Proceedings of the 9th Lunar and Planetary Science Conference, Houston, Texas, Pergamon Press, New York, pp. 2579–2608.
Grieve, R. A. F. (1988). The Haughton impact structure: Summary and synthesis of the results of the HISS project. Meteoritics, Vol. 23, pp. 249–254.
Grieve, R. A. F. (1989). Manson structure implicated. Nature, Vol. 340, pp. 428–429.
Grieve, R. A. F. (2006). Impact Structures in Canada. Geological Association of Canada, GeoText No. 5, 210 pp.
Grieve, R. A. F., and Robertson, P. B. (1976). Variations in shock deformation at the Slate Islands impact structure, Lake Superior, Canada. Contributions to Mineralogy and Petrology, Vol. 58, pp. 37–49.
Grieve, R. A. F., and Floran, R. J. (1978). Manicouagan impact melt, Quebec. 2. Chemical interactions with basement and formational processes. Journal of Geophysical Research, Vol. 83, pp. 2761–2771.
Grieve, R. A. F., and Stöffler, D. (1991). The Sudbury Structure: Controversial or Misunderstood? Journal of Geophysical Research, Vol. 96, pp. 22753–22764.
Grieve, R. A. F., and Ber, T. (1994). Shocked lithologies at the Wanapitei impact structure, Ontario, Canada. Meteoritics, Vol. 29, pp. 621–631.
Grieve, R. A. F., Reny, G., Gurov, E. P., and Ryabenko, A. (1987). The melt rocks of the Boltysh impact crater, Ukraine, U.S.S.R. Contributions to Mineralogy and Petrology, Vol. 96, pp. 56–62.
Grieve, R. A. F., Wood, C. A., Garvin, J. B., Mclaughlin, G., and McHone, J. F. (1988). Astronaut's Guide to Terrestrial Impact Craters. Lunar and Planetary Institute Technical Report, 88-03, Lunar and Planetary Institute, Houston, 89 pp.
Grieve, R. A. F., Garvin, J. B., Coderre, J. M., and Rupert, J. (1989). Test of a geometric model for the modification stage of simple impact crater development. Meteoritics, Vol. 24, pp. 83–88.
Grieve, R. A. F., Kreis, K., Therriault, A. M., and Robertson, P. B. (1998). Impact structures in the Williston Basin. Meteoritics and Planetary Science, Vol. 33, pp. A63–A64.
Grieve, R. A. F., Reimold, W. U., Morgan, J., Riller, U., and Pilkington, M. (2008). Observations and interpretations at Vredefort, Sudbury and Chicxulub: Towards an empirical model of terrestrial impact basin formation. Meteoritics and Planetary Science, Vol. 43, pp. 855–882.
Grosse, E. (1919). Grundlinien der Geologie und Petrographie des östlichen Katanga. Neues Jahrbuch für Mineralogie, Geologie und Palaontologie, Beilage-Band, Vol. 42, pp. 272–419.
Gruber, A., Wessel, B., Martone, M., and Roth, A. (2016). The TanDEM-X DEM mosaicking: Fusion of multiple acquisitions using InSAR quality parameters. IEEE Journal of Selected Topics in Applied Earth Observations and Remote Sensing, Vol. 9, pp. 1047–1057.

Gudlaugsson, S. T. (1993). Large impact crater in the Barents Sea. Geology, Vol. 21, pp. 291–294.

Guillemot, J. (1962). Structures anticlinales circulaires d'origine mal connue. Photo interprétation. Technip Editions, fascicule 4.

Guppy, D. J., Brett, R., and Milton, D. J. (1971). Liverpoool and Strangways craters, Northern Territory: two structures of probable impact origin. Journal of Geophysical Research, Vol. 76, pp. 5387–5393.

Gupta, R. D., Banerjee, A., Goderis, S., Claeys, P., Vanhaecke, F., and Chakrabarti, R. (2017). Evidence for a chondritic impactor, evaporation-condensation effects and melting of the Precambrian basement beneath the 'target' Deccan basalts at Lonar crater, India. Geochimica et Cosmochimica Acta, Vol. 215, pp. 51–75.

Gurov, E., Gurova, E., Chernenko, Y., and Yamnichenko, A. (2009). The Obolon impact structure, Ukraine, and its ejecta deposits. Meteoritics and Planetary Science, Vol. 44, pp. 389–404.

Gurov, E., Nikolaenko, N., Shevchuk, H., and Yamnichenko, A. (2017). Kamenetsk – A new impact structure in the Ukrainian Shield. Meteoritics and Planetary Science, Vol. 52, pp. 2461–2469.

Gurov, E. P. (1988). The Kara-Kul Lake depression in the Pamirs – A Probable Astrobleme (abstract). 8th Soviet-American Microsymposium, Moscow, Russia, pp. 37–39.

Gurov, E. P., and Gurova, E. P. (1995). Impact-melt composition of the Obolon crater; chlorine as a possible indicator of the submarine crater formation. Meteoritics and Planetary Science, Vol. 30, p. 515.

Gurov, E. P., and Gurova, E. P. (1998). The group of Macha craters in western Yakutia. Planetary Space Science, Vol. 46, pp. 323–328.

Gurov, E. P., Gurova, H. P., Rakitskaya, R. B., and Yamnichenko, A. Y. (1993). The Karakul depression in the Pamirs – the first impact structure in central Asia (abstract). 24th Lunar and Planetary Science Conference, Lunar and Planetary Institute, Houston, Texas, pp. 591–592.

Gurov, E. P., Koeberl, C., and Reimold, W. U. (1998). Petrography and geochemistry of target rocks and impactites from the Ilyinets Crater, Ukraine. Meteoritics and Planetary Science, Vol. 33, pp. 1317–1333.

Gurov, E. P., Gurova E. P., and Sokur, T. M. (1999) Impact diamonds in the rocks of the Zapadnaya astrobleme (in Russian). Mineralni resursi Ukraini, Vol. 3, pp. 30–32.

Gurov, E. P., Gurova, E. P., and Sokur, T. M. (2002). Geology and petrography of the Zapadnaya impact crater in the Ukrainian Shield. In: Impacts in Precambrian Shields, Plado, J., and Pesonen, L. J. (eds.), Springer, Berlin Heidelberg, pp. 173–188.

Gurov, E. P., Koeberl, C., Reimold, W. U., Brandstätter, F., and Amare, K. (2005). Shock metamorphism of siliceous volcanic rocks of the El'gygytgyn impact crater (Chukotka, Russia). In: Large Meteorite Impacts III. Kenkmann, T., Hörz, F., and Deutsch, A. (eds.), Geological Society of America Special Paper, Vol. 384, pp. 391–412.

Gurov, E. P., Kelley, S. P., Koeberl, C., and Dykan, N. I. (2006). Sediments and impact rocks filling the Boltysh impact crater. In: Biological Processes Associated with Impact Events, Cockell, C. S., Koeberl, C., and Gilmour, I. (eds.), Springer, Berlin Heidelberg, pp. 335–358.

Gurov, E. P., Koeberl, C., and Yamnichenko, A. (2007). El'gygytgyn impact crater, Russia: Structure, tectonics and morphology. Meteoritics and Planetary Science, Vol. 42, pp. 307–319.

Gutschick, R. C. (1982). Geology of the Kentland structural anomaly, northwestern Indiana – update. Purdue University, Department of Geosciences, and Geological Society of America, North-Sentral Section, Guidebook 4, 38 pp.

Haines, P. W. (1996). Goyder impact structure, Arnhem Land, Northern Territory. AGSO Journal of Australian Geology and Geophysics, Vol. 16, pp. 561–566.

Haines, P. W. (2005) Impact cratering and distal ejecta: the Australian record. Australian Journal of Earth Sciences, Vol. 52, pp. 481–507.

Haines, P. W. (2014). Collaborative scientific drilling at Hickman Crater. In: GSWA 2014 extended abstracts: Promoting the prospectivity of Western Australia. Geological Survey of Western Australia, Record 2014/2, pp. 1–4.

Haines, P. W. (2017). Drilling to basement at Hickman Crater, Western Australia. In: International Workshop on Shock Metamorphism in Terrestrial and Extra-terrestrial Rocks. June 26–29, 2017, Curtin University, Perth.

Haines, P. W., and Rawlings, D. J. (2002). The Foelsche structure, Northern Territory, Australia: an impact crater of probable Neoproterozoic age. Meteoritics and Planetary Science, Vol. 37, pp. 269–280.

Haines, P. W., Therriault, A. M., and Kelly S. P. (1999). Evidence for mid-Cenozoic(?) low-angle multiple impacts in South Australia. Meteoritics and Planetary Sciences, Vol. 34 (Suppl.), pp. A49–A50.

Haines, P. W., Sweet, I. P., and Mitchell, K. (2012). Cleanskin structure, Northern Territory and Queensland, Australia: evidence for an impact origin. Meteoritics and Planetary Science, Abstract. #5176.

Halliday, I., and Griffin, A. A. (1963). Evidence in support of a meteoritic origin for West Hawk Lake, Manitoba, Canada. Journal of Geophysical Research, Vol. 68, pp. 5297–5305.

Halliday, I., and Griffin, A. A. (1967). Summary of drilling at the West Hawk Lake crater. Journal of the Royal Astronomical Society of Canada, Vol. 61, pp. 1–8.

Halls, H. C. (1975). Shock-induced remanent magnetization in Late Precambrian rocks from Lake Superior. Nature, Vol. 255, pp. 692–695.

Halls, H. C., and Grieve, R. A. F. (1976). The Slate Islands: a probable complex meteorite impact structure in Lake Superior. Canadian Journal of Earth Sciences, Vol. 13, pp. 1301–1309.

Hamacher, D. W., and O'Neill, C. (2013). The discovery and history of the Dalgaranga meteorite crater, Western Australia. Australian Journal of Earth Sciences, Vol. 60, pp. 637–646.

Hamann, C., Hecht, L., Ebert, M., and Wirth, R. (2013). Chemical projectile-target interaction and liquid immiscibility in impact glass from the Wabar craters, Saudi Arabia. Geochimica et Cosmochimica Acta, Vol. 121, pp. 291–310.

Hargraves, R. B., Christiansen, P. P., Cullicott, C. E., Deffeyes, K. S., Fiske, P. S., and Hougen, S. (1990). Shatter cones and shocked rocks in southwestern Montana: The Beaverhead impact structure. Geology, Vol. 18, pp. 832–834.

Hargraves, R. B., Kellogg, K. S., Fiske, P. S., and Hougen, S. B. (1994). Allochthonous impact-shocked rocks and superimposed deformations at the Beaverhead site in southwest Montana. Geological Society of America Special Paper, Vol. 293, pp. 225–235.

Harms, J. E., Milton, D. J., Ferguson, J., Gilbert, D. J., Harris, W. K., and Goleby, B. (1980). Goat Paddock cryptoexplosion crater, Western Australia. Nature, Vol. 286, pp. 704–706.

Harper, C. T. (1982). Geology of the Carswell Structure, Central Part. Saskatchewan Geological Survey, Report 214, 6 pp.

Hart, R. J., Andreoli, M. A. G., Tredoux, M., Moser, D., Ashwal, L. D., Eide, E. A., Webb, S. J., and Brandt, D. (1997). Late Jurassic age for the Morokweng impact structure, southern Africa. Earth and Planetary Science Letters, Vol. 147, pp. 25–35.

Hartung, J. B., and Anderson, R. R. (1988). A summary of data on the Manson impact structure (abstract). 19th Lunar and Planetary Science Conference, Lunar and Planetary Institute, Houston, Texas, pp. 455–456.

Hartung, J. B., and Anderson, R. R. (1996). A brief history on investigations of the Manson impact structure. Geological Society of America Special Paper, Vol. 302, pp. 31–43.

Hasch, M. Reimold, W. U., Raschke, U., and Zaag, P. T. (2016). Shatter cones at the Keurusselkä impact structure and their relation to local jointing. Meteoritics and Planetary Science, Vol. 51, pp. 1534–1552.

Hauser, N., Reimold, W. U., Cavosie, A. J., Crósta, A. P., Schwarz, W. H., Trieloff, M., Da Silva Maia de Souza, C., Pereira, L. C., Rodrigues, E. N., and Brown, M. (2019). Linking shock textures revealed by BSE, CL and EBSD with U-Pb data (LA-ICP-MS and SIMS) in zircon from the Araguainha impact structure, Brazil. Meteoritics and Planetary Science, 54, pp. 2386–2311.

Hawke, P. J. (2003). Geophysical investigation of the Wolfe Creek meteorite crater. Geological Survey of Western Australia Record 2003/10, 18 pp.

Hawke, P. J., Buckingham, A. J., and Dentith, M. C. (2006). Modelling source depth and possible origin of magnetic anomalies associated with the Yallalie impact structure, Perth Basin, Western Australia. Exploration Geophysics, Vol. 37, pp. 191–196.

Hecht, L., Reimold, W. U., Sherlock, S., Tagle, R., Koeberl, C., and Schmitt, R. T. (2008). New impact melt rock from the Roter Kamm impact structure, Namibia: further constraints on impact age, melt rock chemistry, and projectile composition. Meteoritics and Planetary Science, Vol. 43, pp. 1201–1218.

Heinrichs, T., Salameh, E., and Khoury, H. (2014). The Waqf as Suwwan crater, Eastern Desert of Jordan: aspects of the deep structure of an oblique impact from reflection seismic and gravity data. International Journal of Earth Sciences, Vol. 103, pp. 233–252.

Heizmann, E. P. J., und Reiff, W. (2002). Der Steinheimer Meteorkrater (in German). Pfeil-Verlag, München, 160 pp.

Hendricks, H. E. (1954). The geology of the Steelville quadrangle, Missouri. Missouri Geological Survey and Water Resources [Report], 2nd Series, Vol. 36, 88 p.

Henkel, H. (1992). Geophysical aspects of meteorite impact craters in eroded shield environment, with special emphasis on electric resistivity. Tectonophysics, Vol. 216, pp. 63–89.

Henkel, H. (2018). Provisional translation of: Deep Drilling in the Puchezh-Katunki Impact Structure (in Russian). Masaitis, V. L., and Pevzner, L. A. (eds.), 1999, VSEGEI Press, St. Petersburg, 185 pp. Published by Henkel, H., ISBN: 978-91-519-0289-0.

Henkel, H., and Pesonen, L. J. (1992). Impact craters and craterform structures in Fennoscandia. Tectonophysics, Vol. 216, pp. 31–40.

Henkel, H., and Reimold, W. U. (1998). Integrated geophysical modeling of a giant, complex impact structure: anatomy of the Vredefort Structure, South Africa. Tectonophysics, Vol. 287, pp. 1–20.

Henkel, H., and Reimold, W. U. (2002). Magnetic model of the central uplift of the Vredefort impact structure, South Africa. Journal of Applied Geophysics, Vol. 51, pp. 43–62.

Henkel, H., Reimold, W. U., and Koeberl, C. (2002). Magnetic and gravity model of the Morokweng impact structure. Journal of Applied Geophysics, Vol. 49, pp. 129–147.

Henkel, H., Ekneligoda, T. C., and Aaro, S. (2010). The extent of impact induced fracturing from gravity modeling of the Granby and Tvären simple craters. Tectonophysics, Vol. 485, pp. 290–305.

Herd, C. D. K., Froese, D. G., Walton, E. L., Kofman, R. S., Herd, E. P. K., and Duke, M. J. M. (2008). Anatomy of a young impact event in central Alberta, Canada: Prospects for the missing Holocene impact record. Geology, Vol. 36, pp. 955–958.

Hergarten, S., and Kenkmann, T. (2015). The number of impact craters on Earth: Any room for further discoveries? Earth and Planetary Science Letters, Vol. 425, pp. 187–192.

Herrick, R. R., and Sharpton, V. L. (1997). Shallow seismic test at Marquez Impact Structure. 28th Lunar and Planetary Science Conference, Lunar and Planetary Institute, Houston, Texas, Part 2, pp. 557–558.

Hervé, G., Gilder, S. A., Marion, C. L., Osinski, G. R., Pohl, J., Petersen, N., and Sylvester, P. J. (2015). Paleomagnetic and rock magnetic study of the Mistastin Lake impact structure (Labrador, Canada): Implications for geomagnetic perturbation and shock effects. Earth and Planetary Science Letters, Vol. 417, pp. 151–163.

Hietala, S., and Moilanen, J. (2004). Keurusselkä – A new impact structure in central Finland. 35th Lunar and Planetary Science Conference, Lunar and Planetary Institute, Houston, Texas, Abstract #1619.

Higgins, M., and Tait, L. (1990). A possible new impact structure near Lac de la Presqu'île, Quebec, Canada. Meteoritics, Vol. 25, pp. 235–236.

Hische, R. (1994). Geology of the Clearwater Impact Structures (English translation). Ph.D. Thesis, Westfälische Wilhelms-Universität Münster (Germany), 200 pp.

Hodge, P. W. (1965). The Henbury meteorite craters. Smithsonian Contributions to Astrophysics, Vol. 8, pp. 199–201.

Hodge, P. W. (1979a). The location of the site of the Haviland meteorite crater. Meteoritics, Vol. 14, pp. 233–234.

Hodge, P. W. (1979b). Meteoritic material in the soil near two meteorite craters. Meteoritics, Vol. 14, pp. 422–423.

Hodge, P. W., and Wright, F. W. (1970) Meteoritic spherules in the soil surrounding terrestrial impact craters. Nature, Vol. 225, pp. 717–718.

Högbom, A. (1910). Minutes of the Geological Society Stockholm 3rd meeting. Geologiska Föreningen i Stockholm Förhandlingar, Vol. 32, pp. 482.

Holliday, V. T., Kring, D. A., Mayer, J. H., and Goble, R. J. (2005). Age and effects of the Odessa meteorite impact, western Texas, USA. Geology, Vol. 33, pp. 945–948.

Holm, S., Alwmark, C., Alvarez, W., and Schmitz, B. (2011). Shock barometry of the Siljan impact structure, Sweden. Meteoritics and Planetary Science, Vol. 46, pp. 1888–1909.

Holm-Alwmark, S., Alwmark, C., Ferrière, L., Lindström, S., Meier, M. M. M., Scherstén, A., Herrmann, M., Masaitis, V. L., Mashchak, M. S., Naumov, M. V., and Jourdan, F. (2019). An Early Jurassic age for the Puchezh-Katunki impact structure (Russia) based on ^{40}Ar-^{39}Ar data and palynology. Meteoritics and Planetary Science, Vol. 54, pp. 1764–1780.

Honda, M., Caffee, M. W., Miura, Y. N., Nagai, H., Nagao, K., and Nishiizumi, K. (2002). Cosmogenic nuclides in the Brenham pallasite. Meteoritics and Planetary Science, Vol. 37, pp. 1711–1728.

Horton, J. W., and Izett, G. E. (2005). Crystalline-rock ejecta and shocked minerals of the Chesapeake Bay impact structure, USGS-NASA Langley core, Hampton, Virginia, with supplemental constraints on the age of impact. In: Studies of the Chesapeake Bay Impact Structure – The USGS-NASA Langley Corehole, Hampton, Virginia and Related Coreholes and Geophysical Surveys, Horton, J. W., Powars, D. S., and Gohn, G. S. (eds.), United States Geological Survey Professional Paper, Vol. 1688, pp. E1–E30.

Horton, J. W., Powars, D. S., and Gohn, D. S. (2005). Studies of the Chesapeake Bay impact structure – Introduction and discussion. In: Studies of the Chesapeake Bay Impact Structure – The USGS-NASA Langley Corehole, Hampton, Virginia and Related Coreholes and Geophysical Surveys, Horton, J. W., Powars, D. S., and Gohn, G. S. (eds.), USGS Professional Paper, Vol. 1688, A1–A24.

Hörz, F., Ostertag, R., and Rainey, D. A. (1983). Bunte Breccia of the Ries: continuous deposits of large impact craters. Review of Geophysics and Space Physics, Vol. 21, pp. 1667–1725.

Hörz, F., See, T. H., Murali, A. V., and Blanchard D. P. (1989). Heterogeneous dissemination of projectile materials in the impact melts from Wabar crater, Saudi Arabia. Proceedings of the 19th Lunar and Planetary Science Conference, Houston, Texas, Cambridge University Press, Cambridge, pp. 697–709.

Hörz, F., Mittlefehldt, D. W., See, T. H., and Galindo, C. (2002). Petrographic studies of the impact melts from Meteor Crater, Arizona, USA. Meteoritics and Planetary Science, Vol. 37, pp. 501–531.

Howard, K. A., Offield, T. W., and Wilshire, H. G. (1972). Structure of Sierra Madera, Texas, as a guide to central peaks of lunar craters. Geological Society of America Bulletin, Vol. 83, pp. 2795–2808.

Hower, J. C., Greb, S. F., Kuehn, K. W., and Eble, C. F. (2009). Did the Middlesboro, Kentucky, bolide impact event influence coal rank? International Journal of Coal Geology, Vol. 79, pp. 92–96.

Hughes, V. H. (1911). Reconnaissance work: Missouri Geological Survey and Mines. Biennial Report for 1909–10, pp. 36–54.

Iasky, R., Mory, A. J., and Blundell, K. A. (2001). The geophysical interpretation of the Woodleigh impact structure, Southern Carnarvon Basin, Western Australia. Western Australia Geological Survey, Report 79, 41 pp.

Innes, M. J. S. (1957). A possible impact crater at Deep Bay, Saskatchewan, Canada. Journal of the Royal Astronomical Society of Canada, Vol. 51, pp. 235–240.

Innes, M. J. S. (1964). Recent advances in meteorite crater research at the Dominion Observatory, Ottawa, Canada. Meteoritics, Vol. 2, pp. 219–241.

Innes, M. J. S., and Dence, M. R. (1965). Nicholson Lake and Pilot Lake craters, N.W.T., Canada. Meteoritical Society Meeting, Odessa, Texas.

Ivanov, B. A. (2005). Numerical modeling of the largest terrestrial meteorite craters. Solar System Research, Vol. 39, pp. 381–409.

Ivanov, B. A., and Petaev, M. I. (1992). Mass and impact velocity of the meteorite formed the Sterlitamak crater in 1990 (abstract). 23rd Lunar and Planetary Science Conference, Lunar and Planetary Institute, Houston, Texas, pp. 573–574.

Ivanov, B. A., Kocharyan, G. G., Kostuchenko, V. N., Kirjakov, A. F., and Pevzner, L. A. (1996). Puchezh-Katunki impact crater: preliminary data on recovered core block structure (abstract). 27th Lunar and Planetary Science Conference, Lunar and Planetary Institute, Houston, Texas, pp. 589–590.

Izett, G. A., Cobban, W. A., Obradovich, J. D., and Kunk, M. J. (1993). The Manson impact structure: ^{40}Ar-^{39}Ar age and its distal impact ejecta in the Pierre shale in southeastern South Dakota. Science, Vol. 262, pp. 729–732.

Izett, G. A., Masaitis, V. L., Shoemaker, E. M., Dalrymple, G. B., and Steiner, M. B. (1994). Eocene age of the Kamensk buried crater of Russia. In: New Developments Regarding the KT Event and Other Catastrophes in Earth History, Lunar and Planetary Institute Contribution No. 825, pp. 55–56.

Jansa, L. F., and Pe-Piper, G. (1987). Identification of an underwater extraterrestrial impact crater. Nature, Vol. 327, pp. 612–614.

Jansa, L. F., Pe-Piper, G., Robertson, P. B., and Friedenreich, O. (1989). Montagnais: A submarine impact structure on the Scotian shelf, eastern Canada. Geological Society of America Bulletin, Vol. 101, pp. 450–463.

Jõeleht, A., Mustasaar, M., Rooni, K., Kalvāns, A., and Popovs, K. (2015). Reflection seismic study of the Dobele impact crater, Latvia. In: Bridging the Gap III: Impact Cratering in Nature, Experiments and Modeling, Freiburg, Germany, Abstract #1105.

Johnson, M. R., Anhaeusser, C. R., and Thomas, R. J. (eds.) (2006). The Geology of South Africa. Geological Society of South Africa, Johannesburg and Council for Geoscience, Pretoria, 691 pp.

Jourdan, F., and Reimold, W. U. (2012). Age of the Siljan impact structure. 75th Annual Meeting of the Meteoritical Society, Cairns, Australia, Abstract #5093.

Jourdan, F., Renne, P. R., and Reimold, W. U. (2008). High-precision ^{40}Ar-^{39}Ar age of the Jänisjärvi impact structure (Russia). Earth and Planetary Science Letters, Vol. 265, pp. 438–449.

Jourdan, F., Moynier, F., Koeberl, C., and Eroglu, S. (2011). ^{40}Ar-^{39}Ar age of the Lonar crater and consequence for the geochronology of planetary impacts. Geology, Vol. 39, pp. 671–674.

Kalleson, E., Dypvik, D., and Naterstad, J. (2007). Post-impact sediments in the Gardnos impact structure, Norway. In: The Sedimentary Record of Meteorite Impacts, Evans, K., Horton Jr., J. W., King Jr. D. T., and Morrow, J. R. (eds.), Geological Society of America Special Paper, Vol. 437, pp. 19–41.

Kalleson, E., Jahren, T., and Dypvik, D. (2008). Excursion Guide to Gardnos Meteorite Crater. Department of Geoscience, University of Oslo, 24 pp.

Kalleson, E., Dypvik, H., Riis, F., and Nilsen, O. (2013). The melt-bearing impactites of the Ritland structure, Norway – Implications for melt formation in small impact craters. Meteoritics and Planetary Science, Vol. 48, pp. 1678–1701.

Karaszewski, W. (1974). Geological studies of "meteorite" craters in Nördlinger Ries (West Germany) and at Morasko (Poland) (in Polish). Przeglad Geologiczny, Vol. 22, pp. 626–627.

Karpoff, R. (1953). The meteorite crater of Talemzane in southern Algeria. Meteoritics, Vol. 1, pp. 31–38.

Kass, M. A., Bedrosian, P. A., Drenth, B. J., Bloss, B. R., McKay, R. M., Liu, H., French, B. M., and Witzke, B. J. (2013). Modeling and inversion results from airborne geophysics over a buried impact structure in Decorah, Iowa, USA. Geological Society of America Abstracts with Programs, Vol. 45, p. 485.

Kastning, H., and Huntoon, P. W. (1996). Cluster of five small Pennsylvanian meteorite impact craters on Sheep Mountain near Douglas, Wyoming. In: NASA's Wyoming Space Grant Fellowship Program 1995–6 Space Science Research: Graduate and Undergraduate Fellowship reports, Ciardi, T. M., and Johnson, P. E. (eds.), Wyoming Space Grant Consortium, University of Wyoming, pp. 57–64.

Keiswetter, D., Black, R., and Steeples, D. (1996). Seismic reflection analysis of the Manson Impact Structure, Iowa. Journal of Geophysical Research, Vol. 101, pp. 5823–5834.

Kelley, S. P., and Spray, J. G. (1997). A late Triassic age for the Rochechouart impact structure, France. Meteoritics and Planetary Science, Vol. 32, pp. 629–636.

Kelley, S. P., and Gurov, E. P. (2002). Boltysh, another end-Cretaceous impact. Meteoritics and Planetary Science, Vol. 37, pp. 1031–1043.

Kellogg, K. S., Snee, L. W., and Unruh, D. M. (2003). The Mesoproterozoic Beaverhead impact structure and its tectonic setting, Montana-Idaho: ^{40}Ar-^{39}Ar and U-Pb isotopic constraints. Journal of Geology, Vol. 111, pp. 639–652.

Kenkmann, T. (2002). Folding within seconds. Geology, Vol. 3, pp. 231–234.

Kenkmann, T. (2003). Dike formation, cataclastic flow, and rock fluidization during impact cratering: an example from the Upheaval Dome structure, Utah. Earth and Planetary Science Letters, Vol. 214, pp. 43–58.

Kenkmann, T., and von Dalwigk, I. (2000). Radial transpression ridges: A new structural feature of complex impact craters. Meteoritics and Planetary Science, Vol. 35, pp. 1189–1201.

Kenkmann, T., and Ivanov, B. A. (2006). Target delamination by spallation and ejecta dragging: An example from the Ries crater's periphery. Earth and Planetary Science Letters, Vol. 252, pp. 15–29.

Kenkmann, T., and Schoenian, F. (2006): Ries and Chicxulub: craters on Earth provide insights for Martian impacts. Meteoritics and Planetary Science, Vol. 41, pp. 1587–1604.

Kenkmann, T., and Poelchau, M. H. (2007). Piccaninny, WA, and Matt Wilson, NT: Two possible complex impact craters in Australia. 17th Annual Goldschmidt Conference, Cologne, Germany. Geochimica et Cosmochimica Acta, Vol. 71 Supplement, A477.

Kenkmann, T., and Poelchau, M. H. (2009). Low-angle collision with Earth: The elliptical impact crater Matt Wilson, Northern Territory, Australia. Geology, Vol. 37, pp. 459–462.

Kenkmann, T., Ivanov, B. A., and Stöffler, D. (2000). Identification of ancient impact structures: low-angle faults and related geological features of crater basements. Lecture Notes in Earth Sciences, Vol. 91, pp. 271–309.

Kenkmann, T., Jahn, A., Scherler, D., and Ivanov, B. A. (2005). Structure and formation of a central uplift: a case study at the Upheaval Dome impact crater, Utah. In: Large Meteorite Impacts III, Kenkmann, T., Hörz, F., and Deutsch, A. (eds.), Geological Society of America Special Paper, Vol. 384, pp. 85–115.

Kenkmann, T., Kiebach, F., Rosenau, M., Raschke, U., Pigowske, A., Mittelhaus, K., and Eue, D. (2007). Coupled effects of impact and orogeny: Is the marine Lockne crater, Sweden, pristine? Meteoritics and Planetary Science, Vol. 42, pp. 1995–2012.

Kenkmann, T., Artemieva, N. A., Wünnemann, K., Poelchau, M. H., Elbeshausen, D., and Nuñéz del Prado, H. (2009a). The Carancas meteorite impact crater, Peru: geologic surveying and modeling of crater formation and atmospheric passage. Meteoritics and Planetary Science, Vol. 44, pp. 985–1000.

Kenkmann, T., Collins, G. S., Wittmann, A., Wünnemann, K., Reimold, W. U., and Melosh, H. J. (2009b). A model for the formation of the Chesapeake Bay impact crater as revealed by drilling and numerical simulation. In: The ICDP-USGS Deep Drilling Project in the Chesapeake Bay Impact Structure: Results from the Eyreville Core Holes, Gohn, G. S., Koeberl, C., Miller, K. G., and Reimold, W. U. (eds.), Geological Society of America Special Paper, Vol. 458, pp. 571–585.

Kenkmann, T., Reimold, W. U., Khirfan, M., Salameh, E., Khoury, H., and Konsul, K. (2010). The complex impact crater Jebel Waqf as Suwwan in Jordan: effects of target heterogeneity and impact obliquity on central uplift formation. In: Large Meteorite Impacts and Planetary Evolution IV. Gibson, R. L. and Reimold W. U. (eds.), Geological Society of America Special Paper, Vol. 465, pp. 471–487.

Kenkmann, T., Vasconcelos, M. A. R., Crósta, A. P., and Reimold, W. U. (2011). The complex impact structure Serra da Cangalha, Tocantins State, Brazil. Meteoritics and Planetary Science, Vol. 46, pp. 875–889.

Kenkmann, T., Poelchau, M. H., and Wulf, G. (2014). Structural geology of impact craters. Journal of Structural Geology, Vol. 62, pp. 156–182.

Kenkmann, T., Afifi, A. M., Stewart, S. A., Poelchau, M. H., Cook, D. J., and Neville, A. S. (2015). Saqqar: A new 34 km diameter impact crater in Saudi Arabia. Meteoritics and Planetary Science, Vol. 50, pp. 1925–1940.

Kenkmann, T., Sturm, S., Krüger, T., Salameh, E., Al-Raggad, M., and Konsul, K. (2017). The structural inventory of a small complex impact crater: Jebel Waqf as Suwwan, Jordan. Meteoritics and Planetary Science, Vol. 52, pp. 1351–1370.

Kenkmann, T., Rae, A. S. P., Cavosie, A. J., Cox, M. A., Timms, N., and Miljkovic, K. (2018). The central uplift of Gosses Bluff, Northern Territory, Australia. 81st Annual Meeting of the Meteoritical Society, Moscow, Russia, Abstract #6077.

Kenkmann, T., Sundell, K. A., and Cook, D. (2018a). Exhumed Paleozoic impact crater strewn field near Douglas, Wyoming, USA: Evidence from microstructural analysis, satellite, and drone imagery. 49th Lunar and Planetary Science Conference, Lunar and Planetary Institute, Houston, Texas, Abstract, #1469.

Kenkmann, T., Sundell, K. A., and Cook, D. (2018b). Evidence for a large Paleozoic impact crater strewn field in the Rocky Mountains. Scientific Reports, Vol. 8, Article #13246.

Kenkmann, T., Haines, P. W., Sweet, I. P., and Mitchell, K. (2019). Shock deformation and faulting in the Cleanskin impact structure, Northern Territory, Australia. Large Meteorite Impacts Conference and Planetary Evolution VI, Brasilia, Abstract #5014.

Kenkmann, T., Wulf, G., and Agarwal, A. (2020). Ramgarh, Rajasthan, India: a 10 km diameter complex impact crater. Meteoritics and Planetary Science, Vol. 55, pp. 936–961.

Kenny, G. G., Whitehouse, M. J., and Kamber, B. S. (2016). Differentiated impact melt sheets may be a potential source of Hadean detrital zircon. Geology, Vol. 44, pp. 435–438.

Kenny, G. G., Schmieder, M., Whitehouse, M. J., Nemchin, A. A., Morales, L. F. G., Buchner, E., Bellucci, J. J., and Snape, J. F. (2019). A new U-Pb age for shock-recrystallised zircon from the Lappajärvi impact crater, Finland, and implications for the accurate dating of impact events. Geochimica et Cosmochimica Acta, Vol. 245, pp. 479–494.

Kettrup, B., and Deutsch, A. (2003). Geochemical variability of the Yucatàn basement: Constraints from crystalline clasts in Chicxulub impactites. Meteoritics and Planetary Science, Vol. 38, pp. 1079–1092.

Khryanina, L. P. (1981). Sobolevskiy meteorite crater (Sikhote Alin' Range). International Geology Review, Vol. 23, pp. 1–10.

Khryanina, L. P. (1982). Indications of meteorite impact in Sobolevka meteorite crater. International Geology Review, Vol. 24, pp. 1019–1026.

Kieffer, S. W., Schaal, R. B., Gibbons, R., Hörz, R., Milton, D. J., and Dube, A. (1976). Shocked basalt from Lonar impact crater, India, and experimental analogues. Proceedings of the 7th Lunar Science Conference, Houston, Texas, Pergamon Press, New York, pp. 1391–1412.

Kiilsgaard, T. H., Heyl, A. V., and Brock, M. R. (1962). The Crooked Creek disturbance, southeast Missouri. U.S. Geological Survey Professional Paper 450-E, pp. E14–E19.

King Jr., D. T., and Petruny, L. W. (1999). Wetumpka impact crater guidebook, Auburn, Alabama, 29 pp.

King Jr., D. T., Neathery, T. L., Petruny, L. W., Koeberl, C., and Hames, W. E. (1999). Evidence confirming meteoritic impact at Wetumpka crater, Alabama, USA. Meteoritics and Planetary Science, Vol. 34, pp. 63–64.

King Jr., D. T., Neathery, T. L., Petruny, L. W., Koeberl, C., and Hames, W. E. (2002). Shallow-marine impact origin of the Wetumpka structure (Alabama, USA). Earth and Planetary Science Letters, Vol. 202, pp. 541–549.

King, D. T., and Ormö, J. (2011). The marine-target Wetumpka impact structure examined in the field and by shallow core-hole drilling. Geological Society of America Special Paper, Vol. 483, pp. 287–300.

Kirschner, C. E., Grantz, A., and Mullen, M. W. (1992). Impact origin of the Avak Structure, Arctic Alaska, and genesis of the Barrow gas fields. The American Association of Petroleum Geologists Bulletin, Vol. 76, pp. 651–679.

Kiselev, N. P., and Korotuschenko, Y. G. (1986) The Bigach astrobleme, Eastern Kazkhstan (in Russian). Meteoritika, Vol. 45, pp. 119–121.

Kjær, K. H., Larsen, N. K., Binder, T., Bjørk, A. A., Eisen, O., Fahnestock, M. A., Funder, S., Garde, A. A., Haack, H., Helm, V., Houmark-Nielsen, M., Kjeldsen, K. K., Khan, S. A., Machguth, H., McDonald, I. Morlighem, M., Mouginot, J., Paden, J. D., Waight, T. E., Weikusat, C., Willerslev, E., and MacGregor, J. A. (2018). A large impact crater beneath Hiawatha Glacier in northwest Greenland. Science Advances, Vol. 4, eaar8173.

Kleesment, A., Konsa, M., Puura, V., Karhu, J., Preeden, U., and Kallaste, T. (2006). Impact induced and diagenetic changes in minerals in the sandy ejecta of the Kärdla crater, NW, Estonia. Proceedings of the Estonian Academy of Science, Geology, Vol. 55, pp. 189–211.

Koeberl, C. (2014). The geochemistry and cosmochemistry of impacts. In: Planets, Asteriods, Comets and The Solar System, Davies, A. M. (ed.), 2nd edition of Treatise on Geochemistry, Elsevier, Vol. 2, pp. 73–118.

Koeberl, C., and Fredriksson, K. (1986). Impact glasses from Zhamanshin crater (U.S.S.R.): chemical composition and discussion of origin. Earth and Planetary Science Letters, Vol. 78, pp. 80–88.

Koeberl, C., and Fiske, P. S. (1991). Beaverhead impact structure, Montana: Geochemistry of impactites and country rock samples (abstract). Meteoritics, Vol. 26, pp. 358–359.

Koeberl, C., and Sharpton, V. L. (1993). Geochemical Study of rocks from the Kentland, Indiana, impact structure: Progress report (abstract). Meteoritics, Vol. 28, pp. 382.

Koeberl, C., and Reimold, W. U. (1995). The Newporte impact structure, North Dakota, USA. Geochimica et Cosmochimica Acta, Vol. 59, pp. 4747–4767.

Koeberl, C., and Anderson, R. R. (eds.) (1996). The Manson impact structure: Anatomy of an impact crater. Geological Society of America Special Paper, Vol. 302, 457 pp.

Koeberl, C., and Reimold, W. U. (2005). Bosumtwi impact crater, Ghana (West Africa): an updated and revised geological map, with explanations. Jahrbuch der Geologischen Bundesanstalt, Vol. 145, Heft 1, pp. 31–70 (one map at 1:50,000).

Koeberl, C., and Ferrière, L. (2019). Libyan Desert Glass area in western Egypt: Shocked quartz in bedrock points to a possible deeply eroded impact structure in the region. Meteoritics and Planetary Science, Vol. 54, pp. 2398–2408.

Koeberl, C., Sharpton, V. L., Murali, A. V., and Burke, K. (1990). Kara and Ust-Kara impact structures (USSR) and their relevance to the K/T boundary event. Geology, Vol. 18, pp. 50–53.

Koeberl, C., Reimold, W. U., and Powell, R. A. (1994). Shocked quartz and impact melt rock at the Ames structure, Oklahoma (abstract). Meteoritics, Vol. 29, p. 483.

Koeberl, C., Reimold, W. U., and Shirey, S. B. (1994). Saltpan impact crater, South Africa: geochemistry of target rocks, breccias and impact glasses, and osmium isotope systematics. Geochimica et Cosmochimica Acta, Vol. 58, pp. 2893–2910.

Koeberl, C., Reimold, W. U., Shirey, S. B., and Le Roux, F. G. (1994). Kalkkop crater, Cape Province, South Africa: confirmation of impact origin using osmium isotope systematics. Geochimica et Cosmochimica Acta, Vol. 58, pp. 1229–1234.

Koeberl, C., Poag, C. W., Reimold, W. U., and Brandt, D. (1996). Impact origin of Chesapeake Bay structure and the source for the North American tektites. Science, Vol. 271, pp. 1263–1266.

Koeberl, C., Reimold, W. U., and Brandt, D. (1996). Red Wing Creek structure, North Dakota: Petrographical and geochemical studies, and confirmation of impact origin. Meteoritics and Planetary Science, Vol. 31, pp. 335–342.

Koeberl, C., Masaitis, V. L., Shafranovsky, G. I., Gilmour, I., Langenhorst, F., and Schrauder, M. (1997). Diamonds from the Popigai impact structure, Russia. Geology, Vol. 25, pp. 967–970.

Koeberl, C., Reimold, W. U., Brandt, D., Dallmeyer, R. D., and Powell, R. A. (1997). Target rocks and breccias from the Ames impact structure, Oklahoma: Petrology, mineralogy, geochemistry and age. In: The Ames Structure and Similar Features, Johnson, K. S., and Campbell, J. A. (eds.), Oklahoma Geological Survey Circular, Vol. 100, pp. 169–198.

Koeberl, C., Reimold, W. U., and Shirey, S. B. (1998). The Aouelloul crater, Mauritania: on the problem of confirming the impact origin of a small crater. Meteoritics and Planetary Science, Vol. 33, pp. 513–517.

Koeberl, C., Reimold, W. U., and Kelley, S. P. (2001). Petrography, geochemistry, and ^{40}Ar-^{39}Ar ages of impact melt rocks and breccias from the Ames Impact Structure, Oklahoma: The Nicor Chestnut 18-4 drill core. Meteoritics and Planetary Science, Vol. 36, pp. 651–671.

Koeberl, C., Reimold, W. U., Cooper, G., Cowan, D., and Vincent, P. (2005). Aorounga and Gweni Fada impact structures, Chad: remote sensing, and petrography and geochemistry of target rocks. Meteoritics and Planetary Science, Vol. 41, pp. 1455–1471.

Koeberl, C., Reimold, W. U., and Plescia, J. (2005). BP and Oasis impact structures, Libya: remote sensing and field studies. In: Impact Tectonics, Impact Studies Series, Koeberl, C., and Henkel, H. (eds.), Springer, Berlin Heidelberg, pp. 161–190.

Koeberl, C., Milkereit, B., and Reimold, W. U. (2007). The ICDP Drilling Project in Bosumtwi, Ghana. Meteoritics and Planetary Science, Vol. 42, pp. 477–896.

Koeberl, C., Milkereit, B., Overpeck, J. T., Scholz, C. A., Amoako, P. Y. O., Boamah, D., Danuor, S. K., Karp, T., Kueck, J., Hecky, R. E., King, J., and Peck, J. A. (2007). An international and multidisciplinary drilling project into a young complex impact structure: the 2004 ICDP Bosumtwi impact crater, Ghana, drilling project – an overview. Meteoritics and Planetary Science, Vol. 42, pp. 483–511.

Koeberl, C., Pittarello, L., Reimold, W. U., Raschke, U., Brigham-Grette, J., Melles, M., and Minyuk, P. (2013). El'gygytgyn impact crater, Chukotka, Arctic Russia: Impact cratering aspects of the 2009 ICDP drilling project. Meteoritics and Planetary Science, Vol. 48, pp. 1108–1129.

Kofman, R. S., Herd, C. D., and Froese, D. G. (2010). The Whitecourt meteorite impact crater, Alberta, Canada. Meteoritics and Planetary Science, Vol. 45, pp. 1429–1445.

Kohman, T. P., and Goel, P. S. (1963). Terrestrial ages of meteorites from cosmogenic ^{14}C. In: Radioactive Dating. International Atomic Energy Agency, Vienna, pp. 395–411.

Kohman, T. P., Lowman, P. D. Jr, and Abdelkhalek, M. L. (1967) Space and aerial photography of the Libyan Desert glass area (abstract). 30th Annual Meeting of The Meteoritical Society Meeting, Moffett Field, California.

Kolesnikov, E. M., Boettger, T., and Kolesnikova, N. V. (1999). Finding of probable Tunguska cosmic body material: isotopic anomalies of carbon and hydrogen in peat. Planetary and Space Science, Vol. 47, pp. 905–916.

Korpikiewicz, H. (1978). Meteoritic shower Morasko. Meteoritics, Vol. 13, pp. 311–326.

Kramers, J. D., Andreoli, M. A. G., Atanasova, M., Belyanin, G. A., Block, D., Franklyn, C., Harris, C., Lekgoathi, M., Montross, C. S., Ntsoane, T., Pischedda, V., Segonyane, P., Viljoen, K. S., and Westraadt, J. E. (2013). Unique chemistry of a diamond-bearing pebble from the Libyan Desert Glass strewnfield, SW Egypt: evidence for a shocked comet fragment. Earth and Planetary Science Letters, Vol. 382, pp. 21–31.

Kraut F. (1969). Über ein neues Impaktitvorkommen im Gebiet von Rochechouart-Chassenon (Department Haute-Vienne und Chartres), Frankreich. Geologica Bavarica, Vol. 61, pp. 428–460.

Kriens, B. J., Shoemaker, E. M., and Herkenhoff, K. E. (1999). Geology of the Upheaval Dome impact structure, southeast Utah. Journal of Geophysical Research, Vol. 104, pp. 18867–18887.

Kring, D. A. (1997). Air blast produced by the Meteor Crater impact event and a reconstruction of the affected environment. Meteoritics and Planetary Science, Vol. 32, pp. 517–530.

Kring, D. A. (2017) Guidebook to the Geology of Barringer Meteorite Crater, Arizona (a.k.a. Meteor Crater), 2nd edition. Lunar and Planetary Institute Contribution No. 2040: https://www.lpi.usra.edu/publications/books/barringer_crater_guidebook

Kring, D. A., Claeys, P., Gulick, S. P. S., Morgan, J. V., Collins, G. S., and the IODP-ICDP Expedition 364 Science Party. (2017). Chicxulub and the exploration of large peak-ring impact craters through scientific drilling. GSA Today, Vol. 27/10, pp. 4–8.

Krinov, E. L. (1959). The main conditions of the fall of a meteoritic shower (in Russian). In: The Sikhote-Alin Iron Meteorite Shower, Fesenkov, V. G., and Krinov, E. L. (eds.), Publishing House of the USSR Academy of Sciences, Vol. 1, pp. 99–156.

Krinov, E. L. (1961). The Kaalijarv meteorite craters on Saarema Island, Estonian SSR. American Journal of Science, Vol. 259, pp. 430–440.

Krinov, E. L. (1971). New studies of the Sikhote-Alin iron meteorite shower. Meteoritics, Vol. 6, pp. 127–138.

Krinov, E. L. (1974). Fragmentation of the Sikhote-Alin meteoritic body. Meteoritics, Vol. 9, pp. 255–262.

Kuivasaari, T., Pesonen, L. J., Elo, S., Plado, J., and Lehtinen, M. (2000). Paasselkä – the ninth meteorite impact structure in Finland. 4th ESF Workshop: Response of the Earth System to Impact Processes (IMPACT), Helsinki, p. 79.

LaFond, C., and Dietz, R. S. (1964). Lonar Crater, India, a meteorite crater? Meteoritics, Vol. 2, pp. 111–116.

Lamali, A., Rochette, P., Merabet, N., Abtout, A., Maouche, S., Gattacceca, J., Ferrière, L., Hamoudi, M., ASTER Team, Meziane, E. H., and Ayache, M. (2016). Geophysical and magneto-structural study of the Maâdna structure (Talemzane, Algeria): Insights on its age and origin. Meteoritics and Planetary Science, Vol. 51, pp. 2249–2273.

Lambert, P. (1977a). The Rochechouart Crater: Shock zoning study. Earth and Planetary Science Letters, Vol. 35, pp. 258–268.

Lambert, P. (1977b). Rochechouart impact crater: Statistical geochemical investigations and meteoritic contamination. In: Impact and Explosion Cratering, Roddy, D. J., Pepin, R. O., and Merrill, R. B. (eds.), Pergamon Press, New York, pp. 449–460.

Lambert, P., and Lamali, A. (2009). Impact structures in Algeria (abstract). In: 1st Arab Impact Cratering and Astrogeology Conference, Amman, pp. 50–53.

Lambert, P., McHone, J. F., Dietz, R. S., and Houfani, M. (1980). Impact and impact-like structures in Algeria, Part 1. Four bowl-shaped depressions. Meteoritics, Vol. 15, pp. 157–179.

Lambert, P., McHone, J. F., Dietz, R. S., Briedj, M., and Djender, M. (1981). Impact and impact-like structures in Algeria, Part 2. Multi-ringed structures. Meteoritics, Vol. 16, pp. 203–227.

Laney, R. T., and Van Schmus, W. R. (1978). A structural study of the Kentland, Indiana, impact site. Proceedings of the 9th Lunar and Planetary Science Conference, Houston, Texas, Pergamon Press, New York, pp. 2609–2632.

LaTouche, T., and Christie, W. (1912). The geology of the Lonar Lake. Records of the Geological Survey of India, Vol. 41, pp. 266–275.

Laurén, L., Lehtovaara, J., Bostrom, R., and Tynni, R. (1978). On the geology and the Cambrian sediments of the circular depression at Söderfjärden, western Finland. Geological Survey of Finland Bulletin, Vol. 297, pp. 1–81.

Layer, P. W. (2000). Argon-40/Argon-39 age of the El'gygytgyn event, Chukotka, Russia. Meteoritics and Planetary Science, Vol. 35, pp. 591–599.

Le Pichon, A., Antier, K., Cansi, Y., Hernandez, B., Minaya, E., Burgoa, B., Drob, D., Evers, L. G., and Vaubaillon, J. (2008). Evidence for a meteoritic origin of the September 15, 2007, Carancas crater. Meteoritics and Planetary Science, Vol. 43, pp. 1797–1809.

Lees, T. C., Murphy, F. C., Tomkins, A. G., and O'Donohue, D. (2019). Displacement of the Proterozoic century ore deposit at the edge of an Ordovician meteorite impact crater, Queensland. Economic Geology, Vol. 114, pp. 427–439.

Lefranc, J. P. (1969). Reconnaissance du cratère météoritique d'Amguid (Mouydir, Sahara Central). Comptes Rendus de l'Académie des Sciences, Paris, Vol. 268, pp. 900–902.

Lehtinen, M. (1976). Lake Lappajärvi, a meteorite impact site in western Finland. Geological Survey of Finland Bulletin, Vol. 282, 92 pp.

Lehtinen, M., Pesonen, L. J., Puranen, R., and Deutsch, A. (1996). Karikkoselkä – a new impact structure in Finland (abstract). 27th Lunar and Planetary Science Conference, Lunar and Planetary Institute, Houston, Texas, pp. 739–740.

Lehtinen, M., Pesonen, L. J., Stehlik, H., and Kuulusa. M. (2002). The Suvasvesi South structure, Central Finland: New evidence for impact. 33rd Lunar and Planetary Science Conference, Lunar and Planetary Institute, Houston, Texas, Abstract #1188.

Lehtovaara, J. J. (1985). ^{40}K-^{40}Ar dating of the Söderfjärden crater, Vaasa, western Finland. Geologiska Föreningen i Stockholm Förhandlingar, Vol. 107, pp. 1–6.

Lehtovaara, J. J. (1992). Söderfjärden: a Cambrian impact crater in western Finland. Tectonophysics, Vol. 216, pp. 157–161.

Lepaulard, C., Gattacceca, J., Swanson-Hysell, N., Quesnel, Y., Demory, F., and Osinski, G. R. (2019). A Paleozoic age for the Tunnunik impact structure. Meteoritics and Planetary Science, Vol. 54, pp. 740–751.

Lightfoot, P. C. (2016). Nickel Sulfide Ores and Impact Melts: Origin of the Sudbury Igneous Complex. Elsevier Publishers, Amsterdam, 662 pp.

Lindström, M., Ekvall, J., Hagenfeldt, S. E., Säwe, B., and Sturkell, E. F. F. (1991). A well-preserved Cambrian impact exposed in Central Sweden. Geologische Rundschau, Vol. 80, pp. 201–204.

Lindström, M., Flodén, T., Grahn, Y., and Kathol, B. (1994). Post-impact deposits in Tvären, a marine Middle Ordovician crater south of Stockholm, Sweden. Geological Magazine, Vol. 131, pp. 91–103.

Lindström, M., Flodén, T., Grahn, Y., Hagenfeldt, S., Ormö, J., Sturkell, E. F. F., and Törnberg, R. (1999). The lower Palaeozoic of the probable impact crater of Hummeln, Sweden. GFF, Vol. 121, pp. 243–252.

Lindström, M., Ormö, J., Sturkell, E. F. F., and von Dalwigk, I. (2005). The Lockne crater: revision and reassessment of structure and impact stratigraphy. In: Impact Tectonics, Koeberl, C. and Henkel, H. (eds.), Springer, Berlin Heidelberg, pp. 357–388.

Lindström, M., Shuvalov V. V., and Ivanov, B. A. (2005). Lockne crater as a result of a marine-target oblique impact. Planetary and Space Science, Vol. 53, pp. 803–815.

Liu, H. P., McKay, R. M., Young, J. N., Witzke, B. J., McVey, K. J., and Liu, X. (2006). A new lagerstätte from the Middle Ordovician St. Peter Formation in northeast Iowa, USA. Geology, Vol. 34, pp. 969–972.

Löffler, J. (2013): The Ritland Impact Structure, southwestern Norway: post-impact Deformation and the Effect of the Crater on Caledonian Thrust Tectonics. Master Thesis (unpublished), University of Freiburg, 127 pp.

Losiak, A., Wild, E. M., Geppert, W. D., Huber, M. S., Jõeleht, A., Kriiska, A., Kulkov, A., Paavel, K., Pirkovic, I., Plado, J., Steier, P., Välja, R., Wilk, J., Wisniowski, T., and Zanetti, M. (2016). Dating a small impact crater: An age of Kaali crater (Estonia) based on charcoal emplaced within proximal ejecta. Meteoritics and Planetary Science, Vol. 51, pp. 681–695.

Lugo, C. M., and Cavosie, A. J. (2014). First report of shocked zircon at the Santa Fe impact structure (USA). 45th Lunar and Planetary Science Conference, Lunar and Planetary Institute, Houston, Texas, Abstract #1839.

Macdonald, F. A., Bunting, J. A., and Cina, S. E. (2003). Yarrabubba – a large, deeply eroded impact structure in the Yilgarn Craton, Western Australia. Earth and Planetary Science Letters, Vol. 213, pp. 235–247.

Macdonald F. A., Mitchell K., and Stewart A. J. (2005). Amelia Creek: a Proterozoic impact structure in the Davenport Ranges, Northern Territory. Australian Journal of Earth Sciences, Vol. 52, pp. 631–640.

Macdonald, F. A., Wingate, M. T. D., and Mitchell, K. (2005). Geology and age of the Glikson impact structure, Western Australia. Australian Journal of Earth Sciences, Vol. 52, pp. 641–651.

Maclagan, E. A., Walton E. L., Herd, C. D. K., and Dence, M. (2018). Investigation of impact melt in allochthonous crater-fill deposits of the Steen River impact structure, Alberta, Canada. Meteoritics and Planetary Science, Vol. 53, pp. 2285–2305.

Mader, M. M., and Osinski, G. R. (2018). Impactites of the Mistastin Lake impact structure: Insights into impact ejecta emplacement. Meteoritics and Planetary Science, Vol. 53, pp. 2492–2518.

Maier, W. D., Andreoli, M. A. G., McDonald, I., Higgins, M. D., Boyce, A. J., Shukolyukov, A., Lugmair, G. W., Ashwal, L. D., Graser, P., Ripley, E. M., and Hart, R. J., (2006). Discovery of a 25-cm asteroid clast in the giant Morokweng impact crater, South Africa. Nature, Vol. 441, pp. 203–206.

Maloof, A. C., Stewart, S. T., Weiss, B. P., Soule, S. A., Swanson-Hysell, N. L., Louzada, K. L., Garrick-Bethell, I., and Poussart, P. M. (2010). Geology of Lonar Crater, India. Bulletin of the Geological Society of America, Vol. 122, pp. 109–126.

Marakushev, A. A. (Ed.) (1981) Impactites (in Russian). Moscow University Press, Moscow, Russia. 240 pp.

Marion, C. L., and Sylvester, P. J. (2010). Composition and heterogeneity of anorthositic impact melt at Mistastin Lake crater, Labrador. Planetary and Space Science, Vol. 58, pp. 552–573.

Mark, D. F., Lindgren, P., and Fallick, A. E. (2014). A high-precision ^{40}Ar-^{39}Ar age for hydrated impact glass from the Dellen impact, Sweden. In: Advances in ^{40}Ar-^{39}Ar Dating: from Archaeology to Planetary Sciences, Jourdan, F., Mark, D. F., and Verati, C. (eds.), Geological Society of London, Special Publications, Vol. 378, pp. 349–366.

Martin, A. J. (1969). Possible impact structure in Southern Cyrenaica, Libya. Nature, Vol. 223, pp. 940–941.

Martins, J. A., Castelo Branco, R. M. G., Castro, N. A., Peaulvast, J.-P., and Lima, Jr. S. B. (2016). Morphological analysis of the São Miguel do Tapuio circular structure, Piauí – Brazil. Revista Brasileira Geomorfologia, Vol. 17, pp. 763–781.

Masaitis, V. L. (1976). Astroblemes in the USSR. International Geology Review, Vol. 18, pp. 1249–1258.

Masaitis, V. L. (1999). Impact structures of northeastern Eurasia: The territories of Russia and adjacent countries. Meteoritics and Planetary Science, Vol. 34, pp. 691–711.

Masaitis, V. L. (2002). The middle Devonian Kaluga impact crater (Russia): new interpretation of marine setting. Deep-Sea Research Part II, Vol. 49, pp. 1157–1169.

Masaitis, V. L., Danilin, A. N., Karpov, G. M., and Raikhlin, A. I. (1976). Karla, Obolon' and Rotmistrovka astroblemes in the European part of the USSR (in Russian). Doklady Akademii Nauk SSSR, Vol. 230, pp. 174–177.

Masaitis, V. L., Sindeev, A. S., and Staritsky, Y. G. (1976). The impactites of the Jänisjärvi astrobleme (in Russian). Meteoritika, Vol. 35, pp. 103–110.

Masaitis, V. L., Danilin, V. N., Mashchak, M. S., Raikhlin A. I., Selivanovskaty. A. V., and Shadenkov, E. M. (1980). The Geology of Astroblemes. Nedra Press, Leningrad, Russia, 231 pp.

Masaitis, V. L., Mashchak, M. S., and Sokolova, I. Y. (1980). High-pressure silica phases in the Ternovka astrobleme (in Russian). Doklady Akademii Nauk SSSR, Vol. 255, pp. 709–713.

Masaitis, V. L., Boiko, Y. I., and Izokh, E. P. (1984). Zhamanshin impact crater (Western Kazakhstan): Additional geological data. 15th Lunar and Planetary Science Conference, Lunar and Planetary Institute, Houston, Texas, pp. 515–516.

Masaitis, V. L., Mashchak, M. S., Naumov, M. V., and Raikhlin, A. I. (1994). Large Astroblemes of Russia. VSEGEI, St. Petersburg, 34 pp.

Masaitis, V. L., Mashchak, M. S., and Naumov, M. V. (2004). Popigai Impact Crater. Guide of Geological Excursions. VSEGEI St. Petersburg, 56 pp.

Mashchak, M. S., and Orlova, J. V. (1986). Shock deformations in lower Proterozoic breccias of the Lake Suav'yarvi region (central Karelia) (in Russian). Meteoritika, Vol. 45, pp. 137–141.

Mashchak, M. S., and Naumov, M. V. (1996). The Suavjärvi structure: An early Proterozoic impact site on the Fennoscandian shield (abstract). 27th Lunar and Planetary Science Conference, Lunar and Planetary Institute, Houston, Texas, pp. 825–826.

Mashchak, M. S., and Naumov, M. V. (2012). The Suavjärvi impact structure, NW Russia. Meteoritics and Planetary Science, Vol. 47, pp. 1644–1658.

Master, S., and Heymann, J. (2000). A possible new impact structure near Gilbués in Piauí Province, Northeastern Brazil. Meteoritics and Planetary Science, Vol. 35, Supplement, p. A105.

Master, S., Rainaud, C., Armstrong, R. A., Phillips, D., and Robb, L. J. (2005). Provenance ages of the Neoproterozoic Katanga Supergroup (Central African Copperbelt), with implications for basin evolution. Journal of African Earth Sciences, Vol. 42, pp. 41–60.

Maziviero, M. V., Vasconcelos, M. A. R., Crósta, A. P., Góes, A. M., Reimold, W. U., and Carneiro, C. C. (2013). Geology and impact features of Riachão structure, northern Brazil. Meteoritics and Planetary Science, Vol. 48, pp. 2044–2058.

Mazur, M. J., Stewart, R. R., and Hildebrand, A. R. (1999). Seismic characterization of buried possible meteorite impact structures. 30th Lunar and Planetary Science Conference, Houston, Texas, Abstract #1393.

Mazur, M., Hildebrand, A., and Stewart, R. (2003). Geophysical Studies of the Steen River Impact Structure, Alberta, Canada. 65th EAGE Conference and Exhibition, Stavanger, Norway, 4 pp.

McCabe, H. R., and Bannatyne, B. B. (1970). Lake St. Martin crypto-explosion crater and geology of the surrounding area. Manitoba Department of Mines and Natural Resources, Mines Branch, Geological Paper, Vol. 3/70, 79 pp.

McElvain, T. H., Read, A. S., Petersen, M. T., Elston, W. E., Newsom, H. E., and Cohen, B. A. (2006). Possible impact structure in the Sangre de Cristo Mountains near Santa Fe, New Mexico: a preliminary report. Geological Society of America Abstracts with Programs, Vol. 38, p. 298.

McGregor, M., McFarlane, C. R. M., and Spray, J. G. (2017). The Nicholson Lake impact structure, Canada: shock features and age of formation. 48th Lunar and Planetary Science Conference, The Woodlands, Texas, Abstract #2151.

McHone, J. F. (1986). Terrestrial Impact Structures: Their Detection and Verification with Two New Examples from Brasil. Ph.D. thesis, University of Illinois at Urbana-Champaign, Urbana, Illinois, 210 pp.

McHone, J. F., and Dietz, R. S. (1976). Tabun Khara Obo crater, Mongolia: Probably meteoritic. Meteoritics, Vol. 11, pp. 332–333.

McHone, J. F., and Dietz, R. S. (1978). Astroblemes in Brazil. Proceedings, 74th Geological Society of America Cordilleran Section Meeting, 116 pp.

McHone, J. F., and Greeley, R. (1987). Talemzane: Algerian impact crater detected on SIR-A orbital imaging radar. Meteoritics, Vol. 22, pp. 253–264.

McHone, J. F., Sargent, M. L., and Nelson, W. J. (1986). Shatter cones in Illinois: Evidence for meteoritic impacts at Glasford and Des Plaines (abstract). Meteoritics, Vol. 21, pp. 446.

Merrill, G. K. (1980). Ordovician conodonts from the Åland Islands, Finland. Geologiska Föreningen i Stockholm Förhandlingar, Vol. 101, pp. 329–341.

Meteor Crater Company: http://www.barringercrater.com.

Meteor Crater Visitor Center: http://www.meteorcrater.com.

Mikhailov, M. V., Shurygin, A. G., and Khariusov, L. S. (1979). Beenchime-Salaaty meteoritic crater (in Russian). Doklady AN SSSR, Vol. 245, pp. 911–914.

Milam, K. A., Deane, B., King, P. L., Lee, P. C., and Hawkins, M. (2006). From the inside of a central uplift: the view from Hawkins impact cave. 37th Lunar and Planetary Science Conference, Lunar and Planetary Institute, Houston, Texas, Abstract #1211.

Miller, J., Evans, K. R., Ethington, R. L., Repetski, J. E., Sandberg, C. A., and Thompson, T. L. (2007). Critical stratigraphic data from reworked Conodonts in impact breccias across Missouri's Ozark Dome (abstract). Geological Society of America South-Central and North-Central Joint Section Meeting, Lawrence, Kansas, Abstracts with Programs, Vol. 39, p. 62.

Miller, R. McG. (2008). Meteorite impact craters and meteorites. In: The Geology of Namibia, Miller, R. McG. (ed.), Ministry of Mines and Energy, Windhoek, Namibia, Chapter 26, pp. 1–8.

Miller, R. McG. (2010). Roter Kamm impact crater of Namibia: new data on rim structure, target rock geochemistry, ejecta, and meteorite trajectory. In: Large Meteorite Impacts and Planetary Evolution IV, Gibson, R. L., and Reimold, W. U. (eds.), Geological Society of America Special Paper, Vol. 465, pp. 489–508.

Millman, P. M., Liberty, B. A., Clark, J. F., Willmore, P., and Innes, M. J. S. (1960). The Brent Crater. Ottawa Dominion Observatory Publications, Vol. 24, 43 pp.

Milstein, R. L. (1988). The Calvin 28 structure: Evidence for impact origin. Canadian Journal of Earth Sciences, Vol. 25, pp. 1524–1530.

Milstein, R. L. (1994). The Calvin Impact Crater, Cass County, Michigan: Identification and Analysis of a Subsurface Ordovician Astrobleme. Ph.D. Thesis, Oregon State University, 114 pp.

Milton, D. J. (1968). Structural Geology of the Henbury Meteorite Craters, Northern Territory, Australia. Contributions to Astrogeology, Geological Survey Professional Paper 599-C, United States Government Printing Office, Washington, 17 pp.

Milton, D. J., and Michel, F. C. (1965). Structure of a ray crater at Henbury, Northern Territory, Australia. U.S. Geological Survey Professional Paper 525-C, C5–C11.

Milton, D. J., and Sutter, J. F. (1987). Revised age for the Gosses Bluff impact structure, Northern Territory, Australia, based on ^{40}Ar-^{39}Ar dating. Meteoritics, Vol. 22, pp. 281–289.

Milton, D. J., and Macdonald, F. A. (2005). Goat Paddock, Western Australia: an impact crater near the simple-complex transition. Australian Journal of Earth Sciences, Vol. 52, pp. 691–698.

Milton, D. J., Barlow, B. C., Brown, A. R., Moss, F. J., Manwaring, E. A., Sedmik, E. C. E., Young, G. A., and Van Son, J. (1996). Gosses Bluff – a latest Jurassic impact structure, central Australia. Part 2: Seismic, magnetic, and gravity studies. AGSO Journal of Australian Geology and Geophysics, Vol. 16, pp. 487–527.

Milton, D. J., Glikson, A. Y., and Brett, R. (1996). Gosses Bluff – a latest Jurassic impact structure, central Australia. Part 1: Geological structure, stratigraphy, and origin. AGSO Journal of Australian Geology and Geophysics, Vol. 16, pp. 453–486.

Minor Planet Center: http://www.minorplanetcenter.net.

Misra, S., Srivastava, P. K., and Arif, M. (2019). Remote sensing, structural and rock magnetic analyses of the Ramgarh structure of SE Rajasthan, Central India – further clues to its impact origin and time of genesis. In: Tectonics and Structural Geology: Indian Context, Mukherjee S. (ed.), Springer Geology, pp. 327–352.

Monod, Th., and Pomerol, C. (1966). Le cratère de Tenoumer et ses laves. Bulletin de la Société Géologique de France, Vol. S7-VIII, pp. 165–172.

Monson, C. C., Sweet, D., Segvic, B., Zanoni, G., Balling, K., Wittmer, J. M., Ganis, G. R., and Cheng, G. (2019). The Late Ordovician (Sandbian) Glasford structure: A marine-target impact crater with a possible connection to the Ordovician meteorite event. Meteoritics and Planetary Science, Vol. 54, pp. 2927–2950.

Moreira, A., Prats-Iraola, P., Younis, M., Krieger, G., Hajnsek, I., and Papathanassiou, K. P. (2013). A tutorial on synthetic aperture radar. IEEE Geoscience and Remote Sensing Magazine, Vol. 1, pp. 6–43.

Morgan, J. W., and Wandless, G. A. (1983). Strangways crater, Northern Territory, Australia: Siderophile element enrichment and lithophile element fractionation. Journal of Geophysical Research, Supplement, Vol. 88, pp. A819–A829.

Morgan J., Warner M., and the Chixculub working group (1997). Size and morphology of the Chicxulub impact crater. Nature, Vol. 390, pp. 472–476.

Morgan, J. V., Warner, M. R., Collins, G. S., Melosh, H. J., and Christeson, G. L. (2000). Peak-ring formation in large impact craters: geophysical constraints from Chicxulub. Earth and Planetary Science Letters, Vol. 183, pp. 347–354.

Morgan, J. V., Gulick, S. P. S., Bralower, T., Chenot, E., Christeson, G., Claeys, P., Cockell, C., Collins, G. S., Coolen, M. J. L., Ferrière, L., Gebhardt, C., Goto, K., Jones, H., Kring, D. A., Le Ber, E., Lofi, J., Long, X., Lowery, C., Mellett, C., Ocampo-Torres, R., Osinski, G. R., Perez-Cruz, L., Pickersgill, A., Poelchau, M. H., Rae, A., Rasmussen, C., Rebolledo-Vieyra, M., Riller, U., Sato, H., Schmitt, D. R., Smit, J., Tikoo, S., Tomioka, N., Urrutia-Fucugauchi, J., Whalen, M., Wittmann, A., Yamaguchi, K. E., and Zylberman, W. (2016). The formation of peak rings in large impact craters. Science, Vol. 354, pp. 878–882.

Mory, A. J., Iasky, R. P., Glikson, A. Y., and Pirajno, F. (2000). Woodleigh, Carnarvon Basin, Western Australia: a new 120 km diameter impact structure. Earth and Planetary Science Letters, Vol. 177, pp. 119–128.

Motuza, G. B., and Gailius, P. (1978). On the supposed astroblemes in Lithuania (in Russian). In: Local Structures in Belarus and the Baltic Region, Mokslas Press, Suveizdis, P. I. (ed.), Geological Research Institute, Vilnius, Lithuania, pp. 91–94.

Müller, N., Hartung, J. B., Jessberger, E. K., and Reimold, W. U. (1990). ^{40}Ar-^{39}Ar ages of Dellen, Janisjärvi, and Sääksjärvi impact craters. Meteoritics, Vol. 25, pp. 1–10.

Mulvany, P. S. (2004). Field Trip II: Geology of the Crooked Creek Ring Structure, Crawford County, Missouri. Association of Missouri Geologists Field Trip Guide Book, 51st Annual Meeting, Rolla, Missouri, pp. 13–30.

Nagera, J. J. (1926). Los hoyos del Campo del Cielo y el meteorito. Direccion General de Minas, Geologia y Hidrologia, Buenos Aires, Argentina.

Naldrett, A. J. (2003). From impact to riches: Evolution of geological understanding as seen at Sudbury, Ontario. Geological Society of America Today, Vol. 13, pp. 4–9.

Nayak, V. K. (1972). Glassy objects (impactite glasses?): A possible new evidence for meteoritic origin of the Lonar Crater, Maharashtra State, India. Earth and Planetary Science Letters, Vol. 14, pp. 1–6.

Neathery, T. L., Bentley, R. D., and Lines, G. C. (1976). Cryptoexplosion structure near Wetumpka, Alabama. Geological Society of America Bulletin, Vol. 87, pp. 567–573. [Reprinted in Alabama Geological Society Guidebook No. 20 (1983) and No. 34c (1997)].

Neville, A. S., Cook D. J., Afifi A. M., and Stewart, S. A. (2014). Five buried crater structures imaged on reflection seismic data in Saudi Arabia. GeoArabia, Vol. 19, pp. 17–44.

Newman, J. D., and Herd, C. D. K. (2015). Mineralogy, petrology, and distribution of meteorites at the Whitecourt Crater, Alberta, Canada. Meteoritics and Planetary Science, Vol. 50, pp. 305–317.

Newman, J. D., and Osinski, G. R. (2016). Geological mapping of the Tunnunik impact structure, Victoria Island, Canadian High Arctic. 47th Lunar and Planetary Science Conference, Lunar and Planetary Institute, Houston, Texas, Abstract #1591.

Nickerson, G. A. J., Spray, J. G., and Mayer, L. A. (2001). Investigation of integrated geologic and geophysical data using GIS; Crooked Creek and Decaturville impact structures, Missouri. Atlantic Geology, Vol. 36, pp. 67–68.

Nininger, H. H., and Figgins, J. D. (1933). The excavation of a meteorite crater near Haviland, Kiowa County, Kansas. American Journal of Science, Vol. 28, pp. 312–313.

Nininger, H. H., and Huss, G. I. (1960). The unique meteorite crater at Dalgaranga, Western Australia. Mineralogical Magazine, Vol. 32, pp. 619–639.

Nördlingen Ries Crater Museum: http://www.riescrater-museum.de.

Northfleet, A. A., Medeiros, R. A., and Muhlmann, H. (1969). Reavaliação dos dados geológicos da Bacia do Paraná. Boletim Técnico da Petrobrás, Vol. 12, pp. 291–346.

Offield, T. W., and Pohn, H. A. (1977). Deformation at the Decaturville impact structure, Missouri. In: Impact and Explosion Cratering, Roddy, D. J., Pepin, R. O., and Merrill, R. B. (eds.), Pergamon Press, New York, pp. 321–341.

Offield, T. W., and Pohn, H. A. (1979). Geology of the Decaturville impact structure, Missouri. U.S. Geological Survey Professional Paper, Vol. 1042, 48 pp.

Öhman, T. (2007). The origin and tectonic modification of the Saarijärvi impact structure, northern Finland. Bridging the Gap II: Effect of target properties on the impact cratering process, Saint-Hubert, Canada, LPI Contribution No. 1360, pp. 85–86.

Öhman, T., Badjukov, D., Raitlala, J., Petrova, T., and Stehlik, H. (2003). Impactites of the Paaselka and Suvasvesi South craters, Finland. 34th Lunar and Planetary Science Conference, Lunar and Planetary Institute, Houston, Texas, Abstract #1571.

Öhman, T., Pesonen, L. J., Elo, S., Uutela, A., Tuiska, P., and Raitala, J. (2003). The origin and evolution of the Saarijärvi impact structure. Meteoritics and Planetary Science, Vol. 38 Supplement, p. A52.

Oliveira, G. J. G. de, Vasconcelos, M. A. R., Crósta, A. P., Reimold, W. U., Góes, A. M., and Kowitz A. (2014). Shatter cones and planar deformation features confirm Santa Marta in Piauí State, Brazil, as an impact structure. Meteoritics and Planetary Science, Vol. 49, pp. 1915–1928.

Oliveira, G. J. G. de, Chamani, M. A. C., Góes, A. M., Crósta, A. P., Vasconcelos, M. A. R., and Reimold, W. U. (2017). Geological investigation of the central portion of the Santa Marta impact structure, Piauí State, Brazil. Brazilian Journal of Geology, Vol. 47, pp. 673–692.

O'Neill, C., Heine, C., Glikson, A. Y., and Haines, P. W. (2005). Reconstructing the Wolfe Creek meteorite impact; deep structure of the crater and effects on target rock. Australian Journal of Earth Sciences, Vol. 52, pp. 699–709.

Ormö, J., and Blomqvist, G. (1996). Magnetic modelling as a tool in the evaluation of impact structures, with special reference to the Tvären Bay impact crater, SE Sweden. Tectonophysics, Vol. 262, pp. 291–300.

Ormö, J., and Lindström, M. (2000). When a cosmic impact strikes the sea bed. Geological Magazine, Vol. 137, pp. 67–80.

Ormö, J., Sturkell, E., Olvak, J. N., Melero-Asensio, I., Frisk, A., and Wikström, T. (2014). The geology of the Målingen structure: A probable doublet to the Lockne marine-target impact crater, central Sweden. Meteoritics and Planetary Science, Vol. 49, pp. 313–327.

Osinski, G. R., and Spray, J. G. (2001). Impact-generated carbonate melts: Evidence from the Haughton structure, Canada. Earth and Planetary Science Letters, Vol. 194, pp. 17–29.

Osinski, G. R., and Spray, J. G. (2005). Tectonics of the Haughton impact event, Devon Island, Canadian High Arctic. Meteoritics and Planetary Science, Vol. 40, pp. 1813–1834.

Osinski, G. R., Grieve, R. A. F., and Spray, J. G. (2004). The nature of the groundmass of surficial suevite from the Ries impact structure, Germany, and constraints on its origin. Meteoritics and Planetary Science, Vol. 39, pp. 1655–1683.

Osinski, G. R., Lee, P., Spray, J. G., Parnell, J., Lim, D., Bunch, T. E., Cockell, C. S., and Glass, B. (2005). Geological overview and cratering model of the Haughton impact structure, Devon Island, Canadian High Arctic. Meteoritics and Planetary Science, Vol. 40, pp. 1759–1776.

Osinski, G. R., Singleton, A. C., Ozaruk, A., and Hanson, J. R. (2012). New investigations of the Gow Lake impact structure, Saskatchewan, Canada: Impact melt rocks, astronaut training, and more. 43rd Lunar and Planetary Science Conference, The Woodlands, Texas, Abstract #2367.

Paiva Filho, A., Andrade, C. A. V. de, and Scheibe, L. F. (1978). Uma janela estratigráfica no oeste de Santa Catarina: o Domo de Vargeão. In: 30° Congresso Brasileiro de Geologia, Recife, Brazil, Proceedings, Vol. 1, pp. 408–412

Papunen, H. (1969). Possible impact metamorphic textures in the erratics of the Lake Saaksjärvi area in southwestern Finland. Bulletin of the Geological Society of Finland, Vol. 41, pp. 151–155.

Partridge, T. C. (ed.) (1999). Investigations into the Origin, Age, and Palaeoenvironments of the Pretoria Saltpan. Council for Geoscience, Memoir 85, 198 pp.

Pati, J. K., Reimold, W. U., Koeberl, C., and Pati, P. (2008). The Dhala structure, Bundelkhand craton, Central India – Eroded remnant of a large Paleoproterozoic impact structure. Meteoritics and Planetary Science, Vol. 43, pp. 1383–1398.

Pati, J. K., Jourdan, F., Armstrong, R., Reimold, W. U., and Prakash, K. (2010). First SHRIMP U-Pb and ^{40}Ar-^{39}Ar chronological results from impact melt breccia from the Paleoproterozoic Dhala impact structure, India. Geological Society of America Special Paper, Vol. 465, pp. 571–591.

Pati, J. K., Qu, W. J., Koeberl, C., Reimold, W. U., Chakarvorty, M., and Schmitt, R. T. (2017). Geochemical evidence of an extraterrestrial component in impact melt breccia from the Paleoproterozoic Dhala impact structure, India. Meteoritics and Planetary Science, Vol. 52, pp. 722–736.

Paya, B. K., Holmes, H., Reimold, W. U., and Farr, J. (1999). Kgagodi Basin: a possible impact structure (abstract). 62nd Annual Meeting of The Meteoritical Society, Johannesburg, South Africa.

Peate, I., Van Soest, M. C., Wartho, J. A., Cabrol, N., Grin, E., Piatek, J., and Chong, G. (2010). A novel application of (U-Th)/He geochronology to constrain the age of small, young meteorite impact craters: A case study of Monturaqui Crater, Chile. 31st Lunar and Planetary Science Conference, Lunar and Planetary Institute, Houston, Texas, Abstract #2161.

Pesonen, L. J. (1998). The Lake Saarijärvi – A new meteorite impact structure in northern Finland. 29th Lunar and Planetary Science Conference, Lunar and Planetary Institute, Houston, Texas, Abstract #1262.

Pesonen, L. J., Järvelä, J., Sarapää, O., and Pietarinen, H. (1996). The Iso-Naakkima meteorite impact structure: Physical properties and paleomagnetism of a drill core (abstract). Meteoritics and Planetary Science, Vol. 31, pp. A105–A106.

Pesonen, L. J., Lehtinen, M., Deutsch, A., Elo, S., and Lukkarinen, H. (1996). New geophysical and petrographic results of the Suvasvesi N impact structure, Finland (abstract). 27th Lunar and Planetary Science Conference, Lunar and Planetary Institute, Houston, Texas, pp. 1021–1022.

Pesonen, L. J., Elo, S., Lehtinen, M., Jokinen, T., Puranen, R., and Kivekäs, L. (1999). Lake Karikkoselkä impact structure, central Finland: New geophysical and petrographic results. In: Large Meteorite Impacts and Planetary Evolution II. Dressler, B. O., and Sharpton, V. L. (eds.), Geological Society of America, Special Paper, Vol. 339, pp. 131–147.

Pesonen, L. J., Kuivasaari, T., Lehtinen, M., and Elo, S. (1999). Paasselkä: A new meteorite impact structure in Eastern Finland. Meteoritics and Planetary Science, Vol. 34 Supplement, p. A90-1.

Pesonen, L. J., Mader, D., Gurov, E. P., Koeberl, C., Kinnunen, K. A., Donadini, F., and Handler, R. (2004). Paleomagnetism and ^{40}Ar-^{39}Ar age determinations of impactites from the Ilyinets structure, Ukraine. In: Cratering in Marine Environments and on Ice, Dypvik, H., Burchell, M. J., and Claeys, P. H. (eds.), Springer, Berlin, Heidelberg, pp. 251–280.

Petaev, M. I., Kisarev, Y. L., Mustafin, S. A., Shakurov, R. K., Pavlov, A. V., and Ivanov B. A. (1991). Meteorite Sterlitamak – A new crater forming fall (abstract). 22nd Lunar and Planetary Science Conference, Lunar and Planetary Institute, Houston, Texas, pp. 1059–1060.

Petrov, V. G. (1974). Peculiarities of the Interior of the Kaluga Circular Structure and its Location in the Moscow Syneclise (in Russian). Unpublished Ph.D. Thesis. Moscow State University, Moscow, Russia.

Philby, H. S. J. (1933). The Empty Quarter. London, Constable and Company Ltd., 433 pp.

Phinney, W. C., Simonds, C. H., Cochran, A., and McGee, P. E. (1978). West Clearwater, Quebec, impact structure, Part II. Petrology. Proceedings of the 9th Lunar and Planetary Science Conference, Houston, Texas, Pergamon Press, New York, pp. 2659–2693.

Pilkington, M., Jansa, L. F., and Grieve, R. A. F. (1995). Geophysical studies of the Montagnais impact crater, Canada. Meteoritics, Vol. 30, pp. 446–450.

Pilkington, M., Pesonen, L. J., Grieve, R. A. F., and Masaitis, V. L. (2002). Geophysics and petrophysics of the Popigai impact structure, Siberia. In: Impacts in Precambrian Shields, Plado, J., and Pesonen, L. J. (eds.), Impact Studies, Springer-Verlag, Berlin-Heidelberg, pp. 87–107.

Pilski, A. S., Wasson, J. T., Muszynski, A., Kryza, R., Karwowski, Ł., and Nowak, M. (2013). Low-Ir IAB irons from Morasko and other locations in central Europe: One fall, possibly distinct from IAB-MG. Meteoritics and Planetary Science, Vol. 48, pp. 2531–2541.

Pirajno, F., Hawke, P., Glikson, A. Y., Haines, P. W., and Uysal, T. (2003). Shoemaker impact structure, Western Australia. Australian Journal of Earth Sciences, Vol. 50, pp. 775–796.

Plado, J. (2012). Meteorite impact craters and possibly impact-related structures in Estonia. Meteoritics and Planetary Science, Vol. 47, pp. 1590–1605.

Plado, J., Pesonen, L. J., Elo, S., Purra. V., and Suuroja, K. (1996). Geophysical research on the Kardla impact structure, Hiiumaa Island, Estonia. Meteoritics and Planetary Science, Vol. 31, pp. 289–298.

Plado, J., Hietala, S., Kreitsmann, T., Lerssi, J., Nenonen, J., and Pesonen, L. (2018). Summanen, a new meteorite impact structure in Central Finland. Meteoritics and Planetary Science, Vol. 53, pp. 2413–2426.

Plescia, J. B. (1996). Gravity investigation of the Manson impact structure, Iowa. Geological Society of America Special Paper, Vol. 302, pp. 89–104.

Plescia, J. B., Shoemaker, E. M., and Shoemaker, C. S. (1994). Gravity survey of the Mount Toondina impact structure, South Australia. Journal of Geophysical Research, Vol. 99E, pp. 13 167–13 179.

Poag, C. W., Plescia, J. B., and Molzer, P. C. (2002). Ancient impact structures on modern continental shelves: The Chesapeake Bay, Montagnais, and Toms Canyon craters, Atlantic margin of North America. Deep-Sea Research Part II: Topical Studies in Oceanography, Vol. 49, pp. 1081–1102.

Poag, C. W., Koeberl, C., and Reimold, W. U. (2004). The Chesapeake Bay crater. Geology and Geophysics of a late Eocene Submarine Impact Structure. Impact Studies, Springer, Berlin Heidelberg, 522 pp.

Poelchau, M. H., and Kenkmann, T. (2008). Asymmetric signatures in simple craters as an indicator for an oblique impact direction. Meteoritics and Planetary Science, Vol. 43, pp. 2059–2072.

Poelchau, M. H., Kenkmann, T., and Kring, D. A. (2009): Rim uplift in simple craters: the effects of target heterogeneities and trajectory obliquity. Journal of Geophysical Research, Vol. 114. doi: 10.1029/2008JE003235.

Pohl, J., Stöffler, D., Gall, H., and Ernstson, K. (1977). The Ries impact crater. In: Impact and Explosion Cratering, Roddy, D. J., Pepin, R. O., and Merrill, R. B. (eds.), Pergamon Press, New York, pp. 343–404.

Pösges, G., and Schieber, M. (2015). The Ries Crater Museum Nördlingen – Museums Guide. Academy Bulletin Nr. 253, Bavarian Academy for Teacher Training, Dillingen, Pfeil-Verlag, München, 128 pp.

Powars, D. S., and Bruce, T. S. (1999). The effects of the Chesapeake Bay impact crater on the geological framework and correlation of hydrogeologic units of the lower York-James peninsula, Virginia: US Geological Survey Professional Paper, Vol. 1612, 82 pp.

Prado, R., Espin Fenoll, I., Miura, G., Ullah, I., Crósta, A. P., Zanon dos Santos, R., Elis, W., Machado, E., Reimold, W. U., Diogo, L., and Riccomini, C. (2018). Geophysical investigation of the Colônia structure, Brazil. Meteoritics and Planetary Science, Vol. 54, pp. 2357–2372.

Pratesi, G., Morelli, M., Rossi, A. P., and Ori, G. G. (2005). Chemical compositions of impact melt breccias and target rocks from the Tenoumer impact crater, Mauritania. Meteoritics and Planetary Science, Vol. 40, pp. 1653–1672.

Puura, V. A., and Suuroja, K. (1992). Ordovician impact crater at Kärdla, Hiiumaa Island, Estonia. Tectonophysics, Vol. 216, pp. 143–156.

Pye, E. G., Naldrett, A. J., and Giblin, P. E. (eds.) (1984). The Geology and Ore Deposits of the Sudbury Structure, Ontario Geological Survey, Special Volume 1, 603 pp.

Rae, A. S. P., Collins, G. S., Grieve, R. A. F., Osinski, G. R., and Morgan, J. V. (2017). Complex crater formation: Insights from combining observations of shock pressure distribution with numerical models at the West Clearwater Lake impact structure. Meteoritics and Planetary Science, Vol. 52, pp. 1330–1350.

Raiskila, S., Salminen, J., Elbra, T., and Pesonen, L. J. (2011). Rock magnetic and palaeomagnetic study of the Keurusselkä impact structure, central Finland. Meteoritics and Planetary Science, Vol. 46, pp. 1670–1687.

Rana, S., and Agarwal, V. (2016). Microscopic evidences for the impact origin of Ramgarh structure, Rajasthan, India. Journal of Indian Geophysical Union, Vol. 20, pp. 544–550.

Raschke, U., Schmitt, R. T., and Reimold, W. U. (2013). Petrography and geochemistry of impactites and volcanic bedrock in the ICDP drill core D1c from Lake El'gygytgyn, NE Russia. Meteoritics and Planetary Science, Vol. 48, pp. 1251–1286.

Raschke, U., Schmitt, R. T., McDonald, I., Reimold, W. U., Mader, D., and Koeberl, C. (2015). Geochemical studies of impact breccias and country rocks from the El'gygytgyn impact structure, Russia. Meteoritics and Planetary Science, Vol. 50, pp. 1071–1088.

Rathbun, K., Ukstins-Peate, I., Drop, S., and Gutierrez, F. (2016). A new geologic map of Monturaqui meteorite impact crater, Chile: Implications for satellite-based geologic mapping of small craters. 47th Lunar and Planetary Science Conference, Lunar and Planetary Institute, Houston, Texas, Abstract #2583.

Raukas, A., Tiirmaa, R., Kaup, E., and Kimmel, K. (2001). The age of the Ilumetsä meteorite craters in southeast Estonia. Meteoritics and Planetary Science, Vol. 36, pp. 1507–1514.

Read, W. F. (1983). Shatter cones at Glover Bluff, Wisconsin. Meteoritics, Vol. 18, pp. 241–243.

Read, W. F. (1984). The circular structure at Glover Bluff: What and where it is (abstract). Meteoritics, Vol. 19, pp. 295–296.

Read, W. F. (1985). Impact bombs in Middle Ordovician marine sediments of southeastern Wisconsin (abstract). 16th Lunar and Planetary Science Conference, Lunar and Planetary Institute, Houston, Houston, Texas, pp. 687–688.

Redeker, H. J., and Stöffler, D. (1988). The allochthonous polymict breccia layer of the Haughton impact crater, Devon Island, Canada. Meteoritics, Vol. 23, pp. 185–196.

Reeves, F., and Chalmers, R. O. (1949). The Wolfe Creek crater. The Australian Journal of Science, Vol. 11, pp. 154–156.

Reidel, S. P., Koucky, F. L., and Stryker, R. J. (1982). The Serpent Mound disturbance, southwestern Ohio. American Journal of Science, Vol. 282, pp. 1343–1375.

Reiff, W. (1977). The Steinheim Basin – An impact structure. In: Impact and Explosion Cratering, Roddy, D. J., Pepin, R. O., and Merrill, R. (eds.), New York. Pergamon Press, pp. 309–320.

Reiff, W. (1979): Guide-book to the Steinheim Basin Impact Crater. Geologisches Landesamt Baden-Württemberg, Stuttgart, 32 pp.

Reimold W. U. (1982). The Lappajärvi meteorite crater, Finland: petrography, Rb-Sr, major and trace element geochemistry of the impact melt and basement rocks. Geochimica et Cosmochimica Acta, Vol. 46, pp. 1203–1225.

Reimold, W. U. (2006). Impact structures in South Africa. In: The Geology of South Africa, Johnson, M. R., Anhaeusser, C. R., and Thomas, R. J. (eds.), Geological Society of South Africa, Johannesburg and Council for Geoscience, Pretoria, pp. 629–649.

Reimold, W. U., and Miller, R. McG. (1989). The Roter Kamm impact crater, SWA/Namibia. Proceedings of the 19th Lunar and Planetary Science Conference, Houston, Cambridge University Press, pp. 711–732.

Reimold, W. U., and Koeberl, C. (2003). Petrography and geochemistry of a deep drill core from the edge of the Morokweng impact structure, South Africa. In: Impact Markers in the Stratigraphic Record. Impact Studies Series, Koeberl, C., and Martinez-Ruiz, F. (eds.), Springer, Berlin Heidelberg, pp. 271–292.

Reimold, W. U., and Gibson, R. L. (2010). Meteorite Impact! The Danger from Space and South Africa's Mega-Impact, the Vredefort Structure (3rd edition), Springer, Berlin Heidelberg, 337 pp.

Reimold, W. U., and Koeberl, C. (2014). Impact structures in Africa: A review. Journal of African Earth Sciences, Vol. 93, pp. 57–175.

Reimold, W. U., Oskierski, W., and Huth, J. (1987). The pseudotachylite from Champagnac in the Rochechouart meteorite crater, France. Journal of Geophysical Research, Vol. 92, pp. E737–E748.

Reimold, W. U., Barr, J. M., Grieve, R. A. F., and Durrheim, R. J. (1990). Geochemistry of the melt and country rocks of the Lake St. Martin impact structure, Manitoba, Canada. Geochimica et Cosmochimica Acta, Vol. 54, pp. 2093–2111.

Reimold, W. U., Koeberl, C., and Reddering, J. S. V. (1998). The Kalkkop impact crater, Eastern Cape Province, South Africa: stratigraphy, petrography, and geochemistry of the 1992 drill core, and age. Journal of African Earth Sciences, Vol. 26, pp. 573–592.

Reimold, W. U., Koeberl, C., and Brandt, D. (1999). The origin of the Pretoria Saltpan crater. In: Investigations into the Origin, Age and Paleoenvironments of the Pretoria Saltpan, Partridge, T. C. (ed.), Memoir 85, Council for Geoscience, Pretoria, pp. 35–54.

Reimold, W. U., Koeberl, C., Hough, R. M., Mcdonald, I., Bevan, A., Amare, K., and French, B. M. (2003). Woodleigh impact structure, Australia: shock petrography and geochemical studies. Meteoritics and Planetary Science, Vol. 38, pp. 1109–1130.

Reimold, W. U., Kelley, S. P., Sherlock, S. C., Henkel, H., and Koeberl, C. (2005). Laser argon dating of melt breccias from the Siljan impact structure, Sweden: Implications for a possible relationship to Late Devonian extinction events. Meteoritics and Planetary Science, Vol. 40, pp. 591–607.

Reimold, W. U., Koeberl, C., Gibson, R. L., and Dressler, B. O. (2005). Economic mineral deposits in impact structures: A review. In: Impact Tectonics, Koeberl, C. and Henkel, H. (eds.), Springer, Berlin Heidelberg, pp. 479–552.

Reimold, W. U., Fischer, L., Müller, J., Kenkmann,T., Schmitt, R. T., Altenberger, U., and Kowitz, A. (2015). Impact generated pseudotachylitic breccia in drill core BH-5 Hättberg, Siljan impact structure, Sweden. GFF 137, Vol. 2, pp. 141–162.

Reimold, W. U., Hauser, N., Hansen, B. T., Thirlwall, M., and Hoffmann, M. (2017). The impact pseudotachylitic breccia controversy: Insights from first isotope analysis of Vredefort PTB and impact melt rock (Vredefort Granophyre). Geochimica et Cosmochimica Acta, Vol. 214, pp. 266–281.

Reimold, W. U., Crósta, A. P., Hasch, M., Kowitz, A., Hauser, N., Sanchez, J. P., Amarante Simões, L. S., de Oliveira, G. J., and Zaag, P. T. (2018). Shock deformation confirms the impact origin for the Cerro do Jarau, Rio Grande do Sul, Brazil, structure. Meteoritics and Planetary Science, Vol. 54, pp. 2384–2397.

Reimold, W. U., Vasconcelos, M. A. R., Jessell, M., Crósta, A. P., Ferrière, L., Hauser, N., Gottwald, M. and Baratoux, D. (2020). Nova Colinas – A new complex impact structure in Maranhão State, Brazil. Contribution to 50th Congress of the Sociedade Brasileira de Geologia (SBG), Brasilia, October 2020.

Reinwald, I. (1928). Bericht über geologische Untersuchungen am Kaalijärv (Krater von Sall) auf Ösel. Loodusuurijate Seltsi Aruanded, Vol. 35, pp. 30–70.

Riccomini, C., Crósta, A. P., Prado, R. L., Ledru, M. P., Turq, B. J., Sant'Anna, L. G., Ferrari, J. A., and Reimold, W. U. (2011). The Colônia structure, São Paulo, Brazil. Meteoritics and Planetary Science, Vol. 46, pp. 1630–1639.

Richard-Molard, J. (1948). Le cratère d'explosion de Tenoumer et l'existence probable d'une grande fracture rectiligne au Sahara occidental. Comptes Rendus de l'Académie des Sciences Paris, Vol. 227, pp. 213–214

Richards, M. A., Alvarez, W., Self, S., Karlstrom, L., Renne, P. R., Manga, M., Sprain, C. J., and Smit, J. (2015). Triggering of the largest Deccan eruptions by the Chicxulub impact. Geological Society of America Bulletin, Vol. 127, pp. 1507–1520.

Riis, F., Kalleson, E., Dypvik, H., Krøgly, S. O., and Nilsen, O. (2011). The Ritland impact structure, southwestern Norway. Meteoritics and Planetary Science, Vol. 46, pp. 748–761.

Riller, U. (2005). Structural characteristics of the Sudbury impact structure, Canada: Impact-induced versus orogenic deformation – A review. Meteoritics and Planetary Science, Vol. 40, 1723–1740.

Riller, U., Poelchau, M. H., Rae, A. S. P., Schulte, F. M., Collins, G. S., Melosh, H. J., Grieve, R. A. F., Morgan, J. V., Gulick, S. P. S., Lofi, J., Diaw, A., McCall, N., Kring, D. A., and IODP–ICDP Expedition 364 Science Party (2018). Rock fluidization during peak-ring formation of large impact structures. Nature, Vol. 562, pp. 511–518.

Robertson, P. B (1968). La Malbaie structure, Quebec. A Paleozoic impact site. Meteoritics, Vol. 4, pp. 89–112.

Robertson, P. B. (1975). Zones of shock metamorphism at the Charlevoix impact structure, Quebec. Geological Society of America Bulletin, Vol. 86, pp. 1630–1638.

Robertson, P. B., and Mason, G. D. (1975). Shatter cones from Haughton Dome, Devon Island, Canada. Nature, Vol. 255, pp. 393–394.

Robertson, P. B., and Grieve, R. A. F. (1977). Shock attenuation at terrestrial impact structures. In: Impact and Explosion Cratering, Roddy, D. J., Pepin, R. O., and Merrill, R. B. (eds.), Pergamon Press, New York, pp. 687–702.

Robertson, P. B., and Grieve, R. A. F. (1978). The Haughton impact structure. Meteoritics, Vol. 13, pp. 615–618.

Robertson, P. B., Dence, M. R., and Vos, M. A. (1968). Deformation in rock forming minerals from Canadian craters. In: Shock Metamorphism of Natural Materials, French, B. M., and Short, N. M. (eds.), Mono Book Corp., Baltimore, pp. 433–452.

Rochette, P., Alac, R., Beck, P., Brocard, G., Cavosie A. J., Debaille, V., Devouard, B., Jourdan, F., Mougel, B., Moustard, F., Moynier, F., Nomade, S., Osinski, G. R., Reynard, B., and Cornec, J. (2019). Pantasma: Evidence for a Pleistocene circa 14 km diameter impact crater in Nicaragua. Meteoritics and Planetary Science, Vol. 54, pp. 880–901.

Roddy, D. J. (1968). The Flynn Creek crater, Tennessee. In: Shock Metamorphism of Natural Materials, French, B. M., and Short, N. M., (eds.), Mono Book Corp., Baltimore, pp. 291–322.

Roddy, D. J. (1977a). Tabular comparisons of the Flynn Creek impact crater, United States, Steinheim impact crater, Germany, and Snowball explosion crater, Canada. In Impact and Explosion Cratering, Roddy, D. J., Pepin, R. A., and Merrill, R. B. (eds.), Pergamon Press, New York, pp. 125–161.

Roddy, D. J. (1977b). Pre-impact conditions and cratering processes at the Flynn Creek crater, Tennessee. In: Impact and Explosion Cratering, Roddy, D. J., Pepin, R. A., and Merrill, R. B. (eds.), Pergamon Press, New York, pp. 277–308.

Roddy, D. J. (1978). Pre-impact geologic conditions, physical properties, energy calculations, meteorite and initial crater dimensions and orientations of joints, faults and walls at Meteor Crater, Arizona. Proceedings of the 9th Lunar and Planetary Science Conference, Houston, Texas, Pergamon Press, New York, pp. 3891–3930.

Roland, N. W. (1976). Die Ringstruktur Aorounga (Borkou, Sud-Sahara). Geologisches Jahrbuch, Vol. 33, pp. 117–131.

Rondot, J. (1966). Geology of La Malbaie area, Charlevoix County. Department of Natural Resources, Quebec. Preliminary report, Vol. 544, 18 pp.

Rothery, D. A., McBride, N., and Gilmour, I. (eds.) (2018). An Introduction to the Solar System. Cambridge University Press, 3rd edition, 440 pp.

Roussell, D. H., and Brown, G. H. (2009). A Field Guide to the Geology of Sudbury, Ontario. Open File Report 6243, Ontario Geological Survey, Sudbury, 222 pp.

Sage, R. P. (1978). Diatremes and shock features in Precambrian rocks of the Slate Islands, northeastern Lake Superior. Geological Society of America, Bulletin 89, pp. 1529–1540.

Sahoui, R., Belhai, D., and Jambon, A. (2013). Impact-generated carbonate melts from the Talemzane structure (Algeria). 44th Lunar and Planetary Science Conference, Lunar and Planetary Institute, Houston, Texas, Abstract #1184.

Salameh, E., Khoury, H., and Schneider, W. (2006). Jebel Waqf as Suwwan, Jordan, a possible impact crater – a first approach. Zeitschrift der deutschen Gesellschaft für Geowissenschaften, Vol. 157, pp. 319–325.

Salameh, E., Khoury, H., Reimold, W. U., and Schneider W. (2008). The first large meteorite impact structure discovered in the Middle East: Jebel Waqf as Suwwan. Meteoritics and Planetary Science. Vol. 43, pp. 1681–1690.

Salminen, J. M., Donadini, F., Pesonen, L. J., Masaitis, V. L., and Naumov, M. V. (2006). Paleomagnetism and petrophysics of the Jänisjärvi impact structure, Russian Karelia. Meteoritics and Planetary Science, Vol. 41, pp. 1853–1870.

Sanchez, J., and Cassidy, W. (1966). A previously undescribed meteorite crater in Chile. Journal of Geophysical Research, Vol. 71, pp. 4891–4895.

Sandbakken, P. T., Langenhorst, F., and Dypvik, H., (2005). Shock metamorphism of quartz at the submarine Mjølnir impact crater, Barents Sea. Meteoritics and Planetary Science, Vol. 40, pp. 1363–1375.

Sawatzky, H. B. (1972). Viewfield – a producing crater? Canadian Society of Exploration Geophysicists Journal, Vol. 8, pp. 22–40.

Sawatzky, H. B. (1975). Astroblemes in the Williston Basin. Journal of the Canadian Society of Exploration Geophysicists, Vol. 10, pp. 23–38.

Sawatzky, H. B. (1976). Two probable late Cretaceous astroblemes in Western Canada – Eagle Butte, Alberta, and Dumas, Saskatchewan. Geophysics, Vol. 41, pp. 1261–1271.

Sawatzky, H. B. (1977). Buried impact craters in the Williston basin and adjacent area. In: Impact and Explosion Cratering, Roddy, D. J., Pepin, R. O., and Merrill, R. B. (eds.), Pergamon Press, New York, pp. 461–480.

Schedl, A. (2006). Applications of twin analysis to studying meteorite impact structures. Earth and Planetary Science Letters, Vol. 244, pp. 530–540.

Scherler, D., Kenkmann, T., and Jahn, A. (2006). Structural record of an oblique impact. Earth and Planetary Science Letters, Vol. 248, pp. 28–38.

Schmidt, G., Palme, H., and Kratz, K.-L. (1997). Highly siderophile elements (Re, Os, Ir, Ru, Rh, Pd, Au) in impact melts from three European impact craters (Sääksjärvi, Mien, and Dellen): Clues to the nature of the impacting bodies. Geochimica et Cosmochimica Acta, Vol. 61, pp. 2977–2987.

Schmieder, M., and Jourdan, F. (2013). The Lappajärvi impact structure (Finland): Age, duration of crater cooling, and implications for early life. Geochimica et Cosmochimica Acta, Vol. 112, pp. 321–339.

Schmieder, M., and Kring, D. A. (2020). Earth's impact events through geologic time: A list of recommended ages for terrestrial impact structures and deposits. Astrobiology, Vol. 20, pp. 1–51, DOI: 10.1089/ast.2019.2085.

Schmieder, M., Moilanen, J., and Buchner, E. (2008). Impact melt rocks from the Paasselkä impact structure (SE Finland): Petrography and geochemistry. Meteoritics and Planetary Science, Vol. 43, pp. 1189–1200.

Schmieder, M., Buchner, E., Schwarz, W., Trieloff, M., and Lambert, P. (2010). A Rhaetian ^{40}Ar-^{39}Ar age for the Rochechouart impact structure (France) and implications for the latest Triassic sedimentary record. Meteoritics and Planetary Science, Vol. 45, pp. 1225–1242.

Schmieder, M., Schwarz, W. H., Buchner, E., Trieloff, M., Moilanen, J., and Ohman, T. A. (2010). A Middle-Late Triassic ^{40}Ar-^{39}Ar age for the Paasselkä impact structure (SE Finland). Meteoritics and Planetary Science, Vol. 45, pp. 572–582.

Schmieder, M., Jourdan, F., Tohver, E., and Cloutis, E. A. (2014). ^{40}Ar-^{39}Ar age of the Lake Saint Martin impact structure (Canada); unchaining the late Triassic terrestrial impact craters. Earth and Planetary Science Letters, Vol. 406. pp. 37–48.

Schmieder, M., Trieloff, M., Schwarz, W. H., Buchner, E., and Jourdan, F. (2014). Supportive comment on: "Morphology and population of binary asteroid impact craters", by K. Miljkovic, G. S. Collins, S. Mannick and P. A. Bland [Earth Planet. Sci. Lett. 363 (2013) 121–132]. An updated assessment. Earth and Planetary Science Letters, Vol. 405, pp. 281–284.

Schmieder, M., Schwarz, W. H., Trieloff, M., Tohver, E., Buchner, E., Hopp, J., and Osinski, G. R. (2015). New ^{40}Ar-^{39}Ar dating of the Clearwater Lake impact structures (Québec, Canada) – Not the binary asteroid impact it seems? Geochimica et Cosmochimica Acta, Vol. 148, pp. 304–324.

Schmieder, M., Jourdan, F., Moilanen, J., Buchner, E., and Öhman, T. (2016). A late Mesoproterozoic $^{40}Ar/^{39}A$ age for a melt breccia from the Keurusselkä impact structure, Finland. Meteoritics and Planetary Science, Vol. 51, pp. 303–322.

Schmieder, M., Schwarz, W. H., Trieloff, M., Buchner, E., Hopp, J., Tohver, E., Pesonen, L. J., Lehtinen, M., Moilanen, J., Werner, S. C., and Öhman, T. (2016). The two Suvasvesi impact structures, Finland: Argon isotopic evidence for a "false" impact crater doublet. Meteoritics and Planetary Science, Vol. 51, pp. 966–980.

Schmieder, M., Shaulis, B. J., Lapen, T. J., Buchner, E., and Kring, D. A. (2019). In situ U-Pb analysis of shocked zircon from the Charlevoix impact structure, Quebec, Canada. Meteoritics and Planetary Science, Vol. 54, pp. 1808–1827.

Schulte, P., Alegret, L., Arenillas, I., Arz, J. A., Barton, P. J., Bown, P. R., Bralower, T. J., Christeson, G. L., Claeys, P., Cockell, C. S., Collins, G. S., Deutsch, A., Goldin, T. J., Goto, K., Grajeles-Nishimura, J. M., Grieve, R. A. F., Gulick, S. P. S., Johnson, K. R., Kiessling, W., Koeberl, C., Kring, D. A., MacLeod, K. G., Matsui, T., Melosh, J., Montanari, A., Morgan, J. V., Neal, C. R., Nichols, D. J., Norris, R. D., Pierazzo, E., Ravizza, G., Rebolledo-Vieyra, M., Reimold, W. U., Robin, E., Salge, T., Speijer, R. P., Sweet, A. R., Urrutia-Fucugauchi, J., Vajda, V., Whalen, M. T., and Willumsen, P. (2010). The Chicxulub asteroid impact and mass extinction at the Cretaceous-Paleogene Boundary. Science, Vol. 327, pp. 1214–1218.

Schultz, P. H., and Lianza, R. E. (1992). Recent grazing impacts on the Earth recorded in the Rio Cuarto crater field, Argentina. Nature, Vol. 355, pp. 234–237.

Schultz, P. H., Koeberl, C., Bunch, T., Grant, J., and Collins, W. (1994). Ground truth for oblique impact processes: new insight from the Rio Cuarto, Argentina, crater field. Geology, Vol. 22, pp. 889–892.

Schultz, P. H., Zarate M. A., Hames, W., Koeberl, C., Bunch, T., Storzer, D., Renne, P. R, and Witke, J. (2004). The Quaternary impact record from the Pampas, Argentina. Earth and Planetary Science Letters, Vol. 219, pp. 211–238.

Schultze, D. S., Jourdan, F., Hecht, L., Reimold, W. U., and Schmitt, R. T. (2016). Tenoumer impact crater, Mauritania: Impact melt genesis from a lithologically diverse target. Meteoritics and Planetary Science, Vol. 51, pp. 323–350.

Schulze, T., Kenkmann, T., Poelchau, M. H., Sundell, K. A., and Cook, D. (2019). Douglas impact crater strewn field, WY, USA: a progress report. Large Meteorite Impact VI, Brasilia, Abstract #5036.

Sergienko, E. S., Kosterov, A., and Kharitonskii, P. V. (2017). Two types of impact melts with contrasting magnetic mineralogy from Jänisjärvi impact structure, Russian Karelia. Geophysical Journal International, Vol. 209, pp. 1080–1094.

Setså, R. (2011). The Ritland Impact Structure: Characteristics and Distribution of the Ejecta Layer and associated Lower Paleozoic sedimentary Succession. Masters Thesis (unpublished), University of Oslo, 114 pp.

Sharma H. S. (1973). Ramgarh structure, India. Nature, Vol. 242, pp. 39–40.

Sharpton, V. L., and Gibson, J. W., Jr. (1990). The Marquez Dome impact structure, Leon County, Texas (abstract). 21st Lunar and Planetary Science Conference, Lunar and Planetary Institute, Houston, Texas, pp. 1136–1137.

Sharpton, V. L., Dressler, B. O., Herrick, R. R., Schnieders, B., and Scott, J. (1996). New constraints on the Slate Islands impact structure, Ontario, Canada. Geology, Vol. 24, pp. 851–854.

Sharpton, V. L., Krochuk, V. R., and Herrick, R. R. (2013). Characterization and morphological reconstruction of the Terny impact structure, central Ukraine. Meteoritics and Planetary Science, Vol. 48, pp. 806–818.

Sherlock, S. C., Kelley, S. P., Glazovskaya, L., and Ukstins Peate, I. (2009). The significance of the contemporaneous Logoisk impact structure (Belarus) and Afro-Arabian flood volcanism. Journal of the Geological Society, Vol. 166, pp. 5–8.

Shkerin, L. M. (1976). Geological structure of the crater-like feature Tabun Khara Obo (south east Mongolia) (in Russian). Meteoritika, Vol. 35, pp. 97–102.

Shmayenok, A. I., and Malakhovskiy, F. B. (1974). Explosion pipe near the southeast shore of Lake Chud' (in Russian). Vestnik Leningrad University, Vol. 24, pp. 97–107.

Shoemaker, E. M., and Chao, E. C. T. (1961). New evidence for the impact origin of the Ries basin, Bavaria, Germany. Journal of Geophysical Research, Vol. 66, pp. 3371–3378.

Shoemaker, E. M., and Kieffer, S. W. (1974). Guidebook to the Geology of Meteor Crater, Arizona. 37th Annual Meeting of The Meteoritical Society, Arizona State University Center for Meteorite Studies, Tempe, Arizona, 66 pp.

Shoemaker, E. M., and Herkenhoff, K. E. (1983). Impact origin of Upheaval Dome, Utah. Eos, Transactions American Geophysical Union, Vol. 44, p. 747.

Shoemaker, E. M., and Shoemaker, C. S. (1985). Impact structures of Western Australia. Meteoritics, Vol. 20, pp. 754–756.

Shoemaker, E. M., and Shoemaker, C. S. (1986). Connolly Basin: a probable eroded impact crater in Western Australia. 17th Lunar and Planetary Science Conference, Houston, Texas, pp. 797–798.

Shoemaker, E. M., and Shoemaker, C. S. (1988). Impact structures of Australia (1987) (Abstract). 19th Lunar and Planetary Science Conference, Houston, Texas, #1547.

Shoemaker, E. M., and Shoemaker, C. S. (1989). Geology of the Connolly Basin impact structure, Western Australia. 20th Lunar and Planetary Science Conference, Houston, Texas, pp. 1008–1009.

Shoemaker, E. M., and Shoemaker, C. S. (1996). The Proterozoic impact record of Australia. AGSO Journal of Australian Geology and Geophysics, Vol. 16, pp. 379–398.

Shoemaker, E. M., and Shoemaker, C. S. (1997a). Glikson, a probable impact structure, Western Australia. 28th Lunar and Planetary Science Conference, Houston, Texas, Abstract #1669.

Shoemaker, E. M., and Shoemaker, C. S. (1997b). Notes on the geology of Liverpool Crater, Northern Territory, Australia. 28th Lunar and Planetary Science Conference, Houston, Texas, Abstract #1662.

Shoemaker, E. M., Roddy, D. J., Shoemaker, C. S., and Roddy, J. K. (1988). The Boxhole Meteorite Crater, Northern Territory, Australia. 19th Lunar and Planetary Science Conference, Houston, Texas, pp. 1081–1082.

Shoemaker, E. M., Shoemaker, C. S., and Plescia, J. B. (1989). Gravity investigation of the Connolly Basin impact structure; Western Australia. 20th Lunar and Planetary Science Conference, Houston, Texas, pp. 1010–1011.

Shoemaker, E. M., Shoemaker, C. S., Nishiizumi, K., Kohl, C. P., Arnold, J. R., Klein, J., Fink, D., Middleton, R., Kubik, P. W., and Sharma, P. (1990). Ages of Australian meteorite craters – A preliminary report (abstract). Meteoritics, Vol. 25, p. 409.

Shoemaker, E. M., Macdonald F. A., and Shoemaker, C. S. (2005). Geology of five small Australian impact craters. Australian Journal of Earth Sciences, Vol. 52, pp. 529–544.

Short, N. M. (1967). Petrographic evidence for an impact origin for the West Hawk Lake Structure, Manitoba, Canada. Eos, Transactions of the American Geophysical Union, Vol. 48, p. 147.

Short, N. M. (1968). Petrographic study of shocked rocks from the Steen River structure, Alberta. In: Shock Metamorphism of Natural Materials, French, B. M., and Short, N. M. (eds.), Mono Book Corp., Baltimore, pp. 374–378.

Short, N. M. (1970). Anatomy of a meteorite impact crater: West Hawk Lake, Manitoba, Canada. Geological Society of America Bulletin, Vol. 81, pp. 609–648.

Shrock, R. R., and Malott, C. A. (1933). The Kentland area of disturbed Ordovician rocks in northwestern Indiana. Journal of Geology, Vol. 41, pp. 337–370.

Shumilova, T. G., Lutoev, V. P., Isaenko, S. I., Kovalchuk, N. S., Makeev, B. A., Lysiuk, A. Yu., Zubov, A. A., and Ernstson, K. (2018). Spectroscopic features of ultrahigh-pressure impact glasses of the Kara astrobleme. Scientific Reports, Vol. 8, No. 6923.

Shuvalov, V. V., and Dypvik, H., (2004). Ejecta formation and crater development of the Mjølnir impact. Meteoritics and Planetary Science, Vol. 39, pp. 467–479.

Shuvalov, V. V., Dypvik, H., and Tsikalas, F. (2002). Numerical simulations of the Mjølnir marine impact crater. Journal of Geophysical Research, Vol. 107, E7.

Shuvalov, V. V., Dypvik H., Kalleson, E., Setsa R., and Riis, F. (2012). Modeling the 2.7 km in diameter, shallow marine Ritland impact structure. Earth Moon Planets, Vol. 108, pp. 175–188.

Sighinolfi, G. P., Sibilia, E., Contini, G., and Martini, M. (2015). Thermoluminescence dating of the Kamil impact crater (Egypt). Meteoritics and Planetary Science, Vol. 50, pp. 204–213.

Simmonds, C. H., Phinney, W., McGee, P., and Cochran, A. (1978). West Clearwater, Quebec, impact structure. Part I: Field geology, structure and bulk chemistry. Proceedings of the 10th Lunar and Planetary Science Conference, Houston, Texas, Pergamon Press, New York, pp. 2633–2658.

Sisodia, M. S., Lashkari G., and Bhandari, N. (2006). Impact origin of the Ramgarh structure, Rajasthan: Some new evidences. Journal of Geological Society of India, Vol. 67, pp. 423–431.

Sokur, T. M. (1999). Shock metamorphism of quartz in rocks from the Zapadnaya impact structure (in Russian). Geologicheskii Zhurnal Vol. 1, pp. 87–93.

Sontheim Meteorkrater-Museum: https://www.steinheim-am-albuch.de/tourismus-a-freizeit/meteorkrater-museum.

Spencer, L. J. (1933). Meteorite craters as topographic features of the earth's surface. Geographic Journal, Vol. 81, pp. 227–248.

Spooner, I., Stevens, G., Morrow, J., Pufahl, P., Grieve, R., Raeside, R., Pilon, J., Stanley, C., Barr, S., and McMullin, D. (2009). Identification of the Bloody Creek structure, a possible impact crater in southwestern Nova Scotia, Canada. Meteoritics and Planetary Science, Vol. 44, pp. 1193–1202.

Spray, J. G. (1997). Superfaults. Geology, Vol. 25, pp. 627–630.

Spray, J. G., Kelley, S. P., and Rowley, D. B. (1998). Evidence for a late Triassic multiple impact event on Earth. Nature, Vol. 392, pp. 171–173.

Spray, J. G., Kelley, S. P., and Dence, M. R. (1999). The Strangways impact structure, Northern Territory, Australia: geological setting and laser probe ^{40}Ar-^{39}Ar geochronology. Earth and Planetary Science Letters, Vol. 172, pp. 199–211.

Spray, J. G., Thompson, L. M., Biren, M. B., and O'Connell-Cooper, C. D. (2010). The Manicouagan impact structure as a terrestrial analogue site for lunar and martian planetary science. Planetary and Space Science, Vol. 58, pp. 538–551.

Stearns, R. G., Wilson, C. W., Jr., Tiedemann, H. A., Wilcox, J. T., and Marsh, P. S. (1968). The Wells Creek structure, Tennessee. In: Shock Metamorphism of Natural Materials, French, B. M. and Short, N. M. (eds.), Mono Book Corp., Baltimore, pp. 323–338.

Stevens, G. R. (1995). Meteor Crater, Nova Scotia. The Journal of the Royal Astronomical Society of Canada, Vol. 89, pp. 111–113.

Stevens, G. R., Spooner, I. S., Morrow, J. R., Pufahl, P., Raeside, R., Grieve, R. A. F., Stanley, C., Barr, S., and McMullin, D. (2007). Physical evidence of a late-glacial (Younger Dryas?) impact event in southwestern Nova Scotia (abstract). Eos Transactions, American Geophysical Union, Fall Meeting Supplement, Vol. 88.

Stewart, A., and Mitchell, K. (1987). Shatter cones at the Lawn Hill circular structure, northwestern Queensland: presumed astrobleme. Australian Journal of Earth Sciences, Vol. 34, pp. 477–485.

Stewart, S. A. (2015). Circular geological structures outcropping in the sedimentary basins of Saudi Arabia. Journal of Asian Earth Sciences, Vol. 106, pp. 95–118.

Stöffler, D. (1977). Research drilling Nördlingen 1973: polymict breccias, crater basement, and cratering model of the Ries impact structure. Geologica Bavarica, Vol. 75, pp. 443–458.

Stöffler, D., Artemieva, N. A., and Pierazzo, E. (2002). Modeling the Ries-Steinheim impact event and the formation of the moldavite strewn field. Meteoritics and Planetary Science, Vol. 37, pp. 1893–1907.

Stöffler, D., Reimold, W. U., Jacob, J., Hansen, B. K., Summerson, I. A. T., Artemieva, N. A., and Wünnemann, K. (2013). Ries crater and suevite revisited – Observations and modeling, Part I: Observations. Meteoritics and Planetary Science, Vol. 48, pp. 515–589.

Stone, D. S. (1999). Cloud Creek: A possible impact structure on the Casper Arch, Wyoming. The Mountain Geologist, Vol. 36, Rocky Mountain Association of Geologists, pp. 212–234.

Stone, D. S., and Therriault, A. M. (2003). The Cloud Creek structure, central Wyoming, U.S.A. – Impact origin confirmed. Meteoritics and Planetary Science, Vol. 38, pp. 445–455.

Storzer, D., Reimold, W. U., and Koeberl, C. (1999). Fission-track age of the Saltpan impact crater. In: Investigations into the Origin, Age and Paleoenvironments of the Pretoria Saltpan, Partridge, T. C. (ed.), Memoir 85, Council for Geoscience, Pretoria, pp. 64–71.

Storzer, D., Sélo, M., Latouche, L., and Fabre, J. (2003). The age of Tenoumer crater, Mauritania, revisited. 34th Lunar and Planetary Science Conference, Lunar and Planetary Institute, Houston, Texas, Abstract #1183.

Sturkell, E. F. F. (1998). The marine Lockne impact structure, Jämtland, Sweden: a review. Geologische Rundschau, Vol. 87, pp. 253–267.

Sturm, S., Wulf, G., Jung, D., and Kenkmann, T. (2013). The Ries impact, a double-layer rampart crater on Earth. Geology, Vol. 41, pp. 531–534.

Suetenko, O. D., and Shkerin, L. M. (1970). A proposed meteorite crater in southeastern Mongolia (in Russian). Astronomicheskii Vestnik, Vol. 4, pp. 261–263.

Suuroja, K., and Suuroja, S. (2000). Neugrund structure – the newly discovered submarine Early Cambrian impact crater. In: Impacts and the Early Earth, Gilmour, I., and Koeberl, C. (eds.), Springer, Berlin Heidelberg, Vol. 91, pp. 389–416.

Suuroja, S., Suuroja, K., and Flodén, T. (2013). A comparative analysis of two Early Palaeozoic marine impact structures in Estonia, Baltic Sea: Neugrund and Kärdla. Bulletin of the Geological Society of Finland, Vol. 85, pp. 83–101.

Svensson, N. B. (1968a). Lake Lappajärvi, Central Finland: a possible meteorite impact crater. Nature, Vol. 217, p. 438.

Svensson, N. B. (1968b). The Dellen lakes: a probable meteorite impact in central Sweden. Geologiska Föreningen i Stockholm Förhandlingar, Vol. 90, pp. 14–316.

Svensson, N. B. (1969). Lake Mien, southern Sweden – a possible astrobleme. Geologiska Föreningen i Stockholm Förhandlingar, Vol. 91, pp. 101–110.

Svensson, N. B. (1971). Probable meteorite impact crater in central Sweden. Nature, Vol. 229, pp. 90–92.

Svensson, N. B. (1993). Lumparn Bay: A meteorite impact crater in the Åland archipelago, southwest Finland. Meteoritics. Vol. 28, pp. 245.

Svensson, N. B., and Wickman, F. E. (1965). Coesite from Lake Mien, southern Sweden. Nature, Vol. 205, pp. 1202–1203.

Sweet, I. P., Haines, P. W., Mitchell, K., and Glikson, A. Y. (2005). Matt Wilson Structure: record of an impact event of possible early Mesoproterozoic age, Northern Territory. Australian Journal of Earth Sciences, Vol. 52, pp. 675–688.

Tagle, R., and Claeys, P. (2005). An ordinary chondrite impactor for the Popigai crater, Siberia. Geochimica et Cosmochimica Acta, Vol. 69, pp. 2877–2889.

Tagle, R., Schmitt, R. T., and Erzinger, J. (2009). Identification of the projectile component in the impact structures Rochechouart, France and Sääksjarvi, Finland; implications for the impactor population for the Earth. Geochimica et Cosmochimica Acta, Vol. 73, pp. 4891–4906.

Therriault, A. M., and Grantz, A. (1995). Planar deformation features in quartz grains from mixed breccia of the Avak structure, Alaska (abstract). 26th Lunar and Planetary Science Conference, Houston, Texas, Lunar and Planetary Institute, pp. 1403–1404.

Therriault, A. M., Fowler, A. D., and Grieve, R. A. F. (2002). The Sudbury Igneous Complex: A differentiated impact melt sheet. Economic Geology, Vol. 97, pp. 1521–1540.

Thomas, M. D., and Innes, M. J. S. (1977). The Gow Lake impact structure, northern Saskatchewan. Canadian Journal of Earth Sciences, Vol. 14, pp. 1788–1795.

Thompson, L. M., and Spray, J. G. (2017). Dynamic interaction between impact melt and fragmented basement at Manicouagan: The suevite connection. Meteoritics and Planetary Science, Vol. 52, pp. 1300–1329.

Tohver, E., Cawood, P. A., Riccomini, C., and Trindade, R. I. F. (2013). Shaking a methane fizz: Seismicity from the Araguainha impact event and the Permian–Triassic global carbon isotope record. Palaeogeography, Palaeoclimatology, Palaeoecology, Vol. 387, pp. 66–75.

Tonkin, P. E. (1973). Discovery of shatter cones at Kelly West near Tennant Creek, Northern Territory, Australia. Geological Society of Australia Journal, Vol. 20, pp. 99–102.

Trepmann, C. A., and Spray, J. G. (2005). Planar microstructure and Dauphiné twins in shocked quartz of the Charlevoix impact structure, Canada. In: Large Meteorite Impacts III, Kenkmann, T., Hörz, F., and Deutsch, A. (eds.), Geological Society of America Special Paper, Vol. 384, pp. 315–328.

Trieloff, M., Deutsch, A., and Jessberger, E. (1998). The age of the Kara impact structure, Russia. Meteoritics and Planetary Science, Vol. 33, pp. 361–372.

Tsikalas, F., and Faleide, J. I. (2007). Post-impact structural crater modification due to sediment loading: An overlooked process. Meteoritics and Planetary Science, Vol. 42, pp. 2013–2029.

Tsikalas, F., Gudlaugsson, S. T., and Faleide, J. I. (1998). The anatomy of a buried complex impact structure: the Mjølnir structure, Barents Sea. Journal of Geophysical Research, Vol. 103, pp. 30469–30483.

Ubide, T. P., Guyett, C., Kenny, G. G., O'Sullivan, E. M., Ames, D. E., Petrus, J. A., Riggs, N., and Kamber, B. S. (2017). Protracted volcanism after large impacts: Evidence from the Sudbury impact basin. Journal of Geophysical Research, Planets, Vol. 122, pp. 701–728.

Ugalde, H., Valenzuela, M., and Milkereit, B. (2007). An integrated geophysical and geological study of the Monturaqui impact crater, Chile. Meteoritics and Planetary Science, Vol. 42, pp. 2153–2163.

Underwood, J. R. (1975). Reconnaissance geology of meteorite impact structures in SE Libya. Abstracts and Programs of the Geological Society of America, Vol. 7, p. 242.

Underwood, J. R. (1976). Impact structures in the Libyan Sahara: some comparisons with Mars. Proceedings of the International Colloquium on Planetary Geology. Geologica Romana, Vol. 15, pp. 337–340.

Underwood, J. R., and Fisk, E. P. (1980). Meteorite impact structures, Southeast Libya. 2nd Symposium on the Geology of Libya, Tripoli, Libya, pp. 893–900.

Urbini, S., Nicolosi, I., Zeoli, A., El Khrepy, S., Lethy, A., Hafez, M., El Gabry, M., El Barkooky, A., Barakat, A., Gomaa, M., Radwan, A., El Sharkawi, M., D'Orazio, M., and Folco, L. (2012). Geological and geophysical investigation of Kamil Crater, Egypt. Meteoritics and Planetary Science, Vol. 47, pp. 1842–1868.

Uutela, A. (2001). Proterozoic and early Palaeozoic microfossils in the Karikkoselkä impact crater, Central Finland. Bulletin of the Geological Society of Finland, Vol. 73, Parts 1–2, pp. 75–85

Uysal, I. T., Golding, S. D., Glikson, A. Y., and Mory, A. J. (2001). K-Ar evidence from illitic clays of a Late Devonian age for the 120 km diameter Woodleigh impact structure, Southern Carnarvon Basin, Western Australia. Earth and Planetary Science Letters, Vol. 192, pp. 281–289.

Valenzuela, M., Rochette, P., Bourlès, D. L., Braucher, R., Faestermann, T., Finkel, R. C., Gattacceca, J., Korschinek, G., Merchel, S., Morata, D., Poutivtsev, J., Rugel, G., and Suavet, C. (2009). The age of Monturaqui impact crater. 72nd Annual Meeting of the Meteoritical Society, Nancy, France, Abstract #5185.

Val'ter, A. A., and Ryabenko, V. A. (1977). Explosion Craters of the Ukrainian Shield (in Russian). Naukova Dumka Press, Kiev, Ukraine. 154 pp.

Val'ter, A. A, Bryansky, V. P., Ryabenko, V. A., and Lazarenko, E. (1976). On the explosion (meteoritic) origin of Zeleny Gai structure on the Ukrainian Shield (in Russian). Doklady ANSSSR, Vol. 229, pp. 160–162.

Val'ter, A. A., Gurov, E. P., and Ryabenko, V. A. (1976). The Obolon' fossil meteorite crater astrobleme on the northeast flank of the Ukrainian Shield (in Russian). Doklady Akademii Nauk SSSR, Vol. 232, pp. 170–173.

Van Lopik, J. R., and Geyer, R. A. (1963). Gravity and magnetic anomalies of the Sierra Madera, Texas, "Dome". Science, Vol. 142, pp. 45–47.

Vasconcelos, M. A. R., Crósta, A. P., and Molina, E. C. (2010). Geophysical characteristics of four possible impact structures in the Parnaíba Basin, Brazil: Comparison and Implications. In: Large Meteorite Impacts and Planetary Evolution IV, Gibson, R. L. and Reimold, W. U. (eds.), Geological Society of America Special Paper, Vol. 465, pp. 201–217.

Vasconcelos, M. A. R., Leite, E. P., and Crósta, A. P. (2012). Contributions of gamma-ray spectrometry to terrestrial impact crater studies: the example of Serra da Cangalha, northeastern Brazil. Geophysical Research Letters, Vol. 39, L04306.

Vasconcelos, M. A. R., Crósta, A. P., Reimold, W. U., Goes, A. M., Kenkmann, T., and Poelchau, M. H. (2013). The Serra da Cangalha impact structure, Brazil: Geological, stratigraphic and petrographic aspects of a recently confirmed impact structure. Journal of South American Earth Sciences, Vol. 45, pp. 316–330.

Vasilyev, N. V. (1998). The Tunguska meteorite problem today. Planetary and Space Science, Vol. 46, pp. 129–150.

Veretennikov, N. V. (2010). The Logoisk impact crater. In: The Sun, the Stars, The Universe and General Relativity: International Conference in Honor of Ya.B. Zeldovich's 95th Anniversary, Minsk, Belarus, Ruffini, R., and Vereshchagin, G. (eds.), AIP Conference Proceedings, Vol. 1205, pp. 185–190.

Veretennikov, N. V., Il'kevich, G., and Makhnacha, S. (1979). Logoisk buried depression – an ancient meteorite crater (in Russian). Doklady Acad. Nauk BSSR, Vol. 23, pp. 156–160.

Vesconi, M. A., Wright, S. P., Spagnuolo, M., Jacob, R., Cerrutti, C., Garcia, L., Fernandez, E., and Cassidy, W. A. (2011). Comparison of four meteorite penetration funnels in the Campo del Cielo crater field, Argentina. Meteoritics and Planetary Science, Vol. 46, pp. 935–949.

Veski, S., Heinsalu, A., Lang, V., Kestlane, Ü., and Possnert, G. (2004). The age of the Kaali meteorite craters and the effect oft he impact on the environment and man: evidence from inside the Kaali craters, island os Saaremaa, Estonia. Vegetation History and Archaeobotany, Vol. 13, pp. 197–206.

Vincent, P. M., and Beauvilain, A. (1996). Découverte d'un nouveau cratère d'impact météoritique en Afrique: l'astroblème de Gweni Fada (Ennedi, Sahara de Tchad). Comptes Rendus de l'Académie des Sciences, Paris, Vol. 323, pp. 987–997.

Vishnevsky, S. A. (1995). The Chykcha impact crater, Taymyr peninsula: Heavily eroded astrobleme of Cretaceous-Paleogene age (abstract). Meteoritics, Vol. 30, p. 591.

Vishnevsky, S. A. (1999). Ragozinka: Eocenian marine impact crater on the middle Urals, Russia, Meteoritics and Planetary Science, Vol. 34 Supplement, pp. A117.

Vishnevsky, S. A. (2007). Shiyli Dome, Kazakhstan: Origin of central uplift by elastic response. Bridging the Gap II: Effect of Target Properties on the Impact Cratering Process, Saint-Hubert, Canada, LPI Contribution No. 1360, pp. 125–126.

Vishnevsky, S. A., and Lagutenko, V. N. (1986). The Ragozinka Astrobleme: An Eocene crater in the central Urals (in Russian). Doklady Akademii Nauk SSSR, Vol. 14, 42 pp.

Vishnevsky, S. A., and Korobkov, V. F. (1989). Chiyli Dome, an eroded impact structure in western Primugodzharye (in Russian). Akademiya Nauk SSSR, pp. 1–51.

Vishnevsky, S. A., Maslov, M. A., Pal'chik, N. A., and Ponomarev, G. Y. (1977). Coesite in rocks of the Kara structure. Akademiya Nauk SSSR Doklady, Vol. 232, pp. 446–448.

Von Engelhardt, W., and Stöffler, D. (1965). Spaltflächen im Quarz als Anzeichen für Einschläge großer Meteoriten. Naturwissenschaften, Vol. 52, pp 489–490.

Von Engelhardt, W., Hörz, F., Stöffler, D., and Bertsch, W. (1968). Observations on quartz deformation in the breccias of West Clearwater Lake, Canada and the Ries Basin, Germany. In: Shock Metamorphism of Natural Materials, French, B. M., and Short, N. M. (eds.), Mono Book Corp., Baltimore, pp. 475–482.

Von Engelhardt, W., Matthai, S. K., and Walzebuck, J. (1992). Araguainha impact crater, Brazil. I. The interior part of the uplift. Meteoritics, Vol. 27, pp. 442–457.

VSEGEI. (2014). State Geological Maps 1:200,000. https://vsegei.ru/en/.

Wacrenier, P., Hudeley, H., and Vincent, P. M. (1958). Carte géologique du Borkou-Ennedi-Tibesti a 1/100000. Brazzaville, Direction des Mines et la Géologie.

Walton, E. L. Sharp, T. G., Hu, J., and Tschauner, O. (2017). Investigating the response of biotite to impact metamorphism: Examples from the Steen River impact structure, Canada. Meteoritics and Planetary Science, Vol. 53, pp. 75–92.

Wartho, J. A., van Soest, M. C., King, D. T., and Petruny, L. W. (2012). An (U-Th)/He age for the shallow-marine Wetumpka impact structure, Alabama, USA. Meteoritics and Planetary Science, Vol. 47, pp. 1243–1255.

Weber, J. C. (2013). Impact geology: Central uplift, Kentland impact structure, Newton County Stone (Kentland) Quarry, Indiana, USA. In: Insights into the Michigan Basin, The Geological Society of America Field Guide, Vol. 31, pp. 1–7.

Werner, S. C., Plado, J., Pesonen, L. J., Janle, P., and Elo, S. (2002). Potential fields and subsurface models of Suvasvesi North impact structure, Finland. Physics and Chemistry of the Earth, Vol. 27, pp. 1237–1245.

Westbroek, H. H., and Stewart, R. R. (1995). The White Valley structure, Saskatchewan: a possible meteorite impact crater. CREWES research report, Chapter 19, 18 pp.

Westbroek, H. H. (1997). Seismic Interpretation of Two Possible Meteorite Impact Craters: White Valley, Saskatchewan and Purple Springs, Alberta. M.Sc.Thesis, University of Calgary, 145 pp.

Wickman, F. E. (1980). The impact center of the Siljan structure, central Sweden. Geologiska Föreningen i Stockholm Förhandlingar, Vol. 102, pp. 105–109.

Wickman, F. E. (1988). Possible impact structures in Sweden. In: Deep Drilling in Crystalline Bedrock. Volume 1, Bodén, A., and Eriksson, K. G. (eds.), Springer, Berlin Heidelberg, pp. 298–327.

Williams, G. E. (1986). The Acraman impact structure: Source of ejecta in Late Precambrian shales, South Australia. Science, Vol. 233, pp. 200–203.

Williams, G. E., and Wallace, M. W. (2003). The Acraman asteroid impact, South Australia: magnitude and implications for the late Vendian environment. Journal of the Geological Society, Vol. 160, pp. 545–554.

Williams, G. E., and Gostin, V. A. (2011). Acraman-Bunyeroo impact event (Ediacaran), South Australia, and environmental consequences: twenty-five years on. Australian Journal of Earth Sciences, Vol. 52, pp. 607–620.

Williams, G. E., Schmidt, P. W., and Boyd, D. M. (1996). Magnetic signature and morphology of the Acraman impact structure, South Australia. AGSO Journal of Australian Geology and Geophysics, Vol. 16, pp. 431–442.

Willman, H. B. (1971). Summary of the Geology of the Chicago Area: Illinois. Geological Survey Circular, Vol. 460, 77 pp.

Wilshire, H. G., and Howard, K. A. (1968). Structural patterns in central uplifts of cryptoexplosion structures as typified by Sierra Madera. Science, Vol. 162, pp. 258–261.

Wilshire, H. G., Offield, T. W., Howard, K. A., and Cummings, D. (1972). Geology of the Sierra Madera cryptoexplosion structure, Pecos County, Texas. U.S. Geological Survey Professional Paper, Vol. 599-H, 42 pp.

Włodarski, W., Papis, J., and Szczuciński, W. (2017). Morphology of the Morasko crater field (western Poland): Influences of pre-impact topography, meteoroid impact processes, and post-impact alterations. Geomorphology, Vol. 295, pp. 586–597.

Wolf, R., Woodrow, A. B., and Grieve, R. A. F. (1980). Meteoritic material at four Canadian impact craters. Geochimica et Cosmochimica Acta, Vol. 44, pp. 1015–1022.

Wong, A. M., Reid, A. M., Hall, S. A., and Sharpton, V. L. (2001). Reconstruction of the subsurface structure of the Marquez impact crater in Leon County, Texas, based on well-log and gravity data. Meteoritics and Planetary Science, Vol. 36, pp. 1443–1455.

Wulf, G., Hergarten, S., and Kenkmann, T. (2018). Combined remote sensing analyses and landform evolution modeling reveal the terrestrial Bosumtwi impact structure as a Mars-like rampart crater. Earth and Planetary Science Letters, Vol. 506, pp. 209–220.

Wünnemann, K., Morgan, J. V., and Jödicke, H. (2005). Is Ries crater typical for its size? An analysis based upon old and new geophysical data and numerical modeling. Geological Society of America Special Paper, Vol. 384, pp. 67–83.

Wynn, J. C., and Shoemaker, E. M. (1998). The day the sands caught fire. Scientific American, Vol. 279, pp. 3645.

Yarmolyuk, V. A. (1951). The Sobolevka crater (in Russian). Priroda, Vol. 6, pp. 40–42.

Yeates, A. N., Crowe, R. W. A., and Towner, R. R. (1976). The Veevers crater: a possible meteoritic feature. BMR Journal of Australian Geology and Geophysics, Vol. 1, pp. 77–78.

Yokoyama, E., Trindade, R. I. F., Lana, C. R., Souza-Filho, C. R., Baratoux, D., Marangoni, Y. R., and Tohver, E. (2012). Magnetic fabric of Araguainha complex impact structure (Central Brazil): Implications for deformation mechanisms and central uplift formation. Earth and Planetary Science Letters, Vol. 331–332, pp. 347–359.

Youles, I. P. (1976). Mount Toondina impact structure. Quarterly Geological Notes, The Geological Survey of South Australia, Vol. 60, pp. 10–11.

Young, K. E., van Soest, M. C., Hodges, K. V., Watson, E. B., Adams, B. A., and Lee, P. (2013). Impact thermochronology and the age of Haughton impact structure, Canada. Geophysical Research Letters, Vol. 40, pp. 3836–3840.

Young, K. E., Mercer, C. M., van Soest, M. C., Hodges, K. V., Wartho, J. A., and Biren, M. B. (2014). An examination of noble gas geochronology and thermochronology in the context of dating impact events. 45th Lunar and Planetary Science Conference, Lunar and Planetary Institute, Houston, Texas, Abstract #2670.

Zink, M. (2015). TanDEM-X mission status. The International Archives of the Photogrammetry, Remote Sensing and Spatial Information Sciences XL-7/W3, pp. 1345–1352.

Zink, M., Bachmann, M., Bräutigam, B., Fritz, T., Hajnsek, I., Krieger, G., Moreira, A., and Wessel, B. (2014). TanDEM-X: The new global DEM takes shape. IEEE Geoscience and Remote Sensing Magazine, Vol. 2, pp. 8–23.

Zumsprekel, H., and Bischoff, L. (2005). Remote sensing and geographic information system analyses of the Strangways impact structure, Northern Territory. Australian Journal of Earth Sciences, Vol. 52, pp. 621–630.

Glossary

Achondrite: Stony differentiated *meteorite* not containing *chondrules* and much nickel-iron. Of all *meteorite* types the achondrites most closely resemble terrestrial rocks.

Airburst: Explosion of a *meteoroid* in the lower atmosphere. This generates a *shock wave* which can produce considerable damage on ground.

Allochthonous: Applies to material that originated from a location other than where it is presently found. In impact cratering this may refer to rock ejected from the crater during its formation or injected into a setting which is different from its point of genesis.

Alluvial: Applies to environments on land which have been formed by the action of rivers and streams.

Amphibole: A family of silicate minerals that have crystal structures with double chains of silica tetrahedra and contain hydroxyl groups in their structure. Upon heating they may decompose to a *pyroxene* mineral with related but single-chain crystal structure under release of the hydroxyl molecule.

Anticline: Arch-shaped fold in rocks, closing upward (in the case of layered rocks) with the oldest rocks in the core. Opposite of *syncline*.

Anorthosite: An *igneous rock* consisting mostly of *plagioclase*. A major constituent of the Lunar Highland crust.

Aphelion: The point in an orbit of a celestial body around the Sun furthest away from the Sun.

Archean: Earlier part of *Precambrian* time, from 4000 until 2500 Myr ago.

Asteroid: Small rocky or metallic body in orbit around the Sun. The majority of *asteroids* is found in the *Asteroid Belt* between Mars and Jupiter.

Autochthonous: Applies to material that originated from a location where it is found. Opposite of *allochthonous*.

Ballen structure: Microscopic thermal alteration feature in silica, i. e., most commonly quartz, *clasts*. Not an established diagnostic shock feature; its occurrence indicates high temperatures.

Basalt: Dark-colored, *mafic igneous*, extrusive rock relatively low in SiO_2 content, composed predominantly of *plagioclase*, *pyroxene*, and olivine.

Bedding: Layering of sheet-like units, called beds or strata, which are the smallest distinguishable units of a layered *sedimentary rock*.

Bedrock: Lithified rock underneath soil, gravel, or other loose surface material of the Earth's crust.

Biotite: Rock-forming sheet silicate mineral.

Bolide: Very bright, large *meteor*; often used as synonymous with (extraterrestrial) projectile.

Bouguer anomaly: Remaining gravity field after taking corrections into account such as local theoretical gravity, elevation and topography correction, or free-air correction. Reflects all variations in density, which cannot be explained by a homogeneous Earth.

Breccia: Coarse-grained rock that is a mixture of angular, broken rock fragments from one (*monomict*) or more (*polymict*) types of rocks included into a fine-grained "matrix" that may be similar to or different from the fragmented material.

Bunte Breccia: Multi-colored (synonymous "variegated") *breccia* from the Ries *impact structure*, which consists of a mixture of all rock types of the target area, but mainly of blocks and fragments of the sedimentary *target rocks*. *Bunte Breccia* is considered material from the *ejecta curtain*.

Calcite: Common rock-forming carbonate mineral. At ambient conditions, it is the stable form of calcium carbonate ($CaCO_3$). The mineral has a great variety of crystal habits and crystallizes in the trigonal system.

Carbonaceous chondrite: Class of *chondrites* representing only about 5 % of all *meteorite* falls. Contains water-bearing minerals and carbon compounds including organic molecules such as amino acids. Carbonaceous chondrites have very primitive compositions similar to that of the protoplanetary nebula, of which the solar system formed.

Cataclastic: Texture of rock caused by compression, extension, or shearing, when brittle minerals have been broken and crushed.

Cenote: Type of *sinkhole* that contains ground water. Typically, on the Yucatán peninsula of Mexico and elsewhere in the Caribbean.

Cenozoic: The third era of the *Phanerozoic* lasting from about 66 Myr ago to the present. It includes the periods Paleogene and Neogene (both formerly comprising the Tertiary), and the Quaternary.

Central peak / Central uplift: *Target rocks* in the center of a *complex impact crater*, uplifted above their original stratigraphic level before the impact due to movements in the crater *modification stage*.

Chondrite: Stony, non-metallic *meteorite*, characterized by the presence of *chondrules*. Chondrites constitute about 86 % of the known *meteorite* falls.

Chondrule: Small spherical grain in chondritic *meteorites*, formed by melting and quenching of silicate material. Considered to represent one of the oldest solid materials in the solar system.

Clast: Fragment broken off a larger rock unit.

Coesite: High-pressure *polymorph* of silica (SiO_2), formed at high pressures and high temperatures.

Comet: Small icy bodies of the *Kuiper belt* or the *Oort cloud* orbiting the Sun in eccentric elliptical orbits. When passing close to the Sun, a comet begins to release gases. Outgassing produces a visible atmosphere or coma, and sometimes also a tail. These phenomena are due to the effects of solar radiation and the solar wind acting upon the nucleus of the comet.

Complex impact crater: Impact crater with a *central uplift*, annular trough, and terraced, down-faulted rim. Formed when the kinetic energy of the impacting body, defined by size and velocity, exceeds a certain value. The resulting crater has a low depth-to-diameter ratio in the order of 0.1. When erosion and tectonic processes have erased much of the complex crater's morphology, it is termed a "complex impact structure". The transition from simple to complex morphologies depends on gravity and on the strength of the target lithologies. On Earth, this transition diameter is at 2–4 km.

Contact and compression stage: First phase in the *impact crater* generation process when the projectile hits the surface with *hypervelocity*, penetrates the *target rocks* and generates a *shock wave* that travels into the target and back into the projectile. The phase ends when the *shock wave* is reflected at a free surface as a release wave.

Country rock: General term for rocks that build up a target prior to impact, or that occur in the surroundings of an *impact structure*.

Crater rim: In pristine craters the crater rim is defined topographically by the crestline of an elevated rim. In eroded craters the apparent crater rim is constrained by concentrically striking rim faults or rim monoclines. Alternative methods to define the crater rim use the extent of geophysical anomalies. The transient crater rim is the rim at the end of the *excavation stage* prior to crater modification.

Craton: Old and stable portions of the continental *lithosphere* of *Proterozoic* age, composed of *shield* regions, and they may include *platform* sediments.

Cretaceous-Paleogene (K-Pg) boundary: Stratigraphic boundary between the Cretaceous and the Paleogene, dated to 66 million years ago. This marks the turn from the *Mesozoic* to the *Cenozoic*. In the past termed "Cretaceous-Tertiary (K-T)" boundary.

Cryptoexplosion structure: Term used in the early 20th century for the description of circular geological structures in North America and elsewhere, where rocks had been observed which suggested a violent explosion without an obvious cause. Many such structures were later identified to be of impact origin.

Crystalline: Applied to *igneous* or *metamorphic rocks*, which were formed by crystallization or recrystallization from liquid or solid precursor material.

Dating: Determination of the age of a geological unit or structure by means of stratigraphic and paleontological data, or radiogenic isotopes. Applied methods may yield a relative timescale or in the case of *radiometric dating*, absolute time.

Diamond: Solid form of carbon with the carbon atoms in a cubic crystal structure. Diamond commonly has octahedral shape and is the hardest natural mineral. Transformation of graphite to diamond can occur upon shock loading.

Diaplectic glass: Glass developed by solid-state shock transformation of minerals – without melting. While the mineral itself is transformed to a quasi-amorphous material, the original forms of the grains and potential inclusions are preserved.

Digital Elevation Model (DEM): Array of regularly spaced elevation values referenced horizontally either to a certain projection or to a geographic coordinate system, usually including features such as vegetation and buildings.

Digital Terrain Model (DTM): Array of regularly spaced elevation values referenced horizontally either to a certain projection or to a geographic coordinate system representing "bare Earth", i. e., the actual terrain/surface topography without features.

Dike: Steeply dipping layer of rock formed in already existing rock by injection or intrusion.

Dip: The attitude of a rock layer perpendicular to its extent (the *strike*). A horizontal layer would have a dip of 0 degrees, whereas a vertical layer one of 90 degrees.

Distal ejecta: Impact ejecta that are found discontinuously at distance from the *impact crater* (typically further than 5 crater radii away).

Doppler history: Sequence of frequency shifts of *radar* pulses backscattered by a target, which is illuminated by the beam of a *radar* antenna that moves with respect to this target.

Ejecta: Material ejected from an *impact crater* during crater formation.

Ejecta blanket: Layer of *ejecta* surrounding an *impact crater*. Composed mainly of rock fragments from crater excavation and may include melt rocks.

Ejecta curtain: Material accelerated during crater excavation to move ballistically on parabolic trajectories. Forms an outward moving conically shaped curtain. Becomes *ejecta blanket* once the ejected material has reached the surface.

Eolian: Applies to processes and phenomena generated by wind.

Excavation stage: Second phase in the *impact crater* formation process that comprises *shock wave* propagation through the *target rocks*, excavation, and *transient cavity* growth. Excavation ends when the *transient cavity* reaches its maximum size and the outward directed flow at the cavity rim reverses to an inward directed flow.

Fabric: Spatial and geometric arrangement of the parts of a rock, such as the alignment of minerals in deformed or *metamorphic rocks*, or the pattern of particles in a *sedimentary rock*.

Fault: *Fracture* in a rock with an approximately planar surface, where a relative displacement between neighboring blocks can be observed.

Feather features (FFs): Type of planar microstructure in shocked grains. Feather Features occur as short, parallel to subparallel lamellae with spacings similar to that of *planar deformation features (PDFs)*. These lamellae branch-off a *planar fracture (PF)* in a comb-like (feather vane-like) manner.

Feldspar: Group of important rock-forming minerals of the general formula $[M(Al,Si)_3O_8]$, where M can be K, Na, Ca, and to a minor degree Ba, Rb, Sr, and Fe. Feldspars form the most abundant group of minerals in the Earth's crust, and basically can occur in all types of rocks.

Felsic: Applies to *igneous rocks* rich in felsic minerals such as *quartz* and *feldspar*. "Felsic" is derived from the terms "feldspar" and "silica".

Fracture: Any separation in a geological formation which forms two pieces of the rock. Fractures have irregular surfaces while *joints* commonly haven even surfaces.

Geomorphology: Scientific study of the forms of the solid terrestrial surface (land and ocean-floor) and the processes that have shaped them.

Gneiss: Type of *metamorphic rock* formed from *igneous rock* or *sedimentary rock* with a well-formed banded structure.

GPa: Abbreviation for Gigapascal (pressure unit, 10^9 Pa; 1 GPa = 10 kbar)

Granite: Coarse-grained, generally light-colored intrusive rock composed of the minerals *quartz* (SiO_2), *feldspar* and some dark minerals, such as *biotite* and/or *amphibole*.

Granitoid: *Igneous* or *metamorphic rocks* composed of the minerals typically occurring in *granite*. It is not necessarily a fresh *igneous rock* and can be metamorphosed.

Granophyre: Rock composed of irregular intergrowths of *quartz* and *feldspar* minerals, on a microscopic scale.

Gravimetry: Measurement of the gravitational acceleration, i. e., the strength of the gravitational field.

Gravity anomaly: see *Bouguer anomaly*.

Hadean: Very first part of the evolution of the Solar System, prior to about 4000 Myr, and best preserved in the Lunar rock record. On Earth, only a few thousand *zircon* grains from comparatively younger *sedimentary rock* have shown this kind of age.

Helix formation: Configuration of the two *radar* satellites of the TanDEM-X mission. The naming indicates that the orbits of both satellites are twisted around each other. This allows safe operations of close spacecraft formations, with minimum collision risk.

High-pressure minerals: Minerals formed at very high pressures. They have specific crystal structures (different from those of their lower pressure *polymorphs*) and become stable under these extreme conditions. So-called "high-pressure polymorphs" are produced in nature either through *hypervelocity* impacts or under static conditions at sufficient depth in the Earth's crust or mantle.

Hillshading: Method of applying light and shadow to elevation data by computational illumination of the scene with a light source, located at a certain elevation and azimuth.

Hugoniot elastic limit: Shock pressure limit below which shock-compressed rocks behave elastically and brittle deformation is dominant. Above this limit rocks undergo a *shock metamorphic* overprint.

Hydrothermal: Applies to processes involving transportation and circulation of hot water in fractured rocks. Hydrothermal processes involve cation and anion exchange between rock and brine and may lead to rock alteration and ore formation.

Hypervelocity: Velocity larger than the speed of sound in a particular medium. In the cratering process the speed of the impacting projectile usually exceeds 11 km/s, the escape velocity at the Earth's surface.

Igneous rock: Rock that has crystallized from a magma.

Impact crater: Commonly circular to sub-circular landform on the solid surface of a planetary body, formed by the *hypervelocity* impact of a projectile. Once the crater's morphology has been considerably modified by erosion or tectonic processes, the term *impact structure* is preferred.

Impact glass: Glass formed from *impact melt* that cooled so quickly that crystals of minerals did not have enough time to form. It is either fully glassy or only semihyaline.

Impact melt: Melt formed by shock-melting of the *target rocks* during an impact with *hypervelocity*.

Impact melt breccia: A *breccia* cemented together by glassy or crystallized *impact melt* derived from the melting of *target rocks*.

Impact melt rock: *Polymict breccia* that is composed of a coherent melt matrix with variable amounts of mineral and lithic *clasts*.

Impact structure: Geological feature on the solid surface of a planetary body formed by the *hypervelocity* impact of a projectile. Particularly used when the circular shape of the fresh crater has been erased by erosion or modified by tectonic processes.

Impactite: Series of lithologies, formed as a direct consequence of the impact event, within and around *impact structures*. Impactites can occur either proximal or distal to an impact site.

Impactor: Projectile that impacts onto a surface with *hypervelocity*.

Interferometry: Measurement techniques where electromagnetic waves are superimposed to cause interference. From interference patterns information such as distance or elevation can be derived (see *Synthetic Aperture Radar Interferometry – InSAR*).

Iron meteorite: Meteorite composed mainly of iron-nickel metal.

Isochron dating: *Radiometric dating* method using daughter and parent isotopes and, in addition for comparison, another isotope of the same element that is not the product of a decay process. Avoids the necessity of knowing the quantity of the initial daughter product in the rock.

Isopleths: Lines having the same vertical displacement relative to a pre-structural configuration.

Joint: Even *fracture* in a rock without significant displacement parallel to the *fracture* but with extension orthogonal to it. Joints develop during exhumation and unloading of stressed rocks or shrinking of rocks.

Kimberlite: An alkaline *igneous rock* that occurs in pipe-like structures. This material is derived from great depths within the Earth's mantle, where it may have picked up *diamonds*.

Kuiper Belt: Region beyond the orbit of Neptune out to a distance of about 50 AU (1 AU = 150 million km) in the outer solar system. The Kuiper Belt objects are composed of mainly ice, frozen volatiles, and had been formed in the early phase of the evolution of the solar system. Named after the Dutch-American astronomer Gerard Kuiper. Sometimes also referred to as "Edgeworth-Kuiper Belt" to honor the astronomer Kenneth E. Edgeworth who also proposed objects beyond Pluto in the 1940s.

Lechatelierite: Fused *silica* glass that has been formed at temperatures above 1713 °C. Lechatelierite can have an impact origin or may have been produced by lightning.

LiDAR: Method exploited in remote sensing meaning "Light Detection and Ranging". Using a laser, pulsed light is emitted. From the backscattered signals the range, i.e., the distance or the velocity of the backscatterer can be derived.

Lithic impact breccia: *Polymict* or *monomict breccia* that contains clastic components only, i.e., the result of mechanical disintegration of rock.

Lithology: Description of macroscopic properties of a rock visible in outcrop, in hand or core samples, or with low magnification microscopy. Physical properties include, e.g., composition, color, *texture*, and grain size.

Long-period comets: *Comets* with a *perihelion* close to the Sun and orbital periods in the order of up to tens of thousands of years. The *Oort Cloud* is considered the origin of the long-period comets.

Lunar highlands: Lunar crust of feldspathic composition, bright in appearance, and older than the dark maria. The highlands have preserved the early lunar impact cratering record.

Mafic: Applies to *igneous rock* or a silicate mineral rich in magnesium and iron.

Magnetic anomaly: Remaining magnetic field after taking corrections into account such as the International Geomagnetic Reference Field.

Maskelynite: *Diaplectic glass* formed by *solid-state transformation* from *plagioclase*. Named after the British geologist M. H. N. Story-Maskelyne.

Mesozoic: The second era of the *Phanerozoic* lasting from about 251 to 66 Myr ago. It includes the periods of the Triassic, Jurassic, and Cretaceous.

Metamorphic rock: Rock that formed through the transition of pre-existing rock in response to increased pressure and temperature.

Meteor: Transient luminous trail produced by a small *meteoroid* when entering the Earth's atmosphere. Very bright meteors are called "fireballs".

Meteorite: Rock or metal from space, found on Earth, which survived the entry of a large *meteoroid* or small *asteroid* into the Earth's atmosphere. Composition can be stony, iron or stony-iron.

Meteoroid: Small solid body with a size of micrometers to about one meter in interplanetary space or in flight through the atmosphere.

Migmatite: Composite rock composed of *igneous* or igneous-looking and/or *metamorphic* materials which are generally distinguishable megascopically.

Modification stage: Third phase in the *impact crater* formation process comprising all processes that modify the *transient cavity*. The driving force is gravity. Crater modification includes *crater rim* failure and the formation of slump terraces, *central uplift* formation, uplift collapse, and the deposition of *allochthonous breccias*.

Monomict: Applies to a brecciated rock, in which the *clasts* are composed of a single rock type.

Multi-ring impact structure: *Impact structure* with at least two concentric rings inside the impact basin. Multi-ring impact structures are known from the Moon and Mars and other planetary bodies. On Earth it has been debated whether the largest craters fall into this category.

Multispectral: Applies to remote sensing methods where electromagnetic radiation is measured in not more than about 10–15 spectral bands. These bands usually cover the wavelength ranges from ultraviolet (UV) to thermal infrared (TIR).

Nappe: Large *thrust* sheet of rock that has moved over neighboring strata, induced by horizontal compressive stresses.

Near Earth Objects (NEOs): *Asteroids* and *comets* with a *perihelion* distance of less than 1.3 times the orbital radius of Earth. In the case of a *comet* its orbital period must not exceed 200 years.

Normal fault: A *fault* in which the hanging wall (that block above the fault plane) appears to have moved downward. Normal faults develop in extensional stress regimes.

Oort Cloud: Hypothetical region surrounding the solar system in interstellar space at distances between 2000 AU (1 AU = 150 million km) and about 1.5 light years, composed of icy objects formed in the early phase of the evolution of the solar system. Named after the Dutch astronomer Jan Oort.

Orogeny: Mountain-building episode, as the result of collision between two lithospheric plates with continental crusts, or continental and oceanic crusts, or a continent and an *island arc*.

Overturned strata: Inverse stratigraphy at the rim, caused by the excavation of debris from the *impact crater*.

Paleozoic: The first era of the *Phanerozoic* lasting from about 542 to 251 Myr ago. It includes the periods Cambrian, Ordovician, Silurian, Devonian, Carboniferous, and Permian.

Parautochthonous: Applies to material with a status between *allochthonous* and *autochthonous*. In impact geology often attributed to rock which has been displaced by moderate slumping but where contact with the source still exists.

Peak ring: Centrally uplifted region of an *impact crater* characterized by a ring of peaks rather than a single peak. A peak ring forms when an overshot *central uplift* collapses under gravity.

Perihelion: The point in an orbit of a celestial body around the Sun closest to the Sun.

Petrography: Study of rock *textures* and mineralogy in thin section or hand samples.

Petrology: General study of rocks including occurrence, origin, field relations, formation conditions and petrographic properties.

Phanerozoic: Geologic time including the *Paleozoic*, *Mesozoic* and *Cenozoic* eras. It covers the time from about 542 Myr ago until the present.

Platform: Part of a *craton* where the basement is overlain by flat-lying sediments.

Potentially Hazardous Objects (PHOs): *Asteroids* and *comets* with a minimum orbit intersection distance with Earth of less than 7.5 Million kilometers and an absolute brightness of 22.0 mag, which can be translated into an approximate diameter of at least 140 meters.

Plagioclase: Important rock-forming mineral belonging to the group of framework silicates, called *feldspars*. Plagioclase substitutes parts of the silicon in the crystal lattice by aluminum and incorporates either Natrium or Calcium, or any mixture of both, in the crystal lattice.

Planetesimal: Small bodies of several kilometer size or more that formed in the protoplanetary disk through coalescence and gravitational attraction.

Planar deformation features (PDFs): Type of planar microstructure in shocked minerals, for example *quartz* or *feldspar*. They occur as narrow-spaced and thin parallel lamellae along certain crystallographic directions. In contrast to *planar fractures* they are not open cracks. Along PDF planes local amorphization of the crystal lattice may take place. PDFs are the most important shock criterion.

Planar fractures (PFs): Type of planar microstructure in shocked grains. Planar fractures occur as parallel sets of multiple (sub)planar cracks in a mineral grain, mostly in *quartz*.

Platinum group elements (PGE): Platinum group elements include platinum (Pt), palladium (Pd), iridium (Ir), osmium (Os), rhodium (Rh) and ruthenium (Ru). Their abundances in *impact melt rocks* may indicate the nature of the impacting projectile.

Polymict: Applies to a brecciated rock composed of fragments of different precursor rock types.

Polymorph: Forms of a mineral with identical chemical composition but different crystal structures. Polymorphs possess different physical properties because of their diverse atomic arrangements (e.g., *diamond* and graphite).

Precambrian: Geologic time preceding the *Phanerozoic* including the *Proterozoic*, *Archean* and *Hadean* eras. It covers the time from about 4600 to 542 Myr ago.

Proterozoic: Most recent period of the *Precambrian* from 2500 Myr to 542 Myr ago.

Proximal ejecta: Impact *ejecta* that are found in the vicinity of the *impact crater* (typically less than 5 crater radii). Proximal ejecta form continuous *ejecta blankets*.

Pseudotachylite: Glassy rock that occurs as dark veins and blind terminating injections into the *country rock*. Pseudotachylite is formed by frictional melting and is known from tectonic shear zones, landslides, and *impact structures*.

Pseudotachylitic breccia: A term applied when the genesis of a material resembling a *pseudotachylite* is not resolved. The different types of rock that might resemble a pseudotachyite of friction origin could be: (ultra) *cataclasite*, ultramylonite, *impact melt rock*, *monomict* or *polymict*, extremely fine-grained *lithic impact breccia*.

Pyroxene: Group of dark-colored silicate minerals of the general formula ($ABSi_2O_6$), where A can be Ca, Na, Mg, Fe^{2+} and B = Mg, Fe^{2+}, Fe^{3+}, Cr, Mn, Al. This group contains important minerals present in many *igneous* and *metamorphic rocks*, and also in *meteorites*.

Quartz: One of the most abundant rock-forming minerals in the Earth's crust (SiO_2).

Radar: Method exploited in remote sensing meaning "Radio Detection and Ranging". Radio or microwaves are transmitted by an antenna. From the backscattered signals the range, i.e., the distance or the velocity of the backscatterer can be derived.

Radiometric dating: Method of *dating* rocks through the measurement of the abundances of daughter and parent isotopes. It compares the concentration of a naturally occurring radioactive isotope within the rock to that of its decay product that forms at a known constant decay rate.

Rarefaction wave: Wave that is formed when the *shock wave* that was reflected back into the projectile upon impact, reaches the back end of the projectile and is reflected forward as a release wave. The rarefaction wave enters the *target rocks* and initiates their decompression.

Raw DEM: Digital elevation data derived for a scene of about 40 × 50 km extent from TanDEM-X acquisitions in single-pass interferometer configuration. All TanDEM-X interferometric data yielded about 570 000 overlapping Raw DEMs. The global TanDEM-X DEM has been processed from this repository.

Reidite: *Zircon* transforms into its high-pressure polymorph *reidite* when exposed to pressures in excess of 20 *GPa*.

RGB: Abbreviation for Red-Green-Blue. The combined RGB-channels yield the visible imagery in *multispectral* remote sensing.

Scattered Disk: Region in the outer solar system populated with small icy bodies having *aphelions* of more than 100 AU (1 AU = 150 million km). *Perihelions* can be as close as 30 AU. The inner region of the Scattered Disk overlaps with the *Kuiper Belt*.

Sedimentary rock: Rock that formed through the solidification of sediment that had been transported and deposited by water, ice or wind.

Shatter cone: Macroscopic deformation features uniquely formed by impact. Shatter cones exhibit a conical shape with striations radiating from its top (apex). Their sizes range from centimeters to several meters.

Shield: Part of a *craton* where *Precambrian crystalline* basement rocks outcrop.

Shock metamorphic effects: Modification in rocks or minerals caused by the passage of the *shock wave* generated upon impact. Shock metamorphic effects can be permanent after shock release or transient during compression.

Shock metamorphism: Changes in a mineral or rock induced by the ultra-high pressures and temperatures generated by the impact of an extraterrestrial projectile at *hypervelocity*.

Shock wave: Abnormal compressional wave radiating outward from the impact point at speeds exceeding the speed of sound in the affected *target rocks* and generates extreme pressures to the rocks. The affected rocks undergo unique deformations triggered by the high-pressure shock wave, are subject to heating, and set into motion.

Shock vaporization: Shock effect in *target rocks* at extremely high shock pressures above 100 *GPa*.

Short-period comets: *Comets* with an *aphelion* inside or near the inner solar system. Their orbital periods are usually less than 200 years. The *Kuiper Belt* and its associated *Scattered Disk* are considered to be the source of the short-period *comets*.

Siderophile elements: Applies to elements soluble in iron such as iridium, osmium, platinum, and palladium. They are relatively common in the metal-rich cores of differentiated solar system bodies and in undifferentiated meteorites but rare on the Earth's surface.

Simple impact crater: Impact crater with a bowl-shaped morphology and raised rim. The cavity is partly filled with *allochthonous* and *parautochthonous breccia*. Formed when the kinetic energy of the impacting body, defined by size and velocity, remains below a certain value so that the *crater modification* adjustments are moderate. The resulting crater has a depth-to diameter ratio in the order of 0.2.

Sinkhole: Karst weathering pit in limestone with steep sidewalls where groundwater is exposed.

Stishovite: *High-pressure polymorph* of *quartz* formed at pressures between about 12–15 *GPa*. Provides mineralogical evidence for impact origin of a suspected structure.

Stony meteorite: *Meteorite* composed mostly, but not only, of silicate minerals.

Stony-iron meteorite: *Meteorite* composed of both silicate minerals and iron-nickel metal in roughly similar proportions.

Stratum: A sheet-like layer of *sedimentary rock*, clearly distinguishable from the layers above and below.

Strike: Orientation defined by the intersection of an inclined bed with a horizontal plane.

Structural uplift: Observed amount of uplift of a lithological marker unit in a *complex impact structure* with respect to its position outside the crater.

Suevite: *Breccia* comprising mineral and lithic *clasts* of variable *shock metamorphic* degree but also including a melt fragment component. Fragments and glassy melt lumps are embedded in a fine-grained clastic matrix. The term "Suevite" originates from the Ries impact where the term originates from the Latin word "Suevia", the region "Schwaben".

Syncline: Where strata form a convex-down trough shape (layers dip towards the fold hinge). Opposite of *anticline*.

Synthetic Aperture Radar (SAR): Method in remote sensing to synthesize a large antenna with a narrow beam by transmitting microwave pulses from a short antenna with a wide beam by exploiting the Doppler shift of the moving transmitter. Processing of SAR data requires the application of complex algorithms and considerable computing effort.

Synthetic Aperture Radar Interferometry (InSAR): Method in remote sensing to derive elevation or deformation information for a surface. It measures the phase differences of the backscattered SAR pulses at either two closely spaced receiving antennas at the same time or with a single receiving antenna at two different times. These phase differences can be translated into linear dimensions under the provision that pulse parameters and antenna trajectories are known with high precision.

Target rocks: The rocks that an *asteroid* or *comet* hits in an impact event.

Tektite: Small, *silica*-rich glass fragment, often of splash-form shape. These small bodies are produced during the early phase of impact events when *target rocks* are melted and spread as droplets which become re-solidified. When dispersed over large distances, they may form extended tektite strewn fields, sometimes hundreds or even thousands of kilometers away from the source crater. Microscopic tektites are termed "Microtektites". They have been found in ocean sediments off some of the large tektite strewn fields.

Texture: General appearance of a rock, in terms of the mutual arrangements of and relationships between the particles or crystals constituting the rock.

Thrust: Fault with a dip of 45° or less, on which the hanging wall, i.e., the block above the fault plane, appears to have moved upwards relative to the footwall. Form in compressional settings.

Transient crater/Transient cavity: Unstable, bowl-shaped cavity which is formed at the end of the *excavation stage*.

X-band: One of the commonly used bands in *radar* remote sensing with a frequency/wavelength of 8–12.5 GHz/$\lambda = 2.5$–4 cm. Other exploited bands are the C-band (3.95–5.8 GHz, $\lambda = 4$–8 cm) and the L-band (1–2.6 GHz, $\lambda = 15$–30 cm).

Zircon: Mineral (zirconium orthosilicate – $ZrSiO_4$), present in the form of small grains or crystals in *igneous* and *metamorphic rock*. Zircon was formed already very early in Earth's history. The oldest dated terrestrial zircon crystals are 4.4 billion years old. Zircon is a very stable mineral resisting weathering.

Chemical Elements

Al	Aluminium
Ar	Argon
Ba	Barium
Be	Beryllium
C	Carbon
Ca	Calcium
Cl	Chlorine
Co	Cobalt
Cr	Chromium
Cu	Copper
Fe	Iron
He	Helium
Ir	Iridium
K	Potassium
Mg	Magnesium
N	Nitrogen
Nd	Neodymium
Ni	Nickel
O	Oxygen
Os	Osmium
Pb	Lead
Pd	Palladium
Pt	Platinum
Rb	Rubidium
Re	Rhenium
Ru	Ruthenium
Si	Silicium
Sr	Strontium
Th	Thorium
U	Uranium
Zr	Zirconium

Abbreviations and Acronyms

AD	Anno Domini
AICAC	Arab Impact Cratering and Astrogeology Conference
ALOS	Advanced Land Observing Satellite
ASTER	Advanced Space-Borne Thermal Emission and Reflection Radiometer
BCE	Before Current Epoch
BP	Before Present
CEA	Central Elevated Area
CFGC	Central Finland Granitoid Complex
CIRIR	Center for International Research and Restitution on Impacts and on Rochechouart
CPRM	Serviço Geológico do Brasil
CTBT	Nuclear-Test Ban Treaty
DEM	Digital Elevation Model
DLR	Deutsches Zentrum für Luft- und Raumfahrt
DRC	Democratic Republic of Congo
EBSD	Electron Back-Scatter Diffraction
EGM96	Earth Gravitational Model 1996
EID	Earth Impact Database
FF	Feather Feature
GPa	GigaPascal (10^9 Pascal)
GSW	Geological Society of Washington
ICDP	International Scientific Continental Drilling Program
IODP	International Ocean Discovery Program
ITCZ	Intertropical Convergence Zone
K-T	Cretaceous-Tertiary
K-Pg	Cretaceous-Paleogene
LDG	Libyan Desert Glass
LiDAR	Light Detection and Ranging
mGal	milliGal (10^{-3} Gal)
MPa	MegaPascal (10^6 Pascal)
Myr	Million years (10^6 years)
NASA	National Aeronautics and Space Administration
NEO	Near Earth Object
NMC	Namaqua Metamorphic Complex
nT	nanoTesla (10^{-9} Tesla)
OCVB	Ochotsk-Chukotsky Volcanic Belt
PASSC	Planetary and Space Science Center
PDF	Planar Deformation Feature
PEMEX	Petróleos Mexicanos
PF	Planar Fracture
PGE	Platinum Group Elements
PHO	Potentially Hazardous Object
PM	Planar Microstructure
PRISM	Panchromatic Remote-sensing Instrument for Stereo Mapping
PTB	Pseudotachylitic Breccia
RGB	Red-Green-Blue
SAR	Synthetic Aperture Radar
SBG	Sociedade Brasileira de Geologia
SHRIMP	Sensitive High-Resolution Ion Microprobe
SIC	Sudbury Igneous Complex
SIMS	Secondary Iron Mass Spectroscopy
SIR-A	Shuttle Imaging Radar
SRTM	Shuttle Radar Topography Mission
TanDEM-X	TerraSAR add-on for Digital Elevation Measurement
TIMS	Thermal Ionization Mass Spectrometry
TNT	Tri-nitro-toluol
U.S.	United States
USGS	United States Geological Survey
UTC	Universal Time Coordinated
UTM	Universal Transverse Mercator
WGS84	World Geodetic System 1984
2D	2-dimensional
3D	3-dimensional

Impact Geology Index

Cartographic Index

Africa

North/Central America

South America

Asia

Australia

Europe

Manfred Gottwald was educated in astronomy and physics at the Ludwig-Maximilians-Universität in Munich. After his Ph.D. in 1983 he worked in high-energy astrophysics for the European Space Agency and the Max-Planck Institute for Extraterrestrial Physics, studying objects far out in our galaxy and beyond. When he joined the German Aerospace Center at Oberpfaffenhofen, our solar system and Earth became the scientific topics of choice. At the Earth Observation Center he was involved in many space-borne missions investigating our atmosphere, our cryosphere and the terrestrial surface. Particularly challenging was his responsibility for the atmospheric science instrument SCIAMACHY on the European ENVISAT platform. Since the International Polar Year 2007/2008 he coordinated the national Earth Observation missions in support of polar science under the auspices of the World Meteorological Organization. This familiarized him with the TanDEM-X radar mission, whose high-resolution digital elevation model allowed Manfred to engage in impact craters, a field where astronomy meets geology. His books, papers and talks, both for scientific communities and the interested public, cover the entire universe, from its distant edge right to the surface of our home planet.

Thomas Kenkmann studied geology and paleontology at the University of Cologne and completed his dissertation in 1997 at the Freie Universität Berlin. Since 2010 he has been Professor of Geology and Structural Geology at the University of Freiburg, Germany. Thomas Kenkmann and his working group are investigating the structure and deformation inventory of impact craters on Earth and other planetary bodies at scales ranging from satellite imagery to micrometers. Thomas is a passionate field geologist, who has studied more than 40 terrestrial impact craters worldwide, and mapped many of them geologically, among them Upheaval Dome in the United States, Jebel Waqf as Suwwan in Jordan, Serra da Cangalha in Brazil, and Matt Wilson, and Gosses Bluff in Australia. He is responsible for a number of crater discoveries and confirmations such as Saqqar in Saudi Arabia, the Douglas crater strewn field in the United States, or Ramgarh in India. He is also experimentalist and initiated and coordinated a multidisciplinary consortium to understand meteorite impact into various rock types by conducting shock and cratering experiments. He has published a multitude of refereed scientific articles in journals and books and edited several special issues and a book on impact craters. In 2012, Thomas Kenkmann won the State Teaching Award of the Ministry of Education, Research and Art of the Federal State of Baden-Württemberg, Germany. Thomas Kenkmann received the Barringer Medal and Award of the Meteoritical Society in 2018 for his key contributions to the areas of impact crater research.

Wolf Uwe Reimold holds M.Sc. (1977) and Ph.D. (1980) degrees in Mineralogy from the University of Münster (Germany). Post-doctoral research at NASA Johnson Space Center in Houston (USA) was followed by research positions and then a Professorship in Mineralogy at the University of the Witwatersrand (Johannesburg, South Africa, 1984–2005). Uwe's main research interests have been multidisciplinary Impact Crater and Shock Metamorphism studies, besides some Economic Geology and Regional Geology exploits in southern Africa and Ethiopia. Impact research was mainly focused on structures in Africa, Europe, and North America, including a major contribution to the understanding of the Vredefort impact structure in South Africa. Other main research targets were Bosumtwi (Ghana), Roter Kamm (Namibia), and various structures in Scandinavia, and in recent years in Brazil. In 2006, Uwe transferred to Humboldt University and the Museum für Naturkunde in Berlin, where he served as Professor of Mineralogy and Petrography, and Head of the Evolution and Geosciences Division of the Museum. In 2018 Uwe took retirement in Berlin, and a new position as Professor Titular at the Institute of Geosciences at the University of Brasília, Brasil. Uwe Reimold was President of the Geological Society of South Africa (2002), and their De Beers Alex L. Du Toit Lecturer in 2006. He was honored for his Impact Cratering Research with the Barringer Medal and Award by The Meteoritical Society in 2009. Uwe Reimold's distinguished scientific output includes authorship of more than 350 refereed scientific articles and chapters in books, as well as author- or editorship of a dozen scientific and popular scientific books.